THE MATHEMATICS FOR ENGINEERS PROBLEM SOLVER®

REGISTERED TRADEMARK

A Complete Solution Guide to Any Textbook

Staff of Research and Education Association
Dr. M. Fogiel, Director

Research and Education Association
61 Ethel Road West
Piscataway, New Jersey 08854

THE MATHEMATICS FOR ENGINEERS
PROBLEM SOLVER ®

Printed in the United States of America

Library of Congress Card Catalog Number 91-67778

International Standard Book Number 0-87891-838-8

PROBLEM SOLVER is a registered trademark of
Research and Education Association, Piscataway, New Jersey

WHAT THIS BOOK IS FOR

To succeed in engineering, students must possess a thorough understanding of the mathematical principles that are applied to solve engineering problems. Solutions to engineering problems may have to be carried out using techniques from a combination of calculus, differential equations, complex variables, linear algebra, numerical analysis, vector analysis, and finite/discrete math. These mathematical subject areas are the tools through which modern engineering problems are solved. Accordingly, students need to develop skills in the use of any one or combination of these mathematical tools.

In recognizing the need to train students in those mathematical subjects, special courses often named "Mathematics for Engineers" have been devised, and numerous textbooks have been published for use in these courses. Despite these training steps that have been taken, however, students remain perplexed due to the numerous conditions that must often be remembered and correlated to solve an engineering problem. Various possible interpretations of terms used in the mathematical and engineering fields have also contributed to much of the difficulties experienced by students.

In a study of the problem, REA found the following basic reasons underlying the difficulties that engineering students experience with the math subjects taught in schools:

(a) No systematic rules of analysis have been developed which students may follow in a step-by-step manner to solve the usual problems encountered. This results from the fact that the numerous different conditions and principles which may be involved in an engineering problem, lead to many possible different methods of solution. To prescribe a set of rules to be followed for each of the possible variations, would involve an enormous number of rules and steps to be searched through by students, and this task would perhaps be more burdensome than solving the problem directly with some accompanying trial and error to find the correct solution route.

(b) Math and engineering textbooks currently available will usually explain a given principle in a few pages written by a professional who has an insight of the subject matter that is not shared by students. The explanations are often written in an abstract manner which leaves the students confused as to the application of the principle. The explanations given are not sufficiently detailed and extensive to make students aware of the wide range of applications and different aspects of the principle being studied.

The numerous possible variations of principles and their applications are usually not discussed, and it is left for the students to discover these for themselves while doing exercises. Accordingly, the average student is expected to rediscover that which has been long known and practiced, but not published or explained extensively.

(c) The examples usually following the explanation of a topic are too few and too simple to enable the student to obtain a thorough grasp of the principles involved. The explanations do not provide sufficient basis to enable students to solve problems that may be subsequently assigned for homework or given on examinations.

The examples are presented in abbreviated form which leaves out much material between steps, and requires that students derive the omitted material themselves. As a result, students find the examples difficult to understand—contrary to the purpose of the examples.

Examples are, furthermore, often worded in a confusing manner. They do not state the problem and then present the solution. Instead, they pass through a general discussion, never revealing what is being solved.

Examples, also, do not always include diagrams/graphs, wherever appropriate, and students do not obtain the training to draw diagrams or graphs to simplify and organize their thinking.

(d) Students can learn the subject only by doing the exercises themselves and reviewing them in class, to obtain experience in applying the principles with their different ramifications.

In doing the exercises by themselves, students find that they are required to devote considerably more time to math subjects than to other subjects of comparable credits, because they are uncertain with regard to the selection and application of the theorems and principles involved. It is also often necessary for students to discover those "tricks" not revealed in their texts (or review books), that make it possible to solve problems easily. Students must usually resort to methods of trial and error to discover these "tricks," and as a result they find that they may sometimes spend several hours solving a single problem.

(e) When reviewing the exercises in classrooms, instructors usually request students to take turns in writing solutions on the boards and explaining them to the class. Students often find it difficult to explain in a manner that holds the interest of the class, and enables the remaining students to follow the material written on the boards. The remaining students seated in the class are, furthermore, too occupied with copying the material from the boards, to listen to the oral explanations and concentrate on the methods of solution.

This book is intended to aid engineering students taking math subjects in overcoming the difficulties described, by supplying detailed illustrations of the solution methods which are usually not apparent to students. The solution methods are illustrated by problems selected from those that are most often assigned for class work and given on examinations. The problems are arranged in order of complexity to enable students to learn and understand a particular topic by reviewing the problems in sequence. The problems are illustrated with detailed step-by-step explanations, to save the students the large amount of time that is often needed to fill in the gaps that are usually found between steps of illustrations in textbooks or review/outline books.

The staff of REA considers that math is best learned by allowing students to view the methods of analysis and solution techniques themselves. This approach to learning the subject matter is similar to that practiced in various scientific laboratories, particularly in the medical fields.

In using this book, students may review and study the illustrated problems at their own pace; they are not limited to the time allowed for explaining problems on the board in class.

When students want to look up a particular type of problem and solution, they can readily locate it in the book by referring to the index which has been extensively prepared. It is also possible to locate a particular type of problem by glancing at just the material within the boxed portions. To facilitate rapid scanning of the problems, each problem has a heavy border around it. Furthermore, each problem is identified with a number immediately above the problem at the right-hand margin.

To obtain maximum benefit from the book, students should familiarize themselves with the section, "How To Use This Book," located in the front pages.

To meet the objectives of this book, staff members of REA have selected problems usually encountered in assignments and examinations, and have solved each problem meticulously to illustrate the steps which are difficult for students to comprehend. Special gratitude is expressed to them for their efforts in this area, as well as to the numerous contributors who devoted their time to this book.

The difficult task of coordinating the efforts of all persons was carried out by Carl Fuchs. His conscientious work deserves much appreciation. He also trained and supervised art and production personnel in the preparation of the book for printing.

Finally, special thanks are due to Helen Kaufmann for her unique talent to render those difficult border-line decisions and constructive suggestions related to the design and organization of the book.

Max Fogiel, Ph.D.
Program Director

HOW TO USE THIS BOOK

This book can be an invaluable aid to engineering students as a supplement to their textbooks. The book is subdivided into 12 chapters, each dealing with a separate topic. The subject matter is developed beginning with differential equations and extending through linear differential equations, applications of differential equations, vectors, multiple integrals, sequence and series, Laplace transforms, and Fourier transforms. Sections on complex variables, special functions, determinants and matrices, probability, and statistics have also been included. An extensive number of applications have been included, since these appear to be most troublesome to students.

TO LEARN AND UNDERSTAND
A TOPIC THOROUGHLY

1. Refer to your class text and read the section pertaining to the topic. You should become acquainted with the principles discussed there. These principles, however, may not be clear to you at that time.

2. Then locate the topic you are looking for by referring to the "Table of Contents" in the front of "The Mathematics for Engineers Problem Solver."

3. Turn to the page where the topic begins and review the problems under each topic, in the order given. For each topic, the problems are arranged in order of complexity, from the simplest to the most difficult. Some problems may appear similar to others, but each problem has been selected to illustrate a different point or solution method.

To learn and understand a topic thoroughly and retain its contents, it will be generally necessary for students to review the problems several times. Repeated review is essential in order to gain experience in recognizing the principles that should be applied and in selecting the best solution technique.

TO FIND A PARTICULAR PROBLEM

To locate one or more problems related to a particular subject matter, refer to the index. In using the index, be certain to note that the numbers given there refer to problem numbers, not to page numbers. This arrangement of the index is intended to facilitate finding a problem more rapidly, since two or more problems may appear on a page.

If a particular type of problem cannot be found readily, it is recommended that the student refer to the "Table of Contents" in the front pages, and then turn to the chapter which is applicable to the problem being sought. By scanning or glancing at the material that is boxed, it will generally be possible to find problems related to the one being sought, without consuming considerable time. After the problems have been located, the solutions can be reviewed and studied in detail. For this purpose of locating problems rapidly, students should acquaint themselves with the organization of the book as found in the "Table of Contents."

In preparing for an exam, it is useful to find the topics to be covered on the exam in the "Table of Contents," and then review the problems under those topics several times. This should equip the student with what might be needed for the exam.

Contents

UNITS CONVERSION FACTORS

This section includes a particularly useful and comprehensive table to aid students and teachers in converting between systems of units.

The problems and their solutions in this book use **SI (International System)** as well as English units. Both of these units are in extensive use throughout the world, and therefore students should develop a good facility to work with both sets of units until a single standard of units has been found acceptable internationally.

In working out or solving a problem in one system of units or the other, essentially only the numbers change. Also, the conversion from one unit system to another is easily achieved through the use of conversion factors that are given in the subsequent table. Accordingly, the units are one of the least important aspects of a problem. For these reasons, a student should not be concerned mainly with which units are used in any particular problem. Instead, a student should obtain from that problem and its solution an understanding of the underlying principles and solution techniques that are illustrated there.

To convert	To	Multiply by	For the reverse, multiply by
acres	square feet	4.356×10^4	2.296×10^{-5}
acres	square meters	4047	2.471×10^{-4}
ampere-hours	coulombs	3600	2.778×10^{-4}
ampere-turns	gilberts	1.257	0.7958
ampere-turns per cm.	ampere-turns per inch	2.54	0.3937
angstrom units	inches	3.937×10^{-9}	2.54×10^8
angstrom units	meters	10^{-10}	10^{10}
atmospheres	feet of water	33.90	0.02950
atmospheres	inch of mercury at 0°C	29.92	3.342×10^{-2}
atmospheres	kilogram per square meter	1.033×10^4	9.678×10^{-5}
atmospheres	millimeter of mercury at 0°C	760	1.316×10^{-3}
atmospheres	pascals	1.0133×10^5	0.9869×10^{-5}
atmospheres	pounds per square inch	14.70	0.06804
bars	atmospheres	9.870×10^{-7}	1.0133
bars	dynes per square cm.	10^6	10^{-6}
bars	pascals	10^5	10^{-5}
bars	pounds per square inch	14.504	6.8947×10^{-2}
Btu	ergs	1.0548×10^{10}	9.486×10^{-11}
Btu	foot-pounds	778.3	1.285×10^{-3}
Btu	joules	1054.8	9.480×10^{-4}
Btu	kilogram-calories	0.252	3.969
calories, gram	Btu	3.968×10^{-3}	252
calories, gram	foot-pounds	3.087	0.324
calories, gram	joules	4.185	0.2389
Celsius	Fahrenheit	(°C × 9/5) + 32 = °F	(°F − 32) × 5/9 = °C

To convert	To	Multiply	For the reverse, multiply by
Celsius	kelvin	°C + 273.1 = K	K − 273.1 = °C
centimeters	angstrom units	1×10^8	1×10^{-8}
centimeters	feet	0.03281	30.479
centistokes	square meters per second	1×10^{-6}	1×10^6
circular mils	square centimeters	5.067×10^{-6}	1.973×10^5
circular mils	square mils	0.7854	1.273
cubic feet	gallons (liquid U.S.)	7.481	0.1337
cubic feet	liters	28.32	3.531×10^{-2}
cubic inches	cubic centimeters	16.39	6.102×10^{-2}
cubic inches	cubic feet	5.787×10^{-4}	1728
cubic inches	cubic meters	1.639×10^{-5}	6.102×10^4
cubic inches	gallons (liquid U.S.)	4.329×10^{-3}	231
cubic meters	cubic feet	35.31	2.832×10^{-2}
cubic meters	cubic yards	1.308	0.7646
curies	coulombs per minute	1.1×10^{12}	0.91×10^{-12}
cycles per second	hertz	1	1
degrees (angle)	mils	17.45	5.73×10^{-2}
degrees (angle)	radians	1.745×10^{-2}	57.3
dynes	pounds	2.248×10^{-6}	4.448×10^5
electron volts	joules	1.602×10^{-19}	0.624×10^{18}
ergs	foot-pounds	7.376×10^{-8}	1.356×10^7
ergs	joules	10^{-7}	10^7
ergs per second	watts	10^{-7}	10^7
ergs per square cm.	watts per square cm.	10^{-3}	10^3
Fahrenheit	kelvin	(°F + 459.67)/1.8	1.8K − 459.67
Fahrenheit	Rankine	°F + 459.67 = °R	°R − 459.67 = °F
faradays	ampere-hours	26.8	3.731×10^{-2}
feet	centimeters	30.48	3.281×10^{-2}
feet	meters	0.3048	3.281
feet	mils	1.2×10^4	8.333×10^{-5}
fermis	meters	10^{-15}	10^{15}
foot candles	lux	10.764	0.0929
foot lamberts	candelas per square meter	3.4263	0.2918
foot-pounds	gram-centimeters	1.383×10^4	1.235×10^{-5}
foot-pounds	horsepower-hours	5.05×10^{-7}	1.98×10^6
foot-pounds	kilogram-meters	0.1383	7.233
foot-pounds	kilowatt-hours	3.766×10^{-7}	2.655×10^6
foot-pounds	ounce-inches	192	5.208×10^{-3}
gallons (liquid U.S.)	cubic meters	3.785×10^{-3}	264.2
gallons (liquid U.S.)	gallons (liquid British Imperial)	0.8327	1.201
gammas	teslas	10^{-9}	10^9
gausses	lines per square cm.	1.0	1.0
gausses	lines per square inch	6.452	0.155
gausses	teslas	10^{-4}	10^4
gausses	webers per square inch	6.452×10^{-8}	1.55×10^7
gilberts	amperes	0.7958	1.257
grads	radians	1.571×10^{-2}	63.65
grains	grams	0.06480	15.432
grains	pounds	$1/_{7000}$	7000
grams	dynes	980.7	1.02×10^{-3}
grams	grains	15.43	6.481×10^{-2}

To convert	To	Multiply	For the reverse, multiply by
grams	ounces (avdp)	3.527×10^{-2}	28.35
grams	poundals	7.093×10^{-2}	14.1
hectares	acres	2.471	0.4047
horsepower	Btu per minute	42.418	2.357×10^{-2}
horsepower	foot-pounds per minute	3.3×10^4	3.03×10^{-5}
horsepower	foot-pounds per second	550	1.182×10^{-3}
horsepower	horsepower (metric)	1.014	0.9863
horsepower	kilowatts	0.746	1.341
inches	centimeters	2.54	0.3937
inches	feet	8.333×10^{-2}	12
inches	meters	2.54×10^{-2}	39.37
inches	miles	1.578×10^{-5}	6.336×10^4
inches	mils	10^3	10^{-3}
inches	yards	2.778×10^{-2}	36
joules	foot-pounds	0.7376	1.356
joules	watt-hours	2.778×10^{-4}	3600
kilograms	tons (long)	9.842×10^{-4}	1016
kilograms	tons (short)	1.102×10^{-3}	907.2
kilograms	pounds (avdp)	2.205	0.4536
kilometers	feet	3281	3.408×10^{-4}
kilometers	inches	3.937×10^4	2.54×10^{-5}
kilometers per hour	feet per minute	54.68	1.829×10^{-2}
kilowatt-hours	Btu	3413	2.93×10^{-4}
kilowatt-hours	foot-pounds	2.655×10^6	3.766×10^{-7}
kilowatt-hours	horsepower-hours	1.341	0.7457
kilowatt-hours	joules	3.6×10^6	2.778×10^{-7}
knots	feet per second	1.688	0.5925
knots	miles per hour	1.1508	0.869
lamberts	candles per square cm.	0.3183	3.142
lamberts	candles per square inch	2.054	0.4869
liters	cubic centimeters	10^3	10^{-3}
liters	cubic inches	61.02	1.639×10^{-2}
liters	gallons (liquid U.S.)	0.2642	3.785
liters	pints (liquid U.S.)	2.113	0.4732
lumens per square foot	foot-candles	1	1
lumens per square meter	foot-candles	0.0929	10.764
lux	foot-candles	0.0929	10.764
maxwells	kilolines	10^{-3}	10^3
maxwells	webers	10^{-8}	10^8
meters	feet	3.28	30.48×10^{-2}
meters	inches	39.37	2.54×10^{-2}
meters	miles	6.214×10^{-4}	1609.35
meters	yards	1.094	0.9144
miles (nautical)	feet	6076.1	1.646×10^{-4}
miles (nautical)	meters	1852	5.4×10^{-4}
miles (statute)	feet	5280	1.894×10^{-4}
miles (statute)	kilometers	1.609	0.6214
miles (statute)	miles (nautical)	0.869	1.1508
miles per hour	feet per second	1.467	0.6818
miles per hour	knots	0.8684	1.152
millimeters	microns	10^3	10^{-3}

To convert	To	Multiply	For the reverse, multiply by
mils	meters	2.54×10^{-5}	3.94×10^4
mils	minutes	3.438	0.2909
minutes (angle)	degrees	1.666×10^{-2}	60
minutes (angle)	radians	2.909×10^{-4}	3484
newtons	dynes	10^5	10^{-5}
newtons	kilograms	0.1020	9.807
newtons per sq. meter	pascals	1	1
newtons	pounds (avdp)	0.2248	4.448
oersteds	amperes per meter	7.9577×10	1.257×10^{-2}
ounces (fluid)	quarts	3.125×10^{-2}	32
ounces (avdp)	pounds	6.25×10^{-2}	16
pints	quarts (liquid U.S.)	0.50	2
poundals	dynes	1.383×10^4	7.233×10^{-5}
poundals	pounds (avdp)	3.108×10^{-2}	32.17
pounds	grams	453.6	2.205×10^{-3}
pounds (force)	newtons	4.4482	0.2288
pounds per square inch	dynes per square cm.	6.8946×10^4	1.450×10^{-5}
pounds per square inch	pascals	6.895×10^3	1.45×10^{-4}
quarts (U.S. liquid)	cubic centimeters	946.4	1.057×10^{-3}
radians	mils	10^3	10^{-3}
radians	minutes of arc	3.438×10^3	2.909×10^{-4}
radians	seconds of arc	2.06265×10^5	4.848×10^{-6}
revolutions per minute	radians per second	0.1047	9.549
roentgens	coulombs per kilogram	2.58×10^{-4}	3.876×10^3
slugs	kilograms	1.459	0.6854
slugs	pounds (avdp)	32.174	3.108×10^{-2}
square feet	square centimeters	929.034	1.076×10^{-3}
square feet	square inches	144	6.944×10^{-3}
square feet	square miles	3.587×10^{-8}	27.88×10^6
square inches	square centimeters	6.452	0.155
square kilometers	square miles	0.3861	2.59
stokes	square meter per second	10^{-4}	10^{-4}
tons (metric)	kilograms	10^3	10^{-3}
tons (short)	pounds	2000	5×10^{-4}
torrs	newtons per square meter	133.32	7.5×10^{-3}
watts	Btu per hour	3.413	0.293
watts	foot-pounds per minute	44.26	2.26×10^{-2}
watts	horsepower	1.341×10^{-3}	746
watt-seconds	joules	1	1
webers	maxwells	10^8	10^{-8}
webers per square meter	gausses	10^4	10^{-4}

CHAPTER 1

DIFFERENTIAL EQUATIONS: TYPES AND METHODS OF SOLUTION

> **Basic Attacks and Strategies for Solving Problems in this Chapter. See pages 2 to 39 for step-by-step solutions to problems.**

A differential equation is an equation that involves at least one derivative of an unknown function. A derivative may be involved either implicitly, or explicitly, through the presence of differentials. Our aim is to solve differential equations using different techniques, by recognizing the "type" of the differential equation, and its appropriate method of analysis.

When an equation involves one derivative with respect to a particular variable, that variable is called an independent variable. A variable is called dependent if its derivative appears in the differential equation. The order of a differential equation is the order of the highest-ordered derivative appearing in the equation.

Methods for the elimination of arbitrary constants vary with the way in which the constants enter the given relation. A method which is efficient for one problem may be poor for another. Since each differentiation yields a new relation, the number of derivatives that need to be used is the same as the number of arbitrary constants to be eliminated.

Polynomials in which all terms are of the same degree are called homogeneous polynomials. The concept of homogeneity is extended so as to homogeneous functions instead of polynomials. If an equation is not exact, it is natural to attempt to make it exact by the introduction of an appropriate factor. By the multiplication of this factor, the variables are separated to obtain an exact equation. For linear equations of the first order, the existence of the integrating factor can be very well demonstrated.

Step-by-Step Solutions to
Problems in this Chapter,
"Differential Equations:
Types and Methods of Solutions"

SEPARABLE DIFFERENTIAL EQUATIONS

PROBLEM 1–1

Solve the differential equations

    a) $y' = y$; b) $y' = 6y$; c) $y' = -5y$.

Give the solution to the differential equation of the general form $y' = ay$ where a is a real number.

<u>Solution</u>: a) $\frac{dy}{dx} = y$. $dx - \frac{1}{y} dy = 0$, for $y \neq 0$. This is in the form $A(x)dx + B(y)dy = 0$. The implicit solution is given by integration:

$\int dx - \int \frac{1}{y} dy = c$, where c is an arbitrary constant of integration. $x = \ln|y| = c$; $x + \ln|y^{-1}| = c$; $|\frac{1}{y}| = e^{c-x}$; $|y| = e^{-c}e^x$ or $y = ke^x$; $(k = \pm e^{-c} \neq 0)$.

b) $\frac{dy}{dx} = 6y$. $6dx - \frac{1}{y} dy = 0$, for $y \neq 0$. Integrating

$6\int dx - \int \frac{1}{y} dy = c$ where c is an arbitrary constant of integration.

$6x - \ln|y| = c$.

$|y| = e^{-c}e^{6x}$ or $y = ke^{6x}$ $(k = \pm e^{-c} \neq 0)$.

c) $\frac{dy}{dx} = -5y$. $5dx + \frac{1}{y} dy = 0$, for $y \neq 0$. Integration yields

$5\int dx + \int \frac{1}{y} dy = c$ where c is an arbitrary constant of integration.

$\ln|y| = c - 5x$; $y = \pm e^c e^{-5x}$; $y = ke^{-5x}$ $(k = \pm r^c \neq 0)$.

Now consider the general case $y' = ay$ or $y' - ay = 0$ where a is any real number. Rewriting in the differential form:

$adx - \frac{1}{y} dy = 0$, for $y \neq 0$.

This is a separable equation of the form $A(x)dx + B(y)dy$, where $A(x) = a$, $B(y) = -1/y$. The solution is found by integrating:

$a\int dx - \int \frac{1}{y} dy = c$, where c is an arbitrary constant of integration,

or $\ln|y^{-1}| = c - ax$. Thus, $y = \pm e^{-c} e^{ax}$ or $y = ke^{ax}$ ($k = \pm e^{-c} \neq 0$).
Since we have $1/y$ in the differential form we must have $y \neq 0$.
This implies, since
$$y = ke^{ax} ,$$
that $k \neq 0$. But the function $y = 0$ is seen to be a solution of the
equation. Thus $k = 0$ implies that $y = 0$ which is a solution.
Hence $y = ke^{ax}$ is a solution for any k which is a real number.

• PROBLEM 1-2

Solve
$$x \, dy + y \, dx = 0 .$$

Solution: An equation is called separable if it is in the form
$F(x)G(y) \, dx + f(x) \, g(y) \, dy = 0$. It can be reduced to an exact equation in the following manner: multiply by the factor
$$\frac{1}{f(x)G(y)} ;$$
this makes the equation $\dfrac{F(x)}{f(x)} \, dx + \dfrac{g(y)}{G(y)} \, dy = 0$. Since
$$\frac{\partial}{\partial y}\left[\frac{F(x)}{f(x)}\right] = 0 = \frac{\partial}{\partial x}\left[\frac{g(y)}{G(y)}\right] ,$$
it is exact.

In the given problem $F(x) = 1$, $f(x) = x$; $G(y) = y$, $g(y) = 1$.
Thus,
$$\frac{1}{f(x)G(y)} = \frac{1}{xy} ,$$
is an integrating factor.
$$x \, dy + y \, dx = 0; \quad \frac{dy}{y} + \frac{dx}{x} = 0 ,$$
for $x \neq 0$ and $y \neq 0$. A solution is
$$\int \frac{dy}{y} + \int \frac{dx}{x} = c ,$$
where c is an arbitrary constant of integration, i.e., $\log|y| +$
$\log |x| = c$. But $\log A + \log B = \log AB$. Hence, $|xy| = e^{c}$;
$xy = \pm e^{c} = k$, where k is an arbitrary nonzero constant.

Note that we excluded the cases where $x = 0$ and where $y = 0$.
Direct substitution in the original equation reveals that $y = 0$ and
$x = 0$ are also solutions:
$$(0)dy + y \, d(0) = 0 + 0 = 0 ,$$
and
$$xd(0) + (0)dx = 0 + 0 = 0 .$$
These last two solutions were lost in the separation process.

• PROBLEM 1-3

Solve the differential equation
$$2(y-1)dx + (x^2 \sin y)dy = 0 .$$

<u>Solution</u>: We rewrite the differential form as an equation in derivative form.

$$\frac{dy}{dx} = \frac{-2(y-1)}{x^2 \sin y} \quad \text{for} \quad x \neq 0 .$$

This equation is separable if we can write it in the form

$$\frac{dy}{dx} = f(x) \ g(y) .$$

Thus

$$\frac{dy}{dx} = \left(\frac{-2}{x^2}\right)\left(\frac{y-1}{\sin y}\right)$$

and the equation is indeed separable.

$$\frac{\sin y}{y-1} \ dy = \frac{-2}{x^2} \ dx \quad \text{for} \quad y \neq 1 .$$

Integration gives

$$\int \frac{\sin y}{y-1} dy = \frac{2}{x} + c$$

where the expression on the left is not an elementary integral and must be evaluated by numerical methods if the limits of integration are given. C is an arbitrary constant of integration.

Note that we excluded the cases where $x = 0$ and where $y = 1$. Substitution in the original equation reveals that $x \equiv 0$ and $y \equiv 1$ are also solutions:

$$2(y-1)d(0) + (0^2 \sin y)dy = 0 + 0 = 0 ,$$

and

$$2(1-1)dx + (x^2 \sin 1)d(1) = 0 + 0 = 0 .$$

These last two solutions were lost in the separation process.

● PROBLEM 1–4

Solve the first-order non-linear differential equation $\frac{dy}{dx} = \frac{1+y^2}{1+x^2}$.

<u>Solution</u>: By inspecting the equation, we suspect that it may be written in the form $A(x)dx + B(y)dy = 0$. Thus

$$\frac{dy}{(1+y^2)} = \frac{dx}{(1+x^2)}$$

and

$$\frac{1}{(1+x^2)} \ dx - \frac{1}{(1+y^2)} \ dy = 0 .$$

Now we proceed to integrate: remembering that

$$\int \frac{1}{(1+t^2)} \ dt = \arctan t + c_1 ,$$

where c_1 is an arbitrary constant of integration. Thus, arctan x - arctan y = c_2, where c_2 is an arbitrary constant of integration. We know that $\tan(\arctan x) = x$

$$y = \tan(\arctan x + c) = \frac{x + \tan c}{1 - x \tan c}$$

We notice that $\frac{x + \tan c}{1 - x \tan c}$ is undefined when $c = \pi/2$. Since it is an indeterminate form, we may evaluate it using L'Hopital's rule: letting c be the variable.

$$\lim_{c \to \pi/2} \frac{x + \tan c}{1 - x \tan c} = \lim_{c \to \pi/2} \frac{\sec^2 c}{-x \sec^2 c} = -\frac{1}{x} .$$

4

This last value of y holds when c is any odd multiple of $\pi/2$, i.e., $c = (k + \frac{1}{2}\pi)$ $k = 0, \pm 1, \ldots$.

● **PROBLEM** 1–5

Find the general solution of: $\frac{dy}{dx} = x^2 y^3$.

Solution: Separating the variables, the given differential equation may be written as:

$$\frac{dy}{y^3} = x^2 dx .$$

The general solution may be obtained by integrating. We obtain:

$$\int \frac{dy}{y^3} = \int x^2 dx + c,$$

or,

$$-\frac{1}{2y^2} = \frac{x^3}{3} + c .$$

This can be rearranged to yield:

$$2y^2 x^2 + cy^2 + 3 = 0.$$

This equation satisfies the given differential equation, for differentiation leads to:

$$\frac{dy}{dx} = x^2 y^3 .$$

VARIABLE TRANSFORMATION: $u = ax + by$

● **PROBLEM** 1–6

Solve $e^x dx - y \, dy = 0$; $y(0) = 1$.

Solution: This is a first-order homogeneous differential equation with an initial condition. Since it is in the form $A(x)dx + B(y)dy$, it is a separable equation whose solution is found by integration. If it had not been in this form, we could have used the fact that a change of variables may sometimes convert a given equation into the form $A(x)dx + B(y)dy = 0$. For example,

$$\frac{dy}{dx} = \cos(x + y)$$

is not separable. But if we let $z = (x + y)$; $\frac{dz}{dx} = 1 + \frac{dy}{dx}$, $\frac{dz}{dx} = \cos(z) + 1$, or $dx + (-1)\frac{dz}{(1+\cos z)} = 0$. This equation is separable.

Returning to our original problem,

$$e^x dx - y \, dy = 0 \quad y(0) = 1$$

$$\int e^x dx + \int (-y)dy = c; \quad y^2 = 2e^x + k, \quad k = -2c,$$

5

where c is an arbitrary constant.

Substituting the initial condition $y(0) = 1$, we find $(1)^2 = 2e^0 + k$; $k = -1$. Thus the solution is

$$y^2 = 2e^x - 1 \quad \text{or} \quad y = \sqrt{2e^x - 1}$$

since y cannot be negative by the initial condition.

● PROBLEM 1-7

Solve the first-order differential equation

$$e^{-y}\left(\frac{dy}{dx} + 1\right) = xe^x . \tag{a}$$

Solution: Given an equation like (a), the best mode of attack is to isolate dy/dx . Thus we have

$$\frac{dy}{dx} = [xe^x \, e^y] - 1.$$

In this form, it is not possible to separate the variables. If we rewrite the equation in the form

$$\frac{dy}{dx} + 1 = xe^x \, e^y = xe^{x+y}$$

and then make the substitution $u = x + y$, it is possible to obtain a separable equation.

$$u = x + y; \quad \frac{du}{dx} = \frac{dx}{dx} + \frac{dy}{dx} = 1 + \frac{dy}{dx} = xe^u .$$

Thus

$$\frac{du}{dx} = xe^u ; \quad \frac{du}{e^u} = x \, dx .$$

Integrating

$$\int \frac{du}{e^u} = \int x \, dx ; \quad -e^{-u} = \frac{x^2}{2} + c_1,$$

where c_1 is an arbitrary constant. We can write this as $e^{-u} + \frac{x^2}{2} = c,$

where $c = -c_1$. Now we substitute $u = x + y$:

$$e^{-(x+y)} + \frac{x^2}{2} = c .$$

Taking logarithms, $-(x + y)\ln e = \ln(c - \frac{x^2}{2})$,

or

$$y = -\ln\left[c - \frac{x^2}{2}\right] - x .$$

● PROBLEM 1-8

Solve the differential equation

$$(x^2 - xy + y^2)dx - xy \, dy = 0 .$$

Solution: The equation is not in the form $f(x)dx + g(y)dy = 0$. Hence it is not separable. But we notice that it is homogeneous of degree two for

$$\frac{dy}{dx} = \frac{x^2 - xy + y^2}{xy}$$

6

and if we assume that xy has degree two, then every term in the expression on the right has the same degree. It is a property of a homogeneous equation that it may be expressed in the form $g(y/x)$, or $g(v)$. Now let $y = vx$, then

$$\frac{dy}{dx} = v + x\frac{dv}{dx} .$$

In this problem

$$v + x\frac{dv}{dx} = \frac{x^2 - x^2v + v^2x^2}{x(vx)} ,$$

or

$$\frac{dv}{dx} = \frac{1}{x}\left(\frac{1-v}{v}\right) .$$

This equation may be separated to give

$$\frac{v\ dv}{1-v} = \frac{dx}{x}$$

or

$$\frac{dx}{x} + \frac{v\ dv}{(v-1)} = 0 .$$

$$\frac{dx}{x} + [1 + \frac{1}{v-1}]\ dv = 0 .$$

Integrating, we obtain

$$\ln|x| + v + \ln|v-1| = \ln|c| , \qquad\qquad (a)$$

where c is a constant of integration. Taking the exponential of both sides of (a):

$$x(v - 1)e^v = c.$$

Substituting $v = y/x$:

$$x(\frac{y}{x} - 1)\ e^{y/x} = c ,$$

or

$$(y - x)\ e^{y/x} = c . \qquad\qquad (b)$$

Since we cannot solve for y explicitly, we call (b) an implicit solution.

EXACT DIFFERENTIAL EQUATIONS

• PROBLEM 1-9

Why is the differential equation

$$\frac{dy}{dx} = -\frac{e^y}{xe^y+2y}$$

exact? What is its solution?

Solution: To check for exactness we must first write the given equation in the differential form

$$e^y dx + (xe^y + 2y)dy = 0.$$

This equation is exact because

$$\frac{\partial}{\partial y}\ e^y = e^y = \frac{\partial}{\partial x}\ (xe^y + 2y).$$

The solution to this equation may be found by the method of grouping.

Write $e^y dx + (xe^y + 2y)dy = 0$ in the form $e^y dx + xe^y dy + 2y dy = 0$. Denoting the process of differentiation (whether partial or regular) by d:

$$d(xe^y) + d(y^2) = d(c)$$

where c is an arbitrary constant.

Combining, $d(xe^y + y^2) = d(c)$ or, $xe^y + y^2 = c$, the required solution.

● **PROBLEM** 1-10

Solve the DE
$$(x + y)y' + (y + 3x) = 0.$$

<u>Solution</u>: Rewriting the equation as
$$(y + 3x)dx + (x + y)dy = 0,$$
we note that the equation is not separable. It is, however, exact, since

$$M_y = \frac{\partial(y+3x)}{\partial y} = 1 = N_x = \frac{\partial(x+y)}{\partial x}$$

There are two methods of solving exact equations. One is by asking oneself which function has its total differential dF, exactly equal to the given differential equation. In symbols, for the present equation,

$$dF = M(x,y)dx + N(x,y)dy$$
$$= (y + 3x)dx + (x + y)dy.$$

Thus we have that $(y + 3x)$ is the partial derivative with respect to x of the required function, F(x,y). Integrating with respect to x gives

$$xy + \frac{3}{2} x^2 + c_1 .$$

Similarly, the integral of $(x + y)$ with respect to y is

$$xy + y^2/2 + c_2 .$$

Combining, $F(x,y) = \frac{3}{2} x^2 + xy + y^2/2 = c.$

A more formal method of solution is as follows: integrate M(x,y) holding y constant. The resulting equation must be equal to the integral of N(x,y) with respect to y holding x constant.

Thus in the given problem,
$$\int(y + 3x)dx = xy + \frac{3}{2} x^2 + \varphi(y).$$

But we have that
$$\frac{\partial}{\partial y}(xy + \frac{3}{2} x^2 + \varphi(y)) = x + y.$$

Therefore, $\varphi'(y) = y$ or $\varphi(y) = \int y \, dy = y^2/2 + c .$

The required function is again
$$xy + \frac{3}{2} x^2 + y^2/2 = c .$$

8

Solve the first order differential equation

$$3x^2 + 6xy^2 + (6x^2 y + 4y^2)y' = 0 .$$

Solution: We first convert the equation into a differential form, i.e.,

$$(3x^2 + 6xy^2)dx + (6x^2 y + 4y^2)dy = 0 .$$

This equation is neither separable (it cannot be expressed as $f(x)dx + g(y)dy$), nor homogeneous $(6x^2 y + 4y^2$ is non-homogeneous). It is, however, exact, since

$$\frac{\partial}{\partial y}(3x^2 + 6xy^2) = 12xy = \frac{\partial}{\partial x}(6x^2 y + 4y^2) .$$

Integrating $3x^2 + 6xy^2$ with respect to x (treating y as a constant),

$$\int (3x^2 + 6xy^2)dx = x^3 + 3x^2 y^2 + \varphi(y).$$

The function that we seek has its partial derivative with respect to y equal to $6x^2 y + 4y^2$. That is,

$$\frac{\partial}{\partial y}(x^3 + 3x^2 y^2 + \varphi(y)) = 6x^2 y + 4y^2 .$$

Therefore, $\varphi'(y) = 4y^2$ and $\varphi(y) = \int 4y^2 dy = \frac{4}{3} y^3 + c.$

Then

$$F(x,y) = x^3 + 3x^2 y^2 + \frac{4}{3} y^3 + c .$$

Solve the equation

$$(3x^2 + 4xy)dx + (2x^2 + 2y)dy = 0.$$

Solution: This is a first-order differential equation as may be seen by rewriting it as:

$$\frac{dy}{dx} = -\frac{(3x^2 + 4xy)}{(2x^2 + 2y)} .$$

Now we examine the original form which is of the type

$$M(x,y)dx + N(x,y)dy = 0.$$

It is exact if

$$\frac{\partial M(x,y)}{\partial y} = \frac{\partial N(x,y)}{\partial x} .$$

Since

$$\frac{\partial}{\partial y}(3x^2 + 4xy) = 4x = \frac{\partial}{\partial x}(2x^2 + 2y) ,$$

the equation is exact. This means that there exists a function $F(x,y)$ whose total differential

$$\frac{\partial f}{\partial x} dx + \frac{\partial f}{\partial y} dy$$

9

is exactly equal to $(3x^2 + 4xy)dx + (2x^2 + 2y)dy$. To find the function, we first find the integral of $M(x,y)dx$, i.e.,

$$\int \left[3x^2 + 4xy\right] dx = x^3 + 2x^2y + \varphi(y).$$

Instead of an arbitrary constant, we have $\varphi(y)$ since we are treating y as a constant.

If $x^3 + 2x^2y + \varphi(y)$ is to be the required solution $F(x,y)$,

$$\frac{\partial}{\partial y} F(x,y) \text{ must equal } 2x^2 + 2y.$$

Hence $\varphi'(y) = 2y$ and $\varphi(y) = y^2 + c$.

The required solution is therefore

$$x^3 + 2x^2y + y^2 + c = 0 .$$

• PROBLEM 1-13

Solve the equation

$$(2x^3 - xy^2 - 2y + 3)dx - (x^2y + 2x)dy = 0 .$$

Solution: The presence of y^2 makes the equation non-linear (an equation is defined to be linear when y and all of the derivatives present in the equation are not raised to any power except 1.) Rewriting as

$$\frac{dy}{dx} = \frac{2x^3 - xy^2 - 2y + 3}{x^2y + 2x}$$

we note that the equation is neither separable nor homogeneous. We therefore check to see if the equation is exact. The test for exactness for the differential form $M(x,y)dx + N(x,y)dy = 0$ is:

$$\frac{\partial M}{\partial y} = \frac{\partial N}{\partial x} .$$

Here $M(x,y) = (2x^3 - xy^2 - 2y + 3)$ and $N(x,y) = (-x^2y - 2x)$.

$\frac{\partial M}{\partial y} = -2xy - 2 = \frac{\partial N}{\partial x}$ which means the equation is exact.

The method of solution for an exact differential form proceeds as follows: first integrate $M(x,y)$ with respect to x holding y constant. The resulting integral will have a function of y as the arbitrary constant. Now take the partial derivative of this integral with respect to y. Since the equation is exact, this derivative is equal to $N(x,y)$. Thus

$$\int M(x,y)dx = \int (2x^3 - xy^2 - 2y + 3)dx .$$

$$\frac{x^4}{2} - \frac{x^2 y^2}{2} - 2xy + 3x + \varphi(y) = F(x,y)$$

$$\frac{\partial F}{\partial y}(x,y) = -x^2 y - 2x + \varphi'(y).$$

But

$$\frac{\partial F}{\partial y}(x,y) = N(x,y) = -x^2 y - 2x .$$

Therefore $\varphi'(y) = 0$; $\varphi(y) = \int \varphi'(y)dy = c$. Then

$$F(x,y) = \frac{x^4}{2} - \frac{x^2 y^2}{2} - 2xy + 3x + c,$$

the solution to the differential equation.

• **PROBLEM** 1–14

Solve $(x + \sin y)dx + (x \cos y - 2y)dy = 0$.

<u>Solution</u>: Rewriting the equation as

$$\frac{dy}{dx} = -\frac{(x + \sin y)}{(x \cos y - 2y)}$$

we see that it is not possible to separate the variables so that $\frac{dy}{dx} = f(x)g(y)$. Nor is the equation homogeneous. But checking for

exactness we find that $\frac{\partial}{\partial y}(x + \sin y) = \cos y = \frac{\partial}{\partial x}(x \cos y - 2y)$.
Therefore the solution of the equation is a function $F(x,y) = c$ such that its total differential is given by $dF = (x + \sin y)dx + (x \cos y - 2y)dy$. To find $F(x,y)$, $\int(x + \sin y)dx$ (treating y as a constant) gives $\frac{x^2}{2} + x \sin y + \varphi(y)$. Now we must have

$$\frac{\partial}{\partial y}(\frac{x^2}{2} + x \sin y + \varphi(y)) = (x \cos y - 2y).$$

Therefore
$\quad x \cos y + \varphi'(y) = x \cos y - 2y$ and $\varphi'(y) = -2y$, $\varphi(y) = -\int 2y \, dy$
$$= -y^2 + c.$$

The solution to the problem is

$$F(x,y) = \frac{x^2}{2} + x \sin y - y^2 = c.$$

This is an implicit solution of y; it is not possible to solve explicitly for y.

Solve the initial value problem:

$$y' = - \frac{(2x \cos y + 3x^2 y)}{x^3 - x^2 \sin y - y} \quad ; \quad y = 2 \quad \text{when} \quad x = 0.$$

Solution: Rewriting as a differential form

$$(2x \cos y + 3x^2 y)dx + (x^3 - x^2 \sin y - y)dy = 0 \ ,$$

we see that $\frac{\partial}{\partial y}(2x \cos y + 3x^2 y) = -2x \sin y + 3x^2$

$$= \frac{\partial}{\partial x} (x^3 - x^2 \sin y - y).$$

The equation is exact which means that we must find F such that

$$\frac{\partial F}{\partial x}(x,y) = 2x \cos y + 3x^2 y$$

and

$$\frac{\partial F}{\partial y}(x,y) = x^3 - x^2 \sin y - y .$$

Thus,

$$F(x,y) = \int (2x \cos y + 3x^2 y)\partial x$$

$$= x^2 \cos y + x^3 y + \varphi(y).$$

Now,

$$\frac{\partial F}{\partial y}(x,y) = -x^2 \sin y + x^3 + \varphi'(y) .$$

Setting this equal to $x^3 - x^2 \sin y - y$ implies that $\varphi'(y) = -y$
or $\varphi(y) = - y^2/2 + c$.

The solution is therefore

$$x^2 \cos y + x^3 y - y^2/2 = c .$$

Now, use the initial condition to find c. $y(0) = 2$; $c = - 4/2 = -2$.
The solution of the initial-value problem is then

$$x^2 \cos y + x^3 y - y^2/2 = -2.$$

The solution cannot be expressed as an explicit equation of y.

Find an integrating factor that makes the differential equation
$(3y + 4xy^2)dx + (2x + 3x^2 y)dy = 0$ exact.

Solution: An equation of the form $M(x,y)dx + N(x,y)dy = 0$ is said to be exact if

$$\frac{\partial M}{\partial y} = \frac{\partial N}{\partial x} .$$

In the given problem,

$$\frac{\partial M}{\partial y} = 3 + 8xy \neq \frac{\partial N}{\partial x} = 2 + 6xy .$$

Hence the equation is not exact.

To make it exact we use an integrating factor, $\mu(x,y)$. Let $\mu(x,y)$ be $x^2 y$. Then

$$x^2 y(3y + 4xy^2)dx + x^2 y(2x + 3x^2 y)dy = 0$$

or

$$(3x^2 y^2 + 4x^3 y^3)dx + (2x^3 y + 3x^4 y^2)dy = 0.$$

Now, with $M(x,y) = (3x^2 y^2 + 4x^3 y^3)$ and $N(x,y) = (2x^3 y + 3x^4 y^2)$ we have the relation

$$\frac{\partial M}{\partial y} = 6x^2 y + 12x^3 y^2 = \frac{\partial N}{\partial x}$$

Hence the equation is now exact.

Solving by the method of grouping,

$$(3x^2 y^2 dx + 2x^3 ydy) + (4x^3 y^3 dx + 3x^4 y^2 dy) = 0$$

or

$$d(x^3 y^2) + d(x^4 y^3) = d(c),$$

noticing that the expressions within the brackets above are the exact differentials of $x^3 y^2$ and $x^4 y^3$.

Thus, $x^3 y^2 + x^4 y^3 = c$ is the solution.

HOMOGENEOUS DIFFERENTIAL EQUATIONS

• PROBLEM 1–17

Solve the differential equation: $(y^2 - x^2)dx - 2xy\, dy = 0$.

Solution: We first identify the equation as homogeneous and then make the substitution $y = vx$ to separate the variables. There are several ways of showing that a given function satisfies the defini- tion of homogeneity. We could rewrite the above equation in the form

$$\frac{dy}{dx} = \frac{y^2 - x^2}{2xy} = f(x,y).$$

We would then proceed to show that $f(tx,ty) = t^n f(x,y)$ where n is the degree of the homogeneous function. Alternatively we can repre- sent the equation as a variant of the differential form

$$M(x,y)dx + N(x,y)dy = 0 .$$

It would then suffice to show that both $M(x,y)$ and $N(x,y)$ are homogeneous functions of degree n. Finally, if, in the equation

$\frac{dy}{dx} = \frac{y^2 - x^2}{2xy}$ we can show that $f(x,y)$ may be expressed in the form

$g(y/x)$, the equation is homogeneous. Using this method,

$$\frac{dy}{dx} = \frac{y^2 - x^2}{2xy} = \frac{1}{2}\frac{y}{x} - \frac{1}{2}\left(\frac{1}{y/x}\right) = g(y/x),$$

as required.

Now let $y = vx$, i.e., $v = y/x$. Then,

$$\frac{dy}{dx} = v + x\frac{dv}{dx} = \frac{1}{2}v - \frac{1}{2}\left(\frac{1}{v}\right) \; ; \; x\frac{dv}{dx} = -\frac{(1 + v^2)}{2v}$$

$$\frac{dx}{x} + \frac{2vdv}{1 + v^2} = 0 .$$

This is a separated equation. Direct integration gives

$\ln|x| + \ln|1 + v^2| = \ln|c|$. Since c is an arbitrary real number we let it equal $\ln|c|$ for convenience. By the laws of logarithms and exponents this may be rewritten as $x(1 + v^2) = c$. Substituting $y/x = v$:

$$x\left(1 + \frac{y^2}{x^2}\right) = c \quad \text{or} \quad x^2 + y^2 = cx .$$

• **PROBLEM** 1–18

Solve $y' = \dfrac{2y^4 + x^4}{xy^3}$.

Solution: This is a homogeneous differential equation of the form

$\dfrac{dy}{dx} = f(x,y)$. This means that $f(tx,ty) = f(x,y)$.

The following method has been found to work with this particular type of equation: let $y = xv$; $y' = v + x \, dv/dx$, its derivative. After simplification, it is found that the resulting differential equation is a separable equation whose solution may be found as follows:

$$v + x\frac{dv}{dx} = \frac{2(xv)^4 + x^4}{x(xv)^3}$$

or

$$x\frac{dv}{dx} = \frac{v^4 + 1}{v^3} \quad \text{which may be written as:}$$

$$\frac{1}{x}\,dx - \frac{v^3}{v^4+1}\,dv = 0 , \quad \text{a separable equation.}$$

The solution is $\ln|x| - \dfrac{1}{4}\ln(v^4+1) = c$. We now substitute $v = y/x$:

$$\ln|x| - \frac{1}{4}\ln\left(\frac{y^4}{x^4} + 1\right) = c.$$

Now let $c = -\ln|k|$; $\ln|x| + \ln|k| = \dfrac{1}{4}\ln\left(\dfrac{y^4}{x^4} + 1\right)$ or

$$\ln(kx)^4 = \ln\left(\frac{y^4}{x^4} + 1\right) ; \text{ that is,}$$

$$k^4x^4 = \frac{y^4}{x^4} + 1. \text{ This may be rewritten as}$$

$$y^4 = cx^8 - x^4 \quad (c = k^4) .$$

[i] Determine the degree of the following functions after verifying for homogeneity.

(a) $x + 2y$; (b) $x^2 e^{y/x}$; (c) $\sqrt{x+y}$; (d) $F(y/x)$; (e) $\dfrac{1}{x-y}$.

[ii] Solve the following equation $\dfrac{dy}{dx} = \sqrt{1 - (y/x)^2} + y/x$.

Solution: A function is said to be homogeneous when all the components of the function have the same degree, k. Suppose a function f(x,y) is given, then by replacing x and y with ax and ay respectively we get a function of degree k namely $f(ax,ay) = a^k f(x,y)$.

(a) $x + 2y$ can be written as $ax + 2ay$ or $a(x+2y)$ is homogeneous with degree 1.

(b) $x^2 e^{y/x} \to (ax)^2 e^{ay/ax} = a^2 x^2 e^{y/x} = a^2 (x^2 e^{y/x})$ is of degree 2.

(c) $\sqrt{x+y}$; substituting ax and ay for x and y respectively $\sqrt{ax+ay} = \sqrt{a(x+y)} = \sqrt{a} \cdot \sqrt{x+y} = a^{\frac{1}{2}} \sqrt{x+y}$.

This function is homogeneous of degree $\frac{1}{2}$.

(d) $F(y/x)$; substituting we get $F(ay/ax) = a^0 F(y/x)$. This function is homogeneous of degree 0.

(e) $\dfrac{1}{x-y}$ can be written as $\dfrac{1}{ax-ay} = \dfrac{1}{a(x-y)} = \dfrac{a^{-1}}{(x-y)}$. The function above is homogeneous of degree -1.

(ii) $\dfrac{dy}{dx} = \sqrt{1 - (y/x)^2} + y/x$.

By the methods outlined above it is apparent that this function is homogeneous with degree zero.

Therefore we can make use of the substitution $y = ux$; therefore,

$\dfrac{dy}{dx} = u + x \dfrac{du}{dx}$. So by substitution

$$u + x \frac{du}{dx} = \sqrt{1 - (ux/x)^2} + ux/x$$

or

$$u + x \frac{du}{dx} = \sqrt{1 - u^2} + u .$$

$$x \frac{du}{dx} = \sqrt{1 - u^2} + u - u$$

or

$$x \frac{du}{dx} = \sqrt{1 - u^2} .$$

Taking the reciprocal on both sides and multiplying throughout by du,

$$\frac{dx}{x} = \frac{du}{\sqrt{1 - u^2}} .$$

This separable equation can be directly integrated to give:

$$c_1 + \ln|x| = \sin^{-1} u$$

or

$$cx = e^{\sin^{-1} u} \qquad \text{where} \quad c = e^{c_1}$$

$$c_1 = \ln c.$$

Substituting the value of u from $y = ux$ so $u = y/x$.

$$cx = e^{\sin^{-1}(y/x)}.$$

Solve: $2xy\,dy = \left(y^2 - x^2\right)dx$.

Solution: This equation has homogeneous coefficients of the second degree. Homogeneous equations are of the form

$$f(xt, yt) = t^n f(x, y)$$

for all $t > 0$. Quantity n is called the degree. For example, if we let $f(x,y) = 2xy$ and $g(x,y) = y^2 - x^2$, then

$$f(xt, yt) = 2xtyt = t^2(2xy) = t^2 f(x,y) \quad \text{and}$$

$$g(xt, yt) = (yt)^2 - (xt)^2 = t^2\left(y^2 - x^2\right) = t^2 g(x,y).$$

Separation of variables appears to be possible if we set $y = vx$ in this kind of equation. If we put $y = vx$, then $dy = v\,dx + x\,dv$. The equation becomes

$$2x \cdot vx\left(v\,dx + x\,dv\right) = \left(v^2 x^2 - x^2\right)dx,$$

or

$$\left(v^2 + 1\right)dx + 2xv\,dv = 0.$$

Separating variables, we get

$$\frac{dx}{x} + \frac{2v\,dv}{v^2 + 1} = 0.$$

Integrating, we have

$$\ln x + \ln\left(v^2 + 1\right) = \ln C,$$

$$\ln\left[x\left(v^2 + 1\right)\right] = \ln C,$$

$$\therefore \quad x\left[\frac{y^2}{x^2} + 1\right] = C, \quad \text{and} \quad y^2 + x^2 = Cx.$$

Transposing Cx to the left-hand side of the general solution and completing the square we obtain

$$y^2 + \left(x - \frac{C}{2}\right)^2 = \frac{C^2}{4},$$

which represents a family of circles centered at $(C/2, 0)$ with radii of $C/2$.

Solve $(x+y)dx - (x-y)dy = 0$.

Solution: We know that in a differential equation $M(x,y)dx + N(x,y)dy = 0$, if $M(x,y)$ and $N(x,y)$ are homogeneous functions of the same degree n, then $M(x,y)dx + N(x,y)dy = 0$ is homogeneous. Since $(x+y)$ and $-(x-y)$ are homogeneous with degree one, the equation is homogeneous and we make the substitution $vx = y$ to obtain a separable equation.

First, however, $(x+y)dx - (x-y)dy = 0$ may be rewritten as

$$\frac{dy}{dx} = \frac{x+y}{x-y} \quad . \tag{1}$$

Let $y = vx$; $y' = v + x\frac{dv}{dx} = \frac{x+vx}{x-vx}$, from (1). Hence $x\frac{dy}{dx} = \frac{1+v}{1-v} - v$;

$$x\frac{dv}{dx} = \frac{1+v^2}{1-v}; \frac{(1-v)}{(1+v^2)} dv = \frac{dx}{x} \quad .$$

The equation has been separated and we perform the suggested integration.

$$\int \frac{1-v}{1+v^2} dv = \int \frac{1}{1+v^2} dv - \int \frac{v}{1+v^2} dv = \int \frac{dx}{x} \quad .$$

Thus, $\tan^{-1}v - \frac{1}{2}\ln|1 + v^2| = \ln|x| + c$. Now,

$$v = y/x; \tan^{-1}\frac{y}{x} - \frac{1}{2}\ln|1 + (y/x)^2| = \ln|x| + c,$$

or,

$$\tan^{-1}\frac{y}{x} = \ln\left|\sqrt{x^2+y^2}\right| + c \quad \text{(since} \quad \ln|x| + \ln\left|\left(\frac{x^2+y^2}{x^2}\right)^{\frac{1}{2}}\right|$$

$$= \ln\left|\sqrt{x^2+y^2}\right| \text{)} \quad .$$

Solve $y' = \frac{2y^4 + x^4}{xy^3}$.

Solution: The above equation can be expressed as $\frac{dy}{dx} = \frac{2y^4 + x^4}{xy^3}$.

We notice that this differential equation is homogeneous and therefore the substitution $x = yu$ can be made enabling us to separate the variables. Rewriting the equation differently by taking the inverse on both sides we get

$$\frac{dx}{dy} = \frac{xy^3}{2y^4 + x^4} \quad .$$

Now we make the substitution $x = yu$. Differentiating the above expression gives

$$\frac{dx}{dy} = u + y\frac{du}{dy} \quad .$$

Substituting the expressions for x and dx/dy in the equation

$$\frac{dx}{dy} = \frac{xy^3}{2y^4 + x^4}$$

we get:

$$u + y \frac{du}{dy} = \frac{(yu)y^3}{2y^4 + (yu)^4}$$

$$= \frac{y^4 u}{2y^4 + y^4 u^4} = \frac{u}{2 + u^4} \, .$$

So $u + y \dfrac{du}{dy} = \dfrac{u}{2 + u^4}$ or $y \dfrac{du}{dy} = \dfrac{u}{2 + u^4} - u$

$$y \frac{du}{dy} = \frac{u - u(2 + u^4)}{2 + u^4} = \frac{u - 2u - u^5}{2 + u^4}$$

$$y \frac{du}{dy} = - \frac{(u + u^5)}{2 + u^4} \, .$$

Taking the inverse on both sides $\dfrac{dy}{y} = - \dfrac{2 + u^4}{u + u^5} du$, or

$$\frac{dy}{y} + \frac{2 + u^4}{u + u^5} du = 0 \, .$$

The above equation is separable and by the use of partial fractions we can write

$$\frac{2 + u^4}{u + u^5} = \frac{2 + u^4}{u(1 + u^4)} = \frac{2}{u} - \frac{u^3}{1 + u^4} \, .$$

So $\qquad \dfrac{dy}{y} + \left(\dfrac{2}{u} - \dfrac{u^3}{1 + u^4} \right) du = 0 \, .$

By direct integration we get

$$\ln|y| + 2 \ln|u| - \tfrac{1}{4} \ln|1 + u^4| = c_1$$

$$\frac{y \cdot u^2}{(1 + u^4)^{\tfrac{1}{4}}} = e^{c_1} \quad \text{or} \quad ky^4 u^8 = 1 + u^4 \quad \text{where} \quad k = -\tfrac{1}{4} \ln k.$$

Since $x = yu$ was used for substitution $u = x/y$ will be substituted back to this resulting answer:

$$k \, y^4 \left(\frac{x}{y} \right)^8 = 1 + \left(\frac{x}{y} \right)^4$$

$$k \frac{x^8}{y^4} = 1 + \frac{x^4}{y^4} \, :$$

or

$$kx^8 = y^4 + x^4 \quad \text{or} \quad y^4 = kx^8 - x^4 \, .$$

• PROBLEM 1–23

Solve the initial-value problem

$$(y + \sqrt{x^2 + y^2})dx - x \, dy = 0 \, ,$$

$$y(1) = 0 \, .$$

Solution: We first check to see whether the differential equation is homogeneous. The equation may be rewritten as:

$$\frac{dy}{dx} = \frac{y + \sqrt{x^2 + y^2}}{x} = f(x,y) \, .$$

You should verify that $f(tx, ty) = f(x, y)$, the homogeneity condition for first order differential equations. This implies that there exists a function $g(y/x)$ which $f(x, y)$ is equivalent to.

$$\frac{dy}{dx} = \frac{y}{x} + \frac{\sqrt{x^2+y^2}}{x} = \frac{y}{x} + \sqrt{1 + (y/x)^2}$$

if we consider $\sqrt{x^2}$ to be x, the positive root, which the initial value one surely indicates. We thus have $y' = g(y/x)$. Let $v = y/x$; then $y = xv$ and $y' = v + x\, dv/dx$. That is,

$$v + x\frac{dv}{dx} = v + \sqrt{1 + v^2}$$

or

$$\frac{dx}{x} - \frac{dv}{\sqrt{1+v^2}} = 0 .$$

This is a separable equation and integration gives

$$\ln|v + \sqrt{v^2+1}| = \ln|x| + \ln|c| = \ln|cx| ,$$

or

$$v + \sqrt{v^2+1} = cx .$$

Replacing v by y/x: $\frac{y}{x} + \sqrt{y^2/x^2 + 1} = cx$ or

$$y + \sqrt{x^2+y^2} = cx^2$$

since $y(1) = 0$,

$$0 + \sqrt{1} = c(1),$$
$$c = 1,$$
$$y + \sqrt{x^2+y^2} = x^2 .$$

• PROBLEM 1-24

Solve

$$y' = \frac{x^2 + y^2}{xy} ; \quad y(1) = -2.$$

Solution: We first convert the given equation into a separable equation. After obtaining the solution we substitute the values $x = 1$, $y(1) = -2$ to get the solution to the initial value problem.

   Since the equation is homogeneous, i.e., $f(tx, ty) = f(x, y)$, we can substitute $y = xv$, which implies $y' = v + x\frac{dv}{dx}$. In the given problem, $y' = \frac{x^2 + y^2}{xy}$, $v + x\frac{dv}{dx} = \frac{x^2 + (xv)^2}{x(xv)}$; simplifying,

$x\frac{dv}{dx} = \frac{1}{v}$ or $\frac{1}{x} dx - v dv = 0$. This is a separable equation whose

solution is given by: $\int \frac{1}{x} dx - \int v\, dv = 0$. We obtain $v^2 = \ln x^2 - 2c$.

At this point we substitute $v = \frac{y}{x}$ to get the solution

$$y^2 = x^2 \ln x^2 + kx^2 \quad (k = -2c) .$$

Finally, "plug in" $x = 1$; $y(1) = -2$,

$$(-2)^2 = (1)^2 \ln(1)^2 + k(1)^2$$

19

which is equivalent to $k = 4$ since $\ln(1)^2 = 0$. Thus,

$$y^2 = x^2 \ln x^2 + 4x^2 \quad \text{or} \quad y = - \sqrt{x^2 \ln x^2 + 4x^2} \ .$$

We take the negative square root since $y(1) = -2$ requires it.

INTEGRATING FACTORS

● PROBLEM 1-25

Solve $(y - xy^2)dx + x \, dy = 0$.

Solution: We can factor out y from the first term to get $y(1 - xy)dx + x \, dy = 0$. This is now in the form $M(x, y) \, dx + N(x, y) \, dy = 0$ where $M = y \, f(xy)$ and $N = x \, g(xy)$. Here $f(xy) = (1 - xy)$ while $g(xy) = 1$. Then an integrating factor is given by:

$$M(x, y) = \frac{1}{x \, M - y \, N} \ . \qquad \text{Here,}$$

$$M(x, y) = \frac{1}{x \, [y(1 - xy)] - yx} = - \frac{1}{(xy)^2}$$

Multiplying the given equation by $- \frac{1}{(xy)^2}$

$$\frac{xy - 1}{x^2 y} \, dx - \frac{1}{xy^2} \, dy = 0.$$

Since $\frac{\partial}{\partial y} \left(\frac{1}{x} - \frac{1}{x^2 y} \right) = \frac{1}{x^2 y^2} = \frac{\partial}{\partial x} \left(- \frac{1}{y^2} \cdot \frac{1}{x} \right)$

the equation is exact.

To solve it, integrate $\frac{1}{x} - \frac{1}{x^2 y}$ with respect to x holding y constant:

$$\int \frac{1}{x} \, dx - \frac{1}{y} \int \frac{1}{x^2} \, dx = \ln |x| + \frac{1}{xy} + \phi(y).$$

Partially differentiating with respect to y,

$$- \frac{1}{xy^2} + \phi'(y) = N(x, y) = - \frac{1}{y^2 x} \ .$$

Hence $\phi'(y) = 0$ or $\phi(y) = c$.

Then $\frac{1}{xy} = - \ln |cx|$ or $y = - \frac{1}{x \ln |cx|}$

which is the solution set (a particular solution is obtained by substituting a value for c).

It may be asked how the integrating factor was derived. The general approach towards obtaining integrating factors is outlined below. The integrating factor is obtained by solving a partial differential equation.

Suppose the equation

$$M(x, y) \, dx + N(x, y) \, dy = 0 \quad \text{is not exact.}$$

Let $\mu(x, y)$ be an integrating factor. Then

$$\mu(x, y) \, M(x, y) \, dx + \mu(x, y) \, N(x, y) \, dy = 0$$

is exact which implies

$$\frac{\partial}{\partial y} [\mu(x, y) \, M(x, y)] = \frac{\partial}{\partial x} [\mu(x, y) \, N(x, y)]$$

$$\frac{\partial \mu}{\partial y}(x, y) \, M(x, y) + \frac{\partial M}{\partial y}(x, y) \mu(x, y)$$

$$= \frac{\partial \mu}{\partial x}(x, y) \, N(x, y) + \frac{\partial N}{\partial x}(x, y) \mu(x, y).$$

or, $\quad N(x, y) \, \dfrac{\partial \mu}{\partial x}(x, y) - M(x, y) \, \dfrac{\partial \mu}{\partial y}(x, y)$

$$= \left[\frac{\partial M}{\partial y}(x, y) - \frac{\partial N}{\partial x}(x, y) \right] \mu(x, y).$$

$\mu(x, y)$ is the unknown integrating factor. Thus

$$N(x, y) \, \frac{\partial \mu}{\partial x} - M(x, y) \, \frac{\partial \mu}{\partial y}$$

$$= \left[\frac{\partial M}{\partial y}(x, y) - \frac{\partial N}{\partial x}(x, y) \right] \mu$$

This is a partial differential equation in μ whose solution gives μ.

• PROBLEM 1–26

Identify and solve the differential equation

$$\frac{dy}{d\theta} + \tan \theta \, y = \cos \theta$$

by choosing an appropriate integrating factor.

Solution: The equation is in the form

$$\frac{dy}{dx} + p(x) \, y = q(x) \quad \text{where } p(x) = \tan \theta \quad \text{and}$$

$q(x) = \cos \theta$. Therefore it is linear.

For a linear differential equation with variable coefficients, the integrating factor is given by $e^{\int p(x)\,dx}$.

Thus the integrating factor for this equation is $e^{\int \tan \theta\,d\theta}$

Now $\int \tan \theta\,d\theta = -\ln |\cos \theta| + c$. Ignoring c for the moment,

$$e^{-\ln |\cos \theta|} = e^{\ln |\sec \theta|} = \sec \theta.$$

Multiplying the equation by $\sec \theta$ to make it exact,

$$\sec \theta \frac{dy}{d\theta} + \sec \theta (\tan \theta\, y) = \cos \theta \sec \theta.$$

Since $\frac{d}{d\theta} \sec \theta = \sec \theta \tan \theta$ and $\cos \theta \sec \theta = 1$,

$(y \sec \theta)' = 1$.

Integrating, $y \sec \theta = \theta + c$

or $$y = (\cos \theta)(\theta + c).$$

● **PROBLEM** 1–27

Find a general solution of

   $y' + 2xy = 0$,

using the integrating factor method.

Solution: The equation is linear. Therefore an integrating factor is given by

$$\exp\left[\int 2x\,dx\right]$$, where 2x is the term

representing p(x) in the general linear equation $\frac{dy}{dx} + p(x)y = q(x)$. The integrating factor is therefore e^{x^2}. Multiplying both sides of the differential equation by e^{x^2}, we obtain

$$e^{x^2} y' + 2xe^{x^2} y = 0.$$

As usual, l·h·s· is the derivative of the product

$ye^{\int p(x)\,dx}$ or, in this case ye^{x^2}.

Thus $(e^{x^2}y)' = 0$. Integrating,

$$e^{x^2}y = c \qquad \text{or} \quad y = ce^{-x^2}.$$

If we had chosen to write the equation in the differential form $(2xy)dx + dy = 0$, we could have used the following method of solution.

An integrating factor is $\frac{1}{v}$. Then the equation is

$2xdx + \frac{dy}{y} = 0$. This is a separable equation:

$$\frac{dy}{y} = -2x \, dx$$

$$\int \frac{dy}{y} = -2 \int x \, dx \, .$$

$$\ln |y| = -x^2 + c; \quad y = e^{-x^2 + c} = ce^{-x^2}$$

which is the same solution as before.

• PROBLEM 1–28

Solve the equation

$$(1 + 3x \sin y)dx - x^2 \cos y \, dy = 0.$$

<u>Solution:</u>   Since the equation is non-linear and inexact, we must find an integrating factor to make it exact.

First, however, note that $\cos y \, dy$ is $\frac{d}{dy} (\sin y)$.
Hence the substitution $\omega = \sin y$ suggests itself.   Then

$$(1 + 3x\omega)dx - x^2 \, d\omega = 0$$

or $\frac{d\omega}{dx} - \frac{3\omega}{x} = \frac{1}{x^2}$.   This is a linear equation whose

integrating factor is given by $\exp \int -\frac{3}{x} \, dx$ or $e^{\ln |x^{-3}|}$

which is $\frac{1}{x^3}$.

Then $x^{-3} \frac{d\omega}{dx} - \frac{3}{x^4} \omega = \frac{1}{x^5}$

Here $x^{-3} \frac{d\omega}{dx} - \frac{3}{x^4} \omega$ is $\frac{d}{dx} (x^{-3} \omega)$

Thus $\frac{d}{dx} (x^{-3} \omega) = \frac{1}{x^5}$

23

Integrating, $\quad x^{-3}\omega = -\frac{1}{4}x^{-4} + \frac{1}{4}c$

or $\qquad\qquad 4x\omega = cx^4 - 1.$

Now insert the original term, $\sin y = \omega$;

$4x \sin y = cx^4 - 1 \quad$ which is the required solution.

• PROBLEM 1–29

Solve the differential equation:

$$\frac{dx}{dt} - (\tan t)x = \sin t,$$

subject to the initial condition: $x(t_0) = x_0$.

<u>Solution:</u>   This is a differential equation with t as the independent variable and x as the dependent variable. It is of the form

$$\frac{dx}{dt} + p(t)x = q(t), \quad \text{i.e., it is linear. Since}$$

the equation is not exact, we must find an integrating factor that will make it exact.

An integrating factor for a linear d·e· is given by $e^{\int p(t)\,dt}$. Since $p(t) = -\tan t$, the integrating factor is $e^{-\int \tan t\,dt}$.

$$e^{-\int \tan t\,dt} = \exp\left(-\int \tan t\,dt\right) = \exp(\ln \cos t).$$

We omit the arbitrary constant c, since it will appear in the solution later.

Exp $(\ln \cos t) = \cos t$. Multiplying the given equation by $\cos t$,

$$(\cos t)x' - (\cos t)(\tan t)\,x = (\cos t)(\sin t)$$

$$(\cos t)x' - (\sin t)x = (\cos t)(\sin t).$$

The expression on the l·h·s· is nothing but $\frac{d}{dt}(x \cos t)$. Therefore $\frac{d}{dt}(x \cos t) = (\cos t)(\sin t)$. Integrating with respect to t,

$$[x \cos t]_{t_0}^{t} = \int_{t_0}^{t} \sin u \cos u \, du$$

$$\int \sin u \cos u \, du = \frac{1}{2} \sin^2 u + c.$$

Therefore $x \cos t - x_0 \cos t_0 = \frac{1}{2} (\sin^2 t - \sin^2 t_0).$

• PROBLEM 1–30

Solve $y' = 2xy - x.$

<u>Solution:</u>   Rewriting the equation as $y' - 2xy = -x$ we see that it is linear. Recalling the method of solving such equations, an integrating factor that will make it exact is

$$e^{\int - 2x \, dx} = e^{-x^2}.$$

We can, however, derive the integrating factor by another method. Rewriting the equation as $(-2xy + x)dx + dy = 0$, we have $M(x, y) = -2xy + x$ and $N(x, y) = 1$. Then

$$\frac{1}{N} \left(\frac{\partial M}{\partial y} - \frac{\partial N}{\partial x} \right) = -2x \quad \text{which is a function of } x \text{ alone.}$$

Therefore, by exponentiation we may obtain the integrating factor, i.e.,

$$e^{\int - 2x \, dx} = e^{-x^2} \quad \text{is an integrating factor.}$$

Multiplying by e^{-x^2},

$$(-2xye^{-x^2} + xe^{-x^2}) \, dx + e^{-x^2} dy = 0 \quad \text{which is}$$

an exact equation.

Integrating $M(x, y)$ with respect to x holding y constant:

$$\int - 2y \, xe^{-x^2} + xe^{-x^2} \, dx = ye^{-x^2} - \frac{1}{2} e^{-x^2} + \phi(y).$$

If this is the solution $F(x, y)$ we must have

$$\frac{\partial}{\partial y} \left[ye^{-x^2} - \frac{1}{2} e^{-x^2} + \phi(y) \right] = e^{-x^2} \quad \text{which}$$

25

implies $\phi'(y) = 0$, $\phi(y) = c$.

Then $ye^{-x^2} = \frac{1}{2} e^{-x^2} + c$

or $\qquad\qquad y = ce^{x^2} + \frac{1}{2}$.

Solve the initial value problem (IVP)

$$\frac{dy}{dx} + 2y = 1 \quad \text{with } y = 0 \text{ at } x = 0.$$

Solution: We recognize the equation as linear, i.e. of the form $\frac{dy}{dx} + p(x)y = q(x)$. In the given equation, $p(x) = 2$, $q(x) = 1$. The solution of the equation requires the use of an integrating factor. An integrating factor is defined to be a function, $\mu(x, y)$, that when multiplied through the given equation makes it exact.

An integrating factor for this problem is e^{2x}.

Then $y'e^{2x} + 2ye^{2x} = e^{2x}$ is exact, as may be verified by writing it in the differential form $M(x,y)dx + N(x,y)dy = 0$. But

$$y'e^{2x} + 2ye^{2x} = (ye^{2x})'.$$

Therefore,

$$(ye^{2x})' = e^{2x}.$$

Integrating,

$$ye^{2x} = \frac{e^{2x}}{2} + c, \qquad \text{or}$$

$$y = ce^{-2x} + \frac{1}{2}.$$

Now, applying the initial condition $y(0) = 0$,

$$0 = ce^{-2(0)} + \frac{1}{2} \quad \text{or} \quad c = -\frac{1}{2}.$$

Thus the solution to the IVP is:

$$y = -\frac{1}{2} e^{-2x} + \frac{1}{2} \quad \text{or} \quad y = \frac{1}{2}(1 - e^{-2x})$$

Solve the linear differential equation

$$\frac{dy}{dx} - 3y = 6$$

by finding an integrating factor.

<u>Solution:</u>   An integrating factor is a function $\mu(x, y)$ that makes the given equation exact. Rewriting in the differential form $(-3y - 6)dx + dy = 0$. This equation is not exact. Multiplying through by $\mu(x)$ (the integrating factor for a linear differential equation depends on x alone).

$$\Big(-3\mu(x)y - 6\mu(x)\Big)dx + \mu(x) \, dy = 0.$$

If this equation is to be exact,

$$\frac{\partial}{\partial y}(-3\mu(x)y) = \frac{\partial}{\partial x}\mu(x).$$

That is, $-3\mu(x) = \frac{d\mu}{dx}$ ($\mu = f(x)$ only).

27

$-3\,dx = \dfrac{d\mu}{\mu}$;   this is a separable equation.

Integrating, $\ln|\mu| = -3x$,   or $\mu = e^{-3x}$.

Multiplying through the equation by e^{-3x},

$$e^{-3x}\dfrac{dy}{dx} - 3ye^{-3x} = 6e^{-3x} .$$

Now $e^{-3x}\dfrac{dy}{dx} - 3ye^{-3x} = \dfrac{d}{dx}(ye^{-3x})$.

Thus $\dfrac{d}{dx}(ye^{-3x}) = 6e^{-3x}$.

Integrating once again

$$ye^{-3x} = -2e^{-3x} + c$$

or $y = ce^{3x} - 2.$

SECOND ORDER HOMOGENEOUS DIFFERENTIAL EQUATIONS WITH CONSTANT COEFFICIENTS

● **PROBLEM** 1–33

Solve $y'' - 5y = 0$.

Solution: This 2nd order linear homogeneous differential equation is of the form $y'' + ay' + by = 0$ where $a = 0$ and $b = -5$. It has exponential solutions i.e. solutions of the form $y = e^{mx}$ for m,a constant.

Thus, assume that $y = e^{mx}$ is a solution. Then $y' = me^{mx}$ and $y'' = m^2 e^{mx}$. Then substituting into the given d.e.

$$m^2 e^{mx} - 5e^{mx} = 0,$$

$$e^{mx}(m^2 - 5) = 0.$$

Since $e^{mx} \neq 0$, this implies $(m^2 - 5) = 0$. Solving for m, $m = \pm\sqrt{5}$.

Therefore, if $y = e^{mx}$ is the solution form then $y = e^{\sqrt{5}x}$ and $y = e^{-\sqrt{5}x}$ are the solutions. The general solution to the problem by the superposition property is:

$$y = c_1 e^{\sqrt{5}x} + c_2 e^{-\sqrt{5}x}$$

What is the superposition property? A proof of its application for the given problem is now presented.

We have that $y_1 = e^{\sqrt{5}x}$ and $y_2 = e^{-\sqrt{5}x}$ are solutions of the d.e. $y'' - 5y = 0$. We must show that the linear combination

$$y = c_1 e^{\sqrt{5}x} + c_2 e^{-\sqrt{5}x} \qquad \text{is a solution for all}$$

values of the constants c_1 and c_2.

Substituting into the original equation the values:

$$y' = \sqrt{5}c_1 e^{\sqrt{5}x} - \sqrt{5}c_2 e^{-\sqrt{5}x}$$

$$y'' = 5c_1 e^{5x} + 5c_2 e^{-5x},$$

$$y'' - 5y = (5c_1 e^{5x} + 5c_2 e^{-5x})$$

$$- 5(c_1 e^{5x} + c_2 e^{-5x})$$

Rearranging so that c_1 and c_2 are explicit multiplicative factors:

$$y'' - 5y = c_1(5e^{5x} - 5e^{5x}) + c_2(5e^{-5x} - 5e^{-5x})$$

Now, each expression within brackets is a solution, i.e. equals zero. Thus the linear combination is also equal to zero. Consequently, it is also a solution.

The theorem on the superposition property asserts that:

If y_1, y_2, ... , y_m are m solutions of a linear homogeneous d·e·, then the linear combination of these solutions,

$$y = c_1 y_1 + c_2 y_2 + \ldots + c_m y_m$$

is also a solution for all choices of the constants c_1, c_2, ..., c_m.

• **PROBLEM** 1–34

Solve $y'' - y' - 2y = 0$.

Solution: This is a 2nd order linear differential equation. It is homogeneous because, if we treat $y^{(i)}$, (the ith derivative) as x_i, we may rewrite $y'' - y' - 2y = 0$ as

$$\sum_{n=0}^{2} c_n x_n = -2x_0 - x_1 + x_2.$$

Now, this is a linear homogeneous equation. In general the 2nd order equation $y'' + ay' + b = 0$ is homogeneous with constant coefficients, a, b.

There is a standard method of solution for these equations. Recall that the first-order linear homogeneous differential equation

$$y' + ky = 0$$

has the solution

$$y = ce^{-kx}.$$

Proceeding by analogy, let us assume that the given equation

$$y'' - y' - 2y = 0$$

has a solution in the form of

$$y = e^{mx}.$$

We must find m, where $m \varepsilon R$. Then $y' = me^{mx}$ and $y'' = m^2 e^{mx}$. Substituting these values into the d.e. $y'' - y' - 2y = 0$

$$m^2 e^{mx} - me^{mx} - 2e^{mx} = e^{mx}(m^2 - m - 2) = 0.$$

Thus, if $y = e^{mx}$ is a solution we must have $(m^2 - m - 2) = 0$ (since $e^{mx} \neq 0$). This quadratic equation may be factored into $(m - 2)(m + 1) = 0$. Thus, $m = 2$ and $m = -1$ are the roots.

We find, therefore, that there are two solutions of the form $y = e^{mx}$, i.e. $y = e^{2x}$ and $y - e^{-x}$. Since the solution to the 2nd order linear homogeneous equation must contain two arbitrary constants, a general solution is given by:

$$y = c_1 e^{2x} + c_2 e^{-x}.$$

• **PROBLEM** 1–35

Solve $\qquad \dfrac{d^2 x}{dt^2} = 0.$

Solution: Since the equation is homogeneous with constant coefficients, we can assume a solution with the form $y = e^{mx}$. Then $y'' = m^2 e^{mx}$ and substituting into $y'' = 0$, we obtain

$$m^2 e^{mx} = 0$$

Since $e^{mx} \neq 0$, this implies that $m^2 = 0$ or $m_1 = m_2 = 0$. Since their roots are equal, it implies that $y = xe^{mx}$ is another solution which satisfies $y'' = 0$. Thus, the general solution is

$$y = c_1 x e^{0x} + c_2 e^{0x},$$

$$y = c_1 x + c_2.$$

• **PROBLEM** 1–36

Solve the equation:

$$\frac{d^2 y}{dx^2} + 2\frac{dy}{dx} + y = 0.$$

Solution: Assume $y = ce^{mx}$ is the solution for the differential equation. Then, since $\frac{dy}{dx} = mce^{mx}$ and $\frac{d^2 y}{dx^2} = m^2 ce^{mx}$, substitution for the given equation gives

$$\frac{d^2 y}{dx^2} + 2\frac{dy}{dx} + y = m^2 ce^{mx} + 2mce^{mx} + ce^{mx} = 0$$

or $\qquad ce^{mx}\left(m^2 + 2m + 1\right) = 0.$

Assuming e^{mx} to be different from zero, and considering $c = 0$ leads to a trivial solution. The only non-trivial solution occurs when

$$m^2 + 2m + 1 = 0,$$

or

$$(m+1)^2 = 0$$

with a repeated root: $m = -1$.

Since a multiple root occurs here, in order to obtain a complete solution, we seek another linearly independent expression that satisfies the obtained auxiliary equation besides ce^{mx}. To get this, we take $y = c_2 x e^{mx}$ and hence:

$$x c_2 e^{mx} \left(m^2 + 2m + 1 \right) + c_1 e^{mx} \left(m^2 + 2m + 1 \right) = 0.$$

Thus
$$y = \left(c_1 + c_2 x \right) e^{mx}$$

is the desired general solution.

Using $m = -1$ as a root, the general solution of the differential equation is

$$y = \left(c_1 + c_2 x \right) e^{-x}.$$

● PROBLEM 1–37

Solve the differential equation: $y'' - 2y' - y = 0$.

Solution: This kind of differential equation is called a second-order, linear, homogeneous equation with constant coefficients.

Setting $y = e^{mx}$ to see if it is a solution of the given equation, and since $y' = m e^{mx}$ and $y'' = m^2 e^{mx}$, substitution gives:

(a) $y'' - 2y' - y = e^{mx} \left(m^2 - 2m - 1 \right) = 0.$

e^{mx} is different from zero in a finite range of x unless m is trivially chosen. Thus, the only way for (a) to hold is for

$$m^2 - 2m - 1 = 0.$$

The roots are m = 1 $\pm\sqrt{2}$.

Thus, either $y = c_1 e^{(1+\sqrt{2})x}$ or $y = c_2 e^{(1-\sqrt{2})x}$
solves the equation. The general solution is the
sum of the two terms.

$$y = c_1 e^{(1+\sqrt{2})x} + c_2 e^{(1-\sqrt{2})x}$$

• PROBLEM 1–38

What are the real solutions of

$y" + 2y' + 5y = 0$?   (a)

Solution: This is a homogeneous linear equation with
constant coefficients. Hence we assume a solution of the
form $y = e^{\lambda x}$. Then $y' = \lambda e^{\lambda x}$, $y" = \lambda^2 e^{\lambda x}$. Substituting
into (a)

$e^{\lambda x} (\lambda^2 + 2\lambda + 5) = 0$ (b)

Thus the characteristic polynomial is

$(\lambda^2 + 2\lambda + 5) = 0$.

Using the quadratic formula,

$\lambda = \dfrac{- 2 \pm \sqrt{4 - 4(5)}}{2} = - 1 \pm 2i$.

Thus the general solution may be written as

$y = c_1 e^{- (1 + 2i)x} + c_2 e^{- (1 - 2i)x}$ (c)

To obtain real solutions we make use of Euler's
formula:

$e^{i\beta x} = \cos \beta x + i \sin \beta x$,

where βx is the imaginary part of the complex number

$z = t + i\beta x$. Then from (c)

$y = e^{(- 1 + 2i)x}$ which is a solution

may be written as

$y = e^{-x} \cdot e^{2ix} = e^{-x} (\cos 2x + i \sin 2x) = \omega_1$.

Similarly, the solution $y = e^{(- 1 - 2i)x}$ equals

$$e^{-x} \cdot e^{-2ix} = e^{-x} \ (\cos \ (- \ 2x) + i \ \sin \ (- \ 2x))$$

$$= e^{-x} \ (\cos \ 2x - i \ \sin \ 2x) = \omega_2 ,$$

since cos x is an even function i.e., cos $(- \ x) = \cos$ x. By contrast sin x is an odd function, i.e., sin $(- \ x) = - \ \sin$ x.

Now form the linear combinations:

$$y_1 (x) = \frac{1}{2} \ \omega_1 + \frac{1}{2} \ \omega_2 \qquad\qquad\qquad\qquad (d)$$

$$y_2 (x) = - \ \frac{i}{2} \ \omega_1 + \frac{i}{2} \ \omega_2 \ . \qquad\qquad\qquad (e)$$

From (d), $y_1 (x) = e^{-x} \cos 2x$. $\qquad\qquad\qquad\qquad$ (f)

From (e), $y_2 (x) = e^{-x} \sin 2x$. $\qquad\qquad\qquad\qquad$ (g)

The general solution is found by forming the linear combination of f and g:

$$y = c_1 e^{-x} \cos 2x + c_2 e^{-x} \sin 2x.$$

Note that this is a real solution.

● PROBLEM 1–39

Find the characteristic equation of the differential equation

$\qquad$ y" - 3y' + 2y = 0 $\qquad\qquad$ and use it to solve

the initial value problem

$\qquad$ y" - 3y' + 2y = 0

y(0) = 1, $\qquad$ y'(0) = 0.

Solution: We must first define what a characteristic equation is. It is known that the 2nd order linear, homogeneous constant coefficients differential equation has exponential solutions, i.e., solutions of the form $y = e^{px}$; p = constant. Thus, given the general equation

$\qquad$ y" + by' + cy = 0, $\quad$ if $y = e^{px}$ is a solution,

$\qquad$ $y' = pe^{px}$, $\quad$ $y" = p^2 e^{px}$ $\qquad$ and substitution of

these values gives

34

$$p^2 e^{px} + bpe^{px} + ce^{px} = 0$$

$e^{px} (p^2 + bp + c) = 0.$ If this equation
is to hold, $(p^2 + bp + c)$ must equal zero since $e^{px} \neq 0.$

We can find the roots of

$$(p^2 + bp + c) = 0$$

by the quadratic formula

$$p = \frac{-b \pm \sqrt{b^2 - 4c}}{2} \quad .$$

Here there are three cases to consider.

(1) $b^2 - 4ac > 0$

The two roots are real and distinct.

(2) $b^2 - 4ac = 0$

There is one real, repeated root.

(3) $b^2 - 4ac < 0$

Here there are two complex roots.

The solution to the differential equation differs
according to the case.

Returning to the given equation,

if $y = e^{mx}$ is a solution,

then $m^2 e^{mx} - 3me^{mx} + 2e^{mx} = 0$

or $e^{mx} (m^2 - 3m + 2) = 0.$

Extracting the roots of $(m^2 - 3m + 2)$,

$(m - 1)(m - 2) = 0$ or,

$m_1 = +1$ and $m_2 = +2.$

Thus the general solution is

$$y = c_1 e^x + c_2 e^{2x} \quad .$$

To find c_1 and c_2 we use the initial conditions
$y(0) = 1, \ y'(0) = 0.$

Then, since $\frac{dy}{dx} = c_1 e^x + 2c_2 e^{2x}$ we have

$$1 = c_1 + c_2$$

$$0 = c_1 + 2c_2$$

from which $c_2 = -1$ and $c_1 = 2$. Using these values for c_1 and c_2 in the solution

$$y = c_1 e^x + c_2 e^{2x}$$

we obtain the solution of the IVP

$$y = 2e^x - e^{2x}$$

or $\quad y = e^x (2 - e^x)$.

• **PROBLEM** 1–40

Solve the equation: $\dfrac{d^2 y}{dx^2} + 16y = 0$, subject to the initial conditions: $y = 0$ and $\dfrac{dy}{dx} = 5$ when $x = 0$.

Solution: This is a second-order, homogeneous differential equation of the form,

$$\frac{d^2 y}{dx^2} + a_1 \frac{dy}{dx} + a_2 y = 0,$$

with $a_1 = 0$ and $a_2 = 16$.

Assuming $y = ke^{mx}$ is the general solution,

and, since $\dfrac{d^2 y}{dx^2} = km^2 e^{mx}$, we obtain:

$$\frac{d^2 y}{dx^2} + 16y = ke^{mx}\left(m^2 + 16 \right) = 0.$$

Taking $k = 0$ leads to trivial solution. A non-trivial solution is found by the auxiliary equation:

$$m^2 + 16 = 0$$

with roots $m = \pm 4i$.

The solution $y_1 = k_1e^{iax}$ or $y_2 = k_2e^{-4x}$ is adequate, but as we are looking for the general solution, the sum of all possible solutions is of importance. Therefore

$$y = y_1 + y_2$$

or
$$y = k_1e^{i4x} + k_2e^{-i4x}$$

is the desired general solution.

Recalling Euler's formula for complex components

$$e^{\pm ix} = \cos x \pm i \sin x,$$

and expressing our general solution in terms of the circular functions, after some algebraic manipulation, we obtain

$$y = c_1 \cos 4x + c_2 \sin 4x$$

where $c_1 = k_1 + k_2$ and $c_2 = i(k_1-k_2)$.

Substituting $y = 0$ and $x = 0$ gives $c_1 = 0$.

$$y = c_2 \sin 4x.$$

Differentiating this, we have

$$\frac{dy}{dx} = 4c_2 \cos 4x;$$

substituting $\frac{dy}{dx} = 5$ and $x = 0$, we get $5 = 4c_2$ or

$c_2 = \frac{5}{4}$. The required particular solution is then

$y = \frac{5}{4} \sin 4x.$

It is frequently convenient to write the solution
$$y = c_1 \cos bx + ic_2 \sin bx,$$

for the case of complex roots, in another form. By trigonometry,

$$c_1 \cos bx + ic_2 \sin bx = c \sin(bx+\alpha) = c \cos(bx+\beta),$$

where $c = \sqrt{c_1^2 + c_2^2}$, $\tan \alpha = c_1/c_2$, $\tan \beta = -c_2/c_1$. The general solution then becomes:

$$y = ce^{ax} \sin(bx+\alpha) = ce^{ax} \cos(bx+\beta).$$

In these forms, c and α, or c and β are the real arbitrary constants.

• **PROBLEM** 1–41

Find a general solution of

$$\frac{d^3y}{dx^3} - 4\frac{d^2y}{dx^2} + \frac{dy}{dx} + 6y = 0 . \tag{a}$$

<u>Solution</u>: This is a third-order linear equation with constant coefficients. By analogy with first and second order linear differential equations with constant co-efficients we suspect that the solution will be an ex-

ponential solution, i.e. $y = e^{mx}$; m = constant.

Thus, let $y = e^{mx}$ be a solution.

Then, $y' = me^{mx}$; $y'' = m^2e^{mx}$; $y''' = m^3e^{mx}$.

Substituting into (a):

$$m^3e^{mx} - 4m^2e^{mx} + me^{mx} + 6e^{mx} = 0$$

$$e^{mx}(m^3 - 4m^2 + m + 6) = 0.$$

If this equation is to be true, we must have

$$(m^3 - 4m^2 + m + 6) = 0 . \tag{b}$$

Notice that m = 2 is a root of this cubic equation because

$$m^3 - 4m^2 + m + 6 = 8 - 16 + 2 + 6 = 0.$$

Dividing the characteristic polynomial by (m - 2), the quotient $m^2 - 2m - 3$ is obtained.

$$(m^3 - 4m^2 + m + 6) = (m - 2)(m^2 - 2m - 3) = 0 .$$

From this, we have $(m^2 - 2m - 3) = 0$ which can be factored to give $(m - 3)(m + 1) = 0$.

Thus, (b) has the roots

$$m_1 = 2; \quad m_2 = 3; \quad m_3 = -1.$$

We originally supposed the existence of a solution of the form $y = e^{mx}$. Since a third order equation with constant coefficients has three arbitrary constants in its solution,

$$y = (c_1 e^{2x} + c_2 e^{3x} + c_3 e^{-x})$$

is a general solution.

• **PROBLEM** 1–42

Solve

$$y^{(4)} + 8y''' + 24y'' + 32y' + 16y = 0. \tag{a}$$

Solution: Since (a) is a fourth order homogeneous equation with constant coefficients, we consider the characteristic equation

$$p^4 + 8p^3 + 24p^2 + 32p + 16 = 0. \tag{b}$$

Using the Binomial Formula, we can factor (b) into

$$(p + 2)^4 = 0 . \tag{c}$$

Equation (c) has a root $p = -2$ of multiplicity four. This gives us four solutions of (a):

$$e^{-2x}, \ xe^{-2x}, \ x^2 e^{-2x}, \ x^3 e^{-2x}.$$

We write the general solution of (a) as

$$y(x) = C_1 e^{-2x} + C_2 xe^{-2x} + C_3 x^2 e^{-2x} + C_4 x^3 e^{-2x},$$

where C_1, C_2, C_3, and C_4 are arbitrary constants.

CHAPTER 2

LINEAR DIFFERENTIAL EQUATIONS

> **Basic Attacks and Strategies for Solving Problems in this Chapter. See pages 41 to 90 for step-by-step solutions to problems.**

Some equations are simple enough that the integrating factor can be found by inspection. The ability to do this depends largely upon recognition of certain common exact differentials and upon experience.

It may be possible by some change of variables to transform the equation into a type which we know how to solve. A natural source of suggestions for useful transformations is the differential equation itself. If a particular function of one or both variables stands out in the equation, then it is worthwhile to examine the equation after that function has been introduced as a new variable.

The general linear differential equation of order n is an equation which can be written as :

$$\sum_{i=0}^{n} b_i(x)\frac{d^i y}{dx^i} = b_0(x)\frac{d^n y}{dx^n} + b_1(x)\frac{d^{n-1}y}{dx^{n-1}} + \ldots b_{n-1}(x)\frac{dy}{dx} + b_n(x)y = R(x)$$

The functions $R(x)$ and $b_i(x)$; $i = 0, 1\ldots, n$, are to be independent of the variable y. If $R(x)$ is identically zero, the equation is said to be linear and homogeneous. if $R(x)$ is not identically zero, the equation is called linear and nonhomogeneous.

Each particular solution of a linear differential equation is a special case of the general solution. The general solution of a nonhomogeneous equation is the sum of the complementary function and any particular solution. There are various methods for obtaining a particular solution. In preparation for the method of undetermined coefficients, it is wise to obtain proficiency in writing a homogeneous differential equation of which a given function of proper form is a solution. It is frequently easy to obtain a particular solution of a nonhomogeneous equation by inspection.

The method of undetermined coefficients applies only to a certain class of differential equations: those for which $R(x)$ itself was a solution of a homogeneous linear equation with constant coefficients. Linear equations with variable coefficients can also be solved by using reduction of order and by variation of parameters.

Step-by-Step Solutions to
Problems in this Chapter,
"Linear Differential Equations"

BY INTEGRATING FACTORS AND
BERNOULLI'S EQUATIONS

• PROBLEM 2–1

Solve $y' + (4/x)y = x^4$.

<u>Solution</u>: This is a first order linear differential equation. Since it is in the form $y' + p(x)y = q(x)$ we can choose

$$e^{\int p(x)dx} = e^{\int 4/x \ dx} = e^{4 \ln x} = x^4$$

as the integrating factor that will make the equation a separable one.

First, consider
$$\left[\frac{4}{x} y - x^4\right]dx + dy = 0 ,$$

the differential form of the given equation. This equation is not exact since

$$\frac{\partial}{\partial y}\left[\frac{4}{x} y - x^4\right] \neq \frac{\partial}{\partial x}[1] .$$

It would be exact only if $4/x$ were zero. Then the equation would reduce to $y' = x^4$, a separable equation. Perhaps there is some factor that will make the differential form an exact equation. Let it be of the form $\mu(x)$. Then

$$\frac{\partial}{\partial y}\left[\mu(x)\{(4/x)y - x^4\}\right] = \frac{\partial}{\partial x}[\mu(x)]$$

i.e.,
$$\mu(x) \frac{4}{x} = \frac{d}{dx} \mu(x)$$

$$\frac{d\mu}{\mu} = 4 \frac{dx}{x} \ , \ \int \frac{d\mu}{\mu} = 4 \int \frac{dx}{x}$$

$$\ln|\mu| = 4 \int \frac{dx}{x} \ ; \ \mu = e^{4\int dx/x} = e^{\ln x^4} = x^4 .$$

41

Another way of obtaining the same result is:
$$\ln|\mu| = \ln|x^4|; \mu = x^4 .$$
Multiplying every term in the given equation by $\mu = x^4$:
$$x^4 y' + 4x^3 y = x^8 .$$
The expression on the left is nothing but $\frac{d}{dx}(x^4 y)$. $\frac{d}{dx}(x^4 y) = x^8$.

Integrating: $yx^4 = \frac{1}{9} x^9 + c$.
$$y = \frac{1}{9} x^5 + \frac{c}{x^4} .$$

● PROBLEM 2–2

Solve the initial-value problem that consists of the differential

equation $\qquad (x^2 + 1)\frac{dy}{dx} + 4xy = x$

and the initial condition

$$y(2) = 1 .$$

<u>Solution</u>: We first rewrite the equation so that it is of the linear form
$$\frac{dy}{dx} + p(x)y = Q(x).$$

Thus, we have
$$\frac{dy}{dx} + \frac{4x}{x^2+1} y = \frac{x}{x^2+1} .$$

The integrating factor is of the form
$$e^{\int p(x)dx} = e^{\int 4x/x^2+1 \, dx} .$$

Now
$$\int \frac{4x}{x^2+1} \, dx = 2\int \frac{2x}{x^2+1} \, dx = 2 \ln|x^2+1| + c$$
$$= \ln(|x^2+1|)^2 + c .$$

Ignoring the arbitrary constant the integrating factor is

$$\mu(x) = e^{\ln(x^2+1)^2} = (x^2+1)^2 , \text{ we have}$$
$$(x^2+1)^2 \frac{dy}{dx} + (x^2+1)^2 \frac{4x}{(x^2+1)} y = (x^2+1)^2 \frac{x}{(x^2+1)} ,$$
or,
$$(x^2+1)^2 \frac{dy}{dx} + 4x(x^2+1)y = (x^2+1)(x) .$$

Then $\frac{d}{dx} [(x^2+1)^2 y] = x^3 + x$. Integrating, $(x^2+1)^2 y = \frac{x^4}{4} + \frac{x^2}{2} + c$.

This is a solution of the given equation. Although implicit it is possible to substitute the initial condition $y(2) = 1$, to evaluate the constant. We have
$$(4+1)^2 (1) = 4 + 2 + c; c = 19.$$

42

The solution of the initial-value problem is, therefore

$$(x^2+1)^2 y = \frac{x^4}{4} + \frac{x^2}{2} + 19 .$$

● PROBLEM 2-3

Solve

$$\frac{dy}{dx} + \frac{(2x + 1)}{x} y = e^{-2x} .$$

Solution: The differential equation may be classified as an ordinary first order linear one. Rewriting it in the differential form

$$\left[\frac{2x+1}{x} y - e^{-2x} \right] dx + dy = 0 ,$$

we note that, since

$$\frac{\partial}{\partial y}\left(\frac{2x+1}{x} y - e^{-2x} \right) \neq \frac{\partial}{\partial x}(1) ,$$

the equation is not exact. We therefore seek an integrating factor.

Let $\mu(x)$ be an integrating factor. Then, in order for it to make the given differential equation exact,

$$\frac{\partial}{\partial y}\left[\mu(x) \frac{2x+1}{x} y - \mu(x)e^{-2x} \right] = \frac{\partial}{\partial x}[\mu(x)]$$

$$\mu(x) \frac{2x+1}{x} = \frac{d\mu}{dx} , \text{ a differential equation in which}$$

μ is an unknown function of x.

$$\frac{d\mu}{\mu} = \left(\frac{2x+1}{x} \right) dx = \left(2 + \frac{1}{x} \right) dx .$$

$$\ln \mu = 2x + \ln\left|x\right| \quad \text{(by integration)} .$$

Finding the antilogarithm, $\mu = e^{2x+\ln\left|x\right|} = xe^{2x}$ which is the required integrating factor. Multiplying the given differential equation by this factor,

$$xe^{2x} \frac{dy}{dx} + e^{2x}(2x+1)y = x.$$

That is, $\frac{d}{dx}(xe^{2x}y) = x.$

We can integrate this: $xe^{2x}y = \frac{x^2}{2} + c$

$$y = \tfrac{1}{2} xe^{-2x} + \frac{c}{x} e^{-2x}$$

where c is an arbitrary constant.

● PROBLEM 2-4

Solve $y' - 2xy = x.$

Solution: This is a case of a linear first order equation and can be solved by the direct application of integrating factors. The general equation is represented by

$$y' - p(x)y = q(x)$$

and the integrating factor for these equations is defined as $e^{\int p(x)dx}$.

Here $p(x) = -2x$ so the integrating factor is $e^{\int -2x dx} = e^{-2x^2/2} = e^{-x^2}$.
Multiplying the original expression throughout by the integrating factor

$$y'e^{-x^2} - 2xe^{-x^2}y = xe^{-x^2}.$$

$\frac{d}{dx}(ye^{-x^2}) = xe^{-x^2}$. Integrating both sides

$$\int\left[\frac{d}{dx} y\ e^{-x^2}\right]dx = \int xe^{-x^2}dx$$

with respect to x. Or

$$ye^{-x^2} = -\tfrac{1}{2}e^{-x^2} + c.$$

Multiply by e^{x^2} on both sides of the equation

$$y = -\tfrac{1}{2} + ce^{x^2}.$$

• PROBLEM 2–5

Solve the first-order differential equation:

$$\frac{dy}{dx} \cos x + y \sin x = 1.$$

Solution: We note that the equation is linear, but neither exact nor separable. We must find an integrating factor, that will make the equation separable.

First, we convert, algebraically,

$$\frac{dy}{dx} \cos x + y \sin x = 1 \quad\text{into}\quad \frac{dy}{dx} = \frac{1 - y \sin x}{\cos x};$$

$\frac{dy}{dx} + y \tan x = \sec x$. Now the equation is in the general linear form

$$\frac{dy}{dx} + p(x)y = q(x).$$

The integrating factor is therefore $e^{\int p(x)dx}$ or $e^{\int \tan x\ dx}$ which is $e^{-\ln|\cos x| + c} = \sec x$ (ignoring the constant e^c).

Then $\sec x \frac{dy}{dx} + y(\tan x \sec x) = \sec^2 x$ or $\qquad\qquad(1)$

$$[(\tan x \sec x)y - \sec^2 x]dx + \sec x\ dy.$$

This is an exact differential equation. From (1), we obtain

$$(y \sec x)' = \sec^2 x$$

$$y \sec x = \int \sec^2 x\ dx$$

$$= \tan x + c$$

$$y \sec x - \tan x = c,$$

or, the solution can also be written as

$$y = \sin x + c \cos x.$$

44

Solve the following differential equations:

a) $\dfrac{dy}{dx} = \dfrac{x^4 + 2y}{x}$

b) $\dfrac{dy}{dx} = \dfrac{2xy}{y^2 - x^2}$.

Solution: a) Rewriting the first equation,

$$\frac{dy}{dx} - \frac{2}{x}\, y = x^3, \text{ and}$$

we note that it is a nonhomogeneous and linear differential equation. Under the general expression:

$$\frac{dy}{dx} + py = Q,$$

where $p = -2/x$ and $Q = x^3$, we obtain the general solution as:

$$y = e^{-\int pdx} \int Q\, e^{\int pdx} dx + c\, e^{-\int pdx}.$$

the integrating factor, $e^{\int pdx}$, in this case is: $e^{\int -2/x\,dx} = e^{-2\ln x} = 1/x^2$. Hence,

$$y = x^2 \int \frac{1}{x^2}\, x^3\, dx + x^2 c$$

$$= \tfrac{1}{2}\, x^4 + x^2 c \; .$$

b) $\qquad\qquad \dfrac{dy}{dx} = \dfrac{2xy}{y^2 - x^2}$.

This is a homogeneous equation of degree zero, where, by substitution of xt for x and yt for y, we have:

$$f(xt, yt) = \frac{2(xt)(yt)}{(yt)^2 - (xt)^2} = \frac{2xy}{y^2 - x^2} = f(x,y) = t^0 f(x,y),$$

The function: dy/dx = f(x,y) is not affected. Under this condition, f(x,y) can be expressed as of a single argument: f(x,y) = f(1+y/x) or f(x/y,1). Now, setting y = vx,

$$v + x\frac{dv}{dx} = \frac{2vx^2}{v^2 x^2 - x^2} = \frac{2v}{v^2 - 1} \; .$$

Separating the variables, the differential equation can be written as

$$\frac{-3(v^2 - 1)}{3v - v^3}\, dv = \frac{3dx}{x} \; .$$

Hence,

$$\ln(3v - v^3) = -3\ln x + \ln c,$$

or,

$$3x^2 y - y^3 = c \; .$$

Solve the first order differential equation:

$$y^2 dx + (3xy - 1)dy = 0.$$

Solution: We first rewrite the equation as $\dfrac{dy}{dx} = \dfrac{y^2}{1 - 3xy}$. But this equation is not linear. Returning to the original formulation which is a particular instance of the differential form

$$M(x,y)dx + N(x,y)dy = 0,$$

we note that the equation is not exact since $\dfrac{\partial M}{\partial y} = 2y \neq \dfrac{\partial N}{\partial x} = 3y$. Further, there is no way of rewriting the original equation in the form $f(x)dx + g(y)dy = 0$, i.e., as a separable equation.

In the differential form, however, either y or x may be regarded as the dependent variable. If we find dx/dy we have,

$$\frac{dx}{dy} = \frac{1 - 3xy}{y^2} \; ; \; \frac{dx}{dy} + \frac{3}{y} x = \frac{1}{y^2} \; .$$

This equation is linear in x. In solving it, however, we must remember that we shall be finding a solution of the form $x = f(y)$ and not $y = f(x)$.

An integrating factor for the equation is given by

$$e^{\int p(y)dy} = e^{\int 3/y \; dy} = e^{\ln |y|^3} = y^3 \; .$$

Then

$$y^3 \frac{dx}{dy} + 3y^2 x = y \; ,$$

or

$$\frac{d}{dy}[y^3 x] = y \; ;$$

Now we may integrate:

$$xy^3 = \frac{y^2}{2} + c; \; x = \frac{1}{2y} + \frac{c}{y^3} \; .$$

This method of solution was possible because of the differential form

$$y^2 dx + (3xy - 1)dy = 0,$$

which made it unnecessary to single out x or y as the only dependent variable.

Solve $\dfrac{dy}{dx} = \dfrac{1}{xy + x^2 y^3}$.

Solution: $\dfrac{dy}{dx} = \dfrac{1}{xy + x^2 y^3}$. Taking the reciprocal on both sides we arrive at

$$\frac{dx}{dy} = xy + x^2 y^3$$

or

$$\frac{dx}{dy} - xy = x^2 y^3 \; .$$

The general form of the Bernoulli equation is

$$x' + p(y)x = q(y)x^n \; .$$

Here n=2 is the value which gives us the necessary substitution to resolve the above nonlinear differential equation to a linear differential equation. The substitution is $z = x^{1-n}$; therefore $z = x^{1-2} = x^{-1}$ or $z = 1/x$. We need to use $x = 1/z$ and $x^1 = z^1/z^2$.

Substituting the above relations in the original equation we get

$$-\frac{z'}{z^2} - (1/z)y = (1/z)^2 \, y^3$$

or

$$-z' - yz = y^3 \ .$$

$$z' + yz = -y^3 \ .$$

Since this is a linear, first order differential equation it can be solved by the use of integrating factors.

$$I(x,y) = e^{\int y \, dy} = e^{y^2/2} \ .$$

Multiplying the equation throughout by $e^{y^2/2}$,

$$z' \, e^{y^2/2} + y e^{y^2/2} z = -y^3 \cdot e^{y^2/2}$$

or

$$\frac{d}{dy}\left(z \, e^{y^2/2}\right) = -y^3 \cdot e^{y^2/2} \ .$$

By integrating both sides of the equation with respect to y.

$$\int \frac{d}{dy}\left(z \, e^{y^2/2}\right) dy = \int -y^3 \, e^{y^2/2} \, dy \ .$$

$$z \, e^{y^2/2} = \int -y^3 \, e^{y^2/2} \, dy \ .$$

Now, the right hand side of the equation can be integrated by the method of integration by parts. $\int -y^3 \, e^{y^2/2} \, dy$ can be written as

$$-\int y^2 \cdot y e^{y^2/2} \, dy \ .$$

$$\int u \, dv = uv - \int v \, du \ .$$

Let $u = y^2$ and $dv = y e^{y^2/2}$ therefore $v = e^{y^2/2}$; $du = 2y$.

$$-\int y^2 \cdot y e^{y^2/2} dy = -y^2 \cdot e^{y^2/2} + \int e^{y^2/2} \cdot 2y \, dy$$

$$= -y^2 \, e^{y^2/2} + e^{y^2/2} \cdot 2 - c_1$$

So $z \, e^{y^2/2} = e^{y^2/2}(2-y^2) - c_1$. Multiplying throughout by $e^{-y^2/2}$,

$$z = (2-y^2) - c_1 e^{-y^2/2} \ ,$$

or

$$1/x = (2-y^2) - c_1 e^{-y^2/2} \ .$$

● PROBLEM 2–9

Solve

$$y' - \frac{3}{x} y = x^4 y^{\frac{1}{3}} \tag{a}$$

<u>Solution</u>: This is a Bernoulli equation with $n = \frac{1}{3}$. Dividing both sides by $y^{\frac{1}{3}}$, we find

$$y^{-\frac{1}{3}} \frac{dy}{dx} - \frac{3}{x} y^{\frac{2}{3}} = x^4 .$$ (b)

We set $z = y^{-n+1} = y^{\frac{2}{3}}$. Then $\frac{dz}{dx} = \frac{2}{3} y^{-\frac{1}{3}} \frac{dy}{dx}$ and (b) becomes

$$\frac{dz}{dx} - (2/x)z = \frac{2}{3} x^4 ,$$ (c)

A first order linear differential equation with integrating factor

$$e^{\int -\frac{2}{x} \, dx} = e^{-\ln x^2} = \frac{1}{x^2} .$$

Multiplying each side of (c) by $\frac{1}{x^2}$ gives us

$$\frac{1}{x^2} \frac{dz}{dx} - \left(\frac{2}{x^3}\right)z = \frac{2}{3} x^2$$

or

$$\frac{d}{dx}\left(\frac{z}{x^2}\right) = \frac{2}{3} x^2$$

or

$$\frac{z}{x^2} = \frac{2}{9} x^3 + c ,$$ (d)

with c an arbitrary constant. Replacing z by $y^{\frac{2}{3}}$ in (d) finally gives the solution

$$y^{2/3} = cx^2 + \frac{2}{9} x^5$$

or, explicitly,

$$y = \left(cx^2 + \frac{2}{9} x^5\right)^{3/2} .$$

• PROBLEM 2–10

Solve $\qquad \dfrac{dy}{dx} + y = xy^3$. $\qquad\qquad$ (a)

<u>Solution</u>: This is a Bernoulli differential equation with $n = 3$. Multiplying through by y^{-3}, we express it in the equivalent form

$$y^{-3} \frac{dy}{dx} + y^{-2} = x.$$

If we let $z = y^{-n+1} = y^{-2}$, then $\frac{dz}{dx} = (-2)y^{-3} \frac{dy}{dx}$, and the preceding differential equation transforms into

$$-\tfrac{1}{2} \frac{dv}{dx} + v = x.$$

Writing this linear equation in the standard form

$$\frac{dv}{dx} - 2v = -2x ,$$ (b)

we see that an integrating factor for this equation is

$$e^{\int(-2)dx} = e^{-2x} .$$

Multiplying (b) by e^{-2x}, we find

$$e^{-2x}\frac{dv}{dx} - 2ve^{-2x} = -2xe^{-2x}$$

or

$$\frac{d}{dx}(e^{-2x}v) = -2xe^{-2x} .$$

Integrating, we find

$$e^{-2x}v = \int (-2x)e^{-2x} dx + c \qquad\qquad (c)$$

Integrating the right hand side of (c) by parts gives

$$e^{-2x}v = \tfrac{1}{2}e^{-2x}(2x+1) + c,$$

or

$$v = x + \tfrac{1}{2} + ce^{2x} ,$$

where c is an arbitrary constant. Since $v = \frac{1}{y^2}$, we obtain the solutions of (a) in the form

$$y^2 = \frac{1}{x+\tfrac{1}{2}+ce^{2x}} .$$

THE WRONSKIAN AND LINEAR INDEPENDENCE

• PROBLEM 2–11

Determine whether the set $\{e^x, e^{-x}\}$ is linearly independent for $-\infty < x < \infty$.

Solution: To solve this problem we require the definition of linear independence. Since the set is a set of functions we frame the definition in terms of functions.

A set of functions $\{f_1,\ldots,f_n\}$ is said to be linearly dependent if there exist some constants $c_i \neq 0$ (i.e., at least one) such that

$$c_1 f_1 + c_2 f_2 + \ldots + c_n f_n = 0 \; . \qquad \text{(a)}$$

If (a) holds only when $c_1 = c_2 \ldots = c_n = 0$ then the set $\{f_1, \ldots, f_n\}$ is said to be linearly independent.

To see whether $\{e^x, e^{-x}\}$ is linearly independent we first form the linear combination

$$c_1 e^x + c_2 e^{-x} \qquad \text{(b)}$$

and set it equal to zero.

$$c_1 e^x + c_2 e^{-x} = 0 \; ,$$

$$c_2 e^{-x} = -c_1 e^x \; ,$$

$$c_2 = -c_1 e^{2x} \; . \qquad \text{(c)}$$

The definition of linear independence stipulated the presence of non-zero constants. But by (c), c_2 is equal to $-c_1 e^{2x}$ which is a variable quantity, for $c_1 \neq 0$. This contradiction suffices to prove that $c_1 = c_2 = 0$ and hence the set $\{e^x, e^{-x}\}$ is linearly independent.

• PROBLEM 2-12

Is the set $\{1-x, 1+x, 1-3x\}$ linearly dependent on $(-\infty, \infty)$?

Solution: The set $\{1-x, 1+x, 1-3x\}$ is linearly dependent if, and only if, there exist some constants c_i, not all equal to zero, such that the linear combination

$$c_1(1-x) + c_2(1+x) + c_3(1-3x) \qquad \text{(a)}$$

is equal to zero.

Rewriting (a) and setting it equal to zero,

$$(-c_1 + c_2 - 3c_3)x + (c_1 + c_2 + c_3) = 0 \qquad \text{(b)}$$

since (b) is a polynomial in x, it will be valid only if

$$-c_1 + c_2 - 3c_3 = 0 \qquad \text{(c)}$$

$$c_1 + c_2 + c_3 = 0 \; . \qquad \text{(d)}$$

From (c) and (d) we obtain

$$c_1 = -2c_3, \; c_2 = c_3 \text{ with } c_3 \text{ an arbitrary constant.}$$

Since it is arbitrary, c_3 does not have to equal zero. Thus, letting $c_3 = 2$, we have the linear combination

$$-4(1-x) + 2(1+x) + 2(1-3x) = 0$$

which means the set of functions is linearly dependent.

50

Are the functions $f_1 = x$, $f_2 = x^2$ linearly independent on the interval $(0,2)$?

Solution: A set of functions $F = \{f_1, \ldots, f_n\}$ is linearly dependent if there exist some constants $c_i \neq 0$, such that

$$c_1 f_1 + c_2 f_2 + \ldots + c_n f_n = 0 \tag{a}$$

is true. If (a) implies $c_1 = c_2 = \ldots = c_n = 0$, the set is linearly independent.

Thus, let
$$c_1 x + c_2 x^2 = 0 . \tag{b}$$

If these functions are linearly dependent, then we can find c_1 and c_2 not both zero such that this equation is true for all x in $(0,2)$.

Since we have two unknowns let us pick two points in the interval $(0,2)$, say a, b; $a \neq b$.

Then
$$c_1 a + c_2 a^2 = 0 \tag{c}$$
$$c_1 b + c_2 b^2 = 0 . \tag{d}$$

Multiplying (c) by b^2 and (d) by a^2 and subtracting (d) from (c),
$$c_1(ab^2 - ba^2) = 0$$

or
$$c_1(ab(b - a)) = 0 . \tag{e}$$

In (e), since $ab \neq 0$ and $(b - a) \neq 0$, it follows that $c_1 = 0$. It may be shown similarly that $c_2 = 0$.

Since (b) is true only if $c_1 = c_2 = 0$, we conclude that the functions $\{x, x^2\}$ are linearly independent on the interval $(0,2)$.

The linear independence may be demonstrated in another way. The Wronskian of the two functions is

$$W(x, x^2) = \begin{vmatrix} x & x^2 \\ \frac{d}{dx}(x) & \frac{d}{dx}(x^2) \end{vmatrix}$$

$$= \begin{vmatrix} x & x^2 \\ 1 & 2x \end{vmatrix} = x^2 .$$

If the Wronskian of a set of functions is never zero, then the functions are linearly independent. Since $x^2 \neq 0$ on the open interval $(0,2)$ we conclude that the set is linearly independent.

The differential equation
$$u'' + \{\tan x - 2 \cot x\}u' = 0$$
has the integrals
$$u_1 = \sin x - \frac{1}{3} \sin 3x \tag{a}$$
and
$$u_2 = \sin 3x . \tag{b}$$
Are these solutions linearly independent?

<u>Solution</u>: To check for the linear independence of two solutions, u_1 and u_2, we compute their Wronskian:

$$W(u_1, u_2) = \begin{vmatrix} u_1 & u_2 \\ u_1' & u_2' \end{vmatrix} = u_1 u_2' - u_1' u_2 .$$

Substituting (a) and (b) for u_1 and u_2, respectively:

$$W(\sin x - \frac{1}{3} \sin 3x, \sin^3 x)$$

$$= \begin{vmatrix} \sin x - \frac{1}{3} \sin 3x & \sin^3 x \\ \cos x - \cos 3x & 3 \sin^2 x \cos x \end{vmatrix}$$

$$= (\sin x - \frac{1}{3} \sin 3x)(3\sin^2 x \cos x)$$
$$\quad - (\cos x - \cos 3x)(\sin^3 x)$$
$$= 2 \sin^3 x \cos x - \sin^2 x(\sin 3x \cos x)$$
$$\quad + \cos 3x \sin^3 x .$$
$$= \sin^2 x(2 \sin x \cos x - \sin 3x \cos x$$
$$\quad + \cos 3x \sin x). \tag{c}$$

Using the identity
$$\sin A \cos B = \frac{1}{2}\{\sin(A - B) + \sin(A + B)\},$$
equation (c) becomes
$$\sin^2(x)\{\sin 2x - \frac{1}{2}(\sin 2x + \sin 4x)$$
$$\quad + \frac{1}{2}(\sin(-2x) + \sin 4x)\} .$$

Since sin(-x) = -sin x, this becomes
$$\sin^2 x(\sin 2x - \sin 2x) = 0 .$$
Since the Wronskian is identically zero, i.e., $W(u_1, u_2) = 0$, the solutions u_1 and u_2 are linearly dependent.

Show that the functions, $y = e^x$ and $y = e^{2x}$ are linearly independent solutions of the differential equation

$$y'' - 3y' + 2y = 0 . \qquad (a)$$

Solution: To solve (a) we note that the equation is linearly homogeneous with constant coefficients. Thus, assuming a solution of the form $y = e^{mx}$ we obtain, upon substitution:

$$e^{mx}(m^2 - 3m + 2) = 0 \qquad (b)$$

The characteristic equation $(m^2 - 3m + 2)$ may be factored into $(m-1)(m-2)$. The equation (a) therefore has two solutions,

$$y = e^x \quad \text{and} \quad y = e^{2x} .$$

We must now show that $y = e^x$ and $y = e^{2x}$ are linearly independent. To do this we use the Wronskian of the two functions $W(e^x, e^{2x})$. The Wronskian of a set of functions, $F = \{f_1, f_2, \ldots, f_n\}$ is defined as

$$W(F) = \begin{vmatrix} f_1 & f_2 & \cdots & f_n \\ f_1' & f_2' & \cdots & f_n' \\ \vdots & & & \\ f_1^{(n-1)} & & \cdots & f_n^{(n-1)} \end{vmatrix} \qquad (c)$$

We are dealing with the set $F = \{e^x, e^{2x}\}$. Applying (c),

$$W(F) = \begin{vmatrix} e^x & e^{2x} \\ e^x & 2e^{2x} \end{vmatrix} = e^{3x} \neq 0 .$$

Since a set of functions is linearly independent if its Wronskian is not equal to zero, we conclude that the solutions $y = e^x$, $y = e^{2x}$ are linearly independent. We can therefore write the general solution of (a):

$$y(x) = c_1 e^x + c_2 e^{2x} ,$$

where c_1 and c_2 are arbitrary constants.

• **PROBLEM 2–16**

Show that the solutions e^x, e^{-x}, and e^{2x} of

$$\frac{d^3y}{dx^3} - 2 \frac{d^2y}{dx^2} - \frac{dy}{dx} + 2y = 0 \tag{a}$$

are linearly independent on every real interval.

<u>Solution</u>: To show that the functions e^x, e^{-x}, and e^{2x} are linearly independent, we compute their Wronskian

$$W(e^x, e^{-x}, e^{2x}) = \begin{vmatrix} e^x & e^{-x} & e^{2x} \\ e^x & -e^{-x} & 2e^{2x} \\ e^x & e^{-x} & 4e^{2x} \end{vmatrix}$$

We compute this determinant by expansion along the first column:

$$e^x \begin{vmatrix} -e^{-x} & 2e^{2x} \\ e^{-x} & 4e^{2x} \end{vmatrix} - e^x \begin{vmatrix} e^{-x} & e^{2x} \\ e^{-x} & 4e^{2x} \end{vmatrix}$$

$$+ e^x \begin{vmatrix} e^{-x} & e^{2x} \\ -e^{-x} & 2e^{2x} \end{vmatrix}$$

$$= e^x(-4e^x - 2e^x) - e^x(4e^x - e^x)$$

$$+ e^x(2e^x + e^x) = -6e^{2x} .$$

Since $-6e^{2x}$ is nonzero for all values of x, we can conclude that the solutions e^x, e^{-x}, and e^{2x} are linearly independent. Furthermore, we may write the general solution of the given third order homogeneous equation (a):

$$y(x) = c_1 e^x + c_2 e^{-x} + c_3 e^{2x} ,$$

where c_1, c_2, and c_3 are arbitrary constants.

• PROBLEM 2–17

Solve the differential equation

$$y'' - y = 0 \qquad\qquad \text{(a)}$$

and show that the solutions obtained, $y_1(x)$ and $y_2(x)$, are linearly independent.

<u>Solution</u>: The equation (a) is linearly homogeneous with constant coefficients. Assuming an exponential solution of the form $y = e^{mx}$, we obtain the characteristic equation

$$m^2 - 1 = 0 . \qquad\qquad \text{(b)}$$

Equation (b) may be factored to give

$$(m+1)(m-1) = 0 \quad \text{or} \quad m = \pm\, 1 .$$

Two solutions of (a) are therefore $y_1 = e^x$ and $y_2 = e^{-x}$.

To show that these solutions are linearly independent, form the Wronskian of (e^x, e^{-x})

$$W(e^x, e^{-x}) = \begin{vmatrix} f_1 & f_2 \\ f_1' & f_2' \end{vmatrix}$$

$$= \begin{vmatrix} e^x & e^{-x} \\ e^x & -e^{-x} \end{vmatrix} = e^x(-e^{-x}) - e^x e^{-x}$$

$$= -2 \neq 0 .$$

Since the Wronskian is not zero for $-\infty < x < \infty$, the two functions are linearly independent for all x.

Given the equation
$$y'' - 2y' + y = x^2 . \tag{a}$$
One solution is $y = x^2 + 4x + 6$ and two solutions of the homogeneous problem
$$y'' - 2y' + y = 0 \tag{b}$$
are e^x and xe^x. Find the general solution.

Solution: The function (a) is inhomogeneous. Its general solution is the sum of the general solution of the corresponding homogeneous differential equation and a particular solution of the inhomogeneous differential equation.

The task is, therefore, to determine whether the two solutions of (b) are linearly independent. A set of functions $F = \{f_1, \ldots, f_n\}$ that are $(n-1)$ times differentiable, are linearly independent, if their Wronskian is nonzero:

$$W(F) = \begin{vmatrix} f_1 & f_2 & \cdots & f_n \\ \vdots & & & \\ f_1^{(n-1)} & \cdots & & f_n^{(n-1)} \end{vmatrix} \neq 0 .$$

The set of functions in the given problem is e^x and xe^x. Their Wronskian is

$$W(F) = \begin{vmatrix} e^x & xe^x \\ e^x & xe^x + e^x \end{vmatrix} = e^{2x} \neq 0 .$$

Thus the two functions (e^x, xe^x) are linearly independent and the general solution to the homogeneous problem is
$$y = c_1 e^x + c_2 xe^x , \tag{c}$$
where c_1 and c_2 are arbitrary constants.

Since the function $y = x^2 + 4x + 6$ is a particular solution of (a) (as may be verified using the method of undetermined coefficients), the general solution of (a) is
$$y = y_h + y_p = c_1 e^x + c_2 xe^x + x^2 + 4x + 6 .$$

METHOD OF UNDETERMINED COEFFICIENTS

● **PROBLEM** 2–19

Solve:

$$y' - 5y = 3e^x - 2x + 1. \qquad \text{(a)}$$

<u>Solution:</u>   We must first solve the associated homogeneous equation

$$y' - 5y = 0.$$

This gives the complementary solution

$$y_c(x) = C e^{5x}, \qquad \text{(b)}$$

where C is an arbitrary constant.

We next assume a particular solution of the form

$$y_p(x) = A e^x + B_1 x + B_0, \qquad \text{(c)}$$

where A, B_1, and B_0 are undetermined coefficients. This is suggested by the inhomogeneous term, $(3e^x - 2x + 1)$. Since no term in the particular solution (c) is included in the complementary solution (b), we may proceed directly with the determination of A, B_1, and B_0.

Substitution of (c) into (a) gives

$$[(Ae^x + B_1) - 5(Ae^x + B_1 x + B_0)] = 3e^x - 2x + 1,$$

or $-4Ae^x - 5B_1 x + B_1 - 5B_0 = 3e^x - 2x + 1.$

Equating coefficients of e^x, x, and x^0 on both sides:

$$-4A = 3,$$

$$-5B_1 = -2,$$

$$B_1 - 5B_0 = 1.$$

Solving this system, we find that $A = -\frac{3}{4}$, $B_1 = \frac{2}{5}$, and $B_0 = -\frac{3}{25}$. Substitution of these values into (c) gives a particular solution

$$y_p(x) = -\frac{3}{4} e^x + \frac{2}{5} x - \frac{3}{25}.$$

Summing this with the complementary solution (b), we obtain the general solution:

$$y(x) = C e^{5x} - \frac{3}{4} e^x + \frac{2}{5} x - \frac{3}{25} \ .$$

• PROBLEM 2-20

Solve

$$y' - 5y = x^2 e^x - xe^{5x} \ . \tag{a}$$

Solution: We must first solve the associated homogeneous equation

$$y' - 5y = 0.$$

This gives the complementary solution

$$y_c(x) = C e^{5x}, \tag{b}$$

where C is an arbitrary constant.

To determine a particular solution of equation (a) by using the method of undetermined coefficients we consider the inhomogeneous term $(x^2 e^x - xe^{5x})$. It is the difference of two terms each in manageable form. For $x^2 e^x$ we assume a solution of the form

$$e^x (A_2 x^2 + A_1 x + A_0), \tag{c}$$

where A_2, A_1, and A_0 are undetermined coefficients. For xe^{5x} we would try initially a solution of the form

$$e^{5x} (B_1 x + B_0) = B_1 xe^{5x} + B_0 e^{5x} \ ,$$

where B_1 and B_0 are undetermined coefficients. But this supposed solution would have, disregarding multiplicative constants, the term e^{5x} in common with $y_c(x)$, as given in (b). We are therefore led to the modified expression

$$xe^{5x} (B_1 x + B_0) = e^{5x} (B_1 x^2 + B_0 x). \tag{d}$$

We now take $y_p(x)$ as the difference of (c) and (d):

$$y_p(x) = e^x (A_2x^2 + A_1x + A_0) - e^{5x} (B_1x^2 + B_0x). \quad (e)$$

Substitution of (e) into (a) gives

$$\{e^x [(- 4A_2)x^2 + (2A_2 - 4A_1)x + (A_1 - 4A_0)]$$
$$+ e^{5x} [(- 2B_1)x - B_0]\}$$
$$= e^xx^2 - e^{5x} x.$$

Equating coefficients of like terms on both sides of the equation:

$$- 4A_2 = 1,$$

$$2A_2 - 4A_1 = 0,$$

$$A_1 - 4A_0 = 0,$$

$$- 2B_1 = - 1,$$

$$- B_0 = 0.$$

Solving this system we find

$$A_2 = - \frac{1}{4}, \qquad A_1 = - \frac{1}{8}, \qquad A_0 = - \frac{1}{32},$$

$$B_1 = \frac{1}{2}, \qquad \text{and} \qquad B_0 = 0.$$

Substitution of these values into (e) gives a particular solution:

$$y_p(x) = - \frac{1}{4} x^2e^x - \frac{1}{8} xe^x - \frac{1}{32} e^x - \frac{1}{2} x^2e^{5x}.$$

Summing this with the complementary solution (b), we obtain the general solution:

$$y(x) = C_1e^{5x} + e^x \left(- \frac{1}{4} x^2 - \frac{1}{8} x - \frac{1}{32}\right) - \frac{1}{2} x^2e^{5x}.$$

● **PROBLEM** 2–21

Solve:

$$y'' - y' - 2y = \sin 2x. \qquad (a)$$

<u>Solution:</u>   We must first solve the associated homogeneous equation:

$$y'' - y' - 2y = 0.$$

Using the characteristic equation

$$p^2 - p - 2 = 0,$$

we find roots at $p_1 = 1$, and $p_2 = -2$. From this we obtain the complementary solution

$$y_c(x) = C_1 e^x + C_2 e^{-2x}, \tag{b}$$

where C_1 and C_2 are arbitrary constants.

Using the method of undetermined coefficients, we assume a particular solution of the form

$$y_p(x) = A \sin 2x + B \cos 2x. \tag{c}$$

This is suggested by the inhomogeneous term (sin 2x) found in equation (a); A and B are the undetermined coefficients. Since no term in our particular solution (c) is contained in the complementary solution (b), we may proceed directly with the determination of A and B.

Substitution of (c) into (a) gives

$$[(- 4 A \sin 2x - 4 B \cos 2x) + (- 2 A \cos 2x + 2 B \sin 2x)$$

$$+ (- 2 A \sin 2x - 2 B \cos 2x)]$$

$$= \sin 2x,$$

or, $[(- 4A + 2B - 2A) \sin 2x + (- 4B - 2A - 2B) \cos 2x]$

$$= \sin 2x.$$

Equating coefficients of sin 2x and cos 2x on both sides of the equation:

$$- 6A + 2B = 1,$$

$$- 2A - 6B = 0.$$

Solving this system, we find that $A = - \frac{3}{20}$ and $B = \frac{1}{20}$. Substitution of these values into (c) gives a particular solution

$$y_p(x) = - \frac{3}{20} \sin 2x + \frac{1}{20} \cos 2x.$$

Summing this with the complementary solution (b), we obtain the general solution

$$y(x) = C_1 e^x + C_2 e^{-2x} - \frac{3}{20} \sin 2x + \frac{1}{20} \cos 2x.$$

Solve by the method of undetermined coefficients:

$$y'' + y' - 6y = 2x^3 + 5x^2 - 7x + 2. \qquad \text{(a)}$$

Solution: The associated homogeneous equation is

$$y'' + y' - 6y = 0$$

and the complementary solution is

$$y_c(x) = c_1 e^{2x} + c_2 e^{-3x} \qquad \text{(b)}$$

where c_1 and c_2 are arbitrary constants.

To determine a particular solution we consider the inhomogeneous term $2x^3 + 5x^2 - 7x + 2$. This suggests a particular solution

$$y_p(x) = Ax^3 + Bx^2 + Cx + D, \qquad \text{(c)}$$

where A, B, C, and D are undetermined coefficients. This is an appropriate form for $y_p(x)$ because none of its terms are contained in the complementary solution (b).

To determine the coefficients A, B, C, and D, we substitute (c) into (a):

$$(6Ax + 2B) + (3Ax^2 + 2Bx + C) - 6(Ax^3 + Bx^2 + Cx + D)$$

$$= 2x^3 + 5x^2 - 7x + 2,$$

or, $-6Ax^3 + (3A - 6B)x^2 + (6A + 2B - 6C)x + (2B + C - 6D)$

$$= 2x^3 + 5x^2 - 7x + 2.$$

Equating coefficients of like terms, we obtain a system of linear equations:

$$- 6A = 2,$$

$$3A - 6B = 5,$$

$$6A + 2B - 6C = - 7,$$

$$2B + C - 6D = 2.$$

Solving, we find $A = -\frac{1}{3}$, $B = -1$, $C = \frac{1}{2}$, and $D = -\frac{7}{12}$. Substitution of these values into (c) gives a particular solution of equation (a)

$$y_p(x) = -\frac{1}{3}x^3 - x^2 + \frac{1}{2}x - \frac{7}{12}.$$

Summing this with the complementary solution (b), we obtain the general solution of (a):

$$y(x) = \left\{ C_1 e^{2x} + C_2 e^{-3x} - \frac{1}{3} x^3 - x^2 + \frac{1}{2} x - \frac{7}{12} \right\} .$$

● **PROBLEM** 2–23

Determine a form for the particular solution of

$$y´´ + 2y´ + 5y = e^{-x} \sin 2x. \tag{a}$$

Solution: The associated homogeneous equation

$$y´´ + 2y´ + 5y = 0$$

has a characteristic equation

$$p^2 + 2p + 5 = 0,$$

which has roots $p_1 = -1 \pm 2i$. The complementary solution is therefore

$$y_c(x) = C_1 e^{-x} \cos 2x + C_2 e^{-x} \sin 2x, \tag{b}$$

where C_1 and C_2 are arbitrary constants.

Although our inhomogeneous term

$$e^{-x} \sin x$$

corresponds to a particular solution

$$y_p(x) = B_1 e^{-x} \cos 2x + B_2 e^{-x} \sin 2x, \tag{c}$$

we must reject this $y_p(x)$ because it is equivalent to the complementary solution (b). If we multiply (c) by x we obtain a particular solution of the form

$$y_p(x) = B_1 x e^{-x} \cos 2x + B_2 x e^{-x} \sin 2x$$

where B_1 and B_2 are constants that have to be determined.

● **PROBLEM** 2–24

Solve, using the method of undetermined coefficients:

$$y´´ - 2y´ + y = (x^2 - 1)e^{2x} + (3x + 4)e^x. \tag{a}$$

<u>Solution:</u>   The associated homogeneous equation is

$$y'' - 2y' + y = 0,$$

and the complementary solution is

$$y_c(x) = C_1 e^x + C_2 x e^x , \qquad (b)$$

where C_1 and C_2 are arbitrary constants.

To determine a particular solution, consider the right side of equation (a):

$$(x^2 - 1)e^{2x} + (3x + 4)e^x.$$

It suggests a particular solution

$$y_p(x) = (A_2 x^2 + A_1 x + A_0)e^{2x} + (B_1 x + B_0)e^x, \qquad (c)$$

where A_2, A_1, A_0, B_1, and B_0 are undetermined coefficients. This, however, is not an appropriate form, for the complementary solution (b) includes, except for a multiplicative constant, the terms xe^x and e^x. We therefore modify (c) by multiplying all e^x terms by x. This gives $(B_1 x^2 + B_0 x)e^x$. This too is inappropriate because of the $B_0 x e^x$ term. We multiply again by x and obtain $(B_1 x^3 + B_0 x^2)e^x$. A particular solution will therefore be of the form

$$y_p(x) = (A_2 x^2 + A_1 x + A_0)e^{2x} + (B_1 x^3 + B_0 x^2)e^x . \qquad (d)$$

To determine the coefficients, we substitute (d) into (a). First we compute the first and second derivatives of $y_p(x)$:

$$y_p'(x) = 2A_2 x^2 e^{2x} + (2A_2 + 2A_1)xe^{2x} + (A_1 + 2A_0)e^{2x}$$
$$+ B_1 x^3 e^x + (3B_1 + B_0)x^2 e^x + 2B_0 x e^x ,$$

$$y_p''(x) = 4A_2 x^2 e^{2x} + (4A_2 + 4A_2 + 4A_1)xe^{2x}$$
$$+ (2A_2 + 2A_1 + 2A_1 + 4A_0)e^{2x}$$
$$+ B_1 x^3 e^x + (6B_1 + B_0)x^2 e^x$$
$$+ (6B_1 + 4B_0)x e^x$$
$$+ 2B_0 e^x .$$

Substituting into (a) and simplifying:

$$x^2 e^{2x} \ (4A_2 - 4A_2 + A_2)$$

$$+ \ xe^{2x} \ (4A_2 + 4A_2 + 4A_1 - 4A_2 - 4A_1 + A_1)$$

$$+ \ e^{2x} \ (2A_2 + 2A_1 + 2A_1 + 4A_0 - 2A_1 - 4A_0 + A_0)$$

$$+ \ x^3 e^x \ (B_1 - 2B_1 + B_1)$$

$$+ \ x^2 e^x \ (6B_1 + B_0 - 6B_1 - 2B_0 + B_0)$$

$$+ \ xe^x \ (6B_1 + 4B_0 - 4B_0)$$

$$+ \ e^x \ (2B_0)$$

$$= \ (x^2 - 1) \ e^{2x} + (3x + 4)e^x.$$

Equating coefficients of like terms:

$$A_2 = 1,$$

$$A_1 + 4A_2 = 0,$$

$$A_0 + 2A_1 + 2A_2 = -1,$$

$$2B_0 = 4,$$

$$6B_1 = 3.$$

Solving we find $A_2 = 1$, $A_1 = -4$, $A_0 = 5$, $B_1 = \frac{1}{2}$, and $B_0 = 2$. Substitution of these values into (d) gives a particular solution:

$$y_p(x) = (x^2 - 4x + 5)e^{2x} + \left[\frac{1}{2} x^3 + 2x^2 \right] e^x .$$

Summing this with the complementary solution (b), we

obtain the general solution:

$$y(x) = C_1 e^x + C_2 xe^x + (x^2 - 4x + 5)e^{2x}$$

$$+ \left[\frac{1}{2} x^3 + 2x^2 \right] e^x .$$

● **PROBLEM 2–25**

Find a particular solution of

$$y''' - 3y'' + 3y' - y = 3e^x. \tag{a}$$

Solution: A complementary solution of (a) is found by solving the associated homogeneous equation

$$y''' - 3y'' + 3y' - y = 0.$$

The characteristic equation of this d.e. is

$$p^3 - 3p^2 + 3p - 1 = 0 \quad \text{with roots } p = 1.$$

The complementary solution is therefore

$$y_c(x) = C_1 e^x + C_2 x e^x + C_3 x^2 e^x , \tag{b}$$

where C_1, C_2, and C_3 are all arbitrary constants.

The inhomogeneous term $(3e^x)$ suggests a particular solution Be^x, but since e^x, xe^x, and $x^2 e^x$ are all contained in the complementary solution (b), we must assume a particular solution of the form

$$y_p(x) = B x^3 e^x, \tag{c}$$

where B is a constant that we must determine. Substitution of (c) into equation (a) gives

$$\left\{ (4Bx\ e^x + 2Bx^2 e^x + 3Bx^2 e^x + Bx^3 e^x + 6Be^x + 6Bxe^x + 6Bxe^x \right.$$

$$+ 3Bx^2 e^x) - 3(2Bx^2 e^x + Bx^3 e^x + 6Bxe^x + 3Bx^2 e^x)$$

$$\left. + 3(Bx^3 e^x + 3Bx^2 e^x) - (Bx^3 e^x) \right\} = 3e^x.$$

Since the sum of the coefficients of all e^x terms must sum to 3, we have

$$6Be^x = 3e^x,$$

or $\quad B = \dfrac{1}{2}$.

Substitution of $B = \dfrac{1}{2}$ into equation (c) gives our particular solution

$$y_p(x) = \frac{1}{2} x^3 e^x.$$

Furthermore, we can now construct the general solution of (a)

$$y(x) = y_c(x) + y_p(x)$$

$$y(x) = C_1 e^x + C_2 xe^x + C_3 x^2 e^x + \frac{1}{2} x^3 e^x.$$

Determine y so that it will satisfy the equation

$$y''' - y' = 4e^{-x} + 3e^{2x} , \qquad (a)$$

with the conditions that when $x = 0$, $y = 0$, $y' = -1$, $y'' = 2$.

Solution: We must first solve the associated homogeneous equation

$$y''' - y' = 0.$$

Using the characteristic equation

$$p^3 - p = p(p^2 - 1) = 0,$$

which has roots $0, \pm 1$, we obtain the complementary solution

$$y_c(x) = C_1 + C_2 e^x + C_3 e^{-x} , \qquad (b)$$

where C_1, C_2, and C_3 are arbitrary constants.

To find a particular solution of (a), we consider the inhomogeneous term $(4e^{-x} + 3e^{2x})$. It suggests a solution of the form

$$y_p(x) = Ae^{-x} + Be^{2x} , \qquad (c)$$

but since e^{-x} is included in the complementary solution (b), we must modify (c) to obtain an appropriate particular solution:

$$y_p(x) = Axe^{-x} + Be^{2x} , \qquad (d)$$

where A and B are undetermined coefficients.

To determine A and B, we substitute (d) into (a):

$$-Axe^{-x} + 3Ae^{-x} + 8Be^{2x} + Axe^{-x} - Ae^{-x} - 2Be^{2x}$$

$$= 4e^{-x} + 3e^{2x},$$

or $2Ae^{-x} + 6Be^{2x} = 4e^{-x} + 3e^{2x} .$

Equating coefficients of e^{-x} and e^{2x}:

$$2A = 4,$$

$$6B = 3.$$

This gives A = 2, and B = $\frac{1}{2}$. Substitution of these values into (d) gives a particular solution:

$$y_p(x) = 2xe^{-x} + \frac{1}{2} e^{2x} .$$

Summing this with (b), the complementary solution, we obtain the general solution

$$y(x) = C_1 + C_2 e^x + C_3 e^{-x} + 2xe^{-x} + \frac{1}{2} e^{2x} . \qquad (e)$$

Having obtained the general solution, we can now solve for the arbitrary constants C_1, C_2, and C_3 by using the initial conditions. From (e) it follows that

$$y'(x) = C_2 e^x - C_3 e^{-x} - 2xe^{-x} + 2e^{-x} + e^{2x} \qquad (f)$$

and $$y''(x) = C_2 e^x + C_3 e^{-x} + 2xe^{-x} - 4e^{-x} + 2e^{2x} \qquad (g)$$

We put x = 0 in each of (e), (f), and (g) to obtain equations for C_1, C_2, and C_3. These are

$$0 = C_1 + C_2 + C_3 + \frac{1}{2} ,$$

$$-1 = C_2 - C_3 + 3,$$

$$2 = C_2 + C_3 - 2.$$

Solving this system, we find $C_1 = -\frac{9}{2}$, $C_2 = 0$, and $C_3 = 4$. Therefore the final result is

$$y(x) = -\frac{9}{2} + 4e^{-x} + 2xe^{-x} + \frac{1}{2} e^{2x} .$$

VARIATION OF PARAMETERS

• PROBLEM 2–27

Use variation of parameters to solve:
$$y' + \frac{4}{x} y = x^4 . \qquad (a)$$

Solution: To solve this inhomogeneous linear equation we must first determine the general solution of the associated homogeneous equation:

$$y' + \frac{4}{x} y = 0 .$$

Solve by using separation of variables:

$$\frac{y'}{y} + \frac{4}{x} = 0 \ .$$

Integrate both sides with respect to x:

$$\int \frac{y'}{y} \, dx + 4 \int \frac{dx}{x} = c \quad \text{(an arbitrary constant)},$$

$$\ln|y| + 4 \ln|x| = c \ ,$$

$$\ln|y| = c - 4 \ln|x| \ ,$$

$$\ln|y| = c + \ln|x^{-4}| \ .$$

Take the exponential of both sides to solve for $y(x)$:

$$y(x) = \ell^{c+\ln|x^{-4}|}$$

$$= \ell^c \, \ell^{\ln|x^{-4}|}$$

$$= \ell^c |x^{-4}| \ .$$

Letting $c_1 = \ell^c$, we rewrite the solution, denoting it by $y_c(x)$ to indicate the complementary solution:

$$y_c(x) = c_1 \, x^{-4} \ .$$

To construct a particular solution to the given inhomogeneous equation, (a) we replace the arbitrary constant c_1 with an undetermined function, $v(x)$, and obtain a trial solution

$$y_p(x) = v(x)x^{-4} \ .$$

In order for $y_p(x)$ to satisfy equation (a), we must have

$$v'(x)(x^{-4}) = x^4 \ . \tag{b}$$

This requirement enables us to determine the appropriate $v(x)$. We solve equation (b) for $v'(x)$

$$v'(x) = x^8 \ .$$

Computing the particular integral we get a solution $v(x)$ that is free of arbitrary constants:

$$v(x) = \int v'(x)dx = \int x^8 dx = \frac{x^9}{9} \ .$$

We may now write our particular solution as

$$y_p(x) = \frac{x^9}{9} \, x^{-4} = \frac{x^5}{9} \ .$$

We can now write the general solution to the given inhomogeneous equation (a) as the sum of the complementary and particular solutions:

$$y(x) = C_1 \, x^{-4} + \frac{x^5}{9} \; .$$

Solve

$$y'' - y' - 2y = e^{3x} \; , \tag{a}$$

by using variation of parameters.

Solution: To obtain the general solution of the inhomogeneous equation (a), variation of parameters is applicable because the equation is linear, of second order, and has constant coefficients.

First the corresponding homogeneous equation,

$$y'' - y' - 2y = 0 \; , \tag{b}$$

is solved, using the characteristic equation

$$p^2 - p - 2 = 0 \; .$$

The roots of this equation are $p_1 = -1$ and $p_2 = +2$. The corresponding solutions are $u_1 = e^{-x}$ and $u_2 = e^{2x}$. The general solution to equation (b):

$$y_c(x) = c_1 e^{-x} + c_2 e^{2x} \; , \tag{c}$$

where c_1 and c_2 are arbitrary constants, is the complementary solution to equation (a) and is used in that equation's general solution.

Assume a particular solution to equation (a) of the form

$$y_p(x) = u_1(x) \, w_1(x) + u_2(x) \, w_2(x) \; . \tag{d}$$

The determination of this solution is possible provided the following system of equations is satisfied simultaneously by the functions $w_1(x)$ and $w_2(x)$:

$$w_1'(x)(e^{-x}) + w_2'(x)(e^{2x}) = 0 \; ,$$

$$w_1'(x)(-e^{-x}) + w_2'(x)(2e^{2x}) = e^{3x} \; .$$

Using Cramer's Rule:

$$w_1' = \cfrac{\begin{vmatrix} 0 & e^{2x} \\[2ex] e^{3x} & 2e^{2x} \end{vmatrix}}{\begin{vmatrix} e^{-x} & e^{2x} \\[2ex] -e^{-x} & 2e^{2x} \end{vmatrix}} = \frac{-e^{4x}}{3} \quad ,$$

$$w_2' = \cfrac{\begin{vmatrix} e^{-x} & 0 \\[2ex] -e^{-x} & e^{3x} \end{vmatrix}}{\begin{vmatrix} e^{-x} & e^{2x} \\[2ex] -e^{-x} & 2e^{2x} \end{vmatrix}} = \frac{e^{x}}{3} \quad .$$

Integrating w_1' and w_2' with respect to x:

$$w_1(x) = \frac{-e^{4x}}{12} \quad \text{and} \quad w_2(x) = \frac{e^x}{3} \quad .$$

Substituting these values into the particular solution (equation (d)):

$$y_p(x) = -\frac{1}{12} e^{4x} e^{-x} + \frac{1}{3} e^x e^{2x}$$

$$y_p(x) = \frac{-1}{12} e^{3x} + \frac{1}{3} e^{3x} = \frac{1}{4} e^{3x} \quad .$$

Summing the complementary and particular solutions the general solution to equation (a) is obtained:

$$y(x) = c_1 e^{-x} + c_2 e^{2x} + \frac{1}{4} e^{3x} \quad .$$

• PROBLEM 2-29

Solve by variation of parameters:

$$y'' - y = e^x . \tag{a}$$

<u>Solution</u>: The general solution of this inhomogeneous equation is the sum of a particular solution, $y_p(x)$, and the complementary solution, $y_c(x)$:

$$y(x) = y_c(x) + y_p(x) ,$$

70

We first determine $y_c(x)$; it is the general solution of the associated homogeneous equation

$$y'' - y = 0 . \tag{b}$$

Its characteristic equation, $p^2 - 1 = 0$, has roots $p_1 = 1$, $p_2 = -1$ so that the general solution of (b) is

$$y_c(x) = c_1 e^x + c_2 e^{-x} ,$$

where c_1 and c_2 are arbitrary constants.

The particular solution $y_p(x)$ is constructed by replacing the constants c_1 and c_2 with functions $w_1(x)$ and $w_2(x)$:

$$y_p(x) = w_1(x)e^x + w_2(x)e^{-x} .$$

We know that $y_p(x)$ exists because the complementary functions e^x and e^{-x} are linearly independent, as evidenced by their nonzero Wronskian:

$$W(e^x, e^{-x}) = \begin{vmatrix} e^x & e^{-x} \\ (e^x)' & (e^{-x})' \end{vmatrix} = \begin{vmatrix} e^x & e^{-x} \\ e^x & -e^{-x} \end{vmatrix}$$

$$= -e^0 - e^0 = -2 \neq 0 .$$

To determine $w_1(x)$ and $w_2(x)$ that make $y_p(x)$ a particular solution of (a) we apply the restrictions:

$$w_1'(x)e^x + w_2'(x)e^{-x} = 0 ,$$
$$w_1'(x)e^x - w_2'(x)e^{-x} = e^x .$$

Solving simultaneously using Cramer's Rule:

$$w_1'(x) = \frac{\begin{vmatrix} 0 & e^{-x} \\ e^x & -e^x \end{vmatrix}}{\begin{vmatrix} e^x & e^{-x} \\ e^x & -e^{-x} \end{vmatrix}} = \frac{-e^0}{-e^0 - e^0} = \frac{-1}{-2} = \frac{+1}{2} .$$

$$w_2'(x) = \frac{\begin{vmatrix} e^x & 0 \\ e^x & e^x \end{vmatrix}}{\begin{vmatrix} e^x & e^{-x} \\ e^x & -e^{-x} \end{vmatrix}} = \frac{e^{2x}}{-e^0 - e^0} = \frac{e^{2x}}{-2} = \frac{-1}{2} e^{2x} .$$

Note that the denominator in each of these expressions is the Wronskian $W(e^x, e^{-x})$. To determine w_1 and w_2 we compute the particular integrals:

$$w_1(x) = \int w_1'(x)\,dx = \int \tfrac{1}{2}\,dx = \frac{x}{2} \ .$$

$$w_2(x) = \int w_2'(x)\,dx = \int -\tfrac{1}{2}\,e^{2x}\,dx = -\tfrac{1}{4}\,e^{2x} \ .$$

We need no constants of integration in either of these expressions because they are particular integrals and therefore free of arbitrary constants.

A particular solution to the inhomogeneous equation (a) is now determined:

$$y_p(x) = \frac{x}{2}(e^x) - \tfrac{1}{4}\,e^{2x}(e^{-x})$$

$$y_p(x) = e^x\left(\frac{x}{2} - \frac{1}{4}\right)$$

and the general solution of the given inhomogeneous equation (a) is therefore obtained as

$$y(x) = \left[c_1 e^x + c_2 e^{-x} + e^x\left(\frac{x}{2} - \frac{1}{4}\right)\right] \ .$$

Letting $c_3 = c_1 - \tfrac{1}{4}$ this simplifies to

$$y(x) = c_3 e^x + c_2 e^{-x} + \frac{x}{2}\,e^x \ .$$

• PROBLEM 2-30

Solve

$$y'' - 2y' + y = \frac{e^x}{x} \ ; \qquad\qquad (a)$$

$$y(1) = 0,\ y'(1) = 1 \ . \qquad\qquad (b)$$

Solution: In order to use the method of variation of parameters, we must first solve the associated homogeneous equation

$$y'' - 2y' + y = 0 \qquad\qquad (c)$$

The characteristic equation corresponding to (c) is

$$p^2 - 2p + 1 = 0 \ ,$$

which has double root $p = 1$. The complementary solution of (a) is therefore

$$y_c(x) = c_1 e^x + c_2 x e^x \ ,$$

where c_1 and c_2 are arbitrary constants.

We now look for a particular solution in the form

$$y_p(x) = v_1(x)e^x + v_2(x)x e^x \ , \qquad\qquad (d)$$

where v_1 and v_2 are functions that we will have to determine.
We do this by imposing the following conditions on their derivatives:

72

$$v_1' e^x + v_2' x e^x = 0 \ ,$$

$$v_1' e^x + v_2' e^x (x + 1) = \frac{e^x}{x} \ .$$

We solve this system of equations by using Cramer's Rule:

$$v_1' = \frac{\begin{vmatrix} 0 & xe^x \\ \dfrac{e^x}{x} & e^x(x+1) \end{vmatrix}}{\begin{vmatrix} e^x & xe^x \\ e^x & e^x(x+1) \end{vmatrix}}$$

$$= -\frac{e^{2x}}{e^{2x}(x+1) - xe^{2x}} = -\frac{e^{2x}}{e^{2x}} = -1 \ , \tag{e}$$

$$v_2' = \frac{\begin{vmatrix} e^x & xe^x \\ e^x & \dfrac{e^x}{x} \end{vmatrix}}{\begin{vmatrix} e^x & xe^x \\ e^x & e^x(x+1) \end{vmatrix}}$$

$$= \frac{e^{2x}/x}{e^{2x}(x+1) - xe^{2x}} = \frac{e^{2x}}{xe^{2x}} = \frac{1}{x} \ . \tag{f}$$

To determine v_1 and v_2 we take the particular integrals of (e) and (f):

$$v_1 = \int v_1' \ dx = \int -1 \ dx = -x \ ,$$

$$v_2 = \int v_2' \ dx = \int \frac{1}{x} \ dx = \ln|x| \ .$$

Substituting these values into (d), we obtain a particular solution:

$$y_p(x) = -xe^x + xe^x \ln|x| \ .$$

The general solution is the sum of the complementary and particular solutions:

$$y(x) = c_1 e^x + c_2 x e^x - xe^x + xe^x \ln|x| \ ,$$

or

$$y(x) = c_1 e^x + c_3 x e^x + xe^x \ln|x| \ , \tag{g}$$

where $c_3 = c_2 - 1$.

We now determine the constants c_1 and c_3 by substituting the initial conditions given in (b). First we compute the derivative of (g):

$$y'(x) = c_1 e^x + c_3 e^x + c_3 x e^x + e^x \ln|x| + xe^x \ln|x| + e^x \tag{h}$$

Applying the first initial condition to (g):

$$y(1) = c_1 e^1 + c_3 (1)e^1 + (1)\, e^1 \ln 1 = 0 \ ,$$

or, since $\ln 1 = 0$,

$$c_1 e + c_3 e = 0 \ . \tag{i}$$

Applying the second initial condition to (h):

$$c_1 e^1 + c_3 e^1 + c_3 (1)e^1 + e^1 \ln 1 + (1)e^1 \ln 1 + e^1 = 1,$$

or

$$c_1 e + 2c_3 e = 1 - e \ . \tag{j}$$

Solving (i) and (j) simultaneously we find

$$c_1 = -c_3 = \frac{e-1}{e} \ .$$

Substituting these values into (g) we obtain the solution of the initial value problem:

$$y = e^{x-1}(e-1)(1-x) + xe^x \ln|x| \ .$$

Find a particular solution of

$$y'' + y = \csc x \ . \tag{a}$$

__Solution__: To find a particular solution of this second order, linear, inhomogeneous, differential equation it is necessary to first determine its complementary solution, which is the general solution of the associated homogeneous equation:

$$y + y = 0 \ . \tag{b}$$

We use the characteristic polynomial $p^2 + 1 = 0$ and obtain complementary functions

$$u_1(x) = \cos x$$

and

$$u_2(x) = \sin x \ .$$

These must be checked for linear independence. To do so, compute their Wronskian:

$$W(u_1, u_2) = \begin{vmatrix} \cos x & \sin x \\ -\sin x & \cos x \end{vmatrix}$$

$$= \cos^2 x + \sin^2 x$$

$$= 1 \neq 0 \ .$$

Since u_1 and u_2 are linearly independent we may express the complementary solution as

$$y_c(x) = c_1 \cos x + c_2 \sin x \qquad\qquad (c)$$

where c_1 and c_2 are arbitrary constants. The linear independence

of u_1 and u_2 guarantees two conditions. First, equation (x)

gives the most general solution of equation (b). Secondly, we may
construct a particular solution using these complementary functions.
We assume the form

$$y_p(x) = w_1(x)\, u_1(x) + w_2(x)\, u_2(x) \; .$$

Provided that $w_1(x)$ and $w_2(x)$ are suitably chosen, we are guar-

anteed that this $y_p(x)$ exists.

To determine the functions $w_1(x)$ and $w_2(x)$ that make $y_p(x)$
a solution of equation (a), we solve the system:

$$w_1'(x)(\cos x) + w_2'(x)(\sin x) = 0 \; ,$$

$$w_1'(x)(-\sin x) + w_2'(x)(\cos x) = \csc x \; .$$

Since $W(u_1,u_2) \neq 0$, we can solve for $w_1'(x)$ and $w_2'(x)$ using
Cramer's Rule.

$$w_1'(x) = \frac{(-\sin x)(\csc x)}{W(\cos x,\, \sin x)} = \frac{-1}{+1} = -1 \; .$$

$$w_2'(x) = \frac{(\cos x)(\csc x)}{W(\cos x,\, \sin x)} = \cot x \; .$$

Integrating these expressions gives

$$w_1(x) = \int -dx = -x \; .$$

$$w_2(x) = \int \cot x \; dx = \ln(\sin x) \; .$$

Now the particular solution to equation (a) may be explicitly
expressed:

$$y_p(x) = (\cos x)(-x) + (\sin x)\, \ln(\sin x) \; .$$

Using superposition we construct the general solution of equation (a)
from the complementary and particular solutions we have determined:

$$y(x) = y_c(x) + y_p(x)$$

$$[y(x) = c_1 \cos x + c_2 \sin x - (x \cos x) + (\sin x)\, \ln(\sin x)]$$

where c_1 and c_2 are arbitrary constants arising from the general
solution of the homogeneous equation (b).

Solve, using variation of parameters:
$$y''' + y' = \sec x .$$
(a)

<u>Solution:</u>   The associated homogeneous equation
$$y''' + y' = 0$$
(b)

must first be solved. The characteristic equation $p^3 + p = 0$ has roots $p_1 = 0$, $p_2 = i$, $p_3 = -i$ so that the general solution of the homogeneous equation (b) is
$$y_c(x) = c_1 + c_2 \cos x + c_3 \sin x ,$$
(c)

where c_1, c_2, c_3 are arbitrary constants. This is the complementary solution of (a) and by replacing the constants c_1, c_2, c_3 with yet unknown functions $w_1(x)$, $w_2(x)$, $w_3(x)$, we will attempt to construct a particular solution that satisfies (a).
$$y_p(x) = w_1(x) + w_2(x)(\cos x) + w_3(x)(\sin x) .$$

The existence of such a solution is guaranteed if the set of solutions of the homogeneous equation is linearly independent. If so, their Wronskian, $w(1, \cos x, \sin x)$ will be identically nonzero:

$$W(1, \cos x, \sin x) = \begin{vmatrix} 1 & \cos x & \sin x \\ 0 & -\sin x & \cos x \\ 0 & -\cos x & -\sin x \end{vmatrix}$$

$$= \sin^2 x + \cos^2 x = 1 \neq 0 .$$

To construct the particular solution $y_p(x)$, we must determine the functions $w_1(x)$, $w_2(x)$, $w_3(x)$. This is done by applying the set of restrictions:

$$w_1'(1) + w_2'(\cos x) + w_3'(\sin x) = 0$$

$$w_1'(1)' + w_2'(\cos x)' + w_3'(\sin x)' = 0$$

$$w_1'(1)'' + w_2'(\cos x)'' + w_3'(\sin x)'' = \sec x .$$

Solving these simultaneously, starting with the last two equations gives

$$w_2'(x) = -1 ,$$

$$w_3'(x) = w_2' \tan x = -\tan x ,$$

$$w_1'(x) = \sec x .$$

Integration gives the solutions

$$w_1(x) = \ln|\sec x + \tan x| \; ,$$

$$w_2(x) = -x \; ,$$

$$w_3(x) = \ln|\cos x| \; .$$

Substituting into the particular solution gives

$$y_p(x) = \ln|\sec x + \tan x| - x \cos x + (\sin x)\ln|\cos x| \; .$$

Adding this to the complementary solution, $y_c(x)$, gives the general solution of the inhomogenous equation (a):

$$y(x) = y_c(x) + y_p(x) \; ,$$

$$y(x) = c_1 + c_2 \cos x + c_3 \sin x + \ln|\sec x + \tan x|$$
$$- x \cos x + (\sin x) \ln|\cos x| \; .$$

• PROBLEM 2-33

Find the general integral of the differential equation:

$$xy'' + y' - \frac{4}{x} y = x + x^3 \; . \tag{a}$$

Solution: Restricting x to the interval $<0,\infty>$, we may rewrite equation (a) as

$$x^2 y'' + xy' - 4y = x(x + x^3) \; . \tag{b}$$

To solve equation (b) by using the method of variation of parameters we must first find the general solution of the associated homogeneous equation

$$x^2 y'' + xy' - 4y = 0 \; . \tag{c}$$

Recognizing (c) as a form of the Euler Equation, we write its indicial polynomial:

$$r(r - 1) + r - 4 = r^2 - 4 \; .$$

Finding roots $r_1 = 2$, $r_2 = -2$, we express the general solution of (c) as

$$y_c(x) = c_1 x^2 + c_2 x^{-2} \; , \tag{d}$$

where c_1 and c_2 are arbitrary constants. We call (d) the complementary solution of the inhomogeneous differential equation (b).

To find a particular solution, we replace the constants c_1 and c_2 in (d) by unknown functions of x:

$$y_p(x) = v_1(x)x^2 + v_2(x)x^{-2} . \qquad (e)$$

To determine the v_1 and v_2 that make (e) a solution of (b), we impose two conditions on the derivatives of v_1 and v_2:

$$v_1'x^2 + v_2'x^{-2} = 0 ,$$

$$v_1'2x + v_2'(-2x^{-3}) = \frac{x(x + x^3)}{x^2} = 1 + x^2 .$$

We solve this system using Cramer's Rule:

$$v_1' = \frac{\begin{vmatrix} 0 & x^{-2} \\ 1+x^2 & -2x^{-3} \end{vmatrix}}{\begin{vmatrix} x^2 & x^{-2} \\ 2x & -2x^{-3} \end{vmatrix}}$$

$$= -\frac{(1+x^2)}{x^2}\left(-\frac{x}{4}\right) = \frac{1+x^2}{4x} , \qquad (f)$$

and

$$v_2' = \frac{\begin{vmatrix} x^2 & 0 \\ 2x & 1+x^2 \end{vmatrix}}{\begin{vmatrix} x^2 & x^{-2} \\ 2x & -2x^{-3} \end{vmatrix}}$$

$$= x^2(1 + x^2)\left(-\frac{x}{4}\right) = -\frac{x^3}{4}(1 + x^2) . \qquad (g)$$

Integration of (f) and (g) gives

$$v_1(x) = \int v_1' \, dx = \tfrac{1}{4} \int \left(\frac{1}{x} + x\right) dx = \tfrac{1}{4}\left(\ln x + \tfrac{1}{2}x^2\right)$$

$$= \tfrac{1}{4}\ln x + \tfrac{1}{8}x^2 , \qquad (h)$$

$$v_2(x) = \int v_2' \, dx = -\tfrac{1}{4} \int (x^3 + x^5) \, dx = -\tfrac{1}{4} \int \left(\tfrac{1}{4}x^4 + \tfrac{1}{6}x^6\right)$$

$$= -\frac{1}{16}x^4 - \frac{1}{24}x^6 . \qquad (i)$$

78

Substituting (h) and (i) into (e), we obtain a particular solution:

$$y_p(x) = \tfrac{1}{4} x^2 \ln x + \frac{1}{8} x^4 - \frac{1}{16} x^2 - \frac{1}{24} x^4 ,$$

or

$$y_p(x) = \tfrac{1}{4} x^2 \ln x + \frac{1}{12} x^4 - \frac{1}{16} x^2 .$$

Summing this with the complementary solution (d), we obtain the general solution of (a):

$$y(x) = c_3 x^2 + \frac{c_2}{x^2} + \tfrac{1}{4} x^2 \ln x + \frac{1}{12} x^4 ,$$

where $c_3 = \left(c_1 - \frac{1}{16} \right) .$

• PROBLEM 2–34

Solve:
$$x^2 y'' - 2xy' + 2y = x^3 \sin x . \tag{a}$$

<u>Solution</u>: We first solve the associated homogeneous equation

$$x^2 y'' - 2xy' + 2y = 0 . \tag{b}$$

If we confine x to the interval $\langle 0, \infty \rangle$ we can solve (b) as a form of the Euler Equation. We can write the indicial polynomial that corresponds to (b):

$$r(r-1) - 2r + 2 = r^2 - 3r + 2 ;$$

its roots are $r_1 = +1, r_2 = +2$. The general solution of (b), the complementary solution, is therefore

$$y_c(x) = c_1 x + c_2 x^2 , \tag{c}$$

where c_1 and c_2 are arbitrary constants

We obtain a particular solution of equation (a) by replacing the constants c_1 and c_2 with unknown functions of x:

$$y_p(x) = v_1(x)x + v_2(x)x^2 . \tag{d}$$

79

We determine v_1 and v_2 by imposing the following conditions on their derivatives:

$$v_1'x + v_2'x^2 = 0 \ ,$$

$$v_1' + v_2' \, 2x = \frac{x^3 \sin x}{x^2} = x \sin x \ .$$

We solve these equations by using Cramer's Rule:

$$v_1'(x) = \frac{\begin{vmatrix} 0 & x^2 \\ x \sin x & 2x \end{vmatrix}}{\begin{vmatrix} x & x^2 \\ 1 & 2x \end{vmatrix}}$$

$$v_1' = - \frac{x^3 \sin x}{2x^2 - x^2} = - x \sin x \ . \tag{e}$$

and

$$v_2' = \frac{\begin{vmatrix} x & 0 \\ 1 & x \sin x \end{vmatrix}}{\begin{vmatrix} x & x^2 \\ 1 & 2x \end{vmatrix}}$$

$$v_2' = \frac{x^2 \sin x}{2x^2 - x^2} = \sin x \ . \tag{f}$$

Taking the particular integrals of (e) and (f):

$$v_1(x) = \int v_1' \, dx = - \int x \sin x \, dx = - \sin x + x \cos x, \tag{g}$$

$$v_2(x) = \int v_2' \, dx = \int \sin x \, dx = -\cos x \ . \tag{h}$$

We do not add constants of integration because we are constructing a particular solution, which must be free of arbitrary constants.

Substituting (g) and (h) into (d) we obtain a particular solution:

$$y_p(x) = - x \sin x + x^2 \cos x - x^2 \cos x = - x \sin x \ .$$

Summing this with the complementary solution (c) we obtain the general solution of (a):

$$y(x) = c_1 x + c_2 x^2 - x \sin x \ .$$

REDUCTION OF ORDER

Solve

$$p^2 - 4y = 0 \tag{a}$$

where $p = \dfrac{dy}{dx}$.

<u>Solution</u>:   We write (a) as

$$y - \frac{1}{4} p^2 = 0$$

and differentiate to obtain

$$\frac{dy}{dx} - \frac{1}{2} p \frac{dp}{dx} = 0$$

or $p - \dfrac{1}{2} p \dfrac{dp}{dx} = 0.$

Factoring,

$$p \left(1 - \frac{1}{2} \frac{dp}{dx}\right) = 0.$$

The solution of this equation is

$$p = 0 \tag{b}$$

or $1 - \dfrac{1}{2} \dfrac{dp}{dx} = 0. \tag{c}$

For $p = 0$ or $\frac{dy}{dx} = 0$ we substitute in (a) to obtain

$$- 4y = 0 \qquad \text{or}$$

$$y = 0$$

which is a solution.

For $\quad 1 - \frac{1}{2} \frac{dp}{dx} = 0$,

$$\frac{dp}{dx} = 2$$

Separating variables and integrating

$$\int dp = 2 \int dx + C$$

where C is an integration constant

$$p = 2x + C.$$

Substituting this in (a)

$$(2x + C)^2 - 4y = 0$$

or $\qquad\qquad\qquad y = \frac{1}{4} (2x + C)^2$

which is also a solution for any value of C.

• PROBLEM 2–36

Solve

$$y - \frac{xp}{2} - \frac{x}{2p} = 0 \qquad\qquad\qquad (a)$$

where $\quad p = \frac{dy}{dx} \neq 0.$

Solution: Differentiate (a)

$$\frac{dy}{dx} - \frac{1}{2} \left(x \frac{dp}{dx} + p \right) - \frac{2p - x \left(2 \frac{dp}{dx} \right)}{(2p)^2} = 0$$

$$p - \frac{1}{2} \left(x \frac{dp}{dx} + p \right) + 2 \left(\frac{x \frac{dp}{dx} - p}{4p^2} \right) = 0,$$

$$2p - \frac{x \, dp}{dx} - p + \frac{x}{p^2} \frac{dp}{dx} - \frac{1}{p} = 0,$$

82

$$2p^3 - xp^2 \frac{dp}{dx} - p^3 + \frac{x \ dp}{dx} - p = 0,$$

$$p^3 + \frac{x \ dp}{dx} (1 - p^2) - p = 0,$$

$$p(p^2 - 1) + (1 - p^2)\left[x \frac{dp}{dx}\right] = 0, \quad \text{and finally,}$$

$$(p^2 - 1)\left[x \frac{dp}{dx} - p\right] = 0 \qquad\qquad (b)$$

Since equation (b) is factored it reduces to

$$p^2 - 1 = 0 \qquad \text{or} \qquad x \frac{dp}{dx} - p = 0.$$

The first factor yields

$$p^2 - 1 = 0,$$

$$\left(\frac{dy}{dx}\right) = \pm 1,$$

$$dy = \pm dx$$

Integrating

$$y = \pm x + C_1 \qquad\qquad (c)$$

where C_1 is the integration constant.

For the second factor

$$x \frac{dp}{dx} = p$$

Separating variables and integrating

$$\int \frac{dp}{p} = \int \frac{dx}{x} + C_2$$

(C_2 being the integration constant).

$$\ln p = \ln x + C_2$$

Exponentiating

$$p = C_3 \ x \qquad\qquad (d)$$

where $C_3 \equiv e^{C_2}$. Thus

$$\frac{dy}{dx} = C_3 \ x$$

$$\int dy = C_3 \int x \, dx + C_4$$

$$y = \frac{C_3 x^2}{2} + C_4 \qquad\qquad (e)$$

Again, C_4 is an integration constant. Thus (c) and (e) constitute our solution.

• **PROBLEM** 2–37

Solve

$$\frac{d^2y}{dx^2} + \frac{1}{x} \frac{dy}{dx} - 2 = 0 \qquad\qquad (a)$$

by reduction of order.

Solution: Let $p = \frac{dy}{dx}$. Then $\frac{d^2y}{dx^2} = \frac{dp}{dx}$ and (a) becomes

$$\frac{dp}{dx} + \frac{1}{x} p = 2. \qquad\qquad (b)$$

We solve this by use of an integrating factor ϕ such that

$$\frac{d}{dx} (\phi p) = \phi \frac{dp}{dx} + \frac{\phi p}{x}$$

or $\quad \phi \frac{dp}{dx} + p \frac{d\phi}{dx} = \phi \frac{dp}{dx} + \frac{\phi p}{x}$.

Subtracting $\phi \frac{dp}{dx}$ and dividing by p yields

$$\frac{d\phi}{dx} = \frac{\phi}{x} .$$

Separating variables and integrating

$$\int \frac{d\phi}{\phi} = \int \frac{dx}{x}$$

$$\ln \phi = \ln x$$

and so $\quad \phi = x$.

We multiply (b) by $\phi = x$ to obtain

$$\frac{d}{dx} (xp) = 2x$$

$$\int d (xp) = 2 \int x \, dx + C_1$$

where C_1 is the integration constant.

$$xp = x^2 + C_1$$

$$p = x + \frac{C_1}{x}$$

or $\quad \frac{dy}{dx} = x + \frac{C_1}{x}$

Separating variables and integrating

$$\int dy = \int \left(x + \frac{C_1}{x}\right) dx + C_2$$

where C_2 is again a constant of integration. Finally,

$$y = \frac{x^2}{2} + C_1 \ln |x| + C_2$$

● **PROBLEM** 2–38

Solve the differential equation:

$$y\frac{d^2y}{dx^2} + \left(\frac{dy}{dx}\right)^2 = 1.$$

<u>Solution:</u> In this kind of differential **equation,**
of the form

$$f\left(y", y', y\right),$$

we substitute $y' = \frac{dy}{dx} = p$ which, in turn, gives an
expression for y", namely $y" = \frac{dp}{dx}$.

But, if we set $\frac{dy}{dx} = p$, using the chain rule,
we can also write, instead, that:

$$y" = \frac{d^2y}{dx} = \frac{dp}{dy}\frac{dy}{dx} = p\frac{dp}{dy},$$

a more useful formulation here, since it does not
involve χ. This gives:

$$yp\frac{dp}{dy} + p^2 = 1.$$

Separating variables:

$$\frac{dy}{y} = \frac{pdp}{1-p^2}.$$

85

Integration yields:

$$\ln y = -\frac{1}{2} \ln\left(1-p^2\right) + \ln c,$$

$$2\ln y + \ln\left(1-p^2\right) = 2\ln c$$

or

$$\ln\left|y^2\left(1-p^2\right)\right| = \ln c^2.$$

Letting $\ln c^2 = c_1$, we have

$$y^2\left(1-p^2\right) = c_1, \text{ or } \frac{dy}{dx} = \frac{\pm \sqrt{y^2-c_1}}{y}.$$

Separating variables again,

$$\pm \frac{ydy}{\sqrt{y^2-c_1}} = dx,$$

and, integrating: $\sqrt{y^2-c_1} = \pm x + c_2,$

or

$$y^2 = \left(\pm x + c_2\right)^2 + c_1.$$

• **PROBLEM** 2–39

Solve the equation: $\dfrac{d^2y}{dx^2} = -y.$

<u>Solution</u>: Let $p = \dfrac{dy}{dx}$, then $\dfrac{d^2y}{dx^2} = \dfrac{dp}{dx} = \dfrac{dp}{dy}\dfrac{dy}{dx} = p\dfrac{dp}{dy}.$

Substitution gives

$$p\frac{dp}{dy} = -y.$$

Separating variables and integrating, we obtain:

$$p^2 = -y^2 + a^2 \text{ or } p = \frac{dy}{dx} = \pm\sqrt{a^2-y^2},$$

where a^2 is a constant of integration.

Separating variables and integrating once more, we have

$$\sin^{-1}(y/a) = \pm x + c,$$

or
$$y = a \sin(\pm x + c).$$

Another approach is the following: Since this equation is a homogeneous, linear and second-order differential equation with constant coefficients we have, in a general form:

$$\frac{d^2y}{dx^2} + a_1 \frac{dy}{dx} + a_2 y = 0$$

where $a_1 = 0$ and $a_2 = 1$.

The auxiliary equation is then

$$m^2 + 1 = 0$$

with roots $m = \pm i$.

Thus, the general solution is $y = c_1 e^{ix} + c_2 e^{-ix}$.

From Euler's formula about complex exponents

$$e^{\pm ix} = \cos x \pm i \sin x.$$

Setting $c_1 + c_2 = k_1$ and $c_1 - c_2 = k_2$, and factoring out the cosine and sine parts, the general solution we obtain will be, in a harmonic form,

$$y = k_1 \cos x + ik_2 \sin x.$$

● **PROBLEM** 2–40

Solve the differential equation:

$$\frac{d^2y}{dx^2} = \cos 2x.$$

Solution: $\frac{dy}{dx} = \frac{1}{2} \sin 2x + a$. Hence

$y = -\frac{1}{4} \cos 2x + ax + b$ is the general solution. This solution is obtained by setting $p = \frac{dy}{dx}$, and hence

$$\frac{dp}{dx} = \frac{d^2y}{dx^2}.$$

If the given equation had been expressed as (y'', y', y), we would (by the chain rule) have used

$$\frac{d^2y}{dx^2} = \frac{dp}{dy} \frac{dy}{dx} = p\frac{dp}{dy}.$$

Substituting $\frac{d^2y}{dx^2} = \frac{dp}{dx}$ in the given differential equation, and integrating yields an expression from which we solve for p. Substituting $\frac{dy}{dx}$ back for p, we obtain the first-order differential equation. Further integration gives the desired general solution. Once we are used to this technique, we can automatically write down the first-order and the general equations.

Solve

$$y'' + k^2y = 0 \qquad\qquad (a)$$

by reduction of order.

<u>Solution:</u>   Let $p = \frac{dy}{dx} = y'$. Then

$$y'' = \frac{dp}{dx} = \frac{dp}{dy} \frac{dy}{dx} = p \frac{dp}{dy} \quad .$$

Substituting in (a) yields

$$p \frac{dp}{dy} + k^2y = 0$$

Separating variables

$$p \, dp = - k^2 \, y \, dy$$

Integrating,

$$\int p \, dp = - k^2 \int y \, dy + C$$

where C is the constant of integration.   Thus,

$$\frac{p^2}{2} = - \frac{k^2 \, y^2}{2} + C$$

or $p = \pm \sqrt{2C - k^2 \, y^2}$

Let $a^2 = \frac{2C}{k^2}$

88

Then $\quad$ $p = \pm k \sqrt{a^2 - y^2}$,

$$\frac{dy}{dx} = \pm k \sqrt{a^2 - y^2} .$$

Separating variables again

$$\frac{dy}{\sqrt{a^2 - y^2}} = \pm k \, dx$$

Integrating

$$\int \frac{dy}{\sqrt{a^2 - y^2}} = \pm k \int dx + A$$

where A is a constant of integration. Since

$$\int \frac{dy}{\sqrt{a^2 - y^2}} = \sin^{-1} \left(\frac{y}{a}\right)$$

we have $\quad$ $\sin^{-1} \left(\frac{y}{a}\right) = \pm kx + A.$

Taking the sin of both sides

$$y = a \sin (\pm kx + A)$$

which is one form of the required solution.

• PROBLEM 2–42

Given that $x = e^{s^2}$ is a solution of

$$x'' - 2sx' - 2x = 0 \tag{a}$$

find a second solution.

Solution: $\qquad$ Let

$$x = e^{s^2} Z(s)$$

where Z(s) is a function to be determined. Then

$$x' = 2se^{s^2} Z + e^{s^2} Z'$$

and $\quad$ $x'' = 2e^{s^2} Z + 2s(2se^{s^2})Z + 2se^{s^2} Z' + 2se^{s^2} Z' + e^{s^2} Z''.$

or $\quad$ $x'' = (2 + 4s^2)e^{s^2} Z + 4se^{s^2} Z' + e^{s^2} Z'')$

89

Substituting these values in (a)

$$e^{s^2}\{(2 + 4s^2)Z + 4sZ' + Z'' - 4s^2Z - 2sZ' - 2Z\} = 0$$

or $\quad e^{s^2}(Z'' + 2sZ') = 0$

Since $e^{s^2} \neq 0$ we must have

$$Z'' + 2sZ' = 0$$

Let $p = Z'$; then

$$p' + 2sp = 0$$

Separating variables

$$\frac{dp}{p} = - 2s \ ds$$

Integrating

$$\int \frac{dp}{p} = - 2 \int s \ ds + C_1$$

where C_1 is a constant of integration.

$$\ln p = - s^2 + C_1$$

Exponentiating

$$p = C_2 e^{-s^2}$$

where $C_2 = e^{C_1}$.

Thus $Z' = C_2 e^{-s^2}$

$$\int \frac{dZ}{ds} \ ds = C_2 \int e^{-s^2} \ ds + C_3$$

where C_3 is a constant of integration.

Finally

$$Z(s) = C_2 \int e^{-s^2} \ ds + C_3.$$

Thus, another solution is

$$x = C_2 e^{s^2} \int e^{-s^2} \ ds + C_3 e^{s^2}.$$

CHAPTER 3

APPLICATIONS OF DIFFERENTIAL EQUATIONS

Basic Attacks and Strategies for Solving Problems in this Chapter. See pages 92 to 140 for step-by-step solutions to problems.

Differential equations arise frequently in physics, engineering, and chemistry, and on occasion in such subjects as biology, physiology, and economics. The solution of differential equation plays an important role in the study of the motions of heavenly bodies such as planets and moons, and for tracking artificial satellites.

Many physical problems involve differential equations of the first order. A classic example is the problem of determining the velocity of a particle projected in a radial direction outward from the earth and acted upon by only one force, the gravitational attraction of the earth.

The current at each point in an electrical network may be determined by solving the equations that result from applying Kirchhoff's laws:

1. The sum of the currents into (or away from) any point is zero, and

2. Around any closed path the sum of the instantaneous voltage drops in a specified direction is zero.

Systems of differential equations occur naturally in the application of Kirchhoff's laws to electric networks.

Step-by-Step Solutions to Problems in this Chapter, "Applications of Differential Equations"

APPLICATIONS OF THE FIRST-ORDER DIFFERENTIAL EQUATIONS

• PROBLEM 3–1

A body falls from rest under the action of gravity. The fall takes place in a viscous medium offering resistance proportional to the velocity. Find expressions for its velocity and distance fallen at any time t.

Solution: Take the origin at the starting point, and let y denote the distance of the body from this origin measured as positive downward. We know that the downward force of the body, with no external force, is its weight mg, where m is the mass and g is the gravitational acceleration. This force is opposed by an upward force which tends to retard any downward motion of the body. This is the resistance force (friction) which, according to the given data, is proportional to the velocity. Thus

$$F_r = kvm \quad \text{or resistance,} \quad R = kv$$

where k is the constant of proportionality.

The net downward force is then

$$m \frac{d^2y}{dt^2} = m(g-kv)$$

or

(a) $$\frac{d^2y}{dt^2} = \frac{dv}{dt} = g - kv,$$

the differential equation of motion.

Separating variables, we obtain:

92

$$\frac{dv}{g-kv} = dt.$$

Integration gives

(b) $-\frac{1}{k} \ln(g-kv) = t + c_1.$

Since the body falls from rest, we have the initial condition v=0 when t=0; substituting these values in (b), we find $c_1 = -1/k \ln g$. Then (b) becomes

$-\frac{1}{k} \ln(g-kv) = t - \frac{1}{k} \ln g,$ or $\ln(g-kv) - \ln g = -kt,$

or

$\ln \frac{g-kv}{g} = -kt,$ hence $\frac{g-kv}{g} = e^{-kt}$

from which we have

(c) $v = \frac{g}{k}\left(1-e^{-kt}\right).$

From this we see that $v \to g/k$ when $t \to \infty$. We therefore call the constant g/k the limiting velocity; from (a) we find the acceleration dv/dt = 0 when v = g/k.

Replacing v by $\frac{dy}{dt}$ and integrating again, we get

$$y = \frac{g}{k} t + \frac{g}{k^2} e^{-kt} + c_2.$$

Using the initial condition y=0 when t=0, we find $c_2 = -g/k^2$. Then

(d) $y = \frac{g}{k} t - \frac{g}{k^2}\left(1-e^{-kt}\right).$

● PROBLEM 3–2

A body falls with an initial velocity of 1000 ft./sec. and is subject to the acceleration of gravity $\left(g \simeq 32 \text{ ft./sec.}^2\right)$. What distance does it fall in 3 sec.?

<u>Solution:</u> Let v = velocity in ft./sec.

$$v = \frac{ds}{dt} = gt + v_0 = 32t + v_0$$

or ds = $\left(32t + v_0\right)$dt (differential-equation form).
Integrating, s = $\int 32t \cdot dt + v_0 \int dt + C$

$$= 32 \cdot \frac{t^2}{2} + v_0 t + C = 16t^2 + v_0 t + C.$$

Now s = 0 when t = 0, which we substitute.

$$0 = 0 + 0 + C \text{ and } C = 0.$$

Therefore, $s = 16t^2 + v_0 t.$

When t = 3 and v_0 = 1,000, as given, then

$s = 16 \cdot (3)^2 + 1,000 \cdot 3 = 3,144\text{-ft. drop in 3 sec.}$

• **PROBLEM** 3-3

A horizontal weightless spring has one end fixed while the other is attached to a mass of m slugs. k is the spring constant. If the mass is moving with velocity v_0 when the spring is unstretched, find v as a function of the stretch.

Solution:

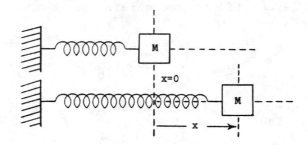

From Hooke's law we know that the spring force is proportional to the displacement of the object. Letting x = 0 be the point where the spring is unstretched, and x the displacement we have the spring force as - kx (it is negative because the force is always towards x = 0).

By Newton's second law,

$$F = m \frac{dv}{dt} .$$

Thus, $- kx = m \dfrac{dv}{dt} .$

But we require v as a function of displacement, i.e. v = f(x).

Now $m \dfrac{dv}{dt} = m \dfrac{dv}{dx} \cdot \dfrac{dx}{dt}$ (by the chain rule).

Since $v = \dfrac{dx}{dt}$ (i.e. velocity is the instantaneous rate of displacement),

$$m \frac{dv}{dx} \cdot \frac{dx}{dt} = mv \frac{dv}{dx} \, .$$

Thus, $- kx = mv \frac{dv}{dx}$ which is a separable equation. Its solution is given by $mv^2 = - kx^2 + c$.

To evaluate c, we use the information that $v = v_0$ when $x = 0$. Then $c = mv_0{}^2$. Thus,

$$mv^2 + kx^2 = mv_0^2 \, .$$

Inserting $\frac{1}{2}$ in the above equation, we obtain

$$\frac{1}{2} mv^2 + \frac{1}{2} kx^2 = \frac{1}{2} mv_0{}^2 = \text{constant} \, .$$

The final equation tells us that the kinetic energy of the mass plus the potential energy stored in the spring is constant.

• PROBLEM 3–4

A 4 lb. weight is attached to the lower end of a coil spring which hangs vertically from a fixed support. The weight comes to rest in its equilibrium position, thereby stretching the spring 6 inches. The weight is then pulled down 3 inches below this equilibrium position and released at t = 0. The medium offers a resistance in pounds numerically equal to a $\frac{dx}{dt}$, where a > 0.

(a) Determine the value of "a" such that the resulting motion would be critically damped and determine the displacement for this critical value of a.

(b) Determine the displacement if "a" is equal to one-half the critical value found in step (a).

(c) Determine the displacement if "a" is equal to twice the critical value found in step (a).

Solution: (a) F = ma

F = 4 (lbs) = m (32 ft/sec^2) so that m = $\frac{1}{8}$(slug) but F=kx, or

F = 4(lbs) = k (1/2 ft) so that k = 8 (lbs/ft).
The initial conditions are:

(1)   x(0) = $\frac{1}{4}$ ft

(2) $\left.\dfrac{dx}{dt}\right|_{t=0} = 0$.

The resulting differential equation is

$$\frac{1}{8} \frac{d^2x}{dt^2} + a \frac{dx}{dt} + 8x = 0$$

or $\dfrac{d^2x}{dt^2} + 8a \dfrac{dx}{dt} + 64x = 0$.

The auxiliary equation is

$$r^2 + 8ar + 64 = 0 \tag{1}$$

or $r = \dfrac{-8a \pm \sqrt{64a^2 - 4 \cdot 64}}{2} = -4a \pm 4\sqrt{a^2 - 4}$.

In order for our system to be critically damped, $a^2 - 4$ must equal 0. Therefore, $a = \pm 2$, but $a > 0$ implies $a = 2$.

The displacement in critically damped motion is given by

$$x = (c_1 + c_2 t)e^{-\frac{a}{2m}t} \qquad . \qquad \text{Thus}$$

$$x = (c_1 + c_2 t)e^{-8t}.$$

To find the constants c_1 and c_2 we must use the initial conditions:

$$x(0) = \frac{1}{4} = c_1 + c_2 (0) e^{-8t} \qquad \text{or} \quad c_1 = \frac{1}{4}$$

$$\frac{dx}{dt} = -8c_1 e^{-8t} + c_2 e^{-8t} - 8c_2 t e^{-8t}$$

$$\left.\frac{dx}{dt}\right|_{t=0} = 0 = -8c_1 + c_2$$

$$0 = (-8)\frac{1}{4} + c_2$$

so that $c_2 = 2$.

Thus, $x(t) = \left[\dfrac{1}{4} + 2t\right] e^{-8t}$.

(b)   If $a = 1$ then $r = -4 \pm 4\sqrt{3}\,i$ by (1) and

$$x(t) = e^{-4t} (c_1 \sin 4\sqrt{3}t + c_2 \cos 4\sqrt{3}t) .$$

Initial conditions give

$$x(0) = \frac{1}{4} = c_2$$

$$\frac{dx}{dt} = -4c_1 e^{-4t} \sin 4\sqrt{3}t + 4\sqrt{3}\, c_1 e^{-4t} \cos 4\sqrt{3}t +$$

$$-4e^{-4t} c_2 e^{-4t} \cos 4\sqrt{3}t + 4\sqrt{3}\, c_2 e^{-4t} \sin 4\sqrt{3}t$$

$$\frac{dx}{dt}\bigg|_{t=0} = 0 = 4\sqrt{3}c_1 - 4c_2 = 4\sqrt{3}c_1 - 1$$

or

$$c_1 = \frac{\sqrt{3}}{12} .$$

This yields

$$x(t) = e^{-4t}\left[\frac{\sqrt{3}}{12}\sin 4\sqrt{3}t + \frac{1}{4}\cos 4\sqrt{3}t\right] .$$

(c) If a = 4 then r = $-16 \pm 4\sqrt{12}$

$$= -16 \pm 8\sqrt{3}$$

or

$$r_1 = -16 + 8\sqrt{3}$$

$$r_2 = -16 - 8\sqrt{3}$$

and $x(t) = c_1 e^{(-16 + 8\sqrt{3})t} + c_2 e^{(-16 - 8\sqrt{3})t}$

so that

$$x(0) = \frac{1}{4} = c_1 + c_2 . \tag{2}$$

$$\frac{dx}{dt} = (-16 + 8\sqrt{3})c_1 e^{(-16+8\sqrt{3})t} + (-16 - 8\sqrt{3})c_2 e^{(-16-8\sqrt{3})t}$$

$$\frac{dx}{dt}\bigg|_{t=0} = 0 = (-16 + 8\sqrt{3})c_1 + (-16 - 8\sqrt{3})c_2 . \tag{3}$$

Solving for c_1 and c_2 yields

$$c_1 = \frac{3 + 2\sqrt{3}}{24} \qquad c_2 = \frac{3 - 2\sqrt{3}}{24}$$

so that

$$x(t) = \frac{3 + 2\sqrt{3}}{24} e^{(-16+8\sqrt{3})t} + \left(\frac{3 - 2\sqrt{3}}{24}\right) e^{(-16-8\sqrt{3})t} .$$

A certain type of glass is such that a slab 1" thick absorbs one-quarter of the light which starts to pass through it. How thin must a pane be made to absorb only 1 per cent of the light?

Solution: From the information given we know that the light passing through a slab is proportional to the thickness of the slab. Let t = thickness of the slab, and x = x(t), the fractional part of entering light which passes through a pane of thickness t. We form the differential equation

$$\frac{dx}{dt} = -kx; \quad x(0) = 1; \quad x(1) = \frac{3}{4}.$$

The equation is solved by separating variables.

In solving x = Ae^{-kt} to find A, recall that x(0) = 1; hence A = 1. Then

$$x = e^{-kt}.$$

Proceeding, $x(1) = \frac{3}{4} = e^{-k}$ or

$$k = -\log \frac{3}{4} = .288.$$

Then x = e$^{-0.288t}$

Now we can find t such that x(t) = .99.

$$.99 = e^{-.288t}; \quad \log .99 = -0.288t;$$

or t = .035".

Thus a pane of thickness .035 inch allows 99% of the light to pass through and therefore absorbs 1 per cent.

● **PROBLEM** 3–6

The population of a certain country is known to increase at a rate proportional to the number of people presently living in the country. If after two years the population has doubled, and after three years the population is 20,000, find the number of people initially living in the country.

Solution: This is a problem in population dynamics. Quite frequently a reasonable assumption concerning the rate of increase of a population is to assume that it is proportional to the present size. Mathematically,

$\frac{dN}{dt} = gN$,   where N is the number of people and g

is the growth rate assumed to be constant. The solution

of this first order separable equation is $N = N_0 e^{gt}$ where

N_0 is the arbitrary constant of integration. To find its value and also the value of g, we note that when t = 0 ,

$N = N_0 e^{g(0)} = N_0$. Combine this with the information that

when t = 2, $N = 2N_0$. Thus $2N_0 = N_0 e^{2g}$ or $g = \frac{\ln 2}{2}$ which is

.347.

We have found the growth rate but we have yet to find the number of people initially living in the country. Utilizing the other piece of information,

$N(3) = 20,000 = N_0 e^{(.347)(3)}$; $N_0 = e^{-1.041} \times (20,000)$

= 7062.

Then $N = 7062\ e^{.347\ t}$.

• **PROBLEM** 3–7

A certain radioactive material loses mass at a rate pro- portional to the mass present. If the material has a half-life of 30 minutes, what percent of the original mass is expected to remain after 0.9 hour?

Solution: If t is the time measured in hours, M_0 is the mass present at time t = 0, and M is the mass present at time t, we can set up the following equation

$\frac{dM}{dt} = kM$ (1)

where k is a constant. The derivative $\frac{dM}{dt}$ is the rate of change of M, given as proportional to M. Since the material is decaying, we expect that M will be a decreasing function of t, and that k will work out to be negative.

Equation (1) separates into

$$\frac{dM}{M} = kT \tag{2}$$

whose solution is

$$\ln M = kT + C' \tag{3}$$

or $\quad M = Ce^{kT} \tag{4}$

where $\quad C = e^{C'}$. Using the initial condition, $M(0) = M_0$, we find that $C = M_0$, leaving

$$\frac{M}{M_0} = e^{kT} \;. \tag{5}$$

Before we can answer the question, we need to evaluate the constant k. For this we must use the other information given in the problem - the half-life of the material. By definition, half-life is the time required for half of the material present to decay. This substance has a half-life of 30 minutes, $t = 0.5$. From (5) we then have

$$\frac{M}{M_0} = \frac{\frac{1}{2} M_0}{M_0} = \frac{1}{2} = e^{0.5\,k} \tag{6}$$

or $\quad 0.5\,k = -\ln 2$

$$k = -\ln 4 \;. \tag{7}$$

The negative sign in equation (7) is as expected by the properties of loragithms. The solution for M(t) is then

$$M = M_0\, e^{-\,t\,\ln 4}$$

$$= M_0\, 4^{-t} \;. \tag{8}$$

We can now determine the percentage remaining after 0.9 hour.

$$M = M_0\, 4^{-\,0.9} \stackrel{\sim}{\sim} 0.29\, M_0 \;. \tag{9}$$

Thus, about 29% of the original quantity remains.

According to Newton's law of cooling, the rate at
which a body loses heat, and therefore the change
in temperature, is proportional to the difference
in temperature between the body and the surrounding
medium:

$$\frac{dT}{dt} = -k(T - T_0),$$

where T is the temperature of the body, T_0 is the
temperature of the surrounding medium, and t is the
time. Show that $T - T_0 = (T_1 - T_0)e^{-kt}$, where T_1
is the value of T when t = 0.

Solution: Dividing the given equation, $\frac{T}{dt} = -k(T-T_0)$,
by $(T-T_0)$ and multiplying by dt, thereby separating
the variables, we obtain

$$\frac{dT}{T-T_0} = -kdt.$$

Integrating, we have

$$\ln(T-T_0) = -kt + \ln C$$

$$\ln(T-T_0) - \ln C = -kt$$

$$\ln\frac{T-T_0}{C} = -kt$$

therefore, $\frac{T-T_0}{C} = e^{-kt}$, or $T-T_0 = Ce^{-kt}$.

We now apply the initial conditions. At t = 0, the
temperature of the body is T_1, from which we obtain
the particular solution with a definite value of C.
Remember that under arbitrary conditions, the integration constant C can assume many parametric values.
Thus we get a family of curves, all of which have
the property determined by the given equation. According to the initial condition,

$$T_1 = T(\text{at } t = 0) = Ce^{-k(0)} + T_0.$$

Therefore $T_1 = C + T_0$ or $C = T_1 - T_0$,

hence $T - T_0 = (T_1-T_0)e^{-kt}.$

As t approaches ∞, T goes to T_0.

If the temperature is constant, the rate of change of the atmospheric pressure at any height is proportional to the pressure at that height:

$$\frac{dp}{dh} = -kp,$$

where p = pressure, h = height.

The minus sign is used since the pressure decreases as the height increases. Express the relationship between p and h.

Solution: Since $\frac{dp}{dh} = -kp$, we wish to find the general solution of the given differential equation. We find, by separating the variables, that:

$$\frac{dp}{p} = -kdh,$$

Integrating,

$$\ln p = -kh + \ln c.$$

Here, c is an arbitrary constant of integration. Hence

$$\frac{p}{c} = e^{-kh},$$

or,

$$p = ce^{-kh}.$$

The pressure at zero elevation (h = 0) is designated p_0, and thus $p_0 = p$ (at h = 0) = $ce^{-k(0)}$ = c . Therefore, $p = p_0 e^{-kh}$. Hence as the height increases, the pressure decreases exponentially. The maximum pressure occurs at zero elevation.

For a first order chemical reaction the law of mass action for a single reacting chemical is given by y' = - ky where k is the rate constant.

The law of mass action for a second order chemical reaction is given by $y' = - ky^2$ where $y(0) = a_0$. Solve the differential equations and evaluate the expression for the half life of the two chemical reactions.

Solution: For a first order reaction (unimolecular) y' = - ky, where y is the concentration of the reactant. By introducing our independent variable to be t where y' = F(t) = - ky, we can write

$\frac{dy}{dt} = - ky$. Notice that this is a separable differential equation (which is the case for nearly all chemical reactions) we can derive $\frac{dy}{y} = - k\ dt$. The minus sign denotes the decrease in the concentration of the reactant as the reaction proceeds.

Integrating, $|\ln y| = - kt + c$

We know that when $t = 0$, $y = y(0)$.

$\therefore \quad |\ln y(0)| = c$.

The final expression becomes

$|\ln y| = - kt + \ln y(0)$

or $\quad y = y(0)\ e^{- kt}$

We must find the half life $(t_{1/2})$ of the reaction, which is defined as the time required for exactly half of the initial reactant to be consumed or in this case for the initial concentration to be halved. Hence,

$y(t_{1/2}) = \frac{y(0)}{2}$.

Substituting this relationship into the final expression

$\frac{y(0)}{2} = y(0)\ e^{- kt_{1/2}}$

$\ln \frac{1}{2} = - kt_{1/2}$

$t_{1/2} = \frac{\ln 2}{2}$.

Note that the half life of a first order reaction does not depend on the initial concentration.

For a second order (bimolecular) reaction

$y' = - ky^2$ or $\frac{dy}{dt} = - ky^2$. This can be identified as a separable differential equation.

$\frac{y'}{y^2} = - k\ dt$.

Integrating directly

$$-\frac{1}{y} = -kt + c \;.$$

When $t = 0$, $y = a_0$, where a_0 represents the initial concentration before the start of the reaction.

$$-\frac{1}{a_0} = 0 + c$$

The final expression is

$$\frac{1}{y} = kt + \frac{1}{a_0} \;;$$

$$y = \frac{a_0}{1 + a_0 \, kt} \;,$$

To calculate the half life we have to find the time when the concentration is $\frac{a_0}{2}$ (initial conc.) By substitution:

$$\frac{a_0}{2} = \frac{a_0}{1 + a_0 \, kt_{1/2}} \;;$$

$$t_{1/2} = \frac{1}{a_0 \, k} \;.$$

In contrast to a first order reaction the half life for a second order reaction does depend on the initial concentrations.

● **PROBLEM 3-11**

What is the time required for one dollar to double when invested at the rate of 5% per annum? Assume that interest is compounded continuously.

Solution: Let P_0 be the amount of money at the initial time t_0. The problem involves the relationship that the change in P with respect to time is proportional to P. More precisely,

$$\frac{dP}{dt} = kP. \quad \text{This is a first order separable equation.}$$

To solve it we proceed as follows:

$$\frac{dP}{P} = k \, dt; \quad \ln |P| = kt + c$$

$$P = e^c \, e^{kt} = P_0 \, e^{kt}.$$

Now in the given problem we have that $P_0 = 1$ and $k = .05$. Substituting these values,

$$P = 1\, e^{.05t} = e^{.05t}.$$

We require that after t years P = 2. Thus

$$2 = e^{.05t}; \quad t = \frac{\log 2}{.05} = 20 \log 2.$$

Since $\log_e 2 = .693147$, the time required to double an initial amount of 1 dollar is 13.86 years.

Note that we were able to set up the problem as a differential equation because of the assumption that interest is compounded continuously. When the compounding process is discrete, a difference-equation is required.

• PROBLEM 3–12

If the marginal cost of producing a certain item is

$$\frac{dy}{dx} = y' = 3 + x + \frac{e^{-x}}{4},$$

what is the cost of producing 1 item if there is a fixed cost of $4?

Solution:
$$y = \int y'\, dx = \int \left(3 + x + \frac{e^{-x}}{4} \right) dx$$
$$= 3x + \frac{x^2}{2} - \frac{e^{-x}}{4} + C.$$

Solving for the constant C, we know that the fixed cost is 4 when x = 0. Therefore, y = 4. Hence,

$$y = 4 = 0 + \frac{0^2}{2} - \frac{1}{4} + C.$$

Thus,
$$C = 4 + \frac{1}{4} = \frac{17}{4},$$

and
$$y = 3x + \frac{x^2}{2} - \frac{e^{-x}}{4} + \frac{17}{4}.$$

when x = 1,
$$y = 3 + \frac{1}{2} - \frac{1}{4e} + \frac{17}{4}$$

$$= \frac{12 + 2 + 17}{4} - \frac{1}{4e}$$

$$= \frac{31}{4} - \frac{1}{4e}$$

$$= \frac{31e - 1}{4e}.$$

105

APPLICATIONS OF SECOND ORDER DIFFERENTIAL EQUATIONS

A 10-kilogram mass is attached to a spring which is thereby stretched 0.7 meters from its natural length. The mass is started in motion from the equilibrium position with an initial velocity of 1 meter/sec in the upward direction. Find the resulting motion if the force due to air resistance is $- 90 \frac{dx}{dt}$ newtons.

Solution: The motion is damped (air resistance) and free (no external force). Our differential equation from Newton's second law is

$$\frac{md^2x}{dt^2} = - 90 \frac{dx}{dt} - kx \qquad (a)$$

where m = the mass (10 kg), and k is the spring constant. Since $k = \frac{\text{force applied}}{\text{length stretched}}$,

$$k = \frac{(9.8)(10)}{.7} = 140 \text{ newtons/meter.}$$

(Note that a mass of 10 kg exerts a force or weight of (9.8) (10) = 98 newtons). Substituting into (a) we have

$$10 \frac{d^2x}{dt^2} + 90 \frac{dx}{dt} + 140 x = 0$$

or $\quad \frac{d^2x}{dt^2} + 9 \frac{dx}{dt} + 14 x = 0. \qquad$ (b)

The characteristic equation is

$$\lambda^2 + 9\lambda + 14 = 0$$

or $\quad (\lambda + 2)(\lambda + 7) = 0$

Thus the roots are

$$\lambda = - 2 \qquad \text{and} \qquad \lambda = - 7.$$

Our general solution is thus

$$x(t) = c_1 e^{-2t} + c_2 e^{-7t} \qquad (c)$$

where c_1 and c_2 are arbitrary constants to be determined by the initial conditions which are

$$x(0) = 0$$

and $\left.\dfrac{dx}{dt}\right|_{t=0} = -1$

(the initial velocity is in the negative x-direction).

Thus $x(0) = c_1 e^0 + c_2 e^0$,

$$0 = c_1 + c_2 \qquad\qquad\qquad\qquad \text{(d)}$$

and $\left.\dfrac{dx}{dt}\right|_{t=0} = -2 c_1 e^0 - 7 c_2 e^0$,

$$-1 = -2 c_1 - 7 c_2$$

or $\quad 2 c_1 + 7 c_2 = 1. \qquad\qquad\qquad\qquad \text{(e)}$

Since by (d) $c_1 = -c_2$, (e) becomes

$$2(-c_2) + 7 c_2 = 1$$

$$c_2 = \frac{1}{5}$$

and $\qquad\qquad\qquad c_1 = -\frac{1}{5}$

Our final solution is

$$x(t) = \frac{1}{5}\,(e^{-7t} - e^{-2t}).$$

● **PROBLEM 3–14**

A mass of 1/4 slug is attached to a spring of force constant k = 1 lb./ft. The mass is set in motion by initially displacing it 2 ft. in the downward direction, and giving it an initial velocity of 2 ft./sec. in the upward direction. Find the subsequent motion of the mass if the force due to air resistance is $-1\,\dfrac{dx}{dt}$ lb.

Solution: The differential equation of this system comes from Newton's second law

$$m\,\frac{d^2x}{dt^2} = -a\,\frac{dx}{dt} - kx \qquad\qquad\qquad \text{(a)}$$

where $-a\,\dfrac{dx}{dt}$ is the force of air resistance and $-kx$ is the spring force. Substituting our values (a) becomes

$$\frac{1}{4} \frac{d^2x}{dt^2} = -1 \frac{dx}{dt} - 1 \, x,$$

or $\quad \dfrac{d^2x}{dt^2} + 4 \dfrac{dx}{dt} + 4x = 0.$ $\hspace{2cm}$ (b)

We obtain the characteristic equation of (b) by assuming a solution of the type $e^{\lambda t}$. Thus

$$\lambda^2 + 4\lambda + 4 = 0$$

which by the quadratic formula yields

$$\lambda = \frac{-4 \pm \sqrt{16 - 4(4)}}{2} = -2.$$

When both roots of the characteristic equation are equal the general solution becomes

$$x(t) = c_1 e^{-2t} + c_2 t e^{-2t}. \hspace{2cm} (c)$$

In other words not only is e^{-2t} a solution, but also te^{-2t}. We must evaluate the coefficients c_1 and c_2 from the initial conditions which are

$$x(0) = 2 \hspace{2cm} (d)$$

and $\quad \dfrac{dx}{dt}\Big|_{t=0} = -2$ $\hspace{2cm}$ (e)

(since the initial velocity is in the negative direction).

From equation (c)

$$x(0) = c_1 e^0 + c_2 (0) e^0,$$

$$2 = c_1$$

by (d). Our solution thus far is

$$x(t) = 2e^{-2t} + c_2 t e^{-2t}$$

$$\frac{dx}{dt} = -4e^{-2t} + c_2 (e^{-2t} - 2te^{-2t}),$$

$$\frac{dx}{dt}\Big|_{t=0} = -4 \, e^0 + c_2 (e^0 - 2(0)e^0),$$

$$-2 = -4 + c_2,$$

$$2 = c_2.$$

Our final solution is

$$x(t) = 2e^{-2t} + 2te^{-2t}$$

$$= 2(1 + t)e^{-2t}$$

A particle is in simple harmonic motion along the y-axis. At

$$t = 0, \quad y = 3 \text{ and } v = \frac{dy}{dt} = 0.$$

Exactly 1/2 second later these values repeat themselves. Find y(t) and v(t).

Solution: A particle is in simple harmonic motion if its position coordinate satisfies the differential equation

$$\frac{d^2y}{dt^2} + w^2y = 0 \qquad\qquad (a)$$

where w is a constant which is the frequency of the motion. An alternative definition of harmonic motion is motion whose position function is of the form

$$y = A \sin wt + B \cos wt \qquad\qquad (b)$$

where A and B are arbitrary constants, and w is again the frequency. Since (b) is a solution of (a) we see that these two definitions are equivalent.

We know that y(0) = 3, thus from (b)

$$y(0) = A \sin 0 + B \cos 0$$

$$3 = B .$$

Further,

$$v(0) = \frac{dy}{dy}\bigg|_{t=0} = 0$$

$$v(t) = w A \cos wt - wB \sin wt$$

$$v(0) = w A \cos 0 - w B \sin 0$$

and $\qquad 0 = wA.$

Therefore, A = 0 or w = 0.

Since w = 0 corresponds to the degenerate case, we take A = 0. Our solution is thus

$$y = B \cos wt \qquad\qquad (c)$$

To find w we note that

$$y\left(\frac{1}{2}\right) = B$$

Thus $\quad y\left(\frac{1}{2}\right) = 3 \cos w \frac{1}{2}$

$$3 = 3 \cos \frac{w}{2}$$

$$1 = \cos \frac{w}{2} . \tag{d}$$

For angle values between 0 and 2π, the solution of

(d) is $\quad \dfrac{w}{2} = 2\pi.$

Thus $\quad w = 4\pi.$

Our final solution is

$$y = 3 \cos 4\pi t$$

and the frequency is 4π.

● **PROBLEM** 3–16

A simple pendulum of length 2 ft is given no initial angular displacement, and an initial angular velocity of 1 rad/sec. Neglecting friction, determine the subsequent motion.

Solution: The differential equation for the angle of displacement is

$$\frac{d^2\theta}{dt^2} + \frac{g}{\ell} \theta = 0 \tag{a}$$

where g = 32 ft/sec/sec. Thus

$$\theta'' + 16\theta = 0 \tag{b}$$

and $\quad \theta(0) = 0, \qquad \theta'(0) = 1 \tag{c}$

where primes denote differentiation. Equation (b) is the harmonic oscillator equation whose solution is seen to be

$$\theta = c_1 \cos 4t + c_2 \sin 4t.$$

Applying (c)

$$\theta(0) = c_1 \cos 0 + c_2 \sin 0$$

$$0 = c_1$$

Thus $\theta(t) = c_2 \sin 4t$

$\quad\quad \theta'(t) = 4c_2 \cos 4t$

$\quad\quad \theta'(0) = 4c_2 \cos 0$

and $\quad\quad 1 = 4c_2$.

Thus $\quad c_2 = \dfrac{1}{4}$

and the final solution is

$$\theta = \frac{1}{4} \sin 4t .$$

The frequency is 4 rad/sec and the amplitude is 1/4 ft.
The period is 2π/frequency which is $\pi/2$ sec.

● **PROBLEM** 3–17

A 2-lb. mass is dropped from a great height. Air resistance is proportional to kv^2 where $k = 1/20,000$. Find the velocity when the mass has fallen 1000 ft.

<u>Solution</u>: From Newton's second law

$$m \frac{d^2y}{dt^2} = mg - kv^2 \tag{a}$$

where y is the vertical distance of the mass measured from the point of the initial position, mg is the weight or force of gravity, and $- kv^2$ is the force of air resistance. Now $v = \dfrac{dy}{dt}$ is the velocity of the particle, so

$$\frac{d^2y}{dt^2} = \frac{dv}{dt} = \frac{dv}{dy}\frac{dy}{dt} = \frac{dv}{dy} \cdot v.$$

The reason for taking this unusual step is that the question asks for v as a function of y (find v for a given y). Changing the independent variable of differentiation accomplishes this neatly as will be seen below.

Substituting in (a)

$$mv \frac{dv}{dy} = mg - kv^2. \tag{b}$$

Since $mg = 2$, and $g = 32$ in the English system of units, $m = \dfrac{1}{16}$. Substituting our values in (b)

111

$$\frac{1}{16} v \frac{dv}{dy} = 2 - \frac{v^2}{20000} \quad .$$

Separating variables,

$$\frac{v\ dv}{40000 - v^2} = \frac{16}{20000}\ dy,$$

$$\int \frac{v\ dv}{40000 - v^2} = \frac{1}{1250} \int dy + c \qquad\qquad \text{(c)}$$

where c is a constant of integration.

Let $p = 40000 - v^2$

$\qquad dp = - 2v\ dv \qquad\qquad$ then

$$\int \frac{v\ dv}{40000 - v^2} = - \frac{1}{2} \int \frac{dp}{p}$$

$$= - \frac{1}{2}\ \ln p$$

$$= - \frac{1}{2}\ \ln (40000 - v^2).$$

Substituting in (c)

$$- \frac{1}{2}\ \ln (40000 - v^2) = \frac{y}{1250} + c$$

$$\ln (40000 - v^2) = - \frac{y}{625} - 2c \quad .$$

Exponentiating

$$40000 - v^2 = Ke^{-y/625}$$

where $K = e^{-2c}$. Since $v = 0$ at $y = 0$,

$\qquad 40000 - 0^2 = Ke^0,$

$$K = 40000.$$

Hence $\qquad\qquad v^2 = 40000 \left(1 - e^{-y/625}\right)$

or $\qquad\qquad v = 200 \sqrt{1 - e^{-y/625}} \quad .$

When $\qquad\qquad y = 1000,$

$$v = 200 \sqrt{1 - e^{-1000/625}}$$

$$v \cong 178.6.$$

A ball is thrown straight up from the ground with velocity
v.   How high will it go?

Solution: Let us have a vertical x-axis along the line
of motion of the ball such that x = 0 at the ground, and
x(t) is the position of the ball at time t.   The force
acting on the ball is gravity which is - mg where m is the
mass of the ball and g is the acceleration due to gravity.
From Newton's second law,

$$m \frac{d^2 x}{dt^2} = - mg$$

or $\frac{d^2 x}{dt^2} + g = 0.$ (a)

We integrate (a):

$$\int \frac{d^2 x}{dt^2} \, dt + g \int dt + C = 0$$

where C is a constant of integration

$$\frac{dx}{dt} + gt + C = 0.$$ (b)

Since $\frac{dx}{dt} = v(t)$, and $v(0) = v$, we may evaluate C in (b) by

$$\frac{dx}{dt} \bigg|_{t=0} + (g \cdot 0) + C = 0$$

$$C = - v.$$

Thus $\frac{dx}{dt} + gt - v = 0.$

Integrating again

$$\int \frac{dx}{dt} \, dt + g \int t \, dt - v \int dt + K = 0$$

where K is again a constant of integration.

$$x(t) + \frac{1}{2} gt^2 - vt + K = 0.$$

Since x(0) = 0, we see that K = 0.   Our final
solution is

$$x(t) = - \frac{1}{2} gt^2 + vt.$$ (c)

To find maximum x(t) we set the first derivative
of (c) equal to zero

$$-gt + v = 0,$$

$$t_{max} = \frac{v}{g} .$$

Since $x''(v/g) = -g < 0$

we indeed do have a max. at $t = v/g$ by the second derivative test.   Hence

$$x\left(\frac{v}{g}\right) = -\frac{1}{2} g \left(\frac{v}{g}\right)^2 + v\left(\frac{v}{g}\right) ,$$

$$x_{max} = \frac{v^2}{2g} .$$

• **PROBLEM** 3–19

A particle accelerates from t=0 in accordance with the law that the acceleration, a=6-3t.
a) Compute the distance that the **particle travels** while the velocity is increasing starting from zero,
b) find the distance traveled by the particle during the interval from t=1 to t=5.

Solution:
a) The velocity increases during the interval in which the acceleration is positive. For this example, this interval begins at t=0 and continues until t=2. (This is obtained from examining the equation for the acceleration.) Hence the distance moved during the interval from t=0 to t=2 is desired. The acceleration is

$$a = \frac{d^2s}{dt^2} = \frac{dv}{dt} = 6 - 3t.$$

Then
$$dv = (6-3t)dt,$$

$$v = 6t - \frac{3}{2}t^2 + c.$$

Since v=0 when t=0, the constant of integration c is zero.

Before evaluating s over the interval from t=0 to t=2, we must see if the motion is in the same direction during the interval. Since the direction of motion may change when v=0, we see from

$$v = 6t - \frac{3}{2}t^2 = \frac{3}{2}(4-t)t = 0$$

that the direction of motion can change only at t=0 or t=4. Hence it does not change between t=0 and t=2. Now, writing s as a definite integral we have

114

$$s = \int^{2}(6t - \frac{3}{2}t^2)dt$$

$$= 3t^2 - \frac{1}{2}t^3 \Big]^2_0 = 12 - 4 = 8 \text{ units.}$$

b) The velocity is positive for $0 < t < 4$, negative for $t > 0$ and zero for $t=0$ and 4. For $t > 4$ the motion of the particle is opposite to the direction in which positive s is measured. To find the total (or absolute) distance traveled we must compute s separately for the intervals $t=1$ to $t=4$ and $t=4$ to $t=5$, and add the two results numerically. Thus

$$s = \int^{4}_{1} \frac{3}{2}(4-t)t dt + \int^{5}_{4} \frac{3}{2}(4-t)t dt = s_1 + s_2.$$

$$s = \int^{4}_{1} \frac{3}{2}(4-t)t dt + \int^{5}_{4} \frac{3}{2}(4-t)t dt = \int^{5}_{1} \left(6t - \frac{3t^2}{2}\right)dt$$

or

$$s = 3t^2 - \frac{t^3}{2} \Big|^5_1 = 75 - \frac{125}{2} - 3 + \frac{1}{2} = 10 \text{ units,}$$

the total distance traveled.

• PROBLEM 3-20

Find the "escape velocity", i.e., the speed with which a particle would have to be projected from the surface of the earth in order never to return, by determining the speed with which a particle would strike the earth's surface if it started from rest at a very great (supposedly infinite) distance and traveled subject only to the earth's gravitation.

Solution: From the law of gravitation,

$$F = - k_1 \frac{mM}{r^2},$$

where k_1 is the gravitational constant, m and M are the masses of the particle and the earth, respectively, and r^2 is the distance separating them.

We can call $\left(k_1 M\right) = k_2$, and

$$F = - \frac{k_2 m}{r^2}.$$

On the surface of the earth, with the radius of the earth $R = r = 3959$ mi., the force is just the weight of the particle, that is:

$$F = - \frac{k_2 m}{R^2} = - mg,$$

and hence, for $r \neq R$,

$$F = -\frac{R^2 gm}{r^2} .$$

Calling $R^2 g = k^2$, we have:

$$\frac{F}{m} = -\frac{k^2}{r^2} = \frac{d^2 r}{dt^2} .$$

To integrate, we set $p = dr/dt$. Then $\dfrac{d^2 r}{dt^2} = \dfrac{dp}{dt} = \dfrac{dp}{dr}\dfrac{dr}{dt} = p\dfrac{dp}{dr}$

$$p\frac{dp}{dr} + \frac{k^2}{r^2} = 0 .$$

Separation of variables and integration yields:

$$\frac{p^2}{2} = \frac{k^2}{r} + c.$$

To find c, we let $r \to \infty$, where the motion starts, and $p = \dfrac{dr}{dt} = 0$. This makes $c = 0$. Furthermore, solving for $p = dr/dt$,

$$\frac{dr}{dt} = k \sqrt{2/r} .$$

Using $k = R\sqrt{g}$, we finally obtain:

$$v = R \sqrt{2g/r} .$$

Since we are interested in finding the velocity of this particle on the earth's surface,

$$v_0 = r \sqrt{2g/R} = \sqrt{2gR} = \left(\frac{2 \times 32.2 \times 3959}{5280}\right)^{1/2}$$

$$= 7 \text{ miles/second.}$$

This value is independent of the mass of the particle.

• **PROBLEM** 3–21

A beam resting on two end supports is called a simple beam (see fig.). Let the length of the beam be ℓ and its weight per foot be w. Find the deflection of the beam.

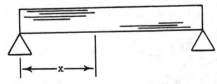

Solution: Take the left end of the beam as the origin of the x-y coordinate system. At point x there is the load force $-wx$, and the reaction force at the right hand support is $\frac{1}{2} w\ell$. The moments of these two forces are

116

$\frac{1}{2}$ wℓx   and $-\frac{1}{2}$ wx^2.

Hence, the differential equation of the deflection, as a function of the moments is

$$EIy'' = \frac{w\ell x}{2} - \frac{wx^2}{2}.$$ (a)

Integrating (a) once we obtain

$$EIy' = \frac{w\ell x^2}{4} - \frac{wx^3}{6} + c_1.$$

Integrating again,

$$EIy = \frac{w\ell x^3}{12} - \frac{wx^4}{24} + c_1 x + c_2$$ (b)

where c_1 and c_2 are constants of integration.

Since $y(0) = 0$ and $y'\left(\frac{\ell}{2}\right) = 0$ we evaluate c_1 and c_2.

$$EI(y(0)) = \frac{w\ell(0)^3}{12} - \frac{w(0)^4}{24} + c_1(0) + c_2$$

$$0 = c_2.$$

Thus $EIy(x) = \frac{w\ell x^3}{12} - \frac{wx^4}{24} + c_1 x$,

$$EIy'\left(\frac{\ell}{2}\right) = \frac{w\ell^3}{16} - \frac{w\ell^3}{48} + c_1$$

$$0 = \frac{3w\ell^3 - w\ell^3}{48} + c_1$$

Thus $c_1 = -\frac{w\ell^3}{24}.$

Our final solution is

$$y = \frac{-w}{24E I} (+ x^4 + 2\ell x^3 + \ell^3 x).$$

• **PROBLEM 3–22**

Chemical A is being converted into chemical B at a reaction rate of $-\frac{1}{2}$ per second. The initial concentration of A is $C_0 = 10$ moles per liter. Determine concentration $C(t)$ at later times if the reaction is assumed to be of first order. Also find the half-life of the reaction.

Solution: A chemical reaction is of first order if it is governed by a first order differential equation which is

$$\frac{dC}{dt} = k\,C \tag{a}$$

where C is the concentration and k is the reaction rate. In our case $k = -\frac{1}{2}$, so (a) becomes

$$\frac{dC}{dt} = -\frac{1}{2}\,C$$

or, separating variables

$$\frac{dC}{C} = -\frac{1}{2}\,dt$$

Integrating,

$$\int \frac{dC}{C} = -\frac{1}{2} \int dt + K'$$

where K' is a constant of integration.

$$\ln C = -\frac{1}{2}\,t + K'\,.$$

Exponentiating,

$$C = K\,e^{-1/2\,t}$$

where $K = e^{K'}$ is a constant to be evaluated from initial condition C(0) = 10.   Thus

$$C(0) = Ke^{-1/2\,0}$$

$$10 = K\,.$$

Thus the final solution is

$$C(t) = 10\,e^{-1/2\,t}\,.$$

The half-life of the reaction is the time at which the concentration decreases to half its initial amount, i.e., when C = 5.   Thus,

$$5 = 10\,e^{-1/2\,t}$$

or   t = 2 ln 2.

In an isolated marine community, barnacles were prey for starfish. Initially, there are $y_0 = 500$ barnacles and $z_0 = 100$ starfish. Determine the population of predator and prey at $t > 0$ if the empirical governing equations are

$$y' = 2y - 5z \tag{a}$$

$$z' = 4y - 2z. \tag{b}$$

Solution: We must solve the coupled system (a) and (b). From (a)

$$z = \frac{2y}{5} - \frac{y'}{5},$$

$$z' = \frac{2y'}{5} - \frac{y''}{5}.$$

Substituting into (b)

$$\frac{2y'}{5} - \frac{y''}{5} = 4y - 2\left(\frac{2y}{5} - \frac{y'}{5}\right),$$

$$2y' - y'' = 20y - 4y + 2y',$$

or $y'' + 16y = 0.$ $\qquad$ (c)

Equation (c) is the equation of the simple harmonic oscillator whose solution may be verified to be

$$y = c_1 \cos 4t + c_2 \sin 4t \tag{d}$$

where c_1 and c_2 are arbitrary constants.

Substituting the value of y from (d) into

$$z = \frac{2y}{5} - \frac{y'}{5}$$

$$z = \frac{2}{5}(c_1 \cos 4t + c_2 \sin 4t)$$

$$- \frac{1}{5}(-4c_1 \sin 4t + 4c_2 \cos 4t)$$

or $z = \frac{2}{5}(c_1 - 2c_2)\cos 4t + \frac{2}{5}(c_2 + 2c_1)\sin 4t.$ $\qquad$ (e)

To evaluate c_1 and c_2 we use the initial conditions

$$y(0) = 500 \qquad \text{and} \qquad z(0) = 100.$$

$$y(0) = c_1 \cos 0 + c_2 \sin 0$$

$$500 = c_1.$$

Then $z(0) = \frac{2}{5}(500 - 2c_2)(1) + \frac{2}{5}(c_2 + 1000)(0)$

$$100 = \frac{2}{5}(500 - 2c_2)$$

or $\quad c_2 = \frac{500}{4}$.

The final solutions are

$$y = 500 \cos 4t + \frac{500}{4} \sin 4t$$

$$z = 250 \cos 4t + 1125 \sin 4t.$$

● **PROBLEM** 3–24

A point P is dragged along the xy-plane by string PT of constant length a. Initially T is at the origin and P is on the x-axis at the point (a, 0). T is then moved to the y-axis. Find the resulting equation for the path of P.

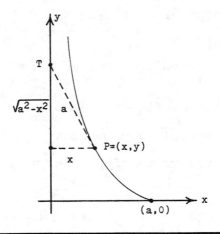

Solution: Since PT is always tangent to the curve generated by P, the definition of slope yields

$$\frac{dy}{dx} = -\frac{\sqrt{a^2 - x^2}}{x} \quad . \quad \text{(see figure.)}$$

Separating variables and integrating

$$\int dy = -\int \frac{\sqrt{a^2 - x^2}}{x} \, dx + C \quad\quad\quad\text{(a)}$$

where C is a constant of integration.   We must now evaluate

$$I \equiv \int \frac{\sqrt{a^2 - x^2}}{x} \, dx .$$

Let us change variables:

$$x = a \cos \theta$$

$$dx = - a \sin \theta \, d\theta .$$

Then $\sqrt{a^2 - x^2} = a \sqrt{1 - \cos^2 \theta}$

$$= \sin \theta .$$

Thus $I = - a \int \dfrac{\sin \theta \, \sin \theta \, d\theta}{\cos \theta}$

But $\sin^2 \theta = 1 - \cos^2 \theta .$ Hence

$$I = - a \left[\int \frac{1}{\cos \theta} \, d\theta - \int \cos \theta \, d\theta \right],$$

$$I = - a \, [\ln \, |\sec \theta + \tan \theta| - \sin \theta].$$

Now, $\sec \theta + \tan \theta = \dfrac{1 + \sin \theta}{\cos \theta}$

where $\cos \theta = \dfrac{x}{a}$

and $\sin \theta = \sqrt{1 - \cos^2 \theta} = \dfrac{\sqrt{a^2 - x^2}}{a}$.

Substituting,

$$I = - a \left[\ln \left| \frac{1 + \sqrt{a^2 - x^2}/a}{x/a} \right| - \frac{\sqrt{a^2 - x^2}}{a} \right],$$

$$I = - a \ln \left| \frac{a + \sqrt{a^2 - x^2}}{x} \right| + \sqrt{a^2 - x^2} .$$

Substituting this value of I in (a) yields

$$y = a \ln \left(\frac{a + \sqrt{a^2 - x^2}}{x} \right) - \sqrt{a^2 - x^2} + C.$$

We drop the absolute value since the expression is always positive.   To evaluate C we note that $y = 0$ when $x = a$, thus $C = 0$.

The final result is

$$y = a \ln \left(\frac{a + \sqrt{a^2 - x^2}}{x} \right) - \sqrt{a^2 - x^2}$$

which is the equation of the tractrix.

ELECTRICAL CIRCUITS

Solve the equation

$$L \frac{dI}{dt} + RI = E_0 \qquad\qquad\qquad (a)$$

for the case in which an initial current I_0 is flowing and a constant emf E_0 is impressed on the circuit at time $t = 0$.

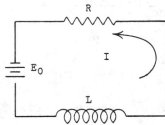

Solution: We consider only $t \geq 0$ as being of interest. We solve (a) by separation of variables:

$$L \frac{dI}{dt} = E_0 - RI ,$$

$$\frac{dI}{E_0 - RI} = \frac{dt}{L} .$$

Integrating,

$$\int \frac{dI}{E_0 - RI} = \int \frac{dt}{L} + C, \qquad\qquad (c)$$

where C is the constant of integration. We use a change of variable:

$$p = E_0 - RI ,$$

$$dp = -R \, dI .$$

Substituting this into (b) yields

$$-\frac{1}{R} \int \frac{dp}{p} = \frac{t}{L} + C ,$$

$$-\frac{1}{R} \ln p = \frac{t}{L} + C ,$$

$$\ln p = -\frac{R}{L} t - RC .$$

Replacing p by $E_0 - RI$ and exponentiating:

$$E_0 - RI = e^{-Rt/L - RC}$$

$$I = \frac{E_0}{R} - e^{-Rt/L - RC} \quad .$$

(c)

Since $I = I_0$ at $t = 0$, (c) becomes

$$I_0 = \frac{E_0}{R} - e^{-RC} \quad ,$$

$$e^{-RC} = \frac{E_0}{R} - I_0 \quad .$$

Substituting this into (c) gives

$$I = \frac{E_0}{R} - e^{-Rt/L}\left(\frac{E_0}{R} - I_0\right)$$

(d)

as the final solution.

The E_0/R term in (d) is called the steady-state part, and the

$\left(I_0 - \frac{E_0}{R}\right)e^{-Rt/L}$ term is called the transient part (which goes to zero

as $t \to \infty$). The solution may thus be written as

$$I = \left\{ \begin{matrix} \text{STEADY-STATE} \\ \text{PART} \end{matrix} \right\} + \left\{ \begin{matrix} \text{TRANSIENT} \\ \text{PART} \end{matrix} \right\} \quad .$$

• PROBLEM 3–26

The RL-circuit:

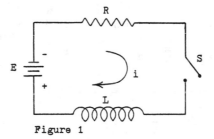

Figure 1

The above circuit has a constant impressed voltage E, a resistor of resistance R, and a coil of impedance L. Find the current $i = i(t)$ flowing in the circuit.

Solution: The differential equation governing this circuit is derived from setting the sum of the voltage drops around the circuit equal to the impressed voltage. Thus,

$$L \frac{di}{dt} + Ri = E \quad ,$$

(a)

where $L \frac{di}{dt}$ is the voltage drop across the coil, and Ri is the vol-

tage drop across the resistor. We may solve equation (a) by the use of an integrating factor. Divide (a) by L:

$$\frac{di}{dt} + \frac{R}{L} i = \frac{E}{L} \quad ,$$

and let $\varphi(t)$ be the function such that

$$\varphi \frac{di}{dt} + \frac{R}{L} \varphi i = \frac{d}{dt} (\varphi i) . \tag{b}$$

Expanding the right-hand side of (b):

$$\varphi \frac{di}{dt} + \frac{R}{L} \varphi i = \varphi \frac{di}{dt} + i \frac{d\varphi}{dt} .$$

Subtract $\varphi \, di/dt$:

$$\frac{R}{L} \varphi i = i \frac{d\varphi}{dt} ,$$

and divide by i to obtain

$$\frac{d\varphi}{dt} = \frac{R}{L} \varphi .$$

This equation may be solved for φ by separation of variables:

$$\frac{d\varphi}{\varphi} = \frac{R}{L} dt .$$

Integrating both sides,

$$\int \frac{d\varphi}{\varphi} = \frac{R}{L} \int dt ,$$

or

$$\ln \varphi = \frac{R}{L} t .$$

Exponentiating both sides,

$$\varphi = e^{\frac{R}{L} t} .$$

This function φ is the integrating factor which will allow us to write (a) as

$$e^{\frac{R}{L} t} \frac{di}{dt} + e^{\frac{R}{L} t} \frac{R}{L} i = \frac{E}{L} e^{\frac{R}{L} t} ,$$

or, using (b),

$$\frac{d}{dt}\left(e^{\frac{R}{L} t} i\right) = \frac{E}{L} e^{\frac{R}{L} t} .$$

We integrate both sides to obtain

$$\int \frac{d}{dt}\left(e^{\frac{R}{L} t} i\right) dt = \frac{E}{L} \int e^{\frac{R}{L} t} dt + C, \tag{c}$$

where C is the constant of integration. Evaluating (c) we have

$$e^{\frac{R}{L} t} i = \frac{E}{L}\left(\frac{L}{R}\right) e^{\frac{R}{L} t} + C .$$

Multiplication by $e^{-\frac{R}{L} t}$ gives

$$i = \frac{E}{R} + Ce^{-\frac{R}{L} t} . \tag{d}$$

To evaluate the constant of integration, we assume that the switch S is closed at time $t = 0$, and the initial current is thus $i(0) = 0$. Using this information and (d) we have

$$i(0) = \frac{E}{R} + Ce^0 ,$$

$$0 = \frac{E}{R} + C ,$$

$$C = \frac{-E}{R} .$$

Our final solution is therefore,

$$i(t) = \frac{E}{R} - \frac{E}{R} e^{-\frac{R}{L} t} ,$$

$$i(t) = \frac{E}{R}\left(1 - e^{-\frac{R}{L} t}\right) . \qquad (e)$$

We see that

$$\lim_{t \to \infty} i(t) = \frac{E}{R} - \frac{E}{R} \lim_{t \to \infty} e^{-\frac{R}{L} t}$$

$$= \frac{E}{R} - \frac{E}{R} (0)$$

$$= \frac{E}{R} .$$

Therefore, $i(t)$ approaches the steady-state value E/R exponentially. The graph of $i(t)$ is shown in Figure 2.

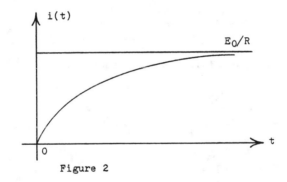

Figure 2

• **PROBLEM** 3–27

An RL-circuit has an emf of 5 volts, a resistance of 50 ohms and an inductance of 1 henry. There is no initial current in the circuit. Find the current in the circuit at any time t.

Solution: A schematic diagram of the above conditions is illustrated.

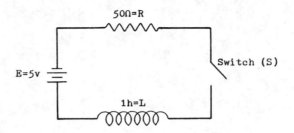

By equating the voltage drops around the circuit with the impressed voltage we have

$$L \frac{di}{dt} + Ri = E$$

or substituting our values

$$1 \frac{di}{dt} + 50i = 5 \; ,$$

$$\frac{di}{dt} + 50i = 5 \; . \tag{a}$$

We may solve (a) by use of an integrating factor, $\varphi(t)$. We want $\varphi(t)$ to be such that

$$\varphi \frac{di}{dt} + 50i\varphi = \frac{d}{dt}(i\varphi) \; , \tag{b}$$

$$\varphi \frac{di}{dt} + 50i\varphi = \varphi \frac{di}{dt} + i \frac{d\varphi}{dt} \; .$$

Subtract $\varphi \, di/dt$ and divide by i to obtain

$$\frac{d\varphi}{dt} = 50\varphi$$

or

$$\frac{d\varphi}{\varphi} = 50 \, dt \; .$$

Integration of this yields

$$\int \frac{d\varphi}{\varphi} = 50 \int dt$$

$$\ln \varphi = 50t + \text{constant} \tag{c}$$

Since we are interested in the simplest function φ which satisfies (b) we choose the integration constant to be zero. If we exponentiate (c) we obtain

$$\varphi = e^{50t} \; .$$

Thus multiply (a) by φ ,

$$e^{50t} \frac{di}{dt} + 50i \, e^{50t} = 5e^{50t} \; .$$

Substituting from (b) yields

$$\frac{d}{dt}\left(i \ e^{50t}\right) = 5e^{50t} \ .$$

Integrating

$$\int \frac{d}{dt}(ie^{50t})dt = 5 \int e^{50t}dt$$

$$ie^{50t} = \frac{5}{50} e^{50t} + C \ ,$$

where C is a new constant. Thus, multiplying by e^{-50t}

$$i = \frac{1}{10} + Ce^{-50t} \ .$$

Since $i(0) = 0$,

$$i(0) = \frac{1}{10} + Ce^{0}$$

$$0 = \frac{1}{10} + C$$

$$C = - \frac{1}{10} \ .$$

The complete solution is thus,

$$i(t) = \frac{1}{10} - \frac{1}{10} \ e^{-50t}$$

or

$$i(t) = \frac{1}{10}\left(1 - e^{-50t}\right) \ .$$

● **PROBLEM** 3–28

The RC-circuit

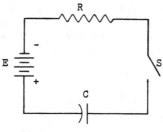

Figure 1

The above circuit has a constant impressed voltage E, a resistance R, a capacitance C and switch S. Find the current $i = i(t)$ flowing in the circuit.

<u>Solution</u>: The differential equation governing the system is obtained by setting the voltage drops around the circuit equal to the impressed voltage. Thus,

$$Ri + \frac{1}{C} q = E \ , \qquad \qquad (a)$$

where Ri is the voltage drop across the resistor and q/C is the voltage drop on the capacitor. The charge, q, is defined by $i = dq/dt$. We differentiate (a) in order to eliminate q:

$$R \frac{di}{dt} + \frac{1}{C} \frac{dq}{dt} = 0 \; ,$$

$$R \frac{di}{dt} + \frac{1}{C} \cdot i = 0 \; ,$$

$$\frac{di}{dt} + \frac{1}{RC} i = 0 \; . \tag{b}$$

We may solve (b) by separation of variables:

$$\frac{di}{dt} = - \frac{1}{RC} i$$

$$\frac{di}{i} = - \frac{1}{RC} dt \; .$$

Integrating,

$$\int \frac{di}{i} = - \frac{1}{RC} \int dt + C_1 \; ,$$

where C_1 is the constant of integration.

$$\ln i = - \frac{1}{RC} t + C_1 \; .$$

Exponentiating,

$$i = e^{-t/RC + C_1} \; ,$$

$$i = e^{C_1} e^{-t/RC} \; .$$

Let $e^{C_1} = C_2$. Then,

$$i = C_2 e^{-t/RC} \; . \tag{c}$$

We may solve for C_2 by using the initial condition that when the switch is closed at $t = 0$ the capacitor has no charge on it. Thus, the voltage drop across it, q/C, is zero. The entire voltage E is applied to R and Ohm's law states that $i(0) = E/R$. Thus, in equation (c) we have

$$i(0) = C_2 e^{0} \; ,$$

$$\frac{E}{R} = C_2 \; .$$

Our general solution is therefore,

$$i(t) = \frac{E}{R} e^{-t/RC} \; . \tag{d}$$

The graph of (d) is Fig. 2.

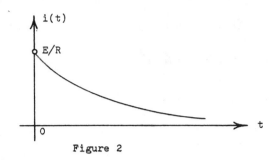

Figure 2

• **PROBLEM** 3–29

A circuit has in series a constant emf of 100 volts, a resistor of 10 ohms, and a capacitor of 2×10^{-4} farads. Find the charge on the capacitor at $t > 0$, if the voltage applied at $t = 0$.

Solution: Let i denote the current and q the charge on the capacitor at time t. The total emf for $t > 0$ is 100 volts. Using the voltage drop laws we see that:

$$E_R = R \frac{dq}{dt} = 10 \frac{dq}{dt} \quad \text{for the resistor}$$

$$E_C = \frac{1}{C} q = 5000 \, q \quad \text{for the capacitor}$$

By Kirchhoff's Law we have

$$R \frac{dq}{dt} + \frac{1}{C} q = E$$

or

$$10 \frac{dq}{dt} + 5000 \, q = 100$$

yielding

$$\frac{dq}{dt} + 500 \, q = 10 \tag{1}$$

The homogeneous part of this constant coefficient differential equation

$$\frac{dq}{dt} + 500 \, q = 0$$

may be solved by assuming $q = e^{mt}$. Then $(m + 500)e^{mt} = 0$ so $m = -500$ and the general solution is $q = Ce^{-500t}$.

For a particular solution since the inhomogeneous term is constant assume $q_p = C_1$; then

$$500 \, C_1 = 10$$

and

$$C_1 = 1/50$$

so

$$q(t) = Ce^{-500t} + 1/50 \, .$$

Now the initial condition must be that $q(t) = 0$ before the voltage is applied. That is

$$q(0) = 0 = C + 1/50$$

so $\qquad C = -1/50$

and $\qquad q(t) = 1/50 \ (1 - e^{-500t})$.

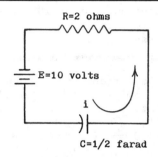

R=2 ohms

E=10 volts

i

C=1/2 farad

In the above circuit the resistance is 2 ohms, the capacitance is $\frac{1}{2}$ farad and the impressed emf is 10 volts. The initial current at $t = 0$ is i_0. Find the current $i = i(t)$ for all $t > 0$.

<u>Solution</u>: The differential equation governing this circuit is obtained by setting the sum of the voltage drops around the circuit equal to the impressed emf. Thus,

$$Ri + \frac{q}{C} = E \ , \qquad\qquad (a)$$

where Ri is the voltage drop across the resistor, and q/C the drop across the capacitor. The variable q represents the charge on the capacitor and is related to the current by $i = dq/dt$.

Differentiating (a) and substituting our values we obtain:

$$2 \frac{di}{dt} + \frac{1}{\frac{1}{2}} i = 0 \ . \qquad\qquad (b)$$

(Note that since E is constant $dE/dt = 0$). Separating variables in (b) we obtain:

$$\frac{di}{i} = -dt \ .$$

Integrating

$$\int \frac{di}{i} = - \int dt + C$$

where C is the constant of integration

$$\ln i = -t + C.$$

130

Exponentiating,
$$i = Ke^{-t}$$
(c)

where
$$K = e^C.$$

Since $i(0) = i_0$, we set $t = 0$ in (c). Thus,

$$i(0) = Ke^0 = i_0 = K.$$

Finally,

$$i(t) = i_0 e^{-t}.$$

• PROBLEM 3–31

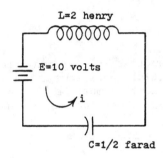

L=2 henry

E=10 volts

C=1/2 farad

In the above circuit the inductance is 2 henrys, the capacitor is $\frac{1}{2}$ farad, and the impressed emf is 10 volts. The initial current in the circuit at $t = 0$ is i_0. Find the current $i = i(t)$ for all $t > 0$.

Solution: The differential equation for this circuit is obtained by setting the sum of the voltage drops around the circuit equal to the impressed emf. Thus,

$$L\frac{di}{dt} + \frac{q}{C} = E$$
(a)

where $L\,di/dt$ is the drop across the inductor and q/C the drop across the capacitor. The charge on the capacitor is q. Since $i = dq/dt$ we differentiate (a) and substitute our values as well as dq/dt. Thus,

$$2\frac{d^2 i}{dt^2} + \frac{1}{\frac{1}{2}}\,i = 0$$
(b)

(Since $E = 5$, $dE/dt = 0$). The differential equation to be solved is

$$\frac{d^2 i}{dt^2} + i = 0$$

or

$$\frac{d^2 i}{dt^2} = -i.$$

A function that has a second derivative equal to the opposite of itself
is

$$i(t) = A \sin t + B \cos t \qquad\qquad (c)$$

is a solution to (b). We verify this by

$$\frac{di}{dt} = A \cos t - B \sin t \; ,$$

$$\frac{d^2 i}{dt^2} = -A \sin t - B \cos t = -(A \sin t + B \cos t)$$

$$= -i \; .$$

In order to evaluate the arbitrary constants A and B in (c) we
need the initial conditions. We are only told that $i(0) = 0$. Thus,

$$i(0) = A \sin(0) + B \cos(0)$$

$$0 = A(0) + B(1)$$

Therefore $\qquad\qquad B = 0$.

Our solution is now

$$i(t) = A \sin t.$$

To evaluate A we need another initial condition. (A second-order
equation requires two initial conditions). $\qquad\qquad\qquad$ (d)
Physically, the second condition must be the initial charge on the
capacitor. Say $q(0) = q_0$. Then looking at the voltage around the
loop at $t = 0$ we obtain

$$10 - L \frac{di(0)}{dt} + \frac{q_0}{C}$$

Thus,

$$\frac{di(0)}{dt} = \frac{10 - q_0/C}{R}$$

for

$$i = A \sin t$$

$$\frac{di}{dt} = A \cos t$$

and

$$\frac{di(0)}{dt} = A$$

Therefore,

$$A = \frac{10 - q_0/C}{R}$$

and finally

$$i(t) = \frac{10 - q_0/C}{R} \sin t \; .$$

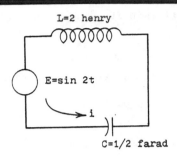

L=2 henry

E=sin 2t

i

C=1/2 farad

In the above circuit, the inductor has inductance 2 henrys, the capacitor has capacitance $\frac{1}{2}$ farad and the impressed voltage is $E = \sin 2t$. The initial current is zero, and the initial charge on the capacitor is 1 coulomb. Find $q(t)$, the charge on the capacitor, for $t > 0$.

<u>Solution</u>: The differential equation for this circuit is obtained by setting the sum of the voltage drops around the circuit equal to the impressed emf. Thus,

$$2 \frac{di}{dt} + \frac{q}{\frac{1}{2}} = \sin 2t \tag{a}$$

where $2\, di/dt$ is the drop across the inductor, and $q/\frac{1}{2}$ is the drop on the capacitor. Since $i = dq/dt$ and $di/dt = d^2 q/dt^2$ we may write (a) as

$$\frac{d^2 q}{dt^2} + q = \frac{1}{2} \sin 2t \; . \tag{b}$$

Equation (b) is an inhomogeneous equation whose solution will be the sum of the homogeneous solution and a particular solution. The homogeneous equation is

$$\frac{d^2 q}{dt^2} + q = 0 \; .$$

This equation is solved by inspection to be

$$q = A \cos t + B \sin t \tag{c}$$

where A and B are the two arbitrary constants needed in the solution of a second-order equation. To get the particular solution we use the method of undetermined coefficients. Assume q_p is a particular solution of the form

$$q_p = U \sin 2t + V \cos 2t \; . \tag{d}$$

(U and V are the undetermined coefficients). Then

$$\frac{d^2 q_p}{dt^2} = -4U \sin 2t - 4V \cos 2t \; .$$

Substituting this into (b)

$$-4(U \sin 2t + V \cos 2t) + (U \sin 2t + V \cos 2t) = \frac{1}{2} \sin 2t,$$

or

$$-3U \sin 2t - 3V \cos 2t = \frac{1}{2} \sin 2t \; .$$

This equation is true only if

$$-3U = \frac{1}{2} \quad \text{and} \quad -3V = 0$$

since sin(t) and cos(t) are linearly independent. Thus, $U = -1/6$ and $V = 0$. Our particular solution is

$$q_p = -1/6 \sin 2t .$$

The general solution is

$$q(t) = A \cos t + B \sin t - 1/6 \sin 2t. \qquad (e)$$

We have only to find A and B from the initial conditions.

(I) $q(0) = 1$: $q(0) = A \cos(0) + B \sin(0) - 1/6 \sin(0)$

$\quad 1 = A .$

(II) $i(0) = 0$:   differentiate (e) since

$$i = \frac{dq}{dt} = -A \sin t + B \cos t - 1/3 \cos 2t,$$

$$i(0) = -A \sin(0) + B \cos(0) - 1/3 \cos(0)$$

$$0 = B - 1/3 \quad \text{or} \quad B = 1/3 .$$

Our final solution is

$$q(t) = \cos t + 1/3 \sin t - 1/6 \sin 2t.$$

● **PROBLEM 3-33**

Show that a simple electrical circuit [consisting of a capacitor, an inductor, a resistor and an electromotive force (either a battery or a generator)] connected in series is governed by an inhomogeneous second-order differential equation with constant coefficients. Also, assuming an initial charge on the capacitor of q_0 coulombs and an initial current in the circuit of I_0 amperes, find initial conditions for the system.

Solution: The first step to this problem would be to draw the electrical circuit involved. Since the four electrical devices are placed in series, the circuit diagram is:

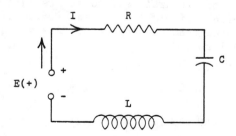

In the above circuit, $E(t)$ is the voltage source which generates the electromotive force in volts, R is the Resistance in ohms, C is the Capacitance in farads, and L is the inductance in henrys.
We will apply Kirchhoff's voltage law on the only loop in the above circuit. Kirchhoff's voltage law states that the algebraic sum of all the voltage drops across each electrical device is zero.

V_R = Voltage drop across resistor = IR

V_C = Voltage drop across capacitor = $\frac{1}{C}$ q.

V_L = Voltage drop across inductor = $L \frac{dI}{dt}$,

and the voltage drop across an emf. = -E(t).

Therefore,

$$IR + \frac{1}{C} q + L \frac{dI}{dt} - E(t) = 0 . \qquad (a)$$

where I = current which is equal to the rate of flow of charge in a given unit time. Written mathematically: I = dq/dt.

$$\frac{dI}{dt} = \frac{d}{dt}\left(\frac{dq}{dt}\right) = \frac{d^2 q}{dt^2} . \qquad (b)$$

Substituting this equality back into equation (a)

$$R \frac{dq}{dt} + \frac{1}{C} q + L \frac{d^2 q}{dt^2} = E(t) . \qquad (c)$$

The general form of an inhomogeneous differential equation with constant coefficients could be represented as:

$$\ddot{x} + a_1 \dot{x} + a_0 x = f(t). \qquad (d)$$

Written a little differently, equation (c) becomes

$$\frac{d^2 q}{dt^2} + \frac{R}{L} \frac{dq}{dt} + \frac{1}{CL} q = \frac{E(t)}{L} . \qquad (e)$$

Comparing (d) with (e) we immediately recognize that a_1 = R/L and a_0 = 1/CL. Both of these physical quantities are constant if the resistance, capacitance, and inductance do not vary. In non-extreme situations this is the case. Also note f(t) = E(t)/L .

Hence, the above circuit is definitely governed by the type of differential equation indicated in the question.

The initial conditions given for q are q(0) = q_0 and $\frac{dq}{dt}\big|_{t=0}$ = I(0) = I_0.

If we need the differential equation governing the current we can differentiate equation (a) to get

$$R \frac{dI}{dt} + L \frac{d^2 I}{dt^2} + \frac{1}{C} \frac{dq}{dt} - \frac{dE(t)}{dt} = 0 .$$

Remembering that I = dq/dt we obtain

$$R \frac{dI}{dt} + L \frac{d^2 I}{dt^2} + \frac{1}{C} I = \frac{dE(t)}{dt} ,$$

or

$$\frac{d^2 I}{dt^2} + \frac{R}{L} \frac{dI}{dt} + \frac{1}{LC} I = \frac{1}{L} \frac{dE(t)}{dt}.$$

Again we have an inhomogeneous second-order differential equation. Now for initial conditions we will need I(0) and $\frac{dI}{dt}\big|_{t=0}$

$$\text{let } I(0) = I_0 .$$

For $\frac{dI}{dt}\big|_{t=0}$ if we rewrite equation (c) as

$$RI + \frac{1}{C} q + L \frac{dI}{dt} = E(t)$$

135

we see

$$\frac{dI}{dt}\Big|_{t=0} = \frac{E(0)}{L} - \frac{R}{L} I_0 - \frac{1}{LC} q_0$$

So in this case we need the initial driving EMF, current, and capacitor charge.

• **PROBLEM** 3-34

An RCL-circuit as shown in Fig. 1 with resistance of 5 ohms, capacitance of 10^{-2} farad, and inductance, L = 1/8 henry has an applied voltage E(t) = sin t. Find the steady state current in the circuit.

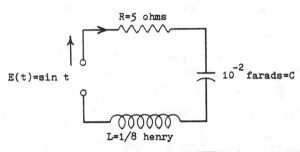

Fig. 1

E(t)=sin t

R=5 ohms

10^{-2} farads=C

L=1/8 henry

Solution: The circuit is shown in the figure. It is known that the voltage drops across a resistor, a capacitor, and an inductor are respectively RI, 1/C q, and L dI/dq (where q is the charge in coulombs on the capacitor, and I is the current). The voltage drop across an EMF is -E(t). From Kirchhoff's law we have:

$$RI + L \frac{dI}{dt} + \frac{1}{C} q - E(t) = 0. \tag{1}$$

Recall that I = dq/dt. Differentiating equation (1) with respect to t and dividing by L yields:

$$\frac{d^2 I}{dt^2} + \frac{R}{L} \frac{dI}{dt} + \frac{1}{LC} I = \frac{1}{L} \frac{dE(t)}{dt}. \tag{2}$$

In this problem R = 5, L = 1/8, C = 10^{-2} and E(t) = sin t. So that equation (2) becomes

$$\ddot{I} + 40\dot{I} + 800I = 8 \cos t \quad (\text{here } \dot{I} = \frac{dI}{dt}) \tag{3}$$

Since we want the steady state current we are only interested in the particular solution; that is to say, we are taking the steady state solution to be that part of the total solution which results due to the driving EMF. The other part of the solution, that of the homogeneous equation, would be the transient or oscillatory solution depending on the nature of the solution. Therefore we use the method of undetermined coefficients.

Letting y_p = a sin t + b cos t we compute successive derivatives $\dot{y}_p$ and $\ddot{y}_p$ and solve $\ddot{y}_p + 40\dot{y}_p + 800y_p = 8 \cos t$ for the constants a and b. Inserting the values for y, $\dot{y}$, and $\ddot{y}$ into this equation and simplifying yields:

136

$(799a - 40b)\sin t + (799b + 40a)\cos t = 8 \cos t$

or equating coefficients yields

$$799a - 40b = 0$$

$$40a + 799b = 8$$

Solving yields

$$a = \frac{320}{640,001}$$

$$b = \frac{6392}{640,001}$$

so that the steady state equation is given by

$$y_p(t) = \frac{1}{640,001} (320 \sin t + 6392 \cos t).$$

● **PROBLEM** 3-35

Set up the differential equation for the general two loop network.

Solution: The basic components of a network are resistors with resistance measured in ohms, capacitors with capacitance measured in farads and inductors with inductance measured in henries. There are also voltage sources measured in volts, and the current measured in amperes.

In our diagram we have resistors R_1, R_2 and R_3; capacitors C_1 and C_2, inductors L_1 and L_2 and voltage source $E(t)$. We may label the current in any part of the circuit by i_1, i_2, $i_3 \ldots$, subject to Kirchhoff's Current Law, which states that the sum of the currents flowing into a point equals the sum of the currents flowing out of the point. A point where the current may proceed along different paths is called a branch point. In our diagram N_1 and N_2 are branch points. Thus, by Kirchhoff's Current Law,

$$i_1 = i_2 + i_3$$

at each of these points.

The differential equation for the network comes from Kirchhoff's Voltage Law which states that the sum of the voltage drop around any loop equals the impressed voltage in that loop. (A loop is any closed path in the circuit. Our circuit has three loops: $AN_1 N_2 DA$, $N_1 BCN_2 N_1$, and $AN_1 BCN_2 DA$). The voltage drop across a resistor is defined as RI where

R is the resistance and I the current passing through it. The voltage drop across an inductor is L dI/dt where L is the inductance
and dI/dt is the rate of change of the current in it. The voltage
drop across a capacitor is Q/C where C is the capacitance and Q
is the charge (measured in coulombs) on the capacitor. The charge
is related to the current I in the branch of the capacitor by
I = dQ/dt. Alternately

$$Q(t) = \int_0^t I(x)\,dx$$

where x is a dummy variable of integration.

We must also follow the convention that if I is the current flowing
from point U to point V in a given circuit, then -I is the current flowing from V to U.

Using Kirchhoff's two laws we must write down as many equations as we
have currents. Thus, in our diagram we need three independent
equations corresponding to i_1, i_2 and i_3. We already have one
equation: $i_1 = i_2 + i_3$. To get a second and third we apply Kirchhoff's voltage law to loops $AN_1 N_2 DA$ and $N_1 BCN_2 N$.

Loop $AN_1 N_2 DA$:

$$R_1 i_1 + \frac{Q_1}{C_1} + L_1 \frac{di_2}{dt} + R_3 i_1 = E \qquad (a)$$

Loop $N_1 BCN_1 N_2$:

$$R_2 i_3 + \frac{Q_2}{C_2} + L_2 \frac{di_3}{dt} - L_1 \frac{di_2}{dt} = 0 \qquad (b)$$

The minus sign for the inductance indicates that we are going against
i_3 in this loop. Since $dQ_1/dt = i_1$ and $dQ_2/dt = i_3$ we differentiate (a) and (b) and substitute:

$$R_1 \frac{di_1}{dt} + \frac{i_1}{C} + L_1 \frac{d^2 i_2}{dt^2} + R_3 \frac{di_1}{dt} = \frac{dE}{dt}$$

and

$$R_2 \frac{di_3}{dt} + \frac{i_3}{C_2} + L_2 \frac{d^2 i_3}{dt^2} - L_1 \frac{d^2 i_1}{dt^2} = 0 .$$

In operator notation these equations become:

$$\left(R_1 D + R_3 D + \frac{1}{C} \right) i_1 + \left(L_1 D^2 \right) i_2 = \frac{dE}{dt} \qquad (c)$$

and

$$\left(-L_1 D^2 \right) i_1 + \left(L_2 D^2 + R_2 D + \frac{1}{C} \right) i_3 = 0 . \qquad (d)$$

Coupling (c) and (d) with

$$i_1 - i_2 - i_3 = 0 \qquad \text{(e)}$$

we have three equations with three unknowns.

The constants of integration which arrive from solving differential equations (c) and (d) would be evaluated initial conditions such as

$$i_1(0) = i_2(0) = i_3(0) = 0,$$

(initial current flow)

$$Q_1(0) = 5 \text{ coulombs}$$

and $\qquad Q_2(0) = 0 \quad$ (initial charge on capacitor).

• PROBLEM 3-36

For the network shown below set up the equations for the currents i_1, i_2, i_3 and the charge q_3. Assume when the switch is closed at $t = 0$ all currents and charges are zero. Find the characteristic polynomial for the matrix equation of the system.

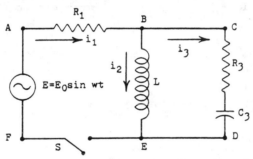

Solution: From Kirchhoff's current law we have that

$$i_1 = i_2 + i_3 . \qquad \text{(a)}$$

By Kirchhoff's voltage law the sum of the voltage drops around any closed loop is equal to the emf in that loop. Thus, around loop ABEFA we have

$$R_1 i_1 + L \frac{di_2}{dt} = E_0 \sin \omega t . \qquad \text{(b)}$$

Around loop ABCDEFA we have

$$R_1 i_1 + R_3 i_3 + \frac{q_3}{C_3} = E_0 \sin \omega t . \qquad \text{(c)}$$

Finally, $\qquad i_3 = \frac{dq_3}{dt} \qquad \text{(d)}$

by definition. Equations (a), (b), (c), (d) constitute four equations in four unknowns i_1, i_2, i_3, q_3 subject to the initial conditions

$$i_1(0) = i_2(0) = i_3(0) = q_3(0) = 0 . \qquad \text{(e)}$$

We use equations (a) and (d) to eliminate i_3 and q_3 as follows.

Differentiate (c) to obtain

$$R_1 \frac{di_1}{dt} + R_3 \frac{di_3}{dt} + \frac{1}{C_3} i_3 = \omega E_0 \cos \omega t \qquad \text{(f)}$$

by (a) $i_3 = i_1 - i_2$ and

$$\frac{di_3}{dt} = \frac{di_1}{dt} - \frac{di_2}{dt} ,$$

Thus (f) becomes

$$R_1 \frac{di_1}{dt} + R_3 \left(\frac{di_1}{dt} - \frac{di_2}{dt} \right) + \frac{1}{C_3}(i_1 - i_2) = \omega E_0 \cos \omega t . \qquad \text{(g)}$$

Solving for di_1/dt we obtain

$$\frac{di_1}{dt} = \frac{1}{R_1 + R_3} \left\{ \omega E_0 \cos \omega t + R_3 \frac{di_2}{dt} - \frac{i_1 - i_2}{C_3} \right\} . \qquad \text{(h)}$$

But by (b)

$$\frac{di_2}{dt} = \frac{E_0}{L} \sin \omega t - \frac{R_1}{L} i_1 \qquad \text{(j)}$$

Substituting this into (h)

$$\frac{di_1}{dt} = -\left(\frac{C_3 R_1 R_3 + L}{C_3 L (R_1 + R_3)} \right) i_1 + \left(\frac{1}{C_3 (R_1 + R_3)} \right) i_2 + \frac{E_0 \omega}{R_1 + R_3} \cos \omega t + \frac{E_0 R_3}{L_2 (R_1 + R_3)} \sin \omega t$$

$$\text{(k)}$$

In the associated homogeneous equation $E = 0$, thus $dE/dt = 0$ and (k) and (j) become

$$\frac{di_1}{dt} = -\left(\frac{C_3 R_1 R_3 + L}{C_3 L (R_1 + R_3)} \right) i_1 + \left(\frac{1}{C_3 (R_1 + R_3)} \right) i_2$$

$$\frac{di_2}{dt} = \left(- \frac{R_1}{L} \right) i_1 + (0) i_2 .$$

The matrix form of this is

$$\frac{d}{dt} \begin{pmatrix} i_1 \\ i_2 \end{pmatrix} = \begin{vmatrix} \dfrac{-(C_3 R_1 R_3 + L}{C_3 L (R_1 + R_3)} & \dfrac{1}{C_3 (R_1 + R_3)} \\[2em] -\dfrac{R}{L} & 0 \end{vmatrix} \begin{pmatrix} i_1 \\ i_2 \end{pmatrix}$$

The characteristic polynomial is thus, $\Delta(m)$ defined by

$$\Delta = \det \begin{vmatrix} -\dfrac{C_3 R_1 R_3 + L}{C_3 L (R_1 + R_3)} - m & \dfrac{1}{C_3 (R_1 + R_3)} \\[2em] -\dfrac{R_1}{L} & -m \end{vmatrix}$$

where det means determinant. Evaluating,

$$\Delta(m) = m^2 + \left(\frac{C_3 R_1 R_3 + L}{C_3 L (R_1 + R_3)} \right) m + \frac{R_1}{C_3 L (R_1 + R_3)} .$$

140

CHAPTER 4

VECTORS

> **Basic Attacks and Strategies for Solving Problems in this Chapter. See pages 142 to 186 for step-by-step solutions to problems.**

Vectors are frequently used in many branches of pure and applied mathematics and in the physical and engineering sciences. The vector analysis applied by the physicist or engineer to problems in their fields differs in many ways from the n-dimensional vector spaces used in pure mathematics. Both types, however, have a common intuitive foundation. The need for a vector concept arose very naturally in mechanics. The force on a body, for example, cannot in general be completely described by a single number. Force has two properties, magnitude and direction, and therefore requires more than a single number for its description. Force is an example of a vector quantity which has both magnitude and direction. A large number of physical quantities are vectors, and, interestingly enough, the same laws of operation apply to all vectors.

Vectors are often represented geometrically by a line with an arrowhead on the end of it. The length of the line indicates the magnitude of the vector, and the arrow denotes its direction.

Step-by-Step Solutions to Problems in this Chapter, "Vectors"

VECTOR PRODUCTS

• **PROBLEM** 4–1

Find vector $\bar{c}$, such that

$$\bar{c} = \bar{a} \times \bar{b} \tag{1}$$

where

$$\bar{a} = (0,2,5)$$
$$\bar{b} = (2,-4,0) \tag{2}$$

$\bar{a} \times \bar{b}$ denotes the vector product of vectors $\bar{a}$ and $\bar{b}$.

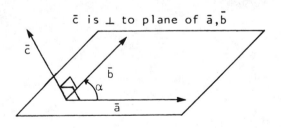

$\bar{c}$ is $\perp$ to plane of $\bar{a},\bar{b}$

Solution: Start with the definition of the vector product of two vectors $\bar{a}$ and $\bar{b}$. Let $\bar{a}$ and $\bar{b}$ be two vectors and α the smallest non-negative angle between them, $0° \leq \alpha \leq 180°$ as shown in the figure. The vector product of $\bar{a}$ and $\bar{b}$ denoted by $\bar{a} \times \bar{b}$ is vector $\bar{c}$

$$\bar{c} = \bar{a} \times \bar{b}$$

such that

1) $\bar{c}$ is perpendicular to $\bar{a}$ and $\bar{b}$

2) the direction of $\bar{c}$ is such that $\bar{a}$, $\bar{b}$ and $\bar{c}$ form a right-handed system

3) $c = ab \sin \alpha$ \hfill (3)

First find the magnitudes of $\bar{a}$ and $\bar{b}$.

$$a = |\bar{a}| = \sqrt{4+25} = \sqrt{29}$$
$$b = |\bar{b}| = \sqrt{4+16} = \sqrt{20}$$

\hfill (4)

From the scalar product, find the angle α

$$\bar{a} \cdot \bar{b} = (0,2,5) \cdot (2,-4,0) = -8$$

Since

$$\bar{a} \cdot \bar{b} = ab \cos \alpha,$$ \hfill (5)

$$\cos \alpha = \frac{\bar{a} \cdot \bar{b}}{ab} = \frac{-8}{\sqrt{29} \cdot \sqrt{20}} = -0.332$$ \hfill (6)

and \qquad\qquad $\alpha = 109°$

The magnitude of $\bar{c}$ is

$$c = ab \sin \alpha = \sqrt{29} \cdot \sqrt{20} \sin 109° = 22.8$$ \hfill (7)

Vector $\bar{c}$ is perpendicular to $\bar{a}$ and $\bar{b}$ and the direction of $\bar{c}$ is such that $\bar{a}$, $\bar{b}$, $\bar{c}$ form a right-handed system.

• PROBLEM 4–2

Vectors $\bar{a} = (2,3,5)$ and $\bar{b} = (1,-3,2)$ are given. Find vector $\bar{c}$

where \quad $\bar{c} = \bar{a} \times \bar{b}$ \hfill (1)

using

1) the component expression for the vector product

2) the determinant expression for the vector product

3) the coordinate expression for the vector product. Remember that for the base vectors

$$\bar{i}_1 \times \bar{i}_1 = \bar{i}_2 \times \bar{i}_2 = \bar{i}_3 \times \bar{i}_3 = \bar{0}$$

$$\bar{i}_1 \times \bar{i}_2 = \bar{i}_3 \qquad\qquad \bar{i}_k \times \bar{i}_1 = -\bar{i}_1 \times \bar{i}_k$$ \hfill (2)

$$\bar{i}_2 \times \bar{i}_3 = \bar{i}_1$$

$$\bar{i}_3 \times \bar{i}_1 = \bar{i}_2$$

<u>Solution</u>: 1) In the component form vector $\overline{c}$

$$\overline{c} = \overline{a} \times \overline{b} \quad \text{is given by}$$

$$\overline{c} = (a_2 b_3 - a_3 b_2, \ a_3 b_1 - a_1 b_3, \ a_1 b_2 - a_2 b_1) \tag{3}$$

Substituting $\overline{a}$ and $\overline{b}$, obtain

$$\overline{c} = (3 \cdot 2 + 5 \cdot 3, \ 5 \cdot 1 - 2 \cdot 2, \ -2 \cdot 3 - 3 \cdot 1)$$
$$= (21, 1, -9) \tag{4}$$

2) In the determinant form, vector $\overline{c}$ can be expressed as

$$\overline{c} = \begin{vmatrix} \overline{i}_1 & \overline{i}_2 & \overline{i}_3 \\ a_1 & a_2 & a_3 \\ b_1 & b_2 & b_3 \end{vmatrix} \tag{5}$$

Substituting $\overline{a}$ and $\overline{b}$ into eq.(5), obtain

$$\overline{c} = \begin{vmatrix} \overline{i}_1 & \overline{i}_2 & \overline{i}_3 \\ 2 & 3 & 5 \\ 1 & -3 & 2 \end{vmatrix} \tag{6}$$

$$= \overline{i}_1 \ 3 \cdot 2 + \overline{i}_2 \ 5 \cdot 1 + \overline{i}_3 \ 2 \cdot (-3) - \overline{i}_1 (-3) \cdot 5 - \overline{i}_2 \ 2 \cdot 2 - \overline{i}_3 \ 3 \cdot 1$$

$$= \overline{i}_1 (6+15) + \overline{i}_2 (5-4) + \overline{i}_3 (-6-3)$$

$$= 21 \ \overline{i}_1 + \overline{i}_2 - 9 \overline{i}_3$$

3) Vectors $\overline{a}$ and $\overline{b}$ can be represented as

$$\overline{a} = 2\overline{i}_1 + 3\overline{i}_2 + 5\overline{i}_3$$
$$\overline{b} = \overline{i}_1 - 3\overline{i}_2 + 2\overline{i}_3 \tag{7}$$

Substituting eq.(7) into

$$\overline{c} = \overline{a} \times \overline{b}$$

and using eq.(2), obtain

$$\overline{c} = \overline{a} \times \overline{b} \tag{8}$$

$$= (2\overline{i}_1 + 3\overline{i}_2 + 5\overline{i}_3) \times (\overline{i}_1 - 3\overline{i}_2 + 2\overline{i}_3)$$

$$= 2\overline{i}_1 \times \overline{i}_1 - 6\overline{i}_1 \times \overline{i}_2 + 4\overline{i}_1 \times \overline{i}_3 + 3\overline{i}_2 \times \overline{i}_1 - 9\overline{i}_2 \times \overline{i}_2 + 6\overline{i}_2 \times \overline{i}_3$$

$$+5\bar{i}_3 \times \bar{i}_1 - 15\bar{i}_3 \times \bar{i}_2 + 10\bar{i}_3 \times \bar{i}_3$$

$$= -6\bar{i}_1 \times \bar{i}_2 - 3\bar{i}_1 \times \bar{i}_2 + 6\bar{i}_2 \times \bar{i}_3 + 15\bar{i}_2 \times \bar{i}_3 - 4\bar{i}_3 \times \bar{i}_1 + 5\bar{i}_3 \times \bar{i}_1$$

$$= -9\bar{i}_1 \times \bar{i}_2 + 21\bar{i}_2 \times \bar{i}_3 + \bar{i}_3 \times \bar{i}_1$$

$$= 21\bar{i}_1 + \bar{i}_2 - 9\bar{i}_3$$

• PROBLEM 4-3

Show that the volume V of the parallelepiped with the vectors $\bar{a}$, $\bar{b}$ and $\bar{c}$ forming adjacent edges as shown in the figure, is given by

$$V = |\bar{a} \cdot (\bar{b} \times \bar{c})| \tag{1}$$

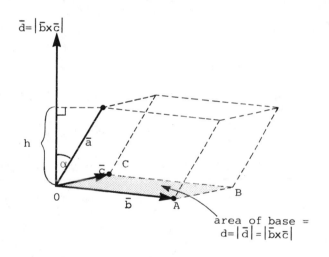

area of base =
$d = |\bar{d}| = |\bar{b} \times \bar{c}|$

<u>Solution</u>: Denote

$$\bar{d} = \bar{b} \times \bar{c}$$
and
$$d = |\bar{b} \times \bar{c}|$$

It was shown previously that d represents the area of the parallelogram OABC. Then

$$V = dh \tag{2}$$

where h is the altitude of the parallelepiped. By definition, $\bar{d}$ is perpendicular to the base, thus

$$h = a|\cos \alpha| \tag{3}$$
and
$$V = dh = ad \ |\cos \alpha| = |\bar{a} \cdot \bar{d}| = |\bar{a} \cdot (\bar{b} \times \bar{c})| \tag{4}$$

The expression
$$\bar{a} \times (\bar{b} \times \bar{c})$$
is called the vector triple product.

Prove the following identities

$$\bar{a} \times (\bar{b} \times \bar{c}) = (\bar{a} \cdot \bar{c})\bar{b} - (\bar{a} \cdot \bar{b})\bar{c} \qquad (1)$$

$$(\bar{a} \times \bar{b}) \times \bar{c} = (\bar{c} \cdot \bar{a})\bar{b} - (\bar{c} \cdot \bar{b})\bar{a} \qquad (2)$$

Solution: To prove eq.(1) compute the components of both sides of eq.(1). From the definition of the vector product, obtain

$$\bar{a} \times (\bar{b} \times \bar{c}) = (a_1, a_2, a_3) \times (b_2c_3 - b_3c_2, \; b_3c_1 - b_1c_3, \; b_1c_2 - b_2c_1)$$

$$= \Big[a_2(b_1c_2 - b_2c_1) - a_3(b_3c_1 - b_1c_3), \; a_3(b_2c_3 - b_3c_2) - a_1(b_1c_2 - b_2c_1),$$
$$a_1(b_3c_1 - b_1c_3) - a_2(b_2c_3 - b_3c_2) \Big] \qquad (3)$$

$$= \Big[b_1(a_2c_2 + a_3c_3) - c_1(a_2b_2 + a_3b_3),$$
$$b_2(a_1c_1 + a_3c_3) - c_2(a_1b_1 + a_3b_3), \; b_3(a_1c_1 + a_2c_2) - c_3(a_1b_1 + a_2b_2) \Big]$$

$$= \Big[b_1(a_1c_1 + a_2c_2 + a_3c_3) - c_1(a_1b_1 + a_2b_2 + a_3b_3), \; b_2(a_1c_1 + a_2c_2 + a_3c_3)$$
$$- c_2(a_1b_1 + a_2b_2 + a_3b_3), \; b_3(a_1c_1 + a_2c_2 + a_3c_3) - c_3(a_1b_1 + a_2b_2 + a_3b_3) \Big]$$

$$= \Big[b_1(\bar{a} \cdot \bar{c}), \; b_2(\bar{a} \cdot \bar{c}), \; b_3(\bar{a} \cdot \bar{c}) \Big] - \Big[c_1(\bar{a} \cdot \bar{b}), \; c_2(\bar{a} \cdot \bar{b}), \; c_3(\bar{a} \cdot \bar{b}) \Big]$$

$$= (\bar{a} \cdot \bar{c})\bar{b} - (\bar{a} \cdot \bar{b})\bar{c}$$

That completes the proof. The same method can be used to prove eq.(2).

The shortest method, however, is to rewrite eq.(1) in the form

$$(\bar{b} \times \bar{c}) \times \bar{a} = (\bar{a} \cdot \bar{b})\bar{c} - (\bar{a} \cdot \bar{c})\bar{b} \qquad (4)$$

then replace
vector $\bar{b}$ by $\bar{a}$

$$\bar{b} \to \bar{a}$$

vector $\bar{c}$ by $\bar{b}$

$$\bar{c} \to \bar{b}$$

vector $\bar{a}$ by $\bar{c}$

$$\bar{a} \rightarrow \bar{c}$$

Then, obtain

$$(\bar{b} \times \bar{c}) \times \bar{a} = (\bar{a} \cdot \bar{b})\, \bar{c} - (\bar{a} \cdot \bar{c})\, \bar{b}$$
$$\downarrow \quad \downarrow \quad \downarrow \quad \downarrow \downarrow \quad \downarrow \quad \downarrow \downarrow \quad \downarrow \qquad (5)$$
$$(\bar{a} \times \bar{b}) \times \bar{c} = (\bar{c} \cdot \bar{a})\, \bar{b} - (\bar{c} \cdot \bar{b})\, \bar{a}$$

That proves identity (2).

• **PROBLEM** 4–5

Find the equation of the line

1) through two given points A and B

2) through a given point A and perpendicular to the plane through three given points B, C and D.

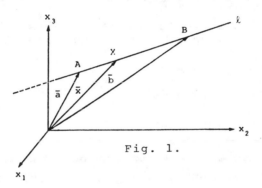

Fig. 1.

Solution: 1) Let $\bar{a}$ and $\bar{b}$ be position-vectors of the given points A and B respectively. Let $\bar{x}$ be the position vector of any point X on the line as shown in Fig. 1.

From Fig. 1,

$$\bar{a} + \overline{AB} = \bar{b} \qquad (1)$$

or

$$\overline{AB} = \bar{b} - \bar{a}$$

Using the result that the equation of the line through a point F whose position vector is $\bar{f}$ and parallel to $\bar{g}$ is

$$(\bar{x} - \bar{f}) \times \bar{g} = 0, \qquad (2)$$

obtain (after substituting eq.(1) into eq.(2))

$$(\bar{x} - \bar{a}) \times (\bar{b} - \bar{a}) = \bar{0} \qquad (3)$$

2) Let A be the given point on the line ℓ and let B, C and D be the given points defining the plane, as shown in Fig. 2.

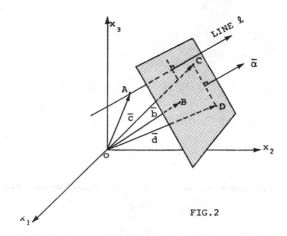

FIG.2

Let $\bar{a}$, $\bar{b}$, $\bar{c}$ and $\bar{d}$ be the position vectors of A, B, C and D. Define vector $\bar{\alpha}$ such that

$$\bar{\alpha} = \bar{b} \times \bar{c} + \bar{c} \times \bar{d} + \bar{d} \times \bar{b} \tag{4}$$

$\bar{\alpha}$ is perpendicular to the plane S. Thus, the equation of the line through a given point A and parallel to $\bar{\alpha}$ is

$$(\bar{x} - \bar{a}) \times \bar{\alpha} = 0 \tag{5}$$

where $\bar{\alpha}$ is given by eq.(4).

● **PROBLEM 4–6**

1) Find the equation of the plane through a given point P and perpendicular to a given vector $\bar{a}$, where $P:(p_1,p_2,p_3)$ and $\bar{a} = (a_1,a_2,a_3)$.

2) Find the equation of the plane that passes through $A = (1,6,2)$ and is perpendicular to the vector $\bar{a} = (2,4,1)$.

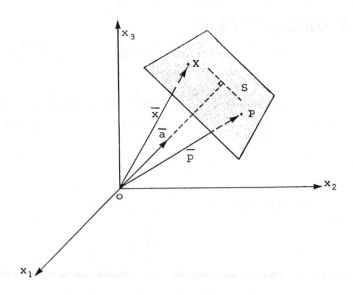

Solution: 1) Vector $\bar{a}$ and point P are shown in the figure.

Let X be any point on S and let $\bar{x}$ and $\bar{p}$ denote the position-vectors of X and P respectively. Vector $\overline{PX}$ is perpendicular to $\bar{a}$, thus

$$\overline{PX} \cdot \bar{a} = 0 \tag{1}$$

and

$$\overline{PX} = \bar{x} - \bar{p} \tag{2}$$

thus

$$(\bar{x} - \bar{p}) \cdot \bar{a} = 0 \tag{3}$$

or in the component form

$$(x_1 - p_1,\ x_2 - p_2,\ x_3 - p_3) \cdot (a_1, a_2, a_3) = 0 \tag{4}$$

2) Vector $\bar{a}$ is given by

$$\bar{a} = (2, 4, 1)$$

and point A is (1,6,2). (A is equivalent to pt. P.)

Substituting these values into eq.(4)

$$(x_1 - 1,\ x_2 - 6,\ x_3 - 2) \cdot (2, 4, 1)$$

$$= 2(x_1 - 1) + 4(x_2 - 6) + (x_3 - 2) = 0 \tag{5}$$

Thus,

$$x_3 = -2(x_1 - 1) - 4(x_2 - 6) + 2$$

Points of the plane can be represented in the form

$$(x_1,\ x_2,\ -2(x_1 - 1) - 4(x_2 - 6) + 2)$$

where x_1 and x_2 are any real parameters.

VECTOR DERIVATIVES

Using the definition of the derivative of a vector function, find

1) $\dfrac{d\overline{r}}{dt}$ where $\overline{r}(t) = (2,\ t,\ t^2)$

2) $\dfrac{d\overline{r}}{dt}$ where $\overline{r}(t) = (t,\ 2t,\ t+1)$

3) $\dfrac{d\overline{u}}{dt}$ where $\overline{u}(t) = (t,\ e^t,\ t^2)$

Solution: The derivative of a vector function $\overline{r}(t)$ is defined as the limit:

$$\frac{d\overline{r}}{dt} = \lim_{\Delta t \to 0} \frac{\overline{r}(t+\Delta t) - \overline{r}(t)}{\Delta t} = \lim_{\Delta t \to 0} \frac{\Delta \overline{r}}{\Delta t} \tag{1}$$

(assume that such a limit exists).

1) To find the derivative of $\overline{r}(t) = (2,t,t^2)$ apply the definition of the derivative given by eq.(1). This direct method of finding the derivative is seldom used in practical calcula- tions, where one deals with functions that are more complicated.

Substituting $\overline{r}(t) = (2,t,t^2)$ into eq.(1), obtain

$$\frac{d\overline{r}}{dt} = \lim_{\Delta t \to 0} \frac{\overline{r}(t+\Delta t) - \overline{r}(t)}{\Delta t} = \lim_{\Delta t \to 0} \frac{\left[2, t+\Delta t, (t+\Delta t)^2\right] - (2,t,t^2)}{\Delta t}$$

$$= \lim_{\Delta t \to 0} \frac{(0, \Delta t, 2t\Delta t + (\Delta t)^2)}{\Delta t} = \lim_{\Delta t \to 0}\ 0, \left(\frac{\Delta t}{\Delta t},\ \frac{2t\Delta t + (\Delta t)^2}{\Delta t}\right)$$

$$= (0,1,2t) \tag{2}$$

2) Here again substitute $\overline{r}(t) = (t,\ 2t,\ t+1)$ into eq.(1) and obtain

$$\frac{d\overline{r}}{dt} = \lim_{\Delta t \to 0} \frac{\left[t+\Delta t, 2(t+\Delta t), t+\Delta t+1\right] - (t,2t,t+1)}{\Delta t}$$

$$= \lim_{\Delta t \to 0} \frac{(\Delta t,\ 2\Delta t,\ \Delta t)}{\Delta t} = (1,2,1) \tag{3}$$

3) Substituting $\bar{u}(t) = (t, e^t, t^2)$ into eq.(1), gives

$$\frac{d\bar{u}}{dt} = \lim_{\Delta t \to 0} \frac{\left[t+\Delta t, e^{t+\Delta t}, (t+\Delta t)^2\right] - (t, e^t, t^2)}{\Delta t}$$

$$= \lim_{\Delta t \to 0} \left[\frac{\Delta t}{\Delta t}, \frac{e^{t+\Delta t} - e^t}{\Delta t}, \frac{2t\Delta t + (\Delta t)^2}{\Delta t}\right] \tag{4}$$

$$= (1, e^t, 2t)$$

● **PROBLEM 4–8**

Let $\bar{r}(t)$ and $\bar{u}(t)$ be given by

$$\bar{r}(t) = (\sin t, t^2, 3)$$
$$\bar{u}(t) = (t+1, \sin t, e^t) \tag{1}$$

evaluate

1) $\frac{d}{dt}(\bar{r} + \bar{u})$

2) $\frac{d}{dt}(\bar{r} \cdot \bar{u})$

3) $\frac{d}{dt}(\bar{r} \cdot \bar{r})$

4) $\frac{d}{dt}(\bar{r} \times \bar{u})$

Solution: First evaluate

$$\frac{d\bar{r}}{dt} \quad \text{and} \quad \frac{d\bar{u}}{dt}$$

$$\frac{d\bar{r}}{dt} = (\cos t, 2t, 0)$$
$$\frac{d\bar{u}}{dt} = (1, \cos t, e^t) \tag{2}$$

1) $\frac{d}{dt}(\bar{r}+\bar{u}) = \frac{d\bar{r}}{dt} + \frac{d\bar{u}}{dt} = (\cos t, 2t, 0) + (1, \cos t, e^t)$

$$= (\cos t + 1, 2t + \cos t, e^t) \tag{3}$$

2) $\frac{d}{dt}(\bar{r}\cdot\bar{u}) = \bar{r}\cdot\frac{d\bar{u}}{dt} + \bar{u}\cdot\frac{d\bar{r}}{dt} = (\sin t, t^2, 3)\cdot(1, \cos t, e^t)$

$$+ (t+1, \sin t, e^t)\cdot(\cos t, 2t, 0) \tag{4}$$

151

$$= \sin t + t^2 \cos t + 3e^t + (t+1)\cos t + 2t \sin t$$

3) $\frac{d}{dt} (\bar{r} \cdot \bar{r}) = \bar{r} \cdot \frac{d\bar{r}}{dt} + \bar{r} \cdot \frac{d\bar{r}}{dt} = 2\bar{r} \cdot \frac{d\bar{r}}{dt}$

$$= 2(\sin t, \, t^2, \, 3) \cdot (\cos t, \, 2t, \, 0) = 2(\sin t \cos t + 2t^3) \tag{5}$$

4) $\frac{d}{dt} (\bar{r} \times \bar{u}) = \bar{r} \times \frac{d\bar{u}}{dt} + \frac{d\bar{r}}{dt} \times \bar{u}$

$$= (\sin t, \, t^2, \, 3) \times (1, \, \cos t, \, e^t) + (\cos t, \, 2t, \, 0) \times (t+1, \, \sin t, \, e^t)$$

$$= (t^2 e^t - 3\cos t, \, 3 - e^t \sin t, \, \sin t \cos t - t^2)$$

$$+ (2te^t, \, -e^t \cos t, \, \sin t \cos t - 2t^2 - 2t) \tag{6}$$

$$= (t^2 e^t - 3\cos t + 2te^t, \, 3 - e^t \sin t - e^t \cos t, \, 2\sin t \cos t - 3t^2 - 2t).$$

Note: The results of problem 4-11 were used in parts 1, 2 and 4 of this problem.

• PROBLEM 4–9

Find $\frac{d}{dt} \left(\bar{a} \cdot \frac{d\bar{b}}{dt} - \frac{d\bar{a}}{dt} \cdot \bar{b} \right)$ (1)

if $\bar{a}(t)$ and $\bar{b}(t)$ are differentiable vector functions of t such that

$$\bar{a} = (e^t, \, 5t^2, \, t+1)$$

$$\bar{b} = (\alpha t^2, \, \sin t, \, t^2) \tag{2}$$

where α is a constant parameter.

Solution: To find the required expression, one can substitute directly $\bar{a}(t)$ and $\bar{b}(t)$ and perform all the operations. However, the faster method would be to first simplify eq.(1) and then substitute into eq.(2).

Thus

$$\frac{d}{dt} \left(\bar{a} \cdot \frac{d\bar{b}}{dt} - \frac{d\bar{a}}{dt} \cdot \bar{b} \right) = \bar{a} \cdot \frac{d^2\bar{b}}{dt^2} + \frac{d\bar{a}}{dt} \cdot \frac{d\bar{b}}{dt} - \frac{d\bar{a}}{dt} \cdot \frac{d\bar{b}}{dt} - \frac{d^2\bar{a}}{dt^2} \cdot \bar{b}$$

$$= \bar{a} \cdot \frac{d^2\bar{b}}{dt^2} - \frac{d^2\bar{a}}{dt^2} \cdot \bar{b} \tag{3}$$

Next, find the first and second derivatives of $\bar{a}(t)$ and $\bar{b}(t)$.

$$\frac{d\bar{a}}{dt} = (e^t, \ 10t, \ 1)$$

$$(4)$$

$$\frac{d^2\bar{a}}{dt^2} = (e^t, \ 10, \ 0)$$

$$\frac{d\bar{b}}{dt} = (2\alpha t, \ \cos t, \ 2t)$$

$$(5)$$

$$\frac{d^2\bar{b}}{dt^2} = (2\alpha, \ -\sin t, \ 2)$$

Substituting eqs.(4) and (5) into eq.(3), gives

$$\frac{d}{dt}\left(\bar{a}\cdot\frac{d\bar{b}}{dt} - \frac{d\bar{a}}{dt}\cdot\bar{b}\right) = \bar{a}\cdot\frac{d^2\bar{b}}{dt^2} - \frac{d^2\bar{a}}{dt^2}\cdot\bar{b}$$

$$= (e^t, 5t^2, t+1)\cdot(2\alpha, -\sin t, 2) - (e^t, 10, 0)\cdot(\alpha t^2, \sin t, t^2) \quad (6)$$

$$= 2\alpha e^t - 5t^2\sin t + 2(t+1) - \alpha t^2 e^t - 10\sin t$$

• PROBLEM 4–10

Find curl $\bar{a}$ at the point $(1,-2,1)$ for

$$\bar{a} = x^2 y^2 \bar{i} + 2xyz \bar{j} + z^2 \bar{k} \qquad (1)$$

Solution: The curl of a vector field $\bar{a}$ is defined as

$$\text{curl } \bar{a} = \left(\frac{\partial a_z}{\partial y} - \frac{\partial a_y}{\partial z}\right)\bar{i} + \left(\frac{\partial a_x}{\partial z} - \frac{\partial a_z}{\partial x}\right)\bar{j} + \left(\frac{\partial a_y}{\partial x} - \frac{\partial a_x}{\partial y}\right)\bar{k} \qquad (2)$$

Since the gradient operator is defined as

$$\nabla \equiv \frac{\partial}{\partial x}\bar{i} + \frac{\partial}{\partial y}\bar{j} + \frac{\partial}{\partial z}\bar{k},$$

definition (2) can be written in the form

$$\text{curl } \bar{a} = \text{rot } \bar{a} \equiv \nabla \times \bar{a} = \left(\frac{\partial}{\partial x}\bar{i} + \frac{\partial}{\partial y}\bar{j} + \frac{\partial}{\partial z}\bar{k}\right) \times (a_x\bar{i} + a_y\bar{j} + a_z\bar{k}) \quad (3)$$

$$= \begin{vmatrix} \bar{i} & \bar{j} & \bar{k} \\ \frac{\partial}{\partial x} & \frac{\partial}{\partial y} & \frac{\partial}{\partial z} \\ a_x & a_y & a_z \end{vmatrix} = \left(\frac{\partial a_z}{\partial y} - \frac{\partial a_y}{\partial z} , \frac{\partial a_x}{\partial z} - \frac{\partial a_z}{\partial x} , \frac{\partial a_y}{\partial x} - \frac{\partial a_x}{\partial y} \right) \quad (4)$$

Note that the notations curl $\bar{a}$, rot $\bar{a}$ and $\nabla \times \bar{a}$ are all equivalent and will be used interchangeably throughout this book. The curl or rotation of a vector plays a very important role in vector differential calculus. It is frequently used in electromagnetism, elasticity and hydrodynamics.

Substituting eq.(1) into eq.(2), we find

$$\text{curl } \bar{a} = \left[\frac{\partial a_z}{\partial y} - \frac{\partial a_y}{\partial z} \right] \bar{i} + \left[\frac{\partial a_x}{\partial z} - \frac{\partial a_z}{\partial x} \right] \bar{j} + \left[\frac{\partial a_y}{\partial x} - \frac{\partial a_x}{\partial y} \right] \bar{k}$$

$$= \left[\frac{\partial}{\partial y} (z^2) - \frac{\partial}{\partial z} (2xyz) \right] \bar{i} + \left[\frac{\partial}{\partial z} (x^2 y^2) - \frac{\partial}{\partial x} (z^2) \right] \bar{j} + \quad (6)$$

$$\left[\frac{\partial}{\partial x} (2xyz) - \frac{\partial}{\partial y} (x^2 y^2) \right] \bar{k}$$

$$= -2xy\bar{i} + (2yz - 2x^2 y)\bar{k}$$

At the point $(1,-2,1)$ curl $\bar{a}$ is equal to

$$\text{curl } \bar{a} \Big|_{(1,-2,1)} = \left(-2xy\bar{i} + (2yz - 2x^2 y)\bar{k} \right) \Big|_{(1,-2,1)}$$
$$= 4\bar{i} \quad (7)$$

Thus
$$\text{curl } \bar{a} \Big|_{(1,-2,1)} = 4\bar{i} = (4,0,0) \quad (8)$$

Given the vectors

$$\bar{a} = 3x^2y \; \bar{i} + 2x \; \bar{j} + yz \; \bar{k} \tag{1}$$

$$\bar{b} = yz^2 \; \bar{i} + xy \; \bar{j} + z^2 \; \bar{k} \tag{2}$$

and the scalar function

$$f = x^2 + yz \tag{3}$$

find

1. $\bar{a} \cdot (\nabla f)$

2. $(\bar{a} \cdot \nabla)f$

3. $(\bar{a} \cdot \nabla)\bar{b}$

4. $\bar{a} \times (\nabla f)$

5. $(\bar{a} \times \nabla)f$

6. $(\bar{a} \times \nabla) \times \bar{b}$

Compare the results of parts 1 and 2, and 4 and 5.

Solution: 1. $\bar{a} \cdot (\nabla f) = (3x^2y, \; 2x, \; yz) \cdot \left(\frac{\partial f}{\partial x}, \; \frac{\partial f}{\partial y}, \; \frac{\partial f}{\partial z} \right)$ (4)

The gradient of the function f is given by

$$\nabla f = \left(\frac{\partial f}{\partial x}, \; \frac{\partial f}{\partial y}, \; \frac{\partial f}{\partial z} \right) = (2x, \; z, \; y) \tag{5}$$

Substituting eq.(5) into eq.(4),

$$\bar{a} \cdot (\nabla f) = (3x^2y, \; 2x, \; yz) \cdot (2x, \; z, \; y) \tag{6}$$

$$= 6x^3y + 2xz + y^2z$$

2. $(\bar{a} \cdot \nabla)f = \left[(3x^2y, \; 2x, \; yz) \cdot \left(\frac{\partial}{\partial x}, \; \frac{\partial}{\partial y}, \; \frac{\partial}{\partial z} \right) \right] (x^2 + yz)$

$$= \left(3x^2y \frac{\partial}{\partial x} + 2x \frac{\partial}{\partial y} + yz \frac{\partial}{\partial z} \right) (x^2 + yz)$$

$$= \left(3x^2y \frac{\partial}{\partial x} (x^2+yz) + 2x \frac{\partial}{\partial y} (x^2+yz) + yz \frac{\partial}{\partial z} (x^2+yz) \right) \tag{7}$$

$$= 3x^2y \, 2x + 2xz + yzy = 6x^3y + 2xz + y^2z$$

Since the results of 1 and 2 are equal, presume there exists an identity such that

$$\bar{a} \cdot (\nabla f) = (\bar{a} \cdot \nabla)f \tag{8}$$

To prove this, find $\overline{a} \cdot (\nabla f)$.

$$\overline{a} \cdot (\nabla f) = (a_1, \ a_2, \ a_3) \cdot \left(\frac{\partial f}{\partial x}, \ \frac{\partial f}{\partial y}, \ \frac{\partial f}{\partial z} \right)$$

$$= a_1 \frac{\partial f}{\partial x} + a_2 \frac{\partial f}{\partial y} + a_3 \frac{\partial f}{\partial z} = \left(a_1 \frac{\partial}{\partial x} + a_2 \frac{\partial}{\partial y} + a_3 \frac{\partial}{\partial z} \right) f$$

$$= (a_1, \ a_2, \ a_3) \cdot \left(\frac{\partial}{\partial x}, \ \frac{\partial}{\partial y}, \ \frac{\partial}{\partial z} \right) f = (\overline{a} \cdot \nabla) f \tag{9}$$

Thus,

$$\overline{a} \cdot (\nabla f) = (\overline{a} \cdot \nabla) f \tag{10}$$

3. $\quad (\overline{a} \cdot \nabla) \overline{b} = \left[(a_1, \ a_2, \ a_3) \cdot \left(\frac{\partial}{\partial x}, \ \frac{\partial}{\partial y}, \ \frac{\partial}{\partial z} \right) \right] (b_1, \ b_2, \ b_3)$

$$= \left[(3x^2 y, \ 2x, \ yz) \cdot \left(\frac{\partial}{\partial x}, \ \frac{\partial}{\partial y}, \ \frac{\partial}{\partial z} \right) \right] (yz^2, \ xy, \ z^2)$$

$$= \left(3x^2 y \frac{\partial}{\partial x} + 2x \frac{\partial}{\partial y} + yz \frac{\partial}{\partial z} \right) (yz^2, \ xy, \ z^2) \tag{11}$$

Now denote by Λ the operator

$$3x^2 y \frac{\partial}{\partial x} + 2x \frac{\partial}{\partial y} + yz \frac{\partial}{\partial z} \equiv \Lambda \tag{12}$$

Eq.(11) can be written in the form

$$(\overline{a} \cdot \nabla) \overline{b} = \Lambda (yz^2, \ xy, \ z^2) \tag{13}$$

$$= (\Lambda yz^2, \ \Lambda xy, \ \Lambda z^2)$$

The operator Λ acts on each component, therefore

$$\Lambda \ yz^2 = \left(3x^2 y \frac{\partial}{\partial x} + 2x \frac{\partial}{\partial y} + yz \frac{\partial}{\partial z} \right) yz^2$$

$$= \left(3x^2 y \frac{\partial}{\partial x}(yz^2) + 2x \frac{\partial}{\partial y}(yz^2) + yz \frac{\partial}{\partial z}(yz^2) \right) \tag{14}$$

$$= 2xz^2 + yz \left(2yz \right) = 2xz^2 + 2y^2 z^2$$

$$\Lambda \ xy = \left(3x^2 y \frac{\partial}{\partial x} + 2x \frac{\partial}{\partial y} + yz \frac{\partial}{\partial z} \right) xy = 3x^2 y^2 + 2x^2 \tag{15}$$

$$\Lambda \ z^2 = 2yz^2 \tag{16}$$

Substituting eqs.(14), (15) and (16) into eq.(13), obtain

$$(\overline{a}\cdot\nabla)\overline{b} = (2xz^2 + 2y^2z^2,\ 3x^2y^2 + 2x^2,\ 2yz^2) \tag{17}$$

4. $\overline{a} \times (\nabla f) = (3x^2y,\ 2x,\ yz) \times (2x,\ z,\ y)$

$$= (2xy-yz^2,\ 2xyz-3x^2y^2,\ 3x^2yz-4x^2) \tag{18}$$

5. $(\overline{a} \times \nabla)f = \left[(3x^2y,\ 2x,\ yz) \times \left(\dfrac{\partial}{\partial x},\ \dfrac{\partial}{\partial y},\ \dfrac{\partial}{\partial z}\right)\right](x^2 + yz)$

$$= \left(2x\dfrac{\partial}{\partial z} - yz\dfrac{\partial}{\partial y},\ yz\dfrac{\partial}{\partial x} - 3x^2y\dfrac{\partial}{\partial z},\ 3x^2y\dfrac{\partial}{\partial y} - 2x\dfrac{\partial}{\partial x}\right)(x^2+yz) \tag{19}$$

$$= (2xy - yz^2,\ 2xyz - 3x^2y^2,\ 3x^2yz - 4x^2)$$

Once again, since the results of 4 and 5 are equal, presume there exists an identity such that

$$\overline{a} \times (\nabla f) = (\overline{a} \times \nabla)f \tag{20}$$

Proving this,

$\overline{a} \times (\nabla f) = (a_1,\ a_2,\ a_3) \times \left(\dfrac{\partial f}{\partial x},\ \dfrac{\partial f}{\partial y},\ \dfrac{\partial f}{\partial z}\right)$

$= \left(a_2\dfrac{\partial f}{\partial z} - a_3\dfrac{\partial f}{\partial y},\ a_3\dfrac{\partial f}{\partial x} - a_1\dfrac{\partial f}{\partial z},\ a_1\dfrac{\partial f}{\partial y} - a_2\dfrac{\partial f}{\partial x}\right)$

$= \left(a_2\dfrac{\partial}{\partial z} - a_3\dfrac{\partial}{\partial y},\ a_3\dfrac{\partial}{\partial x} - a_1\dfrac{\partial}{\partial z},\ a_1\dfrac{\partial}{\partial y} - a_2\dfrac{\partial}{\partial x}\right)f \tag{21}$

$= \left[(a_1,\ a_2,\ a_3) \times \left(\dfrac{\partial}{\partial x},\ \dfrac{\partial}{\partial y},\ \dfrac{\partial}{\partial z}\right)\right]f = (\overline{a} \times \nabla)f$

6. $(\overline{a} \times \nabla) \times \overline{b}$

$= \left[(3x^2y,\ 2x,\ yz) \times (\dfrac{\partial}{\partial x},\ \dfrac{\partial}{\partial y},\ \dfrac{\partial}{\partial z})\right] \times (yz^2,\ xy,\ z^2)$

$= \left(2x\dfrac{\partial}{\partial z} - yz\dfrac{\partial}{\partial y},\ yz\dfrac{\partial}{\partial x} - 3x^2y\dfrac{\partial}{\partial z},\right.$

$$\left. 3x^2y\dfrac{\partial}{\partial y} - 2x\dfrac{\partial}{\partial x}\right) \times (yz^2,\ xy,\ z^2) \tag{22}$$

$= (-6x^2yz - 3x^3y + 2xy,\ 3x^2yz^2 - 4xz,\ -xyz + 6x^2y^2z)$

Evaluate $\frac{\partial}{\partial n}$ (div $\bar{u}$) at the point $(3,4,5)$ where

$$\bar{u} = x^2\bar{i} - 2y^2\bar{j} + z^2\bar{k}$$

and $\bar{n}$ is the unit outer normal vector to the sphere

$$x^2 + y^2 + z^2 = 4.$$

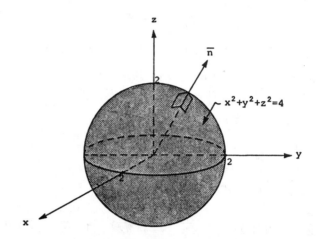

Solution: First compute div $\bar{u}$.

$$\text{div } \bar{u} = \left[\frac{\partial}{\partial x}\bar{i} + \frac{\partial}{\partial y}\bar{j} + \frac{\partial}{\partial z}\bar{k} \right] \cdot \left[x^2\bar{i} - 2y^2\bar{j} + z^2\bar{k} \right] = $$

$$= 2x - 4y + 2z \tag{1}$$

Since div $\bar{u}$ is a scalar function, let

$$f = \text{div } \bar{u} = 2x - 4y + 2z$$

The partial derivative of div $\bar{u}$, or f, with respect to the unit vector $\bar{n}$ is nothing more than the directional derivative of f in the $\bar{n}$ direction, therefore

$$\frac{\partial}{\partial n}(\text{div } \bar{u}) = \frac{\partial f}{\partial n} \equiv \nabla_n f = \nabla f \cdot \bar{n} \tag{2}$$

Substituting eq.(1) into eq.(2), we obtain

$$\frac{\partial}{\partial n}(\text{div } \bar{u}) = \frac{\partial f}{\partial n} = \nabla(2x-4y+2z)\cdot\bar{n} = (2,-4,2)\cdot\bar{n} \tag{3}$$

where $\bar{n}$ is a unit outer vector normal to the sphere

$$x^2 + y^2 + z^2 = 4.$$

Recall that if a surface is given by a function

$$\phi(x,y,z) = \text{const.},$$

158

then the gradient of ϕ, $\nabla\phi$, is an outer normal vector to the surface.

Therefore, for $\phi(x,y,z) = x^2+y^2+z^2 = 4 = $ const., we obtain

$$\nabla(x^2+y^2+z^2) = (2x,\ 2y,\ 2z) \tag{4}$$

Thus the unit outer normal vector is

$$\bar{n} = \frac{(2x,\ 2y,\ 2z)}{\sqrt{4x^2+4y^2+4z^2}} = \frac{(x,\ y,\ z)}{\sqrt{x^2+y^2+z^2}} \tag{5}$$

Substituting eq.(5) into eq.(3), obtain

$$\frac{\partial}{\partial n}(\text{div } \bar{u})\bigg|_{(3,4,5)} = (2,\ -4,\ 2)\cdot\bar{n}\bigg|_{(3,4,5)}$$

$$= \frac{(2,-4,2)\cdot(x,y,z)}{\sqrt{x^2+y^2+z^2}}\bigg|_{(3,4,5)} \tag{6}$$

$$= \frac{2x - 4y + 2z}{\sqrt{x^2+y^2+z^2}}\bigg|_{(3\ 4,5)}$$

$$= \frac{6 - 16 + 10}{\sqrt{9 + 16 + 25}} = 0$$

VECTOR INTEGRALS – ORDINARY, LINE AND DOUBLE INTEGRALS

Evaluate the integral

$$\int_2^4 \overline{a} \times \frac{d^2\overline{a}}{dt^2} \, dt \tag{1}$$

if

$$\overline{a}(4) = (6,7,3), \qquad \left.\frac{d\overline{a}}{dt}\right|_{t=4} = (1,1,1) \tag{2}$$

$$\overline{a}(2) = (3,1,1), \qquad \left.\frac{d\overline{a}}{dt}\right|_{t=2} = (0,0,0)$$

Solution: The value of the vector $\overline{a}$ and its time derivative is given only at the points t=2 and t=4. To evaluate the integral (1), express $\overline{a} \times \frac{d^2\overline{a}}{dt^2}$ as a time derivative of some vector function, $\overline{f}(t)$

$$\overline{a} \times \frac{d^2\overline{a}}{dt^2} = \frac{d}{dt}(\overline{f}(t)) \tag{3}$$

It is now required to find $\overline{f}(t)$. Assume that $\overline{f}(t)$ can be expressed as

$$\overline{f}(t) = \overline{b} \times \overline{c} \tag{4}$$

where $\overline{b}$ and $\overline{c}$ are unknown vector functions. Differentiating eq.(4) results in

$$\frac{d\overline{f}}{dt} = \overline{b} \times \frac{d\overline{c}}{dt} + \frac{d\overline{b}}{dt} \times \overline{c} \tag{5}$$

Now, notice that if we let $\overline{c} = \frac{d\overline{b}}{dt}$, then the second term of the right-hand side of eq.(5) would equal zero. Therefore, if we arbitrarily let

$$\overline{b} = \overline{a} \quad \text{and} \quad \overline{c} = \frac{d\overline{b}}{dt} = \frac{d\overline{a}}{dt}, \tag{6}$$

then eq.(3) becomes

$$\overline{a} \times \frac{d^2\overline{a}}{dt^2} = \frac{d}{dt}\left(\overline{a} \times \frac{d\overline{a}}{dt}\right) \tag{7}$$

160

Now, substituting eq.(7) into eq.(1) results in

$$\int_2^4 \bar{a} \times \frac{d^2\bar{a}}{dt^2} \, dt = \int_2^4 \frac{d}{dt}\left(\bar{a} \times \frac{d\bar{a}}{dt}\right) dt$$

$$= \bar{a} \times \frac{d\bar{a}}{dt}\bigg|_2^4 = \bar{a}(4) \times \frac{d\bar{a}(4)}{dt} - \bar{a}(2) \times \frac{d\bar{a}(2)}{dt}$$

$$(8)$$

$$= (6,7,3)\times(1,1,1)-(3,1,1)\times(0,0,0)$$

$$= (4,-3,-1) = 4\bar{i} - 3\bar{j} - \bar{k}$$

• PROBLEM 4-14

If vector field $\bar{F}$ is given by
$$\bar{F} = xy\bar{i} + x^2\bar{j} + (x-z)\bar{k}, \tag{1}$$
evaluate the line integral
$$\int \bar{F} \cdot d\bar{r} \tag{2}$$
from (0,0,0) to (1,2,4) along

1. the line segment joining these two points

2. the curve given parametrically by

$$x = t^2, \quad y = 2t^3, \quad z = 4t \tag{3}$$

Solution: 1. Begin by writing the parametric equations of the line passing through the points (0,0,0) and (1,2,4). We have,
$$x = t, \quad y = 2t, \quad z = 4t \tag{4}$$

and for this line segment, t is restricted to
$$0 \le t \le 1 \tag{5}$$

Now, note that the integral can be expressed as

$$\int_C \bar{F} \cdot d\bar{r} = \int_C \left[xy\bar{i} + x^2\bar{j} + (x-z)\bar{k}\right] \cdot \left[dx\bar{i} + dy\bar{j} + dz\bar{k}\right]$$

$$(6)$$

$$= \int_C xy\,dx + x^2\,dy + (x-z)\,dz$$

Differentiating eq.(4) gives
$$dx = dt$$

$$dy = 2dt \tag{7}$$

$$dz = 4dt$$

Substituting eq.(7) and eq.(4) into eq.(6) results in

$$\int_C \overline{F} \cdot d\overline{r} = \int_0^1 2t^2 dt + 2t^2 dt + (t-4t)4dt$$

$$= \int_0^1 (4t^2 - 12t)dt = \left(\frac{4}{3}t^3 - 6t^2\right)\Big|_0^1 \qquad (8)$$

$$= \frac{4}{3} - 6 = -\frac{14}{3}$$

At this point, it should be noted that in order to evaluate a line integral we need to know not only the parametric equations of the respective curve, but the orientation as well. In this example, the orientation of the line segment was from the point $(0,0,0)$ to the point $(1,2,4)$.

2. From eq.(3), the differentials dx, dy, dz are found to be
$$dx = 2t \, dt$$

$$dy = 6t^2 dt \qquad (9)$$

$$dz = 4 \, dt$$

Substituting eq.(9) and eq.(2) into the expression (2), we obtain

$$\int_C \overline{F} \cdot d\overline{r} = \int_C \left[xy\overline{i} + x^2\overline{j} + (x-z)\overline{k}\right] \cdot \left[dx\overline{i} + dy\overline{j} + dz\overline{k}\right]$$

$$= \int_C xy\,dx + x^2 dy + (x-z)dz$$

$$= \int_0^1 (t^2)(2t^3)2t\,dt + (t^4)6t^2 dt + (t^2-4t)4dt$$

$$= \int_0^1 (4t^6 + 6t^6 + 4t^2 - 16t)dt \qquad (10)$$

$$= \int_0^1 (10t^6 + 4t^2 - 16t)dt$$

$$= \left(\frac{10t^7}{7} + \frac{4t^3}{3} - 8t^2\right)\Big|_0^1$$

$$= \frac{10}{7} + \frac{4}{3} - 8 = -\frac{110}{21}$$

By reducing the double integral to the iterated integral, evaluate

$$\iint\limits_{D} (x^2+y^2)dxdy \qquad (1)$$

where D is a quarter-circle such that

$$0 \leq x \leq 1 \quad \text{and} \quad 0 \leq y \leq 1.$$

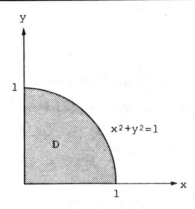

Solution: The following theorem illustrates the method of evaluating double integrals known as the reduction to an iterated integral.

Theorem

Let D be a closed region lying in the xy plane described by the inequalities

$$x_1(y) \leq x \leq x_2(y) \qquad (2)$$

$$y_1 \leq y \leq y_2 \qquad (3)$$

If f(x,y) is a continuous function defined in D, then for $y_1 \leq y \leq y_2$

$$\int_{x_1(y)}^{x_2(y)} f(x,y)dx \qquad (4)$$

is a continuous function of y and

$$\iint\limits_{D} f(x,y)dxdy = \int_{y_1}^{y_2} \int_{x_1(y)}^{x_2(y)} f(x,y)dxdy \qquad (5)$$

The expression on the right side of eq.(5) is called an iterated (double) integral. In the case where region D is described by the inequalities

$$y_1(x) \le y \le y_2(x) \tag{6}$$

$$x_1 \le x \le x_2, \tag{7}$$

the double integral is expressed as

$$\iint_D f(x,y)dxdy = \int_{x_1}^{x_2} \int_{y_1(x)}^{y_2(x)} f(x,y)dydx \tag{8}$$

Now, to return to the problem at hand, notice that region D, as shown in the figure, can be described by the inequalities

$$0 \le x \le 1 \tag{9}$$

$$0 \le y \le \sqrt{1-x^2} \tag{10}$$

and therefore, we can use eq.(9) as a means of evaluation.

For the function $f(x,y) = x^2+y^2$, we have

$$\iint_D (x^2+y^2)dxdy = \int_{x_1}^{x_2} \int_{y_1(x)}^{y_2(x)} (x^2+y^2)dydx$$

$$= \int_0^1 \int_0^{\sqrt{1-x^2}} (x^2+y^2)dydx$$

$$= \int_0^1 \left[\int_0^{\sqrt{1-x^2}} (x^2+y^2)dy \right] dx \tag{11}$$

$$= \int_0^1 \left[\left(x^2 y + \frac{y^3}{3} \right) \Big|_0^{\sqrt{1-x^2}} \right] dx = \int_0^1 \left[x^2\sqrt{1-x^2} + \frac{1}{3}(1-x^2)^{\frac{3}{2}} \right] dx$$

Now, to solve this integral, use the method of trigonometric substitution.

Substituting $x = \sin\alpha$ yields

$$\int_0^1 \left[x^2\sqrt{1-x^2} + \frac{1}{3}(1-x^2)^{\frac{3}{2}} \right] dx$$

$$= \int_0^{\frac{\pi}{2}} \left[\sin^2\alpha\cos\alpha + \frac{1}{3}\cos^3\alpha \right] \cos\alpha d\alpha$$

$$= \int_0^{\frac{\pi}{2}} \sin^2\alpha\cos^2\alpha d\alpha + \int_0^{\frac{\pi}{2}} \frac{1}{3}\cos^4\alpha d\alpha$$

$$= \left[\frac{\alpha}{8} - \frac{1}{32} \sin 4\alpha \right] \Bigg|_0^{\frac{\pi}{2}} + \frac{1}{3} \left[\frac{3}{8} \alpha + \frac{1}{4} \sin 2\alpha + \frac{1}{32} \sin 4\alpha \right] \Bigg|_0^{\frac{\pi}{2}} \qquad (12)$$

$$= \frac{\pi}{8}$$

The above results of eq.(12) may be obtained through the use of a table of integrals.

• PROBLEM 4-16

Using double integrals, evaluate:

1. The area of the triangle S with vertices (0,0), (1,1), (1,0).

2. The volume between the surface

$$z = x^3 + y^3 \qquad (1)$$

and the triangle S on the xy plane whose vertices are (0,0,0), (1,1,0), (1,0,0).

Solution: 1. First, notice that if we take f(x,y)=1, then the double integral may be expressed as

$$\iint_D f(x,y)\,dxdy = \iint_D dxdy = A \qquad (2)$$

where A is the area of region D.

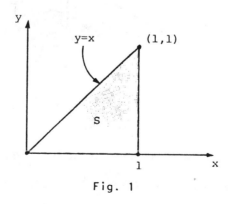

Fig. 1

The triangle S, shown in Fig. 1, can be described by the in-equalities

$$0 \leq y \leq x$$

$$0 \leq x \leq 1 \tag{3}$$

The integral in eq.(2) can then be written

$$\iint_D dxdy = \int_0^1 \int_0^x dydx$$

$$= \int_0^1 \left[y \Big|_0^x \right] dx = \int_0^1 xdx = \frac{1}{2} \tag{4}$$

2. Let $f(x,y)=z$ be the general equation of a surface. The volume between the surface $z=f(x,y)$ and the xy plane is given by

$$V = \iint_D f(x,y)dxdy \tag{6}$$

where D is some general region lying in the xy plane, as shown in Fig. 2. Volumes above the xy plane are considered positive, while volumes below the xy plane are considered negative.

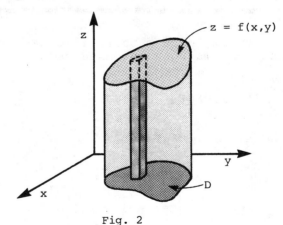

Fig. 2

In this case, the function $z=f(x,y)$ is given by

$$z = x^3 + y^3 \tag{7}$$

The region D can be described by the inequalities

$$0 \leq y \leq x$$

$$0 \leq x \leq 1 \tag{8}$$

Substituting eq.(7) and eq.(8) into eq.(6) results in

166

$$V = \iint\limits_D f(x,y)dxdy$$

$$= \int_0^1 \int_0^x (x^3+y^3)dydx = \int_0^1 \left[\left(x^3 y + \frac{y^4}{4} \right) \Big|_0^x \right] dx \qquad (9)$$

$$= \int_0^1 \left(x^4 + \frac{1}{4} x^4 \right) dx = \frac{5}{4} \int_0^1 x^4 dx$$

$$= \left(\frac{5}{4} \frac{x^5}{5} \right) \Big|_0^1 = \frac{1}{4}$$

• PROBLEM 4–17

Find the volume of the solid bounded by the paraboloid $x^2 + y^2 = 5z$, the xy-plane and the cylinder $x^2 + y^2 = 9$.

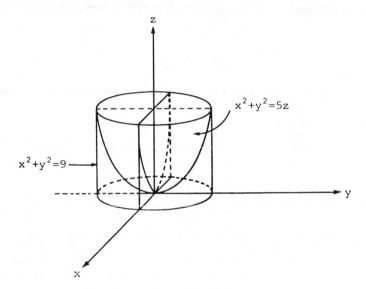

Solution: The solid is shown in the figure.

Now, the volume of the solid is given by

$$V = \iint\limits_D f(x,y)dydx \qquad (1)$$

where $f(x,y) = z$ represents the equation of the surface and D represents the corresponding region in the xy-plane.

In our case

$$z = \frac{x^2+y^2}{5} \qquad (2)$$

and region D is a circle of radius 3

$$x^2 + y^2 = 9 \qquad (3)$$

167

Now, it just so happens that in this case, it is much simpler to compute the volume by first transforming eq.(1) into polar coordinates.

$$V = \iint\limits_{D} \frac{x^2+y^2}{5} \, dydx = \int_{\theta=0}^{2\pi} \int_{r=0}^{3} r \cdot \frac{r^2}{5} \, drd\theta$$

$$= \int_{0}^{2\pi} \left[\frac{r^4}{20} \right] \Big|_{0}^{3} \, d\theta = \frac{81\pi}{10} \tag{4}$$

Note that the region D in the xy-plane is the projection of the intersection of the cylinder $x^2 + y^2 = 9$ with the paraboloid $x^2 + y^2 = 5z$.

● **PROBLEM 4–18**

By transforming into polar coordinates, evaluate the integral

$$\int_{x=0}^{a} \int_{y=0}^{\sqrt{a^2-x^2}} (x^2+y^2) \, dydx \tag{1}$$

Solution: We will not give here a rigorous treatment of the problem of changing the variables in a double integral. Instead, we will only cite the results. Consider the transformation of the x,y coordinates into u,v coordinates.

$$x = f(u,v)$$
$$y = g(u,v) \tag{2}$$

Eq.(2) may be interpreted as the mapping of a region D of the xy-plane into a region G of the uv-plane. Under some restrictions on the functions f and g, the following formula for changing from xy-coordinates to uv-coordinates holds

$$\iint\limits_{D} F(x,y) \, dxdy = \iint\limits_{G} F[f(u,v),g(u,v)] \frac{\partial(x,y)}{\partial(u,v)} \, dudv \tag{3}$$

Here $\frac{\partial(x,y)}{\partial(u,v)}$ denotes the Jacobian of the transformation,

$$\frac{\partial(x,y)}{\partial(u,v)} = \begin{vmatrix} \frac{\partial x}{\partial u} & \frac{\partial x}{\partial v} \\ \frac{\partial y}{\partial u} & \frac{\partial y}{\partial v} \end{vmatrix} \tag{4}$$

For the polar coordinates, we have

$$x = r\cos\theta$$
$$y = r\sin\theta \qquad (5)$$

Thus,

$$\frac{\partial(x,y)}{\partial(r,\theta)} = \begin{vmatrix} \cos\theta & -r\sin\theta \\ \sin\theta & r\cos\theta \end{vmatrix} = r(\cos^2\theta+\sin^2\theta) = r \qquad (6)$$

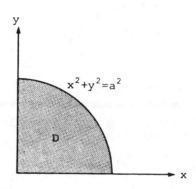

Eq.(3) becomes

$$\iint_D F(x,y)dxdy = \iint_G F(r\cos\theta,r\sin\theta)r\,drd\theta \qquad (7)$$

Applying eq.(7) to eq.(1) and noting that $x^2+y^2 = r^2$ yields

$$\int_{x=0}^{a}\int_{y=0}^{\sqrt{a^2-x^2}} (x^2+y^2)dydx = \int_{\theta=0}^{\frac{\pi}{2}}\int_{r=0}^{a} r^2r\,drd\theta \qquad (8)$$

The region of integration is shown in the figure.

Note that since D is a quarter circle, we have

$$0 \le r \le a$$
$$0 \le \theta \le \frac{\pi}{2} \qquad (9)$$

From eq.(8) we compute

$$\int_{\theta=0}^{\frac{\pi}{2}}\int_{r=0}^{a} r^3drd\theta = \int_{0}^{\frac{\pi}{2}} \frac{a^4}{4}\,d\theta = \frac{\pi a^4}{8} \qquad (10)$$

VECTOR INTEGRALS –
SURFACE AND VOLUME INTEGRALS

Find the surface area of the surface defined by the equations

$$x = \cos u$$

$$y = \sin u \qquad\qquad (1)$$

$$z = v$$

for $0 \le u \le 2\pi$ and $0 \le v \le 1$.

Solution: First, consider a regular surface element described by the equation

$$\overline{r} = \overline{r}(u,v) \qquad\qquad (2)$$

The surface area of a regular surface element is given by

$$S = \iint \left| \frac{\partial \overline{r}}{\partial u} \times \frac{\partial \overline{r}}{\partial v} \right| \, du\, dv \qquad\qquad (3)$$

We often introduce the vector $d\overline{S}$

$$d\overline{S} = \frac{\partial \overline{r}}{\partial u} \, du \times \frac{\partial \overline{r}}{\partial v} \, dv$$

or

$$d\overline{S} = \left(\frac{\partial \overline{r}}{\partial u} \times \frac{\partial \overline{r}}{\partial v} \right) \, du\, dv \qquad\qquad (4)$$

Note that $d\overline{S}$ is a vector normal to the surface, its magnitude $|d\overline{S}| = dS$ is the element of area.

Integral (3) can be written

$$S = \iint \left| \frac{\partial \overline{r}}{\partial u} \times \frac{\partial \overline{r}}{\partial v} \right| \, du\, dv = \iint |d\overline{S}|$$

$$= \iint dS = \iint \overline{n} \cdot d\overline{S} \qquad\qquad (5)$$

where $\overline{n}$ is a unit normal vector in the direction of $d\overline{S}$.

Now, since

$$\overline{r} = (\cos u, \sin u, v)$$

the tangent vectors $\dfrac{\partial \overline{r}}{\partial u}$ and $\dfrac{\partial \overline{r}}{\partial v}$ are

$$\frac{\partial \overline{r}}{\partial u} = (-\sin u, \cos u, 0) \tag{6}$$

$$\frac{\partial \overline{r}}{\partial v} = (0,0,1) \tag{7}$$

Their vector product is

$$\frac{\partial \overline{r}}{\partial u} \times \frac{\partial \overline{r}}{\partial v} = (-\sin u, \cos u, 0) \times (0,0,1)$$

$$= (\cos u, \sin u, 0) \tag{8}$$

Substituting eq.(8) into eq.(4) results in

$$d\overline{S} = (\cos u, \sin u, 0)dudv \tag{9}$$

The area of the surface is given by eq.(5), thus

$$S = \iint |d\overline{S}| = \int_0^1 \int_0^{2\pi} \sqrt{\cos^2 u + \sin^2 u}\ dudv$$

$$= \int_0^1 \left[u \Big|_0^{2\pi} \right] dv \tag{10}$$

$$= 2\pi$$

• **PROBLEM 4–20**

1. Evaluate the integral

$$\iint_D \sqrt{x^2 + y^2}\ dxdy \tag{1}$$

over the region D in the xy plane bounded by $x^2 + y^2 = 25$.

Solution: It is easier to solve this problem in polar co-ordinates. We have

$$x = \rho \cos \alpha$$
$$y = \rho \sin \alpha \tag{2}$$

Thus

$$\sqrt{x^2 + y^2} = \rho \tag{3}$$

The integral in eq.(1) can be written

$$\iint_D \sqrt{x^2 + y^2}\ dxdy = \iint_D \rho\rho\ d\rho\,d\alpha \tag{4}$$

$$= \int_0^{2\pi} \int_0^5 \rho^2\ d\rho\,d\alpha = \int_0^{2\pi} \frac{125}{3} d\alpha = \frac{250\pi}{3}$$

171

Evaluate the integral

$$\iint_S dxdy + dzdx + dydz \tag{1}$$

where S is

$$z = \sqrt{1-x^2-y^2} \tag{2}$$

over the disk $x^2+y^2 \le 1$ (3)

Solution: Integral (1) is the cartesian form of the surface integral. In general, if

$$\overline{F} = P\overline{i} + Q\overline{j} + R\overline{k} \tag{4}$$

then the integral can be written

$$\iint_S \overline{F} \cdot \overline{n} \, dS = \iint_S Pdydz + Qdxdz + Rdxdy \tag{5}$$

Depending on the form of the equation for the surface, we use the appropriate formulas for $\overline{n}$ and dS.

For a surface given by

$$z = f(x,y),$$

as is the case in this problem, the unit normal vector is given by

$$\overline{n} = \frac{-\left(\frac{\partial z}{\partial x}\right)\overline{i} - \left(\frac{\partial z}{\partial y}\right)\overline{j} + \overline{k}}{\sqrt{1 + \left(\frac{\partial z}{\partial x}\right)^2 + \left(\frac{\partial z}{\partial y}\right)^2}} \tag{6}$$

and

$$dS = \sqrt{1 + \left(\frac{\partial z}{\partial x}\right)^2 + \left(\frac{\partial z}{\partial y}\right)^2} \, dxdy \tag{7}$$

Thus, substituting eqs.(6), (7) and (4) into the left-hand side of eq.(5) leads to

$$\iint_S \overline{F} \cdot \overline{n} \, dS = \iint_S (P\overline{i}+Q\overline{j}+R\overline{k}) \cdot \overline{n} \, dS$$

$$= \pm \iint_{D_{xy}} (P\overline{i}+Q\overline{j}+R\overline{k}) \cdot \frac{-\left(\frac{\partial z}{\partial x}\right)\overline{i} - \left(\frac{\partial z}{\partial y}\right)\overline{j} + \overline{k}}{\sqrt{1 + \left(\frac{\partial z}{\partial x}\right)^2 + \left(\frac{\partial z}{\partial y}\right)^2}} \sqrt{1 + \left(\frac{\partial z}{\partial x}\right)^2 + \left(\frac{\partial z}{\partial y}\right)^2} \, dxdy$$

$$= \pm \iint_{D_{xy}} \left(-P\frac{\partial z}{\partial x} - Q\frac{\partial z}{\partial y} + R\right) dxdy \tag{8}$$

The sign + or - is determined by the unit normal vector $\bar{n}$.

For the surface given by eq.(2) we have

$$\frac{\partial z}{\partial x} = \frac{-x}{\sqrt{1-x^2-y^2}} \quad ; \quad \frac{\partial z}{\partial y} = \frac{-y}{\sqrt{1-x^2-y^2}} \tag{9}$$

From eqs.(1) and (5) we see that

$$P = Q = R = 1 \tag{10}$$

Thus

$$\iint_{D_{xy}} \left(\frac{x}{\sqrt{1-x^2-y^2}} + \frac{y}{\sqrt{1-x^2-y^2}} + 1 \right) dxdy \tag{11}$$

Transforming into polar coordinates yields

$$\int_0^{2\pi} \int_0^1 \left(\frac{r\cos\theta}{\sqrt{1-r^2}} + \frac{r\sin\theta}{\sqrt{1-r^2}} + 1 \right) r \, drd\theta$$

$$= \int_0^{2\pi} \int_0^1 \frac{r^2\cos\theta}{\sqrt{1-r^2}} \, drd\theta + \int_0^{2\pi} \int_0^1 \frac{r^2\sin\theta}{\sqrt{1-r^2}} \, drd\theta \tag{12}$$

$$+ \int_0^{2\pi} \int_0^1 r \, drd\theta = \pi$$

• PROBLEM 4-22

1. Evaluate the volume integral

$$\iiint_V x^2 y \, dV \tag{1}$$

where V is the closed region bounded by the coordinate planes and x=1, y=1, z=1.

2. Evaluate the volume integral of

$$\phi(x,y,z) = xz + yz \tag{2}$$

over the closed region V bounded by the coordinate planes and

$$x=2, \quad y=3, \quad z=-x+4 \tag{3}$$

Solution: 1. V is a cube bounded by the planes x=0, x=1, y=0, y=1, z=0, z=1. Thus, we integrate from x=0 to x=1, from y=0 to y=1, from z=0 to z=1. Eq.(1) can be written

173

$$\iiint\limits_V x^2 y \ dV = \int_{x=0}^{1} \int_{y=0}^{1} \int_{z=0}^{1} x^2 y \ dz\,dy\,dx$$

(4)

$$= \int_{x=0}^{1} \int_{y=0}^{1} x^2 y \ dy\,dx$$

$$= \int_{x=0}^{1} \frac{1}{2} x^2 \, dx = \frac{1}{6}$$

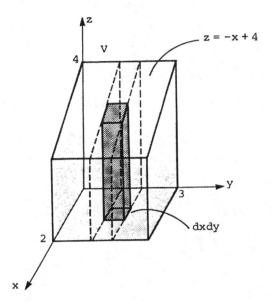

2.   The region V is shown in the figure. It is required to
     find the volume integral

$$\iiint\limits_V (xz+yz)\,dz\,dy\,dx$$

(5)

Consider a point (x,y,z) inside the region V. Letting x and
y be fixed, we can vary z from z=0 to z=-x+4 to obtain a
column, as shown in the figure. Holding y fixed and varying
x from x=0 to x=2, we obtain a slice, as shown. Adding the
slices from y=0 to y=3 gives us the volume integral over the
region V. Thus,

$$\iiint\limits_V (xz+yz)\,dV = \int_{y=0}^{3} \int_{x=0}^{2} \int_{z=0}^{-x+4} (xz+yz)\,dz\,dx\,dy$$

$$= \int_{y=0}^{3} \int_{x=0}^{2} \left[(x+y) \frac{z^2}{2} \right] \Bigg|_{z=0}^{-x+4} dx\,dy$$

(6)

$$= \int_{y=0}^{3} \int_{x=0}^{2} \left[\frac{1}{2} x^3 - 4x^2 + 8x + \frac{1}{2} x^2 y - 4xy + 8y \right] dx\,dy$$

174

$$= \int_{y=0}^{3} \left(\frac{22}{3} + \frac{28}{3} y \right) dy = \left[\frac{22}{3} y + \frac{28}{3} \frac{y^2}{2} \right] \Bigg|_{y=0}^{3}$$

$$= 64$$

● **PROBLEM** 4–23

Find the volume of the region of space bounded by the planes x=0, y=0, z=0, z=3-2x+y and the surface $y=1-x^2$.

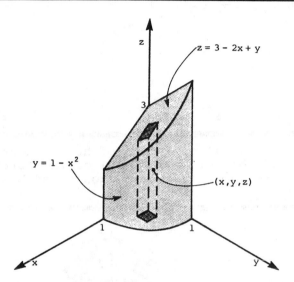

Solution: The region is shown in the figure. The volume integral is given by

$$\iiint_{V} \phi(x,y,z) dV \tag{1}$$

Taking $\phi(x,y,z) \equiv 1$, that is $\phi(x,y,z)$ identically equal to unity, we obtain the expression for the volume of the region V.

$$\text{Volume of } V = \iiint_{V} dxdydz = \iiint_{V} dV \tag{2}$$

where dV = dx dy dz.

Let us now find the limits of integration. Taking an arbitrary point (x,y,z) inside the region and fixing x and y, we find that z can change from z=0 to z=3-2x+y. By the same token, for fixed values of x the coordinate y can vary from y=0 to $y=1-x^2$. Finally, we find that x changes from x=0 to x=1. The volume integral (2) can thus be written

$$\int_{x=0}^{1} \int_{y=0}^{1-x^2} \int_{z=0}^{3-2x+y} dz\ dy\ dx$$

175

$$= \int_{x=0}^{1} \int_{y=0}^{1-x^2} (3 - 2x + y)\,dy\,dx$$

$$= \int_{0}^{1} \left[3y - 2xy + \frac{y^2}{2} \right] \Bigg|_{y=0}^{1-x^2} dx$$

$$= \int_{0}^{1} \left[\frac{1}{2} x^4 + 2x^3 - 4x^2 - 2x + \frac{7}{2} \right] dx$$

$$= \left[\frac{x^5}{10} + \frac{x^4}{2} - \frac{4}{3} x^3 - x^2 + \frac{7}{2} x \right] \Bigg|_{x=0}^{1}$$

$$= \frac{53}{30}$$

• **PROBLEM 4–24**

Consider two coaxial cylinders of height h and radii r_1 and r_2. Find the volume integral of the function $f(x,y,z) = x^2 + y^2$ over the volume contained between the two cylinders.

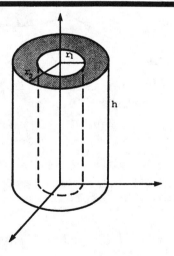

Solution: The cylinders are shown in the figure.

All computations will be performed in cylindrical coordinates:

$$x = \rho \cos\phi$$

$$y = \rho \sin\phi \qquad\qquad (1)$$

$$z = z$$

where $0 \leq z \leq h$.

The function f becomes

$$f = x^2 + y^2$$

$$= (\rho\cos\phi)^2 + (\rho\sin\phi)^2$$

$$= \rho^2(\cos^2\phi + \sin^2\phi) \tag{2}$$

$$= \rho^2$$

or

$$f(\rho,\phi,z) = \rho^2$$

The volume element is given by

$$dV = \rho \; d\rho d\phi dz \tag{3}$$

Thus, the volume integral of $f(\rho,\phi,z)$ is

$$\iiint f(\rho,\phi,z)dV$$

$$= \int_{z=0}^{h} \int_{\phi=0}^{2\pi} \int_{\rho=r_1}^{r_2} \rho^2 \rho d\rho d\phi dz$$

$$= \int_{z=0}^{h} \int_{\phi=0}^{2\pi} \left.\frac{\rho^4}{4}\right|_{r_1}^{r_2} d\phi dz \tag{4}$$

$$= \pi h\left[\frac{r_2^4}{2} - \frac{r_1^4}{2}\right]$$

CURVILINEAR COORDINATES

• PROBLEM 4-25

Consider the system of parabolic cylindrical coordinates

$$x = \frac{1}{2}(u^2 - v^2)$$

$$y = uv$$

$$z = z$$

Express Laplace's equation in terms of these coordinates.

Solution: From eq.(1), we evaluate the scale factors

$$h_u = \left|\frac{\partial \overline{r}}{\partial u}\right| = |(u,v,0)| = \sqrt{u^2+v^2}$$

$$h_v = \left|\frac{\partial \overline{r}}{\partial v}\right| = |(-v,u,0)| = \sqrt{u^2+v^2} \tag{2}$$

$$h_z = 1$$

Laplace's equation is

$$\nabla^2 f = 0 \qquad (3)$$

In parabolic cylindrical coordinates it becomes

$$\frac{1}{h_1 h_2 h_3} \left[\frac{\partial}{\partial u_1}\left(\frac{h_2 h_3}{h_1} \frac{\partial f}{\partial u_1} \right) + \frac{\partial}{\partial u_2}\left(\frac{h_1 h_3}{h_2} \frac{\partial f}{\partial u_2} \right) + \frac{\partial}{\partial u_3}\left(\frac{h_1 h_2}{h_3} \frac{\partial f}{\partial u_3} \right) \right]$$

$$= \frac{1}{u^2 + v^2} \left[\frac{\partial}{\partial u}\left(\frac{\partial f}{\partial u} \right) + \frac{\partial}{\partial v}\left(\frac{\partial f}{\partial v} \right) + \frac{\partial}{\partial z}\left((u^2 + v^2) \frac{\partial f}{\partial z} \right) \right] \qquad (4)$$

$$= \frac{1}{u^2 + v^2} \left[\frac{\partial^2 f}{\partial u^2} + \frac{\partial^2 f}{\partial v^2} \right] + \frac{\partial^2 f}{\partial z^2}$$

and

$$\frac{\partial^2 f}{\partial u^2} + \frac{\partial^2 f}{\partial v^2} + (u^2 + v^2) \frac{\partial^2 f}{\partial z^2} = 0 \qquad (5)$$

• PROBLEM 4–26

If the transformation from (u_1, u_2, u_3) to (x, y, z) is given by

$$x = a_1 u_1 + a_2 u_2 + a_3 u_3$$

$$y = b_1 u_1 + b_2 u_2 + b_3 u_3$$

$$z = c_1 u_1 + c_2 u_2 + c_3 u_3$$

then under what conditions will the (u_1, u_2, u_3) system be orthogonal?

Solution: To determine if the system (u_1, u_2, u_3) is orthogonal, we should first evaluate the vectors $\bar{e}_1$, $\bar{e}_2$, and $\bar{e}_3$.

$$\bar{e}_1 = \frac{\frac{\partial \bar{r}}{\partial u_1}}{\left| \frac{\partial \bar{r}}{\partial u_1} \right|} = \frac{(a_1, b_1, c_1)}{\sqrt{a_1^2 + b_1^2 + c_1^2}} = \frac{a_1 \bar{i} + b_1 \bar{j} + c_1 \bar{k}}{\sqrt{a_1^2 + b_1^2 + c_1^2}} \qquad (2)$$

$$\bar{e}_2 = \frac{\frac{\partial \bar{r}}{\partial u_2}}{\left|\frac{\partial \bar{r}}{\partial u_2}\right|} = \frac{(a_2, b_2, c_2)}{\sqrt{a_2^2 + b_2^2 + c_2^2}} = \frac{a_2\bar{i} + b_2\bar{j} + c_2\bar{k}}{\sqrt{a_2^2 + b_2^2 + c_2^2}} \tag{3}$$

$$\bar{e}_3 = \frac{\frac{\partial \bar{r}}{\partial u_3}}{\left|\frac{\partial \bar{r}}{\partial u_3}\right|} = \frac{(a_3, b_3, c_3)}{\sqrt{a_3^2 + b_3^2 + c_3^2}} = \frac{a_3\bar{i} + b_3\bar{j} + c_3\bar{k}}{\sqrt{a_3^2 + b_3^2 + c_3^2}} \tag{4}$$

The system is orthogonal when

$$\bar{e}_1 \cdot \bar{e}_2 = \bar{e}_1 \cdot \bar{e}_3 = \bar{e}_2 \cdot \bar{e}_3 = 0 \tag{5}$$

The first equation leads to

$$\bar{e}_1 \cdot \bar{e}_2 = \frac{a_1 a_2 + b_1 b_2 + c_1 c_2}{A_{12}} = 0 \tag{6}$$

The denominator A_{12} is immaterial.

$$\bar{e}_1 \cdot \bar{e}_3 = \frac{a_1 a_3 + b_1 b_3 + c_1 c_3}{A_{13}} = 0 \tag{7}$$

$$\bar{e}_2 \cdot \bar{e}_3 = \frac{a_2 a_3 + b_2 b_3 + c_2 c_3}{A_{23}} = 0 \tag{8}$$

The system is orthogonal when

$$a_1 a_2 + b_1 b_2 + c_1 c_2 = a_1 a_3 + b_1 b_3 + c_1 c_3$$
$$= a_2 a_3 + b_2 b_3 + c_2 c_3 = 0 \tag{9}$$

The transformation matrix

$$A = \begin{bmatrix} a_1 & a_2 & a_3 \\ b_1 & b_2 & b_3 \\ c_1 & c_2 & c_3 \end{bmatrix} = \begin{bmatrix} \bar{v}_1, \bar{v}_2, \bar{v}_3 \end{bmatrix} \tag{10}$$

consists of three orthogonal vectors

$$\bar{v}_1 = \begin{bmatrix} a_1 \\ b_1 \\ c_1 \end{bmatrix} \quad \bar{v}_2 = \begin{bmatrix} a_2 \\ b_2 \\ c_2 \end{bmatrix} \quad \bar{v}_3 = \begin{bmatrix} a_3 \\ b_3 \\ c_3 \end{bmatrix} \tag{11}$$

1. Show that the transformation

$$x = 2u_1u_2$$

$$y = u_2^2 - u_1^2 \qquad (1)$$

$$z = u_3$$

   is orthogonal.

2. Evaluate the scale factors.

3. Evaluate div $\overline{F}$ and curl $\overline{F}$ where

$$\overline{F} = u_2\overline{e}_1 + u_3\overline{e}_2 + u_1\overline{e}_3 \qquad (2)$$

<u>Solution</u>: We can start with the evaluation of the vectors

$$\overline{e}_i = \frac{\dfrac{\partial \overline{r}}{\partial u_i}}{\left| \dfrac{\partial \overline{r}}{\partial u_i} \right|} \qquad (3)$$

If $\overline{e}_1, \overline{e}_2, \overline{e}_3$ are orthogonal then (u_1, u_2, u_3) are orthogonal

curvilinear coordinates. The scale factors are obtained from

$$\overline{e}_i = \frac{1}{h_i} \frac{\partial \overline{r}}{\partial u_i} \qquad (4)$$

We have,

$$\overline{e}_1 = \frac{\dfrac{\partial \overline{r}}{\partial u_1}}{\left| \dfrac{\partial \overline{r}}{\partial u_1} \right|} = \frac{(2u_2, -2u_1, 0)}{\sqrt{4u_1^2 + 4u_2^2}} = \frac{u_2\overline{i} - u_1\overline{j}}{\sqrt{u_1^2 + u_2^2}} \qquad (5)$$

$$\overline{e}_2 = \frac{\dfrac{\partial \overline{r}}{\partial u_2}}{\left| \dfrac{\partial \overline{r}}{\partial u_2} \right|} = \frac{(2u_1, 2u_2, 0)}{\sqrt{4u_1^2 + 4u_2^2}} = \frac{u_1\overline{i} + u_2\overline{j}}{\sqrt{u_1^2 + u_2^2}}. \qquad (6)$$

$$\overline{e}_3 = \frac{\dfrac{\partial \overline{r}}{\partial u_3}}{\left| \dfrac{\partial \overline{r}}{\partial u_3} \right|} = \frac{(0, 0, 1)}{1} = \overline{k} \qquad (7)$$

Obviously

$$\overline{e}_1 \cdot \overline{e}_3 = \overline{e}_2 \cdot \overline{e}_3 = 0 \qquad (8)$$

and

$$e_1 \cdot e_2 = \frac{u_1u_2 - u_1u_2}{(u_1^2 + u_2^2)} = 0 \qquad (9)$$

Hence, the system is orthogonal.

The scale factors are

$$h_1 = 2\sqrt{u_1^2 + u_2^2}$$

$$h_2 = 2\sqrt{u_1^2 + u_2^2} \tag{10}$$

$$h_3 = 1$$

Substituting eqs.(10) and (2) into the expression for divergence, we obtain

$$\operatorname{div} \overline{F} = \frac{1}{4(u_1^2 + u_2^2)}\left[\frac{\partial}{\partial u_1}(2u_2\sqrt{u_1^2+u_2^2}) + \frac{\partial}{\partial u_2}(2u_3\sqrt{u_1^2+u_2^2})\right.$$

$$\left. + \frac{\partial}{\partial u_3}(4u_1(u_1^2+u_2^2))\right]$$

$$= \frac{u_2(u_1+u_3)}{2(u_1^2+u_2^2)^{\frac{3}{2}}} \tag{11}$$

The curl of $\overline{F}$ is

$$\nabla \times \overline{F} = \frac{1}{4(u_1^2+u_2^2)}\begin{vmatrix} 2\sqrt{u_1^2+u_2^2}\ \overline{e}_1 & 2\sqrt{u_1^2+u_2^2}\ \overline{e}_2 & \overline{e}_3 \\[2mm] \dfrac{\partial}{\partial u_1} & \dfrac{\partial}{\partial u_2} & \dfrac{\partial}{\partial u_3} \\[2mm] 2u_2\sqrt{u_1^2+u_2^2} & 2u_3\sqrt{u_1^2+u_2^2} & u_1 \end{vmatrix}$$

$$= -\overline{e}_1 - \frac{\overline{e}_2}{2\sqrt{u_1^2+u_2^2}} + \frac{\overline{e}_3}{2(u_1^2+u_2^2)}\left[\frac{u_1u_3-u_2^2}{\sqrt{u_1^2+u_2^2}} - \sqrt{u_1^2+u_2^2}\right] \tag{12}$$

● **PROBLEM 4–28**

For the following orthogonal coordinate system,

$$x = e^{u_3}\cos u_2$$

$$y = e^{u_3}\sin u_2 \tag{1}$$

$$z = u_1$$

evaluate the scale factors and compute $\nabla^2 f$ where

$$f = u_1 u_2 u_3. \tag{2}$$

<u>Solution</u>:   The arc length is

$$ds^2 = dx^2 + dy^2 + dz^2$$

$$= (e^{u_3} \cos u_2 \, du_3 - e^{u_3} \sin u_2 \, du_2)^2$$

$$+ (e^{u_3} \sin u_2 \, du_3 + e^{u_3} \cos u_2 \, du_2)^2 + du_1^2$$

$$= du_1^2 + du_2^2 \left[e^{2u_3} \sin^2 u_2 + e^{2u_3} \cos^2 u_2 \right] \tag{3}$$

$$+ du_3^2 \left[e^{2u_3} \cos^2 u_2 + e^{2u_3} \sin^2 u_2 \right]$$

$$= du_1^2 + e^{2u_3} du_2^2 + e^{2u_3} du_3^2$$

The scale factors are

$$h_1 = 1, \quad h_2 = e^{u_3}, \quad h_3 = e^{u_3} \tag{4}$$

Substituting the scale factors and eq.(2) into the expression for ∇^2 we find

$$\nabla^2 f = \frac{1}{h_1 h_2 h_3} \left[\frac{\partial}{\partial u_1} \left(\frac{h_2 h_3}{h_1} \frac{\partial f}{\partial u_1} \right) + \frac{\partial}{\partial u_2} \left(\frac{h_1 h_3}{h_2} \frac{\partial f}{\partial u_2} \right) \right.$$

$$\left. + \frac{\partial}{\partial u_3} \left(\frac{h_1 h_2}{h_3} \frac{\partial f}{\partial u_3} \right) \right] \tag{5}$$

$$= \frac{1}{e^{2u_3}} \left[\frac{\partial}{\partial u_1} (u_2 u_3 \cdot e^{2u_3}) + \frac{\partial}{\partial u_2} (u_1 u_3) + \frac{\partial}{\partial u_3} (u_1 u_2) \right]$$

$$= 0$$

• PROBLEM 4–29

Let
$$x = u_1 - u_2$$
$$y = u_1 - 2u_3 \tag{1}$$
$$z = u_1 + u_3$$

Express the vector

$$\overline{F} = 2\overline{i} + \overline{j} + 3\overline{k} \tag{2}$$

in terms of its contravariant components and its covariant components.

Solution: Vector $\overline{F}$ can be expressed as a linear combination

$$\overline{F} = \alpha_1 \frac{\partial \overline{r}}{\partial u_1} + \alpha_2 \frac{\partial \overline{r}}{\partial u_2} + \alpha_3 \frac{\partial \overline{r}}{\partial u_3} \tag{3}$$

Here, $\frac{\partial \overline{r}}{\partial u_1}, \frac{\partial \overline{r}}{\partial u_2}, \frac{\partial \overline{r}}{\partial u_3}$ are called the unitary base vectors, and $\alpha_1, \alpha_2, \alpha_3$ are the contravariant components of $\overline{F}$. From eq.(1), we have

$$\frac{\partial \overline{r}}{\partial u_1} = (1,1,1) = \overline{i} + \overline{j} + \overline{k}$$

$$\frac{\partial \overline{r}}{\partial u_2} = (-1,0,0) = -\overline{i} \tag{4}$$

$$\frac{\partial \overline{r}}{\partial u_3} = (0,-2,1) = -2\overline{j} + \overline{k}$$

Substituting eq.(2) and eq.(4) into eq.(3) we obtain

$$(2,1,3) = \alpha_1(1,1,1) + \alpha_2(-1,0,0) + \alpha_3(0,-2,1) \tag{5}$$

The contravariant components of $\overline{F}$ are thus

$$\alpha_1 = \frac{7}{3} \qquad \alpha_2 = \frac{1}{3} \qquad \alpha_3 = \frac{2}{3} \tag{6}$$

Vector $\overline{F}$ can be represented as a linear combination

$$\overline{F} = \beta_1 \nabla u_1 + \beta_2 \nabla u_2 + \beta_3 \nabla u_3 \tag{7}$$

where $\beta_1, \beta_2, \beta_3$ are the covariant components of $\overline{F}$.

From eq.(1), we find

$$u_1 = \frac{y + 2z}{3}$$

$$u_2 = \frac{-3x + y + 2z}{3} \tag{8}$$

$$u_3 = \frac{-y + z}{3}$$

Hence,

$$\nabla u_1 = \left(0, \frac{1}{3}, \frac{2}{3}\right) = \frac{1}{3} \bar{j} + \frac{2}{3} \bar{k}$$

$$\nabla u_2 = \left(-1, \frac{1}{3}, \frac{2}{3}\right) = -\bar{i} + \frac{1}{3} \bar{j} + \frac{2}{3} \bar{k} \qquad (9)$$

$$\nabla u_3 = \left(0, -\frac{1}{3}, \frac{1}{3}\right) = -\frac{1}{3} \bar{j} + \frac{1}{3} \cdot \bar{k}$$

Substituting eqs.(2) and (9) into eq.(7) results in

$$(2,1,3) = \beta_1 \left(0, \frac{1}{3}, \frac{2}{3}\right) + \beta_2 \left(-1, \frac{1}{3}, \frac{2}{3}\right) + \beta_3 \left(0, -\frac{1}{3}, \frac{1}{3}\right) \quad (10)$$

so that $\beta_1 = 6$, $\beta_2 = -2$, $\beta_3 = 1$

Thus, the covariant components of $\bar{F}$ are

$$\beta_1 = 6, \qquad \beta_2 = -2, \qquad \beta_3 = 1.$$

● **PROBLEM 4-30**

The transformation is given by

$$x = u_1 + u_2^2$$

$$y = u_1 + u_3 \qquad\qquad (1)$$

$$z = u_2^2 + u_2 u_3$$

Find g and the Jacobian. Verify that

$$J = \sqrt{g} \qquad\qquad (2)$$

Solution: Let us first find the Jacobian of the transformation.

We have

$$J = \frac{\partial(x,y,z)}{\partial(u_1,u_2,u_3)} = \begin{vmatrix} \dfrac{\partial x}{\partial u_1} & \dfrac{\partial y}{\partial u_1} & \dfrac{\partial z}{\partial u_1} \\[2mm] \dfrac{\partial x}{\partial u_2} & \dfrac{\partial y}{\partial u_2} & \dfrac{\partial z}{\partial u_2} \\[2mm] \dfrac{\partial x}{\partial u_3} & \dfrac{\partial y}{\partial u_3} & \dfrac{\partial z}{\partial u_3} \end{vmatrix}$$

$$= \begin{vmatrix} 1 & 1 & 0 \\ 2u_2 & 0 & 2u_2 + u_3 \\ 0 & 1 & u_2 \end{vmatrix}$$

$$= 2u_2^2 + 2u_2 + u_3 \qquad (3)$$

Now, g is defined as

$$g = \begin{vmatrix} g_{11} & g_{12} & g_{13} \\ g_{21} & g_{22} & g_{23} \\ g_{31} & g_{32} & g_{33} \end{vmatrix} \qquad (4)$$

where

$$g_{ij} = \frac{\partial \bar{r}}{\partial u_i} \cdot \frac{\partial \bar{r}}{\partial u_j} \qquad (5)$$

We have

$$\frac{\partial \bar{r}}{\partial u_1} = (1,1,0)$$

$$\frac{\partial \bar{r}}{\partial u_2} = (2u_2,0,2u_2+u_3) \qquad (6)$$

$$\frac{\partial \bar{r}}{\partial u_3} = (0,1,u_2)$$

Thus,

$$g_{11} = \frac{\partial \bar{r}}{\partial u_1} \cdot \frac{\partial \bar{r}}{\partial u_1} = 2$$

$$g_{12} = 2u_2$$

$$g_{13} = 1 \qquad (7)$$

$$g_{22} = 8u_2^2 + 4u_2 u_3 + u_3^2$$

$$g_{23} = 2u_2^2 + u_2 u_3$$

$$g_{33} = 1 + u_2^2$$

Obviously

$$g_{12} = g_{21}, \quad g_{13} = g_{31}, \quad g_{32} = g_{23}. \qquad (8)$$

We have

$$g = \begin{vmatrix} 2 & 2u_2 & 1 \\ 2u_2 & 8u_2^2+4u_2u_3+u_3^2 & 2u_2^2+u_2u_3 \\ 1 & 2u_2^2+u_2u_3 & 1+u_2^2 \end{vmatrix}$$

$$= 2(1+u_2^2)(8u_2^2+4u_2u_3+u_3^2) + 2u_2(2u_2^2+u_2u_3)$$

$$+ 2u_2(2u_2^2+u_2u_3) - (8u_2^2+4u_2u_3+u_3^2)$$

$$- 2(2u_2^2+u_2u_3)(2u_2^2+u_2u_3) - \left[(1+u_2^2)2u_2\right]2u_2$$

$$= 4u_2^4 + 4u_2^2 + 8u_2^3 + 4u_2u_3 + 4u_2^2u_3 + u_3^2 \qquad (9)$$

Now, note that from eq.(3) we obtain

$$J^2 = (2u_2^2 + 2u_2 + u_3)^2$$

$$= 4u_2^4 + 8u_2^3 + 4u_2^2u_3 + 4u_2^2 + 4u_2u_3 + u_3^2 \qquad (10)$$

which is eq.(9).

Thus, we have verified that

$$J = \sqrt{g} \, .$$

CHAPTER 5

MULTIPLE INTEGRALS

> **Basic Attacks and Strategies for Solving Problems in this Chapter. See pages 188 to 244 for step-by-step solutions to problems.**

The definite integral is really a generalization of the process of summation, and has an elementary application for finding the areas of certain plane regions. However, the definite integral and its generalizations have important applications in the applied sciences of fluid dynamics, electromagnetism, and several others.

The concept of the definite integral was studied by the German mathematician Bernhard Riemann, and is often referred to as the Riemann Integral. The French mathematician Henri Lebesque generalized much of Riemann's work, producing a more general form of the integral bearing Lebesque's name. It is vital to study the various forms of the integral for functions of one, and then several variables, over intervals, and then over certain regions in R^n.

The behavior, relative to integrability, of functions of a single variable that are defined in an interval of infinite length, or are bounded almost everywhere in such an interval, or which become unbounded in an interval, is studied by the integrals of such functions, called improper integrals.

Ideas from vector calculus and the theory of integration are used to produce the concepts of line integral and surface integral. These ideas have significant applications in science and engineering, and are of profound interest in several areas of current and active mathematical research.

Step-by-Step Solutions to
Problems in this Chapter,
"Multiple Integrals"

RIEMANN AND STIELTJES INTEGRALS

Suppose f is defined on [0,2] as follows:

$$f(x) = \begin{cases} 1 \text{ for } 0 \le x < 1 \\ 2 \text{ for } 1 \le x \le 2 \end{cases}.$$

Show that f is Riemann integrable.

<u>Solution</u>: By definition a function which can be integrated according to Riemann's definition is called integrable. Note that a function does not have to be continuous to be integrable (i.e., on [a,b] all continuous functions and certain discontinuous functions are integrable).

For the given problem we need the following theorem: Suppose f is bounded on [a,b] and suppose corresponding to each positive ε there is a partition of [a,b] such that the corresponding upper and lower sums (represented as $U(P,f)$ and $L(P,f)$, respectively) satisfy the inequality

$$U(P,f) - L(P,f) < \varepsilon;$$

then f is integrable. To prove this theorem, we start with the condition

$$U(P,f) < \varepsilon + L(P,f).$$

Since, by definition,

$$\inf_{P \in P[a,b]} U(P,f) \leq U(P,f)$$

and

$$L(P,f) \leq \sup_{P \in P[a,b]} L(P,f) \; ,$$

combining inequalities we have

$$\inf_{P \in P[a,b]} U(P,f) \leq U(P,f) < L(P,f) + \varepsilon \leq$$

$$\sup_{P \in P[a,b]} L(P,f) + \varepsilon$$

or

$$\inf_{P \in P[a,b]} U(P,f) \leq \sup_{P \in P[a,b]} L(P,f) + \varepsilon \qquad .$$

Since this conclusion is valid for every $\varepsilon > 0$, we infer that

$$\inf_{P \in P[a,b]} U(P,f) \leq \sup_{P \in P[a,b]} L(P,f) \; .$$

But, by a lemma, we know that

$$\sup_{P \in P[a,b]} L(P,f) \leq \inf_{P \in P[a,b]} U(P,f)$$

is always true. Thus,

$$\sup_{P \in P[a,b]} L(P,f) = \inf_{P \in P[a,b]} U(P,f)$$

which means f is integrable. This completes the proof.

For the given problem suppose $\varepsilon > 0$. Let h be a positive number such that

$$h < 1 \quad \text{and} \quad h < \frac{\varepsilon}{2} \quad .$$

Consider the partition defined by

$$x_0 = 0, \; x_1 = 1-h, \; x_2 = 1+h, \; x_3 = 2 \; .$$

From fig. 1 it is seen that

$$m_1 = M_1 = 1, \; m_2 = 1, \; M_2 = 2, \; \text{and} \; m_3 = M_3 = 2,$$

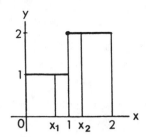

Fig. 1

where m_i and M_i represent the greatest lower bound and least upper bound, respectively, of $f(x)$, defined on the subinterval from x_{i-1} to x_i.

Therefore,

$$L(P,f) =$$

$$m_1(x_1 - x_0) + m_2(x_2 - x_1) + m_3(x_3 - x_2)$$

or

$$L(P,f) =$$

$$1 \cdot (1-h) + 1 \cdot 2h + 2(1-h) = 3 - h \ .$$

Similarly,

$$U(P,f) =$$

$$M_1(x_1 - x_0) + M_2(x_2 - x_1) + M_3(x_3 - x_2)$$

or

$$U(P,f) = 1 \cdot (1-h) + 2 \cdot 2h + 2(1-h) = 3 + h \ .$$

Accordingly

$$U(P,f) - L(P,f) = 2h < \varepsilon$$

since

$$h < \frac{\varepsilon}{2} \ .$$

Thus, by the theorem, f is integrable.

(Note that the function is discontinuous at $x = 1$.)

● **PROBLEM** 5–2

Find the derivatives of

a)

$$\int_1^x t^2 dt \qquad \text{with respect to x.}$$

190

b)

$$\int_1^{t^2} \sin(x^2)\,dx \qquad \text{with respect to t.}$$

Solution: a) Let F(x) denote

$$\int_1^x t^2\,dt.$$

The problem is to find F'(x) where x is a parameter occurring in the limits of integration. To solve this problem, the chain-rule for composite functions of several variables and the theorem on the derivative of an integral which is a function of its upper limit of integration must be used.

Let

$$F(x_1, x_2) = F(x_1(t), x_2(t)),$$

i.e., F is a differentiable function of x_1 and x_2 where x_1 and x_2 are themselves differentiable functions of a parameter t. Then, according to the chain-rule:

$$\frac{dF}{dt} = \frac{\partial F}{\partial x_1}\,\frac{dx_1}{dt} + \frac{\partial F}{\partial x_2}\,\frac{dx_2}{dt} \; .$$

Next, suppose

$$G(y) = \int_a^y g(t)\,dt \; .$$

Then, if g is continuous on [a,y], G'(y) = g(y). To prove this theorem, let $y_0 \in [a,y]$ be a point at which g is continuous. Given $\varepsilon > 0$, choose δ such that

$$|g(y_1) - g(y_0)| < \varepsilon$$

whenever

$$|y_1 - y_0| < \delta \; . \tag{1}$$

Now, G'(y) = g(y) implies that, by the definition of the derivative,

$$\lim_{h \to 0} \left| \frac{G(y_0 + h) - G(y_0)}{h} - g(y_0) \right| = 0. \tag{2}$$

But

$$G(y_0 + h) - G(y_0) = \int_a^{y_0+h} g(t)dt - \int_a^{y_0} g(t)dt$$

$$= \int_{y_0}^{y_0+h} g(t)dt \quad \text{and} \quad g(y_0) = \frac{1}{h}\int_{y_0}^{y_0+h} g(y_0)dt .$$

Therefore,

$$\frac{G(y_0 + h) - G(y_0)}{h} - g(y_0) =$$

$$\frac{1}{h}\int_{y_0}^{y_0+h} [g(t) - g(y_0)]dt \tag{3}$$

Now let $0 < |h| < \delta$. Then the expression under the integral sign on the right side of (3) is in absolute value less than ε, by (1) since

$$|t - y_0| \le |h| < \delta .$$

Accordingly, the entire right side of (3) is less than ε when $0 < |h| < \delta$. This proves (2) and thus proves the theorem. Hence for

$$F(x) = \int_1^x t^2 dt, \quad F'(x) = x^2 .$$

b)

$$\int_1^{t^2} \sin(x^2)dx .$$

Here let $u = t^2$, so that $\frac{du}{dt} = 2t$. Then

$$\frac{dF}{dt} = \frac{dF}{du}\frac{du}{dt} = \sin u^2 \frac{du}{dt} = 2t \sin t^4 .$$

● **PROBLEM 5–3**

Differentiate the integral

192

$$F(y) = \int_0^{y^2} x^5 (y-x)^7 dx$$

Solution: Integrals are differentiated according to Leibniz's rule which states:

Let $f(x,y)$ be an integrable function of x for each value of y. Suppose the partial derivative

$$\frac{\partial f(x,y)}{\partial y}$$

exists and is a continuous function of x and y in some rectangle R then $F(y)$ (i.e.,

$$\left(F(y) = \int_a^b f(x,y) dx \right).$$

has a derivative given by

$$F'(y) = \int_a^b \frac{\partial f(x,y) dx}{\partial y} .$$

The given integral, however, has the parameter y appearing in the limits of integration as well as in the integrand. Since the integral is a function of its upper and lower limits, the following theorem needs to be used: Let

$$F(x) = \int_{x_0}^{x_1} f(t) dt .$$

Then the formula

$$F'(x) = f(x)$$

holds at each point where f is continuous.

Applying the two theorems to the given integral, define

$$G(u,v) = \int_0^u x^5 (v-x)^7 dx$$

and put

$$u = y^2, \quad v = y .$$

Then, treating $G(u,v)$ as an integral which is a function of its upper parameter u and applying the second theorem,

$$\frac{\partial G}{\partial u} = u^5 (v-u)^7 .$$

Next, applying Leibniz's rule to $G(u,v)$ as a function of v:

$$\frac{\partial G}{\partial v} = \int_0^u \frac{\partial G}{\partial v} \, dx = \int_0^u 7x^5 (v-x)^6 dx$$

But $G(u,v)$ is a function of u and v where u and v are themselves functions of y. According to the chain-rule,

$$\frac{dF}{dy} = \frac{\partial G}{\partial u} \frac{du}{dy} + \frac{\partial G}{\partial v} \frac{dv}{dy} .$$

Thus

$$\frac{dF}{dy} = u^5 (v-u)^7 \frac{du}{dy} + \int_0^u 7x^5 (y-x)^6 \, dx \, \frac{dv}{dy} .$$

Since

$$u = y^2; \quad \frac{du}{dy} = 2y;$$

since

$$v = y, \quad \frac{dv}{dy} = 1 .$$

Therefore,

$$F'(y) = y^{10} (y-y^2)^7 \cdot 2y + \int_0^{y^2} 7x^5 (y-x)^6 dx .$$

• **PROBLEM** 5–4

Evaluate the following Stieltjes integrals:

a)

$$\int_a^b f(x) dc ,$$

where c is a constant function

b)

194

$$\int_a^b 1 \, dx$$

c)

$$\int_a^b dx^2$$

d)

$$\int_a^b dg(x)$$

where g(x) is any bounded function.

Solution: The Stieltjes integral is an extension of the Riemann integral. It involves two functions f and g defined on a closed interval [a,b] and its general form is

$$\int_a^b f(x) \, dg(x) \; .$$

The function g is called the integrator.

The Stieltjes integral is defined as the limit of a sum in much the same way as the Riemann integral is defined.

Let

$$P = \{x_0, x_1, \ldots, x_n\}$$

be a partition of [a,b], and let

$$\{z_1, \ldots, z_n\}$$

be respective points in the respective subintervals

$$(x_0, x_1), (x_1, x_2), \ldots, (x_{n-1}, x_n) \; .$$

The mesh of P, denoted by $|P|$ is

$$\max \{(x_1 - x_0), (x_2 - x_1), \ldots, (x_n - x_{n-1})\}$$

(i.e., the length of the largest sub-interval). Then

$$\int_a^b f(x)\,dg(x) = \int_a^b f\,dg = \lim_{|P| \to 0} \sum_{i=1}^{n} f(z_i)[g(x_i) - g(x_{i-1})]$$

if the sum converges to a unique limit for all partitions such that

$$|P| < \delta,$$

where $\delta > 0$ is a given challenge number.

a)

$$\int_a^b f(x)\,dc = \lim_{|P| \to 0} \sum_{i=1}^{n} f(z_i)[c(x_i) - c(x_{i-1})]$$

$$= \lim_{|P| \to 0} \sum_{i=1}^{n} f(z_i)[c-c] = 0 .$$

Thus the Stieltjes integral of any function with respect to a constant function is zero.

b)

$$\int_a^b 1 \cdot dx = \lim_{|P| \to 0} \sum_{i=1}^{n} 1[x_i - x_{i-1}]$$

$$= \lim_{|P| \to 0} \{(x_1-x_0) + (x_2-x_1) + \ldots + (x_n-x_{n-1})\}$$

$$= x_n - x_0 = b-a .$$

c)

$$\int_a^b dx^2 = \lim_{|P| \to 0} \sum_{i=1}^{n} 1[x_i^2 - x_{i-1}^2]$$

$$= \lim_{|P| \to 0} \{(x_1^2 - x_0^2) + (x_2^2 - x_1^2) + \ldots + (x_n^2 - x_{n-1}^2)\}$$

$$= x_n^2 - x_0^2 = b^2 - a^2 .$$

d)

$$\int_a^b dg(x) = \lim_{|P| \to 0} \sum_{i=1}^{n} [g(x_i) - g(x_{i-1})]$$

$$= \left(g(x_1) - g(x_0)\right) + \left(g(x_2) - g(x_1)\right) + \ldots + \left(g(x_n) - g(x_{n-1})\right)$$

$$= g(x_n) - g(x_0) = g(b) - g(a).$$

The Stieltjes integral of any constant function c with respect to an integrator g(x) over [a,b] equals

$$c[g(b) - g(a)] .$$

Let f be a function from [a,b] into R which is continuous at c $\in$ [a,b] and let X_c be the characteristic function of c, i.e.,

$$X_c(x) = \begin{cases} 1 & x = c \\ 0 & x \neq c . \end{cases}$$

Show, using the definition of the Stieltjes integral that

$$\int_a^b f dX_c = \begin{cases} 0 & c \in (a,b) \\ -f(a) & c = a \\ f(b) & c = b . \end{cases}$$

<u>Solution</u>: The definition of the Stieltjes integral is as follows:

Let

$$P = \{(x_{k-1}, x_k) : k = 1, \ldots n\}$$

be a partition of [a,b] and let f,g be defined on [a,b]. Then

$$\int_a^b f \, dg = \lim_{\|P\| \to 0} \sum_{k=1}^n f(z_k)[g(x_k) - g(x_{k-1})]$$

where only partitions with sub-intervals less than any preassigned δ are considered.

Let E $\subset$ R, i.e., E is a subset of the real line. The characteristic function of x with respect to E is

$$X(x) = \begin{cases} 1 & x \in E \\ 0 & x \notin E . \end{cases}$$

197

In the given problem E is the singleton set $\{c\}$.

There are three separate cases to consider:

i) $c \in (a,b)$, ii) $c = a$, the left endpoint of the interval, and iii) $c = b$, the right endpoint of the interval.

i) Suppose $c \in (a,b)$ and that $x_k \neq c$ for

$$k = 0 , \ldots , n .$$

Then

$$S(f,X_c,P) = \sum_{k=1}^{n} f(z_k) [X_c(x_k) - X_c(x_{k-1})]$$

$$= \sum_{k=1}^{n} f(z_k) [0-0] = 0 .$$

On the other hand, if $x_k = c$, i.e., one of the endpoints of the sub-intervals is c, then, since $X_c(x_i) = 1$,

$$S(f,X_c,P) = f(z_i) (X_c(x_i) - X_c(x_{i-1}))$$

$$+ f(z_{i+1}) (X_c(x_{i+1}) - X_c(x_i))$$

$$= f(z_i) (1-0) + f(z_{i+1}) (0-1) = f(z_i) - f(z_{i+1}) .$$

The function f being continuous at c implies that for each $\varepsilon > 0$, there exists $\delta > 0$ such that

$$|f(c) - f(x)| < \varepsilon/2$$

whenever $|c - x| < \delta$. Then for any partition P with $\|P\| < \delta$,

$$|S(f,X_c,P)| = |f(z_i) - f(z_{i+1})|$$

$$\leq |f(z_i) - f(c)| + |f(c) - f(z_{i+1})|$$

$$< \varepsilon/2 + \varepsilon/2 = \varepsilon$$

Hence

$$\int_a^b f \, dX_c = \lim_{\|P\| \to 0} s(f,X_c,P) = 0 .$$

ii) When $c = a$, the left endpoint of the interval $[a,b]$,

$$S(f, X_c, P) = f(z_1)(X_c(x_1) - X_c(x_0))$$

$$= f(z_1)(0-1) = -f(z_1) = -f(a) \ ,$$

and

$$\int_a^b f \, dX_c = \lim_{||P|| \to 0} S(f, X_c, P) = -f(a)$$

iii)   When $c = b$, the right endpoint of the interval $[a,b]$,

$$S(f, X_c, P) = f(z_n)(X_c(x_n) - X_c(x_{n-1}))$$

$$= f(z_n)(1-0) = f(c) = f(b).$$

• **PROBLEM 5–6**

Discuss the existence of

$$\int_0^1 x \, dx^2$$

and find its value.

Solution: The first task in the evaluation of an integral is to show that the integral exists.   The given integral is of the form

$$\int_a^b f(x) \, dg(x)$$

where $g(x)$ is monotonically increasing on $[a,b]$ and $f(x)$ is continuous on $[a,b]$.   A fundamental criterion for integrability is the following: A function $f: [a,b] \to R$ is integrable with respect to a nondecreasing function

$$g: [a,b] \to R$$

if an only if for all $\varepsilon > 0$ there is a partition P_ε of

$[a,b]$ such that if P_ε' is a refinement of P_ε then

$$\sum_{i=1}^n (M_i - m_i)[g(x_i) - g(x_{i-1})] < \varepsilon$$

where

$$M_i = \sup \{f(x): x \in [x_{i-1}, x_i]\}$$

and

$$m_i = \inf \{f(x): x \in [x_{i-1}, x_i]\}$$

for $i = 1, \ldots, n$.

From this result a theorem on the integrability of a continuous function with respect to a monotonically increasing integrator may be deduced.

To prove this latter theorem for the given integral

$$\int_0^1 x\,dx^2,$$

let $f(x) = x$ and $g(x) = x^2$.

Then, since f is uniformly continuous on $[0,1]$, there exists a $\delta(\varepsilon) > 0$ such that

$$|x-y| < \delta(\varepsilon)$$

implies

$$|f(x) - f(y)| < \varepsilon \quad .$$

Let

$$P_\varepsilon = \{w_0, w_1, \ldots, w_k\} \text{ be a partition of } [a,b]$$

such that $\sup\{w_k - w_{k-1}\} < \delta(\varepsilon)$.

If P_ε' is a refinement of P_ε, then

$$\sup \{x_k - x_{k-1}\} < \delta(\varepsilon)$$

and thus $M_i - m_i < \varepsilon$, which implies that

$$\sum_{i=1}^n (M_i - m_i) \{g(x_i) - g(x_{i-1})\} \le \varepsilon (g(b) - g(a)) \quad .$$

Since $\varepsilon > 0$ is arbitrary, the fundamental criterion of integrability, (1), may be applied. Thus

$$\int_0^1 x\,dx^2$$

exists.

The next task is to evaluate this integral. Using the definition of the integral as the limit of a sum,

$$\int_0^1 f(x)\,dg(x) = \lim_{n\to\infty} \sum_{j=1}^n f(x_j)[g(x_j) - g(x_{j-1})] \ ,$$

Let

$$x_j = \frac{j}{n} \ ,$$

then

$$\int_0^1 f(x)\,dg(x) = \lim_{n\to\infty} \sum_{j=1}^n f\left(\frac{j}{n}\right)\left[g(j/n) - g\left(\frac{j-1}{n}\right)\right]$$

or,

$$\int_0^1 x\,dx^2 = \lim_{n\to\infty} \sum_{j=1}^n \frac{j}{n}\left[\frac{j^2}{n^2} - (j-1)^2/n^2\right]$$

$$= \lim_{n\to\infty} \sum_{j=1}^n \frac{j}{n}\,\frac{2j-1}{n}\,\frac{1}{n} = \lim_{n\to\infty} \sum_{j=1}^n \frac{1}{n^3}\,[2j^2 - j]$$

$$= \lim_{n\to\infty} \sum_{j=1}^n \frac{1}{n^3}\,2j^2 - \lim_{n\to\infty} \sum_{j=1}^n \frac{j}{n^3}$$

$$= \lim_{n\to\infty} \frac{1}{n^3}\left[\frac{n(n+1)(2n+1)}{3} - \frac{n(n+1)}{2}\right]$$

$$= \lim_{n\to\infty} \frac{1}{n^3}\left[\frac{4n^3}{6} + \frac{3n^2}{6} - \frac{n}{6}\right]$$

$$= 2/3 \ .$$

Thus

$$\int_0^1 x\,dx^2 = 2/3 \ .$$

LINE INTEGRALS

Evaluate: a) the line integral $\int_C (x^3-y^3)dy$ where C
is the semicircle $y = \sqrt{1-x^2}$ shown in Figure 1.
b) $\int_C \vec{F}\cdot d\vec{C}$ where $\vec{F}(x,y) = (x^2,xy)$ and C is $x=y^2$ between
$(1,-1)$ and $(1,1)$.

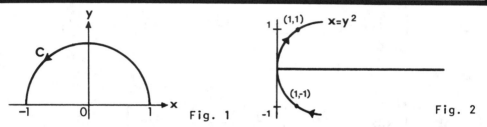

Fig. 1   Fig. 2

Solution: The curve C can be represented parametrically
by $x = \cos\theta$ and $y = \sin\theta$, $0 \le \theta \le \pi$. Therefore the integral
becomes $\int_0^\pi (\cos^3\theta - \sin^3\theta)\cos\theta\,d\theta$ $(dy = \cos\theta\,d\theta)$

$$= \int_0^\pi \cos^4\theta\,d\theta - \int_0^\pi \sin^3\theta\cos\theta\,d\theta. \quad (1)$$

Evaluate the first integral by parts letting $u = \cos^3\theta$,
$dv = \cos\theta\,d\theta$ so that $du = -3\cos^2\theta\sin\theta$ and $v = \sin\theta$. Hence
the first integral in (1) becomes:

$$\cos^3\theta\sin\theta\Big|_0^\pi + 3\int_0^\pi \sin^2\theta\cos^2\theta\,d\theta$$

$$= 0 + 3\int_0^\pi \sin^2\theta\cos^2\theta\,d\theta .$$

Using $2\sin\theta\cos\theta = \sin2\theta$ and $\sin^2\theta = \dfrac{1-\cos2\theta}{2}$ the above

integral is computed to equal $3\int_0^\pi \left(\dfrac{1}{8} - \dfrac{\cos4\theta}{8}\right) d\theta = \dfrac{3\pi}{8}$.

To evaluate the second integral, let $u = \sin\theta$, $du = \cos\theta$,

then $-\int_0^\pi u^3 du = 0$. Therefore $\int_C (x^3-y^3)dy = \dfrac{3\pi}{8}$.

b) Since $\int \vec{F} \cdot d\vec{C}$ is being evaluated, first compute $\vec{F} \cdot d\vec{C} =$ $(x^2, xy) \cdot (dx, dy) = x^2 dx + xy dy$. To parameterize let $y=y$; $x=y^2$ so that $dy=dy$; $dx=2y dy$ and

$$\int_C \vec{F} \cdot d\vec{C} = \int_C x^2 dx + xy dy = \int_{-1}^1 y^4 \cdot 2y dy + y^3 dy$$

$$= \frac{2y^6}{6} + \frac{y^4}{4} \Big|_{-1}^1 = \left(\frac{1}{3} + \frac{1}{4}\right) - \left(\frac{1}{3} + \frac{1}{4}\right) = 0.$$

• PROBLEM 5–8

Evaluate the following line integrals:

a) $\oint_C y^2 dx + x^2 dy$ where C is the triangle with vertices $(1,0), (1,1), (0,0)$ (Figure 1). (1)

b) $\int_C x^2 dx + xy dy$ where C is the straight line segment from $(1,0)$ to $(2,3)$. (2)

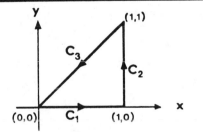

Fig. 1

Solution: a) To compute $\int_C y^2 dx + x^2 dy$ compute $\int_{C_i} y^2 dx + x^2 dy$ along 3 different C_i ($i = 1$ to 3). Then $\oint_C y^2 dx + x^2 dy$ equals the sum of the integrals along each C_i. Along C_1, $y=0$ and $dy=0$; therefore

$$\int_{C_1} y^2 dx + x^2 dy = \int_{C_1} 0 \cdot dx + x^2 \cdot 0 = 0. \qquad (3)$$

Along C_2 use y as the parameter and $x=1$ so that $dx=0$. Therefore

$$\oint_{C_2} y^2 dx + x^2 dy = \int_0^1 y^2 \cdot 0 + dy = \int_0^1 dy = 1. \qquad (4)$$

For C_3 use x as the parameter so that x=x and y=x and dx=dx; dy=dx.

$$\int_{C_3} y^2dx + x^2dy = \int_1^0 x^2dx + x^2dx = \int_1^0 2x^2dx = -\frac{2}{3} \quad . \quad (5)$$

Hence (1) equals the sum of (3), (4), (5) so that

$$\oint_C y^2dx + x^2dy = 0 + 1 - \frac{2}{3} = \frac{1}{3} \quad .$$

b)   To compute (2), note that the line segment is of the form $P = P_1 + t(P_2-P_1)$, $(0 \le t \le 1)$, and can be parameterized accordingly. Since the given line segment goes from (1,0) to (2,3), by substitution:

$$x = 1 + t(2-1) = 1 + t; \quad y = 0 + t(3-0) = 3t \ (0 \le t \le 1).$$

Hence substituting into (2)

$$\oint_C x^2dx + xydy = \int_0^1 (1+t)^2dt + (1+t)3t \cdot 3dt$$

$$= \int_0^1 (1+2t+t^2+9t+9t^2)dt = \int_0^1 (1+11t+10t^2)dt$$

$$= 1 + \frac{11}{2} + \frac{10}{3} = \frac{59}{6} \quad .$$

• PROBLEM 5–9

Let $\vec{F}(x,y) = (x^2y, xy^2)$.

a)   Determine whether this vector field has a potential function.

b)   Evaluate the integral of $\vec{F}$ from O to the point P on Figure 1, along the line segment from (0,0) to $\left(\frac{\sqrt{2}}{2}, \frac{\sqrt{2}}{2}\right)$.

c)   Evaluate the integral of $\vec{F}$ from O to P along the path that consists of the line segment from (0,0) to (1,0) and along the circular arc from (1,0) to P.

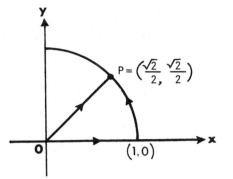

Fig. 1

Solution: a) Since $\vec{F}(x,y) = (x^2y, xy^2)$, $\frac{\partial f}{\partial y} = x^2$ and

$\frac{\partial g}{\partial x} = y^2$. Therefore since $\frac{\partial f}{\partial y} \neq \frac{\partial g}{\partial x}$ there cannot be a potential function.

b) Here integration is over the line segment that is part of the line x=y, from $(0,0)$ to $\left(\frac{\sqrt{2}}{2}, \frac{\sqrt{2}}{2}\right)$. So apply the parameterization x=x, y=x, dx=dx, dy=dx to obtain

$$\int_C x^2y\,dx + xy^2\,dy = \int_0^{\frac{\sqrt{2}}{2}} x^3\,dx + x^3\,dx = \frac{2x^4}{4}\Bigg|_0^{\frac{\sqrt{2}}{2}}$$

$$= \frac{\left(\frac{\sqrt{2}}{2}\right)^4}{2} = \frac{4}{32} = \frac{1}{8}.$$

c) Here C consists of two paths. The first path is the line segment y=0 from $(0,0)$ to $(1,0)$. Therefore use the parameterization x=x, y=0 so that dx=dx, dy=0 and

$$\int_{C_1} 0 \cdot dx + 0 = 0. \tag{1}$$

The second segment is the arc of the circle $x^2+y^2=1$ from the point $(1,0)$ to $\left(\frac{\sqrt{2}}{2}, \frac{\sqrt{2}}{2}\right)$. Here apply the parameterization $x = \cos\theta$, $y = \sin\theta$. To find the range of θ,

$$\tan\theta = \frac{y}{x} \quad \text{so at } (1,0) \quad \tan\theta = \frac{0}{1} = 0; \ \theta = 0$$

$$\text{at } \left(\frac{\sqrt{2}}{2}, \frac{\sqrt{2}}{2}\right) \quad \tan\theta = \left(\frac{\frac{\sqrt{2}}{2}}{\frac{\sqrt{2}}{2}}\right) = 1, \ \theta = \frac{\pi}{4}.$$

205

Thus,

$$\int_{C_2} x^2 y dx + xy^2 dy = \int_0^{\frac{\pi}{4}} \cos^2\theta \sin\theta (-\sin\theta) d\theta + \cos\theta \sin^2\theta \cos\theta d\theta$$

$$= \int_0^{\frac{\pi}{4}} (-\cos^2\theta \sin^2\theta + \cos^2\theta \sin^2\theta) d\theta = \int_0^{\frac{\pi}{4}} 0 d\theta = 0. \qquad (2)$$

Hence $\int_C \vec{F} \cdot d\vec{c}$ equals (1) + (2) = 0 + 0 = 0.

Note that the value of the line integral over the path in part (c) is different from the value found in part (b). Therefore the line integral is not independent of the path and this is because $\vec{F}$ is not exact.

• PROBLEM 5–10

Evaluate the following line integrals:

a) $\int_{(1,-2)}^{(3,4)} \frac{ydx-xdy}{x^2}$ on the line $y = 3x-5$

b) $\int_{(0,2)}^{(1,3)} \frac{3x^2}{y} dx - \frac{x^3}{y^2} dy$ on the parabola $y = 2+x^2$

c) $\int_{(0,0)}^{(2,8)} \vec{\nabla}f \cdot d\vec{c}$ where $\vec{\nabla}f$ is grad f and f is the function $f(x,y) = x^2-y^2$. C is the curve $y = x^3$.

Solution: a) The function $\frac{ydx-xdy}{x^2}$ may be expressed as

$$\vec{F}(x,y) = \vec{F}\left(\frac{y}{x^2}, \frac{-x}{x^2}\right) = (f(x,y), g(x,y)).$$ If there exists

a function $\phi(x,y)$ such that $\vec{\nabla}\phi = \frac{\partial\phi}{\partial x}, \frac{\partial\phi}{\partial y} = (f(x,y),$

$g(x,y))$ then the given vector field has a potential function and

$$\int_{(1,-2)}^{(3,4)} \frac{ydx-xdy}{x^2} = \phi(3,4) - \phi(1,2).$$

Since $\frac{\partial^2\phi}{\partial y\partial x} = \frac{\partial^2\phi}{\partial x\partial y}$, an equivalent way of proving the

existence of the potential function is to show that $\frac{\partial f}{\partial y} =$

$\frac{\partial g}{\partial x}$. For the given vector field, $\vec{F}\left(\frac{y}{x^2} , \frac{-x}{x^2}\right)$, $\frac{\partial f}{\partial y} = \frac{1}{x^2}$

and $\frac{\partial g}{\partial x} = \frac{1}{x^2}$. Therefore $\frac{\partial f}{\partial y} = \frac{\partial g}{\partial x}$. Thus the given vector field has a potential function $\phi(x,y)$ on the given segment of the line ($x \neq 0$, since $1 \leq x \leq 3$). To find this function $\phi(x,y)$ first note that

$$\frac{\partial \phi}{\partial x} = \frac{y}{x^2} \; ; \; \frac{\partial \phi}{\partial y} = \frac{-x}{x^2} = \frac{-1}{x} . \tag{1}$$

Then integrate with respect to x to obtain

$$\phi(x,y) = \int \frac{y}{x^2} dx + H(y) \qquad (H(y) \text{ some function of } y).$$

Then $\phi(x,y) = \frac{-y}{x} + H(y)$. Next find $\frac{\partial \phi}{\partial y}$: \hfill (2)

$$\frac{\partial \phi}{\partial y} = \frac{-1}{x} + H'(y) . \text{ But from (1)} \quad \frac{\partial \phi}{\partial y} = \frac{-1}{x} .$$

Setting $H'(y)$ equal to zero, $\phi(x,y) = \frac{-y}{x}$ (assuming $H(y) = 0$). (3)

Hence $\displaystyle\int_{(1,-2)}^{(3,4)} \frac{ydx-xdy}{x^2} = \phi(3,4) - \phi(1,-2)$

$$= \frac{-4}{3} - 2 = \frac{-10}{3} \text{ (by substitution into (3))}.$$

b) Since $\vec{F} = \left(\frac{3x^2}{y} , \frac{-x^3}{y^2}\right)$, $\frac{\partial f}{\partial y} = \frac{-3x^2}{y^2}$, $\frac{\partial g}{\partial x} = \frac{-3x^2}{y^2}$.

Therefore since $\frac{\partial f}{\partial y} = \frac{\partial g}{\partial x}$, $\vec{F}$ has a potential function

$\phi(x,y)$ such that $\frac{\partial \phi}{\partial x} = \frac{3x^2}{y}$, $\frac{\partial \phi}{\partial y} = \frac{-x^3}{y^2}$. \hfill (4)

(Note: $\vec{F}$ is not defined at $y=0$. However, the line integral is from the points $(0,2)$ to $(1,3)$ along $y = 2+x^2$, so $y \neq 0$ on this path). Integrate $\phi(x,y)$ with respect to x first

to obtain $\phi(x,y) = \int \frac{3x^2}{y} dx + H(y) = \frac{x^3}{y} + H(y)$. \hfill (5)

Then $\frac{\partial \phi}{\partial y} = \frac{-x^3}{y^2} + H'(y)$. But from (4) $\frac{\partial \phi}{\partial y} = \frac{-x^3}{y^2}$.

Thus $H'(y) = 0$ in (5) so that $\phi(x,y) = \frac{x^3}{y}$. \hfill (6)

Then $\int_{(0,2)}^{(1,3)} \frac{3x^2}{y} dx - \frac{x^3}{y^2} dy = \phi(1,3) - \phi(0,2)$

$$= \frac{1}{3} - 0 = \frac{1}{3} \text{ (by substitution into (6))}.$$

c) Given the function $f(x,y) = x^2 - y^2$, since grad f = $=\left(\frac{\partial f}{\partial x}, \frac{\partial g}{\partial y}\right)$, $\vec{\nabla}f = (2x,-2y)$. (Note: gradient makes a vector out of a scalar, therefore the integral is a line integral). Let $\vec{F} = \vec{\nabla}f = (g_1,g_2)$; then $\frac{\partial g_1}{\partial y} = 0$ and $\frac{\partial g_2}{\partial x} = 0$. Therefore there exists a potential function such that

$\frac{\partial \phi}{\partial x} = 2x$ and $\frac{\partial \phi}{\partial y} = -2y$.

Integrating, $\phi(x,y) = \int 2x dx - \int 2y dy$

$$\phi(x,y) = x^2 - y^2. \qquad (7)$$

Then $\int_{(0,0)}^{(2,8)} \vec{\nabla}f \cdot d\vec{c} = \phi(2,8) - \phi(0,0)$

$$= 4 - 64 - 0 = -60 \text{ (by substitution into (7))}.$$

● **PROBLEM 5-11**

Show that the following functions are independent of the path in the xy-plane and evaluate them:

a) $\int_{(1,1)}^{(x,y)} 2xy dx + (x^2-y^2) dy$

b) $\int_{(0,0)}^{(x,y)} \sin y\, dx + x \cos y\, dy.$

<u>Solution</u>: To determine if the line integral is independent of the path apply the following method. First, calculate the following partial derivatives of F(x,y), $\frac{\partial f}{\partial y}$; $\frac{\partial g}{\partial x}$. Then check to see if the condition $\frac{\partial f}{\partial y} = \frac{\partial g}{\partial x}$ is satisfied. If it is, then we know that for this vector field (defined on a simply connected region) there exists some function $\phi(x,y)$ such that $\vec{F} = \text{grad}\phi$ (ϕ is called a potential function). In this case, $\int_{P}^{Q} \vec{F} \cdot d\vec{c} = \phi(Q) - \phi(P),$

which means the value of the line integral depends only
on the endpoints of the interval and not the path taken.
Therefore the line integral is independent of the path.

a) $\vec{F}(x,y) = (2xy, x^2-y^2)$. Computing $\frac{\partial f}{\partial y} = 2x$; $\frac{\partial g}{\partial x} = 2x$
and $\frac{\partial f}{\partial y} = \frac{\partial g}{\partial x}$. Therefore there exists a potential function
$\phi(x,y)$ and the line integral is thus independent of the
path.

To evaluate this line integral, find $\phi(x,y)$ such that
$\frac{\partial \phi}{\partial x} = 2xy$ and $\frac{\partial \phi}{\partial y} = x^2-y^2$.   (1)

Integrating with respect to x first, $\phi(x,y) = \int 2xy\,dx + H(y)$

where $H(y)$ is some function of y. Then

$\phi(x,y) = x^2y + H(y)$.   (2)

From (2) $\frac{\partial \phi}{\partial y} = x^2 + H'(y)$. But from (1) $\frac{\partial \phi}{\partial y} = x^2-y^2$.
This implies that $H'(y) = -y^2$, so let $H(y) = \frac{-y^3}{3}$ and
(2) becomes

$\phi(x,y) = x^2y - \frac{y^3}{3}$.   (3)

Then since $\vec{F} = \text{grad}\phi$

$\int_P^Q \vec{F} \cdot d\vec{c} = \int_{(1,1)}^{(x,y)} 2xy\,dx + (x^2-y^2)\,dy = \phi(x,y) - \phi(1,1)$ (4)

Substituting (3) in (4)

$\int_{(1,1)}^{(x,y)} 2xy\,dx + (x^2-y^2)\,dy = (x^2y - \frac{y^3}{3}) - (1-\frac{1}{3}) = x^2y - \frac{y^3}{3} - \frac{2}{3}$

$= x^2y - \frac{1}{3}(y^3+2)$.

b) $\vec{F}(x,y) = (\sin y, x \cos y)$. Computing, $\frac{\partial f}{\partial y} = \cos y$;
$\frac{\partial g}{\partial x} = \cos y$. Therefore since $\frac{\partial f}{\partial y} = \frac{\partial g}{\partial x}$, there exists a
potential function $\phi(x,y)$ and the line integral is inde-
pendent of the path.

To evaluate this line integral, find a function
$\phi(x,y)$ such that $\frac{\partial \phi}{\partial x} = \sin y$; $\frac{\partial \phi}{\partial y} = x \cos y$.   (5)

Integrating with respect to x first
$\phi(x,y) = \int \sin y\,dx + H(y)$ where $H(y)$ is some function

of y. Then

$$\phi(x,y) = x \sin y + H(y) \tag{6}$$

From (6) $\frac{\partial \phi}{\partial y} = x \cos y + H'(y)$. But from (5) $\frac{\partial \phi}{\partial y} = x \cos y$.

Thus, letting $H(y) = 0$ in (6),

$$\phi(x,y) = x \sin y \tag{7}$$

Hence $\int_P^Q \vec{F} \cdot d\vec{c} = \int_{(0,0)}^{(x,y)} \sin y \, dx + x \cos y dy$

$$= \phi(x,y) - \phi(0,0).$$

Substituting this into (7)

$$\int_{(0,0)}^{(x,y)} \sin y \, dx + x \cos y dy = x \sin y - 0 \cdot \sin 0 = x \sin y.$$

• PROBLEM 5–12

Let $\vec{F}$ be the following vector fields:

a) $\vec{F}_1(x,y,z) = (1-yz, 1-zx, -xy)$

b) $\vec{F}_2(x,y,z) = (yz, xz, xy)$

c) $\vec{F}_3(x,y,z) = (\log(xy), x, y)$

Determine, for each vector field, if $\int \vec{F} \cdot d\vec{c}$ is independent of the path and if so integrate the vector field over the curve from the point $P = (1,6,5)$ to the point $Q = (4,3,2)$.

Solution: To determine if the line integral is independent of the path compute $\frac{\partial f}{\partial y}$, $\frac{\partial g}{\partial x}$, $\frac{\partial f}{\partial z}$, $\frac{\partial h}{\partial x}$, $\frac{\partial g}{\partial z}$, and $\frac{\partial h}{\partial y}$.

If we have the conditions

$$\frac{\partial f}{\partial y} = \frac{\partial g}{\partial x} , \quad \frac{\partial f}{\partial z} = \frac{\partial h}{\partial x} , \quad \frac{\partial g}{\partial z} = \frac{\partial h}{\partial y} \tag{1}$$

then there exists a function ϕ such that $\vec{F} = \text{grad}\phi$. Then $\int_P^Q pdx + qdy = \phi(P) - \phi(Q)$, thus showing that the value of the line integral depends only on the endpoints and not on the path chosen. Hence the line integral is independent of the path.

a) $\vec{F}_1(x,y,z) = (1-yz, 1-zx, -xy)$. Computing

$$\frac{\partial f}{\partial y} = -z = \frac{\partial g}{\partial x} , \quad \frac{\partial f}{\partial z} = -y = \frac{\partial h}{\partial x} , \quad \text{and } \frac{\partial g}{\partial z} = -x = \frac{\partial h}{\partial y} .$$

Therefore since the conditions in (1) are satisfied $\int_C \vec{F}_1 \cdot d\vec{c}$ is independent of the path. To evaluate this line integral look for a function ϕ such that $\vec{F} = \text{grad}\phi$,

i.e. $\dfrac{\partial \phi}{\partial x} = 1 - yz$, $\dfrac{\partial \phi}{\partial y} = 1 - zx$, $\dfrac{\partial \phi}{\partial z} = -xy$. (2)

Integrating with respect to x first

$\phi(x,y,z) = \int (1-yz)\,dx + H(y,z)$ so that

$\phi(x,y,z) = x - yzx + H(y,z)$ (3)

Differentiating (3) with respect to y and z:

$\dfrac{\partial \phi}{\partial y} = -xz + \dfrac{\partial H}{\partial y}$ and $\dfrac{\partial \phi}{\partial z} = -xy + \dfrac{\partial H}{\partial z}$.

But from (2) $\dfrac{\partial \phi}{\partial y} = 1 - xz$ and $\dfrac{\partial \phi}{\partial z} = -xy$.

Take $H(y,z) = y$ so that $\dfrac{\partial H}{\partial y} = 1$ and $\dfrac{\partial H}{\partial z} = 0$.

Therefore from (3), $\phi(x,y,z) = x + y - xyz$

$\int_{(1,6,5)}^{(4,3,2)} (1-yz)\,dx + (1-zx)\,dy - xy\,dz = \phi(4,3,2) - \phi(1,6,5)$

$= -17 - (-23) = 23 - 17 = 6.$

b) $\vec{F}_2(x,y,z) = (yz,xz,xy)$. Computing, $\dfrac{\partial f}{\partial y} = z = \dfrac{\partial g}{\partial x}$;

$\dfrac{\partial f}{\partial z} = y = \dfrac{\partial h}{\partial x}$, $\dfrac{\partial g}{\partial z} = x = \dfrac{\partial h}{\partial y}$. Therefore the conditions in (1) are satisfied and $\int_C \vec{F}_2 \cdot d\vec{c}$ is independent of the path. Thus ϕ is required such that

$\dfrac{\partial \phi}{\partial x} = yz, \dfrac{\partial \phi}{\partial y} = xz, \dfrac{\partial \phi}{\partial z} = xy$ (4)

Integrating with respect to x first

$\phi(x,y,z) = \int yz + H(y,z) = xyz + H(y,z)$. Differentiating with respect to y and z, $\dfrac{\partial \phi}{\partial y} = xz + \dfrac{\partial H}{\partial y}$ and $\dfrac{\partial \phi}{\partial z} = xy + \dfrac{\partial H}{\partial z}$.

But from (4) $\dfrac{\partial \phi}{\partial y} = xz$, $\dfrac{\partial \phi}{\partial z} = xy$. Therefore take $H(y,z) = 0$ and from (5) $\phi(x,y,z) = xyz$. Then

$\int_{(1,6,5)}^{(4,3,2)} yz\,dx + xz\,dy + xy\,dz = \phi(4,3,2) - \phi(1,6,5)$

$= 24 - 30 = -6.$

c) $\vec{F}_3(x,y,z) = (\log(xy),x,y)$. Computing, $\frac{\partial f}{\partial y} = \frac{1}{y}$ and $\frac{\partial g}{\partial x} = 1$. Since $\frac{\partial f}{\partial y} \neq \frac{\partial g}{\partial x}$ there is a condition of (1) that is not satisfied so $\int_C \vec{F}_3 \cdot d\vec{c}$ is not independent of the path.

● **PROBLEM 5–13**

Let C be the ellipse $x^2 + 4y^2 = 4$. Compute

$\oint_C (2x-y)dx + (x+3y)dy$ by Green's Theorem.

Solution: Green's Theorem states: Let p,q be functions on a region R, which is the interior of a closed path C (parameterized counterclockwise). Then

$$\int_C pdx + qdy = \iint_R \left(\frac{\partial q}{\partial x} - \frac{\partial p}{\partial y}\right)dxdy.$$

In the given problem,

$$\oint_C (2x-y)dx + (x+3y)dy \qquad (1)$$

where C is the ellipse $x^2 + 4y^2 = 4$, is the integral. To use Green's Theorem let $p = 2x-y$ and $q = (x+3y)$. Then $\frac{\partial p}{\partial y} = -1$ and $\frac{\partial q}{\partial x} = 1$ so that (1) equals

$$\iint_R (1+1)dxdy = 2\iint_R dxdy = 2 \times (\text{area of the ellipse}). \qquad (2)$$

Hence rewriting the ellipse as $\frac{x^2}{4} + y^2 = 1$, and using the formula Area $= \pi ab$ where $\frac{x^2}{a^2} + \frac{y^2}{b^2} = 1$, the area of the given ellipse is 2π. Thus, (2) becomes $2 \times 2\pi = 4\pi$. Thus the value of the line integral is 4π.

● **PROBLEM 5–14**

Use Green's Theorem to find:

a) $\int_C y^2 dx - xdy$ clockwise around the triangle whose vertices are at $(0,0)$, $(0,1)$, $(1,0)$.

b) The integral of the vector field $\vec{F}(x,y) = (y+3x, 2y-x)$ counterclockwise around the ellipse $4x^2 + y^2 = 4$.

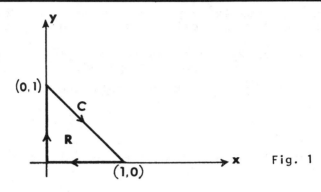

Fig. 1

Solution: Green's Theorem states:

$$\oint_C p\,dx + q\,dy = \iint_R \left(\frac{\partial q}{\partial x} - \frac{\partial p}{\partial y}\right) dx\,dy \qquad (1)$$

where C is the boundary of the region R.

a) From Figure 1 the region lies to the right of the path. Therefore use C^- instead of C so that the region lies to the left. Then use the fact that

$$\int_C \vec{F} = -\int_{C^-} \vec{F}. \quad \text{Hence}$$

$$\int_C y^2 dx - x\,dy = -\int_{C^-} y^2 dx - x\,dy. \qquad (2)$$

Now apply Green's Theorem to the right hand side of equation (2). Let $p = y^2$ and $q = -x$ so that $\frac{\partial q}{\partial x} = -1$ and $\frac{\partial p}{\partial y} = 2y$ to obtain

$$\int_C y^2 dx - x\,dy = -\iint_R (-1-2y)\,dx\,dy.$$

From Figure 1 R is the region bounded by the x-axis, the y-axis, and the line x+y=1. Thus (2) becomes

$$\int_C y^2 dx - x\,dy = -\int_0^1 \int_0^{1-y} -(1+2y)\,dx\,dy$$

$$= \int_0^1 \left(x+2yx \Big|_0^{1-y}\right)dy = \int_0^1 [1-y + 2y(1-y)]\,dy$$

$$= \int_0^1 (1+y-2y^2)\,dy = y + \frac{y^2}{2} - \frac{2y^3}{3}\Big|_0^1 = 1 + \frac{1}{2} - \frac{2}{3}$$

$$= \frac{5}{6}.$$

b) Since $\vec{F}(x,y)$ is being integrated counterclockwise over the ellipse $4x^2 + y^2 = 4$, we have the region R to the left of C with C being a closed curve. Given

$\vec{F}(x,y) = (y+3x, 2y-x)$; if we let $p = y+3x$ and $q = 2y-x$ then Green's Theorem may be applied. First calculate: $\frac{\partial p}{\partial y} = 1$ and $\frac{\partial q}{\partial x} = -1$, then substitute into (1) to obtain

$$\int_C (y+3x)\,dx + (2y-x)\,dy = \iint_R (-1-1)\,dxdy \qquad (3)$$

$$= -2 \iint_R dxdy = -2 \text{ (area of the ellipse)}.$$

If the ellipse is rewritten in the form $\dfrac{x^2}{a^2} + \dfrac{y^2}{b^2} = 1$ we could use the equation for area of the ellipse, $A = \pi ab$. Since the ellipse is given as $4x^2 + y^2 = 4$ rewrite it as $\dfrac{x^2}{1^2} + \dfrac{y^2}{2^2} = 1$ to obtain $A = \pi(1)(2) = 2\pi$. Thus

$$\int_{\text{ellipse}} \vec{F} = -2 \times 2\pi = -4\pi.$$

SURFACE INTEGRALS

• **PROBLEM** 5–15

a) Let $f(x,y)$ be a continuous function in the xy-plane. Find an integral expression for $\iint_{G(R)} f(x,y)\,dxdy$ in polar coordinates using the change of variables formula.

b) Repeat for $f(x,y,z)$ and $\iiint f(x,y,z)\,dxdydz$ represented by spherical coordinates.

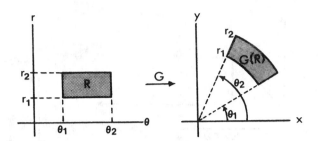

Fig. 1

Solution: a) The change of variables formula in two dimensions is:

$$\iint_{G(R)} f(x,y)\ dxdy = \iint_R f(G(u,v))|\Delta_G(u,v)|\ du\ dv\ .$$

Letting $x = r\cos\theta$ and $y = r\sin\theta$, $G(x,y) = G(r,\theta) = (r\cos\theta, r\sin\theta)$.
Therefore

$$|\Delta_G(r,\theta)| = \left|\frac{\partial(x,y)}{\partial(r,\theta)}\right| = \left\|\begin{matrix} \cos\theta & -r\sin\theta \\ \sin\theta & r\cos\theta \end{matrix}\right\| = r\cos^2\theta + r\sin^2\theta = r.$$

Hence $\displaystyle\iint_{G(R)} f(x,y)\,dx\ dy = \iint_R f(r\cos\theta, r\sin\theta)r\ dr\ d\theta$.

The mapping $G(r,\theta)$ for the rectangle $0 \le r_1 \le r \le r_2$ and
$0 \le \theta_1 \le \theta \le \theta_2 \le 2\pi$ is shown in Figure 1.

b) The change of variables formula in 3 dimensions is:

$$\iiint_{G(R)} f(x,y,z)\,dx\ dy\ dz = \iiint_R f(G(u,v,w))|\Delta_G(u,v,w)|\,du\ dv\ dw\ .$$

Letting $x = r\sin\varphi\cos\theta$, $y = r\sin\varphi\sin\theta$, and $z = r\cos\varphi$

$$G(r,\varphi,\theta) = (r\sin\varphi\cos\theta,\ r\sin\varphi\sin\theta,\ r\cos\varphi)\ .$$

Therefore,

$$|\Delta_G(r,\varphi,\theta)| = \left|\frac{\partial(x,y,z)}{\partial(r,\varphi,\theta)}\right| = \left\|\begin{matrix} \sin\varphi\cos\theta & r\cos\varphi\cos\theta & -r\sin\varphi\sin\theta \\ \sin\varphi\sin\theta & r\cos\varphi\sin\theta & r\sin\varphi\cos\theta \\ \cos\varphi & -r\sin\varphi & 0 \end{matrix}\right\|$$

$$= |\sin\varphi\cos\theta(r^2\sin^2\varphi\cos\theta) + r\cos\varphi\cos\theta(r\sin\varphi\cos\varphi\cos\theta)|$$
$$-\ r\sin\varphi\sin\theta(-r\sin^2\varphi\sin\theta - r\cos^2\varphi\sin\theta)|$$

$$= |r^2\sin\varphi(\sin^2\varphi + \cos^2\theta) + r^2\sin\varphi(\cos^2\varphi\cos^2\theta) - r\sin\varphi\sin\theta(-r\sin\theta)$$

$$= |r^2\sin\varphi(\sin^2\varphi\cos^2\theta + \cos^2\varphi\cos^2\theta + \sin^2\theta)|$$

$$= |r^2\sin\varphi((\cos^2\varphi + \sin^2\varphi)\cos^2\theta + \sin^2\theta)| = |r^2\sin\varphi| = r^2\sin\varphi \text{ since } 0 \le \varphi \le \pi.$$

Hence $\displaystyle\iiint_{G(R)} f(x,y,z)\ dx\ dy\ dz = \iiint_R f(r\sin\varphi\cos\theta, r\sin\varphi\sin\theta,$

$$r\cos\varphi)\ r^2\sin\varphi\ dr\ d\varphi\ d\theta\ .$$

The mapping $G(r,\varphi,\theta)$ for $0 \le r$, $0 \le \varphi \le \pi$, $0 \le \theta \le 2\pi$ is shown
in Figure 2.

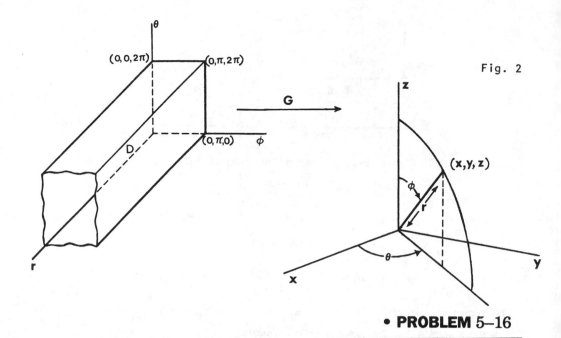

Fig. 2

• **PROBLEM** 5–16

Compute the area of the paraboloid given by the equation $z = x^2 + y^2$,
with $0 \leq z \leq 2$.

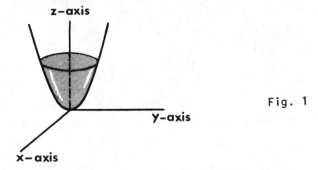

Fig. 1

Solution: Let a surface be defined by the relation $z = f(x,y)$
where (x,y) are in some region R of the plane. Let $(x_0,y_0) \in R$
and consider the curve on this surface defined by $x = x_0 + s$,
 $y = y_0$, $z = f(x_0 + s, y_0)$ for sufficiently small s . The tangent

vector is

$$\left(\frac{dx}{ds}, \frac{dy}{ds}, \frac{dz}{ds}\right) = \left(1, 0, \frac{\partial f}{\partial x}(x_0,y_0)\right)$$

at the point (x_0,y_0,z_0) . Similarly by considering the curve
 $x = x_0$, $y = y_0 + s$, $z = f(x_0,y_0 + s)$ one gets another tangent vector

 $(0, 1, \frac{\partial f}{\partial y}(x_0,y_0))$. These vectors are linearly independent and
if one considers an infinitesimal length of arcs along these two
directions and writing these vectors as

216

$$(dx, 0, \frac{\partial f}{\partial x} dx), \quad (0, dy, \frac{\partial f}{\partial y} dy)$$

then the area of the "parallelogram" spanned by these vectors is

$$\left\| (1, 0, \frac{\partial f}{\partial x}) \times (0, 1, \frac{\partial f}{\partial y}) \right\| dx\, dy = \left\| (-\frac{\partial f}{\partial x}, -\frac{\partial f}{\partial y}, 1) \right\| dx\, dy.$$

This small patch of area lies on the surface. Summing over all such patches one sees that the total area (in magnitude) of the surface is

$$A = \iint_R \sqrt{1 + \left(\frac{\partial f}{\partial x}\right)^2 + \left(\frac{\partial f}{\partial y}\right)^2}\; dx\, dy \tag{1}$$

Having motivated the discussion, one takes (1) as the definition of the area of a surface defined by $z = f(x,y)$.

In this problem the region R is the disc of radius $\sqrt{2}$. Therefore the area of the surface $= \iint_R \sqrt{1 + (2x)^2 + (2y)^2}\; dx\, dy$.

To solve this integral, use polar coordinates letting $x = r \cos\theta$, $y = r \sin\theta$ and $dx\, dy = r\, dr\, d\theta$. Then by the change of variables formula

$$\text{area} = \int_0^{2\pi} \int_0^{\sqrt{2}} \sqrt{1 + 4(r^2 \cos^2\theta + r^2 \sin^2\theta)}\; r\, dr\, d\theta$$

$$= \int_0^{2\pi} \int_0^{\sqrt{2}} \sqrt{1 + 4r^2}\; r\, dr\, d\theta$$

Substitute, letting $w = 1 + 4r^2$ and $dw = 8r\, dr$, with w ranging from 1 to 9.

$$\text{area} = \frac{1}{8} \int_0^{2\pi} \int_0^9 w^{\frac{1}{2}}\, dw\, d\theta$$

$$= \frac{1}{8} \left[\int_0^{2\pi} d\theta \int_0^9 w^{\frac{1}{2}}\, dw \right]$$

$$= \frac{2\pi}{8} \int_0^9 w^{\frac{1}{2}}\, dw$$

$$= \frac{\pi}{4} \left[\frac{2}{3} w^{3/2} \Big|_1^9 \right] = \frac{\pi}{4} \left[\frac{54}{3} - \frac{2}{3} \right] = \frac{13\pi}{3}$$

● PROBLEM 5-17

Find the integral of the function $f(x,y,z) = x$ over the surface $z = x^2 + y$ with x,y satisfying the inequalities $0 \le x \le 1$ and $-1 \le y \le 1$.

Solution: By definition, the integral of a function over a surface is given by the formula

$$\iint_S f\, dA = \iint_R f(X(t,u)) \left\| \frac{\partial X}{\partial t} \times \frac{\partial X}{\partial u} \right\| dt\, du . \tag{1}$$

In this problem notice that the surface is of the form $z = f(x,y)$. Parametrize this surface by $X(x,y) = (x,y,f(x,y))$ and find

217

$\frac{\partial X}{\partial x} = (1,0,\frac{\partial f}{\partial x})$ and $\frac{\partial X}{\partial y} = (0,1,\frac{\partial f}{\partial y})$. Thus $\frac{\partial X}{\partial x} \times \frac{\partial X}{\partial y} = (-\frac{\partial f}{\partial x}, -\frac{\partial f}{\partial y}, 1)$

and

$$\left\| \frac{\partial X}{\partial x} \times \frac{\partial X}{\partial y} \right\| = \sqrt{1 + \left(\frac{\partial f}{\partial x}\right)^2 + \left(\frac{\partial f}{\partial y}\right)^2} \ . \tag{2}$$

Since $z = x^2 + y$, $\frac{\partial f}{\partial x} = 2x$ and $\frac{\partial f}{\partial y} = 1$. Hence, substituting (2) into (1),

$$\iint_S f \ dA = \iint_R x \sqrt{1 + (2x)^2} + 1 \ dx \ dy$$

$$= \int_{-1}^1 \int_0^1 x \sqrt{2 + 4x^2} \ dx \ dy$$

$\left(\text{since} \ \ 0 \leq x \leq 1 \ \ \text{and} \ \ -1 \leq y \leq 1\right)$

$$= \int_{-1}^1 dy \int_0^1 x \sqrt{2 + 4x^2} \ dx$$

$$= 2 \int_0^1 x (4x^2 + 2)^{\frac{1}{2}} \ dx \ .$$

To evaluate this integral let $u = 2 + 4x^2$ and $du = 8x \ dx$. Therefore,

$$2 \int_0^1 x(4x^2 + 2)^{\frac{1}{2}} \ dx = \frac{2}{8} \int_0^1 (8x)(4x^2 + 2)^{\frac{1}{2}} \ dx$$

which upon the above substitution gives

$$\iint_S x \ dA = \frac{1}{4} \int_2^6 u^{\frac{1}{2}} \ du = \frac{1}{4}\left(\frac{2}{3} \ u^{3/2} \Big|_2^6\right)$$

$$= \frac{1}{6}\left(6^{3/2} - 2^{3/2}\right) \ .$$

• PROBLEM 5–18

Compute the integral of the vector field $\vec{F}(x,y,z) = (y,-x,z^2)$ over the paraboloid $z = x^2+y^2$ with $0 \leq z \leq 1$.

Solution: Given that $X(t,u)$ parametrizes a surface, and letting $\vec{n}$ represent the outward normal unit vector to the surface, the integral of a vector field over a surface is defined by

$$\iint_S \vec{F} \cdot \vec{n} \ dA = \iint_R \vec{F} \cdot \vec{n} \left\| \frac{\partial X}{\partial t} \times \frac{\partial X}{\partial u} \right\| dt \ du$$

where $\vec{F}$ is a vector field in some open set in R^3. Furthermore, using the fact that $\frac{\vec{N}}{\|\vec{N}\|} = \vec{n}$, where $\vec{N}$ is any normal in the same direction as $\vec{n}$ (unit normal) and $\|\vec{N}\|$ is its norm,

$$\vec{n} \left\| \frac{\partial X}{\partial t} \times \frac{\partial X}{\partial u} \right\| = \pm \frac{\partial X}{\partial t} \times \frac{\partial X}{\partial u} \ .$$

Thus

$$\iint_S \vec{F} \cdot \vec{n} \, dA = \iint_R \vec{F}(X(t,u)) \cdot \left[\pm \left(\frac{\partial X}{\partial t} \times \frac{\partial X}{\partial u} \right) \right] dt \, du \ . \tag{1}$$

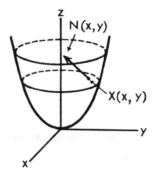

Fig. 1

In this problem the surface $z = x^2 + y^2$ with $0 \leq z \leq 1$ is given. Parametrize the surface by $X(x,y) = (x,y,x^2+y^2)$. Computing

$$\vec{N}(x,y) = \left(\frac{\partial x}{\partial x}, \frac{\partial y}{\partial x}, \frac{\partial z}{\partial x} \right) \times \left(\frac{\partial x}{\partial y}, \frac{\partial y}{\partial y}, \frac{\partial z}{\partial y} \right)$$

we find $\vec{N}(x,y) = (1,0,2x) \times (0,1,2y)$

$$= \begin{vmatrix} \hat{i} & \hat{j} & \hat{k} \\ 1 & 0 & 2x \\ 0 & 1 & 2y \end{vmatrix} = -2x\hat{i} - 2y\hat{j} + \hat{k}$$

$$= (-2x, -2y, 1) \ .$$

Note that $\vec{N}(x,y)$ points inward and hence the outward normal direction is that of $-\vec{N}$. Next compute

$$\vec{F}(X(x,y)) \cdot \vec{N}(x,y) = (y,-x,z^2) \cdot (-2x,-2y,1)$$
$$= -2xy + 2xy + z^2 = z^2$$
$$= (x^2 + y^2)^2 \ .$$

Substituting this value into (1) along with the negative sign due to the inward orientation

$$\iint_S \vec{F} \cdot \vec{n} \, dA = -\iint_R (x^2+y^2)^2 \, dx \, dy$$

where R is the disc $x^2 + y^2 \leq 1$ in the xy-plane. To evaluate this integral change to polar coordinates, letting $x = r \cos \theta$, $y = r \sin \theta$ and $dx \, dy = r \, dr \, d\theta$ to obtain

$$-\int_0^{2\pi} \int_0^1 (r^2 \cos^2 \theta + r^2 \sin^2 \theta)^2 \, r \, dr \, d\theta$$

219

$$= - 2\pi \int_0^1 r^4 r \, dr = -2\pi \int_0^1 r^5 \, dr$$

$$= - 2\pi \left. \frac{r^6}{6} \right|_0^1 = - \frac{2\pi}{6} = - \frac{\pi}{3} \, .$$

• PROBLEM 5-19

Using the divergence theorem compute the integral $\iint_S \vec{F} \cdot \vec{n} \, dA$ where $\vec{F}$ is the vector field $\vec{F}(x,y,z) = (xy, yz, x)$ and S is the boundary of the domain $x^2 + y^2 < z$, $0 < z < 1$.

Solution: The Divergence Theorem states: Let S be a closed surface bounding an open set U and let $\vec{n}$ be the outward unit normal vector to S. Then $\iint_S \vec{F} \cdot \vec{n} \, dA = \iiint_U \text{div} \, \vec{F} \, dV$ where $\vec{F}$ is a vector field on an open set containing U and S. Next, to use the divergence theorem compute the divergence of $\vec{F}$. Since $\text{div} \, \vec{F} = \frac{\partial F_1}{\partial x} + \frac{\partial F_2}{\partial y} + \frac{\partial F_3}{\partial z}$, $\text{div} \, \vec{F} = (y+z)$ for the given vector field.

Therefore,

$$\iint_S \vec{F} \cdot \vec{n} \, dA = \int_0^1 \iint_{x^2+y^2 \leq z} (y+z) dx \, dy \, dz \qquad (1)$$

To compute this triple integral first compute the inner double integral $\iint_{x^2+y^2 \leq z} (y+z) dx \, dy$. $\qquad (2)$

To do this change to polar coordinates letting $x = r \cos\theta$; $y = r \sin\theta$; and $z = z$. Thus (2) equals $\int_0^{2\pi} \int_0^{\sqrt{z}} (r \sin\theta + z) r \, dr \, d\theta$

$$= \int_0^{2\pi} \int_0^{\sqrt{z}} r^2 \sin\theta \; dr \; d\theta + z \int_0^{2\pi} \int_0^{\sqrt{z}} r \; dr \; d\theta$$

$$= \frac{(\sqrt{z})^3}{3} \int_0^{2\pi} \sin\theta \; d\theta + z \frac{(\sqrt{z})^2}{2} \int_0^{2\pi} d\theta$$

$$= 0 + \frac{z^2}{2} \cdot 2\pi = \pi z^2 \; .$$

Hence (1) equals $\pi \int_0^1 z^2 \; dz = \pi \frac{z^3}{3} \Big|_0^1 = \frac{\pi}{3} \; .$

• **PROBLEM 5–20**

Using the divergence theorem find the integral of the vector field $\vec{F} = \vec{F}(x,y,z) = (x,y,z)$ taken over the sphere of radius a.

Solution: The divergence theorem states: Let U be a region in 3-space, forming the inside of a smooth surface S. Let $\vec{F}$ be a vector field defined on an open set containing U and S. Let $\vec{n}$ be the unit outward normal vector to S. Then

$$\iint_S \vec{F} \cdot \vec{n} \; d\sigma = \iiint_U \text{div} \; \vec{F} \; dV \; .$$

First compute the divergence of $\vec{F}$ and then use the divergence theorem

$$\iint_S \vec{F} \cdot \vec{n} \; da = \iiint_B \text{div} \; \vec{F} \; dV \; .$$

Writing $\vec{F}$ as (F_1, F_2, F_3)

$$\text{div} \; \vec{F} = \frac{\partial F_1}{\partial x} + \frac{\partial F_2}{\partial y} + \frac{\partial F_3}{\partial z} \; .$$

In this case, div $\vec{F} = \frac{\partial x}{\partial x} + \frac{\partial y}{\partial y} + \frac{\partial z}{\partial z} = 3$. Therefore,

221

$$\iint_S \vec{F} \cdot \vec{n} \; da = \iiint_B 3 \; dV = 3 \iiint_B dV \; .$$

But $\iiint_B dV$ is just the volume of the sphere of radius a, which

is $4/3 \; \pi \; a^3$. So $\iint \vec{F} \cdot \vec{n} \; da = 3 \cdot 4/3 \; \pi \; a^3 = 4\pi a^3$.

● PROBLEM 5–21

Verify Stokes's Theorem for the vector field
$$\vec{F}(x,y,z) = (z-y, \; x+z, \; -(x+y))$$
and the surface bounded by the paraboloid $z = 4 - x^2 - y^2$ and the plane $z = 0$.

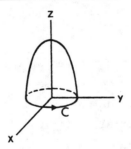

Fig. 1

Solution: Stokes's Theorem states that under routine assumptions of continuity and differentiability,

$$\iint_S (\text{curl } \vec{F}) \cdot \vec{n} \; dA = \int_C \vec{F} \cdot \vec{ds}.$$

First compute $\int_C \vec{F} \cdot \vec{ds}$ where C is the circle: $x^2 + y^2 = 4$.

Parametrizing the circle as $x = 2 \cos\theta$, $y = 2 \sin\theta$, $z = 0$.

$\vec{F} = (z-y, \; x+z, \; -(x+y)) = (-2 \sin\theta, \; 2 \cos\theta, \; -(2 \cos\theta + 2 \sin\theta))$

and
$$\vec{ds} = (dx, dy, dz) = (-2 \sin\theta \; d\theta, \; 2 \cos\theta \; d\theta, 0) \; .$$
$$\vec{F} \cdot \vec{ds} = (4 \sin^2\theta + 4 \cos^2\theta) \; d\theta = 4 \; d\theta$$

so
$$\int_C \vec{F} \cdot \vec{ds} = \int_0^{2\pi} 4 \; d\theta = 8\pi \; .$$

222

Now compute $\int_S \nabla \times \vec{F} \cdot \vec{n} \, dA$ where $(\nabla \times \vec{F})$ is the curl of f, and S is the portion of the paraboloid above the plane $z = 0$. $\nabla \times \vec{F}$ can be found by expanding the following determinant along the top row

$$\begin{vmatrix} \hat{i} & \hat{j} & \hat{k} \\ \frac{\partial}{\partial x} & \frac{\partial}{\partial y} & \frac{\partial}{\partial z} \\ z-y & x+z & -x-y \end{vmatrix} = \left(\frac{\partial}{\partial y}(-x-y) - \frac{\partial}{\partial z}(x+z), \frac{\partial}{\partial z}(z-y) - \frac{\partial}{\partial x}(-x-y), \right.$$

$$\left. \frac{\partial}{\partial x}(x+z) - \frac{\partial}{\partial y}(z-y) \right)$$

$$= (-2, 2, 2).$$

When the surface is given by $g(x,y,z) = z - f(x,y) = 0$ an expression for the normal is $\frac{\nabla g}{|\nabla g|}$. This becomes for $f = -x^2 - y^2 + 4$

$$\frac{(2x, 2y, 1)}{\| (2x, 2y, 1) \|} = \frac{1}{(4x^2 + 4y^2 + 1)^{\frac{1}{2}}} (2x, 2y, 1). \quad \text{Therefore} \int \nabla \times \vec{F} \cdot \vec{n} \, dA =$$

$$\iint_O \left(-4x + 4y + 2 \right) dx \, dy \quad \text{where } O \text{ is the disc of radius two. To}$$

evaluate this notice that

$$\iint -4x \, dx \, dy = -4 \int_0^2 \int_0^{2\pi} r \cos\theta \, r \, dr \, d\theta$$

$$= -4 \int_0^2 r^2 \, dr \int_0^{2\pi} \cos\theta \, d\theta = 0$$

since $\int_0^{2\pi} \cos\theta \, d\theta = 0$. Similarly $\iint 4y \, dx \, dy = 0$ and all that

remains is $\iint 2 \, dx \, dy = 2 \iint dx \, dy = 2 \cdot \text{area of disc}$

$$= 2 \cdot \pi(2)^2 = 8\pi$$

which agrees with $\int_C \vec{F} \cdot \vec{ds}$.

● PROBLEM 5–22

Show that the 2-form

$$\sigma = \frac{x \, dy \, dz + y \, dz \, dx + z \, dx \, dy}{(x^2 + y^2 + z^2)^{3/2}}$$

satisfies $d\sigma = 0$ but that σ is not exact. Do this by proving that $\iint_S \sigma$, where S is the unit sphere, is not zero.

Solution: For an arbitrary 2-form

$$\omega = f(x,y,z)dy\ dz + g(x,y,z)dz\ dx + h(x,y,z)dx\ dy,$$

by definition $d\omega = \left(\dfrac{\partial f}{\partial x} + \dfrac{\partial g}{\partial y} + \dfrac{\partial h}{\partial z}\right) dx\ dy\ dz$. In this case

$$f = x(x^2 + y^2 + z^2)^{-3/2}$$

$$g = y(x^2 + y^2 + z^2)^{-3/2}$$

$$h = z(x^2 + y^2 + z^2)^{-3/2}$$

$$\frac{\partial f}{\partial x} = (x^2+y^2+z^2)^{-3/2} + x(-3/2)(x^2+y^2+z^2)^{-5/2}(2x)$$

$$\frac{\partial g}{\partial y} = (x^2+y^2+z^2)^{-3/2} + y(-3/2)(x^2+y^2+z^2)^{-5/2}(2y)$$

$$\frac{\partial h}{\partial z} = (x^2+y^2+z^2)^{-3/2} + z(-3/2)(x^2+y^2+z^2)^{-5/2}(2z)\ .$$

Adding these three equations

$$\frac{\partial f}{\partial x} + \frac{\partial g}{\partial y} + \frac{\partial h}{\partial z} = 3(x^2+y^2+z^2)^{-3/2} - 3x^2(x^2+y^2+z^2)^{-5/2}$$

$$- 3y^2(x^2+y^2+z^2)^{-5/2} - 3z^2(x^2+y^2+z^2)^{-5/2}$$

$$= 3(x^2+y^2+z^2)^{-3/2} - 3(x^2+y^2+z^2)(x^2+y^2+z^2)^{-5/2}$$

$$= 0.$$

To compute $\iint_S \sigma$ parametrize:

$$x = \sin\varphi\ \cos\theta \qquad 0 \le \varphi \le \pi$$
$$y = \sin\varphi\ \sin\theta \qquad 0 \le \theta \le 2\pi$$
$$z = \cos\varphi$$

then by definition

$$dx\ dy = \frac{\partial(x,y)}{\partial(\varphi,\theta)}\ d\varphi\ d\theta$$

$$dz\ dx = \frac{\partial(z,x)}{\partial(\varphi,\theta)}\ d\varphi\ d\theta$$

and

$$dy\ dz = \frac{\partial(y,z)}{\partial(\varphi,\theta)}\ d\varphi\ d\theta\ .$$

Thus

$$dx\ dy = \begin{vmatrix} \dfrac{\partial x}{\partial \varphi} & \dfrac{\partial x}{\partial \theta} \\[2mm] \dfrac{\partial y}{\partial \varphi} & \dfrac{\partial y}{\partial \theta} \end{vmatrix} d\varphi\ d\theta$$

$$= \begin{vmatrix} \cos\varphi\ \cos\theta & -\sin\varphi\ \sin\theta \\ \cos\varphi\ \sin\theta & \sin\varphi\ \cos\theta \end{vmatrix} d\varphi\ d\theta$$

$$= (\cos^2\theta\ \cos\varphi\ \sin\varphi + \sin^2\theta\ \sin\varphi\ \cos\varphi)d\varphi\ d\theta$$

$$= \cos\varphi\ \sin\varphi\ d\varphi\ d\theta\ .$$

Similarly $dz\ dx = \sin^2\varphi\ \sin\theta\ d\varphi\ d\theta$ and $dy\ dz = \sin^2\varphi\ \cos\theta\ d\varphi\ d\theta$.

Now

$$\iint \frac{x\,dy\,dz + y\,dz\,dx + z\,dx\,dy}{(x^2 + y^2 + z^2)^{3/2}}$$

becomes, with the above substitutions,

$$\int_0^{2\pi}\int_0^{\pi}\Big[(\sin\varphi\,\cos\theta)(\sin^2\varphi\,\cos\theta) + (\sin\varphi\,\sin\theta)(\sin^2\varphi\,\sin\theta)$$
$$+ (\cos\varphi)(\cos\varphi\,\sin\varphi)\Big]d\varphi\,d\theta$$

$$= \int_0^{2\pi}\int_0^{\pi}\Big[\sin^3\varphi\,\cos^2\theta + \sin^3\varphi\,\sin^2\theta + \cos^2\varphi\,\sin\varphi\Big]d\varphi\,d\theta$$

$$= \int_0^{2\pi}\int_0^{\pi}\sin\varphi\,d\varphi\,d\theta$$

$$= 2\pi\int_0^{\pi}\sin\varphi\,d\varphi$$

$$= 2\pi\Big(-\cos\varphi\Big|_0^{\pi}\Big) = 2\pi(2) = 4\pi \neq 0 .$$

Therefore σ is not exact.

IMPROPER INTEGRALS

● **PROBLEM** 5–23

Determine if the following integrals are proper or improper. If an integral is improper, determine if it is of the first, second or third kind.

a) $\displaystyle\int_{-\infty}^{-1} \frac{dx}{x(x-1)}$ b) $\displaystyle\int_0^{\infty} \frac{e^{-at}-e^{-bt}}{t}\,dt$ c) $\displaystyle\int_0^{\frac{1}{2}} \frac{dx}{x(x-1)}$

d) $\displaystyle\int_4^8 \frac{x\,dx}{(x-3)^2}$ e) $\displaystyle\int_0^{\infty} \frac{e^{-x}}{\sqrt{x}}\,dx .$

Solution: Given that a function is bounded, and the intervals or regions of integration are bounded, if this function is integrable, the integral is then said to be a proper integral.

An integral is improper if it is an integral over an unbounded interval or region or if it is the integral of an unbounded function or both. Furthermore, the integral $\int_a^b f(x)\,dx$ is:

(i) an improper integral of the first kind if $a = -\infty$ or $b = \infty$ or both.

(ii) an improper integral of the second kind if $f(x)$ is unbounded at one or more points of $a \le x \le b$. These points are termed singularities of $f(x)$.

(iii) an improper integral of the third kind if both conditions (i) and (ii) are satisfied.

a) $\int_{-\infty}^{-1} \dfrac{dx}{x(x-1)}$ This is an improper integral of the first kind, since

one of the integration limits is infinite and the integrand is bounded.

b) $\int_{0}^{\infty} \dfrac{e^{-at} - e^{-bt}}{t}\, dt$. This is an improper integral of the first kind,

for $\lim\limits_{t \to 0} \dfrac{e^{-at} - e^{-bt}}{t} = b - a$. That is, as $t \to 0$, the integrand ap-

proaches a finite limit. Hence, by condition (i), the improper integral
is of the first kind.

c) $\int_{0}^{\frac{1}{2}} \dfrac{dx}{x(x-1)}$. This is an improper integral of the second kind since

the integral is unbounded at $x = 0$.

d) $\int_{4}^{8} \dfrac{x\, dx}{(x-3)^2}$. This is a proper integral since, even though the

integrand becomes unbounded at $x = 3$, this value for x is outside
the interval of integration $4 \le x \le 8$.

e) $\int_{0}^{\infty} \dfrac{e^{-x}}{\sqrt{x}}\, dx$. This is an improper integral of the third kind by

condition (iii) since one of the limits of integration is infinite
and the integrand is unbounded at $x = 0$.

● PROBLEM 5–24

Determine if the following improper integrals of the first kind
converge or diverge:

a) $\int_{1}^{\infty} \dfrac{1}{x^p}\, dx$ b) $\int_{0}^{\infty} e^{-rx}\, dx$ c) $\int_{0}^{\infty} \sin x\, dx$.

Solution: Let $f(x)$ be continuous in the interval $a \le x < \infty$.

Then $\int_{a}^{\infty} f(x)\, dx$ converges if and only if $\lim\limits_{R \to \infty} \int_{a}^{R} f(x)\, dx = A$

exists and is finite; in such a case one defines $\int_{a}^{\infty} f(x)\, dx = A$.

If $\lim\limits_{R \to \infty} \int_{a}^{R} f(x)\, dx$ does not exist, then $\int_{a}^{\infty} f(x)\, dx$ is said to

diverge. Applying these definitions we have:

a) For $\int_{1}^{\infty} \dfrac{1}{x^p}\, dx$, $\lim\limits_{R \to \infty} \int_{1}^{R} \dfrac{1}{x^p}\, dx = \lim\limits_{R \to \infty} \left. \dfrac{x^{1-p}}{1-p} \right|_{1}^{R}$ $(p \ne 1)$

$$= \lim\limits_{R \to \infty} \left[\dfrac{R^{1-p}}{1-p} - \dfrac{1}{1-p} \right] ,$$

which exists and equals $1/(1-p)$ for $p > 1$ and does not exist for
$p < 1$. At $p = 1$ we have

$$\lim\limits_{R \to \infty} \int_{1}^{R} \dfrac{dx}{x} = \lim\limits_{R \to \infty} \left[\ln R - \ln 1 \right]$$

which does not exist. Hence, $\int_{1}^{\infty} \dfrac{1}{x^p}\, dx$ converges if and only if

$p > 1$ and diverges if $p \leq 1$.

b) For $\int_0^\infty e^{-rx}\, dx$,

$$\lim_{R \to \infty} \int_0^R e^{-rx}\, dx = \lim_{R \to \infty} \left. \frac{-e^{-rx}}{r} \right|_0^R \qquad (r \neq 0)$$

$$= \lim_{R \to \infty} \left[\frac{-e^{-Rr}}{r} + \frac{1}{r} \right]$$

which exists and equals $1/r$ if $r > 0$ and does not exist if $r < 0$. At $r = 0$

$$\lim_{R \to \infty} \int_0^R dx = \lim_{R \to \infty} R \quad \text{which does not exist. Therefore}$$

$\int_0^\infty e^{-rx}\, dx$ converges if $r > 0$ and diverges if $r \leq 0$.

c) For $\int_0^\infty \sin x\, dx$,

$$\lim_{R \to \infty} \int_0^R \sin x\, dx = \lim_{R \to \infty} (1 - \cos R)$$

which does not exist. Thus $\int_0^\infty \sin x\, dx$ diverges.

● **PROBLEM 5–25**

Test for convergence:

a) $\int_2^\infty \dfrac{x^2\, dx}{\sqrt{x^7+1}}$ b) $\int_2^\infty \dfrac{x^3\, dx}{\sqrt{x^7+1}}$ c) $\int_2^\infty \dfrac{x^2+4x+4}{(\sqrt{x}-1)^3 \sqrt[3]{x^3-1}}\, dx$.

Solution: To test the given integrals for convergence, apply the following theorem (Comparison test):

Let $\int_a^\infty f(x)\, dx$ and $\int_b^\infty g(x)\, dx$ be two integrals of the first kind with nonnegative integrands. Suppose that $f(x) \leq g(x)$ for all values of x beyond a certain point $x = c$. Then if $\int_b^\infty g(x)\, dx$ is convergent, so is $\int_a^\infty f(x)\, dx$, and if $\int_a^\infty f(x)\, dx$ is divergent, so is $\int_b^\infty g(x)\, dx$.

a) $\int_2^\infty \dfrac{x^2\, dx}{\sqrt{x^7+1}}$. By the theorem, to prove that this integral is divergent we must find an integrand which is less than the one given whose integral we know to diverge; to prove it is convergent we must find an integrand greater than the one given whose integral we know to converge. Proceeding, for $2 \leq x < \infty$, we know that $x^4 x^3 < x^7 + 1$ so that $x^4/(x^7+1) < 1/x^3$ and, taking square roots,

227

$$0 < \frac{x^2}{\sqrt{x^7+1}} < \frac{1}{x^{3/2}} \quad . \tag{1}$$

Furthermore, $\int_2^\infty \frac{1}{x^{3/2}}$ converges since it has previously been shown

that $\int_2^\infty \frac{1}{x^p}$ converges for $p > 1$. Hence, by the comparison test and

by (1), since $\int_2^\infty \frac{1}{x^{3/2}}$ converges, $\int_2^\infty \frac{x^2 dx}{\sqrt{x^7+1}}$ converges.

b) $\int_2^\infty \frac{x^3}{\sqrt{x^7+1}}\, dx$. Since $\frac{x^3}{\sqrt{x^7+1}} = \frac{1}{\sqrt{x^7+1}/\sqrt{x^6}} = \frac{1}{\sqrt{x+x^{-6}}} = \frac{1}{\sqrt{x}\sqrt{1+x^{-7}}} \geq$

$\frac{1}{\sqrt{x}\sqrt{1+2^{-7}}}$ for $2 \leq x < \infty$, and $\frac{1}{\sqrt{1+2^{-7}}} \int_2^\infty \frac{1}{\sqrt{x}}\, dx$ diverges $\bigl($since

$\int_0^\infty \frac{1}{x^p}\, dx$ diverges for $p \leq 1\bigr)$, by the comparison test, $\int_2^\infty \frac{x^3}{\sqrt{x^7+1}}\, dx$

diverges.

c) $\int_2^\infty \frac{x^2+4x+4}{(\sqrt{x}-1)^3\sqrt{x^3-1}}\, dx$. Since $\frac{x^2+4x+4}{(\sqrt{x}-1)^3\sqrt{x^3-1}} \geq \frac{x^2+4x+4}{(\sqrt{x})^3\sqrt{x^3}} = \frac{x(x+4+4/x)}{(\sqrt{x})^3 x\sqrt{x}}$

$= \frac{x+4+4/x}{x^2} = \frac{1+4/x+4/x^2}{x} \geq \frac{1}{x}$ for $2 \leq x < \infty$ and $\int_2^\infty \frac{1}{x}\, dx$ diverges,

by the comparison test, the given integral diverges.

• PROBLEM 5–26

Test for convergence:

a) $\int_1^\infty \frac{x^2 dx}{2x^4-x+1}$ b) $\int_1^\infty \frac{x\, dx}{3x^4+6x^2+1}$.

Solution: a) To test this integral for convergence, use the following test, known as the Quotient test: Suppose $\int_a^\infty f(x)\, dx$ and $\int_b^\infty g(x)\, dx$ are integrals of the first kind with nonnegative integrands:

i) If the limit $\lim_{x\to\infty} \frac{f(x)}{g(x)} = L$ exists (finite) and is not zero,

then either both integrals are convergent or both are divergent.

ii) If $L = 0$ in (1) and $\int_b^\infty g(x)\, dx$ converges, then $\int_a^\infty f(x)\, dx$ converges.

iii) If $L = \infty$ in (i) and $\int_b^\infty g(x)\, dx$ diverges, then $\int_a^\infty f(x)\, dx$ diverges.

To apply the test, first observe that for large values of x the integrand is approximately $1/2x^2$, since $2x^4 - x + 1$ will be approximately equal to $2x^4$. Specifically, taking

228

$$f(x) = \frac{x^2}{2x^4 - x + 1}, \quad g(x) = \frac{1}{x^2}$$

we find $\lim\limits_{x \to \infty} \frac{f(x)}{g(x)} = \lim\limits_{x \to \infty} \frac{x^4}{2x^4 - x + 1} = 1/2.$

Now the integral $\int_1^\infty \frac{1}{x^2}\, dx$ is convergent since $\int_1^\infty \frac{1}{x^p}\, dx$ is

convergent for $p > 1$. Therefore, since both integrals are either convergent or divergent by (i), and because it has been shown that

$\int_1^\infty \frac{1}{x^2}\, dx$ is convergent, then both are convergent. Thus,

$$\int_1^\infty \frac{x^2\, dx}{2x^4 - x + 1} \text{ is convergent.}$$

b) $\int_1^\infty \frac{x}{3x^4 + 6x^2 + 1}\, dx.$ This integral will be tested for convergence

by two methods.

Method 1: The Comparison Test.

For large x, the integrand is approximately equal to
$$x/3x^4 = 1/3x^3.$$

Since $\frac{x}{3x^4 + 6x^2 + 1} \le \frac{1}{3x^3}$ and $\frac{1}{3}\int_1^\infty \frac{1}{x^3}\, dx$ converges (because $\int_1^\infty \frac{1}{x^p}\, dx$

converges for $p > 1$), by the comparison test $\int_1^\infty \frac{x}{3x^4 + 6x^2 + 1}\, dx$

converges.

Method 2: The Quotient Test.

Let $f(x) = \frac{x}{3x^4 + 6x^2 + 1}$, $g(x) = \frac{1}{x^3}$. Then $\lim\limits_{x \to \infty} \frac{f(x)}{g(x)} = \lim\limits_{x \to \infty} \frac{x^4}{3x^4 + 6x^2 + 1} = \frac{1}{3}.$

Now, the integral $\int_1^\infty \frac{1}{x^3}\, dx$ is convergent (p integral with $p = 3$).

Therefore, since $\lim\limits_{x \to \infty} \frac{f(x)}{g(x)} = \frac{1}{3}$ and $\int_1^\infty g(x)\, dx$ converges,

$\int_1^\infty f(x)\, dx = \int_1^\infty \frac{x}{3x^4 + 6x^2 + 1}\, dx$ converges by the quotient test.

• PROBLEM 5–27

Test the following integrals of the first kind for convergence:

a) $\displaystyle\int_0^\infty (1/t) \sin t\, dt$ (1)

b) $\displaystyle\int_0^\infty \sin u^2\, du.$ (2)

Solution: a) $\int_0^\infty (1/t) \sin t\, dt$. At first glance the integrand shows

a singularity at t = 0 so that it is an improper integral of the third kind. However, closer inspection of the integral shows this to be untrue since by L'Hospital's rule,

$$\lim_{x \to 0} \frac{\sin x}{x} = 1 \, .$$

Therefore, there is no singularity at t = 0 and the integral is of the first kind. To test this integral for convergence the following theorem is necessary: let

$$\int_a^\infty \varphi(t) f(t) dt$$

be an improper integral of the first kind. Given that the functions φ and f satisfy the following conditions:

    (i) $\varphi'(t)$ is continuous, $\varphi'(t) \le 0$ and $\lim_{t \to \infty} \varphi(t) = 0$,

    (ii) f(t) is continuous, and bounded if the integral F(x) =

$\int_a^x f(t) dt$ is bounded for all $x \ge a$, then the integral $\int_a^\infty \varphi(t) f(t) dt$ is convergent.

To apply this theorem, take $\varphi(t) = 1/t$, $f(t) = \sin t$ (from (1)).

Then $\varphi'(t) = -1/t^2 \le 0$, $\lim_{t \to \infty} 1/t = 0$ and $F(x) = \int_0^x \sin t\, dt = 1 - \cos x$,

so that $0 \le F(x) \le 2$. (F(x) is bounded for all $x \ge a$). Therefore, since the conditions of the theorem are satisfied, the given integral (1) is convergent.

b) $\int_0^\infty \sin u^2\, du$. To test this integral for convergence, apply

the theorem mentioned in part a) of this problem. However, to do this, first make the following change of variable:

$$u = \sqrt{t} \quad \text{so that} \quad du = \frac{dt}{2\sqrt{t}} \, .$$

Then

$$\int_0^x \sin u^2\, du = \frac{1}{2} \int_0^{x^2} \frac{\sin t}{\sqrt{t}}\, dt \, . \tag{3}$$

Therefore, letting $x \to \infty$ in (2), we have

$$\int_0^\infty \sin u^2\, du = \frac{1}{2} \int_0^\infty \frac{\sin t}{\sqrt{t}}\, dt \, . \tag{4}$$

(Note there is no singularity at t = 0 since $\lim_{t \to 0} \frac{\sin t}{\sqrt{t}} = 0$, by

L'Hospital's rule).

    Now to apply the theorem to the integral on the right side of equation (4) take

$$\varphi(t) = t^{-1/2}, \quad f(t) = \sin t \, .$$

Then $\varphi'(t) = -\dfrac{1}{2t^{3/2}} \le 0$ for $0 \le t < \infty$, $\lim_{t \to \infty} 1/\sqrt{t} = 0$ and

$$F(x) = \int_0^x \sin t\, dt = 1 - \cos x$$

so that $0 \leq F(x) \leq 2$. Thus, by the theorem, $\frac{1}{2} \int_0^\infty \frac{\sin t}{\sqrt{t}} dt$ is

convergent, so that by (4), the integral (2) is convergent.

● PROBLEM 5-28

Determine if the following improper integrals of the second kind are convergent:

a) $\int_{-1}^{4} \frac{dx}{(x+1)^{1/3}}$, b) $\int_{-1}^{1} \frac{dx}{x}$, c) $\int_0^1 \ln\left(\frac{1}{1-x}\right) dx$.

<u>Solution</u>: An improper integral of the second kind is an integral of the form $\int_a^b f(x)dx$ where $f(x)$ has one or more singularities in the interval $a \leq x \leq b$.

a) $\int_{-1}^{4} \frac{dx}{(x+1)^{1/3}}$. To determine if this integral is convergent, the following information is needed:

(i) if $f(x)$ has a singularity at the end point $x = a$ of the interval $a \leq x \leq b$, then by definition

$$\int_a^b f(x)dx = \lim_{\epsilon \to 0^+} \int_{a+\epsilon}^b f(x)dx \qquad (1)$$

Furthermore, if the limit on the right of (1) exists, then the integral on the left is convergent (divergent otherwise).
Similarly, if $f(x)$ has a singularity at the end point $x = b$ in the same interval, then

$$\int_a^b f(x)dx = \lim_{\epsilon \to 0^+} \int_a^{b-\epsilon} f(x)dx . \qquad (2)$$

Again, if the limit exists in (2), the integral on the left is convergent (divergent otherwise).

(ii) If $f(x)$ has a singularity at an interior point $x = x_0$ of the interval $a \leq x \leq b$, then

$$\int_a^b f(x)dx = \lim_{\epsilon_1 \to 0^+} \int_a^{x_0 - \epsilon_1} f(x)dx + \lim_{\epsilon_2 \to 0^+} \int_{x_0 + \epsilon_2}^b f(x)dx . \qquad (3)$$

As before, if the limits exist in (3), the integral on the left is convergent (otherwise it is divergent).
Note, however, that if the limits on the right of (3) do not exist, it is possible that if we choose $\epsilon_1 = \epsilon_2 = \epsilon$ in (3), the limits do exist.
If this is the case, this value is the Cauchy principal value of the integral on the left side of (3).

For the given integral, the intergrand has a singularity at $x = -1$, which is one of the end points of the interval. Hence, applying (i),

$$\int_{-1}^{4} \frac{dx}{(x+1)^{1/3}} = \lim_{\epsilon \to 0^+} \int_{-1+\epsilon}^{4} \frac{dx}{(x+1)^{1/3}} = \lim_{\epsilon \to 0^+} \frac{(x+1)^{2/3}}{2/3} \Big|_{-1+\epsilon}^{4}$$

$$= \lim_{\epsilon \to 0^+} \frac{5^{2/3}}{2/3} - \frac{3}{2} \epsilon^{2/3} = \frac{3}{2} 5^{2/3} .$$

Therefore, since the limit exists, the integral converges and equals $\left(\frac{3}{2}\right)5^{2/3}$.

b) $\int_{-1}^{1} \frac{dx}{x}$. The integrand has a singularity at $x = 0$. Hence

$$\int_{-1}^{1} \frac{dx}{x} = \lim_{\epsilon_1 \to 0^+} \int_{-1}^{-\epsilon_1} \frac{dx}{x} + \lim_{\epsilon_2 \to 0^+} \int_{\epsilon_2}^{1} \frac{dx}{x}$$

$$= \lim_{\epsilon_1 \to 0} (\ln|-\epsilon_1| - \ln|-1|) + \lim_{\epsilon_2 \to 0} (\ln(1) - \ln(\epsilon_2)) .$$

However, these limits do not exist, so the integral does not converge in the usual sense. For the Cauchy value we have

$$\lim_{\epsilon \to 0^+} \left(\int_{-1}^{-\epsilon} \frac{dx}{x} + \int_{\epsilon}^{1} \frac{dx}{x} \right) = \lim_{\epsilon \to 0^+} (\ln(-\epsilon) - \ln(-1) + \ln(1) - \ln(\epsilon))$$

$$= \lim_{\epsilon \to 0^+} (\ln(-\epsilon/\epsilon) - \ln(-1) + \ln(1))$$

$$= \lim_{\epsilon \to 0^+} (\ln(-1) - \ln(-1) + \ln(1)) = 0$$

Thus the Cauchy value of the integral is zero.

c) $\int_{0}^{1} \ln\left(\frac{1}{1-x}\right) dx$. The integrand has a singularity at $x = 1$, therefore this improper integral is of the second kind. To determine if this integral converges, first set $y = (1-x)^{-1}$ so that $dx = y^{-2}dy$ and the integral becomes

$$\int_{1}^{\infty} \frac{\ln y}{y^2} dy , \tag{1}$$

which is an improper integral of the first kind (i.e., the improper integral of the second kind was transformed into an improper integral of the first kind).

Now, by theorem, given two integrals of the first kind with positive integrands such that

$$\lim_{x \to \infty} \frac{f(x)}{g(x)} = L = 0 ,$$

then if

$$\int_{b}^{\infty} g(x)dx$$

is convergent, so is

$$\int_{a}^{\infty} f(x)dx .$$

For the given integral, let $g(y) = y^{-3/2}$. Then $\int_{b}^{\infty} g(y)dy = \int_{1}^{\infty} y^{-3/2}dy,$

which is convergent (p integral, with $p > 1$). In addition,

$$\lim_{y \to \infty} \frac{f(y)}{g(y)} = \lim_{y \to \infty} \frac{\ln y}{y^2} \cdot y^{3/2} = \lim_{y \to \infty} \frac{\ln y}{y^{1/2}}$$

$$= \lim_{y \to \infty} \frac{2}{y^{1/2}} = 0 \quad \text{by L'Hospital's rule.}$$

Thus (1) is convergent by the theorem and consequently, so is $\int_0^1 \ln\left(\frac{1}{1-x}\right)dx$, since this integral equals (1).

• PROBLEM 5-29

Test the following improper integrals of the second kind for convergence:

a) $\int_0^1 \dfrac{dx}{x^{\frac{1}{2}}(x+2x^2)^{\frac{1}{3}}}$

b) $\int_1^2 \dfrac{8-x^3}{(2x-x^2)^2} \, dx$.

Solution: For improper integrals of the second kind, there is a comparison test analogous to that for integrals of the first kind. Here we give the test for the case where $f(x)$ has a singularity at $x = a$ (the test is similar if $f(x)$ has a singularity at $x = b$ or $x = x_0$ for $a < x_0 < b$).

If $g(x) \geq 0$ for $a < x \leq b$ and if $\int_a^b g(x)dx$ converges, then if $0 \leq f(x) \leq g(x)$ for $a < x \leq b$, $\int_a^b f(x)dx$ converges. However, if $\int_a^b g(x)dx$ diverges, then if $f(x) \geq g(x)$ for $a < x \leq b$, $\int_a^b f(x)dx$ diverges.

Before applying this test, however, we make note that the basic, standard, reference integrals for improper integrals of the second kind are

$$\int_a^b \frac{dx}{(b-x)^p} \, , \quad \int_a^b \frac{dx}{(x-a)^p} \, . \tag{1}$$

These integrals are convergent for $p < 1$ divergent if $p \geq 1$, and if $p \leq 0$ they are proper integrals, since there would be no singularities in the integrand.

a) $\int_0^1 \dfrac{dx}{x^{\frac{1}{2}}(x+2x^2)^{\frac{1}{3}}}$. For this integral, $0 < \dfrac{1}{x^{\frac{1}{2}}(x+2x^2)^{\frac{1}{3}}} \leq \dfrac{1}{x^{5/6}}$ for

$x > 0$. Since the singularity is at $x = 0$, use the second integral in (1). Therefore,

$$\int_0^1 \frac{dx}{x^{5/6}}$$

is an integral of the second form in (1) with $a = 0$, $b = 1$ and $p = 5/6$; hence, it is convergent. Thus, by the comparison test,

$$\int_0^1 \frac{dx}{x^{\frac{1}{2}}(x+2x^2)^{\frac{1}{3}}}$$

is convergent.

b) $\int_1^2 \dfrac{8-x^3}{(2x-x^2)^2} dx$. Here, the integrand is nonnegative for

$1 \leq x < 2$ (the singularity is at $x = 2$). Now, using the first integral in (1) we have

$$\int_1^2 \dfrac{dx}{(2-x)}$$

is divergent (i.e., $b = 2$, $a = 1$, $p = 1$). In addition

$$\dfrac{8-x^3}{(2x-x^2)^2} = \dfrac{1}{x^2} \dfrac{(2-x)(4+2x+x^2)}{(2-x)^2} = \dfrac{(4+2x+x^2)}{x^2(2-x)} = \dfrac{\left(\dfrac{4}{x^2}\right)+\left(\dfrac{2}{x}\right)+1}{(2-x)}$$

$$\geq \dfrac{1}{(2-x)} \quad \text{for } 1 \leq x < 2.$$

Hence, by the comparison test, since $f(x) \geq g(x)$ and $\int_a^b g(x)dx$ diverges, the integral

$$\int_1^2 \dfrac{8-x^3}{(2x-x^2)^3} dx$$

diverges.

● **PROBLEM 5-30**

Determine if the following improper integrals of the second kind are convergent:

a) $\int_0^1 \dfrac{dx}{(1-x^3)^{\frac{1}{3}}}$ 　 b) $\int_0^\pi \dfrac{\sin x}{x^3} dx$ 　 c) $\int_0^1 \ln\left(\dfrac{1}{1-x}\right) dx$.

<u>Solution</u>: a) $\int_0^1 \dfrac{dx}{(1-x^3)^{\frac{1}{3}}}$. For this integral, we use a quotient test

for integrals with positive integrands analogous to that for improper integrals of the first kind. Here, assuming a singularity at $x = a$ (for singularities at $x = b$ or $x = x_0$, $a < x_0 < b$ the test is similar), if $f(x) \geq 0$ and $g(x) \geq 0$ for $a < x \leq b$ and if

$\lim\limits_{x \to a} \dfrac{f(x)}{g(x)} = L \neq 0$ or ∞, then the two integrals $\int_a^b f(x)dx$ and

$\int_a^b g(x)dx$ either both converge or both diverge. Furthermore, if

$L = 0$ and $\int_a^b g(x)dx$ converges, then $\int_a^b f(x)dx$ converges, and if

$L = \infty$ and $\int_a^b g(x)dx$ diverges then $\int_a^b f(x)dx$ diverges.

To apply this test to the given integral let $f(x) = \dfrac{1}{(1-x^3)^{\frac{1}{3}}}$ and

since the integrand has a singularity at the upper limit, let $g(x)$ be an integrand of the form $\dfrac{1}{(b-x)^P}$ because it is known that

$\int_a^b \dfrac{dx}{(b-x)^p}$ converges for $p < 1$ and diverges for $p \geq 1$. Let

$g(x) = \dfrac{1}{(1-x)^{\frac{1}{3}}}$. Now,

$$\lim_{x \to b^-} \frac{f(x)}{g(x)} = \lim_{x \to 1^-} \frac{(1-x)^{\frac{1}{3}}}{(1-x^3)^{\frac{1}{3}}} \ . \tag{1}$$

However, since

$$\frac{1}{(1-x^3)^{\frac{1}{3}}} = \frac{1}{(1-x)^{\frac{1}{3}}(1+x+x^2)^{\frac{1}{3}}} \ ,$$

(1) becomes

$$\lim_{x \to 1^-} \frac{1}{(1+x+x^2)^{\frac{1}{3}}} = (\tfrac{1}{3})^{\frac{1}{3}} \ .$$

In addition,

$$\int_0^1 \frac{dx}{(1-x)^{\frac{1}{3}}}$$

is convergent since $p = \frac{1}{3} < 1$. Thus, by the quotient test, since $L \neq 0$ or ∞, both integrals are either convergent or divergent. However, since $\int_a^b g(x)dx$ is convergent, $\int_a^b f(x)dx$ (the given integral) is also convergent.

b) $\int_0^{\pi} \dfrac{\sin x}{x^3} \, dx$. The test for convergence of this integral involves making use of the following test: If $f(x)$ has a singularity at the lower limit, let $\lim\limits_{x \to a^+} (x-a)^p f(x) = L$ (for the case of a singularity at the upper limit b, let $\lim\limits_{x \to b^-} (b-x)^p f(x) = L$); then

(i) $\int_a^b f(x) \, dx$ converges if $p < 1$ and L is finite.

(ii) $\int_a^b f(x)dx$ diverges if $p \geq 1$ and $L \neq 0$ (however, L may be infinite).

To apply this test, note that the given integral has a singularity at the lower limit, $x = 0$. Hence, choosing $p = 2$, $\lim\limits_{x \to 0^+} x^2 \dfrac{\sin x}{x^3} = \lim\limits_{x \to 0^+} \dfrac{\sin x}{x} = 1$ by L'Hospital's rule. Now, since $p > 1$, by (ii) the given integral diverges.

c) $\int_0^1 \ln\left(\frac{1}{1-x}\right)dx$. This integral was proven convergent in a previous problem by transforming it to an improper integral of the first kind. However, by the test given in part (b) of this problem, we can determine the convergence or divergence of this integral directly. To do this let $f(x) = \ln\left(\frac{1}{1-x}\right)$ and since the integral has a singularity at the upper limit, $x = 1$, let

$$\lim_{x \to b^-} (b-x)^P f(x) = \lim_{x \to 1^-} (1-x)^{\frac{1}{2}}\ln\left(\frac{1}{1-x}\right) = \lim_{x \to 1^-} \frac{\ln\left(\frac{1}{1-x}\right)}{(1-x)^{-\frac{1}{2}}} = \lim_{x \to 1^-} \frac{-\ln(1-x)}{(1-x)^{-\frac{1}{2}}} .$$

Then by L'Hospital's rule this equals

$$\lim_{x \to 1^-} \frac{1/(1-x)}{-\frac{1}{2}(1-x)^{-3/2}} = \lim_{x \to 1^-} -2(1-x)^{\frac{1}{2}} = 0 .$$

Therefore, by (i) since $p = \frac{1}{2} < 1$ and since L is finite, the given integral converges.

• PROBLEM 5–31

Determine for what values of x the integral

$$\int_0^1 e^{-t} t^{x-1} dt \qquad (1)$$

is a) proper; b) improper, but convergent.

Solution: a) The integral (1) is proper for $x \geq 1$. This results from the fact that for these values, the exponent $x-1$ is not negative, and thus t^{x-1} has no singularity at $t = 0$. However, if $x < 1$, the integral is improper, for the exponent $x-1$ would now be negative and t^{x-1} would have a singularity at $t = 0$.

b) For $x < 1$ the integral has a singularity at its lower limit, $t = 0$. To determine for what values of x this improper integral is convergent, use the following limit test:

Let $\lim_{x \to a^+} (x-a)^P f(x) = L$. Then

(i) $\int_a^b f(x)dx$ converges if $p < 1$ and L is finite.

(ii) $\int_a^b f(x)dx$ diverges if $p \geq 1$ and $L \neq 0$ (L may be infinite).

For the given problem, to apply this test consider

$$\lim_{t \to 0^+} t^P e^{-t} t^{x-1} . \qquad (2)$$

Then, at $x = 0$, (2) becomes $\lim_{t \to 0^+} \frac{t^P e^{-t}}{t} .$ \qquad (3)

Then letting $p = 1$ in (3) yields

$$\lim_{t \to 0^+} \frac{t e^{-t}}{t} = \lim_{t \to 0^+} e^{-t} = 1 .$$

Hence, by (ii) the integral diverges at $x = 0$. At $0 < x < 1$, let $p = 1-x$ so that in this interval, $p < 1$ always. (2) now becomes

$$\lim_{t \to 0^+} t^{1-x} e^{-t} t^{x-1} = \lim_{t \to 0^+} e^{-t} = 1.$$ Thus, by (i) the integral converges

for $0 < x < 1$. Finally, for $x < 0$, let $p = 1-x$ again so that in this interval $p > 1$. As before, $\lim_{t \to 0^+} t^{1-x} e^{-t} t^{x-1} = 1.$ However,

here $p > 1$, so that the integral diverges for $x < 0$ by (ii).

Thus, summarizing the results we have, the integral

$$\int_0^1 e^{-t} t^{x-1} dt$$

is 1) proper for $x \geq 1$.

2) improper but convergent for $0 < x < 1$.

3) improper but divergent for $x \leq 0$.

• PROBLEM 5–32

Determine if the following improper integrals converge absolutely:

a) $\int_1^\infty \frac{\sin x}{x^2} dx$ b) $\int_1^\infty \frac{\cos x}{x} dx$ c) $\int_0^\infty \frac{\sin x}{x} dx$.

Solution: By definition, the integral $\int_a^\infty f(x) dx$ is absolutely convergent if $\int_a^\infty |f(x)| dx$ converges. If $\int_a^\infty f(x) dx$ converges and

$\int_a^\infty |f(x)| dx$ diverges, then $\int_a^\infty f(x) dx$ is said to be conditionally convergent.

a) $\int_1^\infty \frac{\sin x}{x^2} dx$. By definition this integral is absolutely convergent

if $\int_1^\infty \left| \frac{\sin x}{x^2} \right| dx$ \hfill (1)

is convergent. Since $\frac{|\sin x|}{x^2} \leq \frac{1}{x^2}$, $\int_1^\infty \frac{|\sin x|}{x^2} dx \leq \int_1^\infty \frac{dx}{x^2}$, which

converges (p integral with $p = 2$). Hence, by the comparison test the integral (1) converges and thus the given integral is absolutely convergent.

b) $\int_1^\infty \frac{\cos x}{x} dx$. For this integral one method would be to use the

series test which states: for integrals with nonnegative integrands, the integral

$$\int_a^\infty f(x) dx$$

converges or diverges according to whether or not Σa_n, where $a_n = f(n)$, converges or diverges, respectively. However, we use another method. First, note that

$$\int_1^\infty \left|\frac{\cos x}{x}\right| dx \geq \sum_{n=1}^\infty \int_{2\pi n-\pi/3}^{2\pi n+\pi/3} \frac{\cos x}{x}\, dx \ .$$

Then, between $2\pi n-\pi/3$ and $2\pi n+\pi/3$,

$$\frac{\cos x}{x} \geq (2\pi m + \pi/3)^{-1}\, \tfrac{1}{2}\ ,$$

(since $\cos \pi/3 = \tfrac{1}{2}$ and in the given interval $\cos x \geq \tfrac{1}{2}$ and $x \leq 2\pi m + \pi/3$). Thus,

$$\int_1^\infty \left|\frac{\cos x}{x}\right| dx \geq \sum_{n=1}^\infty \frac{1}{2} \cdot \frac{1}{(2\pi n+\pi/3)} \cdot \frac{2\pi}{3} = \infty$$

i.e., the series diverges. Therefore, the integral diverges and the given integral is not absolutely convergent.

c) $\displaystyle \int_0^\infty \frac{\sin x}{x} dx = \int_0^\pi \frac{\sin x}{x} dx + \int_\pi^{2\pi} \frac{\sin x}{x} dx + \ldots + \int_{n\pi}^{(n+1)\pi} \frac{\sin x}{x} dx + \ldots$

$$= \sum_{n=0}^\infty \int_{n\pi}^{(n+1)\pi} \frac{\sin x}{x} dx \ .$$

Therefore,

$$\int_0^\infty \left|\frac{\sin x}{x}\right| dx = \sum_{n=0}^\infty \int_{n\pi}^{(n+1)\pi} \left|\frac{\sin x}{x}\right| dx \ . \tag{2}$$

Then, letting $x = z + n\pi$, (2) becomes

$$\int_0^\infty \left|\frac{\sin x}{x}\right| dx = \sum_{n=0}^\infty \left|(-1)^n\right| \left|\int_0^\pi \frac{\sin z}{z+n\pi} dz\right| = \sum_{n=0}^\infty \int_0^\pi \frac{\sin z}{z+n\pi} dz \ . \tag{3}$$

Since $\dfrac{1}{z+n\pi} \geq \dfrac{1}{(n+1)\pi}$ for $0 \leq z \leq \pi$, the integral

$$\int_0^\pi \frac{\sin z}{z+n\pi} dz \geq \frac{1}{(n+1)\pi} \int_0^\pi \sin z\, dz = \frac{1-(-1)}{(n+1)\pi}$$

(the absolute value sign can be removed since $\dfrac{\sin z}{z+n\pi}$ is positive in the interval from $z = 0$ to $z = \pi$). Hence,

$$\sum_{n=0}^\infty \int_0^\pi \frac{\sin z}{z+n\pi} dz \geq \sum_{n=0}^\infty \frac{2}{(n+1)\pi}$$

which diverges. Thus, the series $\displaystyle \sum_{n=0}^\infty \int_0^\pi \frac{\sin z}{z+n\pi} dz$ diverges, so that by (3) and the comparison test, the integral $\displaystyle \int_0^\infty \left|\frac{\sin x}{x}\right| dx$ diverges.

This shows that the given integral is not absolutely convergent. Note: the integral is convergent; therefore given this, we have shown the integral to be conditionally convergent.

Show that the integral $\int_0^\infty e^{-xy} \dfrac{\sin x}{x} dx$ (1)

converges uniformly on $[0,\infty]$ by:
  a) Dirichlet's test for uniform convergence.
  b) Abel's test.

Solution: a) Dirichlet's test for uniform convergence states: If $g(x,y)$ is continuous on $\{(x,y) | c \le x, a \le y \le b\}$ and

$$\left| \int_c^R g(x,y)dx \right| < k$$

(k a constant) for all $R \ge c$ and all y on $[a,b]$, if $f(x,y)$ is a decreasing function of x for $x \ge c$ and each fixed y on $[a,b]$, and if $f(x,y)$ approaches zero uniformly in y as $x \to \infty$, then

$$\int_c^\infty f(x,y)g(x,y)dx$$

converges uniformly on $a \le y \le b$. To apply this to the integral (1), let $g(x,y) = e^{-xy} \sin x$ and $f(x,y) = x^{-1}$. Now, $f(x,y) = x^{-1}$ is a decreasing function of x for each fixed y and for $x \ge 0$, and $\lim_{x \to \infty} x^{-1} = 0$. In addition, $g(x,y)$ is continuous on $\{(x,y) | 0 \le x, 0 \le y\}$.
Then $\left| \int_c^R g(x,y)dx \right| = \left| \int_0^R e^{-xy} \sin x\, dx \right|$.

However, using the theorem

$$\left| \int_{n\pi}^R g(x)\sin x\, dx \right| \le g(n\pi) \int_{n\pi}^{(n+1)\pi} |\sin x| dx = 2g(n\pi),$$ (2)

let $n = 0$ and $g(x) = e^{-xy}$ in (1) to obtain

$$\left| \int_0^R e^{-xy} \sin x\, dx \right| \le g(0) \int_0^\pi |\sin x| dx = 2g(0) = 2.$$

Hence,
$$\left| \int_0^R e^{-xy} \sin x\, dx \right| \le 2 \quad \text{if } R \ge 0,\ y \ge 0.$$

Thus, by Dirichlet's test, the integral (1) converges uniformly on $y \ge 0$ (i.e., $[0,\infty]$).

b) Abel's test states: If $g(x,y)$ is continuous on $[c,\infty] \times [a,b]$ and $\int_c^\infty g(x,y)dx$ converges uniformly on $[a,b]$, if $f(x,y)$ is a decreasing function of x for $x \ge c$ and each fixed y on $[a,b]$, and if $f(x,y)$ is bounded on $[c,\infty] \times [a,b]$, then

$$\int_c^\infty f(x,y)g(x,y)dx$$

converges uniformly on $[a,b]$.
To apply this test, again let
$$g(x,y) = e^{-xy} \sin x \quad \text{and} \quad f(x,y) = x^{-1}.$$

Then $f(x,y)$ is a decreasing function for $x \ge c$ and since

$\lim x^{-1} = 0$, $f(x,y)$ is bounded on
$x \to \infty$

$[0,\infty] \times [0,\infty]$. Now, all that is left to show is that $\int_0^\infty e^{-xy} \sin x \, dx$

converges uniformly on $[0,\infty]$. To do this, apply the Weierstrass M-test. Since e^{-xy} is positive for all values of x and y, we have

$|e^{-xy} \sin x| = e^{-xy} |\sin x| \le e^{-xy}$ for $y > 0$, $x > 0$. Hence, let

$M(x) = e^{-xy}$ so that

$$\int_0^\infty M(x) dx = \int_0^\infty e^{-xy} \, dx = \lim_{R \to \infty} \int_0^R e^{-xy} \, dx$$

$$= \lim_{R \to \infty} \left(\frac{-e^{-Ry}}{y} + \frac{1}{y} \right) = \frac{1}{y}$$

for $y > 0$. Hence, since $\int_0^\infty M(x) dx$ converges for $y > 0$ and

$|f(x,y)| \le M(x)$ for $y > 0$, $x > 0$, the integral $\int_0^\infty e^{-xy} \sin x \, dx$

converges uniformly on $[0,\infty]$. Hence, by Abel's theorem,

$$\int_c^\infty g(x,y) f(x,y) dx = \int_0^\infty e^{-xy} \frac{\sin x}{x} dx \quad \text{converges uniformly on}$$

$[0,\infty]$.

• **PROBLEM 5-34**

Show that

a) $\int_0^\infty x^{-1} \sin x \, dx = (1/2)\pi$ (1)

b) $\int_0^\infty \frac{e^{-ax} - e^{-bx}}{x \sec rx} dx = \frac{1}{2} \ln \frac{b^2 + r^2}{a^2 + r^2}$ (2)

where $a, b > 0$.

Solution: a) To show (1) let

$$F(u) = \int_0^\infty e^{-xu} \frac{\sin x}{x} dx \, . \quad u \ge 0 \qquad (3)$$

This converges for all $u \ge 0$ and $F(0)$ is the value to be found. Now $F(u)$ is uniformly convergent for all u with $0 \le u < \infty$. To prove this note that by definition, the integral $F(\alpha) = \int_a^\infty f(x,\alpha) dx$

is uniformly convergent in $[a,b]$ if for each $\epsilon > 0$ there is an N depending on ϵ but not on α such that $|F(\alpha) - \int_a^R f(x,\alpha) dx| < \epsilon$ for

all $R > N$ and all α in $[a,b]$. This can be rewritten by noting that

$|F(\alpha) - \int_a^R f(x,\alpha) dx| = |\int_R^\infty f(x,\alpha) dx|$. Hence, using integration by parts

we have (integrating $\sin x$ and differentiating the remaining factor):

$$\int_R^\infty f(x,u) dx = \int_R^\infty e^{-xu} x^{-1} \sin x \, dx = [-e^{-xu} x^{-1} \cos x]_R^\infty - \int_R^\infty \frac{(1+xu)e^{-xu} \cos x}{x^2} dx$$

The integrand in the second integral is always less than $1/x^2$ in absolute value, for any $u \geq 0$. Hence,

$$\left|\int_R^\infty e^{-xu}x^{-1}\sin x\, dx\right| \leq \frac{e^{-uR}|\cos R|}{R} + \int_R^\infty x^{-2}dx \leq \frac{e^{-uR}}{R} + \frac{1}{R} \leq \frac{1}{R} + \frac{1}{R}$$

$$\leq \frac{2}{R} < \epsilon \quad \text{for} \quad R > N > \frac{2}{\epsilon}.$$

Therefore

$$\left|F(u) - \int_0^R e^{-xu}\frac{\sin x}{x}dx\right| = \left|\int_R^\infty e^{-xu}\frac{\sin x}{x}\,dx\right| < \epsilon$$

so that $F(u)$ is uniformly convergent for all u with $0 \leq u < \infty$. Consequently,

$$\lim_{u\to\infty}\int_0^\infty e^{-xu}\frac{\sin x}{x}\,dx = \int_0^\infty \lim_{u\to\infty}\frac{e^{-xu}\sin x}{x}\,dx = \int_0^\infty 0\,dx = 0 \qquad (4)$$

and

$$\lim_{u\to0}\int_0^\infty e^{-xu}\frac{\sin x}{x}\,dx = \int_0^\infty \lim_{u\to0}\frac{e^{-xu}\sin x}{x}\,dx = \int_0^\infty \frac{\sin x}{x}\,dx. \qquad (5)$$

Now, differentiating (3) yields

$$F'(u) = \int_0^\infty \frac{\partial}{\partial u}\left(e^{-xu}\frac{\sin x}{x}\right)dx = -\int_0^\infty e^{-xu}\sin x\, dx \qquad (6)$$

which may be integrated by parts twice, to yield $F'(u) = -(1+u^2)^{-1}$. This is valid for all u with $u > 0$ since the integrand in (6) is dominated by e^{-xu} and the integral of this is uniformly convergent for all u with $\emptyset \leq u < \infty$ and any $\emptyset > 0$. Integrating yields $F(u) = C - \arctan u$, for all $u > 0$. Now, let u increase, so that from (4), $\lim_{u\to\infty} F(u) = 0$, and

$$0 = C - \lim_{u\to\infty} \arctan u = C - \tfrac{1}{2}\pi \quad \text{and} \quad C = \tfrac{1}{2}\pi.$$

Hence, $F(u) = \pi/2 - \arctan u$, so that $F(0) = \int_0^\infty x^{-1}\sin x\, dx = \tfrac{1}{2}\pi$.

b) Start with the integral $\int_0^\infty e^{-\varphi x}\cos rx\, dx$. $\qquad (7)$

Then, $\lim_{M\to\infty}\int_0^M e^{-\varphi x}\cos rx\, dx = \lim_{M\to\infty} e^{-\varphi x}\frac{(r \sin rx - \varphi \cos rx)}{\varphi^2 + r^2}\Big|_0^M = \frac{\varphi}{\varphi^2 + r^2}$

(integration by parts). Furthermore, this integral converges uniformly for $a \leq \varphi \leq b$, where $0 < a < b$ and any r (use the Weierstrass M-test, taking $M(x) = e^{-\varphi x}$ since

$$\left|e^{-\varphi x}\cos rx\right| \leq e^{-\varphi x} \quad \text{and} \quad \int_0^\infty e^{-\varphi x}\, dx$$

converges). Now by theorem, if $f(t,x)$ is continuous for $t \geq a$, and $c \leq x \leq d$ and if $\int_a^\infty f(t,x)dt$ is uniformly convergent for $c \leq x \leq d$, then we can integrate $F(x) = \int_a^\infty f(t,x)dt$ with respect to x from c to d to obtain

241

$$\int_c^d F(x)dx = \int_c^d \left\{ \int_a^\infty f(t,x)dt \right\} dx = \int_a^\infty \left\{ \int_c^d f(t,x)dx \right\} dt .$$

Applying this to (7) yields

$$\int_{x=0}^\infty \left\{ \int_{\varphi=a}^b e^{-\varphi x} \cos rx \, d\varphi \right\} dx = \int_{\varphi=a}^b \left\{ \int_{x=0}^\infty e^{-\varphi x} \cos rx \, dx \right\} d\varphi$$

or

$$\int_{x=0}^\infty \frac{e^{-\varphi x}\cos rx}{-x} \Big|_{\varphi=a}^b dx = \int_{\varphi=a}^b \frac{\varphi}{\varphi^2 + r^2} d\varphi ,$$

which is equivalent to

$$\int_0^\infty \frac{e^{-ax} - e^{-bx}}{x \sec rx} dx = \frac{1}{2} \ln \frac{b^2 + r^2}{a^2 + r^2} .$$

This is (2), the desired result.

• **PROBLEM** 5–35

Verify the relation

$$B(x,y) = \frac{\Gamma(x)\Gamma(y)}{\Gamma(x+y)} , \quad x,y > 0 . \tag{1}$$

where $B(x,y)$ is the beta function and $\Gamma(x)$ is the gamma function of x.

Solution: The beta function is defined by

$$B(x,y) = \int_0^1 t^{x-1}(1 - t)^{y-1} dt . \tag{2}$$

If $x \geq 1$ and $y \geq 1$, the integral (1) is proper. However, if $x > 0$ and $y > 0$ and either $x < 1$ or $y < 1$ (or both), the integral is improper, but convergent.
The gamma function is defined by

$$\Gamma(x) = \int_0^\infty t^{x-1} e^{-t} dt \tag{3}$$

This improper integral is convergent for $x > 0$.
To verify (1), begin by letting $z = t^2$ in (3), to yield

$$\Gamma(x) = \int_0^\infty z^{x-1} e^{-z} dz = 2\int_0^\infty t^{2x-1} e^{-t^2} dt.$$

Similarly, letting $v = s^2$ in (3),

$$\Gamma(y) = \int_0^\infty v^{y-1} e^{-v} dv = 2\int_0^\infty s^{2y-1} e^{-s^2} ds .$$

Consequently,

$$\Gamma(x)\Gamma(y) = 4\left[\int_0^\infty t^{2x-1} e^{-t^2} dt\right]\left[\int_0^\infty s^{2y-1} e^{-s^2} ds\right]$$

$$= 4\int_0^\infty \int_0^\infty t^{2x-1} s^{2y-1} e^{-(t^2+s^2)} dt\, ds \ . \tag{4}$$

Now transform to polar coordinates, letting $t = r \cos \theta$, $s = r \sin \theta$, so that (4) becomes

$$\Gamma(x)\Gamma(y) = 4\int_{\theta=0}^{\pi/2} \int_{r=0}^\infty (r \cos \theta)^{2x-1} (r \sin \theta)^{2y-1} e^{-r^2} r\, dr\, d\theta$$

$$= 4 \int_{\theta=0}^{\pi/2} \int_{r=0}^\infty rr^{2(x+y-1)} e^{-r^2} \cos^{2x-1}\theta \, \sin^{2y-1}\theta \, dr\, d\theta$$

$$= 4\left[\int_0^\infty rr^{2(x+y-1)} e^{-r^2} dr\right]\left[\int_0^{\pi/2}\cos^{2x-1}\theta \, \sin^{2y-1}\theta \, d\theta\right]. \tag{5}$$

Let $p = r^2$ in the first integral in (5) to give

$$= \left[2\int_0^\infty \rho^{(x+y)-1} e^{-\rho} d\rho\right]\left[\int_0^{\pi/2} \cos^{2x-1}\theta \, \sin^{2y-1}\theta \, d\theta\right]$$

or

$$\Gamma(x)\Gamma(y) = 2\Gamma(x+y) \int_0^{\pi/2}\cos^{2x-1}\theta \, \sin^{2y-1}\theta \, d\theta \ . \tag{6}$$

Now, let $t = \sin^2\theta$ in (2) to give,

$$B(x,y) = \int_0^1 t^{x-1}(1-t)^{y-1} dt = \int_0^{\pi/2}(\sin^2\theta)^{x-1}(\cos^2\theta)^{y-1} 2 \sin \theta \cos \theta \, d\theta.$$

$$B(x,y) = 2\int_0^{\pi/2}\sin^{2x-1}\theta \, \cos^{2y-1}\theta \, d\theta \ .$$

Hence, (6) becomes

$$\Gamma(x)\Gamma(y) = \Gamma(x+y) \, B(y,x) \ . \tag{7}$$

However, using the transformation $t = 1-s$ in (2) yields

$$B(x,y) = \int_0^1 t^{x-1}(1-t)^{y-1}dt = -\int_1^0 (1-s)^{x-1} s^{y-1} ds = \int_0^1 (1-s)^{x-1}s^{y-1}ds$$

$$= B(y,x).$$

Thus (7) becomes

$$\Gamma(x)\Gamma(y) = \Gamma(x+y)B(x,y)$$

or

$$B(x,y) = \frac{\Gamma(x)\Gamma(y)}{\Gamma(x+y)} \ , \quad x,y > 0 \ .$$

● **PROBLEM** 5–36

Prove

$$\int_0^{\pi/2} \sin^p\theta \, d\theta = \int_0^{\pi/2} \cos^p\theta \, d\theta \tag{1}$$

a) $= \dfrac{1\cdot 3\cdot 5\ldots(p-1)}{2\cdot 4\cdot 6\ldots p}\,\dfrac{\pi}{2}$ if p is an even positive integer.

b) $= \dfrac{2\cdot 4\cdot 6\ldots(p-1)}{1\cdot 3\cdot 5\ldots p}$ if p is an odd positive integer.

c) Evaluate $\int_0^{\pi/2} \cos^6\theta \, d\theta$.

<u>Solution</u>: Using the fact that

243

$$B(x,y) = 2\int_0^{\pi/2} \sin^{2x-1}\theta \, \cos^{2y-1}\theta \, d\theta = \frac{\Gamma(x)\Gamma(y)}{\Gamma(x+y)} \tag{2}$$

take $p = 2x-1$, $2y-1 = 0$ so that $x = \frac{1}{2}(p+1)$, $y = \frac{1}{2}$, to yield

$$\int_0^{\pi/2} \sin^p\theta \, d\theta = \frac{\Gamma[\frac{1}{2}(p+1)]\Gamma(\frac{1}{2})}{2\Gamma[\frac{1}{2}(p+2)]} \tag{3}$$

To prove that $\int_0^{\pi/2}\sin^p\theta \, d\theta = \int_0^{\pi/2}\cos^p\theta \, d\theta$, let $\theta = \pi/2 - \varphi$.

Then $\sin\theta = \sin(\frac{\pi}{2} - \varphi) = \sin\frac{\pi}{2}\cos\varphi - \cos\frac{\pi}{2}\sin\varphi = \cos\varphi$. Hence,

$$\int_0^{\pi/2}\sin^p\theta \, d\theta = -\int_{\pi/2}^0 \cos^p\varphi \, d\varphi = \int_0^{\pi/2}\cos^p\varphi \, d\varphi = \int_0^{\pi/2}\cos^p\theta \, d\theta \, .$$

a) Case 1: P is an even positive integer. Let $p = 2r$, so that (3) becomes

$$\int_0^{\pi/2}\sin^p\theta \, d\theta = \frac{\Gamma(r+\frac{1}{2})\Gamma(\frac{1}{2})}{2\Gamma(r+1)}$$

$$= \frac{(r-\frac{1}{2})(r-3/2)\ldots\frac{1}{2}\,\Gamma(\frac{1}{2})\cdot\Gamma(\frac{1}{2})}{2r(r-1)\ldots 1} \, .$$

Using the fact that $\Gamma(\frac{1}{2}) = \sqrt{\pi}$, we have

$$\int_0^{\pi/2}\sin^p\theta \, d\theta = \frac{(2r-1)(2r-3)\ldots 1}{2r(2r-2)\ldots 2}\frac{\pi}{2}$$

$$= \frac{1\cdot3\cdot5\ldots(2r-1)}{2\cdot4\cdot6\cdot8\ldots 2r}\frac{\pi}{2} = \frac{1\cdot3\cdot5\ldots(p-1)}{2\cdot4\cdot6\cdot8\ldots p}\frac{\pi}{2} \, .$$

b) Case 2: P is an odd positive integer. Let $p = 2r+1$; from (3) we have

$$\int_0^{\pi/2}\sin^p\theta \, d\theta = \frac{\Gamma(r+1)\Gamma(\frac{1}{2})}{2\Gamma(r+3/2)}$$

$$= \frac{r(r-1)\ldots 1\cdot\sqrt{\pi}}{2(r+\frac{1}{2})(r-\frac{1}{2})\ldots\frac{1}{2}\sqrt{\pi}} = \frac{2r(2r-2)\ldots 2}{(2r+1)(2r-1)\ldots 1}$$

$$= \frac{2\cdot4\cdot6\ldots 2r}{1\cdot3\cdot5\ldots(2r+1)} = \frac{2\cdot4\cdot6\ldots(p-1)}{1\cdot3\cdot5\ldots p}$$

c) $\int_0^{\pi/2}\cos^6\theta \, d\theta$. Take $2x-1 = 0$, $2y-1 = 6$ in (2), so that $x = \frac{1}{2}$ and $y = 7/2$, to yield

$$\int_0^{\pi/2}\cos^6\theta \, d\theta = \frac{\Gamma(\frac{1}{2})\Gamma(7/2)}{2\Gamma(4)} = \frac{5/2 \cdot 3/2 \cdot 1/2 \sqrt{\pi}\cdot\sqrt{\pi}}{2\cdot 3!} = \frac{5\pi}{32}$$

or by part (a) of this problem, the integral equals

$$\frac{1\cdot3\cdot5}{2\cdot4\cdot6}\frac{\pi}{2} = \frac{5\pi}{32} \, .$$

CHAPTER 6

SEQUENCES AND SERIES

> **Basic Attacks and Strategies for Solving Problems in this Chapter. See pages 246 to 359 for step-by-step solutions to problems.**

The concepts of infinite sequences and infinite series play a vital role in higher analyses and their applications. An infinite series of constants is the most fundamental form of the infinite series, and uses ideas concerning sequences of real numbers. It would be helpful to have certain standard series that converge, and others that diverge, so that they may be used for comparison to others whose convergence or divergence are to be established. Various tests have been formulated to test the convergence of different kinds of series as they compare to the standard series.

Several important, analytical questions can be answered by using the concept of uniform convergence. Once these analytical properties are presented, their applications to interesting physical problems can be considered.

Let $f(x)$ satisfy the following conditions:

1. $f(x)$ is defined in the interval $c < x < c + 2\ell$

2. $f(x)$ and $f^1(x)$ are sectionally continuous in $c < x < c + 2\ell$

3. $f(x + 2\ell) = f(x)$, i.e., $f(x)$ is periodic with period 2ℓ.

Then at every point of continuity, we have

$$f(x) = \frac{a_0}{2} + \sum_{n=1}^{\infty} \left(a_n \cos \frac{n\pi x}{\ell} + b_n \sin \frac{n\pi x}{\ell} \right) \text{ where:}$$

$$a_n = \frac{1}{\ell} \int_c^{c+2\ell} f(x) \cos \frac{n\pi x dx}{\ell}$$

$$b_n = \frac{1}{\ell} \int_c^{c+2\ell} f(x) \sin \frac{n\pi x dx}{\ell}$$

The series given for $f(x)$ is called the former series of $f(x)$. In the fourier series corresponding to an odd function, only sine terms are present. In the fourier series corresponding to an even function, only cosine terms are present.

In complex notation, the former series can be written as:

$$f(x) = \sum_{n=\infty}^{\infty} c_n e^{in\pi x/\ell} \text{ where, taking } c = -\ell, \ c_n = \frac{1}{2\ell} \int_{-\ell}^{\ell} f(x) e^{-in\pi x/\ell} dx$$

Step-by-Step Solutions to
Problems in this Chapter,
"Sequences and Series"

INFINITE SEQUENCES

• **PROBLEM** 6–1

Show that:

a)
$$\lim_{n \to \infty} \frac{n^4 + n^3 - 1}{(n^2 + 2)(n^2 - n - 1)} = 1$$

b)
$$\lim_{n \to \infty} \left(1 + \frac{C}{n^2}\right)^n = 1 \ ,$$

C a constant.

<u>Solution</u>:   By definition,

$$\lim_{n \to \infty} P_n = p$$

if for all $\epsilon > 0$, there exists N such that $n \geq N$ implies that

$$|P_n - p| < \epsilon \ .$$

a)
$$\lim_{n \to \infty} \frac{n^4 + n^3 - 1}{(n^2 + 2)(n^2 - n - 1)}$$

$$= \lim_{n \to \infty} \frac{n^4(1 + 1/n - 1/n^4)}{n^2(1 + 2/n^2)n^2(1 - 1/n - 1/n^2)}$$

$$= \lim_{n \to \infty} \frac{(1 + 1/n - 1/n^4)}{(1 + 2/n^2)(1 - 1/n - 1/n^2)}$$

$$= \frac{1 + 0 - 0}{(1 + 0)(1 - 0 - 0)} = 1$$

since

$$\lim_{n \to \infty} \frac{1}{n^p} = 0 \text{ for } p > 0 .$$

b)   We want to show that

$$\lim_{n \to \infty} \left(1 + \frac{c}{n^2}\right)^n = 1 .$$

However, from the definition of a convergent sequence, this is equivalent to showing that

$$\left|\left(1 + \frac{c}{n^2}\right)^n - 1\right| < \varepsilon$$

where $\varepsilon > 0$, and $n \geq N$.

To do this we make use of the binomial theorem,

$$(a+b)^m = \sum_{r=0}^{m} \binom{m}{r} a^{m-r} b^r ,$$

where

$$\binom{m}{r} = \frac{m!}{r!(m-r)!} .$$

Hence,

$$\left|\left(1 + \frac{c}{n^2}\right)^n - 1\right| = \left| \sum_{r=0}^{n} \binom{n}{r}\left(\frac{c}{n^2}\right)^r - 1 \right|$$

$$= \left| \sum_{r=1}^{n} \binom{n}{r}\left(\frac{c}{n^2}\right)^r \right|$$

$$\leq \sum_{r=1}^{n} \frac{n^r}{r!} \left(\frac{|c|}{n^2}\right)^r \qquad \text{since } \binom{n}{r} \leq \frac{n^r}{r!}$$

$$\leq \sum_{r=1}^{n} \left(\frac{|c|}{n}\right)^r \qquad \text{since } \frac{n^r}{r!n^r} < 1 .$$

But

$$\sum_{r=1}^{n} \left(\frac{|c|}{n}\right)^r = \frac{1 - \left(\frac{|c|}{n}\right)^{n+1}}{1 - \frac{|c|}{n}} - 1 = \frac{\frac{|c|}{n} - \left(\frac{|c|}{n}\right)^{n+1}}{1 - \frac{|c|}{n}}$$

$$= \frac{\frac{|c|}{n}\left(1 - \left(\frac{|c|}{n}\right)^n\right)}{1 - \frac{|c|}{n}}$$

$$< \frac{\frac{|c|}{n}}{1 - \frac{|c|}{n}}$$

$$\left(\text{if } n > |c| \quad \text{since} \quad 1 - \left(\frac{|c|}{n}\right)^n < 1\right).$$

Therefore,

$$\left|\left(1 + \frac{c}{n^2}\right)^n - 1\right| < \frac{\frac{|c|}{n}}{1 - \frac{|c|}{n}} = \frac{|c|}{n - |c|}$$

which $\to 0$ as $n \to \infty$.

Thus,

$$\lim_{n \to \infty} \left(1 + \frac{c}{n^2}\right)^n = 1 \;.$$

• PROBLEM 6–2

Find

$$\lim_{x \to \infty} (x\sqrt{x^2 + 1} - x^2).$$

Solution: We can find this limit by three different methods. For the first method, let

$$\lim_{x \to \infty} (x\sqrt{x^2 + 1} - x^2) = \lim_{x \to \infty} (x\sqrt{x^2 + 1} - x^2)\left(\frac{x\sqrt{x^2 + 1} + x^2}{x\sqrt{x^2 + 1} + x^2}\right)$$

$$= \lim_{x \to \infty} \frac{x^2}{x\sqrt{x^2 + 1} + x^2}$$

$$= \lim_{x \to \infty} \frac{1}{\left(\sqrt{1 + \frac{1}{x^2}} + 1\right)} = \frac{1}{2}$$

since
$$\lim_{n \to \infty} \frac{1}{n^2} = 0 \ .$$

For the second method, we use the following theorem:
Let
$$f(x),\ g(x) \in C^{n+1} \quad \text{for } a \leq x \leq b \ .$$
In addition, let $f^{(k)}(a) = g^{(k)}(a) = 0$ for $k = 0, 1, \ldots, n$ and let $g^{(n+1)}(a) \neq 0$. Then,

$$\lim_{x \to a^+} \frac{f(x)}{g(x)} = \frac{f^{(n+1)}(a)}{g^{(n+1)}(a)} \ .$$

To be able to apply this theorem to the given problem, we must replace x by $\frac{1}{y}$ and let y approach zero. Then

$$\lim_{x \to \infty} (x\sqrt{x^2 + 1} - x^2) = \lim_{y \to 0} \left(\frac{1}{y}\sqrt{\frac{1}{y^2} + 1} - \frac{1}{y^2} \right) \qquad (1)$$

$$= \lim_{y \to 0} \left(\frac{\sqrt{y^2 + 1} - 1}{y^2} \right) \ .$$

Now, let $f(y) = (y^2 + 1)^{1/2} - 1$ and $g(y) = y^2$
Then

$$f'(y) = y(y^2 + 1)^{-1/2} \ , \quad g'(y) = 2y$$

$$f''(y) = (y^2 + 1)^{-1/2} - y^2(y^2 + 1)^{-3/2} \qquad g''(y) = 2$$

Here, $n = 1$ since

$$f(0),\ g(0),\ f'(0),\ g'(0)$$

each equal zero. Hence,

$\lim_{y \to 0} \frac{f(y)}{g(y)} = \frac{f''(a)}{g''(a)}$ where a is equal to 0

hence,

$$\lim_{y \to 0} \frac{f(y)}{g(y)} = \frac{f''(0)}{g''(0)} = \frac{1}{2} \ .$$

Thus, by (1),

$$\lim_{x \to \infty} (x\sqrt{x^2 + 1} - x^2) = \frac{1}{2} \ .$$

For the third method expand the function in powers of $1/x$. Hence,

$$\lim_{x \to \infty} x\sqrt{x^2 + 1} - x^2 = \lim_{x \to \infty} x^2 \left[\left(1 + \frac{1}{x^2}\right)^{\frac{1}{2}} - 1 \right]$$

(by the binomial theorem)

$$= \lim_{x \to \infty} x^2 \left[\frac{1}{2x^2} - \frac{1}{8x^4} + \ldots \right]$$

$$= \lim_{x \to \infty} \left[\frac{1}{2} - \frac{1}{8x^2} + \ldots \right] = \frac{1}{2}$$

Since $\lim_{n \to \infty} \dfrac{1}{n^p} = 0$ for $p > 0$.

and

$$\lim_{n \to \infty} (S_{n_1} + S_{n_2} + \ldots) = S_1 + S_2 + \ldots$$

given that

$$\lim_{n \to \infty} S_{n_i} = S_i \quad .$$

• PROBLEM 6–3

Show that the following sequences are convergent:

a)
$$a_n = \frac{1 \cdot 3 \cdot 5 \, \ldots \, (2n - 1)}{2 \cdot 4 \cdot 6 \, \ldots \, (2n)} \qquad n = 1, \, 2, \, 3, \, \ldots$$

b)
$$a_n = \frac{1}{1!} + \frac{1}{2!} + \ldots + \frac{1}{n!} \qquad n = 1, \, 2, \, 3, \, \ldots$$

Solution: To solve this problem, we first must define a monotonically decreasing sequence, a monotonically increasing sequence, and a bounded sequence.

A sequence $\{a_n\}$ of real numbers is said to be:

(i) monotonically increasing if

$$a_n \leq a_{n+1} \ (n = 1, \, 2, \, 3, \, \ldots);$$

if $a_n < a_{n+1}$, it is called strictly increasing.

(ii) monotonically decreasing if

$$a_n \geq a_{n+1} \quad (n = 1, 2, 3, \ldots);$$

if $a_n > a_{n+1}$, it is called strictly decreasing.

If $a_n \leq M$ for $n = 1, 2, 3, \ldots$ where M is a constant, then the sequence $\{a_n\}$ is bounded above and M is called an upper bound. If

$$a_n \geq m \quad \text{for } n = 1, 2, 3, \ldots \quad , \text{ where } m \text{ is a}$$

constant, the sequence is bounded below and m is a lower bound. A sequence is called bounded if it is bounded both above and below (i.e., $m \leq a_n \leq M$, for $n = 1, 2, 3, \ldots$).

We note an important theorem which states that a monotonic sequence is convergent (has a limit) provided that it is bounded.

a)
$$a_n = \frac{1 \cdot 3 \cdot 5 \ldots (2n - 1)}{2 \cdot 4 \cdot 6 \ldots (2n)}$$

By substituting $n = 1, 2, 3, \ldots$ successively, we can write out a few terms of the sequence. Hence,

$$a_1 = \frac{1}{2} , \quad a_2 = \frac{1 \cdot 3}{2 \cdot 4} = \frac{3}{8} , \quad a_3 = \frac{1 \cdot 3 \cdot 5}{2 \cdot 4 \cdot 6} = \frac{5}{16} , \quad \ldots$$

observing the first few terms, we see that

$$a_2 = \frac{3}{4} a_1 , \quad a_3 = \frac{5}{6} a_2 , \quad a_4 = \frac{7}{8} a_3 .$$

Therefore, the sequence is monotonically decreasing. In general,

$$a_{n+1} = \frac{2n + 1}{2n + 2} a_n < a_n .$$

Now, since all the terms of $\{a_n\}$ are positive, we have

$0 < a_n \leq a_1 = \frac{1}{2}$. Therefore, the sequence is bounded, and since it is monotonically decreasing, by the theorem, the sequence is convergent (i.e., has a limit).

b)
$$a_n = \frac{1}{1!} + \frac{1}{2!} + \frac{1}{3!} + \ldots + \frac{1}{n!} :$$

The first few terms of $\{a_n\}$ are

$$a_1 = 1, \quad a_2 = 1 + \frac{1}{1 \cdot 2} = \frac{3}{2} , \quad a_3 = 1 + \frac{1}{1 \cdot 1} + \frac{1}{1 \cdot 2 \cdot 3} = \frac{5}{3} ,$$

etc. ...
Note that,

$$a_2 = a_1 + \frac{1}{2!} \;, \; a_3 = a_2 + \frac{1}{3!} \;, \; \ldots \;, \; a_{n+1} = a_n + \frac{1}{(n+1)!} \;, \; \ldots$$

...

Hence, $a_n < a_{n+1}$ so that the sequence is monotonically increasing. Therefore to show that the sequence is convergent, all that is needed is to show that it is bounded. Now, if

$$n > 2, \quad \frac{1}{n!} = \frac{1}{1 \cdot 2 \cdot 3 \ldots n} < \frac{1}{2^{n-1}}$$

since

$$2^{n-1} < 1 \cdot 2 \cdot 3 \ldots n.$$

Consequently for $n > 2$,

$$a_n < 1 + \frac{1}{2} + \frac{1}{2^2} + \ldots + \frac{1}{2^{n-1}} \; . \tag{1}$$

Now, using the formula for the sum of a geometric progression

$$\sum_{k=0}^{n-1} ar^k = a \left(\frac{1-r^n}{1-r} \right) \;, \text{ we have}$$

$$1 + \frac{1}{2} + \ldots + \frac{1}{2^{n-1}} = \sum_{k=0}^{n-1} 1 \cdot \left(\frac{1}{2} \right)^k = 1 \left(\frac{1 - \left(\frac{1}{2} \right)^n}{1 - \frac{1}{2}} \right)$$

$$= 2 \left[1 - \left(\frac{1}{2} \right)^n \right] < 2 \; . \tag{2}$$

Therefore, by (1) and (2), $a_n < 2$ for all n. Hence,

$$1 \le a_n < 2 \;,$$

and so, by the theorem stated in this problem, the sequence $\{a_n\}$ is convergent. The limit of the sequence plus one is conventionally denoted by e.

Note that we have shown these sequences to be convergent without determining what their limits are.

Find the limit of the sequence defined by

$$x_1 = \frac{2}{3}$$

and

$$x_{n+1} = \frac{(x_n + 1)}{(2x_n + 1)} \ .$$

Solution: We write the first four terms of the sequence

$$\left\{ \frac{2}{3} , \ \frac{5}{7} , \ \frac{12}{17} , \ \frac{29}{41} , \ \cdots \right\} \ .$$

Note that we pass from a/b to

$$\frac{\frac{a}{b} + 1}{\frac{2a}{b} + 1} = (a+b)/(2a+b)$$

from one term of the sequence to the next. To find the limit of this sequence, apply Banach's fixed point theorem, which states for the case of one variable:

Let S be a closed nonempty subset of R. Let f be a contraction mapping on S; f maps S into S such that for some k, $0 < k < 1$, and all x and y in S,

$$|f(x) - f(y)| \leq k |x - y| \ . \tag{1}$$

Then, there is one and only one point x in S for which $f(x) = x$. In addition, if $x_1 \in S$ and $x_{n+1} = f(x_n)$ for all

n, then $x_n \to x$ as $n \to \infty$.

To apply this theorem, first note that for $x \geq 0$,

$$\frac{1}{2} \leq f(x) = \frac{x + 1}{2x + 1} \leq 1$$

because

$$1 - \frac{x + 1}{2x + 1} = \frac{x}{2x + 1} \geq 0 \ .$$

Therefore, with

$$S = \left[\frac{1}{2} \ ; \ 1 \right] \ ,$$

f maps S to S and is a contraction. To prove this, note that

$$f(x) - f(y) = f'(z)(x - y)$$

where $x < z < y$, by the Mean Value Theorem. Hence,

$$|f(x) - f(y)| = |f'(z)| |x-y| \leq k |x-y|$$

if $|f'(z)| \leq k$. Therefore, to prove that f is a contraction mapping it suffices to show that $|f'(x)| \leq k < 1$ where $x \in S$. Since

$$|f'(x)| = \frac{1}{(2x+1)^2} \leq \frac{1}{4} ,$$

(1) is satisfied. Consequently, for x, $y \in S$ we have

$$|f(x) - f(y)| \leq \frac{1}{4} |x-y|$$

Therefore, by the theorem, $x_n \to x$, where

$$x = f(x) = \frac{x+1}{2x+1} \quad \text{and} \quad \frac{1}{2} \leq x \leq 1 .$$

Hence, since

$$x = \frac{x+1}{2x+1} ,$$

we have

$$2x^2 + x = x + 1$$

or

$$x^2 = \frac{1}{2} \Longrightarrow x = \left(\frac{1}{2}\right)^{\frac{1}{2}} \quad \left(\frac{1}{2} \leq x \leq 1\right) .$$

Thus,

$$\lim_{n \to \infty} x_n = x = \left(\frac{1}{2}\right)^{\frac{1}{2}} .$$

• PROBLEM 6–5

a) Find

$$\lim_{n \to \infty} \frac{(n!)^{1/n}}{n} .$$

b) Show that

$$\frac{(n+p)!}{n!} \sim n^p$$

as $n \to \infty$, $p = 1, 2, 3, \ldots$ (The symbol $\sim$ is here read as "is asymptotic to").

<u>Solution</u>: To solve this problem, we make use of Stirling's formula, which states that if n is an integer, for large n, n! is approximately equal to

$$\sqrt{2\pi n}\ \ n^n\ e^{-n}\ .$$

This can also be written as

$$\lim_{n\to\infty}\frac{(n/e)^n\sqrt{2\pi n}}{n!} = 1\ . \tag{1}$$

Now, for n large,

$$n!\ \underset{\sim}{\sim}\ \left(\frac{n}{e}\right)^n\sqrt{2\pi n}$$

so that

$$\frac{(n!)^{1/n}}{n}\ \underset{\sim}{\sim}\ \frac{(2\pi)^{1/2n}\ n^{1/2n}}{e}\ .$$

However,

$$\lim_{n\to\infty}(2\pi)^{1/2n} = 1\quad\text{and}\quad\lim_{n\to\infty}n^{1/2n} = 1\ .$$

Hence,

$$\lim_{n\to\infty}\frac{(n!)^{1/n}}{n} = \frac{1}{e}\ .$$

b) Given two sequences a_n, b_n , by definition, $a_n \sim b_n$ (a_n is asymptotic to b_n) as $n \to \infty$ if and only if

$$\lim_{n\to\infty}\frac{a_n}{b_n} = 1\ .$$

Hence, we wish to show that

$$\lim_{n\to\infty}\frac{(n+p)!}{n!\,n^p} = 1\quad\text{for } p = 1, 2, 3, \ldots$$

This can be done by two different methods. For the first method, use Stirling's Formula. However, first note that

$$\lim_{n\to\infty}\frac{(n+p)!}{n!\,n^p} = \lim_{n\to\infty}\frac{n^n\sqrt{2\pi n}}{n!\,e^n}\ \frac{(n+p)!\,e^n}{n^n n^p\sqrt{2\pi n}}$$

$$= \lim_{n\to\infty}\frac{n^n\sqrt{2\pi n}}{n!\,e^n}\ \frac{(n+p)!\,e^{n+p}}{(n+p)^{n+p}\sqrt{2\pi}}\ \cdot\ \frac{e^n\ (n+p)^{n+p}}{\sqrt{n}\ e^{n+p}\ n^{n+p}}$$

$$= \lim_{n\to\infty}\frac{n^n\sqrt{2\pi n}}{n!\,e^n}\ \frac{(n+p)!\,e^{n+p}}{(n+p)^{n+p}\sqrt{2\pi}\ \sqrt{n+p}}\ \frac{(n+p)^{n+p+\frac12}}{e^p\ n^{n+p+\frac12}}$$

$$= \lim_{n \to \infty} \frac{n^n \sqrt{2\pi n}}{n!\, e^n} \quad \frac{(n+p)!\ e^{n+p}}{(n+p)^{n+p}\ \sqrt{2\pi(n+p)}} \quad \frac{\left(1 + \frac{p}{n}\right)^{n+p+\frac{1}{2}}}{e^p} \qquad (2)$$

Now, by (1) we see that

$$\lim_{n \to \infty} \frac{n^n \sqrt{2\pi n}}{e^n\, n!} = 1 \quad \text{and} \quad \lim_{n \to \infty} \frac{(n+p)!\ e^{n+p}}{(n+p)^{n+p}\ \sqrt{2\pi(n+p)}} = 1 \ . \qquad (3)$$

Additionally,

$$\lim_{n \to \infty} \frac{(1 + p/n)^{n+p+\frac{1}{2}}}{e^p} = \frac{1}{e^p}\ \lim_{n \to \infty}\ (1 + p/n)^n (1 + p/n)^{p+\frac{1}{2}}$$

However,

$$\lim_{n \to \infty}\ (1 + p/n)^n = e^p \quad \text{and} \quad \lim_{n \to \infty}\ (1 + p/n)^{p+\frac{1}{2}} = 1 \ .$$

Therefore,

$$\lim_{n \to \infty} \frac{(1 + p/n)^{n+p+\frac{1}{2}}}{e^p} = \frac{e^p}{e^p} = 1 \ . \qquad (4)$$

Thus, by (2), (3), and (4),

$$\lim_{n \to \infty} \frac{(n+p)!}{n!\, n^p} = 1 \ ,$$

which is what we wanted to show.

For the second method, observe that

$$\frac{(n+p)!}{n!} = (n+p)(n+p-1) \ \dots \ (n+1)$$

(note that we have p terms on the right).

Therefore,

$$\lim_{n \to \infty} \frac{(n+p)!}{n!\, n^p} = \lim_{n \to \infty} \frac{(n+p)(n+p-1) \ \dots \ (n+1)}{n^p}$$

$$= \lim_{n \to \infty} \left(\frac{n+p}{n}\right) \left(\frac{n+p-1}{n}\right) \ \dots \ \left(\frac{n+1}{n}\right)$$

$$= \lim_{n \to \infty}\ (1 + p/n)\left(1 + \frac{p-1}{n}\right) \dots \left(1 + \frac{1}{n}\right) = 1$$

since each of the p factors on the right approaches
1 as $n \to \infty$.

Consequently,

$$\lim_{\to \infty} \frac{(n+p)!}{n!\, n^p} = 1 \ .$$

Thus,

$$\frac{(n+p)!}{n!} \sim n^p \text{ as } n \to \infty , \quad p = 1, 2, \ldots$$

Find:

a)

$$\lim_{n \to \infty} \left[\frac{(3n)!}{n^{3n}} \right]^{1/n} .$$

b)   for the sequence of vectors

$$\left(\frac{m}{m+1} , 2^{-m} , \left(1 + \frac{1}{m} \right)^m \right) , \quad m = 1, 2, \ldots \text{ in } R^3$$

the limit as $m \to \infty$.

Solution: a)   We want to find

$$\lim_{n \to \infty} \left[\frac{(3n)!}{n^{3n}} \right]^{1/n} .$$

To do this, first note that

$$\frac{(3n)!}{n^{3n}} = \frac{(3n)! \ 3^{3n}}{3^{3n} \ n^{3n}} = \frac{(3n)! \ 3^{3n}}{(3n)^{3n}} .$$

Now, the 3nth root of this expression is

$$\left(\left[\frac{(3n)!}{(3n)^{3n}} \right]^{1/3n} \right) 3 = \frac{[(3n)!]^{1/3n}}{3n} \cdot 3 .$$

However, in a previous problem, we have shown that

$$\lim_{k \to \infty} \frac{(k!)^{1/k}}{k} = \lim_{k \to \infty} \left(\frac{k!}{k^k} \right)^{1/k} = \frac{1}{e} .$$

Consequently, we take k=3n so that

$$\left[\frac{(3n)!}{n^{3n}} \right]^{1/3n} = \left[\frac{3^k \ k!}{k^k} \right]^{1/k} = 3 \left[\frac{k!}{k^k} \right]^{1/k} .$$

Thus,

$$\lim_{n \to \infty} \left[\frac{(3n)!}{n^{3n}} \right]^{1/3n} = \frac{3}{e} .$$

Hence, since

$$\lim_{n \to \infty} \left[\frac{(3n)!}{n^{3n}} \right]^{1/n} = \lim_{n \to \infty} \left[\frac{(3n)!}{n^{3n}} \right]^{1/3n} \left[\frac{(3n)!}{n^{3n}} \right]^{1/3n} \left[\frac{(3n)!}{n^{3n}} \right]^{1/3n} ,$$

we have

$$\lim_{n \to \infty} \left[\frac{(3n)!}{n^{3n}} \right]^{1/n} = \frac{27}{e^3} .$$

b)

$$\lim_{m \to \infty} \left(\frac{m}{m+1} , \ 2^{-m} , \ \left(1 + \frac{1}{m} \right)^m \right) , \qquad m = 1, 2, \ldots$$

By definition, a sequence of vectors in R^n is a sequence $U_1 , \ldots , U_m , \ldots$, each element of which is a vector of R^n . To determine if a sequence of vectors in R^n converges we have the following theorem:

Let $\vec{U}_m$ (m = 1, 2, ...) be a sequence of vectors in R^n . Let

$$\vec{U}_m = (U_{m1} , \ldots , U_{mn}), \ \vec{U} = (U_1 , \ldots , U_n) .$$

Then

$$\lim_{m \to \infty} \vec{U}_m = \vec{U} \quad \text{if and only if}$$

$$\lim_{m \to \infty} U_{m1} = U_1 , \ \lim_{m \to \infty} U_{m2} = U_2 , \ \ldots \ \lim_{m \to \infty} U_{mn} = U_n .$$

For the given sequence,

$$\lim_{m \to \infty} \frac{m}{m+1} = \lim_{m \to \infty} \left(\frac{1}{1 + \frac{1}{m}} \right) = 1, \ \lim_{m \to \infty} 2^{-m} = 0, \ \lim_{m \to \infty} \left(1 + \frac{1}{m} \right)^m = e .$$

Hence,

$$\lim_{m \to \infty} \left(\frac{m}{m+1} , \ 2^{-m} , \ \left(1 + \frac{1}{m} \right)^m \right) = (1, 0, e) .$$

● **PROBLEM** 6-7

Evaluate the limit

$$\lim_{x \to \infty} \frac{\sqrt{\log x} \ \log(\log x)}{e^{\sqrt{x}}} .$$

Solution: To evaluate this limit, we make use of the notion of orders of infinity. Let $f(x)$ and $g(x)$ be two functions which become positively infinite as the variable x approaches a finite limit or becomes infinite. Then by definition,

$$f(x) < g(x)$$

(read $f(x)$ is a lower order infinity than $g(x)$) if and only if

$$\lim_{x \to \infty} \frac{f(x)}{g(x)} = 0 \ .$$

Now to apply this definition, we make use of the following arrangement of infinities, written in the order of increasing strength:

$$\ldots < \log(\log x) < \log x < x < e^x < e^{e^x} . \tag{1}$$

Note that the order of infinity is increased by raising it to a power $p > 1$, and is decreased if $0 < p < 1$.

For the given limit, from (1) we see that the infinity in the denominator is the strongest of the three infinities,

$$\left(\log(\log x) < \sqrt{\log x} < e^{\sqrt{x}} \right) .$$

Now, since $x^2 < e^{\sqrt{x}}$ and $\log(\log x) < \sqrt{\log x} < e^{\sqrt{x}}$ (2)

we have,

$$\lim_{x \to \infty} \frac{\sqrt{\log x} \ \log(\log x)}{e^{\sqrt{x}}} = \lim_{x \to \infty} \frac{(\sqrt{\log x}/x)[\log(\log x)/x]}{e^{\sqrt{x}}/x^2} . \tag{3}$$

However,

$$\lim_{x \to \infty} \frac{\sqrt{\log x}}{x} = 0 \ , \quad \lim_{x \to \infty} \frac{\log(\log x)}{x} = 0$$

and

$$\lim_{x \to \infty} \frac{e^{\sqrt{x}}}{x^2} = \infty \quad \text{by (2)} \ .$$

Thus, by (3)

$$\lim_{x \to \infty} \frac{\sqrt{\log x} \ \log(\log x)}{e^{\sqrt{x}}} = \frac{0 \cdot 0}{\infty} = 0 \ .$$

Let $\{S_n\}$ be a sequence of real numbers. Define the limit superior and limit inferior of $\{S_n\}$. Then find the limit superior and limit inferior of the following sequences:

a) $\{S_n\} = 1, 0, -1, 2, 0, -2, 3, 0, -3, \ldots$

b) the sequence containing all rationals.

c)
$$\{S_n\} = \frac{1}{2}, -\frac{1}{3}, \frac{1}{4}, -\frac{1}{5}, \ldots, (-1)^{n-1}/(n+1), \ldots$$

d)
$$\{S_n\} = -\frac{1}{2}, \frac{2}{3}, -\frac{3}{4}, \frac{4}{5}, \ldots, (-1)^n/[1 + (1/n)], \ldots$$

Solution: Let G be the set of numbers x (in the extended real number system) such that $\left(S_{n_k}\right)$ converges to x for some subsequence $\left(S_{n_k}\right)$. Then this set G contains all subsequential limits (i.e., limits of all the convergent subsequences) plus possibly the numbers $+\infty$ and $-\infty$. Then, the limit superior and limit inferior of $\{S_n\}$ are respectively,

$$\lim_{n\to\infty} \text{Sup } S_n = \text{Sup } G, \quad \lim_{n\to\infty} \inf S_n = \text{Inf } G$$

(also written as $\overline{\lim} S_n$ or $\underline{\lim} S_n$, respectively).

That is, the limit superior is the least upper bound of the set of subsequential limits, and the limit inferior is the greatest lower bound of the set of subsequential limits.

In other words, $x = \text{Sup } G$ if $\beta \leq x$ for every $\beta \in G$ and if $\beta < x$, then β is not an upper bound of G. Similarly, $y = \inf G$ if $\beta \geq y$ for every $\beta \in G$ and if $\beta > y$, then β is not a lower bound of G. (Note that the terms least upper bound and greatest bound of a sequence were also defined in Chapter 1).

An alternate definition is: A number $\overline{L}$ is the limit superior of the sequence $\{S_n\}$ if infinitely many terms of the sequence are greater than $\overline{L}-\epsilon$ (where $\epsilon > 0$), while only a finite number are greater than $\overline{L} + \epsilon$. A number $\underline{L}$ is the limit inferior of the sequence $\{S_n\}$ if infinitely many terms of the sequence are less than $\underline{L} + \epsilon$ (where $\epsilon > 0$) while only finitely many are less than $\underline{L} - \epsilon$.

Note: If the sequence $\{S_n\}$ has no upper bound, then

$$\lim_{n\to\infty} \text{Sup } S_n = +\infty \; ;$$

if there is no lower bound then

$$\lim_{n\to\infty} \inf S_n = -\infty .$$

In addition, if

$$\lim_{n\to\infty} S_n = +\infty \quad \text{then} \quad \lim_{n\to\infty} \text{Sup } S_n = \lim_{n\to\infty} \inf S_n = +\infty$$

and if

$$\lim_{n\to\infty} S_n = -\infty \quad \text{then} \quad \lim_{n\to\infty} \text{Sup } S_n = \lim_{n\to\infty} \inf S_n = -\infty .$$

Furthermore, a sequence $\{S_n\}$ converges if and only if

$$\lim_{n\to\infty} \text{Sup } S_n = \lim_{n\to\infty} \inf S_n \text{ is finite.}$$

a) $\{S_n\} = 1, 0, -1, 2, 0, -2, 3, 0, -3 :$

This sequence has only one finite limit point, 0. However, it is not bounded above and it is not bounded below; therefore,

$$\lim_{n\to\infty} \text{Sup } S_n = +\infty \quad \text{and} \quad \lim_{n\to\infty} \inf S_n = -\infty .$$

b) Sequence containing all rationals: Here, every real number is a sequential limit; therefore,

$$\lim_{n\to\infty} \text{Sup } S_n = +\infty \quad \text{and} \quad \lim_{n\to\infty} \inf S_n = -\infty .$$

c) $\{S_n\} = \frac{1}{2} , -\frac{1}{3} , \frac{1}{4} , -\frac{1}{5} , \ldots , (-1)^{n-1}/(n+1) , \ldots :$

Here, the greatest lower bound is $-\frac{1}{3}$ and the least upper bound is $\frac{1}{2}$. However,

$$\lim_{n\to\infty} \frac{(-1)^{n-1}}{(n+1)} = 0,$$

therefore

$$\lim_{n\to\infty} \text{Sup } S_n = \lim_{n\to\infty} \inf S_n = 0 .$$

d) $\{S_n\} = -\frac{1}{2} , \frac{2}{3} , -\frac{3}{4} , \frac{4}{5} , \ldots , (-1)^n/[1 + (1/n)] , \ldots :$

This sequence is bounded above by 1 and bounded below by -1. Here,

$$\lim_{n\to\infty} \text{Sup } S_n = 1 \quad \text{and} \quad \lim_{n\to\infty} \inf S_n = -1$$

For any sequence $\{a_n\}$ of positive numbers, prove:

a)
$$\lim_{n\to\infty} \text{Sup } a_n^{1/n} \leq \lim_{n\to\infty} \text{Sup } \frac{a_{n+1}}{a_n}$$

b)
$$\lim_{n\to\infty} \inf \frac{a_{n+1}}{a_n} \leq \lim_{n\to\infty} \inf a_n^{1/n} .$$

Solution: a) Let $\alpha = \lim_{n\to\infty} \text{Sup } \frac{a_{n+1}}{a_n}$

If $\alpha = +\infty$, clearly
$$\lim_{n\to\infty} \text{Sup } a_n^{1/n} \leq \alpha .$$

Therefore, for α finite, choose $\lambda > \alpha$. There exists an integer N such that

$$\frac{a_{n+1}}{a_n} \leq \lambda$$

for $n \geq N$. In particular, for any $k > 0$,

$$a_{N+p+1} \leq \lambda a_{N+p} \quad (p = 0, 1, \ldots, k-1).$$

Thus,

$$a_{N+1} \leq \lambda a_N$$

$$a_{N+2} \leq \lambda a_{N+1}$$

$$\vdots$$

$$a_{N+p} \leq \lambda a_{N+p-1}$$

Multiplying these inequalities yields

$$(a_{N+1})(a_{N+2}) \cdots (a_{N+k}) \leq \lambda^k (a_N)(a_{N+1}) \cdots (a_{N+k-1})$$

which is equivalent to

$$a_{N+k} \leq \lambda^k a_N$$

or

$$a_n \leq a_N \lambda^{-N} \cdot \lambda^n \quad (n \geq N)$$

(to see this, let k = n - N).

Hence,

$$\sqrt[n]{a_n} \le \sqrt[n]{a_N \lambda^{-N}} \cdot \lambda \ .$$

However, since

$$\lim_{n\to\infty} \sqrt[n]{p} = 1 \quad \text{if} \quad p > 0, \text{ we have}$$

$$\lim_{n\to\infty} \text{Sup} \sqrt[n]{a_n} \le \lambda \tag{1}$$

(because

$$\lim_{n\to\infty} \sqrt[n]{a_N \lambda^{-N}} = 1).$$

Since this is true for all $\lambda > \alpha$, we have

$$\lim_{n\to\infty} \text{Sup} \sqrt[n]{a_n} \le \alpha \ = \ \lim_{n\to\infty} \text{Sup} \frac{a_{n+1}}{a_n} \ .$$

The inequality

$$\lim_{n\to\infty} \text{Sup} \sqrt[n]{a_n} \le \alpha$$

can be shown to be valid as follows:

Assume that

$$\lim_{n\to\infty} \text{Sup} \sqrt[n]{a_n} > \alpha \ .$$

Then

$$\lim_{n\to\infty} \text{Sup} \sqrt[n]{a_n} = \alpha + \varepsilon \ ,$$

where $\varepsilon > 0$. Now, since the inequality (1) must be true for all $\lambda > \alpha$, we can let $\lambda = \alpha + \beta$ where $\beta > 0$ and where $\beta < \varepsilon$. Then

$$\lim_{n\to\infty} \text{Sup} \sqrt[n]{a_n} = \alpha + \varepsilon > \alpha + \beta \ . \quad \text{So}$$

$$\lim_{n\to\infty} \text{Sup} \sqrt[n]{a_n} > \lambda \ .$$

However, this contradicts our previous result that

$$\lim_{n\to\infty} \text{Sup} \sqrt[n]{a_n} \le \lambda \ ,$$

Therefore,

$$\lim_{n\to\infty} \text{Sup} \sqrt[n]{a_n} \leq \alpha .$$

b) Here, let

$$\alpha = \lim_{n\to\infty} \text{inf} \frac{a_{n+1}}{a_n}$$

For $\alpha = -\infty$, there is nothing to prove. For α finite, choose, $\lambda < \alpha$. There exists an integer N such that

$$\frac{a_{n+1}}{a_n} \geq \lambda$$

for $n \geq N$. Hence, for any $k > 0$,

$$a_{N+p+1} \geq \lambda a_{N+p} \qquad (p = 0, 1, \ldots , k-1) .$$

Multiplying these inequalities yields

$$a_{N+k} \geq \lambda^k a_N$$

or

$$a_n \geq a_N \lambda^{-N} \cdot \lambda^n \qquad (n \geq N) .$$

Consequently,

$$\sqrt[n]{a_n} \geq \sqrt[n]{a_N \lambda^{-N}} \cdot \lambda$$

so that

$$\lim_{n\to\infty} \text{inf} \sqrt[n]{a_n} \geq \lambda$$

$$\lim_{n\to\infty} \sqrt[n]{a_N \lambda^{-N}} = 1 .$$

Since this inequality is true for every $\lambda < \alpha$, we have

$$\lim_{n\to\infty} \text{inf} \sqrt[n]{a_n} \geq \alpha = \lim_{n\to\infty} \text{inf} \frac{a_{n+1}}{a_n}$$

● **PROBLEM 6–10**

Show that

$$\lim_{n\to\infty} \frac{n}{(n!)^{1/n}} = e .$$

<u>Solution</u>: Since we know that:

(i)
$$\lim_{n \to \infty} \inf \frac{a_{n+1}}{a_n} \leq \lim_{n \to \infty} \inf a_n^{1/n}$$

(ii)
$$\lim_{n \to \infty} \sup a_n^{1/n} \leq \lim_{n \to \infty} \sup \frac{a_{n+1}}{a_n}$$

(iii)
$$\lim_{n \to \infty} \inf \frac{a_{n+1}}{a_n} = \lim_{n \to \infty} \sup \frac{a_{n+1}}{a_n}$$

is finite if

$$\lim_{n \to \infty} \frac{a_{n+1}}{a_n}$$

exists, we can see that given these conditions,

$$\lim_{n \to \infty} a_n^{1/n}$$

also exists and this limit equals

$$\lim_{n \to \infty} \frac{a_{n+1}}{a_n} \ .$$

For the given limit, take $a_n = \dfrac{n^n}{n!}$

so that

$$\lim_{n \to \infty} \frac{n}{(n!)^{1/n}} = \lim_{n \to \infty} a_n^{1/n} \ . \tag{1}$$

Now,

$$\lim_{n \to \infty} \frac{a_{n+1}}{a_n} = \lim_{n \to \infty} \frac{(n+1)^{n+1}}{(n+1)!} \ \frac{n!}{n^n} = \lim_{n \to \infty} \frac{(n+1)^{n+1}}{(n+1) n^n}$$

$$\lim_{n \to \infty} \left(\frac{n+1}{n} \right)^n = \lim_{n \to \infty} \left(1 + \frac{1}{n} \right)^n = e \ .$$

Therefore,

$$\lim_{n \to \infty} a_n^{1/n} = e \quad \text{and by (1)}$$

$$\lim_{n \to \infty} \frac{n}{(n!)^{1/n}} = e \ .$$

Given that

$$e = \sum_{n=0}^{\infty} \frac{1}{n!}$$

prove that

$$\lim_{n\to\infty} \left(1 + \frac{1}{n}\right)^n = e .$$

Solution: Let

$$S_n = \sum_{k=0}^{n} \frac{1}{k!}$$

and

$$t_n = \left(1 + \frac{1}{n}\right)^n .$$

By the binomial theorem $\left((a+b)^m = \sum_{r=0}^{m} \binom{m}{r} a^{m-r} b^r , \right.$

$\left. m \text{ a positive integer}\right)$

$$t_n = 1 + 1 + \frac{1}{2!}\left(1 - \frac{1}{n}\right) + \frac{1}{3!}\left(1 - \frac{1}{n}\right)\left(1 - \frac{2}{n}\right) + \cdots$$

$$+ \frac{1}{n!}\left(1 - \frac{1}{n}\right)\left(1 - \frac{2}{n}\right) \cdots \left(1 - \frac{n-1}{n}\right) .$$

Since

$$S_n = 1 + 1 + \frac{1}{2!} + \frac{1}{3!} + \cdots + \frac{1}{n!} ,$$

we have $t_n \leq S_n$. Now by a theorem, if

$$t_n \leq S_n \quad \text{for} \quad n \geq N,$$

(where N is fixed), then

$$\lim_{n\to\infty} \sup t_n \leq \lim_{n\to\infty} \sup S_n .$$

Therefore

$$\lim_{n\to\infty} \sup t_n \leq \lim_{n\to\infty} \sup S_n = e . \tag{1}$$

Now if $n \geq m$,

$$t_n \geq 1 + 1 + \frac{1}{2!}\left(1 - \frac{1}{n}\right) + \cdots + \frac{1}{m!}\left(1 - \frac{1}{n}\right) \cdots \left(1 - \frac{m-1}{n}\right) .$$

Keeping m fixed, let $n \to \infty$ to yield

$$\lim_{n \to \infty} \inf t_n \geq 1 + 1 + \frac{1}{2!} + \ldots + \frac{1}{m!} \, ,$$

so that

$$S_m \leq \lim_{n \to \infty} \inf t_n \, .$$

Letting $m \to \infty$ then gives

$$e \leq \lim_{n \to \infty} \inf t_n \, . \tag{2}$$

Hence, from (1) and (2) we have

$$e \leq \lim_{n \to \infty} \inf t_n \leq \lim_{n \to \infty} \sup t_n \leq e \, .$$

Thus

$$\lim_{n \to \infty} \left(1 + \frac{1}{n} \right)^n = e \, .$$

• PROBLEM 6–12

Find

$$\lim_{n \to \infty} f_n(x) \qquad \text{where}$$

a)

$$f_n(x) = \frac{x}{1 + nx^2} \qquad \text{for } -1 \leq x \leq 1 \, .$$

Also find,

$$\lim_{n \to \infty} f_n'(x) \, .$$

b)

$$f_n(x) = x^n \qquad \text{for } 0 \leq x \leq 1 \, .$$

Solution: For this problem we start with a definition.

Suppose $\{f_n\}$, $n = 1, 2, 3, \ldots$, is a sequence of functions defined on a set E, and suppose that the sequence of numbers $\{f_n(x)\}$ converges for every $x \in E$. Then a function f can be defined by the following:

$$f(x) = \lim_{n \to \infty} f_n(x) \qquad (x \in E) \, .$$

267

Given this, we say that $\{f_n\}$ converges on E and that f is the limit, or the limit function of $\{f_n\}$.

Though the process of finding the limit of a sequence of functions is similar to that of finding the limit of a sequence, an important problem still arises: If the functions f_n are continuous, differentiable, or integrable, will the same be true of the limit function f? In addition, is it always true that

$$f'(x) = \lim_{n \to \infty} f_n'(x)$$

or that

$$\int_a^b f(x)\,dx = \lim_{n \to \infty} \int_a^b f_n(x)\,dx \ ?$$

Furthermore, f is continuous at x means that

$$\lim_{t \to x} f(t) = f(x).$$

However, to say that the limit of a sequence of continuous functions is continuous means that

$$\lim_{t \to x} \lim_{n \to \infty} f_n(t) = \lim_{n \to \infty} \lim_{t \to x} f_n(t)$$

a)
$$f_n(x) = \frac{x}{1 + nx^2} \qquad \text{for } -1 \le x \le 1 \ .$$

Here,
$$\lim_{n \to \infty} f_n(x) = \frac{x}{1 + nx^2} = 0$$

for all x in [-1, +1]. That is, for $-1 \le x \le 1$,

$\lim\limits_{n \to \infty} f_n(x) = f(x) = 0$. Observe that

$$f_n'(x) = \frac{1-nx^2}{(1+nx^2)^2} \qquad \text{and for}$$

$x = 0$, $f_n'(x) = 1$ so that

$$\lim_{n \to \infty} f_n'(0) = f'(0) = 1 \ .$$

For $x \ne 0$,

$$\lim_{n\to\infty} f_n'(x) = \lim_{n\to\infty} \frac{1-nx^2}{(1+nx^2)^2} = \lim_{n\to\infty} \frac{1}{(1+nx^2)^2} - \lim_{n\to\infty} \frac{nx^2}{(1+nx^2)^2}$$

Now, immediately we see that

$$\lim_{n\to\infty} \frac{1}{(1+nx^2)^2} = 0 \ .$$

However,

$$\lim_{n\to\infty} \frac{nx^2}{(1+nx^2)^2} = \lim_{n\to\infty} \frac{nx^2}{1+n^2x^4+2nx^2} = \lim_{n\to\infty} \frac{x^2}{1/n + nx^4 + 2x^2} = 0 \ .$$

Therefore,

$$\lim_{n\to\infty} f_n'(x) = \begin{cases} 1 & \text{if} \quad x = 0 \\ 0 & \text{if} \quad x \neq 0 \ . \end{cases}$$

Consequently, since $f(x) = 0$, we have

$$f'(0) = 0 \neq \lim_{n\to\infty} f_n'(0) = 1 \ .$$

b)
$f_n(x) = x^n$ for $0 \leq x \leq 1$. We want to find

$$\lim_{n\to\infty} f_n(x) = \lim_{n\to\infty} x^n \quad \text{for} \quad 0 \leq x \leq 1 \ .$$

Now, for $x = 1$,

$$\lim_{n\to\infty} f_n(1) = \lim_{n\to\infty} 1^n = 1 \ .$$

For $0 \leq x < 1$,

$$\lim_{n\to\infty} x^n = 0 . \quad \text{Thus}$$

$$\lim_{n\to\infty} f_n(x) = f(x) = \begin{cases} 1 & x = 1 \\ 0 & 0 \leq x < 1 \end{cases} .$$

● **PROBLEM** 6–13

For $m = 1, 2, 3, \ldots$ let

$$f_m(x) = \lim_{n\to\infty} (\cos m! \, \pi x)^{2n} \ .$$

Find

$$\lim_{m\to\infty} f_m(x) \ .$$

269

Solution: Let
$$f(x) = \lim_{m \to \infty} f_m(x) = \lim_{m \to \infty} \lim_{n \to \infty} (\cos m! \pi x)^{2n}$$

so that the problem is to find $f(x)$ (i.e., the limit function of $\{f_m\}$). Now, when $m!x$ is an integer,

$$\cos m! \pi x = \begin{cases} 1 & m!x = 0 \\ 1 & m!x \text{ even} \\ -1 & m!x \text{ odd} \end{cases} .$$

Hence, $(\cos m! \pi x)^{2n} = 1$ for all $n \in Z$ and $f_m(x) = 1$.

For all other values of x, $|\cos m! \pi x| < 1$ so that

$$f_m(x) = \lim_{n \to \infty} (\cos m! \pi x)^{2n} = 0 .$$

This is equivalent to the following:

$$f_m(x) = \begin{cases} 1 & m!x \text{ integer} \\ 0 & m!x \text{ noninteger} \end{cases} .$$

Then for irrational x, $f_m(x) = 0$ for every m, and consequently

$$f(x) = \lim_{m \to \infty} f_m(x) = 0 .$$

For x rational, (that is, let $x = s/t$, where s and t are integers), we have
$$m!x = \frac{m!s}{t} .$$

Therefore, $m!x$ is an integer if $m \geq t$. (To see this let t be some integer that is equal to $m-r$, where r is some positive integer. Then

$$\frac{m!s}{t} = \frac{m(m-1)(m-2) \ldots (m-r) \ldots 1 \cdot s}{(m-r)}$$

$$= m(m-1) \ldots (m-(r-1)(m-(r+1)) \ldots 1 \cdot s$$

which is an integer value). Therefore,
$$f(x) = \lim_{m \to \infty} 1 = 1 .$$

Hence,
$$f(x) = \lim_{m \to \infty} \lim_{n \to \infty} (\cos m! \pi x)^{2n} = \begin{cases} 0 & (x \text{ irrational}). \\ 1 & (x \text{ rational}) \end{cases}$$

We note that this function $f(x)$ is an everywhere discontinuous function which is not Riemann-integrable.

a) Let

$$f_n(x) = n^2 x \, (1-x^2)^n \quad (0 \le x \le 1, \; n = 1, 2, 3, \ldots).$$

Show that

$$\lim_{n \to \infty} \int_0^1 f_n(x) \, dx \ne \int_0^1 \left[\lim_{n \to \infty} f_n(x) \right] dx \; .$$

b) Repeat for $f_n(x) = nx(1-x^2)^n$.

Solution: a) For $x = 0$, $f_n(0) = 0$ and for $0 < x \le 1$, we have

$$\lim_{n \to \infty} f_n(x) = \lim_{n \to \infty} n^2 x (1-x^2)^n = 0.$$

This by the theorem that if $p > 1$, α real, then

$$\lim_{n \to \infty} \frac{n^\alpha}{p^n} = 0$$

and by the fact that since $x > 0$, we have $1 - x^2 < 1$.

Therefore,

$$\lim_{n \to \infty} f_n(x) = 0 \qquad (0 \le x \le 1).$$

Now,

$$\int_0^1 x(1-x^2)^n \, dx = \frac{1}{2} \int_0^1 u^n du \qquad (\text{where } u = 1 - x^2)$$

$$= \frac{1}{2n + 2} \; .$$

Hence,

$$\lim_{n \to \infty} \int_0^1 f_n(x) \, dx = \lim_{n \to \infty} \int_0^1 n^2 x (1-x^2)^n \, dx = \lim_{n \to \infty} \frac{n^2}{2n+2} = +\infty \; .$$

However,

$$\int_0^1 \left[\lim_{n \to \infty} f_n(x) \right] dx = \int_0^1 0 \, dx = 0 \; .$$

Thus

$$\lim_{n \to \infty} \int_0^1 n^2 x (1-x^2)^n dx \neq \int_0^1 \left[\lim_{n \to \infty} n^2 x (1-x^2)^n \right] dx \ ,$$

$$(0 \leq x \leq 1) \ .$$

b)
$$f_n(x) = nx(1-x^2)^n \qquad (0 \leq x \leq 1, \ n = 1, 2, 3, \ldots) :$$

Here again $f_n(0) = 0$ and for $0 < x \leq 1$ we have

$$\lim_{n \to \infty} f_n(x) = 0 \ .$$

Hence,

$$\lim_{n \to \infty} f_n(x) = 0 \qquad (0 \leq x \leq 1)$$

as before. Now,

$$\int_0^1 f_n(x) \, dx = \int_0^1 nx(1-x^2)^n \, dx = \frac{n}{2n+2}$$

so that

$$\lim_{n \to \infty} \int_0^1 f_n(x) \, dx = \lim_{n \to \infty} \frac{n}{2n+2} = \frac{1}{2}$$

whereas

$$\int_0^1 \left[\lim_{n \to \infty} f_n(x) \right] dx = 0 \ .$$

Thus these two problems show that the limit of the integral need not be equal to the integral of the limit, even if both have finite values as in part (b).

● **PROBLEM 6–15**

a) Show that $f_n(x) = \dfrac{x^n}{n}$ converges uniformly on the interval $0 \leq x \leq 1$. Then show that

$$\left(\text{for } f_n(x) = \frac{x^n}{n} \right) \ ,$$

$$\lim_{n \to \infty} \int_0^1 f_n(x) \, dx = \int_0^1 \lim_{n \to \infty} f_n(x) \, dx$$

272

and that for $0 \leq x < 1$

$$\lim_{n \to \infty} f'_n(x) = f'(x) .$$

b) Show that

$$f_n(x) = \frac{x}{2(n+1)}$$

converges uniformly on the interval $0 \leq x \leq 1$.

c) Show that $f_n(x) = x^n$ does not converge uniformly on the interval $0 < x < 1$.

Solution: a)   For $0 \leq x \leq 1$,

$$\lim_{n \to \infty} f_n(x) = \lim_{n \to \infty} \frac{x^n}{n} = 0 ,$$

so that $f_n(x)$ converges to 0. We wish to show that this convergence is uniform on this interval. That is, for each $\varepsilon > 0$ there corresponds some integer N such that for every x in the interval,

$$|f_n(x) - f(x)| < \varepsilon$$

if $n \geq N$. Of importance is that N is to be independent of x for convergence to be uniform.

Here,

$$|f_n(x) - f(x)| = \left|\frac{x^n}{n}\right| \leq \left|\frac{1}{n}\right|$$

since $x^n \leq 1$.

Then

$$\left|\frac{x^n}{n}\right| < \varepsilon \quad \text{if} \quad \frac{1}{\varepsilon} < n .$$

Therefore, N can be chosen as the smallest integer greater than $\frac{1}{\varepsilon}$. Note that this choice is independent of x. Thus $f_n(x)$ converges uniformly to 0 in this interval.

Now

$$\lim_{n \to \infty} \int_0^1 f_n(x)\,dx = \lim_{n \to \infty} \int_0^1 \frac{x^n}{n}\,dx = \lim_{n \to \infty} \frac{1}{n(n+1)} = 0 .$$

Hence

$$\lim_{n \to \infty} \int_0^1 f_n(x)\,dx = \int_0^1 \lim_{n \to \infty} f_n(x)\,dx = \int_0^1 0\,dx = 0 .$$

In addition,

$$\lim_{n \to \infty} f_n'(x) = \lim_{n \to \infty} x^{n-1} = 0 \quad \text{since } 0 \le x < 1 .$$

Therefore,

$$\lim_{n \to \infty} f_n'(x) = f'(x) = 0 .$$

b) We have that

$$\lim_{n \to \infty} f_n(x) = \lim_{n \to \infty} \frac{x}{2(n+1)} = 0 .$$

Therefore, since $0 \le x \le 1$,

$$\left| f_n(x) - f(x) \right| = \left| \frac{x}{2n+2} \right| \le \left| \frac{1}{2n+2} \right| < \left| \frac{1}{n} \right| < \varepsilon$$

if $n > \frac{1}{\varepsilon}$.

Hence, choose N as the smallest integer greater than $\frac{1}{\varepsilon}$. Again this choice is independent of x.

c) Here,

$$\lim_{n \to \infty} f_n(x) = \lim_{n \to \infty} x^n = 0 \qquad \text{if } 0 \le x < 1$$

and
$$\lim_{n \to \infty} x^n = 1 \qquad \text{if } x = 1.$$

Suppose that $0 < \varepsilon < 1$ and $0 < x < 1$. Then

$$\left| f_n(x) - f(x) \right| < \varepsilon$$

is the same as $x^n < \varepsilon$.

Therefore,

$$n \log x < \log \varepsilon$$

which can be rewritten

$$\log\left(\frac{1}{\varepsilon} \right) < n \log\left(\frac{1}{x} \right)$$

since $\log x < 0$ and $\log\left(\frac{1}{x} \right) > 0$ if $0 < x < 1$.

Then

$$\frac{\log\left(\frac{1}{\varepsilon}\right)}{\log\left(\frac{1}{x}\right)} < n \quad .$$

Now to have this true for all $n \geq N$ it is necessary to choose the integer N large enough so that

$$N > \frac{\log\left(\frac{1}{\varepsilon}\right)}{\log\left(\frac{1}{x}\right)} \quad .$$

Here, we see that N depends both on ε and x. That is, as $\varepsilon \to 0$,

$$\log\left(\frac{1}{\varepsilon}\right) \to +\infty \quad \text{so that } N \to \infty \quad .$$

If ε is remained fixed, as $x \to 1^-$, $\log\left(\frac{1}{x}\right) \to 0^+$

and consequently $N \to \infty$. Thus, there is no value of N such that the given inequality holds simultaneously for all values of x in the interval $0 < x < 1$.

INFINITE SERIES

• **PROBLEM** 6–16

Test the following series for convergence:

a)

$$\sum_{n=2}^{\infty} \frac{1}{n(\log n)^2}$$

b)

$$\sum_{n=2}^{\infty} \frac{1}{n(\log n)}$$

c)

$$\sum_{n=4}^{\infty} \frac{1}{n(\log n)[\log(\log n)]} \quad .$$

<u>Solution</u>: a) To test this series for convergence, apply the integral test. Since the series is

$$\sum_{n=2}^{\infty} \frac{1}{n(\log n)^2} \quad ,$$

set up the integral

$$\int_2^b \frac{dx}{x(\log x)^2} \quad .$$

Then, if the limit of the integral as $b \to \infty$ exists, it follows that the series converges. If the limit of the integral as $b \to \infty$, goes to infinity, then the series diverges. Therefore, to solve

$$\int_2^b \frac{dx}{x(\log x)^2}$$

let $u = \log x$, $du = \frac{dx}{x}$. This yields

$$\int_{\log 2}^{\log b} \frac{du}{u^2} = -u^{-1} \Big|_{\log 2}^{\log b} = \frac{1}{\log 2} - \frac{1}{\log b}$$

Hence

$$\lim_{b \to \infty} \int_2^b \frac{dx}{x(\log x)} = \lim_{b \to \infty} \left(\frac{1}{\log 2} - \frac{1}{\log b} \right) = \frac{1}{\log 2}$$

and thus the series converges.

b)   To test the series

$$\sum_{n=2}^{\infty} \frac{1}{n(\log n)} \quad ,$$

again use the integral test. Therefore set up the integral

$$\int_2^b \frac{dx}{x(\log x)} \, dx \quad .$$

To solve this integral let $u = \log x$, $du = \frac{dx}{x}$.
This yields

276

$$\int_{\log 2}^{\log b} \frac{du}{u} = \log u \Bigg|_{\log 2}^{\log b} = \log (\log b) - \log (\log 2) \ .$$

Hence

$$\lim_{b \to \infty} \int_{2}^{b} \frac{dx}{x (\log x)} = \lim_{b \to \infty} \Big(\log (\log b) - \log (\log 2) \Big)$$

which is unbounded and thus the series diverges.

c)   To test the series,

$$\sum_{n=4}^{\infty} \frac{1}{n \log n \, [\log (\log n)]^2}$$

also use the integral test. Therefore to solve

$$\int_{4}^{b} \frac{dx}{x (\log x)[\log (\log x)]^2} \ , \text{ let } u = \log (\log x), \ du = \frac{dx}{x \log x} \cdot$$

Since $\displaystyle\int \frac{du}{u^2} = -u^{-1}$, this yields

$$- \frac{1}{\log (\log x)} \Bigg|_{4}^{b} = - \frac{1}{\log (\log b)} + \frac{1}{\log (\log 4)} \ .$$

Hence

$$\lim_{b \to \infty} \int_{4}^{b} \frac{dx}{x \log x \, [\log (\log x)]^2}$$

$$= \lim_{b \to \infty} \left(\frac{-1}{\log (\log b)} + \frac{1}{\log (\log 4)} \right)$$

$$= \frac{1}{\log (\log 4)} \ ,$$

and the series converges.

Determine if the series

$$\frac{1}{2} + \frac{1}{3} + \frac{1}{2^2} + \frac{1}{3^2} + \frac{1}{2^3} + \frac{1}{3^3} + \ldots$$

is convergent or divergent.

Solution: The series can be rewritten as the sum of the sequence of numbers given by

$$a_n = \begin{cases} \dfrac{1}{2^{(n+1)/2}} & \text{if } n \text{ is odd } (n > 0) \\[3mm] \dfrac{1}{3^{n/2}} & \text{if } n \text{ is even } (n > 0) \end{cases}$$

Now the ratio test states: If $a_k > 0$ and $\lim\limits_{k \to \infty} \dfrac{a_{k+1}}{a_k} = \ell < 1$,

then

$$\sum_{k=1}^{\infty} a_k$$

converges. Similarly, if

$$\lim_{k \to \infty} \frac{a_{k+1}}{a_k} = \ell \ (1 < \ell \leq \infty) \qquad \text{then}$$

$$\sum_{k=1}^{\infty} a_k \quad \text{diverges.}$$

If $\ell = 1$, the test fails. Therefore applying this test gives:

If a_n is odd,

$$\lim_{n \to \infty} \frac{a_{n+1}}{a_n}$$

$$= \lim_{n \to \infty} \frac{\dfrac{1}{3^{n/2}}}{\dfrac{1}{2^{(n+1)/2}}} = \lim_{n \to \infty} \frac{2^{(n+1)/2}}{3^{n/2}} = \lim_{n \to \infty} \left(\frac{2}{3}\right)^{n/2} 2^{\frac{1}{2}} = 0 \ .$$

If a_n is even,

$$\lim_{n\to\infty} \frac{a_{n+1}}{a_n} = \lim_{n\to\infty} \frac{\dfrac{1}{2^{(n+1)/2}}}{\dfrac{1}{3^{n/2}}} = \lim_{n\to\infty} \left(\frac{3}{2}\right)^{n/2} 2^{\frac{1}{2}}$$

and no limit exists.

Hence, the ratio test gives two different values, one < 1 and the other > 1, therefore the test fails to determine if the series is convergent. Thus, another test, known as the root test is now applied. This test states: Let

$$\sum_{k=1}^{\infty} a_k$$

be a series of nonnegative terms, and let

$$\lim_{n\to\infty} \left(\sqrt[n]{a_n}\right) = S, \text{ where } 0 \leq S \leq \infty \quad . \text{ If:}$$

1) $0 \leq S < 1$,   the series converges

2) $1 < S \leq \infty$,   the series diverges

3) $S = 1$, the series may converge or diverge.

Applying this test yields, if a_n is odd

$$\lim_{n\to\infty} \sqrt[n]{a_n} = \lim_{n\to\infty} \sqrt[n]{\frac{1}{2^{(n+1)/2}}} = \lim_{n\to\infty} \sqrt[n]{\frac{1}{2^{n/2}}} \sqrt[n]{\frac{1}{2^{\frac{1}{2}}}}$$

$$= \lim_{n\to\infty} \frac{1}{\sqrt{2}} \frac{1}{2^{1/2n}} = \frac{1}{\sqrt{2}} < 1 \quad .$$

If a_n is even

$$\lim_{n\to\infty} \sqrt[n]{a_n} = \lim_{n\to\infty} \sqrt[n]{\frac{1}{3^{n/2}}} = \frac{1}{\sqrt{3}} < 1 \quad .$$

Thus, since for both cases, the

$$\lim_{n\to\infty} \sqrt[n]{a_n} < 1 \quad ,$$

the series converges.

Determine if the series:

a)

$$\sum_{k=1}^{\infty} \frac{(k + 1)^{\frac{1}{2}}}{(k^5 + k^3 - 1)^{1/3}}$$

converges or diverges.

b)

$$\sum_{k=1}^{\infty} \frac{k \log k}{7 + 11k - k^2}$$

converges or diverges.

Solution: a) To determine if the given series converges or diverges, the following test called the limit test for convergence is used. This test states:

If

$$\lim_{k \to \infty} k^p U_k = A \text{ for } p > 1,$$

then

$$\sum_{k=1}^{\infty} U_k$$

converges absolutely.

To apply this test let $p = \frac{7}{6} > 1$. Then since

$$U_k = \frac{(k + 1)^{\frac{1}{2}}}{(k^5 + k^3 - 1)^{1/3}},$$

$$\lim_{k \to \infty} k^p U_k = \lim_{k \to \infty} \frac{k^{7/6}(k + 1)^{\frac{1}{2}}}{(k^5+k^3-1)^{1/3}} = \lim_{k \to \infty} \frac{k^{7/6} k^{\frac{1}{2}} (1 + 1/k)^{\frac{1}{2}}}{(k^5+k^3-1)^{1/3}}$$

$$= \lim_{k \to \infty} \frac{k^{5/3}(1 + 1/k)^{\frac{1}{2}}}{(k^5 + k^3 - 1)^{1/3}} = \lim_{k \to \infty} \frac{(1 + 1/k)^{\frac{1}{2}}}{\left[k^{-5}(k^5+k^3-1)\right]^{1/3}}$$

$$= \lim_{k \to \infty} \frac{(1 + 1/k)^{\frac{1}{2}}}{(1 + 1/k^2 - 1/k^5)^{1/3}} = \frac{(1)^{\frac{1}{2}}}{(1)^{1/3}} = 1$$

Therefore, the series converges absolutely.

b) For the series

$$\sum_{k=1}^{\infty} \frac{k \log k}{7 + 11k - k^2} ,$$

the following test called the limit test for divergence

is used. This test states, If

$$\lim_{k \to \infty} k \, U_k = A \neq 0 \text{ (or } \pm \infty)$$

then

$$\sum_{k=1}^{\infty} U_k$$

diverges. If $A = 0$, the test fails.

Since

$$U_k = \frac{k \log k}{7 + 11k - k^2} ,$$

$$\lim_{k \to \infty} k \, U_k = \lim_{k \to \infty} \frac{k^2 \log k}{7 + 11k - k^2} = \lim_{k \to \infty} \frac{k^2 \log k}{k^2 \left(\frac{7}{k^2} + \frac{11}{k} - 1 \right)}$$

$$= \lim_{k \to \infty} \frac{\log k}{\frac{7}{k^2} + \frac{11}{k} - 1} .$$

As $k \to \infty$, $\log k \to \infty$

while

$$\frac{7}{k^2} + \frac{11}{k} - 1 \to -1 .$$

Thus

$$\sum_{k=1}^{\infty} \frac{k \log k}{7 + 11k - k^2} \quad \text{diverges.}$$

Determine if the following series are absolutely convergent, conditionally convergent or divergent.

a)

$$\sum_{n=1}^{\infty} \frac{(-1)^{n+1}}{n}$$

b)

$$\sum_{n=2}^{\infty} (-1)^n \left(\frac{n}{1 + n^2}\right)^n$$

c)

$$\sum_{n=1}^{\infty} \frac{(-1)^n 2^n}{n!}$$

Solution: a) To determine if the series

$$\sum_{n=1}^{\infty} \frac{(-1)^{n+1}}{n}$$

is convergent or divergent, the following test called the alternating series test is used. This test states:

An alternating series

$$a_1 - a_2 + a_3 - a_4 + \ldots = \sum_{n=1}^{\infty} (-1)^{n+1} a_n , \ a_n > 0 ,$$

converges if the following two conditions are satisfied:

    i) its terms are decreasing in absolute value:

$$|a_{n + 1}| \leq |a_n| \text{ for } n = 1, 2, \ldots$$

   ii)
$$\lim_{n \to \infty} a_n = 0$$

For this series, the terms are decreasing in absolute value since

$$1 > \frac{1}{2} > \frac{1}{3} \ldots \ .$$

Also, the nth term approaches zero so

$$\lim_{n\to\infty} a_n = 0 .$$

Hence, the series converges. Next, if $\Sigma |a_n|$ converges also, then the series Σa_n is absolutely convergent. But the series of absolute values is the harmonic series

$$\sum_{n=1}^{\infty} \frac{1}{n}$$

which is known to diverge. Hence,

$\sum_{n=1}^{\infty} |a_n|$ does not converge and the series $\sum_{n=1}^{\infty} \frac{(-1)^{n+1}}{n}$

is conditionally convergent.

b)   For the series

$$\sum_{n=2}^{\infty} (-1)^n \left[\frac{n}{n^2 + 1} \right]^n$$

use the root test, which states: Let a series

$$\sum_{n=1}^{\infty} a_n$$

be given and let

$$\lim_{n\to\infty} \sqrt[n]{|a_n|} = R$$

Then if $R < 1$, the series is absolutely convergent. If $R > 1$, the series diverges. If $R = 1$, the test fails.

For this series

$$\lim_{n\to\infty} \sqrt[n]{|a_n|} = \lim_{n\to\infty} n \sqrt[n]{\left(\frac{n}{1+n^2}\right)^n} = \lim_{n\to\infty} \frac{n}{1+n^2} = \lim_{n\to\infty} \frac{1}{\frac{1}{n} + n} = 0$$

Therefore the series converges absolutely.

c)   For the series

$$\sum_{n=1}^{\infty} \frac{(-1)^n 2^n}{n!} ,$$

use the ratio test.   This states that if $a_n \neq 0$ for $n = 1, 2, \ldots$ and

$$\lim_{n \to \infty} \left| \frac{a_n + 1}{a_n} \right| = L$$

then if $L < 1$,

$$\sum_{n=1}^{\infty} a_n$$

is absolutely convergent, if $L = 1$, the test fails, if $L > 1$,

$$\sum_{n=1}^{\infty} a_n \qquad \text{is divergent.}$$

Here

$$\lim_{n \to \infty} \left| \frac{a_n + 1}{a_n} \right| = \lim_{n \to \infty} \frac{2^{n+1}}{(n+1)!} \cdot \frac{n!}{2^n} = \lim_{n \to \infty} \frac{2^{n+1} \ 2^{-n} \ (1 \cdot 2 \cdot 3 \ \cdots \ n)}{1 \cdot 2 \cdot 3 \cdot n \cdot (n+1)}$$

$$= \lim_{n \to \infty} \frac{2}{n + 1} = 0 \ .$$

Hence $L = 0$ and the series converges absolutely.

• PROBLEM 6–20

Determine if the series

$$1 - \frac{1}{2} + \frac{1 \cdot 3}{2 \cdot 4} - \frac{1 \cdot 3 \cdot 5}{2 \cdot 4 \cdot 6} + \ \cdots \ + (-1)^n \frac{1 \cdot 3 \ \cdots \ (2n-1)}{2 \cdot 4 \ \cdots \ 2n} + \ \cdots$$

is absolutely convergent, conditionally convergent, or divergent.

Solution: To determine if the given series is absolutely convergent, the following test, known as Raabe's test is used. The test states, let

$$t = \lim_{n \to \infty} n \left(1 - \left| \frac{U_{n+1}}{U_n} \right| \right) .$$

Then the series

$$\sum_{n=1}^{\infty} U_n$$

is absolutely convergent if $t > 1$, and is divergent or conditionally convergent if $t < 1$. If $t = 1$ the test fails.

Since

$$U_n = \frac{(-1)^n \, 1 \cdot 3 \, \ldots \, (2n-1)}{2 \cdot 4 \, \ldots \, 2n}$$

$$\frac{U_{n+1}}{U_n} = \frac{1 \cdot 3 \, \ldots \, (2n-1)(2n+1)}{2 \cdot 4 \cdot 6 \, \ldots \, 2n(2n+2)} \cdot \frac{2 \cdot 4 \, \ldots \, 2n}{1 \cdot 3 \, \ldots \, (2n-1)}$$

$$= \frac{2n + 1}{2n + 2}$$

Then

$$t = \lim_{n \to \infty} n\left(1 - \frac{2n+1}{2n+2}\right) = \lim_{n \to \infty} n\left(\frac{1}{2n + 2}\right)$$

i.e.,

$$t = \lim_{n \to \infty} \frac{n}{2n + 2} = \frac{1}{2}$$

Therefore, the series is not absolutely convergent. To determine if the series is conditionally convergent or divergent apply the alternating series test. Here, it is seen

$$|U_{n + 1}| \leq |U_n|$$

since

$$1 > \frac{1}{2} > \frac{1 \cdot 3}{2 \cdot 4} > \ldots \; .$$

Then, for the series to converge,

$$\lim_{n \to \infty} U_n$$

must be 0; that is, it must be shown that

$$\lim_{n \to \infty} \frac{1 \cdot 3 \, \ldots \, (2n - 1)}{2 \cdot 4 \, \ldots \, 2n} = 0 \; . \tag{1}$$

To prove (1) let

$$C_n = \frac{1 \cdot 3 \, \ldots \, (2n - 1)}{2 \cdot 4 \, \ldots \, 2n}$$

Then

$$c_n < \frac{2 \cdot 4 \ \dots \ 2n}{3 \cdot 5 \ \dots \ (2n-1)(2n+1)}$$

since

$$\frac{1 \cdot 3 \cdot 5 \ \dots \ (2n-1)}{2 \cdot 4 \ \dots \ 2n} < \frac{2 \cdot 4 \ \dots \ 2n}{3 \cdot 5 \ \dots \ (2n+1)}$$

follows from

$$(1 \cdot 3)(3 \cdot 5)(5 \cdot 7) \ \dots \ (2n-1)(2n+1)$$

$$< (2 \cdot 2)(4 \cdot 4)(6 \cdot 6) \ \dots \ (2n)(2n) \ .$$

But

$$\frac{2 \cdot 4 \ \dots \ 2n}{3 \cdot 5 \ \dots \ (2n+1)} = \frac{1}{c_n} \ \frac{1}{2n+1}$$

so that

$$c_n^2 < \frac{1}{2n+1} \quad \text{or} \quad c_n < \frac{1}{\sqrt{2n+1}} \ .$$

Then

$$\lim_{n \to \infty} c_n = 0$$

since

$$\lim_{n \to \infty} \frac{1}{\sqrt{2n+1}} = 0 \ .$$

Therefore, since both conditions of the alternating series test are satisfied, and since Raabe's test proved the series to be not absolutely convergent, the series is conditionally convergent.

• PROBLEM 6-21

a) Show that

$$\sum_{k=1}^{\infty} \frac{\cos kx}{k^2}$$

converges uniformly in the interval $-R \leq x \leq R$, where R is any real number.

b) Repeat for

$$\sum_{k=1}^{\infty} \frac{\cos kx}{2^k}$$

in the interval $- \infty < x < \infty$

Solution: One important test that can be applied to a large number of series to determine uniform convergence is the Weierstrass M-test. This test states:

Let

$$\sum_{k=1}^{\infty} U_k(x)$$

be a series of functions all defined in some interval $a \leq x \leq b$. If there is a convergent series of constants

$$\sum_{k=1}^{\infty} M_k \; ,$$

such that $\left| U_k(x) \right| \leq M_k$ for $a \leq x \leq b$ and

$k = 1, 2, \ldots$, then the series

$$\sum_{k=1}^{\infty} U_k(x)$$

converges absolutely for each x in the interval $a \leq x \leq b$ and is uniformly convergent in this same interval. Note that if a series converges uniformly by this test, it also converges absolutely. However, not all uniformly convergent series are absolutely convergent. This shows that for some uniformly convergent series, the sequence M_k required for the Weierstrass test

cannot be found and therefore this test cannot be applied.

a) $\sum_{k=1}^{\infty} \frac{\cos kx}{k^2}$.

Now, to apply the test, it is needed to find a series of constants such that

$$\sum_{k=1}^{\infty} M_k < \infty$$

and

287

$$\left| U_k(x) \right| \leq M_k .$$

For the given series choose $M_k = k^{-2}$. Then $\sum\limits_{k=1}^{\infty} \dfrac{1}{k^2} < \infty$ because of the fact that

$$\sum_{k=1}^{\infty} \frac{1}{k^p}$$

is convergent for $p > 1$.

Next,

$$\left| \frac{\cos kx}{k^2} \right| \leq \frac{1}{k^2}$$

because $|\cos kx| \leq 1$ for all values of x. Thus, since the conditions of the test are satisfied,

$$\sum_{k=1}^{\infty} \frac{\cos kx}{k^2}$$

converges uniformly in the interval $- R \leq x \leq R$, where R is any number. (i.e., the series converges uniformly in the interval $-\infty < x < \infty$).

b)   To apply the test to the series

$$\sum_{k=1}^{\infty} \frac{\cos kx}{2^k}$$

choose

$$M_k = \frac{1}{2^k} .$$

Then using the root test,

$$\lim_{k \to \infty} \left(\sqrt[k]{\frac{1}{2^k}} \right) = \frac{1}{\sqrt{2}} < 1 .$$

Therefore, the series

$$\sum_{k=0}^{\infty} \frac{1}{2^k}$$

is covergent. Now

$$\left| \frac{\cos kx}{2^k} \right| \leq \frac{1}{2^k}$$

because $|\cos kx| \leq 1$ for all values of x. Thus, by the Weierstrass M-test, the series is uniformly convergent for $-\infty < x < \infty$.

Show that the series

$$\frac{\cos x}{1} + \frac{\cos 3x}{3} + \frac{\cos 5x}{5} + \ldots$$

is convergent if x is not one of the values 0, $\pm\pi$, $\pm 2\pi$, ...

Solution: To determine if the series, rewritten as

$$\sum_{n=1}^{\infty} \frac{\cos n x}{n} ,$$

is convergent, the following theorem (known as Dirichlet's test) is used:

Consider a series of the form

$$a_0 b_0 + a_1 b_1 + a_2 b_2 + \ldots + a_n b_n + \ldots \qquad (1)$$

which satisfies the following conditions:

(i) the terms b_n are positive, decreasing in value (i.e., $b_{n+1} \le b_n$) and

$$\lim_{n \to \infty} b_n = 0 .$$

(ii) there is some constant M independent of n such that

$$|a_0 + a_1 + \ldots + a_n| \le M$$

for all values of n.

Then the series (1) is convergent.

Hence, to apply this theorem, take

$$a_0 = 0, \ a_1 = \cos x, \ a_2 = \cos 3x, \ a_3 = \cos 5x, \ \ldots ,$$

Let a_n be the sequence

$$a_n = \frac{(-1)^{n+1}}{n} .$$

Then

$$\sum_{n=1}^{\infty} \frac{(-1)^{n+1}}{n}$$

is the alternating harmonic series. Given that this series is convergent and that its sum is denoted by S, determine if the sums of the following sequences are convergent, and if so, find their sum in terms of S.

a)

$$b_n = \frac{(-1)^{n+1}}{2n} \qquad n = 1, 2, \ldots$$

b) The sequence formed by

$$c_{2j-1} = 0, j = 1, 2, \ldots$$

$$c_{2j} = b_j , \quad j = 1, 2, \ldots$$

c)

$$d_n = a_n + c_n \qquad n = 1, 2, \ldots$$

Solution: a) It is given that the series

$$\sum_{n=1}^{\infty} a_n = \sum_{n=1}^{\infty} \frac{(-1)^{n+1}}{n}$$

is convergent to S. Therefore

$$\sum_{n=1}^{\infty} b_n = \sum_{n=1}^{\infty} \frac{(-1)^{n+1}}{2n} = \frac{1}{2} \sum_{n=1}^{\infty} \frac{(-1)^{n+1}}{n} ,$$

which converges to $\frac{S}{2}$.

b) The series

$$\sum_{n=1}^{\infty} c_n$$

of the sequence formed by

$$c_{2j-1} = 0 , \; j = 1, 2, \ldots , \quad c_{2j} = b_j , \quad j = 1, 2, \ldots ,$$

is not the same series as

$$\sum_{n=1}^{\infty} b_n , \text{ but since}$$

$$\sum_{k=1}^{2m} c_k = \sum_{k=1}^{m} b_k \qquad m = 1, 2, \ldots$$

and

$$\sum_{k=1}^{2m+1} c_k = \sum_{k=1}^{m} b_k \qquad m = 1, 2, \ldots \text{ (because } c_{2j-1} = 0)$$

the series

$$\sum_{n=1}^{\infty} c_n$$

must also be convergent because as $m \rightarrow \infty$, the series

$$\sum_{k=1}^{\infty} b_k$$

converges to $\frac{S}{2}$. Therefore, as $m \rightarrow \infty$,

$$\sum_{k=1}^{2m} c_k$$

converges to $\frac{S}{2}$ so that the series

$$\sum_{n=1}^{\infty} c_n$$

is convergent to $\frac{S}{2}$.

c) The last sequence is formed by taking

$$d_n = a_n + c_n \qquad n = 1, 2, \ldots$$

Then for $\qquad j = 0, 1, 2, \ldots$,

$$d_{4j+1} = a_{4j+1} + c_{4j+1} = \frac{(-1)^{4j+2}}{4j+1} + 0$$

(4j+2 is always even)

$$= \frac{1}{4j+1} + 0 = \frac{1}{4j+1} \qquad .$$

$$d_{4j+2} = a_{4j+2} + c_{4j+2} = a_{4j+2} + b_{2j+1}$$

$$= \frac{(-1)^{4j+3}}{4j+2} + \frac{(-1)^{2j+2}}{2(2j+1)} = \frac{-1}{4j+2} + \frac{1}{4j+2} = 0$$

$$d_{4j+3} = a_{4j+3} + c_{4j+3}$$

$$= \frac{1}{4j+3} + 0 = \frac{1}{4j+3}$$

$$d_{4j+4} = a_{4j+4} + c_{4j+4} = a_{4j+4} + b_{2j+2}$$

$$= \frac{-1}{4j+4} - \frac{1}{2(2j+2)} = \frac{-1}{2j+2}$$

Hence, the numbers in the range of the sequence given by

$$\{d_n / n = 1, 2, \ldots \}$$

include the reciprocals of all the odd positive integers, the negatives of the reciprocals of all the even positive integers, and the number zero repeated infinitely often. Therefore, forming a new series by dropping all the zero terms in this sequence, this new series is a rearrangement of the alternating harmonic series. This because, by definition, the series

$$\sum_{n=1}^{\infty} d_n$$

is a rearrangement of the series

$$\sum_{n=1}^{\infty} a_n$$

provided there exists a one-to-one function g of the positive integers onto the positive integers such that for

$$n = 1, 2, \ldots , \qquad d_n = a_{g(n)} \quad ;$$

and for every positive integer n there exists a positive integer m > n such that g(m) ≠ m .

i.e.

$$\sum_{n=1}^{\infty} \frac{(-1)^{n+1}}{n} = 1 - \frac{1}{2} + \frac{1}{3} - \frac{1}{4} + \ldots - \frac{1}{2j+2} + \ldots$$

$$+ \frac{1}{4j+1} - \frac{1}{4j+2} + \frac{1}{4j+3} - \frac{1}{4j+4} + \ldots$$

$$d_n = 1 - \frac{1}{3} - \frac{1}{4} - \frac{1}{2} + \ldots$$

$$+ \frac{1}{4j+1} + \frac{1}{4j+3} - \frac{1}{2j+2} + \frac{1}{4j+5} - \cdots$$

Hence, since

$$\sum_{k=1}^{n} d_k = \sum_{k=1}^{n} a_k + \sum_{k=1}^{n} c_k, \qquad n = 1, 2, 3, \ldots,$$

this rearrangement of the alternating harmonic series must be convergent and its sum is $3S/2$.

• **PROBLEM 6–24**

Given that

$$\log (1 + x) = x - \frac{1}{2} x^2 + \frac{1}{3} x^3 - \cdots$$

$$\cdots + (-1)^{n-1} \frac{1}{n} x^n + \cdots \qquad (1)$$

Prove that

$$\frac{3}{2} \log 2 = 1 + \frac{1}{3} - \frac{1}{2} + \frac{1}{5} + \frac{1}{7} - \frac{1}{4} + \frac{1}{9} + \frac{1}{11} - \frac{1}{6} + \cdots$$

Solution: To do this problem let $x = 1$ in equation
(1). This yields

$$\log 2 = 1 - \frac{1}{2} + \frac{1}{3} - \frac{1}{4} + \frac{1}{5} - \frac{1}{6} + \frac{1}{7} - \quad \cdots \qquad (2)$$

From which

$$\frac{1}{2} \log 2 = \frac{1}{2} - \frac{1}{4} + \frac{1}{6} - \frac{1}{8} + \frac{1}{10} - \frac{1}{12} + \frac{1}{14} - \quad \cdots \qquad (3)$$

Thus in (3), zero terms may be inserted without affecting the value. Hence (3) becomes

$$\frac{1}{2} \log 2 = 0 + \frac{1}{2} + 0 - \frac{1}{4} + 0 + \frac{1}{6} + 0 - \cdots \qquad (4)$$

Now, it is known that if m and n are the values of two convergent series (i.e.,

$$m = r_1 + r_2 + \ldots + r_n + \ldots,$$

$$n = t_1 + t_2 + \ldots t_n, \ldots),$$

these two series may be combined by adding corresponding terms, and the result will be a new convergent series whose value is m + n. That is,

$$m + n = U_1 + U_2 + \ldots + U_n + \ldots$$

where

$$U_1 = r_1 + t_1 \; , \; U_2 = r_2 + t_2 \; , \; \ldots \; , \; U_n = r_n + t_n \; , \; \ldots \; .$$

Applying this fact to the problem at hand lets us add
(2) and (4) term by term. Therefore, adding gives

$$\frac{3}{2} \log 2 = 1 + 0 + \frac{1}{3} - \frac{1}{2} + \frac{1}{5} + 0 + \frac{1}{7} - \frac{1}{4} + \frac{1}{9} - \ldots \; .$$

Then, the zero terms may be deleted without affecting
the value. This yields

$$\frac{3}{2} \log 2 = 1 + \frac{1}{3} - \frac{1}{2} + \frac{1}{5} + \frac{1}{7} - \frac{1}{4} + \frac{1}{9} + \ldots$$

which is exactly what was required to be proved.

Note: a different value of the series was reached on
rearrangement of the terms because the series is conditionally convergent. Also note that if a series is
conditionally convergent, rearrangement of the terms
may result in a divergent series. However, if a series
is absolutely convergent with sum S, any rearrangement
of terms will also be convergent with sum S.

● **PROBLEM 6–25**

Given that the series

$$\sum_{k=0}^{\infty} x^k = \frac{1}{1 - x}$$

for $-1 < x < 1$, determine the value of the differentiated
series.

Solution: The solution of this problem is facilitated
by the use of the following theorem:

A convergent series can be differentiated term by term,
provided that the functions of the series have continuous
derivatives and that the series of derivatives is uniformly convergent; that is if

    (i)

$$U_k{}'(x) = \frac{dU_k}{dx}$$

is continuous for $a \leq x \leq b$,

(ii) the series

$$\sum_{k=1}^{\infty} U_k(x)$$

converges for $a \leq x \leq b$ to $f(x)$ $\left(\text{i.e.,} \right.$

$$\sum_{k=1}^{\infty} U_k(x) = f(x) \Big) \quad ,$$

and

(iii) the series

$$\sum_{k=1}^{\infty} U_k'(x)$$

converges uniformly for $a \leq x \leq b$,

then

$$f'(x) = \sum_{k=1}^{\infty} U_k'(x) \quad , \quad a \leq x \leq b \quad .$$

(The derivatives at a and b are understood as right-handed and left-handed derivatives respectively).

In this problem the functions of the series have continuous derivatives in $-1 < x < 1$. Since

$$f(x) = \frac{1}{1 - x} = \sum_{k=0}^{\infty} x^k$$

the derived series is

$$\sum_{k=0}^{\infty} k x^{k-1} \quad . \tag{1}$$

To determine if (1) converges uniformly, apply the Weierstrass M-test. Hence, take

$$M_k = k a^{k-1} \quad ,$$

because

$$|k x^{k-1}| \leq k a^{k-1}$$

for $|x| \le a$.

Then, using the ratio test for

$$\sum_{k=0}^{\infty} ka^{k-1} ,$$

$$\lim_{k \to \infty} \left| \frac{a_{k+1}}{a_k} \right| = \lim_{k \to \infty} \left| \frac{(k+1) a^k}{k \, a^{k-1}} \right| = \lim_{k \to \infty} \left| \frac{(k+1) a}{k} \right| = |a|$$

Therefore, the series

$$\sum_{k=0}^{\infty} k_a^{k-1}$$

is convergent for $|a| < 1$.

Hence (1) is uniformly convergent because

$$|U_k(x)| < M_k$$

and

$$\sum_{k=0}^{\infty} M_k < \infty \qquad \text{for} \quad -a \le x \le a , \quad a < 1 .$$

Thus, since the conditions of the theorem (stated at the beginning of this problem) are satisfied, we have

$$\sum_{k=0}^{\infty} U_k'(x) = f'(x) ,$$

which yields

$$\sum_{k=0}^{\infty} kx^{k-1} = \frac{1}{(1 - x)^2}$$

for $|x| \le a, \ a < 1$.

● **PROBLEM 6–26**

a) Show that $\sum_{k=0}^{\infty} x^k \log x$ converges uniformly to

$\dfrac{\log x}{1 - x}$ for $0 < x \le a$ $(0 < a < 1)$.

b)   Given that

$$\frac{\pi^2}{6} = 1 + \frac{1}{2^2} + \frac{1}{3^2} + \ldots$$

Show

$$\int_{0+}^{1} \frac{\log x}{1 - x} \, dx = \frac{-\pi^2}{6}$$

using the results from (a).

Solution: The partial sum of the series is

$$S_n(x) = \sum_{k=0}^{n} x^k \log x = \log x + x \log x + x^2 \log x + \ldots$$

$$+ x^n \log x$$

which can be rewritten

$$S_n(x) = (\log x)(1 + x + x^2 + \ldots + x^n) \, . \tag{1}$$

Multiplying each side of equation (1) by (1-x) yields

$$(1-x) \, S_n(x) = (\log x)(1-x)(1 + x + x^2 + \ldots + x^n)$$

$$(1-x) \, S_n(x) = (\log x)(1 + x + x^2 + \ldots + x^n - x - x^2 - x^3$$

$$- \ldots - x^{n+1})$$

$$S_n(x) = \frac{(\log x)(1 - x^{n+1})}{(1 - x)}$$

Then

$$S(x) = \lim_{n \to \infty} S_n(x) = \lim_{n \to \infty} \frac{(\log x)(1 - x^{n+1})}{(1 - x)} \quad ,$$

$$S(x) = \frac{\log x}{(1-x)}$$

for $0 < x \leq a$; $0 < a < 1$.

Now to show that the series converges uniformly to $S(x)$, apply the Weierstrass M-test. Take

$$M_k = a^k \log a$$

because

$$|x^k \log x| < a^k |\log a| \quad \text{for} \quad |x| < a .$$

Then using the ratio test for

$$\sum_{k=0}^{\infty} a^k |\log a| \quad , \tag{2}$$

$$\lim_{k \to \infty} \left| \frac{a_{k+1}}{a_k} \right| = \lim_{k \to \infty} \left| \frac{a^{k+1} \log a}{a^k \log a} \right| = |a|$$

Therefore, the series (2) is convergent for $|a| < 1$.

Hence, the series

$$\sum_{k=0}^{\infty} x^k \log x$$

is uniformly convergent for $0 < x \le a$; $0 < a < 1$.

Also, at the point $x = 1$ the series has sum 0 and therefore is convergent here also.

b) Since the series is uniformly convergent, term-by-term integration is justified. This is because of a theorem which states: A uniformly convergent series of continuous functions can be integrated term by term. That is, if

$$U_k(x) \in C \quad \text{for} \quad a \le x \le b , \quad n = 1, 2, \dots$$

and if

$$f(x) = \sum_{k=1}^{\infty} U_k(x)$$

uniformly in $a \le x \le b$ then

$$\int_a^b f(x)\, dx = \sum_{k=1}^{\infty} \int_a^b U_k(x)\, dx$$

$$= \int_a^b U_1(x)\, dx + \int_a^b U_2(x)\, dx + \dots + \int_a^b U_k(x)\, dx + \dots$$

Therefore, since $\dfrac{\log x}{1 - x} = \displaystyle\sum_{k=0}^{\infty} x^k \log x,$

298

$$\int_{0+}^{1} \frac{\log x}{(1-x)} \, dx = \sum_{k=0}^{\infty} \int_{0+}^{1} x^k \log x \, dx \, . \tag{3}$$

Integrating by parts, letting $u = \log x$; $du = \frac{dx}{x}$

and

$$dv = x^k \, dx \text{ so that } v = \frac{x^{k+1}}{k+1}$$

transforms the right side of equation (3) into

$$\sum_{k=0}^{\infty} \left[(\log x) \frac{x^{k+1}}{k+1} \Bigg|_{0+}^{1} - \int_{0+}^{1} \frac{x^k}{k+1} \, dx \right] \, . \tag{4}$$

Then noticing that $x^{k+1} \to 0$ faster than $\log x \to -\infty$ as $x \to 0$, the left hand portion of (4) evaluated at the limits is seen to be 0. The integral is then evaluated and equals

$$\frac{-1}{(k+1)^2}$$

Therefore (3) can be rewritten as

$$\int_{0+}^{1} \frac{\log x}{(1-x)} \, dx = \sum_{k=0}^{\infty} \frac{-1}{(k+1)^2}$$

$$= - \sum_{k=1}^{\infty} \frac{1}{k^2} \qquad \text{(by substituting } k = k + 1 \text{)}$$

$$= - \left(1 + \frac{1}{2^2} + \frac{1}{3^2} + \dots \right) = \frac{-\pi^2}{6}$$

as was given.

Hence,

$$\int_{0+}^{1} \frac{\log x}{(1-x)} \, dx = \frac{-\pi^2}{6} \qquad .$$

● **PROBLEM** 6–27

Evaluate the series

$$\sum_{n=1}^{\infty} \frac{n+1}{n \cdot 2^n}$$

with an error less than 0.037 .

Solution: To determine the value of the given series
with an error less than that specified, the following
theorem is needed:

If

$$\left| \frac{a_{n+1}}{a_n} \right| \leq r < 1 \quad \text{for} \quad n > n_1$$

(i.e., the series Σa_n converges by the ratio test),

then

$$|R_n| \leq \frac{|a_{n+1}|}{1-r} = K_n \ , \quad n \geq n_1 \ ;$$

the sequence K_n is monotonously decreasing and converges to 0.

In addition, if

$$\lim_{n \to \infty} \left| \frac{a_{n+1}}{a_n} \right| = L < 1 \ ,$$

then r will be at least equal to L .

   For this problem

$$a_n = \frac{n+1}{n \cdot 2^n} \quad .$$

Consequently,

$$\left| \frac{a_{n+1}}{a_n} \right| = \left| \frac{(n+2)}{(n+1) \, 2^{n+1}} \cdot \frac{n \, 2^n}{(n+1)} \right|$$

$$= \left| \frac{n^2 + 2n}{n^2 + 2n + 1} \cdot \frac{1}{2} \right|$$

Then

$$\lim_{n \to \infty} \left| \frac{a_{n+1}}{a_n} \right| = \lim_{n \to \infty} \left| \frac{n^2 + 2n}{n^2 + 2n + 1} \cdot \frac{1}{2} \right|$$

$$= \lim_{n \to \infty} \left| \frac{1 + 2/n}{1 + 2/n \cdot 1/n^2} \cdot \frac{1}{2} \right| = \frac{1}{2} \quad .$$

(i.e., the series converges by the ratio test since

$$L = \frac{1}{2} < 1).$$

Hence, since it is needed that

$$\left| \frac{a_{n+1}}{a_n} \right| \leq r < 1 , \quad r = \frac{1}{2} \quad \text{can be used.}$$

Accordingly, by the theorem

$$|R_n| \leq \frac{\left| a_{n+1} \right|}{1-r} = K_n$$

$$R_n \leq \frac{\left| \dfrac{(n+2)}{(n+1) 2^{n+1}} \right|}{\dfrac{1}{2}} = K_n \quad .$$

Thus, for an error less than .037, 5 terms are sufficient. This is because

$$|R_5| \leq \left| \frac{7}{6 \cdot 2^6} \cdot 2 \right| = .03645 < .037 .$$

That is

$$R_5 = \frac{7}{6 \cdot 2^6} + \frac{8}{7 \cdot 2^7} + \dots < \frac{7}{6 \cdot 2^6} + \frac{7}{6 \cdot 2^7} + \frac{7}{6 \cdot 2^8} + \dots.$$

$$R_5 < \frac{7}{6 \cdot 2^6} \left(1 + \frac{1}{2} + \frac{1}{4} + \dots \right) = \frac{7}{6 \cdot 2^6} \quad (2) \quad .$$

Since $\displaystyle \sum_{n=0}^{\infty} \frac{1}{2^n} = 2$ (geometric series),

$$R_5 < \frac{7}{192} = 0.037.$$

Finally,

$$\sum_{n=1}^{\infty} \frac{n+1}{n \cdot 2^n} \approx 1 + \frac{3}{2 \cdot 2^2} + \frac{4}{3 \cdot 2^3} + \frac{5}{4 \cdot 2^4} + \frac{6}{5 \cdot 2^5} = 1.6573 \quad .$$

301

Evaluate the series

$$\sum_{n=2}^{\infty} \frac{1}{(\log n)^n}$$

with an error less than 0.06.

<u>Solution</u>: For the solution to the given problem the following theorem is needed:

If $\sqrt[n]{|a_n|} \leq r < 1$ for $n > n_1$,

(i.e., the series Σa_n converges by the root test), then

$$|R_n| \leq \frac{r^{n+1}}{1-r} = K_n \text{ , for } n \geq n_1 .$$

In addition, if

$$\lim_{n\to\infty} \sqrt[n]{|a|} = R < 1 ,$$

then r will be at least equal to R. If

$$1 > |a_{n+1}|^{1/(n+1)} \geq |a_{n+2}|^{1/(n+2)}$$

for $n \geq n_1$, then

$$|R_n| \leq \frac{|a_{n+1}|}{1 - |a_{n+1}|^{1/(n+1)}} = K_n^* \text{ , for } n \geq n_1$$

Here, the sequences K_n and K_n^* are both monotonously decreasing and converge to zero.

To apply this theorem to the series, first note that

$$\sqrt[n]{|a_n|} = \frac{1}{\log n}$$

which is decreasing and less than 1 for $n = 3, 4, \ldots$
Then

$$\lim_{n\to\infty} \sqrt[n]{|a_n|} = \lim_{n\to\infty} \frac{1}{\log n} = 0 = R < 1 .$$

However, for this problem

$$1 > |a_{n+1}|^{1/(n+1)} = \left|\frac{1}{\log (n+1)}\right| \geq |a_{n+2}|^{1/(n+2)} = \left|\frac{1}{\log (n+2)}\right|$$

for $n \geq n_1$; then, using $n = 5$ yields

$$|R_n| \leq \frac{|a_{n+1}|}{1 - |a_{n+1}|^{1/(n+1)}} = \frac{\frac{1}{(\log 6)^6}}{1 - \frac{1}{\log 6}} = 0.06 = K_n^*$$

which gives the value of the series with an error less than 0.06 as

$$\sum_{n=2}^{\infty} \frac{1}{(\log n)^n} = \frac{1}{(\log 2)^2} + \frac{1}{(\log 3)^3} + \frac{1}{(\log 4)^4} + \frac{1}{(\log 5)^5}$$

$$= 3.816 .$$

• PROBLEM 6–29

For the series

$$\sum_{n=1}^{\infty} \frac{(-1)^{n+1}}{n}$$

show that

a) $-\frac{1}{2} < R_1 < 0$, $0 < R_2 < \frac{1}{3}$, $-\frac{1}{4} < R_3 < 0$.

b) If 3 terms are used, use the theorem to find two values the sum is between.

c) How many terms are needed to compute the sum with an error less than 0.01 .

Solution: a) The solution to the problem requires the following theorem:

If the series

$$a_1 - a_2 + a_3 - a_4 + \ldots = \sum_{n=1}^{\infty} (-1)^{n+1} a_n , \quad a_n > 0$$

converges by the alternating series test, then

$$0 < |R_n| < a_{n+1} = K_n .$$

Therefore, $N(\varepsilon)$ can be chosen as the smallest integer such that

303

$$a_{n+1} < \varepsilon$$

That is, when a series converges by the alternating series test, the error made in stopping at n terms is in absolute value less than the first term neglected.

To apply this theorem to the given problem, first determine if the series converges by the alternating series test. Here

$$a_{n+1} \leq a_n \quad \text{since} \quad \frac{1}{n+1} \leq \frac{1}{n}$$

(i.e., $1 > \frac{1}{2} > \frac{1}{3} > \frac{1}{4} > \dots$)

and

$$\lim_{n\to\infty} a_n = \lim_{n\to\infty} \frac{1}{n} = 0 \ .$$

Hence, the series converges by the alternating series test. Then, by the theorem

$$0 < |R_1| < a_2 \quad \text{or} \quad 0 < |R_1| < \frac{1}{2}$$

$$0 < |R_2| < a_3 \quad \text{or} \quad 0 < |R_2| < \frac{1}{3}$$

$$0 < |R_3| < a_4 \quad \text{or} \quad 0 < |R_3| < \frac{1}{4} \ .$$

The above can be written more precisely as:

$$-\frac{1}{2} < R_1 < 0 \quad \text{since} \quad n = 2 \quad \text{gives} \quad -\frac{1}{2}$$

$$0 < R_2 < \frac{1}{3} \quad \text{since} \quad n = 3 \quad \text{gives} \quad +\frac{1}{3}$$

$$-\frac{1}{4} < R_3 < 0 \quad \text{since} \quad n = 4 \quad \text{gives} \quad -\frac{1}{4} \ .$$

b) The partial sums are

$$S_1 = 1 \ , \ S_2 = 1 - \frac{1}{2} = \frac{1}{2} \ , \ S_3 = \frac{5}{6} \ , \ S_4 = \frac{7}{12} \ , \ \dots \ .$$

If 3 terms are used, the sum is between $\frac{5}{6}$ and $\frac{5}{6} - \frac{1}{4} = \frac{7}{12}$

since $\qquad -\frac{1}{4} < R_3 < 0 \ .$

c) To compute the sum with an error less than 0.01, one would need 100 terms, for the 101st term: 1/101 is the first term less than 0.01 .

Show that

a)   the series $1 + 0 - 1 + 1 + 0 - 1 + 1 + 0 - \ldots$

is summable (C, 1) to $\frac{2}{3}$.

b)   the series

$$\sum_{k=1}^{\infty} (-1)^k k$$

is not summable (C, 1) but is summable (C, 2).

Solution: a)   The specified series has the partial sums

$S_1 = 1; \; S_2 = 1; \; S_3 = 0; \; S_4 = 1, \; S_5 = 1, \; S_6 = 0, \; S_7 = 1;$

$$S_8 = 1; \; S_9 = 0; \; \ldots \; .$$

Therefore, letting

$$t_n = \sum_{k=1}^{n} S_k \quad \text{yields}$$

$t_1 = 1, \; t_2 = 2, \; t_3 = 2, \; t_4 = 3, \; t_5 = 4, \; t_6 = 4, \; t_7 = 5,$

$$t_8 = 6, \; t_9 = 6, \; \ldots \; .$$

From which

$$\sigma_n = \frac{t_n}{n} \qquad \text{gives the sequence}$$

$$\sigma_n = 1, \; 1, \frac{2}{3} \; , \; \frac{3}{4} \; , \; \frac{4}{5} \; , \; \frac{2}{3} \; , \; \frac{5}{7} \; , \; \frac{3}{4} \; , \; \frac{2}{3} \; , \; \ldots \qquad (1)$$

Accordingly, (1) can be rewritten as

$$\sigma_{3n+1} = \frac{2n + 1}{3n + 1} \; , \qquad \sigma_{3n+2} = \frac{2n + 2}{3n + 2} \; ,$$

$$\sigma_{3n+3} = \frac{2n + 2}{3n + 3}$$

By definition, the series is summable (C, 1) to A if and only if

$$\lim_{n \to \infty} \sigma_n = A \; .$$

For this problem,

305

$$\lim_{n\to\infty} \sigma_{3n+1} = \lim_{n\to\infty} \sigma_{3n+2} = \lim_{n\to\infty} \sigma_{3n+3} = \lim_{n\to\infty} \sigma_n = \frac{2}{3} \ .$$

Consequently, the series is summable (C, 1) to $\frac{2}{3}$.

b) The series

$$\sum_{k=1}^{\infty} (-1)^k \, k = -1 + 2 - 3 + 4 - 5 + 6 \ldots \tag{2}$$

has the partial sums:

$$S_1 = -1, \ S_2 = 1, \ S_3 = -2, \ S_4 = 2, \ S_5 = -3, \ S_6 = 3, \ \ldots \ .$$

Letting $\quad t_n = \sum\limits_{k=1}^{n} S_k$

yields

$$t_1 = -1, \ t_2 = 0, \ t_3 = -2, \ t_4 = 0, \ t_6 = -3, \ t_6 = 0,$$

$$\ldots \ \text{or} \ \ t_{2n-1} = -n \ , \ t_{2n} = 0 \quad n = 1, 2, \ \ldots \ .$$

Letting

$$\sigma_n = \frac{t_n}{n}$$

furnishes the following sequence

$$\sigma_n = -1, \ 0, \ \frac{-2}{3} \ , \ 0, \ \frac{-3}{5} \ , \ 0, \ \ldots$$

or similarly written as

$$\sigma_{2n} = 0 \ , \quad \sigma_{2n-1} = \frac{-n}{2n-1} \qquad n = 1, \ 2, \ \ldots$$

from which

$$\lim_{n\to\infty} \sigma_{2n} = 0 \ , \quad \lim_{n\to\infty} \sigma_{2n-1} = -\frac{1}{2} \ .$$

This shows the series is not summable (C, 1) due to the different values in the two limits. Hence a more powerful method of summation is needed. This second method states:

$$\text{If} \qquad \lim_{n\to\infty} \frac{2}{n(n+1)} \ \sum_{k=1}^{n} t_k = A \tag{3}$$

then

$$A = \sum_{k=1}^{\infty} U_k \qquad\qquad (C, \ 2)$$

(i.e., the series is summable (C, 2) to A). Therefore letting

$$C_n = \sum_{k=1}^{n} t_k$$

yields the sequence

$$C_n: \ -1, \ -1, \ -3, \ -3, \ -6, \ -6, \ \ldots$$

or,

$$C_n = \frac{\frac{-(n+1)}{2}\left(\frac{(n+1)}{2} + 1\right)}{2}$$

for $n = 1, \ 3, \ 5, \ 7, \ \ldots$

and

$$C_n = \frac{\frac{-n}{2}\left(\frac{n}{2} + 1\right)}{2}$$

for $n = 2, \ 4, \ 6, \ 8, \ \ldots$.

Hence,

$$\lim_{n\to\infty} \frac{2}{n(n+1)} \sum_{k=1}^{n} U_k = \lim_{n\to\infty} \frac{2}{n(n+1)} \left[\frac{\frac{-(n+1)}{2}\left(\frac{(n+1)}{2} + 1\right)}{2}\right]$$

$$(\text{for } n \text{ odd})$$

$$= \lim_{n\to\infty} \frac{-\left[\frac{(n+1)}{4} + \frac{1}{2}\right]}{n} = \lim_{n\to\infty} -\left(\frac{n + 1 + 2}{4n}\right)$$

$$= \lim_{n\to\infty} -\left(\frac{1 + \frac{1}{n} + \frac{2}{n}}{4}\right) = -\frac{1}{4} \qquad (\text{for } n \text{ odd})$$

and

$$\lim_{n\to\infty} \frac{2}{n(n+1)} \left[\frac{-\frac{n}{2}\left(\frac{n}{2} + 1\right)}{2}\right] \qquad (\text{for } n \text{ even})$$

$$= \lim_{n\to\infty} \frac{-\left(\frac{n^2 + 2n}{4}\right)}{n^2 + n} = \lim_{n\to\infty} -\frac{\left(1 + \frac{2}{n}\right)}{4\left(1 + \frac{1}{n}\right)} = -\frac{1}{4} \ .$$

Thus by (3) the series (2) is summable (C, 2) to $-\frac{1}{4}$.

Prove that the infinite product:

a)

$$\prod_{k=2}^{\infty} \left(1 + \frac{1}{k^2-1}\right)$$

converges and has the value 2,

b)

$$\prod_{k=1}^{\infty} \left(1 - \frac{1}{k+1}\right)$$

diverges.

Solution: An infinite product is a product of the form

$$\prod_{k=1}^{\infty} (1 + U_k) = (1 + U_1)(1 + U_2)(1 + U_3) \cdots$$

where $U_k \neq -1$ for $k = 1, 2, 3, \ldots$). To determine if an infinite product converges or diverges, let

$$P_n = \prod_{k=1}^{n} (1 + U_k) \,,$$

then if there exists a number $P \neq 0$ such that

$$\lim_{n \to \infty} P_n = P \,,$$

the infinite product, given by

$$\prod_{k=1}^{\infty} (1 + U_k)$$

converges to P. An infinite product that does not converge is said to diverge.

We note that theorems on infinite products often depend upon theorems for infinite series.

a)

$$\prod_{k=2}^{\infty} \left(1 + \frac{1}{k^2-1}\right) = \prod_{k=2}^{\infty} \frac{k^2}{k^2-1}$$

Now

$$P_n = \prod_{k=2}^{n} \frac{k^2}{k^2-1} = \left(\frac{2^2}{2^2-1}\right)\left(\frac{3^2}{3^2-1}\right) \cdots \left(\frac{n^2}{n^2-1}\right)$$

$$= \left(\frac{4}{3}\right)\left(\frac{9}{8}\right)\left(\frac{16}{15}\right) \cdots \left(\frac{n^2}{n^2-1}\right) = \frac{2n}{n+1} \quad .$$

Then

$$\lim_{n\to\infty} P_n = \lim_{n\to\infty} \frac{2n}{n+1} = 2 \quad .$$

Therefore

$$\prod_{k=2}^{\infty} \left(1 + \frac{1}{k^2-1}\right)$$

converges to 2.

b)

$$\prod_{k=1}^{\infty} \left(1 - \frac{1}{k+1}\right) = \prod_{k=1}^{\infty} \frac{k}{k+1} \quad .$$

Then

$$P_n = \prod_{k=1}^{n} \frac{k}{k+1} = \left(\frac{1}{2}\right)\left(\frac{2}{3}\right)\left(\frac{3}{4}\right) \cdots \left(\frac{n-1}{n}\right)\left(\frac{n}{n+1}\right) = \frac{1}{n+1} \quad .$$

Hence

$$\lim_{n\to\infty} P_n = \lim_{n\to\infty} \frac{1}{n+1} = 0 \quad .$$

Consequently,

$$\prod_{k=1}^{\infty} \left(1 - \frac{1}{k+1}\right)$$

diverges.

POWER SERIES

● **PROBLEM** 6–32

Find the radius of convergence of the following power series:

a) $\displaystyle\sum_{n=1}^{\infty} \frac{x^n}{n}$

b)

$$\sum_{n=1}^{\infty} \frac{(2n)!}{(n!)^2} x^n$$

c)

$$\sum_{n=1}^{\infty} \frac{(3n)!}{(n!)^2} x^n$$

Solution: a)

$$\sum_{n=1}^{\infty} \frac{x^n}{n} = x + \frac{x^2}{2} + \frac{x^3}{3} + \cdots \ .$$

Here we let

$$U_n = \frac{x^n}{n}$$

so that

$$\lim_{n \to \infty} \left| \frac{U_{n+1}}{U_n} \right| = \lim_{n \to \infty} \left| \frac{x^{n+1}}{n+1} \cdot \frac{n}{x^n} \right| = \lim_{n \to \infty} |x| \frac{n}{n+1} = |x| \ .$$

Therefore, $R = 1$ and the series converges for $-1 < x < 1$. Note that the series may or may not converge at the end points $x = \pm 1$. That is, it may converge at both $x = \pm 1$, at just one, or neither. For example, given the endpoints ± 1, we have at $x = 1$ the series,

$$\sum_{n=1}^{\infty} \frac{1}{n} \ ,$$

which is the divergent harmonic series. For $x = -1$ we have the alternating series,

$$\sum_{n=1}^{\infty} \frac{(-1)^n}{n}$$

where

$$\left| \frac{(-1)^{n+1}}{n+1} \right| < \left| \frac{(-1)^n}{n} \right|$$

and

$$\lim_{n \to \infty} \frac{(-1)^n}{n} = 0$$

Hence this series is convergent. Thus, $R = 1$ and the series converges absolutely for $-1 \le x < 1$.

b)

$$\sum_{n=1}^{\infty} \frac{(2n)!}{(n!)^2} x^n = \frac{2!}{(1)^2} x + \frac{4!}{(2!)^2} x^2 + \frac{6!}{(3!)^2} x^3 + \dots .$$

Here we have

$$a_n = \frac{(2n)!}{(n!)^2} \quad \text{and} \quad U_n = \frac{(2n)!}{(n!)^2} x^n$$

$$\lim_{n \to \infty} \left| \frac{U_{n+1}}{U_n} \right| = \lim_{n \to \infty} \left| \frac{(2n+2)!}{[(n+1)!]^2} x^{n+1} \cdot \frac{(n!)^2}{(2n)! x^n} \right| =$$

$$= \lim_{n \to \infty} |x| \frac{(2n+2)(2n+1)}{(n+1)^2} = 4|x| .$$

Therefore $R = \frac{1}{4}$ since we want $4|x| < 1$.

c)

$$\sum_{n=1}^{\infty} \frac{(3n)!}{(n!)^2} x^n = 3! x + \frac{6!}{(2!)^2} x^2 + \frac{9!}{(3!)^2} x^3 + \dots .$$

Here,

$$a_n = \frac{(3n)!}{(n!)^2} \quad \text{and} \quad U_n = \frac{(3n)!}{(n!)^2} x^n$$

$$\lim_{n \to \infty} \left| \frac{U_{n+1}}{U_n} \right| = \lim_{n \to \infty} \left| \frac{(3n+3)! \, x^{n+1}}{[(n+1)!]^2} \cdot \frac{(n!)^2}{(3n)! \, x^n} \right|$$

$$= \lim_{n \to \infty} x \frac{(3n+3)(3n+2)(3n+1)}{(n+1)^2} = \infty$$

Therefore, the series does not converge for any value of x except x = 0. This means that

$$R = 0 \left(\text{i.e., } R = \frac{1}{\infty} \right) .$$

● **PROBLEM 6-33**

Prove that

$$f(x) = \sum_{n=0}^{\infty} a_n x^n$$

and

$$f'(x) = \sum_{n=0}^{\infty} na_n x^{n-1}$$

have the same radius of convergence.

Solution:

$$f(x) = \sum_{n=0}^{\infty} a_n x^n = a_0 + a_1 x + a_2 x^2 + a_3 x^3 + \ldots \tag{1}$$

$$f'(x) = \sum_{n=0}^{\infty} na_n x^{n-1} = a_1 + 2a_2 x + 3a_3 x^2 + 4a_4 x^3 + \ldots \ . \tag{2}$$

Let R and R' denote the radius of convergence of (1) and (2), respectively. Suppose $|x| < R$, and choose x_0 so that

$$|x| < |x_0| < R \ .$$

Then (1) is convergent with $x = x_0$ and consequently $a_n x_0^n \to 0$ as $n \to \infty$, since for a convergent series

$$\sum_{n=0}^{\infty} U_n$$

we have

$$\lim_{n \to \infty} U_n = 0 \ .$$

Therefore a number $A > 0$ may be chosen such that

$$\left| a_n x_0^n \right| \leq A$$

for all n. Then

$$na_n x^{n-1} = \frac{x_0^n}{x_0^n} \ n a_n x^{n-1} = \frac{n}{x_0} a_n x_0^n \left(\frac{x}{x_0} \right)^{n-1}$$

Therefore

$$\left| na_n x^{n-1} \right| \ \leq \ \frac{A}{|x_0|} \ n \ \left| \left(\frac{x}{x_0} \right)^{n-1} \right| \tag{3}$$

or

$$\left| na_n x^{n-1} \right| \ \leq \ \frac{A}{|x_0|} \ nr^{n-1} \tag{3}$$

where

$$r = \frac{|x|}{|x_0|} < 1$$

The series

$$\sum_{n=0}^{\infty} \frac{A}{|x_0|} \, nr^{n-1}$$

is convergent since

$$\lim_{n \to \infty} \left| \frac{U_{n+1}}{U_n} \right| = \lim_{n \to \infty} \left| \frac{(n+1)r^n}{n \, r^{n-1}} \right| = \lim_{n \to \infty} \left| \frac{n+1}{n} \right| r = r < 1 \; ;$$

here

$$U_n = \frac{A}{|x_0|} \, nr^{n-1}$$

Thus, by (3), the series

$$\sum_{n=0}^{\infty} na_n x^{n-1} \, ,$$

which is the series (2), is also convergent. Hence, (2) converges if $|x| < R$. It follows that the radius of convergence of (2) is not less than that of (1). That is $R' \geq R$ and if $R = \infty$ this means that $R' = \infty$.

Now we assume that $R' > R$ and we choose x so that $R < |x| < R'$. Then for this x, the series (2) is absolutely convergent, but the series (1) is divergent. Now

$$|a_n x^n| = |n \, a_n x^{n-1}| \, \left| \frac{x}{n} \right| < |na_n x^{n-1}|$$

as soon as $n > |x|$. Hence, this shows that the series (1) must be convergent by the comparison test. However this is a contradiction. Therefore $R' \not> R$, and since we have already shown that $R' \geq R$ (i.e., $R' \not< R$), this means that $R' = R$. This completes the proof.

● PROBLEM 6-34

Find the radius of convergence of the power series

$$\sum_{n=1}^{\infty} \frac{x^n}{n^2} \, .$$

Then determine if the convergence is uniform for $-R \leq x \leq R$.

313

<u>Solution</u>: Here the ratio gives

$$\lim_{n\to\infty} \left| \frac{U_{n+1}}{U_n} \right| = \lim_{n\to\infty} |x| \frac{n^2}{(n+1)^2} = |x| \ ,$$

so that R = 1. That is, the series converges absolutely for $-1 < x < 1$ and diverges for $|x| > 1$. Note we could have also found this result by the relation,

$$R = \frac{1}{\alpha} \quad \text{where} \quad \alpha = \lim_{n\to\infty} \sup \sqrt[n]{|a_n|}$$

where $a_n = \dfrac{1}{n^2}$ which yields

$$\alpha = \lim_{n\to\infty} \sup \sqrt[n]{|1/n^2|} = \lim_{n\to\infty} \sup \sqrt[n]{\frac{1}{n^2}}$$

$$= \lim_{n\to\infty} \sup \frac{1}{n^{2/n}} = \lim_{n\to\infty} \sup \frac{1}{e^{(2/n) \log n}}$$

$$= \frac{1}{e^{\displaystyle \lim_{n\to\infty} \sup(2/n) \log n}} = \frac{1}{e^0} = 1 \ ,$$

therefore,

$$\frac{1}{R} = 1 \ ,$$

so that as before R = 1. For $x = \pm 1$ the series converges by comparison with the harmonic series of order 2, that is

$$\left| \frac{(\pm 1)^n}{n^2} \right| \le \frac{1}{n^2} \quad .$$

Hence the series converges for $-1 \le x \le 1$. To determine if the convergence is uniform in this interval, we need the Weierstrass M-test for uniform convergence. That is, we must find a convergent series of constants

$$\sum_{n=1}^{\infty} M_n$$

such that

$$\left| \frac{x^n}{n^2} \right| \le M_n$$

for all x in $-1 \le x \le 1$ if we are to determine that

314

$$\sum_{n=1}^{\infty} \frac{x^n}{n^2}$$

is uniformly convergent in this interval.

Since $-1 \le x \le 1$ we have

$$\left| \frac{x^n}{n^2} \right| \le \frac{1}{n^2}$$

for all x in the range.

Therefore since

$$\sum_{n=1}^{\infty} M_n$$

converges, this shows the given power series converges uniformly on $-1 \le x \le 1$.

• **PROBLEM 6-35**

Derive the series expansion

$$\sin^{-1}x = x + \frac{1}{2} \frac{x^3}{3} + \frac{1}{2} \cdot \frac{3}{4} \frac{x^5}{5} + \frac{1 \cdot 3 \cdot 5}{2 \cdot 4 \cdot 6} \frac{x^7}{7} + \cdots .$$

<u>Solution</u>: To derive the given series for $\sin^{-1}x$ we need the following theorem:

Let f be a function defined by

$$f(x) = \sum_{n=0}^{\infty} a_n x^n$$

where $R \ne 0$ for the power series. Then f is continuous in the open interval of convergence of the series. Moreover, if a and b are points of this interval,

$$\int_a^b f(x)\,dx = \sum_{n=0}^{\infty} a_n \frac{b^{n+1} - a^{n+1}}{n+1} . \tag{1}$$

That is, the integral of the function is equal to the series obtained by integrating the original power series term by term, i.e.,

$$\int_a^b f(x)\,dx = \sum_{n=0}^{\infty} a_n \int_a^b x^n\,dx .$$

315

To apply this theorem we start with the fact that

$$\sin^{-1}x = \int_0^x \frac{dt}{\sqrt{1-t^2}} \ . \tag{2}$$

Now (2) is valid if $|x| \leq 1$; however the integral is improper if $x = \pm 1$ since the integrand becomes infinite at $t = \pm 1$. Now by the binomial theorem,

$$(1+x)^r = 1 + rx + \frac{r(r-1)}{2!} x^2 + \ldots + \frac{r(r-1) \ \ldots \ (r-n+1)}{n!} x^n + \ldots$$

when $|x| < 1$, and where r is any real number.

We replace x by $-t^2$ and let $r = -\frac{1}{2}$ to yield

$$(1-t^2)^{-\frac{1}{2}} = 1 - \frac{1}{2}(-t^2) + \frac{\left(-\frac{1}{2}\right)\left(-\frac{3}{2}\right)}{2!}(-t^2)^2 +$$

$$+ \frac{\left(-\frac{1}{2}\right)\left(-\frac{3}{2}\right)\left(-\frac{5}{2}\right)}{3!}(-t^2)^3 + \ldots$$

$$= 1 + \frac{1}{2} t^2 + \frac{1 \cdot 3}{2 \cdot 4} t^4 + \frac{1 \cdot 3 \cdot 5}{2 \cdot 4 \cdot 6} t^6 + \ldots \ ;$$

where this result is valid if $|t| < 1$. Therefore the series has radius of convergence $R = 1$. Thus, by the theorem we may integrate the series from $t = 0$ to $t = x$ if $|x| < 1$. That is

$$\sin^{-1}x = \int_0^x \frac{dt}{\sqrt{1-t^2}} = \int_0^x dt + \int_0^x \frac{t^2}{2} dt + \int_0^x \frac{1 \cdot 3t^4}{2 \cdot 4} dt + \ldots$$

$$= x + \frac{1}{2} \frac{x^3}{3} + \frac{1 \cdot 3}{2 \cdot 4} \frac{x^5}{5} + \frac{1 \cdot 3 \cdot 5}{2 \cdot 4 \cdot 6} \frac{x^7}{7} + \ldots$$

if $|x| < 1$. This is the desired expansion.

● **PROBLEM 6–36**

a) Find an expansion in powers of x of the function

$$f(x) = \int_0^1 \frac{1 - e^{-tx}}{t} dt \ .$$

b) Use the result from part (a) to find $f\left(\frac{1}{2}\right)$ approximately.

Solution: a) Using the fact that for all values x the series representation for e^x is

$$e^x = \sum_{n=0}^{\infty} \frac{x^n}{n!} = 1 + x + \frac{x^2}{2!} + \ldots + \frac{x^n}{n!} + \ldots$$

we have

$$e^{-tx} = 1 - tx + \frac{t^2 x^2}{2!} - \frac{t^3 x^3}{3!} + \ldots \; .$$

Hence,

$$1 - e^{-tx} = tx - \frac{t^2 x^2}{2!} + \frac{t^3 x^3}{3!} - \ldots$$

so that

$$\frac{1 - e^{-tx}}{t} = x - \frac{tx^2}{2!} + \frac{t^2 x^3}{3!} - \ldots + (-1)^{n-1} \frac{t^{n-1} x^n}{n!} + \ldots \; .$$

Now this series representation is valid for all values of x and t. In addition, the radius of convergence of the power series in t is $R = \infty$. This is because,

$$\lim_{n \to \infty} \left| \frac{a_{n+1}}{a_n} \right| = \lim_{n \to \infty} |tx| \frac{1}{n+1} = 0$$

so that $R = \infty$.

Therefore, we can integrate the series term by term (this by the theorem in the previous problem) to obtain,

$$f(x) = \left. xt - \frac{t^2 x^2}{2 \cdot 2!} + \frac{t^3 x^3}{3 \cdot 3!} - \ldots + (-1)^{n-1} \frac{t^n x^n}{n \cdot n!} + \ldots \right|_0^1$$

or

$$f(x) = x - \frac{x^2}{2 \cdot 2!} + \frac{x^3}{3 \cdot 3!} - \ldots + (-1)^{n-1} \frac{x^n}{n \cdot n!} + \ldots \; . \qquad (1)$$

b) From (1) we have,

$$f\left(\frac{1}{2}\right) = \int_0^1 \frac{1 - e^{-t/2}}{t} \, dt = \frac{1}{2} - \frac{1}{2 \cdot 2!} \left(\frac{1}{2}\right)^2 + \frac{1}{3 \cdot 3!} \left(\frac{1}{2}\right)^3 - \ldots$$

which approximately equals 1.13 .

● **PROBLEM 6–37**

a) Prove that

$$\tan^{-1} x = x - \frac{x^3}{3} + \frac{x^5}{5} - \frac{x^7}{7} + \ldots$$

317

where the series is uniformly convergent in $-1 \leq x \leq 1$.

b) Prove that

$$\frac{\pi}{4} = 1 - \frac{1}{3} + \frac{1}{5} - \frac{1}{7} + \ldots \quad .$$

Solution: By the geometric series

$$\sum_{n=1}^{\infty} ar^{n-1} = a + ar + ar^2 + \ldots$$

which converges to $\frac{a}{(1-r)}$ if $|r| < 1$ (it diverges for $r \geq 1$), we have, with $r = -x^2$ and $a = 1$, that

$$\frac{1}{1 + x^2} = 1 - x^2 + x^4 - x^6 + \ldots \quad -1 < x < 1 . \quad (1)$$

Now

$$\tan^{-1}x = \int_0^x \frac{dx}{1 + x^2} \quad -1 < x < 1 .$$

Therefore by (1),

$$\tan^{-1}x = \int_0^x (1 - x^2 + x^4 - x^6 + \ldots) dx$$

or

$$\tan^{-1}x = x - \frac{x^3}{3} + \frac{x^5}{5} - \frac{x^7}{7} + \ldots \quad -1 < x < 1 \quad (2)$$

since, if $\Sigma V_n(x)$ converges uniformly to the sum $S(x)$ in $[a,b]$ and if $V_n(x)$, $n = 1, 2, 3, \ldots$, are continuous in $[a,b]$ then

$$\int_a^b S(x)\,dx = \sum_{n=1}^{\infty} \int_a^b V_n(x)\,dx$$

Now, at $x = 1$, we have on the right of (2). the series

$$1 - \frac{1}{3} + \frac{1}{5} - \frac{1}{7} + \ldots \quad .$$

However, this series is a convergent alternate series. Similarly, for $x = -1$ we also have a convergent alternate series. Therefore by Abel's Theorem we know that the interval of uniform convergence includes the endpoints $x = \pm 1$. Hence, the series is uniformly convergent for $-1 \leq x \leq 1$ and in this interval

$$\tan^{-1}x = x - \frac{x^3}{3} + \frac{x^5}{5} - \frac{x^7}{7} + \dots \quad . \qquad (3)$$

b) From part (a), using (3) we have,

$$\lim_{x\to1^-}\tan^{-1}x = \lim_{x\to1^-}(x - \frac{x^3}{3} + \frac{x^5}{5} - \frac{x^7}{7} + \dots)$$

which means, since $\tan\frac{\pi}{4} = 1$, that

$$\frac{\pi}{4} = 1 - \frac{1}{3} + \frac{1}{5} - \frac{1}{7} + \dots \quad .$$

This is the desired result.

● **PROBLEM** 6–38

Find a power series in x for:

a)   tan x

b) $\dfrac{\sin x}{\sin 2x}$ $(x \neq 0)$

Solution: To find the power series of the given functions, we need the following theorem:

Given the two power series

$$\sum_{n=o}^{\infty} a_n x^n = a_0 + a_1 x + a_2 x^2 + \dots + a_n x^n + \dots$$

and

$$\sum_{n=0}^{\infty} b_n x^n = b_0 + b_1 x + b_2 x^2 + \dots + b_n x^n + \dots ,$$

where $b_0 \neq 0$, and where both of the series are convergent in some interval $|x| < R$, let f be a function defined by

$$f(x) = \frac{a_0 + a_1 x + a_2 x^2 + \dots + a_n x^n + \dots}{b_0 + b_1 x + b_2 x^2 + \dots + b_n x^n + \dots} \quad .$$

Then for sufficiently small values of x the function f can be represented by the power series

$$f(x) = c_0 + c_1 x + c_2 x^2 + \dots + c_n x^n + \dots ,$$

319

where the coefficients c_0 , c_1 , c_2 , $\ldots$, c_n , $\ldots$ are found by long division or equivalently by solving the following relations successively for each c_i $(i = 0$ to ∞):

$$b_0 \ c_0 = a_0$$

$$b_0 \ c_1 + b_1 \ c_0 = a_1$$

$$\vdots$$

$$b_0 \ c_n + b_1 \ c_{n-1} + \ldots + b_n \ c_0 = a_n$$

$$\vdots$$

a) To find the power series expansion of tan x we need the Taylor's series for sin x and cos x. That is

$$\sin x = x - \frac{x^3}{3!} + \frac{x^5}{5!} - \ldots$$

and

$$\cos x = 1 - \frac{x^2}{2!} + \frac{x^4}{4!} - \ldots$$

Then,

$$\tan x = \frac{\sin x}{\cos x} = \frac{x - \frac{x^3}{3!} + \frac{x^5}{5!} - \ldots}{1 - \frac{x^2}{2!} + \frac{x^4}{4!} - \ldots} \qquad (1)$$

Therefore, by the theorem we can find the power series expansion of tan x by dividing the numerator by the denominator on the right side of (1). Hence, using long division we have

$$
\require{enclose}
\begin{array}{r}
x + \frac{1}{3} x^3 + \frac{2}{15} x^5 + \ldots \\[4pt]
1 - \frac{1}{2} x^2 + \frac{1}{24} x^4 - \ldots \enclose{longdiv}{\; x - \frac{1}{6} x^3 + \frac{1}{120} x^5 - \ldots} \\
\end{array}
$$

$$x - \frac{1}{2} x^3 + \frac{1}{24} x^5 - \ldots$$

$$\frac{1}{3} x^3 - \frac{1}{30} x^5 + \ldots$$

$$\frac{1}{3} x^3 - \frac{1}{6} x^5 + \ldots$$

$$\frac{2}{15} x^5 - \ldots$$

$$\frac{2}{15} x^5 - \ldots$$

Thus

$$\tan x = x + \frac{1}{3} x^3 + \frac{2}{15} x^5 + \ldots$$

b) Since
$$\sin x = x - \frac{x^3}{3!} + \frac{x^5}{5!} - \ldots$$

we have
$$\sin (2x) = 2x - \frac{(2x)^3}{3!} + \frac{(2x)^5}{5!} - \ldots \quad ,$$

so that

$$\frac{\sin x}{\sin 2x} = \frac{x - \frac{x^3}{3!} + \frac{x^5}{5!} - \ldots}{2x - \frac{(2x)^3}{3!} + \frac{(2x)^5}{5!} - \ldots} . \qquad (2)$$

Now multiplying the numerator and denominator on the right side of (2) by 1/x yields

$$\frac{\sin x}{\sin 2x} = \frac{1 - \frac{x^2}{6} + \frac{x^4}{120} - \ldots}{2 - \frac{4}{3} x^2 + \frac{4}{15} x^4 - \ldots} .$$

Now by long division

$$2 - \frac{4}{3} x^2 + \frac{4}{15} x^4 \overline{\Big)} \frac{\frac{1}{2} + \frac{1}{4} x^2 + \frac{5}{48} x^4 + \ldots}{1 - \frac{1}{6} x^2 + \frac{1}{120} x^4 - \ldots}$$

$$\frac{1}{2} - \frac{4}{6} x^2 + \frac{4}{30} x^4 - \ldots$$

$$\frac{1}{2} x^2 - \frac{15}{120} x^4 + \ldots$$

$$\frac{1}{2} x^2 - \frac{1}{3} x^4 + \ldots$$

$$\frac{25}{120} x^4 - \ldots$$

$$\frac{25}{120} x^4 - \ldots$$

Thus, for $x \neq 0$,
$$\frac{\sin x}{\sin (2x)} = \frac{1}{2} + \frac{1}{4} x^2 + \frac{5}{48} x^4 + \ldots \quad .$$

● **PROBLEM** 6–39

Starting from the power series expansion for $\tan^{-1} x$ find the power series expansion for $\tan x$. That is, derive the series expansion by a different method from the one used in the previous problem.

Solution: We are given that

$$\tan^{-1} x = x - \frac{x^3}{3} + \frac{x^5}{5} - \frac{x^7}{7} + \ldots + (-1)^{n-1} \frac{x^{2n-1}}{2n-1} + \ldots$$

for $-1 \leq x \leq 1$ and are asked to derive the power series expansion for tan x. To do this we need the following theorem:

If given the power series

$$y = f(x) = \sum_{n=0}^{\infty} c_n x^n$$

and if

$$|x| < R_0 \quad \text{and } c_1 \neq 0 ,$$

then there is an inverse function

$$x = g(y) = \sum_{n=1}^{\infty} b_n y^n$$

where $\quad |y| < R_1$, $R_1 > 0$.

In addition, the coefficients b_n are determined from the identity

$$x \equiv \sum_{n=1}^{\infty} b_n \left[\sum_{m=1}^{\infty} c_m x^m \right]^n .$$

For the given problem, we have

$$y = \tan^{-1} x = x - \frac{x^3}{3} + \frac{x^5}{5} - \frac{x^7}{7} + \ldots + (-1)^{n-1} \frac{x^{2n-1}}{2n-1} + \ldots$$

Therefore

$$x = \tan y \equiv \sum_{n=1}^{\infty} b_n y^n = \sum_{n=1}^{\infty} b_n \left(x - \frac{x^3}{3} + \frac{x^5}{5} + \ldots \right)^n$$

or

$$x \equiv b_1 \left(x - \frac{x^3}{3} + \ldots \right) + b_2 \left(x - \frac{x^3}{3} + \ldots \right)^2 +$$

$$+ b_3 \left(x - \frac{x^3}{3} + \ldots \right)^3 + \ldots ,$$

or

$$x \equiv b_1 x + b_2 x^2 + x^3 \left(-\frac{1}{3} b_1 + b_3 \right) + \ldots$$

Hence $\quad b_1 = 1, \; b_2 = 0, \; b_3 - \frac{1}{3} b_1 = 0 , \; \ldots ,$

Thus

$$x = \tan y = y + \frac{y^3}{3} + \ldots .$$

Show that

$$\int_0^1 \frac{\log (1-t)}{t} \, dt = -\left(\frac{1}{1^2} + \frac{1}{2^2} + \frac{1}{3^2} + \dots + \frac{1}{n^2} + \dots \right)$$

Solution: We first observe that the integral is improper at $t = 1$, but not at $t = 0$, since, by L'Hospital's rule, we have the integrand approaching the finite limit -1 as $t \to 0$.

Now to show that the integral is equal to the given power series, we need to start from Taylor's formula with remainder for

$$\log (1+x) = x - \frac{1}{2} x^2 + \frac{1}{3} x^3 - \dots + (-1)^{n-1} \frac{1}{n} x^n + R_{n+1}$$

if $1 > x > -1$ and where

$$\left| R_{n+1} \right| \le \begin{cases} \dfrac{|x|^{n+1}}{n+1} & \text{if} \quad 0 \le x \le 1 \\[3mm] \dfrac{|x|^{n+1}}{1+x} & \text{if} \quad -1 < x \le 0 \end{cases} \qquad (1)$$

However (1) shows that $R_{n+1} \to 0$ as $n \to \infty$ when x is limited as indicated. Therefore

$$\log (1+x) = x - \frac{1}{2}x^2 + \frac{1}{3}x^3 - \dots + (-1)^{n-1} \frac{1}{n} x^n + \dots \quad . \qquad (2)$$

Now letting $x = -t$ in (2) yields

$$\log (1-t) = -t - \frac{1}{2} t^2 - \frac{1}{3} t^3 - \dots - \frac{1}{n} t^n - \dots \quad .$$

Dividing by t gives,

$$\frac{\log (1-t)}{t} = -1 - \frac{1}{2} t - \frac{1}{3} t^2 - \dots - \frac{1}{n} t^{n-1} - \dots \quad .$$

Note that the series diverges at $t = 1$. However, we can still integrate from 0 to x. This yields,

$$\int_0^x \frac{\log (1-t)}{t} \, dt = \int_0^x \left(-1 - \frac{1}{2} t - \frac{1}{3} t^2 - \quad \dots \right) dt$$

$$= -x - \frac{1}{2^2} x^2 - \frac{1}{3^2} x^3 - \dots - \frac{1}{n^2} x^n - \dots \quad . \qquad (3)$$

However, this series converges when x = 1. That is, the series

$$-1 - \frac{1}{2^2} - \frac{1}{3^2} - \ldots - \frac{1}{n^2} - \ldots$$

$$= -\left(1 + \frac{1}{2^2} + \frac{1}{3^2} + \ldots + \frac{1}{n^2} + \ldots\right) = -\sum_{n=1}^{\infty} \frac{1}{n^2} \, ,$$

which we know to be convergent. Therefore, as a special case of (3) we have,

$$\int_0^1 \frac{\log (1-t)}{t} \, dt = -\left(\frac{1}{1^2} + \frac{1}{2^2} + \frac{1}{3^2} + \ldots + \frac{1}{n^2} + \ldots\right) .$$

• PROBLEM 6–41

Find the sum of the series

$$\sum_{n=1}^{\infty} n^2(x + 3)^n .$$

Solution: To find the sum of the given power series, we first need to determine its interval of convergence. By the ratio test, if $U_n = n^2 (x+3)^n$

$$\lim_{n \to \infty} \left| \frac{U_{n+1}}{U_n} \right| = \lim_{n \to \infty} \left| \frac{(n+1)^2 (x+3)^{n+1}}{n^2 (x+3)^n} \right| = |x+3| \lim_{n \to \infty} \frac{n^2 + 2n+1}{n^2}$$

$$= |x + 3| .$$

Therefore, the series converges for $|x + 3| < 1$ (i.e., $-4 < x < -2$) and diverges for $|x + 3| > 1$.

Now to proceed further, we consider the geometric series

$$\sum_{n=0}^{\infty} x^n ,$$

which converges if $|x| < 1$. That is $R = 1$ and

$$\sum_{n=0}^{\infty} x^n = \frac{1}{(1-x)}$$

where $|x| < 1$. By differentiating this series, which we can do since a power series can be differentiated term by term over any interval lying entirely within the interval of convergence, we obtain

$$\sum_{n=1}^{\infty} nx^{n-1} = \frac{1}{(1-x)^2} \qquad -1 < x < 1 \qquad\qquad (1)$$

Differentiating again yields

$$\sum_{n=2}^{\infty} (n^2 - n)x^{n-2} = \frac{2}{(1-x)^3}$$

or

$$\sum_{n=2}^{\infty} n^2 x^{n-2} - \sum_{n=2}^{\infty} nx^{n-2} = \frac{2}{(1-x)^3} \ . \qquad\qquad (2)$$

Multiplying each side of (2) by x^2 gives

$$\sum_{n=2}^{\infty} n^2 x^n - \sum_{n=2}^{\infty} nx^n = \frac{2x^2}{(1-x)^3} \qquad -1 < x < 1$$

or

$$\sum_{n=2}^{\infty} n^2 x^n - \sum_{n=1}^{\infty} nx^n + x = \frac{2x^2}{(1-x)^3}$$

which by (1) means

$$\sum_{n=1}^{\infty} n^2 x^n - \frac{x}{(1-x)^2} = \frac{2x^2}{(1-x)^3} \qquad -1 < x < 1$$

Hence

$$\sum_{n=1}^{\infty} n^2 x^n = \frac{2x^2}{(1-x)^3} + \frac{x}{(1-x)^2} \qquad -1 < x < 1$$

Then substituting x by x+3 we obtain

$$\sum_{n=1}^{\infty} n^2 (x+3)^n = \frac{2(x+3)^2}{(-x-2)^3} + \frac{(x+3)}{(-x-2)^2} \qquad -1 < x + 3 < 1$$

or

$$\sum_{n=1}^{\infty} n^2 (x+3)^n = -\ \frac{x^2 + 7x + 12}{(x+2)^3} \qquad -4 < x < -2 \ .$$

● **PROBLEM 6–42**

Using Power series, show that

$$\log 2 = 1 - \frac{1}{2} + \frac{1}{3} - \frac{1}{4} + \cdots \ .$$

Solution: Starting with the fact that

$$e^x = 1 + x + \frac{x^2}{2!} + \frac{x^3}{3!} + \ldots$$

we have

$$e^x \to \infty \quad \text{as } x \to \infty \quad \text{and}$$

$$e^{-x} = \frac{1}{e^x} \to 0 \quad \text{as } x \to \infty \;.$$

Therefore e^x increases from 0 to ∞ as x increases from $-\infty$ to ∞. Now by the implicit function theorem we know that $x = e^y$ defines y as a continuous, differentiable function of x, increasing from $-\infty$ to ∞ as x increases from 0 to ∞. This means $y = \log x$ if and only if $x = e^y$. In addition

$$\frac{d(\log x)}{dx} = \frac{1}{x}$$

and so for $x > -1$ we have

$$\log (1 + x) = \int_0^x \frac{dt}{1+t}$$

$$= \int_0^x \sum_{n=0}^{\infty} (-1)^n t^n \, dt$$

if $|x| < 1$. This because, for

$$\sum_{n=0}^{\infty} (-1)^n t^n$$

we have

$$S_n = 1 - t + t^2 - t^3 + \ldots + (-1)^n t^n$$

$$tS_n = t - t^2 + t^3 - t^4 + \ldots + (-1)^n t^{n+1}$$

Adding yields

$$S_n + tS_n = 1 + (-1)^n t^{n+1}$$

or

$$S_n = \frac{1 + (-1)^n t^{n+1}}{1 + t}$$

and

$$\lim_{n \to \infty} S_n = \frac{1}{1+t}$$

326

for $|t| < 1$.

Thus,

$$\log (1+x) = x - \frac{x^2}{2} + \frac{x^3}{3} - \ldots$$

for $|x| < 1$.

Abel's limit theorem states that if

$$\sum_{n=0}^{\infty} a_n$$

converges, then

$$\sum_{n=0}^{\infty} a_n x^n$$

converges uniformly on $[-r,1]$ if $0 \leq r < 1$. In particular

$$\lim_{\substack{x \to 1 \\ x<1}} \sum_{n=0}^{\infty} a_n x^n = \sum_{n=0}^{\infty} \lim_{\substack{x \to 1 \\ x<1}} a_n x^n = \sum_{n=0}^{\infty} a_n \; .$$

Therefore, since

$$\log (1+x) = \sum_{n=1}^{\infty} \frac{(-1)^{n-1}}{n} x^n$$

and since the alternating series

$$\sum_{n=1}^{\infty} \frac{(-1)^{n-1}}{n}$$

converges, we have

$$\lim_{\substack{x \to 1 \\ x<1}} \sum_{n=1}^{\infty} \frac{(-1)^{n-1}}{n} x^n = \sum_{n=1}^{\infty} \frac{(-1)^{n-1}}{n}$$

Thus,

$$\log 2 = \sum_{n=1}^{\infty} \frac{(-1)^{n-1}}{n} = 1 - \frac{1}{2} + \frac{1}{3} - \frac{1}{4} + \ldots \; .$$

Approximate the value of the integral

$$\int_0^1 \frac{1 - e^{-x^2}}{x^2} \, dx$$

Solution: Starting from the fact that

$$e^a = 1 + a + \frac{a^2}{2!} + \frac{a^3}{3!} + \frac{a^4}{4!} + \cdots$$

for $-\infty < a < \infty$,

and upon setting $a = -x^2$ gives

$$e^{-x^2} = 1 - x^2 + \frac{x^4}{2!} - \frac{x^6}{3!} + \frac{x^8}{4!} - \cdots$$

for $-\infty < x < \infty$.

Therefore

$$1 - e^{-x^2} = x^2 - \frac{x^4}{2!} + \frac{x^6}{3!} - \frac{x^8}{4!} + \cdots$$

which means that

$$\frac{1 - e^{-x^2}}{x^2} = 1 - \frac{x^2}{2!} + \frac{x^4}{3!} - \frac{x^6}{4!} + \cdots$$

Now this series converges for all x and in addition converges uniformly for $0 \le x \le 1$.

This because

$$\frac{1 - e^{-x^2}}{x^2} = \sum_{n=1}^{\infty} \frac{(-1)^{n+1} x^{2n-2}}{n!} \qquad (1)$$

and since

$$\left| \frac{x^{2n-2}}{n!} \right| \le \left| \frac{1}{n!} \right|$$

for $0 \le x \le 1$,

from the fact that

$$\sum_{n=1}^{\infty} \frac{1}{n!}$$

converges, it follows that the series (1) converges

328

uniformly for $0 \leq x \leq 1$ by the Weierstrass M-test.
Integrating both the sides of (1) from 0 to 1 gives

$$\int_0^1 \frac{1 - e^{-x^2}}{x^2} \, dx = x - \frac{x^3}{3 \cdot 2!} + \frac{x^5}{5 \cdot 3!} - \frac{x^7}{7 \cdot 4!} + \dots \Bigg|_0^1$$

$$= 1 - \frac{1}{3 \cdot 2!} + \frac{1}{5 \cdot 3!} - \frac{1}{7 \cdot 4!} + \dots$$

$$= 1 - 0.166666 + 0.033333 - 0.005952 + \dots$$

$$\approx 0.8607$$

● **PROBLEM 6–44**

Show that

$$\sum_{n=0}^{\infty} \frac{a_n x^n}{1-x} = \sum_{n=0}^{\infty} (a_0 + a_1 + \dots + a_n) x^n . \qquad (1)$$

Then use the result to find the function represented
by the following series:

a)

$$\sum_{n=0}^{\infty} \left(1 + \frac{1}{1!} + \dots + \frac{1}{n!} \right) x^n$$

b)

$$\sum_{n=1}^{\infty} \left(0 + 1 - \frac{1}{2} + \frac{1}{3} - \dots + (-1)^{n+1} \frac{1}{n} \right) x^n .$$

Solution: Starting from the right side of equation (1)
it follows that

$$\sum_{n=0}^{\infty} (a_0 + a_1 + \dots + a_n) x^n = \sum_{n=0}^{\infty} (a_0 x^n + a_1 x^n + \dots + a_n x^n)$$

$$= a_0 + a_0 x + a_1 x + a_0 x^2 + a_1 x^2 + a_2 x^2$$

$$+ a_0 x^3 + a_1 x^3 + a_2 x^3 + a_3 x^3 + \dots$$

or

$$\sum_{n=0}^{\infty} (a_0 + a_1 + \dots + a_n) x^n$$

$$= a_0 (1 + x + x^2 + x^3 + \dots) + a_1 (x + x^2 + x^3 + \dots)$$

$$+ a_2 (x^2 + x^3 + x^4 + \dots) + a_3 (x^3 + x^4 + x^5 + \dots)$$

$$+ \dots . \qquad (2)$$

329

Now multiplying (2) by (1-x) gives

$$(1-x) \sum_{n=0}^{\infty} (a_0 + \ldots + a_n)x^n =$$

$$(1-x)\left[a_0(1 + x + x^2 + \ldots) + a_1(x + x^2 + x^3 + \ldots) \right.$$
$$+ a_2(x^2 + x^3 + x^4 + \ldots)$$

$$\left. + a_3(x^3 + x^4 + x^5 + \ldots) + \ldots \right] - \left[a_0(x + x^2 + x^3 + \ldots) \right.$$

$$\left. + a_1(x^2 + x^3 + x^4 + \ldots) + a_2(x^3 + x^4 + x^5 + \ldots) + \ldots \right]$$

$$= a_0 + a_1 x + a_2 x^2 + a_3 x^3 + \ldots$$

This means that

$$(1-x) \sum_{n=0}^{\infty} (a_0 + a_1 + \ldots + a_n)x^n = \sum_{n=0}^{\infty} a_n x^n$$

or

$$\sum_{n=0}^{\infty} \frac{a_n x^n}{1-x} = \sum_{n=0}^{\infty} (a_0 + a_1 + \ldots + a_n)x^n$$

a)

$$\sum_{n=0}^{\infty} \left(1 + \frac{1}{1!} + \ldots + \frac{1}{n!}\right)x^n .$$

For this series

$$a_0 = 1, \quad a_1 = \frac{1}{1!}, \ldots, \quad a_n = \frac{1}{n!}$$

Therefore by equation (1),

$$\sum_{n=0}^{\infty} \left(1 + \frac{1}{1!} + \ldots + \frac{1}{n!}\right)x^n = \sum_{n=0}^{\infty} \frac{x^n}{n!} \cdot \frac{1}{1-x} = \frac{1}{1-x} \sum_{n=0}^{\infty} \frac{x^n}{n!}$$

However,

$$\sum_{n=0}^{\infty} \frac{x^n}{n!} = e^x ,$$

so the function represented by the series is

$$\frac{e^x}{1 - x}$$

b)

$$\sum_{n=1}^{\infty} \left(0 + 1 - \frac{1}{2} + \frac{1}{3} - \dots + (-1)^{n+1} \frac{1}{n} \right) x^n \; .$$

By (1) this series equals

$$\sum_{n=1}^{\infty} (-1)^{n+1} \frac{x^n}{n} \cdot \frac{1}{1-x} \; .$$

Since

$$\log (1+x) = \sum_{n=1}^{\infty} (-1)^{n+1} \frac{x^n}{n} \; ,$$

this means that the function represented by the series is

$$\frac{\log (1+x)}{1 - x} \; .$$

FOURIER SERIES

Consider the infinite trigonometric series

$$\frac{a_0}{2} + \sum_{n=1}^{\infty} (a_n \cos nx + b_n \sin nx) \text{ and assume that it converges}$$

uniformly for all $x \in (-\pi, \pi)$. It can then be considered as a function f of x with period 2π, i.e.

$$f(x) = \frac{a_0}{2} + \sum_{n=1}^{\infty} a_n \cos nx + b_n \sin nx. \tag{1}$$

Determine the values of a_n, b_n in terms of f(x).

Solution: It is this computation which leads to the definition of the Fourier Series of a given function f(x). First multiply both sides of (1) by cos mx where m is a positive integer which we will vary later. This yields

$$f(x) \cos mx = \frac{a_0}{2} \cos mx + \sum_{n=1}^{\infty} a_n \cos nx \cos mx$$

$$+ \sum_{n=1}^{\infty} b_n \sin nx \cos mx. \qquad (2)$$

The next step is to integrate both sides of equation (2) from $-\pi$ to π. In order to integrate the two series on the right term by term these two series would have to be uniformly convergent, but since this exercise is intended only to motivate a definition, we will simply assume that termwise integration is valid. Thus, (2) becomes

$$\int_{-\pi}^{\pi} f(x) \cos mx dx = \frac{a_o}{2} \int_{-\pi}^{\pi} \cos mx dx + \sum_{n=1}^{\infty} \left(a_n \int_{-\pi}^{\pi} \cos nx \cos mx dx \right)$$

$$+ \sum_{n=1}^{\infty} \left(b_n \int_{-\infty}^{\infty} \sin nx \cos mx dx \right) \qquad (3)$$

This rather formidable expression yields useful information if one recalls the trigonometric identities

$$\sin nx \sin mx = \frac{1}{2} \cos(n-m)x - \frac{1}{2} \cos(n+m)x \qquad (4)$$

$$\cos nx \cos mx = \frac{1}{2} \cos(n+m)x + \frac{1}{2} \cos(n-m)x \qquad (5)$$

$$\sin nx \cos mx = \frac{1}{2} \sin(n+m)x + \frac{1}{2} \sin(n-m)x. \qquad (6)$$

Using these three identities, the following equations may be verified by carrying out the integrations:

$$\int_{-\pi}^{\pi} \sin nx \cos mx dx = 0 \qquad \text{(for all } n, m > 0\text{)} \qquad (7)$$

$$\int_{-\pi}^{\pi} \cos nx \cos mx dx = \begin{cases} 0 & \text{(if } n \neq m\text{)} \\ \pi & \text{(if } n = m\text{)} \end{cases} \qquad (8)$$

$$\int_{-\pi}^{\pi} \sin nx \sin mx dx = \begin{cases} 0 & \text{(if } n \neq m\text{)} \\ \pi & \text{(if } n = m\text{)} \end{cases} \qquad (9)$$

For instance, using the identity (4) in the integral of equation (9) yields

$$\int_{-\pi}^{\pi} \sin nx \sin mx \, dx = \frac{1}{2} \int_{-\pi}^{\pi} \cos(n-m)x dx - \frac{1}{2} \int_{-\pi}^{\pi} \cos(n+m)x dx. \qquad (10)$$

If $n \neq m$, then

$$\int_{-\pi}^{\pi} \cos(n-m)x\,dx = \left.\frac{\sin(n-m)x}{n-m}\right|_{-\pi}^{\pi} = 0$$

and if n = m, then

$$\int_{-\pi}^{\pi} \cos(n-n)x\,dx = \int_{-\pi}^{\pi} dx = 2\pi.$$

Also,

$$\int_{-\pi}^{\pi} \cos(n+m)x\,dx = \left.\frac{\sin(n+m)x}{n+m}\right|_{-\pi}^{\pi} = 0 \qquad \text{(for all } n,m>0\text{)}.$$

Using these results in (10) yields the result quoted in (9) and the other formulas are established in a similar fashion. These formulas are called the orthogonality properties of sin and cos.

Returning to the series in (3), it is seen that all terms in the second sum are zero (by equation (7)) and that for any m, only one term in the first sum is nonzero by equation (8). That is, for m>0,

$$\int_{-\pi}^{\pi} \cos mx\,dx = 0$$

so that (3) gives

$$\int_{-\pi}^{\pi} f(x)\cos mx\,dx = a_m\pi \qquad (m>0). \tag{11}$$

The coefficients b_n are treated similarly, that is the expansion (1) is multiplied by sin mx and integrated. Again the orthogonality properties (7)-(9) are employed to yield

$$\int_{-\pi}^{\pi} f(x)\sin mx\,dx = b_m\pi. \tag{12}$$

Finally, to obtain a_0, simply integrate the expansion (1) as it stands from $-\pi$ to π. This results in

$$\int_{-\pi}^{\pi} f(x)\,dx = a_0\pi. \tag{13}$$

The results of equations (11), (12), (13) may be summarized as

$$a_n = \frac{1}{\pi} \int_{-\pi}^{\pi} f(x) \cos nx \, dx \qquad\qquad (n \geq 0) \qquad\qquad (14)$$

$$b_n = \frac{1}{\pi} \int_{-\pi}^{\pi} f(x) \sin nx \, dx \qquad\qquad (n > 0). \qquad\qquad (15)$$

Thus, it has been proved that if a function f is representable by a uniformly convergent trigonometric series then that series must have the coefficients of equations (14) and (15).

● **PROBLEM 6–46**

Find the Fourier series of the function $f(x) = e^x$, $-\pi < x < \pi$.

Solution: The given function is defined on the familiar interval $(-\pi, \pi)$. A good habit to acquire is to recall the most general definition of a Fourier series of a function and then use the particular values given in the problem in this definition. Thus, recall that the Fourier series of a function $f(x)$ which is periodic of period 2c and defined on some domain D of the real numbers, $f:D \rightarrow R$, is given by

$$f(x) \sim \frac{a_o}{2} + \sum_{n=1}^{\infty} \left[a_n \cos\left(\frac{\pi n x}{c}\right) + b_n \sin\left(\frac{\pi n x}{c}\right) \right] \qquad (1)$$

where the coefficients are given by

$$a_n = \frac{1}{c} \int_a^{a+2c} f(x) \cos\left(\frac{\pi n x}{c}\right) dx \qquad (2)$$

$$b_n = \frac{1}{c} \int_a^{a+2c} f(x) \sin\left(\frac{\pi n x}{c}\right) dx , \qquad (3)$$

where a is any number such that the interval $(a, a+2c)$ is contained in D. In the case at hand, $f(x) = e^x$, $x \in (-\pi, \pi)$ is only defined on an interval of length 2π. Hence, (1) represents the periodic extension of f over the whole real axis. (Note that f is not periodic within that interval so 2π is the only period that can be assigned any meaning here.) Therefore, in equations (1)-(3) the substitutions to be made for this particular function are

334

$$c = \pi; \quad a = -\pi; \quad f(x) = e^x. \tag{4}$$

Using (4), the coefficients become

$$a_o = \frac{1}{\pi} \int_{-\pi}^{\pi} e^x dx = \frac{2}{\pi} \frac{e^\pi - e^{-\pi}}{2} = \frac{2}{\pi} \sinh \pi.$$

$$a_n = \frac{1}{\pi} \int_{-\pi}^{\pi} e^x \cos nx \, dx = \frac{1}{\pi} \left[e^x \frac{(\cos nx + n\sin nx)}{n^2 + 1} \right]_{-\pi}^{\pi}$$

$$= (-1)^n \frac{e^\pi - e^{-\pi}}{(n^2+1)\pi} = (-1)^n \frac{2 \sinh \pi}{(n^2+1)\pi}.$$

$$b_n = \frac{1}{\pi} \int_{-\pi}^{\pi} e^x \sin nx \, dx = \frac{1}{\pi} \left[e^x \frac{(\sin nx - n\cos nx)}{n+1} \right]_{-\pi}^{\pi}$$

$$= (-1)^{n+1} \frac{e^\pi - e^{-\pi}}{(n^2+1)\pi} n$$

$$= - \frac{(-1)^n 2n \sinh \pi}{(n^2+1)\pi}.$$

Substituting these values back into the Fourier expression, (1), with the substitutions of (3) gives

$$f(x) \sim \frac{a_o}{2} + \sum_{n=1}^{\infty} \left(a_n \cos nx + b_n \sin nx \right)$$

$$= \frac{\sinh \pi}{\pi} + \sum_{n=1}^{\infty} \left[\frac{(-1)^n 2 \sinh \pi}{(n^2+1)\pi} \cos nx \right.$$

$$\left. - \frac{(-1)^n 2n \sinh \pi}{(n^2+1)\pi} \sin nx \right]$$

$$= \frac{\sinh \pi}{\pi} \left\{ 1 + 2 \sum_{n=1}^{\infty} \left[\frac{(-1)^n}{n^2+1} (\cos nx - n \sin nx) \right] \right\}. \tag{5}$$

Recall that the symbol $\sim$ is used in (5) since no determination of the convergence of the series to the function has yet been made. This will be done later and it will be shown that $\sim$ may be replaced by =.

Determine the Fourier series of the function given by

$$\left\{\begin{array}{l} f(x) = x^2, \ x \in (-\pi, \pi) \\ f(x+2\pi) = f(x), \ \text{all} \ x \end{array}\right\} .$$
(1)

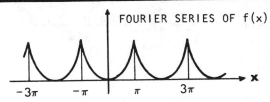

FOURIER SERIES OF f(x)

Solution: The most general definition of the Fourier series of a function f which is periodic with period 2c and defined on the interval D is given by

$$f(x) \sim \frac{a_o}{2} + \sum_{n=1}^{\infty} \left[a_n \cos\left(\frac{\pi n x}{c}\right) + b_n \sin\left(\frac{\pi n x}{c}\right) \right]$$
(2)

where

$$a_n = \int_a^{a+2c} f(x) \cos\left(\frac{\pi n x}{c}\right) dx$$
(3)

$$b_n = \int_a^{a+2c} f(x) \sin\left(\frac{\pi n x}{c}\right) dx .$$
(4)

Here, a is any number such that $(a, a+2c) \subseteq D$. For the function in (1), D is the whole real axis, $2c = 2\pi$ and since any a can be used, the most convenient is $a = -\pi$. Then

$$a_o = \frac{1}{\pi} \cdot \int_{-\pi}^{\pi} x^2 dx = \frac{2\pi^2}{3}$$
(5)

$$a_n = \frac{1}{\pi} \int_{-\pi}^{\pi} x^2 \cos nx \, dx$$
(6)

Integrating (6) by parts twice gives

$$a_n = \frac{1}{\pi} \left[\frac{x^2}{n} \sin nx \right]_0^{\pi} - \frac{2}{n\pi} \int_{-\pi}^{\pi} x \sin nx \, dx$$

$$= 0 + \frac{2}{\pi n^2} \left[x \cos nx \right]_{-\pi}^{\pi} - \frac{2}{\pi n^2} \int_{-\pi}^{\pi} \cos nxdx$$

$$= \frac{4}{n^2} \left[\cos n\pi \right] - \frac{2}{\pi n^3} \left(\sin nx \right)_{-\pi}^{\pi}$$

$$= \frac{4}{n^2} (-1)^n . \qquad\qquad (n > 0) \qquad (7)$$

Finally,

$$b_n = \frac{1}{\pi} \int_{-\pi}^{\pi} x^2 \sin nxdx.$$

Note that the integrand, $F(x) = x^2 \sin nx$, is an odd function of x, that is

$$F(-x) = (-x)^2 \sin(-nx) = -x^2 \sin nx = -F(x). \quad \text{Thus}$$

$$b_n = \frac{1}{\pi} \int_{-\pi}^{0} x^2 \sin nxdx + \frac{1}{\pi} \int_{0}^{\pi} x^2 \sin nxdx$$

$$= \frac{-1}{\pi} \int_{0}^{\pi} y^2 \sin nydy + \frac{1}{\pi} \int_{0}^{\pi} x^2 \sin nxdx = 0 \qquad (8)$$

where the change of variables $x = -y$ was made in the first integral of (8). Using these values of a_n and b_n in (2) gives

$$f(x) \sim \frac{\pi^2}{3} + \sum_{n=1}^{\infty} (-1)^n \left(\frac{4}{n^2} \right) \cos nx. \qquad (9)$$

The graph of this series is shown in the Figure. Note that even if f had been defined only on $(-\pi, \pi)$, it would have the same Fourier series (9) and this series would have the same graph. Thus, the series would represent a periodic extension of the values of x^2 in the interval $(-\pi, \pi)$ if it converged (it does by the Weierstrass M-test, but convergence questions will be handled more generally later on in this chapter).

• PROBLEM 6–48

Determine the Fourier series of the function given by

337

$$f(x) = \begin{cases} -1 & x \in (-\pi, 0) \\ 1 & x \in [0, \pi] \end{cases}. \tag{1}$$

Solution: This is the familiar square wave used in elec-tronics (see Figure 1). First notice that this function is discontinuous at 0 but it will still have a Fourier series since the Fourier coefficients can be determined for any integrable function. The Fourier series of a function with period 2c defined on a domain D is given by

$$f(x) \sim \frac{a_o}{2} + \sum_{n=1}^{\infty} \left[a_n \cos\left(\frac{\pi n x}{c}\right) + b_n \sin\left(\frac{\pi n x}{c}\right) \right] \tag{2}$$

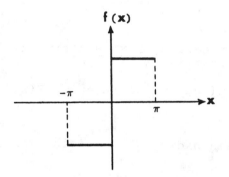

Fig. 1

FOURIER SERIES OF f

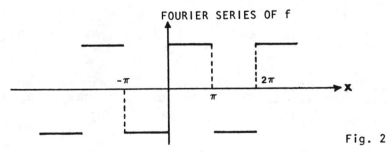

Fig. 2

where

$$a_n = \frac{1}{c} \int_a^{a+2c} f(x) \cos\left(\frac{\pi n x}{c}\right) dx \tag{3}$$

$$b_n = \frac{1}{c} \int_a^{a+2c} f(x) \sin\left(\frac{\pi n x}{c}\right) dx.$$

Here a is any number such that the interval (a, a+2c) is in D. But since f is only defined on $(-\pi, \pi)$, $c = 2\pi$ and a is $-\pi$ by necessity. Thus, the Fourier coefficients are

$$a_o = \frac{1}{\pi} \int_{-\pi}^{0} (-1)dx + \frac{1}{\pi} \int_{0}^{\pi} (+1)dx = -1 + 1 = 0$$

$$a_n = \frac{1}{\pi} \int_{-\pi}^{0} -\cos nx dx + \frac{1}{\pi} \int_{0}^{\pi} \cos nx dx = 0 + 0 = 0.$$

$$b_n = \frac{1}{\pi} \int_{-\pi}^{0} (-\sin nx)dx + \frac{1}{\pi} \int_{0}^{\pi} (\sin nx)dx.$$

But

$$\frac{1}{\pi} \int_{-\pi}^{0} (-\sin nx)dx = \frac{1}{n\pi}[1 - \cos n\pi] = \frac{1}{\pi} \int_{0}^{\pi} \sin nx dx.$$

Thus,

$$b_n = \frac{2}{\pi} \int_{0}^{\pi} \sin nx dx = \left\{ \begin{array}{ll} \frac{4}{n\pi} & (n = odd), \\ 0 & (n = even) \end{array} \right\}$$

And so the Fourier series reads

$$f(x) \sim \frac{4}{\pi} \sum_{n=odd}^{\infty} \frac{\sin nx}{n}. \tag{4}$$

It can be proven by Dirichlet's test that this series converges for all x, but convergence questions will be attended to later. It is particularly interesting to note that at x = 0 the series does not converge to f(x), and this will also be examined later. For now, note that the series in (4) represents a periodic extension of the values of f in the interval $(-\pi, \pi)$ as graphed in Figure 2. Thus, (4) is a representation of the square wave used in electronics.

• PROBLEM 6-49

Find the Fourier series of the function $f(x) = |x|$, $-\pi < x \le \pi$.

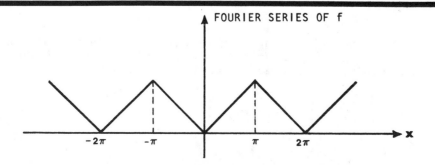

FOURIER SERIES OF f

Solution: The Fourier series of a function f with period 2c (or defined only on the interval 2c) is given by

$$f(x) \sim \frac{a_o}{2} + \sum_{n=1}^{\infty} \left[a_n \cos\left(\frac{\pi n x}{c}\right) + b_n \sin\left(\frac{\pi n x}{c}\right) \right] \tag{1}$$

where

$$a_n = \frac{1}{c} \int_a^{a+2c} f(x) \cos\left(\frac{\pi n x}{c}\right) dx \tag{2}$$

$$b_n = \frac{1}{c} \int_a^{a+2c} f(x) \sin\left(\frac{\pi n x}{c}\right) dx. \tag{3}$$

Here a is any number such that (a, a+2c) is contained within the domain of definition of f. In the case at hand, 2c = 2π, and a = -π by necessity.

In solving the given problem we make use of the notions of odd functions and even functions. A function f(x), is said to be even if f(-x) = f(x). Some examples of even functions are y = cos x, y = x^2, and y = |x|. One characteristic of such functions is that they are symmetric with respect to the y - axis.

A function f(x) is called odd if f(-x) = -f(x). Examples of odd functions are y = sin x, y = x, y = x^3. We note further that the product of two even functions is itself an even function while the product of an odd with an even function is an odd function. Finally, if f(x) is even on the interval -A ≤ x ≤ A, then

$$\int_{-A}^{A} f(x)\,dx = 2 \int_0^A f(x)\,dx,$$

and if f(x) is odd over the same interval,

$$\int_{-A}^{A} f(x)\,dx = 0.$$

In the given problem, f(x) = |x| is an even function. The evaluation of the Fourier coefficients is now shown to be considerably simplified. First, consider

$$a_n = \frac{1}{c} \int_{-c}^{c} |x| \cos \frac{n \pi x}{c}\,dx$$

for $-\pi \leq x \leq \pi$ and $a = -c$.   Since $|x| = x$ for $0 \leq x \leq \pi$, this may be written as

$$a_n = \frac{2}{c} \int_0^c x \cos \frac{n\pi x}{c} \, dx.$$

Next, since $\sin x$ is odd, the coefficients

$$b_n = \frac{1}{c} \int_{-c}^c f(x) \sin \frac{n\pi x}{c} \, dx, \quad (n = 1, 2, 3, \ldots)$$

are all equal to zero.   Hence we need consider only the a_n in the series (a).

Now,

$$a_n = \frac{2}{\pi} \int_0^\pi x \cos \frac{n\pi x}{\pi} \, dx = \frac{2}{\pi} \int_0^\pi x \cos nx \, dx.$$

Integrating by parts,

$$a_n = \frac{2}{\pi} \left[\frac{\cos nx}{n^2} + \frac{x \sin nx}{n} \right]\Bigg|_0^\pi$$

$$= \frac{2}{\pi} \left[\frac{\cos n\pi - 1}{n^2} \right] = \frac{2}{\pi} \left[\frac{(-1)^n - 1}{n^2} \right]$$

(since $\cos n\pi = -1$ for odd n and 1 for even n).

$$\frac{2}{\pi} \left[\frac{(-1)^n - 1}{n^2} \right] = \begin{cases} \dfrac{-4}{\pi n^2}, & n \text{ odd} \\[2mm] 0, & n \text{ even} \end{cases} \qquad n = 1, 2, 3, \ldots \ .$$

The above integration and evaluation is not valid for $n = 0$. To find a_0,

$$a_0 = \frac{2}{\pi} \int_0^\pi f(x) \cos(0) \frac{\pi x}{\pi} \, dx = \frac{2}{\pi} \int_0^\pi f(x) \, dx = \frac{2}{\pi} \int_0^\pi x \, dx$$

$$= \frac{x^2}{\pi} \Bigg|_0^\pi = \pi.$$

Thus, the required series is

$$\frac{\pi}{2} - \frac{4}{\pi} \sum_{\substack{n=1 \\ (n\ \text{odd})}}^{\infty} \frac{\cos nx}{n^2} = \frac{\pi}{2} - \frac{4}{\pi} \sum_{n=1}^{\infty} \cos \frac{(2n-1)x}{(2n-1)^2} \quad .$$

That is,

$$f \sim \frac{\pi}{2} - \frac{4}{\pi} \sum_{n=1}^{\infty} \frac{\cos(2n-1)x}{(2n-1)^2}$$

is the Fourier series of f. The graph of this series is shown in the Figure and it can be seen that the triangular waveform there represents a periodic extension of the values of $|x|$ in the interval $[-\pi, \pi]$.

• PROBLEM 6–50

Find the Fourier series for the sawtooth waveform shown in Fig. 1.

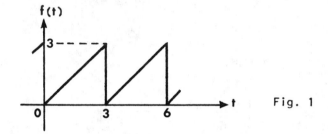

Fig. 1

<u>Solution</u>: We note that this waveform repeats every 3 seconds. Thus the period $\tau = 3s$ and since $\omega_o = \frac{2\pi}{\tau}$, $\omega_o = \frac{2\pi}{3}$.

A periodic function has a Fourier series in the form

$$f(t) \sim a_o + \sum_{n=1}^{\infty} (a_n \cos n\omega_o t + b_n \sin n\omega_o t) \qquad (1)$$

where a_o is the average value of the function f(t) and is defined as

$$a_o = \frac{1}{\tau} \int_0^{\tau} f(t)\,dt. \qquad (2)$$

f(t) represents one period of the entire periodic function, thus $f(t) = t; \ 0 < t < 3$.

Substituting these values yields:

$$a_o = \frac{1}{3} \int_0^3 t \, dt$$

$$a_o = \frac{1}{3} \left(\frac{t^2}{2} \right) \Big|_0^3$$

$$a_o = \frac{1}{3} \left(\frac{9}{2} - 0 \right) = \frac{3}{2} \, .$$

The coefficients b_n and a_n are defined as

$$b_n = \frac{2}{\tau} \int_0^\tau f(t) \sin n\omega_o t \, dt \qquad\qquad (3)$$

$$a_n = \frac{2}{\tau} \int_0^\tau f(t) \cos n\omega_o t \, dt. \qquad\qquad (4)$$

Hence,

$$b_n = \frac{2}{3} \int_0^3 t \sin n \, \frac{2\pi}{3} t \, dt$$

$$b_n = \frac{2}{3} \left[\frac{1}{\left(\frac{n2\pi}{3} \right)^2} \sin n \left(\frac{2\pi}{3} \right) t - \frac{t}{\frac{n2\pi}{3}} \cos n \, \frac{2\pi}{3} t \right] \Bigg|_0^3$$

$$b_n = \frac{2}{3} \left[\frac{1}{\left(\frac{n2\pi}{3} \right)^2} \sin n2\pi - \frac{9}{n2\pi} \cos n2\pi \right]$$

$$b_n = \frac{3}{n^2 2\pi^2} \sin n \, 2\pi - \frac{3}{n\pi} \cos n \, 2\pi.$$

The sine term is zero for all n since any multiple of 2π in the sine term is zero.

Hence

$$b_n = - \left[\frac{3}{\pi} \frac{1}{n} \right] \qquad\qquad n = 1,2,3....$$

since the cos term is 1 for any multiple of 2π.

$$a_n = \frac{2}{3} \int_0^3 t \cos n\omega_o t \, dt$$

$$a_n = \frac{2}{3} \left[\frac{1}{\left(\frac{n2\pi}{3}\right)^2} \cos \frac{n2\pi}{3} t + \frac{t}{\frac{n2\pi}{3}} \sin \frac{n2\pi}{3} t \right] \Big|_0^3$$

$$a_n = \frac{2}{3} \left[\frac{9}{n^2 4\pi^2} \cos n2\pi - \frac{9}{n^2 4\pi^2} \cos 0 \right]$$

But $\cos n2\pi = \cos 0$ for all n; therefore, $a_n = 0$ for all n.

The Fourier representation of this waveform is written

$$f(t) \sim \frac{3}{2} - \frac{3}{\pi} \sum_{n=1}^{\infty} \frac{1}{n} \sin n \frac{2\pi}{3} t$$

$$f(t) \sim \frac{3}{2} - \frac{3}{\pi} \left(\sin \frac{2\pi}{3} t + \frac{1}{2} \sin \frac{4\pi}{3} t + \frac{1}{3} \sin \frac{6\pi}{3} + \ldots \right).$$

• PROBLEM 6–51

The Fourier series for $f(x) = |x|$ $-\pi < x \le \pi$
and $f(x+2\pi) = f(x)$ is

$$\frac{\pi}{2} - \frac{4}{\pi} \sum_{n=1}^{\infty} \frac{\cos(2n-1)x}{(2n-1)^2} . \tag{1}$$

Without computing any Fourier coefficients, find the
Fourier series for

$$g(x) = \begin{cases} -1 & -\pi < x < 0 \\ +1 & 0 \le x \le \pi \end{cases}$$

and $g(x+2\pi) = g(x)$.

Solution: The function f may be written as

$$f(x) = \begin{array}{c} -x \\ x \end{array} \begin{cases} -\pi < x < 0 \\ 0 \le x \le \pi \end{cases} , \quad f(x+2\pi) = f(x).$$

In the intervals $(-\pi,0)$ and $(0,\pi)$, f is continuous and
$f'(x) = -1$ and $+1$ respectively, so that f' is piecewise

continuous in each interval (in fact it is continuous).
Therefore, by the Uniform Convergence Theorem the Fourier
series of f converges to f uniformly in these intervals
and termwise differentiation is therefore valid, i.e.,

$$f'(x) = \sum_{n=1}^{\infty} (f_n')$$

where f_n is the nth term in the Fourier expansion. But
note that on $(-\pi,0)\cup(0,\pi)$, $f'(x) = g(x)$ so that

$$g(x) = \frac{d}{dx} \frac{\pi}{2} - \frac{4}{\pi} \sum_{n=1}^{\infty} \frac{d}{dx} \frac{\cos(2n-1)x}{(2n-1)^2} \quad \begin{array}{c} x\in(-\pi,0), \\ x\in(0,\pi) \end{array}$$

or

$$g(x) = \frac{4}{\pi} \sum_{n=1}^{\infty} \frac{\sin(2n-1)x}{2n-1} , \quad x\in(-\pi,0)\cup(0,\pi). \qquad (2)$$

At the point 0 (and hence at all points $0+2n\pi = 2n\pi$) the
Piecewise Convergence Theorem states that the Fourier
series must converge to

$$\frac{1}{2} (g(0+) + g(0-)) = \frac{1}{2} (1 - 1) = 0$$

and the series of (2) satisfies this. The same condition
must hold at all points $(2n+1)\pi$, that is the Fourier
series of g must converge to

$$\frac{1}{2} (g[(2n+1)\pi+] + g[(2n+1)\pi-]) = \frac{1}{2} (1-1) = 0$$

and again (2) satisfies this condition. Thus, the series
in (2) is the Fourier series of g and the equality sign is
valid for all points such that $x \neq n\pi$.

• PROBLEM 6–52

The Fourier series of the following functions have been
found in previous problems but no convergence questions
were discussed. Determine now which of these functions
has Fourier series which are (i) pointwise convergent
(ii) uniformly convergent and (iii) convergent in the mean:

(a) $f(x) = e^x \qquad (-\pi < x < \pi)$

(b) $f(x) = \begin{cases} x & , \ 0 < x < \pi \\ 0 & , \ \pi < x < 2\pi \end{cases}$

(c) $f(x) = x^2 \ , \ x \in (-\pi,\pi)$
         $f(x+2\pi) = f(x)$, all x

345

(d) $f(x) = \begin{cases} -1 & x \in (-\pi, 0) \\ 1 & x \in (0, \pi) \end{cases}$

     $f(x+2\pi) = f(x)$ all x

(e) $f(x) = |x|$ $-\pi < x < \pi$

(f) $f(t) = t$, $0 < t < 3$
     $f(t+3) = f(t)$ all t.

Solution: (a)(i) The function f is continuous on $(-\pi,\pi)$ (and therefore, of course, piecewise continuous) and has right and left hand derivatives at every point in $(-\pi,\pi)$ (in fact $f'_-(x_o) = f'_+(x_o) = f'(x_o)$) so that for all $x_o \in (-\pi,\pi)$ the Fourier series converges pointwise to $\frac{1}{2}\{f(x_o+) + f(x_o-)\}$, i.e.

$$\frac{1}{2}\{f(x_o+) + f(x_o-)\} = \frac{1}{2} a_o + \sum_{n=1}^{\infty} \left[a_n \cos\left(\frac{\pi n x_o}{c}\right) \right.$$

$$\left. + b_n \sin\left(\frac{\pi n x_o}{c}\right) \right] \tag{1}$$

where $c = \pi$ in this case.

(ii) Since f_1 is continuous on $(-\pi,\pi)$ and $f'(x) = e^x$ is continuous on $(-\pi,\pi)$ (and hence piecewise continuous), the Fourier series of f converges uniformly to f.

(iii) The function is continuous and hence piecewise continuous. Therefore, its Fourier series converges in the mean to f.

(b)(i) The function is piecewise continuous on $(0,2\pi)$ and has right and left hand derivatives at all $x_o \in (0,2\pi)$ so that its Fourier series converges pointwise to $\frac{1}{2}\{f(x_o+) + f(x_o-)\}$ as in (1).   (ii) The function is discontinuous at $x = \pi$ and hence does not have a uniformly convergent Fourier series in $(0,2\pi)$. However, in each of the intervals $(0,\pi)$ and $(\pi,2\pi)$, f is continuous and f' is piecewise continuous (in fact, continuous) so that f's Fourier series converges uniformly to f in each of these intervals.   (iii) Since f is piecewise continuous in $(0,2\pi)$, the Fourier series of f converges in the mean to f.

(c)(i) Since f is piecewise continuous on R and has right and left hand derivatives at all $x \in R$ the Fourier series of f at x_o converges to $\frac{1}{2}\{f(x_o+) + f(x_o-)\}$ for all x.

(ii) f is discontinuous at all points $(2n+1)\pi$ so that its Fourier series is not uniformly convergent on R.

However, on any interval $((2n-1)\pi, (2n+1)\pi)$ f is continuous and f' is piecewise continuous (in fact, continuous) so that the Fourier series of f converges uniformly to f on any such interval. (iii) f is piecewise continuous on R so the Fourier series of f converges in the mean to f on any interval in R.

(d)(i) Pointwise convergence for all $x \in R$; (ii) Uniform convergence in any interval $(n\pi, (n+1)\pi)$; (iii) convergence in the mean on any interval in R.

(e)(i) Pointwise convergence on $(-\pi, \pi)$; (ii) Uniform convergence on $(-\pi, \pi)$; (iii) Convergence in the mean on $(-\pi, \pi)$.

(f) (i) Pointwise convergence on R; (ii) Uniform convergence on any interval $(3n, 3(n+1))$; (iii) Convergence in the mean on any interval.

• PROBLEM 6–53

Find the Fourier sine series of $f(x) = x^2$ over the interval $(0,1)$.

<u>Solution</u>: The Fourier sine series of a function defined on an interval $(0,c)$ is given by

$$f(x) \sim_s \sum_{n=1}^{\infty} b_n \sin\left(\frac{n\pi x}{c}\right) \tag{1}$$

where

$$b_n = \frac{2}{c} \int_0^c f(x) \sin\left(\frac{n\pi x}{c}\right) dx \tag{2}$$

We now turn to the given problem. Here $f(x) = x^2$, and $c = 1$. Thus we obtain the Fourier series

$$x^2 \sim_s \sum_{n=1}^{\infty} b_n \sin n\pi x \tag{3}$$

where

$$b_n = 2 \int_0^1 x^2 \sin n\pi x \, dx.$$

We must evaluate the b_n. Using integration by parts, we obtain

$$b_n = 2 \int_0^1 x \sin n\pi x \, dx$$

$$= 2 \left\{ \left[\frac{-x^2}{n\pi} \cos n\pi x \right] \Bigg|_0^1 + \frac{2}{n\pi} \int_0^1 x \cos n\pi x \, dx \right\}$$

$$= 2 \left\{ \frac{-(-1)^n}{n\pi} + \frac{2}{n^2 \pi^2} x \sin n\pi x \Big|_0^1 \right.$$

$$\left. - \int_0^1 \frac{2}{n^2 \pi^2} \sin n\pi x \, dx \right\}$$

$$= 2 \left\{ \frac{(-1)^{n+1}}{n\pi} + \frac{2}{n^3 \pi^3} [(-1)^n - 1] \right\}. \qquad (4)$$

Substituting (4) for b_n in (3), the required Fourier sine series over $0 < x < 1$ is

$$x^2 \sim_s 2 \sum_{n=1}^{\infty} \left[\frac{(-1)^{n+1}}{n\pi} - \frac{2\{1 - (-1)^n\}}{n^3 \pi^3} \right] \sin n\pi x.$$

● **PROBLEM 6–54**

Find the Fourier sine series for the function defined by

$f(x) = 0 \qquad\qquad 0 \le x < \pi/2$

$f(x) = 1 \qquad\qquad \pi/2 < x \le \pi.$

Solution: The Fourier sine series of a function defined on $0 \le x \le L$

$$f(x) \sim_s \sum_{n=1}^{\infty} b_n \sin \frac{n\pi x}{L} \qquad (n = 1, 2, \ldots) \qquad (1)$$

where

$$b_n = \frac{2}{L} \int_0^L \sin nx \, dx. \qquad (2)$$

Recall that the Fourier sine series is equivalent to finding the Fourier trigonometric series of an odd function, i.e., a function such that $f(-x) = -f(x)$.

In the given problem, since $f(x) = 0$ for $0 \le x \le \pi/2$ we need find its series development only for the interval $\pi/2 < x \le \pi$. Taking $L = \pi$,

$$b_n = \frac{2}{\pi} \int_{\pi/2}^{\pi} \sin nx \, dx = -\frac{2}{n\pi} \left(\cos nx \Big|_{\pi/2}^{\pi} \right)$$

$$= \frac{2}{n\pi} \left[\cos \frac{n\pi}{2} - \cos n\pi \right] = \frac{2}{n\pi} \left[\cos \frac{n\pi}{2} + (-1)^{n+1} \right]. \quad (3)$$

Substituting (3) into (1),

$$f(x) \sim_s \frac{2}{\pi} \left[\frac{\sin x}{1} - \frac{2\sin 2x}{2} + \frac{\sin 3x}{3} + \frac{\sin 5x}{5} \right.$$

$$\left. - \frac{2\sin 6x}{6} + \dots \right].$$

● PROBLEM 6–55

Find the exponential Fourier series of the periodic wave-form from the graph shown below.

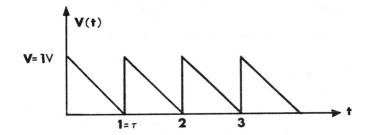

Solution: This is a typical example of the way Fourier series are used to represent functions in electrical engineering. The function to be represented here is v(t) = the voltage at a certain point in a circuit at time t, where v is periodic in time with period $\tau = 2 \frac{\tau}{2}$ = 1. The complex form of the Fourier series of such a function is given by

$$v(t) \sim \sum_{-\infty}^{\infty} c_n e^{i\left(\frac{n2\pi t}{\tau}\right)}, \quad c_n = \frac{1}{\tau} \int_a^{a+\tau} v(t) e^{-i\left(\frac{n2\pi t}{\tau}\right)} dt. \quad (1)$$

It is convenient to define $\omega = \frac{2\pi}{\tau}$, take the limits of integration from 0 to τ and let j stand for the number whose square is -1, (i.e. j = i). This is convenient in electrical applications since i is usually used for current. In this case (1) becomes

$$v(t) \sim \sum_{-\infty}^{\infty} c_n e^{jn\omega t}, \quad c_n = \frac{1}{\tau} \int_0^{\tau} v(t) e^{-jn\omega t} dt. \quad (2)$$

where

$$v(t) = 1 - t, \quad t \in (0, \tau)$$

$$v(t+\tau) = v(t) \quad \text{all } t \quad .$$

First evaluate c_0 by

$$c_0 = \frac{1}{\tau} \int_0^\tau v(t)\,dt$$

(note that this may be interpreted as the average value of the voltage v over the interval $(0,\tau)$). This gives

$$c_0 = \frac{1}{\tau} \int_0^\tau (1-t)\,dt = \left[\frac{1}{\tau}\, t - \frac{t}{2} \right]\Bigg|_0^\tau$$

and evaluating yields

$$c_0 = \frac{1}{\tau}\left[\tau - \frac{\tau^2}{2} \right] = 1 - \frac{\tau}{2} = \frac{1}{2} \tag{3}$$

since $\tau = 1$ second.

Similarly, evaluate equation c_n, $n \neq 0$:

$$c_n = \frac{1}{\tau} \int_0^\tau (1-t) e^{-jn\omega t}\,dt,$$

$$c_n = \frac{1}{\tau} \int_0^\tau e^{-jn\omega t}\,dt - \frac{1}{\tau} \int_0^\tau t e^{-jn\omega t}\,dt$$

but

$$\int e^{ax}\,dx = \frac{e^{ax}}{a}$$

and, using integration by parts,

$$\int x e^{ax}\,dx = \frac{x e^{ax}}{a} - \frac{e^{ax}}{a^2} \quad .$$

Finally, substitute equations (3), (5) and (11) into equation (2):

$$v(t) = \frac{1}{2} + \sum_{n=-\infty}^{\infty} \left(\frac{1}{2\pi n}\right) e^{-j\frac{\pi}{2}} e^{j2\pi nt} \quad .$$

This can be rewritten as

$$v(t) = \frac{1}{2} + \sum_{n=-\infty}^{\infty} \left(\frac{1}{2\pi n}\right) e^{j(2\pi nt - \frac{\pi}{2})} \quad \text{volts}$$

since $e^a e^b = e^{a+b}$.

Find the Fourier series for the waveform shown in the Figure. What is this series if the origin is shifted to O'?

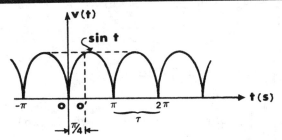

Solution: This is an example of the way Fourier series are used to represent functions in electrical engineering. It is interesting to note that this waveform is called a full-wave rectified wave and is a primitive form of the voltage output of say the rectifier which is attached between a battery which is being recharged and a wall socket. The function to be represented here is v(t) = the voltage at a certain point in a circuit at time t where v is periodic in time with period $\tau = \pi$. The Fourier series of such a function is given by

$$v(t) \sim \frac{a_0}{2} + \sum_{n=1}^{\infty} \left[a_n \cos \frac{n2\pi t}{\tau} + b_n \sin \frac{n2\pi t}{\tau} \right] \tag{1}$$

where

$$a_n = \frac{2}{\tau} \int_a^{a+\tau} v(t) \cos\left(\frac{n2\pi t}{\tau}\right) dt \tag{2}$$

$$b_n = \frac{2}{\tau} \int_a^{a+\tau} v(t) \sin\left(\frac{n2\pi t}{\tau}\right) dt \ . \tag{3}$$

It is convenient to define $\omega = \frac{2\pi}{\tau} = 2$ and take the limits of integration from 0 to $\tau = \pi$. In this case, (1), (2) and (3) become

$$v(t) \sim \frac{a_0}{2} + \sum_{n=1}^{\infty} (a_n \cos 2nt + b_n \sin 2nt) \tag{4}$$

$$a_n = \frac{2}{\pi} \int_0^{\pi} v(t) \cos 2nt \ dt \tag{5}$$

$$b_n = \frac{2}{\pi} \int_0^{\pi} v(t) \sin 2nt. \tag{6}$$

Now the function v(t) is given by

$$v(t) = \sin(t) \qquad\qquad 0 \le t < \pi$$

$$v(t+\pi) = v(t) \qquad\qquad \text{all } t$$

so that from (5) and (6)

$$a_o = \frac{2}{\pi} \int_0^\pi \sin t\, dt = \frac{4}{\pi}$$

$$b_n = \frac{2}{\pi} \int_0^\pi \sin t \sin 2nt\, dt$$

$$= \frac{2}{\pi} \int_0^\pi \left[\frac{1}{2} \cos(1-2n)t - \frac{1}{2}\cos(1+2n)t \right] dt$$

$$= \frac{1}{\pi} \frac{\sin(1-2n)t}{(1-2n)} \Big|_0^\pi - \frac{1}{\pi} \frac{\sin(1\ 2n)t}{(1+2n)} \Big|_0^\pi$$

$$= 0 - 0 = 0.$$

And finally

$$a_n = \frac{2}{\pi} \int_0^\pi \sin t \cos 2nt\, dt.$$

Again, using trigonometric identities, this integral is equivalent to

$$a_n = \frac{2}{\pi} \int_0^\pi \left[\frac{1}{2}\sin(1+2n)t + \frac{1}{2}\sin(1-2n)t \right] dt$$

$$= \frac{1}{\pi} \left[-\frac{1}{1+2n}\cos(1+2n)t - \frac{1}{1-2n}\cos(1-2n)t \right] \Big|_0^\pi$$

$$= \frac{1}{\pi} \left[-\frac{1}{1+2n}\cos(2n+1)\pi - \frac{1}{1-2n}\cos(1-2n)\pi \right]$$

$$\qquad - \frac{1}{\pi} \left[-\frac{1}{1+2n}\cos 0 - \frac{1}{1-2n}\cos 0 \right]$$

$$a_n = \frac{1}{\pi} \left[\frac{4}{1-4n^2} \right]$$

The expression for $v(t)$ is, thus,

$$v(t) \sim \frac{2}{\pi} \left[1 + \sum_{n=1}^\infty \frac{2}{1-4n^2} \cos 2nt \right] \qquad\qquad \text{or since}$$

$$\frac{1}{1-4n^2} = -\frac{1}{4n^2-1}$$

$$v(t) \sim \frac{2}{\pi} [1 - \frac{2}{3} \cos 2t - \frac{2}{15} \cos 4t - \frac{2}{35} \cos 6t ....]$$

Now if the origin is shifted to O' as indicated, the time functions are shifted a quarter period. The maximum value of $f(t)$ will now occur at $\frac{\pi}{4}$ which is $\frac{\pi}{4}$ less than the original peak. Hence each term in the series must be shifted a like amount, and

$$v(t) \sim \frac{2}{\pi} [1 - \frac{2}{3} \cos(2t + \frac{\pi}{4}) - \frac{2}{15} \cos(4t + \frac{\pi}{4}) - ...].$$

● PROBLEM 6–57

The electrical circuit shown in the Figure is driven by a variable electromotive force E(t) which is periodic (but not necessarily sinusoidal) in time. The response of the system is the current I(t) and is known from electromagnetic theory to satisfy

$$L \frac{d^2 I}{dt^2} + R \frac{dI}{dt} + \frac{1}{C} I = \frac{dE}{dt}. \tag{1}$$

Find I(t) in terms of R, L. C and E(t).

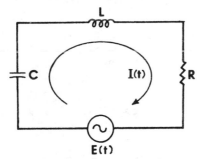

Solution: The solution of a differential equation such as (1) is the sum of a transient term (the general solution to the homogeneous equation $L \frac{d^2 I}{dt^2} + R \frac{dI}{dt} + \frac{1}{C} I = 0$) and a steady state term (a particular solution of (1)). The solution of the homogeneous equation is straightforward, does not require Fourier methods, and dies out rapidly with time. Therefore, assuming that enough time has elapsed so that steady state conditions prevail, (1) will be solved and the solution will give I(t) quite accurately.

Now, under steady state conditions the function I(t) is also periodic with the same period T as for E(t).

353

Let us assume that $E(t)$ and $I(t)$ possess Fourier expansions given in complex form by

$$E(t) = \sum_{n=-\infty}^{\infty} E_n e^{in\omega t} \quad , \quad I(t) = \sum_{n=-\infty}^{\infty} C_n e^{in\omega t} \qquad (2)$$

where $\omega = \frac{2\pi}{T}$. Furthermore, let us assume that the series may be differentiated the necessary number of times (these assumptions will be discussed later). Then

$$\frac{dE}{dt} = \sum_{n=-\infty}^{\infty} in\omega E_n e^{in\omega t} \qquad (3)$$

$$\frac{dI}{dt} = \sum_{n=-\infty}^{\infty} in\omega C_n e^{in\omega t} \qquad (4)$$

$$\frac{d^2 I}{dt^2} = \sum_{n=-\infty}^{\infty} (-n^2\omega^2) C_n e^{in\omega t} . \qquad (5)$$

Now substitute (2), (3), (4), and (5) into (1) and note that the coefficients with the same exponential $e^{in\omega t}$ may be equated due to the fact that the set $\{e^{in\omega t}\}_n$ is orthogonal over $(\frac{-T}{2} , \frac{T}{2})$. Recall that orthogonality of complex functions is defined by $\int f_n(x) f^*_m(x) dx = 0$ for $n \neq m$. That is

$$\int_{-\frac{T}{2}}^{\frac{T}{2}} e^{in\omega t} e^{-im\omega t} dt = \left\{ \begin{array}{ll} 0 & n \neq m \\ 2L & n = m \end{array} \right\} \quad \text{so to prove that we}$$

may equate the coefficients of $e^{in\omega t}$, multiply through the equation

$$L \sum_{n=-\infty}^{\infty} (-n^2\omega^2) C_n e^{in\omega t} + R \sum_{n=-\infty}^{\infty} in\omega C_n e^{in\omega t} + \frac{1}{C} \sum_{n=-\infty}^{\infty} C_n e^{in\omega t}$$

$$= \sum_{n=-\infty}^{\infty} in\omega E_n e^{in\omega t} \qquad (6)$$

by $e^{-in\omega t}$ and integrate from $\frac{-T}{2}$ to $\frac{T}{2}$. Then all terms with $m \neq n$ will vanish and all coefficients of $e^{in\omega t}$ will be multiplied by 2L. Thus, the equality will be proven. Hence, equating like terms in (6) gives

$$(-n^2\omega^2 L + in\omega R + \frac{1}{C})C_n = in\omega E_n .$$

Therefore

$$C_n = \frac{i(n\frac{\omega}{L})}{(\omega_o^2 - n^2\omega^2) + 2\alpha n\omega i} E_n \qquad (7)$$

where $\omega_o^2 = \frac{1}{LC}$ is called the natural frequency of the circuit and $2\alpha = \frac{R}{L}$ is called the attenuation factor of the circuit. Thus the problem is solved for a given $E(t)$ since (2) gives the Fourier expansion of $I(t)$ where C_n are given in terms of the Fourier coefficients of $E(t)$ which may be computed. I.e.,

$$C_n = \frac{i(n\frac{\omega}{L})}{(\omega_o^2 - n^2\omega^2) + 2\alpha n\omega i} \frac{1}{T} \int_{-\frac{T}{2}}^{\frac{T}{2}} E(t)e^{-in\omega t} dt. \qquad (8)$$

Once the C_n are computed the series in (2) may be converted into a real Fourier series

$$I(t) = \frac{a_o}{2} + \sum_{n=1}^{\infty} \left(a_n \cos n\omega t + b_n \sin n\omega t \right) \qquad (9)$$

by recalling the relations

$$C_n = \begin{cases} \frac{1}{2}(a_n - ib_n) & n \neq 0 \\ \frac{1}{2}a_o & n = 0 \end{cases} .$$

A few terms of (9) may then be used to make good approximations of I at any time t.

Finally, the validity of term by term differentiation was assumed valid. But this procedure has only been shown to be valid for uniformly convergent series so far and this is a fairly stringent requirement. However, as will be discussed in the chapter on transform methods, this procedure can be shown to be valid for a wide range of functions, including almost any physically significant function, by using the theory of distributions. In any case, when solving such problems it is usually best to assume such procedures valid and then justify them later after the solution has been developed.

A beam supported at each of its ends is shown in the Figure. It is uniformly loaded by a load q per unit length and the deflection of the beam y(x) is sought. If we choose the direction of the y-axis as downward as indicated in the Figure, the function y(x) is known to satisfy the equation

$$\frac{d^4y}{dx^4} = \frac{1}{EI} q(x) \tag{1}$$

where q(x) is the load per unit length at point x (q = constant in our case) and $\frac{1}{EI}$ is the rigidity of the beam (also constant). Find y(x).

Solution: Note that since the function y(x) must vanish at x = 0 and x = L, it may be conveniently expanded into a Fourier sine series

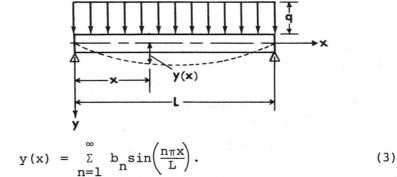

$$y(x) = \sum_{n=1}^{\infty} b_n \sin\left(\frac{n\pi x}{L}\right). \tag{3}$$

Assuming the validity of the fourfold term-by-term differentiation, (3) yields

$$\frac{d^4y(x)}{dx^4} = \sum_{n=1}^{\infty} \left(\frac{n\pi}{L}\right)^4 b_n \sin\left(\frac{n\pi x}{L}\right). \tag{4}$$

(This is a reasonable assumption since on physical grounds one would expect y(x) to be "infinitely smooth" in which case each of the series for $y^{(n)}(x)$ would converge uniformly, justifying term-by-term differentiations.) Also expand q(x) = q into the Fourier sine series

$$q = \sum_{n=1}^{\infty} q_n \sin\left(\frac{n\pi x}{L}\right). \tag{5}$$

where

$$q_n = \frac{2}{L} \int_0^L q \sin\left(\frac{n\pi x}{L}\right) dx = \begin{cases} \dfrac{4q}{n\pi} & (n=\text{odd}) \\[2ex] 0 & (n=\text{even}) \end{cases} . \qquad (6)$$

Substitute both series, (4) and (5), into (1) and note that due to the orthogonality of the set of functions $\{\sin \frac{n\pi x}{L}\}_{n=1}^{\infty}$ we may equate the coefficients of terms with the same n. I.e. since

$$\int_0^L \sin\left(\frac{n\pi x}{L}\right) \sin\left(\frac{m\pi x}{L}\right) dx = \begin{cases} 0 & n \neq m \\[2ex] \dfrac{L}{2} & n = m \end{cases}$$

we may multiply through the resulting equation

$$\sum_{n=1}^{\infty} \left(\frac{n\pi}{L}\right)^4 b_n \sin\left(\frac{n\pi x}{L}\right) = \frac{1}{EI} \sum_{\substack{n=1 \\ n=\text{odd}}}^{\infty} \frac{4q}{n\pi} \sin\left(\frac{n\pi x}{L}\right) \qquad (7)$$

by $\sin\left(\dfrac{n\pi x}{L}\right)$ and integrate from 0 to L. Then all terms with $m \neq n$ will vanish while the coefficients of the nth terms will be multiplied by $\dfrac{L}{2}$. When this is done, we find

$$\frac{L}{2}\left(\frac{n\pi}{L}\right)^4 b_n = \begin{cases} \dfrac{L}{2} \dfrac{1}{EI} \dfrac{4q}{n\pi} & n \text{ odd} \\[2ex] 0 & n \text{ even} \end{cases}$$

or

$$b_n = \begin{cases} \dfrac{4qL^4}{EI\pi^5} \dfrac{1}{n^5} & n \text{ odd} \\[2ex] 0 & n \text{ even} \end{cases} \qquad (8)$$

so that

$$y(x) = \frac{4qL^4}{EI\pi^5} \left[\sum_{n=1,3,5,\ldots}^{\infty} \frac{1}{n^5} \sin\frac{n\pi x}{L} \right]. \qquad (9)$$

A practical advantage of (5) is the rapid convergence of the series due to the fifth power of n in the denominator. for instance, at $x = \dfrac{L}{2}$ (maximum deflection) the second term in the series represents only $\dfrac{1}{3^5} \approx .00412$ or .4% of the first term. Thus, the first term may be used for calculations with about 99% accuracy.

Finally, since the series for q(x) (and, therefore,

for $\dfrac{d^4y}{dx^4}$) is not uniformly convergent, the validity of
the procedure remains in doubt. However it is easily
justified by the theory of distributions which will be
discussed in the section on integral transforms.

• PROBLEM 6–59

Use Parseval's equality and the results of previous problems concerned with calculation of Fourier coefficients
of particular functions to find

(a) $\displaystyle\sum_{n=1}^{\infty} \frac{1}{(2n-1)^2} = 1 + \frac{1}{3^2} + \frac{1}{5^2} + \ldots$

(b) $\displaystyle\sum_{n=1}^{\infty} \frac{1}{(2n-1)^4} = 1 + \frac{1}{3^4} + \frac{1}{5^4} + \ldots$

Solution: Recall that Parseval's equality holds for any
periodic function f with period 2c whose Fourier series
converges in the mean to f. Since all piecewise continuous functions have convergent Fourier series, Parseval's
equality holds for all such functions and is written

$$\frac{1}{c} \int_a^{a+2c} [f(x)]^2 dx = \frac{a_o^2}{2} + \sum_{k=1}^{\infty} (a_k^2 + b_k^2) \qquad (1)$$

where a_o, a_k, b_k are the Fourier coefficients of f and a
is any number such that $(a, a+2c)$ is in the domain of
definition of f.

(a) Earlier it was found that the Fourier series of the
piecewise continuous function

$$f(x) = \begin{cases} -1 & -\pi < x < 0 \\ +1 & 0 < x < \pi \end{cases}$$

$$f(x+2\pi) = f(x) \qquad \text{all } x$$

is

$$f \sim \frac{4}{\pi} \sum_{n=1}^{\infty} \frac{1}{(2n-1)} \sin(2n-1)x .$$

Now $\dfrac{1}{\pi} \displaystyle\int_{-\pi}^{\pi} [f]^2 dx = \dfrac{1}{\pi} \displaystyle\int_{-\pi}^{0} (-1)^2 dx + \dfrac{1}{\pi} \displaystyle\int_{0}^{\pi} (1)^2 dx$

$$= \frac{x}{\pi}\Big|_{-\pi}^{0} + \frac{x}{\pi}\Big|_{0}^{\pi} = 2$$

so that Parseval's equality gives

$$2 = \frac{a_o^2}{2} + \sum_{n=1}^{\infty} a_k^2 + b_k^2 = \sum_{n=1}^{\infty} \left(\frac{4}{\pi(2n-1)}\right)^2$$

so that

$$\frac{\pi^2}{8} = \sum_{n=1}^{\infty} \frac{1}{(2n-1)^2} = 1 + \frac{1}{3^2} + \frac{1}{5^2} + \ldots$$

(b)   Earlier it was found that the Fourier series of the piecewise continuous function

$$f(x) = |x| \qquad -\pi < x < \pi$$

is

$$f \qquad \frac{\pi}{2} - \frac{4}{\pi} \sum_{n=1}^{\infty} \frac{\cos(2n-1)x}{(2n-1)^2}$$

Now

$$\frac{1}{\pi} \int_{-\pi}^{\pi} [f]^2 dx = \frac{1}{\pi} \int_{-\pi}^{\pi} x^2 dx = \left(\frac{1}{\pi} \times \frac{x^3}{3}\right)\Big|_{-\pi}^{\pi} = \frac{2\pi^2}{3}$$

so that Parseval's equality gives

$$\frac{2\pi^2}{3} = \frac{\pi^2}{2} + \sum_{n=1}^{\infty} \left(\frac{-4}{\pi(2n-1)^2}\right)^2$$

$$= \frac{\pi^2}{2} + \frac{16}{\pi^2} \sum_{n=1}^{\infty} \frac{1}{(2n-1)^4}$$

so that

$$\sum_{n=1}^{\infty} \frac{1}{(2n-1)^4} = \frac{\pi^4}{96} \quad .$$

CHAPTER 7

LAPLACE TRANSFORMS AND FOURIER TRANSFORMS

> **Basic Attacks and Strategies for Solving Problems in this Chapter. See pages 362 to 466 for step-by-step solutions to problems.**

Let $f(t)$ be any function such that the integration encountered may be legitimately performed on $f(t)$. The Laplace transform of $f(t)$ is denoted by $L\{f(t)\}$ and is defined by:

$$L\{f(t)\} = \int_0^\infty e^{-st} f(t)dt$$

This integral is a function of the parameter s. We may write

$$L\{f(t)\} = \int_0^\infty e^{-st} f(t)dt = F(s)$$

It is customary to refer to $F(s)$, as well as to the symbol $L\{f(t)\}$, as the Laplace transform of $f(t)$.

Suppose the function $f(t)$ is to be determined from a differential equation with initial conditions. The Laplace operator L is used to transform the original problem into a new problem from which the transform $F(s)$ is to be found. If the Laplace transformation is to be effective, the new problem must be simpler than the original problem. We first find $F(s)$ and then obtain $f(t)$ from $F(s)$.

It is therefore desirable to develop methods for finding the object function $f(t)$ when its transform $F(s)$ is known.

If $L\{f(t)\} = F(s)$, we say that $f(t)$ is the inverse Laplace transform of $F(s)$ and we write: $f(t) = L^{-1}\{F(s)\}$

Let $f(x)$ satisfy the following conditions:

1. $f(x)$ satisfies the Dirichlet conditions in every finite interval $-\ell \leq x \leq \ell$

2. $\int_{-\infty}^\infty |f(x)|dx$ converges, i.e., $f(x)$ is absolutely integrable in $-\infty < x < \infty$.

Then Fourier's integral theorem states that

$$f(x) = \int_0^\infty \{A(\lambda) \cos \lambda x + \beta(\lambda)x\} d\lambda, \text{where:}$$

$$A(\lambda) = \frac{1}{\pi} \int_{-\infty}^\infty f(x) \cos \lambda x dx$$

$$B(\lambda) = \frac{1}{\pi} \int_{-\infty}^\infty f(x) \sin \lambda x dx.$$

In complex notation, the fourier integral can be written as:

$$f(x) = \frac{1}{2\pi} \int_{-\infty}^\infty e^{i\lambda x} dx \int_{-\infty}^\infty f(u) e^{-i\lambda u} du = \frac{1}{2\pi} \int_{-\infty}^\infty \int_{-\infty}^\infty f(u) e^{i\lambda x - u} du dx.$$

$$\text{If } F(x) = \int_{-\infty}^\infty e^{-i\lambda u} f(u) du \text{ then } f(u) = \frac{1}{2\pi} \int_{-\infty}^\infty e^{i\lambda u} F(\lambda) dx$$

The function of $F(\lambda)$ is called the Fourier transform of $f(x)$ and is sometimes written as $F(\lambda) = F\{f(x)\}$.

The function $f(x)$ is the inverse Fourier transform of $F(\lambda)$ and is written as $f(x) = F^{-1}\{F(\lambda)\}$.

Note that the constants preceding the integral signs can be any constants whose product is $\frac{1}{2\pi}$. If they are each taken as $\frac{1}{\sqrt{2\pi}}$ we obtain the so-called symmetric form.

Step-by-Step Solutions to
Problems in this Chapter,
"Laplace Transforms and
Fourier Transforms"

LAPLACE TRANSFORMS

Determine those values of s for which the improper integral

$$G(s) = \int_0^\infty e^{-st} \, dt$$

converges, and find the Laplace transform of f(t) = 1.

<u>Solution</u>: By definition, the improper integral in the problem is a limit of a proper integral

$$G(s) = \lim_{R \to \infty} \int_0^R e^{-st} \, dt \ . \tag{1}$$

For s = 0 this becomes

$$G(s) = \lim_{R \to \infty} \int_0^R e^{-(0)t} \, dt = \lim_{R \to \infty} \int_0^R dt$$

$$= \lim_{R \to \infty} t \Big|_{t=0}^R = \lim_{R \to \infty} R = \infty \ .$$

Hence, the integral G(s) diverges for s = 0. For s ≠ 0 equation (1) becomes

$$G(s) = \lim_{R \to \infty} \left. -\left(\frac{1}{s} e^{-st}\right) \right|_{t=0}^{R}$$

$$= \lim_{R \to \infty} \frac{1}{s} \left(1 - e^{-Rs}\right) = \frac{1}{s} - \lim_{R \to \infty} \frac{e^{-Rs}}{s} . \tag{2}$$

If $s < 0$, then $-Rs > 0$ for positive R; hence, e^{-Rs} approaches infinity as R approaches infinity, and the integral diverges. If $s > 0$, then $-Rs < 0$ for positive R; hence, e^{-Rs} approaches zero as R approaches infinity, and the integral converges to $1/s$.

Extending the domain of the Laplace transform to complex values of s, we evaluate the expression of e^{-Rs} by Euler's formula:

$$e^{-Rs} = e^{-R(Re\{s\}) - iR(Im\{s\})}$$

$$= e^{-R(Re\{s\})} \left[\cos[-R(Im\{s\})] + i \sin[-R(Im\{s\})]\right], \tag{3}$$

where $Re\{s\}$ is the real part of s, $Im\{s\}$ is the imaginary part of s, and $i \equiv \sqrt{-1}$ is the imaginary constant. The cosine and sine functions are bounded; hence expression (3) diverges, as $R \to \infty$, for $Re\{s\} < 0$ and converges to zero for $Re\{s\} > 0$. In the case where $s \neq 0$ and $Re\{s\} = 0$, we have

$$e^{-R(Re\{s\})} = 1;$$

hence,

$$e^{-Rs} = \cos[-R(Im\{s\})] + i \sin[-R(Im\{s\})],$$

which is a nonconstant periodic function, so e^{-Rs} does not converge to any value as $R \to \infty$.

Since e^{-Rs} is the only expression that varies with R in equality (2), its convergence properties for $R \to \infty$ determine the convergence properties of $G(s)$. Thus, in general, $G(s)$ converges to $1/s$ for $Re\{s\} > 0$ and diverges otherwise.

Using L as the Laplace transform operator, with

$$L\{f(t)\} = F(s) \equiv \int_0^{\infty} e^{-st} f(t) \, dt ,$$

we take the Laplace transform of $f(t)$ for $t \geq 0$:

363

$$L\{1\} = \int_0^\infty (e^{-st} \cdot 1)dt = \int_0^\infty e^{-st}\, dt \ ,$$

which is the same integral as G(s) in the problem. As we already showed, this integral converges to 1/s when Re{s} > 0 and diverges otherwise. Thus, the required Laplace transform is

$$L\{1\} = \frac{1}{s} \ , \text{ for } Re\{s\} > 0 \ .$$

Note that the function f(t) of which we take a Laplace transform needs only be defined for positive real values of its argument t since the integral,

$$\int_0^\infty e^{-st} f(t)dt \ ,$$

is in a region in which t ≥ 0 .

Find the Laplace transform of

$$f(t) = t^n \ ,$$

where n is a positive integer.

Solution: Using L as the Laplace transform operator with

$$L\{f(t)\} = \int_0^\infty e^{-st} f(t)\, dt \ ,$$

and first considering real s, we find

$$L\{t^n\} = \int_0^\infty (e^{-st})(t^n)dt = \lim_{R\to\infty} \int_0^R t^n e^{-st}\, dt \ .$$

For s = 0 ,

$$L\{t^n\} = \lim_{R\to\infty} \int_0^R t^n e^{0t}\, dt = \lim_{R\to\infty} \int_0^R t^n\, dt$$

$$= \lim_{R\to\infty} \frac{R^{n+1}}{n+1} = \infty \ ;$$

hence,

$$L\{t^n\}$$

364

does not exist for s = 0. For s $\neq$ 0 ,

integrating by parts,

$$L\{t^n\} = \lim_{R \to \infty} \left\{ \left. \frac{-t^n e^{-st}}{s} \right|_{t=0}^{R} + \frac{n}{s} \int_0^R e^{-st} t^{n-1} dt \right\}$$

$$= \lim_{R \to \infty} \left\{ \frac{-R^n e^{-sR}}{s} + \frac{n}{s} \int_0^R e^{-st} t^{n-1} dt \right\}$$

$$= \lim_{R \to \infty} \left(\frac{-R^n e^{-sR}}{s} \right) + \frac{n}{s} \lim_{R \to \infty} \left(\int_0^R e^{-st} t^{n-1} dt \right)$$

$$= \lim_{R \to \infty} \left(\frac{-R^n e^{-sR}}{s} \right) + \frac{n}{s} L\{t^{n-1}\} \quad . \tag{1}$$

For s $\leq$ 0 the argument of the limit in expression (1)
diverges as R$\to\infty$; hence,

$$L\{t^n\}$$

does not exist. For s > 0 rewrite (1) as

$$L\{t^n\} = \lim_{R \to \infty} \left(\frac{-R^n}{se^{sR}} \right) + \frac{n}{s} L\{t^{n-1}\} \quad . \tag{2}$$

Since both the numerator and denominator in the argument of
the limit in equation (2) approach ∞ as R$\to\infty$, we can apply
L'Hospital's rule:

$$\lim_{R \to \infty} \left(\frac{-R^n}{se^{sR}} \right) = \lim_{R \to \infty} \left[\frac{d/dR(-R^n)}{d/dR(se^{sR})} \right] = \lim_{R \to \infty} \left(\frac{-nR^{n-1}}{s^2 e^{sR}} \right) \tag{3}$$

As long as the numerator and denominator of the argument of
our limit approach ∞ as R$\to\infty$, we can iteratively apply
L'Hospital's rule to equality (3):

$$\lim_{R \to \infty} \left(\frac{-R^n}{se^{sR}} \right) = \lim_{R \to \infty} \left(\frac{-nR^{n-1}}{s^2 e^{sR}} \right) = \lim_{R \to \infty} \left[\frac{-n(n-1)R^{n-2}}{s^3 e^{sR}} \right]$$

$$= \ldots = \lim_{R \to \infty} \left[\frac{-n(n-1)(n-2)\ldots(2)(1)R^{n-n}}{s^{n+1} e^{sR}} \right]$$

$$= (-1) \lim_{R \to \infty} \left[\frac{n!}{s^{n+1} e^{sR}} \right] = 0 \; .$$

Substituting this result in equation (2),

$$L\{t^n\} = \frac{n}{s} \, L\{t^{n-1}\} \quad \text{for } s > 0 \; . \tag{4}$$

Substituting (n-1) for n in equation (4), we find

$$L\{t^{n-1}\} = \frac{n-1}{s} \, L\{t^{n-2}\} \; .$$

Substituting this result back into equation (4),

$$L\{t^n\} = \frac{n}{s} \left(\frac{n-1}{s} \, L\{t^{n-2}\} \right) = \frac{n(n-1)}{s^2} \, L\{t^{n-2}\} \; .$$

By iterating this process, we obtain

$$L\{t^n\} = \frac{n(n-1)(n-2)\ldots(2)(1)}{s^n} \, L\{t^0\} = \frac{n!}{s^n} \, L\{1\}$$

$$= \frac{n!}{s^n} \lim_{R \to \infty} \int_0^R e^{-st}(1) dt \; ,$$

which converges, since s > 0 , to

$$L\{t^n\} = \frac{n!}{s^n} \lim_{R \to \infty} \left(\frac{1 - e^{-sR}}{s} \right) = \frac{n!}{s^n} \cdot \frac{1}{s}$$

$$= \frac{n!}{s^{n+1}} \; , \quad \text{for } s > 0 \; . \tag{5}$$

If s is complex, a similar computation will give the same result, (5), except that the condition on s will now be Re s > 0 .

● **PROBLEM 7–3**

Find the Laplace transforms of

(a) $f(t) = e^{kt}$,

where k is a complex constant of the form

$$k = Re\{k\} + i \, Im\{k\}$$

with Re{k} the real part of k, Im{k} the imaginary part of k, and

$$i \equiv \sqrt{-1} \ .$$

Use this Laplace transform to find the Laplace transforms of

$$f(t) = e^{-kt} \quad \text{and} \quad f(t) = 1 \ .$$

(b) $f(t) = \sin kt$ where k is a real constant.

Solution: (a)   Using L as the Laplace transform operator with

$$L\{f(t)\} = \int_0^{\infty} e^{-st} f(t)dt \ ,$$

for complex s of the form $s = \text{Re}\{s\} + i\text{Im}\{s\}$, where $\text{Re}\{s\}$ is the real part of s and $\text{Im}\{s\}$ is the imaginary part of s,

$$L\{e^{kt}\} = \int_0^{\infty} e^{-st}(e^{kt})dt = \lim_{R\to\infty} \int_0^R e^{(k-s)t} dt \ .$$

We chose to solve the complex case in this problem because complex numbers are necessary when we determine

$$L\{\cos kt\} \quad \text{and} \quad L\{\sin kt\}$$

from $L\{e^{kt}\}$.

For $s = k$,

$$L\{e^{kt}\} = \lim_{R\to\infty} \int_0^R e^{(k-k)t} dt = \lim_{R\to\infty} \int_0^R dt$$

$$= \lim_{R\to\infty} \left(t \bigg|_{t=0}^R \right) = \lim_{R\to\infty} R = \infty \ ;$$

hence

$$L\{e^{kt}\}$$

does not exist for $s = k$.   For $s \neq k$,

$$L\{e^{kt}\} = \lim_{R\to\infty} \int_0^R e^{(k-s)t} dt$$

$$= \lim_{R\to\infty} \left[\frac{e^{(k-s)t}}{k-s} \bigg|_{t=0}^R \right]$$

$$= \lim_{R\to\infty} \left[\frac{e^{(k-s)R} - 1}{k-s} \right]$$

367

$$= \frac{1}{s-k} + \frac{1}{k-s} \lim_{R \to \infty} e^{(k-s)R} \; ,$$

which diverges for

$$\text{Re}\{s\} \le \text{Re}\{k\}$$

and converges to

$$\frac{1}{s-k}$$

for

$$\text{Re}\{s\} > \text{Re}\{k\} \; .$$

Thus, the Laplace transform of e^{kt} (for $t > 0$) is

$$L\{e^{kt}\} = \frac{1}{s-k} \quad \text{for Re}\{s\} > \text{Re}\{k\} \; . \tag{1}$$

Using the constant $(-k)$ in place of k in formula (1), we obtain a new formula:

$$L\{e^{-kt}\} = \frac{1}{s-(-k)} = \frac{1}{s+k}$$

for

$$\text{Re}\{s\} > -\text{Re}\{k\} \; .$$

If the special case $k = 0$ is used, equation (1) gives

$$L\{1\} = L\{e^{0t}\} = \frac{1}{s-0} = \frac{1}{s} \; , \quad \text{Re}\{s\} > 0 \; .$$

(b) Using L as the Laplace transform operator with

$$L\{f(t)\} = \int_0^\infty e^{-st} f(t) \; dt,$$

and first considering real s,

$$L\{\sin kt\} = \int_0^\infty (e^{-st})(\sin kt) dt = \lim_{R \to \infty} \int_0^R e^{-st} \sin kt \; dt \; .$$

In the above equality, we substitute the exponential formula for the sine function:

$$\sin kt = \frac{e^{ikt} - e^{-ikt}}{2i} \; ,$$

where $i \equiv \sqrt{-1}$, so that

$$L\{\sin kt\} = \lim_{R \to \infty} \int_0^R e^{-st} \left(\frac{e^{ikt} - e^{-ikt}}{2i} \right) dt$$

$$= \frac{1}{2i} \lim_{R \to \infty} \left[\int_0^R e^{(ik-s)t} dt - \int_0^R e^{(-ik-s)t} dt \right] . \tag{2}$$

For k = 0 equation (2) gives us

$$L\{\sin kt\} = \frac{1}{2i} \lim_{R \to \infty} \left[\int_0^R e^{(0-s)t} dt - \int_0^R e^{(0-s)t} dt \right]$$

$$= \frac{1}{2i} \lim_{R \to \infty} [0] = 0 .$$

When $k \neq 0$, s cannot equal $\pm ik$, since $\pm ik$ is nonzero imaginary; thus, equation (2) gives us

$$L\{\sin kt\} = \frac{1}{2i} \lim_{R \to \infty} \left[\frac{e^{(ik-s)R}}{ik-s} - \frac{1}{ik-s} + \frac{e^{(-ik-s)R}}{ik+s} - \frac{1}{ik+s} \right]$$

$$= \frac{1}{2i} \lim_{R \to \infty} \left[\frac{(ik+s)e^{(ik-s)R} - (ik+s) + (ik-s)e^{(-ik-s)R} - (ik-s)}{(ik-s)(ik+s)} \right]$$

$$= \frac{1}{2i} \lim_{R \to \infty} \left[\frac{(ik+s)e^{(ik-s)R} + (ik-s)e^{(-ik-s)R} - 2ik}{-s^2 - k^2} \right]$$

$$= \frac{k}{s^2+k^2} + \frac{1}{2i} \lim_{R \to \infty} \left[\frac{(ik+s)e^{(ik-s)R} + (ik-s)e^{(-ik-s)R}}{-s^2 - k^2} \right] . \tag{3}$$

When the real part of (ik-s) is negative (i.e., when -s < 0) and the real part of (-ik-s) is negative (i.e., when -s < 0), the argument of the limit in expression (3) approaches zero as $R \to \infty$; hence

$$L\{\sin kt\} = k/(s^2 + k^2) \text{ for } s > 0 .$$

Otherwise, the argument of the limit diverges, and the Laplace transform of sin kt does not exist.

If s is complex, the result

$$L\{\sin kt\} = \frac{k}{s^2 + k^2}$$

still holds, but the condition on s becomes Re s > 0 .

Use the Laplace transform of

$$f(t) = e^{kt} , \tag{1}$$

where k is a complex constant of the form

k = Re{k} + i Im{k}

with Re{k} the real part of k, Im{k} the imaginary part of k, and

$$i \equiv \sqrt{-1} ,$$

to find the Laplace transforms of

f(t) = cosh kt, sinh kt, cos kt, and sin kt .

Solution:

$$L\{e^{kt}\} = \frac{1}{s-k} \quad \text{for} \quad \text{Re}\{s\} > \text{Re}\{k\} . \tag{2}$$

This result could also be looked up in a table of Laplace transforms. In either case, we use the definitions

$$\cosh kt \equiv \frac{e^{kt} + e^{-kt}}{2} ,$$

and

$$\sinh kt \equiv \frac{e^{kt} - e^{-kt}}{2} ,$$

and the additional formula

$$L\{c_1 f_1(t) + c_2 f_2(t)\} =$$

$$c_1 L\{f_1(t)\} + c_2 L\{f_2(t)\} , \tag{3}$$

to find

$$L\{\cosh kt\} = L\left\{\frac{e^{kt} + e^{-kt}}{2}\right\}$$

$$= \frac{1}{2}\left(L\{e^{kt}\} + L\{e^{-kt}\}\right)$$

$$= \frac{1}{2}\left(\frac{1}{s-k} + \frac{1}{s+k}\right) = \frac{1}{2}\left(\frac{2s}{s^2-k^2}\right)$$

$$= \frac{s}{s^2-k^2} \; , \; \text{for} \; \; \text{Re}\{s\} > |\text{Re}\{k\}| \quad , \tag{4}$$

and

$$L\{\sinh kt\} = L\left\{\frac{e^{kt} - e^{-kt}}{2}\right\}$$

$$= \frac{1}{2}\left(L\{e^{kt}\} - L\{e^{-kt}\}\right)$$

$$= \frac{1}{2}\left(\frac{1}{s-k} - \frac{1}{s+k}\right) = \frac{1}{2}\left(\frac{2k}{s^2-k^2}\right)$$

$$= \frac{k}{s^2-k^2} \; , \; \text{for} \; \text{Re}\{s\} > |\text{Re}\{k\}| \; . \tag{5}$$

The condition $\text{Re}\{s\} > |\text{Re}\{k\}|$ in formulas (4) and (5) comes from the fact that we derived those formulas for $L\{e^{kt}\}$ and $L\{e^{-kt}\}$, which require $\text{Re}\{s\} > \text{Re}\{k\}$ and $\text{Re}\{s\} > -\text{Re}\{k\}$, respectively. To insure that both $\text{Re}\{s\} > \text{Re}\{k\}$ and $\text{Re}\{s\} > -\text{Re}\{k\}$, it is necessary that $\text{Re}\{s\}$ be greater than the greater of $\text{Re}\{k\}$ and $-\text{Re}\{k\}$. Since one of these ($\text{Re}\{k\}$ or $-\text{Re}\{k\}$) must be positive and the other negative, the greater of the two is the positive one, which is equal to the absolute value of $\text{Re}\{k\}$.

Using the exponential formulas for the cosine and sine functions

$$\cos kt = \frac{e^{ikt} + e^{-ikt}}{2} \; ,$$

and

$$\sin kt = \frac{e^{ikt} - e^{-ikt}}{2i} \; ,$$

and again the addition formula (3), we find

$$L\{\cos kt\} = L\left\{\frac{e^{(ik)t} + e^{-(ik)t}}{2}\right\}$$

$$= \frac{1}{2} (L\{e^{(ik)t}\} + L\{e^{-(ik)t}\}) \ ,$$

and a similar expression holds for $L\{\sin kt\}$.

By substituting (ik) for k in formulas (1) and (2),

$$L\{\cos kt\} = \frac{1}{2} \left(\frac{1}{s-ik} + \frac{1}{s+ik} \right)$$

$$= \frac{1}{2} \left(\frac{2s}{s^2+k^2} \right)$$

$$= \frac{s}{s^2+k^2} \ , \tag{6}$$

and

$$L\{\sin kt\} = L \left\{ \frac{e^{(ik)t} - e^{-(ik)t}}{2i} \right\}$$

$$= \frac{1}{2i}(L\{e^{(ik)t}\} - L\{e^{-(ik)t}\})$$

$$= \frac{1}{2i} \left(\frac{1}{s-ik} - \frac{1}{s+ik} \right) = \frac{1}{2i} \left(\frac{2ik}{s^2+k^2} \right)$$

$$= \frac{k}{s^2+k^2} \ . \tag{7}$$

Laplace transforms (6) and (7) are both subject to the same two existence conditions from the Laplace transforms of e^{kt} and e^{-kt} (which were the base of (6) and (7)). Since we used ik instead of k, the conditions are

$$Re\{s\} > Re\{ik\} \quad ,$$

and

$$Re\{s\} > -Re\{ik\} \quad .$$

Combining these two conditions as we did for the cosh and sinh Laplace transforms,

$$Re\{s\} > |Re\{ik\}| \quad .$$

But

$$|Re\{ik\}| = |Re\{i(Re\{k\} + iIm\{k\})\}|$$

$$= |Re\{-Im\{k\} + iRe\{k\}\}|$$

$$= |-Im\{k\}| = |Im\{k\}| \quad ;$$

hence, the condition for the existence of Laplace transforms (6) and (7) is

$$Re\{s\} > |Im\{k\}| \quad ,$$

which, for s and k real, is equivalent to $s > 0$.

• **PROBLEM** 7–5

Find the Laplace transform, $L\{f(t)\} = F(s)$, of

(a) $f(t) = 2 \sin t + 3 \cos 2t$

(b) $g(t) = \dfrac{1 - e^{-t}}{t}$.

<u>Solution:</u> (a) We shall use the addition formula

$$L\{c_1 f_1(t) + c_2 f_2(t)\} = c_1 L\{f_1(t)\} + c_2 L\{f_2(t)\} \quad , \tag{1}$$

where c_1 and c_2 are constants, and the formulas which were derived in the previous problems:

$$L\{\sin kt\} = \frac{k}{s^2+k^2} \quad \text{(for } s > 0\text{)} \tag{2}$$

$$L\{\cos kt\} = \frac{s}{s^2+k^2} \quad \text{(for } s > 0\text{)} \quad , \tag{3}$$

where k is a real constant and s is a real variable.

By formula (1),

$$L\{2 \sin t + 3 \cos 2t\} = 2L\{\sin t\} + 3L\{\cos 2t\} \quad .$$

Applying formulas (2) and (3) to the above equality.

$$L\{2 \sin t + 3 \cos 2t\} = 2 \left(\frac{1}{s^2+1^2}\right) + 3\left(\frac{s}{s^2+2^2}\right)$$

$$= \frac{2}{s^2+1} + \frac{3s}{s^2+4} \quad , \quad \text{for } s > 0 \quad .$$

(b) Expanding e^{-t} as an infinite series,

$$e^{-t} = \sum_{n=0}^{\infty} \frac{(-1)^n t^n}{n!} \quad ;$$

hence,

$$\frac{1 - e^{-t}}{t} = \frac{1 - \left[\sum\limits_{n=0}^{\infty} \frac{(-1)^n t^n}{n!}\right]}{t}$$

$$= \frac{1 - \left[1 + \sum\limits_{n=1}^{\infty} \frac{(-1)^n t^n}{n!}\right]}{t}$$

$$= -\frac{1}{t} \sum\limits_{n=1}^{\infty} \frac{(-1)^n t^n}{n!} = \sum\limits_{n=1}^{\infty} \frac{(-1)^{n-1} t^{n-1}}{n!} \quad .$$

Letting $k = n-1$ in the summation, this becomes

$$\frac{1 - e^{-t}}{t} = \sum\limits_{k=0}^{\infty} \frac{(-1)^k t^k}{(k+1)!} \quad ;$$

thus

$$L\left\{\frac{1 - e^{-t}}{t}\right\} = L\left\{\sum\limits_{k=0}^{\infty} \frac{(-1)^k t^k}{(k+1)!}\right\} \tag{4}$$

Since L is a linear operator (i.e., since

$$L\{c_1 f_1(t) + c_2 f_2(t) + \ldots\} = c_1 L\{f_1(t)\}$$
$$+ c_2 L\{f_2(t)\} + \ldots \ , \ c_j = \text{constant}),$$

equality (4) becomes

$$L\left\{\frac{1 - e^{-t}}{t}\right\} = \sum\limits_{k=0}^{\infty} \frac{(-1)^k}{(k+1)!} L\{t^k\} \quad . \tag{5}$$

From the results of a previous problem, or from a table of Laplace transforms, we find that

$$L\{t^k\} = \frac{k!}{s^{k+1}} \ , \quad s > 0 \ ,$$

where k is a nonnegative integer. Substituting this result into equality (5), we obtain

$$L\left\{\frac{1 - e^{-t}}{t}\right\} = \sum\limits_{k=0}^{\infty} \frac{(-1)^k}{(k+1)!} \left(\frac{k!}{s^{k+1}}\right)$$

$$= \sum_{k=0}^{\infty} \frac{(-1)^k}{(k+1)} \left(\frac{1}{s}\right)^{k+1} \quad .$$

This last summation is the infinite series for the natural logarithm of $(1 + 1/s)$, where $|1/s| < 1$, i.e., where $|s| > 1$. Thus,

$$L\left\{\frac{1 - e^{-t}}{t}\right\} = \log\left(1 + \frac{1}{s}\right) .$$

An alternative method of solving this problem is to use the formula

$$L\left\{\frac{1}{t} f(t)\right\} = \int_{s}^{\infty} F(\sigma) d\sigma ,$$

where $F(s) = L\{f(t)\}$ (we obtained it in problem 7). Here

$$f(t) = 1 - e^{-t} ,$$

and

$$L\{1 - e^{-t}\} = L\{1\} - L\{e^{-t}\} = \frac{1}{s} - \frac{1}{s+1}$$

since

$$L\{e^{kt}\} = \frac{1}{s-k} \qquad \text{(for } k = 0, \ e^{kt} \equiv 1).$$

Then

$$L\left\{\frac{1 - e^{-t}}{t}\right\} = \int_{s}^{\infty} \left(\frac{1}{\sigma} - \frac{1}{\sigma+1}\right) d\sigma$$

$$= \left(\log |\sigma| - \log |\sigma + 1|\right)\Big|_{s}^{\infty}$$

$$= \log \left|\frac{\sigma}{\sigma+1}\right| \Big|_{s}^{\infty} = \log 1 - \log \frac{s}{s+1}$$

$$= \log \frac{s+1}{s} \quad \text{(we have } s > 0) .$$

In fact, the condition $|s| > 1$, or $s > 1$ (since $s > 0$ for $L\{t^k\}$ to exist) is too restrictive, and the second solution shows that for any $s > 0$

$$L\left\{\frac{1 - e^{-t}}{t}\right\}$$

exists.

Use the derivative property of Laplace transforms to solve
the differential equation

$$y' - y = e^{-x} \qquad (1)$$

where $y(0) = 0$ is the initial value of y.

Solution: First multiply equation (1) by e^{-sx} and
integrate from 0 to ∞ to get

$$\int_0^\infty e^{-sx} y'(x)\,dx - \int_0^\infty e^{-sx} y(x)\,dx = \int_0^\infty e^{-sx} \cdot e^{-x}\,dx \qquad (2)$$

or

$$L\{y'(x)\} - L\{y(x)\} = L\{e^{-x}\} \quad . \qquad (3)$$

Now the fact that $L\{e^{-ax}\} = \dfrac{1}{s+a}$ can be obtained from
a previous problem in this chapter or can be looked up on a
table of Laplace transforms and is valid for s > a. Also,
the derivative property states that

$$L\{y'(x)\} = sL\{y(x)\} - y(0) \text{ , so that}$$

denoting

$$Y(s) = L\{y(x)\}$$

and using (3), one obtains

$$[s\, Y(x) - y(0)] - Y(s) = \frac{1}{s+1}$$

or since

$$y(0) = 0 \text{ , } (s-1)\, Y(s) = \frac{1}{s+1} \text{ ,}$$

and

$$Y(s) = \frac{1}{(s+1)(s-1)} = \frac{1}{2}\frac{1}{s-1} - \frac{1}{2}\frac{1}{s+1} \text{ ,} \qquad (4)$$

where we have used rational fraction decomposition of

$$\frac{1}{(s+1)(s-1)} \quad .$$

The equality in (4) is valid only for $s > 1$, but this will turn out to be insignificant since we wish to invert (4) anyway. Thus, recall that

$$L\{e^x\} = \frac{1}{s-1} \quad , \quad L\{e^{-x}\} = \frac{1}{s+1}$$

so that (4) reads

$$L\{y(x)\} = \frac{1}{2}L\{e^x\} - \frac{1}{2} L\{e^{-x}\}$$

$$= L\ \{\frac{1}{2}\ e^x - \frac{1}{2}\ e^{-x}\} \qquad\qquad (5)$$

since L is a linear operator. Thus $y(x)$ and

$$\frac{1}{2}\ e^x - \frac{1}{2}\ e^{-x}$$

have the same Laplace transforms. As will be discussed later, under very general conditions, if two functions have the same Laplace transform, they are identical. This is not true for all functions, but is true for a class of almost all functions of any applicability, and we assume $y(x)$ and

$$\frac{1}{2}\ e^x - \frac{1}{2}\ e^{-x}$$

to be such functions. Thus (5) implies that

$$y(x) = \frac{1}{2}\ e^x - \frac{1}{2}\ e^{-x} = \sinh x$$

is the solution to the problem. Questions concerning the validity of our "inversion" of the Laplace transform will be discussed more rigorously later in the chapter.

● PROBLEM 7–7

Find the Laplace transforms of

(a)
$$g(t) = e^{-2t} \sin 5t ,$$

(b)
$$h(t) = e^{-t} t \cos 2t .$$

Solution: (a) We shall use the formula

$$L\{\sin kt\} = \frac{k}{s^2+k^2} , \qquad\qquad (1)$$

where k is a real constant and s is a real variable. We also use the theorem that states that if $f(t)$ is defined for all nonnegative t, is piecewise continuous on every closed interval [0,b] for b > 0, and is of exponential order $e^{\alpha t}$, then

$$L\{e^{at}f(t)\} = F(s-a) ,\tag{2}$$

for s > α + a, where a is a real constant and

$$F(s) = L\{f(t)\} .$$

Using formula (1), with f(t) = sin 5t, we find

$$f(s) = L\{\sin 5t\} = \frac{5}{s^2+5^2} = \frac{5}{s^2+25} .$$

In order to use the theorem associated with formula (2), we must first demonstrate that the function f(t) = sin 5t is of exponential order; i.e., there exists a real constant α and positive real constants M and t_0 such that

$$e^{-\alpha t}|f(t)| < M$$

for all t > t_0 . For α > 0 ,

$$\lim_{t\to\infty} e^{-\alpha t} = 0 ,$$

and for α = 0 ,

$$\lim_{t\to\infty} e^{-\alpha t} = 1 .$$

Thus, for $\alpha \geq 0$,

$$\lim_{t\to\infty} e^{-\alpha t} < 2 .\tag{3}$$

Also note that

$$|\sin 5t| \leq 1 ,\tag{4}$$

for all t. Multiplying inequalities (3) and (4) together, we obtain

$$\left(\lim_{t\to\infty} e^{-\alpha t}\right)|\sin 5t| < 2 ,$$

for $\alpha \geq 0$. It follows from the definition of a limit that there exists a positive constant t_0 such that, for $t > t_0$,

$$e^{-\alpha t}|\sin 5t| < 2 ,$$

which, taking $M = 2$, shows that $f(t) = \sin 5t$ is of exponential order $e^{\alpha t}$, for $\alpha \geq 0$.

Since $f(t) = \sin 5t$ is defined for all nonnegative t, is piecewise continuous on every closed interval $[0,b]$ for $b > 0$, and is of exponential order $e^{\alpha t}$ for $\alpha \geq 0$, we can use formula (1) in formula (2);

$$L\{g(t)\} = L\{e^{-2t} f(t)\}$$

$$= F(s-a)$$

$$= \frac{5}{(s+2)^2 + 25} , \text{ for } s > -2 .$$

(b)   Using the property that

$$L\{e^{bt}f(t)\} = F(s-b) ,$$

where b is a constant and $F(s) = L\{f(t)\}$, provided that $L\{f(t)\}$ exists, we obtain

$$L\{e^{-t} t \cos 2t\} = G(s+1) , \tag{5}$$

where $G(s) = L\{t \cos 2t\}$.

We could find, in a table of Laplace transforms, that

$$L\{t \cos kt\} = \frac{s^2 - k^2}{(s^2 + k^2)^2} , s > 0 .$$

However, this result may be established easily without consulting such a table.   Using the fact that

$$L\{\cos kt\} = \frac{s}{s^2 + k^2}$$

which was established in a previous problem and the theorem which states that

$$L\{t^n f(t)\} = (-1)^n \frac{d^n}{ds^n} L\{f(t)\} ,$$

it is apparent that

379

$$L\{t \cos kt\} = (-1)\frac{d}{ds} L\{f(t)\}$$

$$= (-1) \frac{d}{ds} \left(\frac{s}{s^2 + k^2}\right)$$

$$= \frac{s^2 - k^2}{(s^2 + k^2)^2} \quad .$$

The region of validity of this formula is the same as that for $L\{\cos kt\}$, i.e., $s > 0$ for real s.

Hence, taking k = 2, we find

$$L\{t \cos 2t\} = \frac{s^2 - 4}{(s^2 + 4)^2} \quad .$$

Thus,

$$G(s) = \frac{s^2 - 4}{(s^2 + 4)^2} \quad ,$$

and, from equality (5),

$$L\{e^{-t} t \cos 2t\} = \frac{(s+1)^2 - 4}{[(s+1)^2 + 4]^2} \quad .$$

● **PROBLEM** 7-8

Find the Laplace transform of

(a)
$$g(t) = te^{4t}$$

(b)
$$f(t) = t^{7/2}$$

Solution: (a) We shall use the formula

$$L\{e^{kt}\} = \frac{1}{s-k} \text{, for } s > k, \tag{1}$$

where k is a real constant, and s is a real variable. We also use the theorem that states that if f(t) is defined for all nonnegative t, is piecewise continuous on every closed interval [0,b] for b > 0, and is of exponential order $e^{\alpha t}$, then

$$L\{t^n f(t)\} = (-1)^n \frac{d^n}{ds^n} L\{f(t)\} \text{,} \tag{2}$$

380

for $s > \alpha$, where n is a positive integer.

Using formula (1), we find that

$$L\{e^{4t}\} = \frac{1}{s-4} , \text{ for } s > 4 .$$

In order to use the theorem associated with formula (2), we must first demonstrate that the function

$$f(t) = e^{4t}$$

is of exponential order; i.e., there exists a real constant α and positive real constants M and t_0 such that

$$e^{-\alpha t}|f(t)| < M \text{ for all } t > t_0 .$$

For $\alpha = 4$,

$$\lim_{t \to \infty} e^{-\alpha t}|e^{4t}| = \lim_{t \to \infty} e^{(4-\alpha)t} = 1.$$

For $\alpha > 4$,

$$\lim_{t \to \infty} e^{-\alpha t}|e^{4t}| = \lim_{t \to \infty} e^{(4-\alpha)t} = 0 .$$

Thus, for $\alpha \geq 4$,

$$\lim_{t \to \infty} e^{-\alpha t}|e^{4t}| < 2 .$$

It follows from the definition of a limit that there exists a positive constant t_0 such that, for $t > t_0$,

$$e^{-\alpha t}|e^{4t}| < 2,$$

which, taking M = 2, shows that $f(t) = e^{4t}$ is of exponential order $e^{\alpha t}$, for $\alpha \geq 4$.

Since $f(t) = e^{4t}$ is defined for all nonnegative t, is piecewise continuous on every closed interval [0,b] for b > 0, and is of exponential order $e^{\alpha t}$ for $\alpha \geq 4$, we can substitute formula (1) for $L\{f(t)\}$ in formula (2):

$$L\{g(t)\} = L\{te^{4t}\}$$

$$= (-1) \frac{d}{ds} L\{e^{4t}\} = - \frac{d}{ds} \left(\frac{1}{s-4}\right)$$

$$= \frac{1}{(s-4)^2} \text{ , for } s > 4 \text{ .}$$

(b)   Using the property that

$$L\{t^n f(t)\} = (-1)^n \frac{d^n}{ds^n} L\{f(t)\} \text{ ,}$$

where n is a positive integer constant, provided that $L\{f(t)\}$ exists, we obtain

$$L\{t^{7/2}\} = L\{t^3 \sqrt{t}\}$$

$$= (-1)^3 \frac{d^3}{ds^3} L\{\sqrt{t}\}$$

$$= - \frac{d^3}{ds^3} L\{\sqrt{t}\} \text{ .} \tag{3}$$

Now $L\{\sqrt{t}\}$ may be found using the Γ function whose values are as well tabulated as, say, log x so that an answer expressible in terms of $\Gamma(x)$ is very useful.   The definition of the Γ function is

$$\Gamma(k) = \int_0^\infty t^{(k-1)} e^{-t} dt$$

so that

$$\Gamma(k+1) = \int_0^\infty t^k e^{-t} dt \text{ .} \tag{4}$$

Now make the substitution t = sx where s is some constant so that dt = sdx, $e^{-t} = e^{-sx}$, and $t^k = s^k x^k$.   Then (d) becomes

$$\Gamma(k+1) = \int_0^\infty s^k x^k e^{-sx} sdx$$

$$= s^{k+1} \int_0^\infty x^k e^{-sx} dx \tag{5}$$

$$= s^{k+1} L\{x^k\}$$

so that

382

$$L\{x^k\} = \frac{\Gamma(k+1)}{s^{k+1}} \ , \ k > -1, \ s > 0 \ , \qquad (6)$$

where the restrictions on s and k are imposed to insure the convergence of the integrals in (4) and (5) respectively. Therefore, using

$k = \frac{1}{2}$ in (6) gives

$$L(t^{\frac{1}{2}}) = L(\sqrt{t}) = \Gamma(3/2)s^{-3/2} .$$

Now look up $\Gamma\{3/2\}$ in a table of $\Gamma(x)$ to find $\Gamma(3/2) = \sqrt{\pi}/2$ and

$$L(\sqrt{t}) = \frac{1}{2}\sqrt{\pi}\, s^{-3/2} .$$

Substituting this result in equality (3),

$$L\{t^{7/2}\} = -\frac{d^3}{ds^3}\left(\frac{1}{2}\sqrt{\pi}s^{-3/2}\right)$$

$$= -\frac{d^2}{ds^2}\left(-\frac{3}{2}\cdot\frac{1}{2}\sqrt{\pi}\,s^{-5/2}\right)$$

$$= -\frac{d}{ds}\left[\left(-\frac{5}{2}\right)\left(-\frac{3}{2}\right)\cdot\frac{1}{2}\sqrt{\pi}\,s^{-7/2}\right]$$

$$= -\left(-\frac{7}{2}\right)\left(-\frac{5}{2}\right)\left(-\frac{3}{2}\right)\cdot\frac{1}{2}\sqrt{\pi}\,s^{-9/2}$$

$$= \frac{105}{16}\sqrt{\pi}\,s^{-9/2} .$$

● **PROBLEM 7-9**

Find the Laplace transform

$$L\left\{\frac{\sin 3t}{t}\right\}.$$

<u>Solution</u>: Using the property that

$$L\left\{\frac{1}{t}f(t)\right\} = \int_s^\infty F(x)\,dx ,$$

where $F(s) = L\{f(t)\}$, provided that $L\{f(t)\}$ and

$$\lim_{\substack{x \to 0 \\ x > 0}} [f(x)/x^r]$$

exist, for some $r > 0$, we obtain

$$L\left\{\frac{\sin 3t}{t}\right\} = \int_s^\infty G(x)\,dx \ , \tag{1}$$

where $G(s) = L\{\sin 3t\}$. Now either recall the results of a previous problem or look in a table of Laplace transforms to find that

$$L\{\sin bt\} = \frac{b}{s^2 + b^2} \ , \ s > 0 \ ;$$

hence, taking b = 3, we find

$$G(s) = L\{\sin 3t\} = \frac{3}{s^2 + 3^2} \ .$$

Thus,

$$G(x) = \frac{3}{x^2 + 3^2} \ .$$

Substituting this result into equality (1),

$$L\left\{\frac{\sin 3t}{t}\right\} = \int_s^\infty \left(\frac{3}{x^2 + 3^2}\right) dx$$

$$= \lim_{R \to \infty} \int_s^R \left(\frac{3}{x^2 + 3^2}\right) dx$$

$$= \lim_{R \to \infty} 3 \int_s^R \frac{dx}{x^2 + 3^2}$$

$$= \lim_{R \to \infty} 3 \left(\frac{1}{3} \operatorname{Arctan} \frac{x}{3}\right)\Bigg|_{x=s}^R$$

$$= \lim_{R \to \infty} \left(\operatorname{Arctan} \frac{R}{3} - \operatorname{Arctan} \frac{s}{3}\right)$$

$$= \frac{\pi}{2} - \operatorname{Arctan} \frac{s}{3} \ ,$$

384

where the capitalized A on the Arctangent function indicates that the range of that function is the open interval $(-\pi/2, \pi/2)$.

Find the Laplace transform $L\{g(t)\}$, where

$$g(t) = \begin{cases} 0, & t < 4 \\ \\ (t-4)^2, & t \geq 4 \end{cases} .$$

<u>Solution</u>: The function $g(t)$ can be expressed as $(t-4)^2 \times \alpha(t-4)$, where α is the unit step function, defined as follows:

$$\alpha(x) = \begin{cases} 0, & x < 0 \\ \\ 1, & x \geq 0 \end{cases}$$

so that

$$\alpha(t-4) = \begin{cases} 0, & t < 4 \\ \\ 1, & t \geq 4 \end{cases} .$$

Using the shifting property,

$$L\{f(t-c)\alpha(t-c)\} = e^{-cs} L\{f(t)\} ,$$

where c is a nonnegative constant, provided that $L\{f(t)\}$ exists, we obtain (taking $c = 4$)

$$L\{g(t)\} = L\{(t-4)\}^2 \alpha(t-4)\}$$

$$= e^{-4s} L\{t^2\} . \tag{1}$$

By recalling the result of a previous problem or by looking in a table of Laplace transforms, we find that

$$L\{t^n\} = \frac{n!}{s^{n+1}} , \quad s > 0 ,$$

where n is a nonnegative integer constant; hence, taking $n = 2$, we find

$$L\{t^2\} = \frac{2}{s^3} \; .$$

Substituting this result into equality (1),

$$L\{g(t)\} = e^{-4s} \left(\frac{2}{s^3}\right) = \frac{2e^{-4s}}{s^3}$$

• PROBLEM 7–11

Find the Laplace transform of the function f(t) shown in the accompanying figure and defined by

$$f(t) = \begin{cases} t \, , & 0 < t < 4 \\ \\ 5 \, , & t > 4 \; . \end{cases}$$

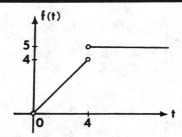

Solution: It is possible to handle some step-function problems without using the shifting property of the Laplace integral in the problem is fairly easy to solve. Thus, note that f(t) need not be defined at t = 0 or at t = 4 since such information is not essential in performing the integration to determine the Laplace transform.

Using L as the Laplace transform operator, with

$$L\{f(t)\} = \int_0^\infty e^{-st} f(t)\,dt,$$

and considering real s, we find

$$L\{f(t)\} = \lim_{R\to\infty} \int_0^R f(t)\, e^{-st}\, dt$$

$$= \int_0^4 te^{-st}\, dt + \lim_{R\to\infty} \int_4^R 5e^{-st}\, dt \; .$$

For s = 0,

386

$$L\{f(t)\} = \int_0^4 te^{0t}\, dt + \lim_{R \to \infty} \int_4^R 5e^{0t}\, dt$$

$$= \int_0^4 t\, dt + \lim_{R \to \infty} \int_4^R 5\, dt$$

$$= \frac{4^2}{2} + \lim_{R \to \infty} 5(R-4) = \infty \quad ;$$

hence $L\{f(t)\}$ does not exist for $s = 0$. For $s \neq 0$,

$$L\{f(t)\} = \int_0^4 te^{-st}\, dt + \lim_{R \to \infty} \int_4^R 5e^{-st}\, dt$$

$$= \int_0^4 te^{-st}\, dt + \lim_{R \to \infty} \left(\frac{-5e^{-sR} + 5e^{-4s}}{s} \right)$$

$$= \int_0^4 te^{-st}\, dt + \frac{5e^{-4s}}{s} - \frac{5}{s} \lim_{R \to \infty} e^{-sR} \quad .$$

Integrating by parts,

$$L\{f(t)\} = \left[-\frac{t}{s} e^{-st} - \frac{1}{s^2} e^{-st} \right] \Bigg|_{t=0}^{4} + \frac{5e^{-4s}}{s} - \frac{5}{s} \lim_{R \to \infty} e^{-sR}$$

$$= -\frac{4}{s} e^{-4s} - \frac{1}{s^2} e^{-4s} + \frac{1}{s^2} + \frac{5}{s} e^{-4s} - \frac{5}{s} \lim_{R \to \infty} e^{-sR} \quad .$$

For $s > 0$, the above expression converges to

$$L\{f(t)\} = \frac{e^{-4s}}{s} - \frac{e^{-4s}}{s^2} + \frac{1}{s^2} \quad ;$$

otherwise, $L\{f(t)\}$ does not exist.

Thus, the required Laplace transform is

$$L\{f(t)\} = \frac{e^{-4s}}{s} - \frac{e^{-4s}}{s^2} + \frac{1}{s^2} \quad , \text{ for } s > 0 \ .$$

Find the Laplace transform $L\{f(t)\}$ of the function shown in Figure 1 and defined by

$$f(t) = \begin{cases} t^2, & 0 < t < 2 \\ \\ 6, & t > 2 \end{cases}.$$

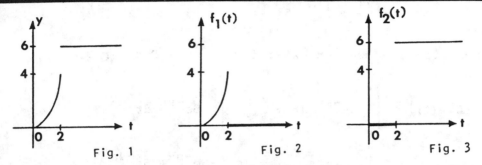

Fig. 1   Fig. 2   Fig. 3

<u>Solution</u>: We wish to express $f(t)$ in terms of the unit step function $\alpha(t)$, defined as

$$\alpha(t) = \begin{cases} 0, & t < 0 \\ \\ 1, & t \geq 0. \end{cases}$$

We can then use the shifting property which states that $L\{g(t-c)\alpha(t-c)\} = e^{-cs}L\{g(t)\}$, where c is a nonnegative constant, provided that $L\{g(t)\}$ exists.

    Let us build $f(t)$ out of continuous functions and unit step functions. First, we need a function $f_1(t)$ that is equal to t^2 when $0 < t < 2$ and zero when $t > 2$. The function $t^2\alpha(t-2)$ is equal to t^2 when $t > 2$ and zero elsewhere, so that if we subtract it from t^2, we obtain the desired function (see Fig. 2):

$$f_1(t) = t^2 - t^2\alpha(t-2).$$

Second, we need a function $f_2(t)$ that is equal to 6 when $t > 2$ and zero when $0 < t < 2$. We obtain this by multiplying the constant 6 by $\alpha(t-2)$, since $\alpha(t-2) = 0$ when $t < 2$ and $= 1$ when $t \geq 2$. Thus (see Fig. 3),

$$f_2(t) = 6\alpha(t-2).$$

The function f(t) is obtained by adding together $f_1(t)$ and $f_2(t)$:

$$f(t) = f_1(t) + f_2(t)$$
$$= t^2 - t^2\alpha(t-2) + 6\alpha(t-2)$$
$$= t^2 + (6-t^2)\alpha(t-2). \tag{1}$$

This is still not the form we require in order to use the property mentioned at the beginning of this solution. We need to express $(6-t^2)$ as a function of $(t-2)$. We know that

$$(t-2)^2 = t^2 - 4t + 4;$$

hence,

$$6 - t^2 = -(t^2 - 4t + 4) - 4t + 10$$
$$= -(t-2)^2 - 4t + 10$$
$$= -(t-2)^2 - 4t + 8 + 2$$
$$= -(t-2)^2 - 4(t-2) + 2.$$

Substituting this last expression into equality (1),

$$f(t) = t^2 + [-(t-2)^2 - 4(t-2) + 2]\alpha(t-2)$$
$$= t^2 - (t-2)^2\alpha(t-2) - 4(t-2)\alpha(t-2) + 2\alpha(t-2).$$

Taking the Laplace transform of this,

$$L\{f(t)\} = L\{t^2 - (t-2)^2\alpha(t-2) - 4(t-2)\alpha(t-2) + 2\alpha(t-2)\}.$$

Since L is a linear operator,

$$L\{f(t)\} = L\{t^2\} - L\{(t-2)^2\alpha(t-2)\} - 4L\{(t-2)\alpha(t-2)\} + 2L\{(1)\alpha(t-2)\}.$$

To the above equality we apply the property mentioned at the beginning of this solution:

$$L\{f(t)\} = L\{t^2\} - e^{-2s}L\{t^2\} - 4e^{-2s}L\{t\} + 2e^{-2s}L\{1\}. \tag{2}$$

Now recall the result of a previous problem or use a table of Laplace transforms to find that

$$L\{t^n\} = \frac{n!}{s^{n+1}}, \quad s > 0,$$

where n is a nonnegative integer constant. Taking $n = 0$, 1, and 2, respectively, we obtain

$$L\{1\} = \frac{1}{s}, \quad L\{t\} = \frac{1}{s^2},$$

and

$$L\{t^2\} = \frac{2}{s^3}.$$

Substituting these results into equality (2), we obtain

$$L\{f(t)\} = \frac{2}{s^3} - \frac{2e^{-2s}}{s^3} - \frac{4e^{-2s}}{s^2} + \frac{2e^{-2s}}{s} \ .$$

● PROBLEM 7-13

Find the Laplace transform $L\{f(t)\}$, where $f(t)$ is the function shown in figures (a) and (b).

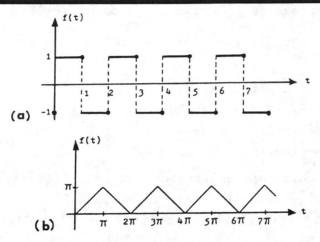

(a)

(b)

Solution: (a) Note that $f(t)$ is a periodic function with period P = 2; i.e., $f(t+2) = f(t)$ for all t. On the interval $0 < t \leq 2$, $f(t)$ can be expressed in analytic form:

$$f(t) = \begin{cases} 1, & 0 < t \leq 1 \\[2mm] -1, & 1 < t \leq 2 \ . \end{cases}$$

Using the property that

$$L\{g(t)\} = \frac{\displaystyle\int_0^P e^{-st} g(t)dt}{1 - e^{-Ps}} \ ,$$

where $g(t)$ is periodic with period P, we obtain

$$L\{f(t)\} = \frac{\displaystyle\int_0^2 e^{-st} f(t)dt}{1 - e^{-2s}}$$

$$= \frac{\displaystyle\int_0^1 e^{-st}(1)dt + \int_1^2 e^{-st}(-1)dt}{1 - e^{-2s}}$$

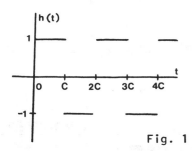
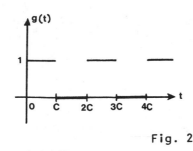

Fig. 1   Fig. 2

Solution: (a) The crucial observation here is that

$$h(t) = 2g(t) - 1$$

for all t. Since L is a linear operator,

$$L\{h(t)\} = L\{2g(t) - 1\}$$

$$= 2L\{g(t)\} - L\{1\} \quad . \tag{3}$$

From a table of Laplace transforms, we find that

$$L\{t^n\} = \frac{n!}{s^{n+1}} \quad ,$$

where n is a nonnegative integer constant; hence, taking n = 0,

$$L\{1\} = 1/s \quad .$$

Substituting this result and equality (1) into equality (3) we find that

$$L\{h(t)\} = 2\left[\frac{1}{s(1+e^{-cs})}\right] - \frac{1}{s}$$

$$= \frac{1}{s}\left(\frac{2}{1 + e^{-cs}} - 1\right)$$

$$= \frac{1}{s}\left[\frac{2 - (1 + e^{-cs})}{1 + e^{-cs}}\right]$$

$$= \frac{1}{s}\left(\frac{1 - e^{-cs}}{1 + e^{-cs}}\right) \quad .$$

Multiplying the numerator and denominator by $e^{cs/2}$,

$$L\{h(t)\} = \frac{1}{s}\left(\frac{e^{cs/2} - e^{-cs/2}}{e^{cs/2} + e^{-cs/2}}\right)$$

394

$$= \frac{\dfrac{-e^{-st}}{s}\bigg|_{t=0}^{1} + \dfrac{e^{-st}}{s}\bigg|_{t=1}^{2}}{1 - e^{-2s}}$$

$$= \frac{\dfrac{1}{s}(1 - 2e^{-s} + e^{-2s})}{1 - e^{-2s}} \quad .$$

Factoring the numerator and denominator of this last fraction,

$$L\{f(t)\} = \frac{\dfrac{1}{s}(1 - e^{-s})^2}{(1+e^{-s})(1-e^{-s})}$$

$$= \frac{1}{s}\left(\frac{1 - e^{-s}}{1 + e^{-s}}\right) \quad .$$

Multiplying the numerator and denominator by $e^{s/2}$,

$$L\{f(t)\} = \frac{1}{s}\left(\frac{e^{s/2} - e^{-s/2}}{e^{s/2} + e^{-s/2}}\right)$$

$$= \frac{1}{s}\tanh\frac{s}{2} \quad .$$

(b) Note that f(t) is a periodic function with period P = 2π; i.e., f(t+2π) = f(t) for all t. On the interval 0 ≤ t ≤ 2π, f(t) can be expressed in analytic form:

$$f(t) = \begin{cases} t, & 0 \le t \le \pi \\ 2\pi - t, & \pi < t \le 2\pi \end{cases} \quad .$$

Using the property that

$$L\{g(t)\} = \frac{\displaystyle\int_0^P e^{-st}g(t)\,dt}{1 - e^{-Ps}} \quad ,$$

where g(t) is periodic with period P, we obtain

$$L\{f(t)\} = \frac{\displaystyle\int_0^{2\pi} e^{-st}f(t)\,dt}{\rule{2cm}{0.4pt}}$$

391

$$= \frac{\int_0^\pi e^{-st}(t)\,dt + \int_\pi^{2\pi} e^{-st}(2\pi - t)\,dt}{1 - e^{-2\pi s}}$$

$$\frac{1 - e^{-2\pi s}}$$

$$= \frac{\int_0^\pi t e^{-st}\,dt - \int_\pi^{2\pi} t e^{-st}\,dt + 2\pi \int_\pi^{2\pi} e^{-st}\,dt}{1 - e^{-2\pi s}}.$$

Integrating by parts

$$L\{f(t)\} = \frac{\left(\frac{-te^{-st}}{s} - \frac{e^{-st}}{s^2}\right)\Big|_{t=0}^{\pi} - \left(\frac{-te^{-st}}{s} - \frac{e^{-st}}{s^2}\right)\Big|_{t=\pi}^{2\pi} - \left(\frac{2\pi e^{-st}}{s}\right)\Big|_{t=\pi}^{2\pi}}{1 - e^{-2\pi s}}$$

$$= \frac{\left(\frac{-\pi e^{-\pi s}}{s} + \frac{1 - e^{-\pi s}}{s^2}\right) + \left(\frac{2\pi e^{-2\pi s} - \pi e^{-\pi s}}{s} + \frac{e^{-2\pi s} - e^{-\pi s}}{s^2}\right)}{1 - e^{-2\pi s}}$$

$$+ \frac{\left(\frac{2\pi e^{-\pi s} - 2\pi e^{-2\pi s}}{s}\right)}{1 - e^{-2\pi s}}$$

$$= \frac{1 - 2e^{-\pi s} + e^{-2\pi s}}{s^2(1 - e^{-2\pi s})}$$

$$= \frac{\left(1 - e^{-\pi s}\right)^2}{s^2(1 + e^{-\pi s})(1 - e^{-\pi s})}$$

$$= \frac{1}{s^2}\left(\frac{1 - e^{-\pi s}}{1 + e^{-\pi s}}\right).$$

Multiplying both the numerator and denominator in the last fraction by $e^{\pi s/2}$,

$$L\{f(t)\} = \frac{1}{s^2}\left(\frac{e^{\pi s/2} - e^{-\pi s/2}}{e^{\pi s/2} + e^{-\pi s/2}}\right)$$

$$= \frac{1}{s^2}\tanh\frac{\pi s}{2}.$$

Find the Laplace transform $L\{h(t)\}$, where

$$h(t) = \begin{cases} 1, & 0 < t < c \\ -1, & c < t < 2c \end{cases}$$

and $h(t+2c) = h(t)$ for all t, with c a constant in the following two ways. (See Figure 1.)

(a) Use the fact that

$$L\{g(t)\} = \frac{1}{s(1 + e^{-cs})}, \tag{1}$$

where

$$g(t) = \begin{cases} 1, & 0 < t < c \\ 0, & c < t < 2c \end{cases}$$

and

$$g(t+2c) = g(t), \text{ for all } t .$$

(see Figure 2.)

(b) Use the result developed in the previous problem, i.e., that for antiperiodic functions of the form

$$f(t+c) = -f(t) ,$$

the Laplace transform of f is given by

$$L\{f(t)\} = \frac{\int_0^c f(t)e^{-st}\,dt}{1 + e^{-cs}}. \tag{2}$$

$$= \frac{1}{s} \tanh \left(\frac{cs}{2} \right) .$$

(b)   In this case the relation (2) takes the form

$$L\{h(t)\} = \frac{\displaystyle\int_0^c h(t) e^{-st} \, dt}{1 + e^{-cs}} = \frac{\displaystyle\int_0^c 1 \cdot e^{-st} \, dt}{1 + e^{-cs}}$$

since $h(t) = 1$ in $(0,c)$.   Hence

$$L\{h(t)\} = \frac{\left. \dfrac{-1}{s} e^{-st} \right|_0^c}{1 + e^{-cs}} = \frac{\dfrac{1}{s} (1 - e^{-cs})}{1 + e^{-cs}} .$$

Again, multiply numerator and denominator by $e^{cs/2}$ to get

$$L\{h(t)\} = \frac{1}{s} \left(\frac{e^{cs/2} - e^{-cs/2}}{e^{cs/2} + e^{-cs/2}} \right) = \frac{1}{s} \tanh \left(\frac{cs}{2} \right) .$$

INVERSE LAPLACE TRANSFORMS

● **PROBLEM** 7–15

Find the inverse Laplace transforms

(a)
$$L^{-1} \left\{ \frac{2s}{(s^2 + 1)^2} \right\},$$

(b)
$$L^{-1} \left\{ \frac{1}{\sqrt{s}} \right\}$$

Solution: (a) In a table of Laplace transforms we find

$$L\{t \sin bt\} = \frac{2bs}{(s^2 + b^2)^2} ,$$

where b is a constant; hence, taking b = 1, we find

$$L\{t \sin t\} = \frac{2s}{(s^2 + 1)^2} .$$

Therefore, by definition of the inverse Laplace transform,

$$L^{-1}\left\{\frac{2s}{(s^2 + 1)^2}\right\} = t \sin t \ .$$

(b) In a table of Laplace transforms, we find

$$L\left\{\frac{1}{\sqrt{t}}\right\} = \frac{\sqrt{\pi}}{\sqrt{s}} \quad .$$

Thus, by definition of the inverse Laplace transform,

$$L^{-1}\left\{\frac{\sqrt{\pi}}{\sqrt{s}}\right\} = \frac{1}{\sqrt{t}}$$

Since L^{-1} is a linear operator,

$$L^{-1}\left\{\frac{1}{\sqrt{s}}\right\} = \frac{1}{\sqrt{\pi}} L^{-1}\left\{\frac{\sqrt{\pi}}{\sqrt{s}}\right\} = \frac{1}{\sqrt{\pi}} \cdot \frac{1}{\sqrt{t}}$$

● **PROBLEM** 7–16

Find and sketch the function g(t) which is the inverse Laplace transform

$$g(t) = L^{-1}\left\{\frac{3}{s} - \frac{4e^{-s}}{s^2} + \frac{4e^{-3s}}{s^2}\right\} \ .$$

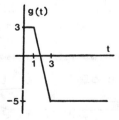

Solution: L^{-1} is a linear operator under the very weak restrictions on g(t) given in an earlier problem. Since these conditions are assumed to hold for all functions dealt with in this book,

$$g(t) = 3L^{-1}\left\{\frac{1}{s}\right\} - 4L^{-1}\left\{\frac{e^{-s}}{s^2}\right\} + 4L^{-1}\left\{\frac{e^{-3s}}{s^2}\right\} . \qquad (1)$$

Recalling the property that

$$L^{-1}\{e^{-ks}F(s)\} = f(t-k)\alpha(t-k) ,$$

where $f(t) = L^{-1}\{f(s)\}$, k is nonnegative constant, and α is the unit step function

$$\alpha(x) = \begin{cases} 0 , & x < 0 \\ \\ 1 , & x \geq 0 \end{cases} ,$$

provided that $L^{-1}\left\{e^{-ks}F(s)\right\}$ and $L^{-1}\left\{F(s)\right\}$ exist, equality (1) becomes

$$g(t) = \left[3L^{-1}\left\{\frac{1}{s}\right\}\bigg|_{t} -4L^{-1}\left\{\frac{1}{s^2}\right\}\bigg|_{t-1} \cdot \alpha(t-1) + 4L^{-1}\left\{\frac{1}{s^2}\right\}\bigg|_{t-3} \right.$$

$$\left. X \alpha(t-3) \right]. \qquad (2)$$

From a table of Laplace transforms, or by recalling a previous problem, we find that

$$L^{-1}\left\{\frac{1}{s^n}\right\}\bigg|_{x} = \frac{x^{n-1}}{(n-1)!} ,$$

where n is a positive integer; hence,

$$L^{-1}\left\{\frac{1}{s}\right\}\bigg|_{t} = 1, \quad L^{-1}\left\{\frac{1}{s^2}\right\}\bigg|_{t-1} = t-1 ,$$

and

$$L^{-1}\left\{\frac{1}{s^2}\right\}\bigg|_{t-3} = t-3 .$$

Substituting these results into equality (2),

$$g(t) = 3-4(t-1)\alpha(t-1) + 4(t-3)\alpha(t-3) . \qquad (3)$$

To remove the α function from expression (3), consider first the case t < 1. In that case,

$$\alpha(t-1) = \alpha(t-3) = 0 ,$$

so that

$$g(t) = 3-4(t-1)(0) + 4(t-3)(0) = 3 . \qquad (4)$$

When $1 \le t < 3$, we have $\alpha(t-1) = 1$ and $\alpha(t-3) = 0$, so that

$$g(t) = 3-4(t-1)(1) + 4(t-3)(0) = 7-4t . \qquad (5)$$

When $t \ge 3$, we have $\alpha(t-1) = \alpha(t-3) = 1$, so that

$$g(t) = 3-4(t-1)(1) + 4(t-3)(1) = -5 . \qquad (6)$$

Grouping together results (4), (5), and (6), we have

$$g(t) = \begin{cases} 3 , & t < 1 \\ 7-4t , & 1 \le t < 3 \\ -5 , & t \ge 3 \end{cases} .$$

The graph of g(t) is shown in the figure.

● **PROBLEM** 7–17

Find the inverse Laplace transform

$$f(t) = L^{-1}\left\{ \log \frac{s+1}{s-1} \right\} , \quad s > 1 .$$

Solution: From a table of infinite series, we can find

$$\log \frac{1+x}{1-x} = 2 \sum_{n=0}^{\infty} \frac{x^{2n+1}}{2n+1} , \qquad (1)$$

for $|x| < 1$. We wish to put $\log[(s+1)/(s-1)]$ into some form for which the series (1) will be useful. Dividing the numerator and denominator by s does not change the value:

$$\log \frac{s+1}{s-1} = \log \frac{1+1/s}{1-1/s} \ .$$

Now, $|1/s| < 1$, since $s > 1$. Thus,

$$\log \frac{1+1/s}{1-1/2} = 2 \sum_{n=0}^{\infty} \frac{(1/s)^{2n+1}}{2n+1} \ ,$$

and

$$L^{-1}\left\{\log \frac{s+1}{s-1}\right\} = L^{-1}\left\{2 \sum_{n=0}^{\infty} \frac{(1/s)^{2n+1}}{2n+1}\right\}.$$

Since L^{-1} is a linear operator, under the very weak restrictions on $f(t)$ given in an earlier problem and which are assumed here,

$$L^{-1}\left\{\log \frac{s+1}{s-1}\right\} = 2 \sum_{n=0}^{\infty} \left(\frac{1}{2n+1}\right) L^{-1}\left\{\left(\frac{1}{s}\right)^{2n+1}\right\} . \qquad (2)$$

From a table of Laplace transforms or from the result of a previous problem, it is found that

$$L^{-1}\left\{\left(\frac{1}{s}\right)^{k}\right\} = \frac{t^{k-1}}{(k-1)!} \ ,$$

where k is a positive integer. Substitution $(2n+1)$ for k in the above formula,

$$L^{-1}\left\{\left(\frac{1}{s}\right)^{2n+1}\right\} = \frac{t^{2n}}{(2n)!} \ ,$$

where $(2n+1)$ is a positive integer; i.e., where n is a nonnegative integer. Substituting this result into equality (2),

$$f(t) = L^{-1}\left\{\log \frac{s+1}{s-1}\right\} = 2 \sum_{n=0}^{\infty} \left(\frac{1}{2n+1}\right) \frac{t^{2n}}{(2n)!}$$

$$= 2 \sum_{n=0}^{\infty} \frac{t^{2n}}{(2n+1)!}$$

$$= \frac{2}{t} \sum_{n=0}^{\infty} \frac{t^{2n+1}}{(2n+1)!} = \frac{2}{t} \sinh t \ .$$

An alternative way of solving this problem is to use the equality

$$L\{tf(t)\} = -\frac{d}{ds} L\{f(t)\}$$

Differentiating our function,

$$\frac{d}{ds}\left(\log \frac{s+1}{s-1}\right) = -\frac{2}{s^2-1} \quad .$$

If we find $h(t)$ such that

$$L\{h(t)\} = -\frac{2}{s^2-1} \quad ,$$

then

$$L\{h(t)\} = L\{t \cdot \frac{h(t)}{t}\} = -\frac{d}{ds} L\left\{\frac{h(t)}{t}\right\}$$

$$= -\frac{2}{s^2-1} = \frac{d}{ds}\left(\log\frac{s+1}{s-1}\right) \quad ,$$

and

$$L\left\{-\frac{h(t)}{t}\right\} = \log \frac{s+1}{s-1} + \text{const.}$$

Constant we will show to be equal zero. We know

$$L\{\sinh t\} = \frac{1}{s^2-1} \quad ,$$

so

$$L^{-1}\left\{-\frac{2}{s^2-1}\right\} = -2 \sinh t \quad ,$$

and we have

$$L\left\{-\left(-\frac{2 \sinh t}{t}\right)\right\} = L\left\{\frac{2 \sinh t}{t}\right\} = \log \frac{s+1}{s-1} + \text{const.}$$

Now, as $s \to +\infty$ both

$$L\left\{\frac{2 \sinh t}{t}\right\} \quad \text{and} \quad \log \frac{s+1}{s-1} \to 0,$$

so const = 0 and

$$L^{-1}\left\{\log\frac{s+1}{s-1}\right\} = \frac{2\sinh t}{t} \quad .$$

Find the inverse Laplace transforms

(a)

$$L^{-1}\left\{\frac{1}{s^2 - 2s + 9}\right\},$$

(b)

$$L^{-1}\left\{\frac{s + 1}{s^2 + 6s + 25}\right\} \quad .$$

<u>Solution</u>: (a) Completing the square in the denominator $(s^2 - 2s + 9)$,

$$\frac{1}{s^2 - 2s + 9} = \frac{1}{(s-1)^2 + (\sqrt{8})^2} = \frac{1}{\sqrt{8}}\left[\frac{\sqrt{8}}{(s-1)^2 + (\sqrt{8})^2}\right] \quad .$$

Thus,

$$L^{-1}\left\{\frac{1}{s^2 - 2s + 9}\right\} = L^{-1}\left\{\frac{1}{\sqrt{8}} \cdot \frac{\sqrt{8}}{(s-1)^2 + (\sqrt{8})^2}\right\} \quad . \qquad (1)$$

Noting the property that

$$L^{-1}\{F(s-k)\} = e^{kt} L^{-1}\{F(s)\}$$

for any function F and constant k, provided that $L^{-1}\{F(s-k)\}$ and $L^{-1}\{F(s)\}$ exist, equality (1) becomes

$$L^{-1}\left\{\frac{1}{s^2 - 2s + 9}\right\} = e^t L^{-1}\left\{\frac{1}{\sqrt{8}} \cdot \frac{\sqrt{8}}{s^2 + (\sqrt{8})^2}\right\} \quad .$$

Since L^{-1} is a linear operator under the very weak restrictions on f(t) given in an earlier problem and which are assumed here,

$$L^{-1}\left\{\frac{1}{s^2 - 2s + 9}\right\} = \frac{1}{\sqrt{8}} e^t L^{-1}\left\{\frac{\sqrt{8}}{s^2 + (\sqrt{8})^2}\right\} \quad . \qquad (2)$$

From a table of Laplace transforms or from the results of a previous problem we find that

$$L^{-1}\left\{\frac{b}{s^2 + b^2}\right\} = \sin bt,$$

where b is a constant; hence, $b = \sqrt{8}$, we find

$$L^{-1}\left\{\frac{\sqrt{8}}{s^2 + (\sqrt{8})^2}\right\} = \sin (\sqrt{8}t) .$$

Substituting this result into equality (2),

$$L^{-1}\left\{\frac{1}{s^2 - 2s + 9}\right\} = \frac{1}{\sqrt{8}} e^t \sin (\sqrt{8}t) .$$

(b)   Completing the square in the denominator $(s^2 + 6s + 25)$,

$$\frac{s+1}{s^2 + 6s + 25} = \frac{s+1}{(s+3)^2 + 4^2} = \frac{(s+3)-2}{(s+3)^2 + 4^2}$$

$$= \frac{(s+3)}{(s+3)^2 + 4^2} - \frac{1}{2}\left[\frac{4}{(s+3)^2 + 4^2}\right] .$$

Thus, since L^{-1} is a linear operator   under the restrictions mentioned earlier

$$L^{-1}\left\{\frac{s+1}{s^2 + 6s + 25}\right\} = L^{-1}\left\{\frac{(s+3)}{(s+3)^2 + 4^2}\right\} - \frac{1}{2} L^{-1}\left\{\frac{4}{(s+3)^2 + 4^2}\right\} .$$

(3)

Noting the property that

$$L^{-1}\{F(s-k)\} = e^{kt} L^{-1}\{F(s)\}$$

for any function F and constant k, provided that $L^{-1}\{F(s-k)\}$

and $L^{-1}\{F(s)\}$ exist, equality (3) becomes

$$L^{-1}\left\{\frac{s+1}{s^2 + 6s + 25}\right\} = e^{-3t} L^{-1}\left\{\frac{s}{s^2 + 4^2}\right\} - \frac{1}{2} e^{-3t} L^{-1}\left\{\frac{4}{s^2 + 4^2}\right\} .$$

(4)

From a table of Laplace transforms or from the results of a previous problem   we find that

$$L^{-1}\left\{\frac{s}{s^2 + b^2}\right\} = \cos bt, \text{ and } L^{-1}\left\{\frac{b}{s^2 + b^2}\right\} = \sin bt,$$

where b is a constant; hence, taking b = 4, we find

$$L^{-1}\left\{\frac{s}{s^2 + 4^2}\right\} = \cos 4t, \text{ and } L^{-1}\left\{\frac{4}{s^2 + 4^2}\right\} = \sin 4t \ .$$

Substituting these results into equality (4),

$$L^{-1}\left\{\frac{s+1}{s^2 + 6s + 25}\right\} = e^{-3t}\cos 4t - \frac{1}{2} e^{-3t}\sin 4t \ .$$

• PROBLEM 7-19

Find the inverse Laplace transform

$$f(t) = L^{-1}\{F(s)\} = L^{-1}\left\{\frac{3s^2 + 17s + 47}{(s + 2)(s^2 + 4s + 29)}\right\} \ .$$

Solution: Since the degree of the numerator polynomial is less than the degree of the denominator polynomial, factor the denominator and expand the rational fraction by partial fractions.

Thus,

$$F(s) = \frac{3s^2 + 17s + 47}{(s + 2)(s + r_1)(s + r_2)}$$

where

$$r_1 = \frac{-4 + \sqrt{16-116}}{2} \ ; \quad r_2 = \frac{-4 - \sqrt{16-116}}{2}$$

$$r_1 = -2 + j5 \quad ; \quad r_2 = -2 - j5$$

and

$$j = \sqrt{-1} \ .$$

Expanding by partial fractions gives

$$F(s) = \frac{K_1}{s+2} + \frac{K_2}{s+2+j5} + \frac{K_2*}{s+2-j5} \tag{1}$$

or, finding a common denominator for the right hand side of (1)

$$F(s) = \frac{K_1(s+2+j5)(s+2-j5) + K_2(s+2)(s+2-j5) + K_2^*(s+2)(s+2+j5)}{(s+2)(s+2+j5)(s+2-j5)}$$

But

$$F(s) = \frac{3s^2 + 17s + 47}{(s+2)(s+2+j5)(s+2-j5)}$$

and using this in (2) gives

$$3s^2 + 17s + 47 = K_1(s+2+j5)(s+2-j5) + K_2(s+2)(s+2-j5)$$

$$+ K_2^*(s+2)(s+2+j5) . \qquad (3)$$

Now (3) must hold for all s. In particular, when s = -2, (3) yields

$$\left. (3s^2 + 17s + 47) \right|_{s=-2} = \left. K_1(s+2+j5)(s+2-j5) \right|_{s=-2}$$

so that

$$K_1 = \left. \frac{3s^2 + 17s + 47}{s^2 + 4s + 29} \right|_{s=-2} = \frac{25}{25} = 1 .$$

Similarly at s = -2 -j5, (3) must hold so that

$$K_2 = \left. \frac{3s^2 + 17s + 47}{(s+2)(s+2-j5)} \right|_{s=-2-j5} = \frac{3(-21 + j20) - 34 - j85 + 47}{(-j5)(-j10)}$$

$$= \frac{-50 - j25}{-50} = 1 + j(0.5)$$

and

$$K_2^* = 1 - j(0.5) .$$

Substituting K_1 , K_2 , and K_2^* into the partial fraction expansion gives

$$F(s) = \frac{1}{s+2} + \frac{1+j0.5}{s+2+j5} + \frac{1-j0.5}{s+2-j5} .$$

Taking the inverse Laplace transform of F(s) gives

$$f(t) = e^{-2t} + (1+j0.5)e^{-2t-j5t} + (1-j0.5)e^{-2t+j5t}.$$

Multiplying out:

$$f(t) = e^{-2t} + e^{-2t}e^{-j5t} + j0.5e^{-2t}e^{-j5t} + e^{-2t}e^{j5t}$$

404

$$-j0.5e^{-2t}e^{j5t} \, ,$$

and factoring to obtain the form:

$$\cos \omega t = \frac{e^{j\omega t} + e^{-j\omega t}}{2} \, , \quad \sin \omega t = \frac{e^{j\omega t} - e^{-j\omega t}}{2j} \, ,$$

$$f(t) = e^{-2t} + 2e^{-2t} \left| \frac{e^{j5t} + e^{-j5t}}{2} \right| + (2j)(-j0.5e^{-2t})$$

$$\times \left| \frac{e^{j5t} - e^{-j5t}}{2j} \right|$$

$$f(t) = [e^{-2t} + 2e^{-2t} \cos 5t + e^{-2t} \sin 5t] \, .$$

• PROBLEM 7-20

Use partial fractions to decompose

(a)
$$\frac{1}{(s+1)(s^2+1)} \, ;$$

(b)
$$\frac{1}{(s^2+1)(s^2+4s+8)} \, .$$

Solution: (a) From the linear factor (s+1) in the denominator, we obtain A/(s+1), where A is a constant, undetermined as yet.

From the quadratic factor (s^2+1), we obtain $(Bs+C)/(s^2+1)$, where B and C are constants, undetermined as yet. Now set

$$\frac{1}{(s+1)(s^2+1)} = \frac{A}{s+1} + \frac{Bs+C}{s^2+1} \, . \tag{1}$$

Multiplying both sides of equation (1) by

$$(s+1)(s^2+1) \, ,$$

we obtain

$$1 = A(s^2+1) + (Bs+C)(s+1).$$

Collecting like terms,

$$(0)s^2 + (0)s + 1 = (A+B)s^2 + (B+C)s + (A+C) \, .$$

Equating coefficients of like powers of s on both sides of the above equation,

405

$$0 = A + B , \qquad\qquad (2)$$

$$0 = B + C , \qquad\qquad (3)$$

and

$$1 = A + C . \qquad\qquad (4)$$

From equation (2), A = -B. From equation (3), C = -B.
Substituting these two results into equation (4),

$$1 = (-B) + (-B)$$

$$= -2B .$$

Thus, $B = -\frac{1}{2}$. From this, it follows that $A = -(-\frac{1}{2}) = \frac{1}{2}$,
and

$$C = -(-\frac{1}{2}) = \frac{1}{2} .$$

Substituting these results into equation (1),

$$\frac{1}{(s+1)(s^2+1)} = \frac{(\frac{1}{2})}{s+1} + \frac{(-\frac{1}{2})s+(\frac{1}{2})}{s^2+1} .$$

(b) From the quadratic factors (s^2+1) and (s^2+4s+8) in
the denominator, we obtain $(As+B)/(s^2+1)$ and
$(Cs+D)/(s^2+4s+8)$, respectively, where A,B,C, and D are
constants, undetermined as yet. Now set

$$\frac{1}{(s^2+1)(s^2+4s+8)} = \frac{As+B}{s^2+1} + \frac{Cs+D}{s^2+4s+8} . \qquad\qquad (5)$$

Multiplying both sides of equation (5) by

$$(s^2+1)(s^2+4s+8) ,$$

we obtain

$$1 = (As+B)(s^2+4s+8) + (Cs+D)(s^2+1) .$$

Collecting like terms,

$$(0)s^3+(0)s^2+(0)s+1 = (A+C)s^3 + (4A+B+D)s^2 + (8A+4B+C)s$$
$$+ (8B+D) .$$

Equating coefficients of like powers of s on both sides of
the above equation,

$$0 = A + C , \qquad\qquad (6)$$

$$0 = 4A + B + D , \qquad\qquad (7)$$

$$0 = 8A + 4B + C , \tag{8}$$

and

$$1 = 8B + D . \tag{9}$$

The system of equations (6) through (9) can be solved by Cramer's rule, yielding

$$A = -4/65, \ B = 7/65, \ C = 4/65, \ \text{and} \ D = 9/65.$$

Substituting these results into equation (5),

$$\frac{1}{(s^2+1)(s^2+4s+8)} = \frac{(-4/65)s+(7/65)}{s^2+1} + \frac{(4/65)s+(9/65)}{s^2+4s+8}$$

• PROBLEM 7-21

Find the inverse Laplace transforms

(a)

$$L^{-1}\left\{\frac{s+3}{(s-2)(s+1)}\right\} ,$$

(b)

$$L^{-1}\left\{\frac{8}{s^3(s^2-s-2)}\right\} .$$

Solution: (a) We use the method of partial fractions to decompose $(s+3)/[(s-2)(s+1)]$. Setting

$$\frac{s+3}{(s-2)(s+1)} = \frac{A}{s-2} + \frac{B}{s+1}$$

for all s, with A and B constant, we find $A = 5/3$ and $B = -2/3$ by the following method. Multiply through this equation by $(s-2)(s+1)$ giving

$$s+3 = A(s+1) + B(s-2) .$$

Now since this equation must be true for all s, it is true for $s = 2$ in particular. Substituting $s = 2$ gives

$$2 + 3 = A(2+1) + B(2-2) = 3A$$

or

$$A = 5/3.$$

Substituting $s = -1$ gives

$$-1 + 3 = A(-1 + 1) + B(-1 -2) = -3B$$

or

$$B = -2/3 .$$

Thus,

$$L^{-1}\left\{\frac{s+3}{(s-2)(s+1)}\right\} = L^{-1}\left\{\frac{5}{3}\left(\frac{1}{s-2}\right) - \frac{2}{3}\left(\frac{1}{s+1}\right)\right\} .$$

Since L^{-1} is a linear operator under the very weak restrictions imposed on

$$f(t) = L^{-1}\left\{\frac{s+3}{(s-2)(s+1)}\right\}$$

in a previous problem and which are assumed throughout this chapter,

$$L^{-1}\left\{\frac{s+3}{(s-2)(s+1)}\right\} = \frac{5}{3}L^{-1}\left\{\frac{1}{s-2}\right\} - \frac{2}{3}L^{-1}\left\{\frac{1}{s+1}\right\} . \qquad (1)$$

From a table of Laplace transforms, or from a previous problem, we find that

$$L^{-1}\left\{\frac{1}{s-a}\right\} = e^{at} ,$$

where a is a constant; hence,

$$L^{-1}\left\{\frac{1}{s-2}\right\} = e^{2t}$$

and

$$L^{-1}\left\{\frac{1}{s+1}\right\} = e^{-t} .$$

Substituting these results into equality (1),

$$L^{-1}\left\{\frac{s+3}{(s-2)(s+1)}\right\} = \frac{5}{3}e^{2t} - \frac{2}{3}e^{-t} .$$

(b) We use the method of partial fractions, employing only real roots, to decompose

$$\frac{8}{s^3(s^2-s-2)} , \quad \text{or} \quad \frac{8}{s^3(s-2)(s+1)} .$$

Setting

$$\frac{8}{s^3(s^2-s-2)} = \frac{A}{s} + \frac{B}{s^2} + \frac{C}{s^3} + \frac{D}{s-2} + \frac{E}{s+1}$$

for all s, with A,B,C,D, and E constant, we find, using the method of the previous problem,

$$A = -3, \; B = 2, \; C = -4, \; D = 1/3, \; and \; E = 8/3 \; .$$

Thus,

$$L^{-1}\left\{\frac{8}{s^3(s^2-s-2)}\right\} = L^{-1}\left\{\frac{-3}{s} + \frac{2}{s^2} - \frac{4}{s^3} + \frac{1/3}{s-2} + \frac{8/3}{s+1}\right\} \; .$$

Since L^{-1} is a linear operator under the conditions stated earlier,

$$L^{-1}\left\{\frac{8}{s^3(s^2-s-2)}\right\} = \left[-3L^{-1}\left\{\frac{1}{s}\right\} + 2L^{-1}\left\{\frac{1}{s^2}\right\} -4L^{-1}\left\{\frac{1}{s^3}\right\} \right.$$

$$\left. + \frac{1}{3} L^{-1}\left\{\frac{1}{s-2}\right\} + \frac{8}{3} L^{-1}\left\{\frac{1}{s+1}\right\} \right] \; . \tag{2}$$

From a table of Laplace transforms, we find that

$$L^{-1}\left\{\frac{1}{s^n}\right\} = \frac{t^{n-1}}{(n-1)!} \; ,$$

where n is a positive integer constant; hence,

$$L^{-1}\left\{\frac{1}{s}\right\} = 1, \; L^{-1}\left\{\frac{1}{s^2}\right\} = t \; ,$$

and

$$L^{-1}\left\{\frac{1}{s^3}\right\} = \frac{t^2}{2} \; .$$

Substituting these results into equality (2),

$$L^{-1}\left\{\frac{8}{s^3(s^2-s-2)}\right\} = \left[-3+2t-2t^2 + \frac{1}{3} L^{-1}\left\{\frac{1}{s-2}\right\} \right.$$

$$\left. + \frac{8}{3} L^{-1}\left\{\frac{1}{s+1}\right\} \right] \; . \tag{3}$$

From a table of Laplace transforms, we find that

$$L^{-1}\left\{\frac{1}{s-a}\right\} = e^{at} \text{ ,}$$

where a is a constant; hence,

$$L^{-1}\left\{\frac{1}{s-2}\right\} = e^{2t}$$

and

$$L^{-1}\left\{\frac{1}{s+1}\right\} = e^{-t} \text{ .}$$

Substituting these results into equality (3),

$$L^{-1}\left\{\frac{8}{s^3(s^2-s-2)}\right\} = -3+2-2t^2 + \frac{1}{3}e^{2t} + \frac{8}{3}e^{-t} \text{ .}$$

• **PROBLEM** 7–22

(a) Define the convolution of two functions f(t) and g(t).

(b) State the convolution theorem for Laplace transforms.

(c) Find the inverse Laplace transform

$$f(t) = L^{-1}\{F(s)\} = L^{-1}\left\{\frac{1}{(s^2+c^2)^2}\right\} \text{ ,}$$

(c = constant) .

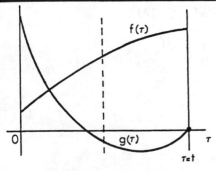

Fig. 1

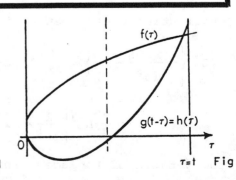

Fig. 2

410

Solution: (a) A convolution of the two functions f(t) and g(t) is denoted by (f*g) and is defined by

$$(f*g) \equiv \int_0^t f(\tau)g(t-\tau)d\tau \quad . \tag{1}$$

Thus (f*g) is the integral of the product $f(\tau)h(\tau)$ where $h(\tau)$ is the reflection of $g(\tau)$ with respect to the line $\tau = t/2$ (see Figures 1 and 2). It can be verified that

(f*g) = (g*f) by making the change of variable $u = t - \tau$.

(b) The convolution theorem states that if f(t) and g(t) are piecewise continuous and of exponential orders α and β (see the first problem of this chapter) then

$$L\{f*g\} = F(s)G(s) \quad [\text{Re } s > \max (\alpha,\beta)] \tag{2}$$

where $F(s) = L\{f(t)\}$ and $G = L\{g(t)\}$. This theorem is very important in the application of Laplace transform technique to differential equations since at the end of a problem one might be faced with inverting F(s)G(s), i.e., finding

$$L^{-1}\{F(s)G(s)\} \quad \text{where} \quad L^{-1}\{F(s)\} = f(t)$$

and

$$L^{-1}\{G(s)\} = g(t)$$

are known functions. But by the convolution theorem,

$$L^{-1}\{F(s)G(s)\} = (f*g) = \int_0^t f(\tau)g(t-\tau)d\tau \quad . \tag{3}$$

(c) Since the inverse transform of

$$\frac{1}{s^2+c^2}$$

is known to be

$$L^{-1}\left\{\frac{1}{s^2+c^2}\right\} = \frac{1}{c} \sin ct$$

from the previous problem or a table of Laplace transforms, the inverse of

411

$$F(s) = \left(\frac{1}{s^2+c^2}\right)^2$$

can be found by the convolution theorem. Thus letting

$$F(s) = \frac{1}{s^2+c^2} = G(s)$$

and

$$f(t) = \frac{1}{c} \sin ct = g(t)$$

in (3) yields

$$L^{-1}\left\{\frac{1}{s^2+c^2} \cdot \frac{1}{s^2+c^2}\right\} = \int_0^t \frac{1}{c} \sin c\tau \frac{1}{c} \sin c(t-\tau)d\tau \quad . \qquad (4)$$

Using the trigonometric identity

$$\sin(ct-c\tau) = \sin ct \cos c\tau - \cos ct \sin c\tau$$

to evaluate the integral in (4) gives

$$\int_0^t \sin c\tau \sin c(t-\tau)d\tau = \sin ct \int_0^t \sin c\tau \cos c\tau \, d\tau$$

$$- \cos ct \int_0^t \sin^2 c\tau \, d\tau$$

$$= \sin ct \left(\frac{1-\cos 2ct}{4c}\right) - \cos ct \left(\frac{2ct - \sin 2ct}{4c}\right)$$

Using this result in (4) yields

$$L^{-1}\left\{\frac{1}{(s^2+c^2)^2}\right\} = \frac{1}{2c^3}(\sin ct - ct \cos ct) \quad .$$

Solve the following problems by employing the convolution theorem:

(a)   Find the inverse Laplace transform

$$x_2(t) = L^{-1}\{X_2(s)\} = L^{-1}\left\{\frac{F(s)}{s^2 + 2\lambda s + \omega_0^2}\right\}$$

which arises in the solution of the motion of a forced damped harmonic oscillator.

(b)   Find the Laplace transform

$$L\left\{\int_0^t \sinh 2x \, dx\right\} \quad .$$

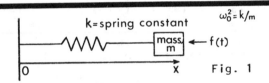

k=spring constant $\omega_0^2 = k/m$

mass m ← f(t)

0   x   Fig. 1

Solution: (a) A brief sketch of the way in which this problem arises is as follows. The equation of motion of a damped   harmonic oscillator which is subject to a time dependent force f(t) (see Fig. 1) is given by

$$\ddot{x} + 2\lambda\dot{x} + \omega_0^2 x = f(t) \qquad \frac{dx}{dt} = \dot{x} \ , \ \frac{d^2x}{dt^2} = \ddot{x} \qquad (1)$$

with initial conditions $x(0) = x_0$, $\dot{x}(0) = v_0$.   The Laplace transform method calls for taking the transform of (1) giving (using the derivative property of L)

$$s^2 X(s) - sx_0 - v_0 + 2\lambda s X(s) - 2\lambda x_0 + \omega_0^2 X(s) = F(s) \ ,$$

where

$F(s) = L\{f(t)\}$ is known and $X(s) = L\{x(t)\}$

is sought. When solved for $X(s)$, this equation becomes

$$X(s) = \frac{2\lambda x_0 + v_0 + sx_0}{s^2 + 2\lambda s + \omega_0{}^2} + \frac{F(s)}{s^2 + 2\lambda s + \omega_0{}^2}$$

$$= X_1(s) + X_2(s) \quad . \tag{2}$$

The first term is inverted by completing the square in the denominator so that

$$\frac{2\lambda x_0 + v_0 + sx_0}{s^2 + 2\lambda s + \omega_0{}^2} = \frac{x_0(s+\lambda)}{(s+\lambda)^2 + (\omega_0{}^2 - \lambda^2)} + \frac{v_0 + \lambda x_0}{(s+\lambda)^2 + (\omega_0{}^2 - \lambda^2)} \quad ,$$

which, upon term by term inversion and making use of a table of Laplace transforms, yields

$$x_1(t) = L^{-1}\{X_1(s)\} = L^{-1}\left\{\frac{x_0(s+\lambda)}{(s+\lambda)^2 + (\omega_0{}^2 - \lambda^2)}\right\}$$

$$+ L^{-1}\left\{\frac{v_0 + \lambda x_0}{(s+\lambda)^2 + (\omega_0{}^2 - \lambda^2)}\right\}$$

$$= x_0 e^{-\lambda t} \cos \omega t + \frac{v_0 + \lambda x_0}{\omega} e^{-\lambda t} \sin \omega t \,(\omega = \sqrt{\omega_0{}^2 - \lambda^2}) \quad .$$

This brings us to the problem at hand, i.e., that of inverting

$$\frac{F(s)}{s^2 + 2\lambda s + \omega_0{}^2}$$

where

$$F(s) = L\{f(t)\} \quad .$$

The relevant pieces of information to be used here are

$$L^{-1}\{F(s)\} = f(t),$$

$$L^{-1}\left\{\frac{1}{s^2 + 2\lambda s + \omega_0^2}\right\} = \frac{1}{\omega} e^{-\lambda t} \sin \omega t,$$

where

$$\omega = \sqrt{\omega_0^2 - \lambda^2},$$

(this can be deduced by writing

$$\frac{1}{s^2 + 2\lambda s + \omega_0^2} = \frac{1}{(s+\lambda)^2 + (\omega_0^2 - \lambda^2)} = \frac{1}{(s+\lambda)^2 + \omega^2},$$

recalling the fact that

$$L\{\sin \omega t\} = \frac{\omega}{s^2 + \omega^2}$$

and using the attentuation rule)   and the convolution theorem

$$L^{-1}\{F(s)G(s)\} = \int_0^t f(\tau)g(t-\tau)d\tau \quad . \tag{3}$$

Thus, using

$$G(s) = \frac{1}{s^2 + 2\lambda s + \omega_0^2}$$

so that

$$g(t) = L^{-1}\{G(s)\} = \frac{1}{\omega} e^{-\lambda t} \sin \omega t,$$

(3) becomes

$$L^{-1}\left\{\frac{F(s)}{s^2 + 2\lambda s + \omega_0^2}\right\} = \int_0^t \frac{1}{\omega} e^{-\lambda(t-\tau)}$$

$$\times \sin \omega(t-\tau) f(\tau)d\tau \quad . \tag{4}$$

Thus, given $f(t)$, if the integral in (4) can be evaluated the problem can be solved and the complete solution reads

$$x(t) = x_0 e^{-\lambda t} \cos \omega t + \frac{v_0 + \lambda x_0}{\omega} e^{-\lambda t} \sin \omega t$$

$$+ \frac{1}{\omega} \int_0^t e^{-\lambda(t-\tau)} \sin \omega(t-\tau) f(\tau) d\tau \qquad .$$

(b) Recall that the convolution theorem states that if f and g are piecewise continuous and of exponential orders α and β respectively, then

$$L\{(f*g)\} = L\left\{ \int_0^t f(\tau)g(t-\tau)d\tau \right\}$$

$$= L\{f(t)\}L\{g(t)\} \ , \ s > \max \ (\alpha, \beta) \ . \tag{5}$$

In particular, if

$$L\{g(t)\} = \frac{1}{s}$$

then g = 1 and (5) yields

$$L\left\{ \int_0^t f(\tau)d\tau \right\} = \frac{1}{s} L\{f(t)\} = \frac{1}{s} F(s) \ , \tag{6}$$

where $F(s) = L\{f(t)\}$. Using this property in the given problem yields

$$L\left\{ \int_0^t \sinh 2x \ dx \right\} = \frac{1}{s} L\{\sinh 2t\} \qquad . \tag{7}$$

Now either look in a table of Laplace transforms or a previous problem in this chapter to find that

$$L\{\sinh bt\} = \frac{b}{s^2-b^2} \ , \ s > |b| \qquad ;$$

Hence, taking b = 2 yields

$$L\{\sinh 2t\} = \frac{2}{s^2-4} \ , \ s > 2 \ .$$

Substituting this result into (7).

$$L\left\{ \int_0^t \sinh 2x \ dx \right\} = \frac{1}{s} \cdot \frac{2}{s^2-4} = \frac{2}{s(s^2-4)} \ .$$

APPLICATIONS OF LAPLACE TRANSFORMS

Find the solution to the initial value problem

$$y''(t) + 4y'(t) + 8y(t) = \sin t, \qquad (1)$$

where

$$y(0) = 1 \text{ and } y'(0) = 0 .$$

<u>Solution</u>: This type of differential equation arises quite often in physical problems. One case is that of a damped harmonic oscillator subject to a sinusoidal force sin t (Fig. 1).

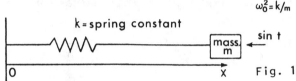

Fig. 1

The differential equation governing the motion of the mass is known to be (with initial conditions)

$$x''(t) + 2\lambda x'(t) + \omega_0^2 x = \sin t, \ x(0) = x_0 , \ x'(0) = v_0 ,$$

and where λ is the damping constant. It is seen that (1) is a special case of this. Equations such as (1) also arise in electric circuit theory. In fact, the equation governing the charge q on the capacitor in Figure 2 is known to be

$$L \frac{d^2 q}{dt} + R \frac{dq}{dt} + \frac{q}{C} = \sin t$$

where L,R,C are constants and $E(t) = \sin t$ is a sinusoidally

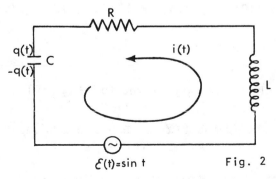

Fig. 2

varying voltage. Thus, equation (1) is also a particular case of this situation. Now to the solution.

Taking the Laplace transform of both sides of equation (1),

$$L\{y''(t) + 4y'(t) + 8y(t)\} = L\{\sin t\}. \qquad (2)$$

From a table of Laplace transforms, we find that

$$L\{\sin bt\} = \frac{b}{s^2 + b^2} \quad , \qquad s > 0 \ ,$$

where b is a constant; hence, taking b = 1, we obtain

$$L\{\sin t\} = \frac{1}{s^2 + 1} \quad , \qquad s > 0 \ .$$

Substituting this result into equation (b),

$$L\{y''(t) + 4y'(t) + 8y(t)\} = \frac{1}{s^2 + 1} \quad .$$

Since L is a linear operator,

$$L\{y''(t)\} + 4L\{y'(t)\} + 8L\{y(t)\} = \frac{1}{s^2 + 1} \quad . \qquad (3)$$

Using the properties that

$$L\{y'(t)\} = sL\{y(t)\} - y(0) \ ,$$

and

$$L\{y''(t)\} = s^2 L\{y(t)\} - sy(0) - y'(0) \ ,$$

provided that y'(t), y''(t), and $L\{y(t)\}$ exist, equation (3) becomes

$$[s^2 L\{y(t)\} - sy(0) - y'(0)] + 4[sL\{y(t)\} - y(0)]$$

$$+ 8L\{y(t)\} = \frac{1}{s^2 + 1} \quad .$$

Solving the above equation for $L\{y(t)\}$,

$$L\{y(t)\} = \frac{(s^2 + 1)(s + 4)y(0) + (s^2 + 1)y'(0) + 1}{(s^2 + 1)(s^2 + 4s + 8)} \quad , \qquad (4)$$

and since y(0) = 1 and y'(0) = 0, equation (4) becomes

$$L\{y(t)\} = \frac{(s^2 + 1)(s + 4) + 1}{(s^2 + 1)(s + 4s + 8)}$$

$$= \frac{s^3 + 4s^2 + s + 5}{(s^2 + 1)(s^2 + 4s + 8)} \qquad (5)$$

Inverting equation (5),

$$y(t) = L^{-1}\left\{\frac{s^3 + 4s^2 + s + 5}{(s^2 + 1)(s^2 + 4s + 8)}\right\}. \qquad (6)$$

We use the method of partial fractions to decompose

$$\frac{(s^3 + 4s^2 + s + 5)}{[(s^2 + 1)(s^2 + 4s + 8)]}.$$

Setting

$$\frac{s^3 + 4s^2 + s + 5}{(s^2 + 1)(s^2 + 4s + 8)} = \frac{As + B}{s^2 + 1} + \frac{Cs + D}{s^2 + 4s + 8},$$

for all s, with A, B, C, and D constant, we find upon multiplying through by

$$(s^2 + 1)(s^2 + 4s + 8)$$

that

$$s^3 + 4s^2 + s + 5 = (As + B)(s^2 + 4s + 8) + (Cs + D)(s^2 + 1).$$

Now choose any 4 values for s and substitute them in this equation to obtain four equations in four unknowns (A,B,C,D) Solving these equations will yield

$$A = -\frac{4}{65}, \quad B = \frac{7}{65}, \quad C = \frac{69}{65}, \quad \text{and } D = \frac{269}{65}.$$

Thus equation (6) becomes

$$y(t) = L^{-1}\left\{\frac{-\dfrac{4}{65}s + \dfrac{7}{65}}{s^2 + 1} + \frac{\dfrac{69}{65}s + \dfrac{269}{65}}{s^2 + 4s + 8}\right\}. \qquad (7)$$

Completing the square in the right-hand denominator in equation (4),

$$y(t) = L^{-1}\left\{\frac{-\dfrac{4}{65}s + \dfrac{7}{65}}{s^2 + 1} + \frac{\dfrac{69}{65}s + \dfrac{269}{65}}{(s + 2)^2 + 2^2}\right\}$$

$$= L^{-1} \left\{ \frac{-\dfrac{4}{65} s + \dfrac{7}{65}}{s^2 + 1} + \frac{\dfrac{69}{65} (s + 2) + \dfrac{131}{130}}{(s + 2)^2 + 2^2} \right\} \cdot \quad (2)$$

Since L^{-1} is a linear operator under the conditions assumed in this chapter,

$$y(t) = \left[-\frac{4}{65} L^{-1} \left\{ \frac{s}{s^2 + 1} \right\} + \frac{7}{65} L^{-1} \left\{ \frac{1}{s^2 + 1} \right\} \right.$$

$$+ \frac{69}{65} L^{-1} \left\{ \frac{(s + 2)}{(s + 2)^2 + 2^2} \right\}$$

$$\left. + \frac{131}{130} L^{-1} \left\{ \frac{2}{(s + 2)^2 + 2^2} \right\} \right] \cdot \quad (8)$$

Using the property that

$$L^{-1}\{f(s - k)\} = e^{kt} L^{-1}\{f(s)\} \quad ,$$

where k is a constant, provided that $L^{-1}\{f(s)\}$ exists, equation (8) becomes

$$y(t) = \left[-\frac{4}{65} L^{-1} \left\{ \frac{s}{s^2 + 1} \right\} + \frac{7}{65} L^{-1} \left\{ \frac{1}{s^2 + 1} \right\} \right.$$

$$+ \frac{69}{65} e^{-2t} L^{-1} \left\{ \frac{s}{s^2 + 2^2} \right\}$$

$$\left. + \frac{131}{130} e^{-2t} L^{-1} \left\{ \frac{2}{s^2 + 2^2} \right\} \right] \cdot \quad (9)$$

From a table of Laplace transforms, we find that

$$L^{-1} \left\{ \frac{s}{s^2 + c_1{}^2} \right\} = \cos c_1 t \ ,$$

and

$$L^{-1} \left\{ \frac{c_2}{s^2 + c_1{}^2} \right\} = \sin c_2 t \ ,$$

where c_1 and c_2 are constants; hence, taking $c_1 = 1$ and 2 and $c_2 = 1$ and 2, respectively, we obtain

420

$$L^{-1}\left\{\frac{s}{s^2+1}\right\} = \cos t, \quad L^{-1}\left\{\frac{s}{s^2+2^2}\right\} = \cos 2t,$$

$$L^{-1}\left\{\frac{1}{s^2+1}\right\} = \sin t, \quad L^{-1}\left\{\frac{2}{s^2+2^2}\right\} = \sin 2t.$$

Substituting these results into equation (9),

$$y(t) = -\frac{4}{65}\cos t + \frac{7}{65}\sin t + \frac{69}{65}e^{-2t}\cos 2t$$

$$+ \frac{131}{130}e^{-2t}\sin 2t. \tag{10}$$

Since we made some assumptions about existence of $y(t)$, $y'(t)$, $y''(t)$, and $L\{y(t)\}$, we must check our result (10) against the original problem. Differentiating twice, we obtain

$$y'(t) = \frac{7}{65}\cos t + \frac{4}{65}\sin t + \frac{e^{-2t}}{65}(-7\cos 2t - 269\sin 2t),$$

$$y''(t) = \frac{4}{65}\cos t - \frac{7}{65}\sin t + \frac{e^{-2t}}{65}$$

$$\times (-524\cos 2t + 552\sin 2t).$$

Checking in equation (1),

$$y''(t) + 4y'(t) + 8y(t)$$

$$= \left[\frac{4}{65}\cos t - \frac{7}{65}\sin t\right.$$

$$+ \frac{e^{-2t}}{65}(-524\cos 2t + 552\sin 2t)\bigg]$$

$$+ 4\left[\frac{7}{65}\cos t + \frac{4}{65}\sin t\right.$$

$$+ \frac{e^{-2t}}{65}(-7\cos 2t - 269\sin 2t)\bigg]$$

$$+ 8\left[-\frac{4}{65}\cos t + \frac{7}{65}\sin t + \frac{69}{65}e^{-2t}\cos 2t\right.$$

$$+ \frac{131}{130}e^{-2t}\sin 2t\bigg]$$

$$= \sin t,$$

$$y(0) = -\frac{4}{65} \cos 0 + \frac{7}{65} \sin 0 + \frac{69}{65} e^0 \cos 0 + \frac{131}{130} e^0 \sin 0$$

$$= 1,$$

$$y'(0) = \frac{7}{65} \cos 0 + \frac{4}{65} \sin 0 + \frac{e^0}{65} (-7 \cos 0 - 269 \sin 0)$$

$$= 0;$$

hence, equation (10) is the solution.

• PROBLEM 7-25

A beam of rigidity EI is clamped at one end and is loaded as shown in the figure; the weight of the beam is neglected. Find the deflection of the beam, y(x), where, for the coordinate system shown, the following differential relations are known to hold.

(a)

$$\frac{d^2 y(x)}{dx^2} = -\frac{1}{EI} m(x) ,$$

where y(x) is the deflection of the beam at point x, and m(x) is the bending moment [counterclockwise torque of all (external) forces to the right of point x],

(b)

$$\frac{dm(x)}{dx} = t(x),$$

where t(x) is the shearing force (resultant of all vertical forces to the right of point x),

(c)

$$\frac{dt(x)}{dx} = -q(x),$$

where q(x) is the load per unit length at point x.

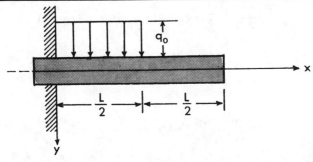

<u>Solution</u>: From these relations it follows that

$$EI \frac{d^4 y(x)}{dx^4} = q(x)$$

Since the beam is clamped, we have $y(0) = 0$, $y'(0)$. Also from the definition of t, since the weight, q_0, per unit length from 0 to L/2 is the only vertical force, it is seen that

t(0) = resultant or "sum" of all vertical forces to the right of

$$0 = q_0 \frac{units\ of\ forces}{unit\ length} \times \frac{L}{2} \text{ units of length.}$$

That is,

$$t(0) = q_0 (L/2).$$

Finally, to find m(0) note the following relations for any small interval dx

$$q_0 \cdot dx = \text{force at points in } (x, x+dx), \text{ where}$$
$$x \in (0, L/2),$$

$$(q_0 \cdot dx) \cdot x = \text{torque about 0 due to load at point x.}$$

Therefore,

$$m(0) = \text{total counterclockwise torque} = -\int_0^{L/2} q_0 x\ dx$$

$$= - q_0 \frac{x^2}{2} \Big|_0^{L/2} = - \frac{q_0 L^2}{8}.$$

Using these results for m(0) and t(0) in the conditions (a) and (b) gives

$$y''(0) = q_0 L^2 / 8EI, \qquad y'''(0) = - q_0 L / 2EI.$$

Now, take the Laplace transform of the equation (1) and use the derivative property

$$L\{y^{(n)}(t)\} = s^n L\{y(t)\} - \sum_{k=1}^{n} s^{k-1} y^{(n-k)}(0)$$

to find

$$EI[s^4 L\{y(x)\} - y^{(3)}(0) - sy''(0) - s^2 y'(0) - s^3 y(0)] =$$

423

$$L\{q(x)\}$$

or, letting $Y(s) = L\{y(t)\}$,

$$EI\left[s^4 Y(s) + \frac{q_0 L}{2EI} - \frac{s q_0 L^2}{8EI}\right] = L\left\{q_0[1 - \alpha(x - L/2)]\right\}$$

$$= L\{q_0\} - L\{q_0 \alpha(x - L/2)\} \tag{2}$$

where

$$\alpha(x) = \begin{cases} 1, & x \geq 0 \\ \\ 0, & x < 0 \end{cases}$$

is the unit step function so that

$$1 - \alpha\{x - L/2\} = \begin{cases} 0, & x \geq L/2 \\ \\ 1, & x < L/2 \end{cases}$$

From a table of Laplace transforms or the results of previous problems it can be found that

$$L\{1\} = \frac{1}{s}, \quad s > 0$$

$$L\{\alpha(t-k)\} = \frac{e^{-ks}}{s}, \qquad s > 0 .$$

Using these results with $k = L/2$ and rearranging (2) gives

$$s^4 Y(s) - s\frac{q_0 L^2}{8EI} + \frac{q_0 L}{2EI} = \frac{q_0}{EI}\frac{1 - e^{-sL/2}}{s} ,$$

so that

$$Y(s) = \frac{q_0}{EI}\frac{1}{s^5} - \frac{q_0}{EI}\frac{e^{-sL/2}}{s^5} + \frac{q_0 L^2}{8EI}\frac{1}{s^3} - \frac{q_0 L^2}{2EI}\frac{1}{s^4} .$$

Inversion yields

$$y(x) = L^{-1}\{Y(s)\} = L^{-1}\left\{\frac{q_0}{EI}\frac{1}{s^5}\right\} - L^{-1}\left\{\frac{q_0}{EI}\frac{e^{-sL/2}}{s^5}\right\}$$

$$+ L^{-1}\left\{\frac{q_0 L^2}{8EI}\frac{1}{s^3}\right\} - L^{-1}\left\{\frac{q_0 L}{2EI}\frac{1}{s^4}\right\}$$

or

$$y(x) = \frac{q_0}{24EI} x^4 - \frac{q_0}{24EI} \left(x - \frac{L}{2} \right)^4 \alpha \left(x - \frac{L}{2} \right) + \frac{q_0 L^2}{16EI} x^2 - \frac{q_0 L}{12EI} x^3 .$$

It is more convenient to rewrite this solution in the form

$$y(x) = \begin{cases} \frac{q_0}{EI} \left(\frac{x^4}{24} - \frac{Lx^3}{12} + \frac{L^2 x^2}{16} \right) & \left(0 \leq x < \frac{L}{2} \right) \\\\ \frac{q_0}{EI} \left(\frac{L^3 x}{48} - \frac{L^4}{384} \right) & \left(\frac{L}{2} < x \leq L \right) \end{cases}$$

from which, for instance, it is clearly seen that the right half of the beam will remain straight, a fact anticipated on physical grounds.

● **PROBLEM** 7–26

Solve the initial value problem

$$y''(t) + 2y'(t) + 5y(t) = H(t) \tag{1}$$

$$y(0) = y'(0) = 0,$$

where

$$H(t) = \begin{cases} 1, & 0 \leq t < \pi \\\\ 0, & t \geq \pi, \end{cases}$$

as shown in the accompanying graph.

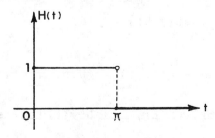

Solution: The function $H(t)$ can be expressed as

$$H(t) = 1 - \alpha(t - \pi),$$

where α is the unit step function

425

$$\alpha(x) = \begin{cases} 0, & x < 0, \\ 1, & x \geq 0 . \end{cases}$$

Thus, equation (1) is equivalent to

$$y''(t) + 2y'(t) + 5y(t) = 1 - \alpha(t - \pi).$$

Taking the Laplace transform of both sides,

$$L\{y''(t) + 2y'(t) + 5y(t)\} = L\{1 - \alpha(t - \pi)\} .$$

Since L is a linear operator,

$$L\{y''(t)\} + 2L\{y'(t)\} + 5L\{y(t)\} = L\{1\} - L\{\alpha(t - \pi)\} . \qquad (2)$$

From a table of Laplace transforms, we find that

$$L\{1\} = \frac{1}{s} , \qquad\qquad s > 0,$$

and

$$L\{\alpha(t - k)\} = \frac{e^{-ks}}{s} , \qquad s > 0,$$

where k is a nonnegative constant. Taking $k = \pi$, the latter result becomes

$$L\{\alpha(t - \pi)\} = \frac{e^{-\pi s}}{s} , \qquad s > 0.$$

Substituting these results into equation (2),

$$L\{y''(t)\} + 2L\{y'(t)\} + 5L\{y(t)\} = \frac{1}{s} - \frac{e^{-\pi s}}{s} . \qquad (3)$$

Using the properties that

$$L\{y'(t)\} = sL\{y(t)\} - y(0)$$

and

$$L\{y''(t)\} = s^2 L\{y(t)\} - sy(0) - y'(0),$$

provided that $y'(t)$, $y''(t)$, and $L\{y(t)\}$ exist, equation (3) becomes

$$[s^2 L\{y(t)\} - sy(0) - y'(0)] + 2[sL\{y(t)\} - y(0)]$$

$$+ 5L\{y(t)\} = \frac{1 - e^{-\pi s}}{s} .$$

Solving for L{y(t)},

$$L\{y(t)\} = \frac{1 - e^{-\pi s} + s(s + 2)y(0) + sy'(0)}{s(s^2 + 2s + 5)},$$

and since

$$y(0) = y'(0) = 0,$$

$$L\{y(t)\} = \frac{1 - e^{-\pi s}}{s(s^2 + 2s + 5)}.$$

Inverting the above equation,

$$y(t) = L^{-1}\left\{\frac{1 - e^{-\pi s}}{s(s^2 + 2s + 5)}\right\}.$$

Since L^{-1} is a linear operator under the conditions assumed in this chapter

$$y(t) = L^{-1}\left\{\frac{1}{s(s^2 + 2s + 5)}\right\} - L^{-1}\left\{\frac{e^{-\pi s}}{s(s^2 + 2s + 5)}\right\}.$$

Let

$$G(t) = L^{-1}\left\{\frac{1}{s(s^2 + 2s + 5)}\right\}.$$

Thus,

$$y(t) = G(t) - L^{-1}\left\{\frac{e^{-\pi s}}{s(s^2 + 2s + 5)}\right\}. \qquad (4)$$

Using the property that

$$L^{-1}\left\{e^{-ks} F(s)\right\} = f(t - k)\alpha(t - k),$$

where k is a nonnegative constant and $f(t) = L^{-1}\{F(s)\}$, provided that $L^{-1}\{F(s)\}$ exists, we find (with $k = \pi$) that

$$L^{-1}\left\{\frac{e^{-\pi s}}{s(s^2 + 2s + 5)}\right\} = G(t - \pi)\alpha(t - \pi);$$

hence, equation (4) is equivalent to

$$y(t) = G(t) - G(t - \pi)\alpha(t - \pi). \qquad (5)$$

We now evaluate the function G(t). We use the method of partial fractions to decompose

$$\frac{1}{[s(s^2 + 2s + 5)]} .$$

Setting

$$\frac{1}{s(s^2 + 2s + 5)} = \frac{A}{s} + \frac{Bs + C}{s^2 + 2s + 5} ,$$

for all s, with A, B, and C constant, we find

$$A = \frac{1}{5} , \qquad B = -\frac{1}{5} , \qquad \text{and } C = -\frac{2}{5} .$$

Thus,

$$G(t) = L^{-1} \left\{ \frac{1}{5}\left(\frac{1}{s}\right) - \frac{1}{5}\left(\frac{s + 2}{(s^2 + 2s + 5)}\right) \right\} .$$

Completing the square in the denominator of the fraction

$$\frac{(s + 2)}{(s^2 + 2s + 5)}$$

above,

$$G(t) = L^{-1} \left\{ \frac{1}{5}\left(\frac{1}{s}\right) - \frac{1}{5}\left[\frac{s + 2}{(s + 1)^2 + 2^2}\right] \right\}$$

$$= L^{-1} \left\{ \frac{1}{5}\left(\frac{1}{s}\right) - \frac{1}{5}\left[\frac{s + 1}{(s + 1)^2 + 2^2}\right] \right.$$

$$\left. - \frac{1}{10}\left[\frac{2}{(s + 1)^2 + 2^2}\right] \right\} .$$

Since L^{-1} is a linear operator,

$$G(t) = \frac{1}{5} L^{-1}\left\{\frac{1}{s}\right\} - \frac{1}{5} L^{-1}\left\{\frac{s + 1}{(s + 1)^2 + 2^2}\right\}$$

$$- \frac{1}{10} L^{-1}\left\{\frac{2}{(s + 1)^2 + 2^2}\right\} . \qquad (6)$$

Using the property that

$$L^{-1}\{f(s - b)\} = e^{bt} L^{-1}\{f(s)\} ,$$

428

where b is a constant, provided that $L^{-1}\{f(s)\}$ exists, equation (6) becomes (with b = -1)

$$G(t) = \frac{1}{5} L^{-1} \left\{ \frac{1}{s} \right\} - \frac{1}{5} e^{-t} L^{-1} \left\{ \frac{s}{s^2 + 2^2} \right\}$$

$$- \frac{1}{10} e^{-t} L^{-1} \left\{ \frac{2}{s^2 + 2^2} \right\} . \tag{7}$$

From a table of Laplace transforms, we find that

$$L^{-1} \left\{ \frac{1}{s} \right\} = 1, \quad L^{-1} \left\{ \frac{s}{s^2 + c^2} \right\} = \cos ct,$$

and

$$L^{-1} \left\{ \frac{c}{s^2 + c^2} \right\} = \sin ct,$$

where c is a constant. Taking c = 2, the last two of the above results become

$$L^{-1} \left\{ \frac{s}{s^2 + 2^2} \right\} = \cos \ 2t,$$

and

$$L^{-1} \left\{ \frac{2}{s^2 + 2^2} \right\} = \sin \ 2t .$$

Substituting these results into equation (7),

$$G(t) = \frac{1}{5} - \frac{1}{5} e^{-t} \cos 2t - \frac{1}{10} e^{-t} \sin 2t, \tag{8}$$

and thus,

$$G(t - \pi) = \frac{1}{5} - \frac{1}{5} e^{-(t-\pi)} \cos 2(t - \pi)$$

$$- \frac{1}{10} e^{-(t-\pi)} \sin 2(t - \pi) . \tag{9}$$

Note that

$$\cos 2(t - \pi) = \cos (2t - 2\pi) = \cos 2t,$$

and

$$\sin 2(t - \pi) = \sin (2t - 2\pi) = \sin 2t;$$

429

therefore, equation (9) is equivalent to

$$G(t - \pi) = \frac{1}{5} - \frac{1}{5} e^{-(t-\pi)} \cos 2t - \frac{1}{10} e^{-(t-\pi)} \sin 2t . \quad (10)$$

Substituting equations (8) and (10) into equation (5),

$$y(t) = \frac{1}{5} - \frac{1}{5} e^{-t} \cos 2t - \frac{1}{10} e^{-t} \sin 2t$$

$$- \left[\frac{1}{5} - \frac{1}{5} e^{-(t-\pi)} \cos 2t - \frac{1}{10} e^{-(t-\pi)} \sin 2t \right] \alpha(t - \pi) .$$

$$(11)$$

Since we made some assumptions about existence of $y(t), y'(t), y''(t)$ and $L(y(t))$, we must check our result (10) against the original problem. Differentiating twice, we obtain

$$y'(t) = \frac{1}{2} \left[1 - e^{\pi} \alpha(t - \pi) \right] e^{-t} \sin 2t$$

$$y''(t) = \frac{1}{2} \left[1 - e^{\pi} \alpha(t - \pi) \right] e^{-t} (- \sin 2t + 2 \cos 2t).$$

Checking in equation (1),

$$y''(t) + 2y'(t) + 5y(t)$$

$$= \frac{1}{2} [1 - e^{\pi} \alpha(t - \pi)] e^{-t} (- \sin 2t + 2 \cos 2t)$$

$$+ 2 \cdot \frac{1}{2} [1 - e^{\pi} \alpha(t - \pi)] e^{-t} \sin 2t$$

$$+ 5 \left[\frac{1}{5} - \frac{1}{5} e^{-t} \cos 2t - \frac{1}{10} e^{-t} \sin 2t \right.$$

$$\left. - \left(\frac{1}{5} - \frac{1}{5} e^{-(t-\pi)} \cos 2t - \frac{1}{10} e^{-(t-\pi)} \sin 2t \right) \alpha(t-\pi) \right]$$

$$= 1 - \alpha(t - \pi) = H(t),$$

$$y(0) = \frac{1}{5} - \frac{1}{5} e^{0} \cos 0 - \frac{1}{10} e^{0} \sin 0$$

$$- \left[\frac{1}{5} - \frac{1}{5} e \cos 0 - \frac{1}{10} e^{\pi} \sin 0 \right] \alpha(0 - \pi)$$

$$= 0$$

$$y'(0) = \frac{1}{2} [1 - e^{\pi} \alpha(0 - \pi)] e^{0} \sin 0 = 0;$$

hence, equation (11) is the solution.

Two circuits are coupled magnetically, as shown in Fig. 1. Find the currents $i_1(t)$ and $i_2(t)$ after the switch S is closed, given that the currents are known to obey the differential equations

$$L_i \frac{di_1}{dt} + M \frac{di_2}{dt} + R_1 i_1 = e_0 S(t), \tag{1}$$

$$M \frac{di_1}{dt} + L_2 \frac{di_2}{dt} + R_2 i_2 = 0, \tag{2}$$

where

$$S(t) = \begin{cases} 1, & t > 0 \\ 0, & t < 0 \end{cases}$$

is the unit step function and the initial currents are assumed to be zero.

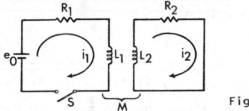

Fig. 1

Solution: The first step is to take the Laplace transform of each of equations (1) and (2) to get

$$L\left\{ L_1 \frac{di_1}{dt} + M \frac{di_2}{dt} + R_1 i_1 \right\} = L\{e_0 S(t)\} \tag{3}$$

$$L\left\{ M \frac{di_1}{dt} + L_2 \frac{di_2}{dt} + R_2 i_2 \right\} = L\{0\}. \tag{4}$$

Since L is a linear operator (3) and (4) can be written as

$$L_1 L\left\{ \frac{di_1}{dt} \right\} + ML\left\{ \frac{di_2}{dt} \right\} + R_1 L\{i_1\} = e_0 L\{S(t)\}. \tag{5}$$

$$ML\left\{ \frac{di_1}{dt} \right\} + L_2 L\left\{ \frac{di_2}{dt} \right\} + R_2 L\{i_2\} = L\{0\}. \tag{6}$$

Note the derivative property states that

$$L\{f'(t)\} = sL\{f(t)\} - f(0) .$$

Using this in (5) and (6) yields

$$L_1\left[sL\{i_1\} - i_1(0)\right] + M\left[sL\{i_2\} - i_2(0)\right]$$

$$+ R_1L\{i_1\} = e_0 \cdot \frac{1}{s} \tag{7}$$

$$M\left[sL\{i_1\} - i_1(0)\right] + L_2\left[sL\{i_2\} - i_2(0)\right]$$

$$+ R_2L\{i_2\} = 0 \tag{8}$$

where a table of Laplace transforms may be consulted to find

$$L\{S(t)\} = 1/s$$

and

$$L\{0\} = 0 .$$

Using the fact that

$$i_1(0) = i_2(0) = 0$$

and letting

$$I_1(s) = L\{i_1(t)\} , \quad I_2(s) = L\{i_2(t)\} , \quad (7) \text{ and } (8)$$

become a system of two equations in two unknowns I_1, I_2 :

$$(L_1s + R_1)I_1 + MsI_2 = e_0/s,$$

$$MsI_1 + (L_2s + R_2)I_2 = 0 . \tag{9}$$

The determinant of the system is

$$\Delta = \det \begin{pmatrix} L_1s + R_1 & Ms \\ \\ Ms & L_2S + R_2 \end{pmatrix}$$

$$= (L_1s + R_1)(L_2s + R_2) - (Ms)(Ms)$$

$$= (L_1L_2 - M^2)s^2 + (L_1R_2 + L_2R_1)s + R_1R_2 .$$

It is known from electromagnetic theory that $L_1L_2 \geq M^2$. The usual condition in practice is $L_1L_2 > M^2$. Thus we have, using Cramer's rule on the system (9),

$$I_1(s) = \frac{\begin{vmatrix} e_0/s & Ms \\ 0 & L_2s + R_2 \end{vmatrix}}{\Delta} = \frac{e_0(L_2S + R_2)}{s \cdot \Delta}$$

$$I_2(s) = \frac{\begin{vmatrix} L_1s + R_1 & e_0/s \\ Ms & 0 \end{vmatrix}}{\Delta} = -\frac{Me_0}{\Delta}$$

which may be written in the form

$$I_1(s) = \frac{e_0}{(L_1L_2 - M^2)} \frac{L_2s + R_2}{s(s - r_1)(s - r_2)}, \tag{10}$$

$$I_2(s) = -\frac{Me_0}{(L_1L_2 - M^2)} \frac{1}{(s - r_1)(s - r_2)}, \tag{11}$$

where r_1 and r_2 are the roots of

$$\Delta = (L_1L_2 - M^2)s^2 + (L_1R_2 + L_2R_1)s + R_1R_2 = 0 . \tag{12}$$

Since the discriminant of (12) is

$$(L_1R_2 + L_2R_1)^2 - 4R_1R_2(L_1L_2 - M^2) = (L_1R_2 - R_1L_2)^2$$

$$+ 4R_1R_2M^2 > 0 ,$$

both roots are real.. Moreover, they are both negative since

$$\frac{L_1R_2 + L_2R_1}{L_1L_2 - M^2} > 0$$

and

$$\frac{R_1R_2}{L_1L_2 - M^2} > 0 .$$

Rational fraction decomposition yields

$$I_1(s) = \frac{e_0}{L_1L_2 - M^2} \left(\frac{R_2}{r_1r_2s} + \frac{L_2r_1 + R_2}{r_1(r_1 - r_2)(s - r_1)} \right.$$

$$+ \frac{L_2r_2 + R_2}{r_2(r_2 - r_1)(s - r_2)} \Bigg), \tag{13}$$

$$I_2(s) = - \frac{Me_0}{L_1L_2 - M^2} \cdot \frac{1}{r_1 - r_2} \left(\frac{1}{s - r_1} - \frac{1}{s - r_2} \right). \tag{14}$$

Inverting [using $r_1r_2 = R_1R_2/(L_1L_2 - M^2)$], we find that since L^{-1} is a linear operator under the assumptions made in this chapter, (13) and (14) become

$$i_1(t) = L^{-1}\{I_1(s)\} = \frac{e_0}{L_1L_2 - M^2} \left[\frac{R_2(L_1L_2 - M^2)}{R_1R_2} L^{-1}\left\{\frac{1}{s}\right\} \right.$$

$$+ \frac{L_2r_1 + R_2}{r_1(r_1-r_2)} L^{-1}\left\{\frac{1}{s-r_1}\right\} + \frac{Lr_2 + R_2}{r_2(r_2-r_1)} L^{-1}\left\{\frac{1}{s-r_2}\right\}$$

$$i_2(t) = \frac{-Me_0}{L_1L_2 - M^2} \frac{1}{r_1-r_2} \left[L^{-1}\left\{\frac{1}{s-r_1}\right\} - L^{-1}\left\{\frac{1}{s-r_2}\right\} \right].$$

Recalling that

$$L^{-1}\left\{\frac{1}{s-c}\right\} = e^{ct} \, ,$$

$$i_1(t) = \frac{e_0}{R_1} + \frac{e_0}{L_1L_2 - M^2} \left(\frac{L_2r_1 + R_2}{r_1(r_1-r_2)} e^{r_1t} \right.$$

$$\left. - \frac{L_2r_2 + R_2}{r_2(r_1-r_2)} e^{r_2t} \right) \, ,$$

$$i_2(t) = - \frac{Me_0}{(L_1L_2 - M^2)} \frac{1}{(r_1 - r_2)} (e^{r_1t} - e^{r_2t}) \, .$$

It is seen that $i_1 \rightarrow e_0/R$ and $i_2 \rightarrow 0$ as $t \rightarrow \infty$ as expected on physical grounds. Also, $i_1(0) = 0$ and $i_2(0) = 0$.

FOURIER TRANSFORMS

Find the Fourier transform, $F(k) = \Phi\{f(x)\}$ of the Gaussian probability function

$$f(x) = Ne^{-\alpha x^2} \qquad (N, \alpha = \text{constant}). \qquad (1)$$

Show directly that $f(x)$ is retrievable from the inverse transform. I.e., show that

$$f(x) = \frac{1}{\sqrt{2\pi}} \int_{-\infty}^{\infty} F(k)e^{-ikx}\, dx = \Phi^{-1}\left\{\Phi\{f(x)\}\right\} \quad .$$

Solution: The defintion of the Fourier transform of $f(x)$ is

$$F(k) = \Phi\{f(x)\} = \frac{1}{\sqrt{2\pi}} \int_{-\infty}^{\infty} f(x)e^{ikx}\, dx . \qquad (2)$$

Using the Gaussian function of (1) yields

$$F(k) = \frac{1}{\sqrt{2\pi}} \int_{-\infty}^{\infty} Ne^{-\alpha x^2}e^{ikx}\, dx = \frac{N}{\sqrt{2\pi}} \int_{-\infty}^{\infty} e^{(-\alpha x^2 + ikx)}\, dx . \qquad (3)$$

It is convenient to complete the square in the integrand of (3) to get

$$-\alpha x^2 + ikx = -\left(x\sqrt{\alpha} - \frac{ik}{2\sqrt{\alpha}}\right)^2 - \frac{k^2}{4\alpha} \quad . \qquad (4)$$

Now make the change of variables

$$x\sqrt{\alpha} - \frac{ik}{2\sqrt{\alpha}} = u \text{ in (4) to obtain}$$

$$-\alpha x^2 + ikx = -u^2 - \frac{k^2}{4\alpha} \;,\; dx = \frac{1}{\sqrt{\alpha}}\, du \qquad (5)$$

and substitute (5) into (3) to find

$$F(k) = \frac{N}{\sqrt{2\pi\alpha}} e^{-k^2/4\alpha} \int_{-\infty}^{\infty} e^{-u^2} du = N \frac{1}{\sqrt{2\alpha}} e^{-k^2/4\alpha} . \qquad (6)$$

Note that F(k) is also a Gaussian probability function with a peak (the mean) at x = 0. Also note that if f(x) is sharply peaked due to a large α, then F(k) is broadened and vice versa (see Figure 1). This has important applications in Quantum physics where

$$|f(x)|^2$$

represents the probability of finding a (one-dimensional

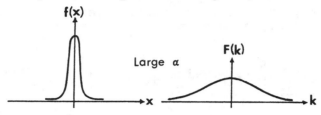

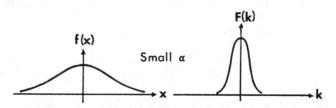

Fig. 1

particle at point x and $|F(k)|^2$ represents the probability of finding this particle with momentum p = $\hbar$k where $\hbar$ is a known constant. Thus, the better one is able to predict the location of the particle (narrow f(x)), the harder it is to predict its momentum (broad F(k)) and vice versa. This is the basic idea behind what is known as the Heisenberg's uncertainty principle and the fact that "narrow" functions have "broad" Fourier transforms and vice versa is a property of all Fourier transforms.

The inverse Fourier integral is given by

$$\Phi^{-1}\{F(k)\} = \frac{1}{\sqrt{2\pi}} \int_{-\infty}^{\infty} F(k) e^{-ikx} dx = \frac{1}{\sqrt{2\pi}} \frac{N}{\sqrt{2\alpha}}$$

$$X \int_{-\infty}^{\infty} e^{-k^2/4\alpha} e^{-ikx} dk \qquad (7)$$

and it is desired to see whether this equals f(x) as one would expect from the theory. The integral in (7) is calculated in the same way as that in (3). In fact, as a short-cut to that calculation, set

$$\alpha^1 = \frac{1}{4\alpha} \quad \text{and} \quad x^1 = -x$$

to deduce

$$\frac{1}{\sqrt{2\pi}} \int_{-\infty}^{\infty} e^{-\alpha^1 k^2} e^{ix^1 k} \, dk = \frac{1}{\sqrt{2\alpha^1}} e^{-(x^1)^2/4\alpha^1}$$

$$= \sqrt{2\alpha} \, e^{-\alpha x^2}$$

so that from (7)

$$\Phi^{-1}\{F(k)\} = \frac{1}{\sqrt{2\pi}} \int_{-\infty}^{\infty} F(k) e^{-ikx} \, dk = \frac{N}{\sqrt{2\alpha}} \sqrt{2\alpha} \, e^{-\alpha x^2}$$

$$= Ne^{-\alpha x^2}$$

or

$$\Phi^{-1}\{F(k)\} = f(x)$$

as expected.

Find the Fourier transform, $F(k) = \Phi\{f(x)\}$, of the function

$$f(x) = \frac{a}{x^2 + a^2} \qquad (a > 0)$$

by using the residue theorem of complex analysis.

Solution: The Fourier transform of a function is defined by

$$F(k) = \Phi\{f(x)\} = \frac{1}{\sqrt{2\pi}} \int_{-\infty}^{\infty} f(x) \, e^{ikx} \, dx$$

and for the given function this definition becomes

$$F(k) = \int_{-\infty}^{\infty} \frac{a}{\sqrt{2\pi}} \frac{e^{ikx}}{a^2 + a^2} \, dx = \frac{a}{\sqrt{2\pi}} \int_{-\infty}^{\infty} \frac{e^{ikx}}{(x + ai)(x - ai)} \, dx \qquad (1)$$

where the denominator has been factored to make clearer
the applicability of the calculus of residues to the
problem. First, for the case where $k \geq 0$, make use of
the contour shown in Figure 1 and consider the integral

$$J = \oint g(z) \, dz = \oint_C \frac{a}{\sqrt{2\pi}} \frac{e^{ikx}}{z^2 + a^2} \, dz$$

$$= \int_{-R}^{R} \frac{a}{\sqrt{2\pi}} \frac{e^{ikx}}{(x + ai)(x - ai)} \, dx + \int_{C_R} \frac{a}{\sqrt{2\pi}} \frac{e^{ikz}}{z^2 + a^2} \, dz \,. \qquad (2)$$

The residue theorem states that

$$J = 2\pi i \sum_{j=1}^{n} \mathrm{Res} \; g(b_j) \qquad (3)$$

where there are n isolated singularities b_j within the
given contour and $\mathrm{Res} \; g(b_j)$ is the residue of g at b_j .
In this case $n = 1$ and $b_1 = ai$. The residue of g at ai

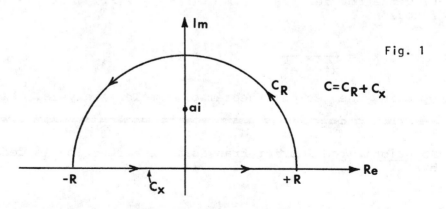

Fig. 1

$C = C_R + C_X$

is easy to calculate in this case since ai is a simple
pole of g and we may use the formula

$$\mathrm{Res} \; g(ai) = \lim_{z \to ai} (z - ai) g(z)$$

438

so that

$$\text{Res } g(ai) = \lim_{z \to ai} (z-ai) \frac{a}{\sqrt{2\pi}} \frac{e^{ikz}}{(z^2 + a^2)} =$$

$$= \lim_{z \to ai} (z-ai) \frac{a}{\sqrt{2\pi}} \frac{e^{ikz}}{(z-ai)(z+ai)}$$

$$= \frac{a}{\sqrt{2\pi}} \frac{e^{-ka}}{2ai}$$

and using (3) gives

$$J = 2\pi i \frac{a}{\sqrt{2\pi}} \frac{e^{-ka}}{2ai} = \sqrt{\frac{\pi}{2}} \ e^{-ka} \qquad (k \geq 0). \qquad (4)$$

Now, to show that the second integral in (2) goes to zero as $R \to \infty$, make the estimates

$$\left| \int_{C_R} \frac{e^{iz} \ dz}{z^2 + a^2} \right| \leq \int_{C_R} \frac{|e^{iz} \ dz|}{|z^2 + a^2|} \leq \max \frac{1}{|z^2 + a^2|}$$

$$X \int_{C_R} e^{-y}(dz)$$

$$= \max \frac{1}{|z^2 + a^2|} \cdot \int_0^\pi e^{-\cos\theta} \ Rd\theta \leq \max \frac{1}{|z^2 + a^2|} \int_0^\pi eRd\theta$$

$$= eR\pi \max \frac{1}{|z^2 + a^2|} < \frac{eR\pi}{R^2 - a^2} \qquad (5)$$

where

$$\max \ \frac{1}{|z^2 + a^2|}$$

is the maximum value of $\dfrac{1}{z^2 + a^2}$ over the contour C_R .

Equation (5) shows that

$$\lim_{R\to\infty} \int_{C_R} \frac{e^{iz}\,dz}{z^2 + a^2} = \lim_{R\to\infty} \frac{eR\pi}{R^2 - a^2} = 0$$

so that taking the limit as $R \to \infty$ in (2) and recalling (4) gives

$$\lim_{R\to\infty} J = \lim_{R\to\infty} \sqrt{\frac{\pi}{2}}\ e^{-ka} = \int_{-\infty}^{\infty} \frac{a}{\sqrt{2\alpha}}\frac{e^{ikx}\,dx}{x^2 + a^2} + 0$$

$$F(k) = \int_{-\infty}^{\infty} \frac{a}{\sqrt{2\pi}}\frac{e^{ikx}}{x^2 + a^2}\,dx = \sqrt{\frac{\pi}{2}}\ e^{-ka} \qquad (k \ge 0) \ . \qquad (6)$$

For the case where $k < 0$, make use of the contour shown in Figure 2. Again the residue theorem is employed via equation (2) except that C_R is now the curve shown in Figure 2, and equation (3) is now written as

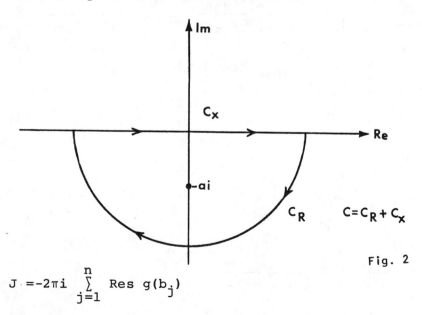

Fig. 2

$$J = -2\pi i \sum_{j=1}^{n} \text{Res } g(b_j)$$

(since C is taken as clockwise) with $n=1$, $b_1 = -ai$.
Using the formula

$$\text{Res } g(-ai) = \lim_{z \to -ai} (z + ai) g(z)$$

yields

$$\text{Res } g(ai) = \lim_{z \to -ai} (z + ai) \frac{a}{\sqrt{2\pi}} \frac{e^{ikz}}{(z+ai)(z-ai)}$$

$$= \frac{a}{\sqrt{2\pi}} \frac{e^{ka}}{(-2ai)}$$

so that from (7)

$$J = -2\pi i \frac{a}{\sqrt{2\pi}} \frac{e^{ka}}{2} = \sqrt{\frac{\pi}{2}} e^{ka}$$

for $k < 0$. The proof that the integral over C_R goes to

0 as $R \to \infty$ is completely analogous to that for the C_R of Figure 1 so that (2) yields

$$\lim_{R \to \infty} J = \lim_{R \to \infty} \sqrt{\frac{\pi}{2}} e^{ka} = \int_{-\infty}^{\infty} \frac{a}{\sqrt{2\pi}} \frac{e^{ikx}}{x^2 + a^2} dx + 0$$

so that

$$F(k) = \int_{-\infty}^{\infty} \frac{a}{\sqrt{2\pi}} \frac{e^{ikx}}{x^2 + a^2} dx = \sqrt{\frac{\pi}{2}} e^{ka} \qquad (k < 0) \qquad (8)$$

and equations (6) and (8) may be combined to yield

$$F(k) = \sqrt{\frac{\pi}{2}} e^{-|k|a} \qquad \text{(all k)}. \qquad (9)$$

As in the previous problem, if f(x) has a sharp peak
(small a) then F(k) is broadened and vice versa, a general
feature of Fourier transforms.

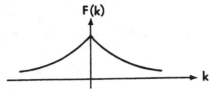

F(k)

k

● **PROBLEM** 7–30

Find the Fourier transforms, $F_i(k) = \Phi\{f_i(t)\}$, of the
functions whose graphs are shown in the accompanying
figure.

<u>Solution:</u> The Fourier transform of a function f(t) is
defined by

$$F(k) = \frac{1}{\sqrt{2\pi}} \int_{-\infty}^{\infty} f(t) e^{-jkt} \, dt \qquad (1)$$

where $j = \sqrt{-1}$.

a) The "box" function graphed in Figure 1(a) may be
written analytically as

$$f_a(t) = \begin{cases} 1 & |t| \le d \\ \\ 0 & |t| > d \end{cases} \qquad (d > 0).$$

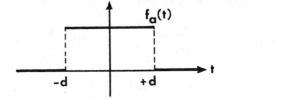

$f_a(t)$

-d   +d   t

Fig. 1(a)

441

Using (1), its Fourier transform is found to be

$$F_a(k) = \frac{1}{\sqrt{2\pi}} \int_{-d}^{d} e^{jkt} \, dt = \frac{1}{\sqrt{2\pi}} \frac{e^{jkd} - e^{-jkd}}{jk} \, . \qquad)2)$$

Using the formulas

$$e^{j\theta} = \cos\theta + j \sin\theta$$
$$e^{-j\theta} = \cos\theta - j \sin\theta$$

in (2) yields

$$F_a(k) = \sqrt{\frac{2}{\pi}} \, \frac{\sin dk}{k} \qquad (3)$$

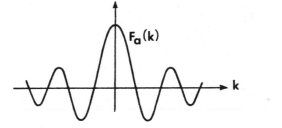

Fig. 2

$F_a(k)$ is graphed in Figure 2. Note the property that as f_a becomes narrower (small d) F_a becomes "broader" and vice versa.

b) The function $f_b(t)$ of Figure 1 can be written in analytic form as

$$f_b(t) = \begin{cases} \dfrac{-V_0}{b} (t-b) \; ; & 0 \le t \le b \\ \\ 0 & ; \quad t < 0, \, t > b \end{cases} \; .$$

Hence, from (1)

$$F_b(k) = \frac{-V_0}{b\sqrt{2\pi}} \int_0^b (t-b) e^{-jkt} \, dt$$

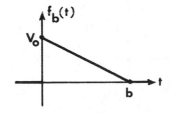

Fig.1(b)

442

$$= \frac{-V_0}{b\sqrt{2\pi}} \int_0^b t e^{-jkt} \, dt \; \frac{V_0}{\sqrt{2\pi}} \int_0^b e^{-jkt} \, dt \; .$$

Using integration by parts on the first integral yields

$$F_b(k) = \left[\frac{-V_0}{b\sqrt{2\pi}} \left(\frac{t}{-jk} e^{-jkt} \right) \Big|_0^b - \frac{V_0}{b\sqrt{2\pi}} \int_0^b \frac{1}{jk} e^{-jkt} \, dt \right]$$

$$+ \left(\frac{V_0}{\sqrt{2\pi}} \; \frac{1}{-jk} \; e^{-jkt} \right) \Big|_0^b$$

$$= \frac{V_0}{b\sqrt{2\pi}} \left(\frac{t}{-jk} e^{-jkt} - \frac{1}{(jk)^2} e^{-jkt} \right) \Big|_0^b$$

$$+ \left[\frac{V_0}{\sqrt{2\pi}} \; \frac{1}{-jk} \; e^{-jkb} + \frac{1}{jk} \right]$$

$$\frac{V_0}{\sqrt{2\pi}} \; \frac{1}{jk} \; e^{-jkb} - \frac{V_0}{\sqrt{2\pi} \; bk^2} e^{-jkb} +$$

$$\frac{V_0}{\sqrt{2\pi} \; bk^2} - \frac{V_0}{\sqrt{2\pi} \; jk} e^{-jkb} + \frac{V_0}{\sqrt{2\pi} \; jk}$$

$$= \frac{V_0}{\sqrt{2\pi}} \left[\frac{-1}{bk^2} e^{-jkb} + \frac{1}{bk^2} + \frac{1}{jk} \right] \; .$$

c) For the functions in Fig. 1(c) use ω as the transfer parameter instead of k. Then

$$f_c(t) = \sin t; \qquad -\pi \le t \le \pi \; .$$

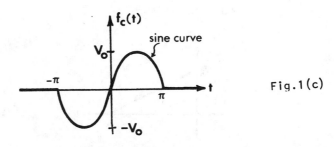

Fig.1(c)

Hence,

$$F_c(\omega) = V_0 \int_{-\pi}^{\pi} \sin t \; e^{-j\omega t} \, dt$$

$$= V_0 \int_{-\pi}^{\pi} \left(\frac{e^{jt} - e^{-jt}}{2j} \right) e^{-j\omega t} \, dt$$

$$= V_0 \int_{-\pi}^{\pi} \frac{e^{jt(1-\omega)} - e^{-jt(1+\omega)}}{2j} \, dt$$

$$= V_0 \left[\frac{e^{jt(1-\omega)}}{2j(j(1-\omega))} + \frac{e^{-jt(1+\omega)}}{2j(j(1+\omega))} \right] \Bigg|_{-\pi}^{\pi}$$

$$= V_0 \left[-\frac{e^{jt(1-\omega)}}{2(1-\omega)} - \frac{e^{-jt(1+\omega)}}{2(1+\omega)} \right] \Bigg|_{-\pi}^{\pi}$$

$$= V_0 \; - \left[\frac{e^{j\pi(1-\omega)}}{2(1-\omega)} + \frac{e^{-j\pi(1+\omega)}}{2(1+\omega)} - \frac{e^{-j\pi(1-\omega)}}{2(1-\omega)} \right.$$

$$\left. - \; e^{\frac{j\pi(1+\omega)}{2(1+\omega)}} \right]$$

$$= \frac{V_0 j}{(1-\omega)} \left[\frac{-e^{j\pi(1-\omega)} + e^{-j\pi(1-\omega)}}{2j} \right]$$

$$+ \frac{V_0 j}{(1+\omega)} \left[\frac{e^{j\pi(1+\omega)} - e^{-j\pi(1+\omega)}}{2j} \right] \; .$$

$$= - \frac{j V_0}{1-\omega} \sin \pi(1-\omega) + \frac{j V_0}{1+\omega} \sin \pi(1+\omega)$$

This can be simplified by combining the two terms

$$F_c = \frac{-jV_0(1+\omega)\sin\pi(1-\omega) + jV_0(1-\omega)\sin\pi(1+\omega)}{(1-\omega)(1+\omega)}$$

$$= \frac{-jV_0(1+\omega)\sin\pi\omega - jV_0(1-\omega)\sin\pi\omega}{1-\omega^2}$$

or

$$F_c(\omega) = \frac{-j2V_0\sin\pi\omega}{1-\omega^2} \quad .$$

Prove the following properties of Fourier transforms:

a)   The Fourier transform of f(x) exists if f is absolutely integrable over $(-\infty,+\infty)$,

b)   If f(x) is real valued then

$$F(-k) = F*(k)$$

where F*(k) is the complex conjugate of F(k).

Solution: By definition, if f is absolutely integrable over $(-\infty,+\infty)$ then

$$\int_{-\infty}^{\infty} |f(x)|\,dx \tag{1}$$

exists. The Fourier transform of f is given by

$$F(k) = \frac{1}{\sqrt{2\pi}}\int_{-\infty}^{\infty} f(x)\,e^{ikx}\,dx \quad . \tag{2}$$

Now, recalling that $|e^{iy}| = 1$ for all y we have that

$$|f(x)\,e^{ikx}| = |f(x)|$$

445

so that

$$\frac{1}{\sqrt{2\pi}} \int_{-\infty}^{\infty} \left| f(x)\ e^{ikx} \right| dx = \frac{1}{\sqrt{2\pi}} \int_{-\infty}^{\infty} |f(x)|\ dx\ ,$$

and we know that this second integral exists. Thus,

$$f(x)e^{ikx}$$

is absolutely integrable over $(-\infty,\infty)$ and is therefore integrable over $(-\infty,\infty)$. Hence, $F(k)$ exists.

b) This proof is immediate. From (2)

$$F(-k) = \frac{1}{\sqrt{2\pi}} \int_{-\infty}^{\infty} f(x)\ e^{-ikx}\ dx \tag{3}$$

and, recalling the identity $e^{iy} = \cos y + i\,\sin y$,

$$F^*(k) = \left(\frac{1}{\sqrt{2\pi}} \int_{-\infty}^{\infty} f(x)\ e^{ikx}\ dx \right)^*$$

$$= \left(\frac{1}{\sqrt{2\pi}} \int_{-\infty}^{\infty} f(x)\ \cos kx\ dx + i\,\frac{1}{\sqrt{2\pi}} \int_{-\infty}^{\infty} f(x)\ \sin kx\ dx \right)^*$$

$$= \frac{1}{\sqrt{2\pi}} \int_{-\infty}^{\infty} f(x)\ \cos kx\ dx - i\,\frac{1}{\sqrt{2\pi}} \int_{-\infty}^{\infty} f(x)\ \sin kx\ dx \qquad (f\ \text{real})$$

$$= \frac{1}{\sqrt{2\pi}} \int_{-\infty}^{\infty} f(x)\ \cos(-kx)\,dx + i\,\frac{1}{\sqrt{2\pi}} \int_{-\infty}^{\infty} f(x)\ \sin(-kx)\,dx$$

$$= \frac{1}{\sqrt{2\pi}} \int_{-\infty}^{\infty} f(x)\ (\cos(-kx) + i\,\sin(-kx))\,dx$$

$$= \frac{1}{\sqrt{2\pi}} \int_{-\infty}^{\infty} f(x) \ e^{-ikx} \ dx \ . \tag{4}$$

In the third step the facts that $\cos(-y) = \cos y$ and $\sin(-y) = -\sin y$ were used. Equating (3) and (4) yields

$$F(-k) = F^*(k) \ , \text{for real } f(x) \ . \qquad \bullet \ \textbf{PROBLEM } 7\text{--}32$$

a) Prove the attentuation property of Fourier transforms:

$$\Phi\{f(x) \ e^{ax}\} = F(k-ai)$$

where

$$F(k) = \Phi\{f(x)\} \quad .$$

b) Prove the shifting property of Fourier transforms:

$$\Phi\{f(x-a)\} = e^{ika} \ F(k) \ .$$

c) Prove the derivative properties of Fourier transforms:

$$\Phi\{f'(x)\} = -ik\Phi\{f(x)\}$$

$$\Phi\{f''(x)\} = -k^2\Phi\{f(x)\} \quad .$$

Solution: a) The Fourier transform of $f(x)$ is defined by

$$F(k) = \Phi\{f(x)\} = \frac{1}{\sqrt{2\pi}} \int_{-\infty}^{\infty} f(x) \ e^{ikx} \ dx \ . \tag{1}$$

The Fourier transform of $g(x) = f(x) \ e^{ax}$ is then

$$G(k) = \Phi\{f(x) \ e^{ax}\} = \frac{1}{\sqrt{2\pi}} \int_{-\infty}^{\infty} f(x) \ e^{ax} \ e^{ikx} \ dx$$

$$= \frac{1}{\sqrt{2\pi}} \int_{-\infty}^{\infty} f(x) \ e^{(a+ik)x} \ dx$$

447

$$= \frac{1}{\sqrt{2\pi}} \int_{-\infty}^{\infty} f(x) \; e^{i(k-ia)x} \; dx \; . \qquad (2)$$

Now make the change of variable r = k -ia in (2) to give

$$\Phi\{f(x) \; e^{ax}\} = \frac{1}{\sqrt{2\pi}} \int_{-\infty}^{\infty} f(x) \; e^{irx} \; dx = F(r) = F(k-ia) \; .$$

b) Suppose f(x) is shifted a length a to the right.
Then the Fourier transform of this new function f(x-a)
is

$$\Phi\{f(x-a)\} = \frac{1}{\sqrt{2\pi}} \int_{-\infty}^{\infty} f(x-a) \; e^{ikx} \; dx \qquad (3)$$

Now make the substitution x' = x-a. Then (3) becomes

$$\Phi\{f(x-a)\} = \frac{1}{\sqrt{2\pi}} \int_{-\infty}^{\infty} f(x') \; e^{ik(x'+a)} \; dx'$$

$$= e^{ika} \frac{1}{\sqrt{2\pi}} \int_{-\infty}^{\infty} f(x') \; e^{ikx'} \; dx'$$

$$= e^{ika} \; F(k) \; .$$

c) Suppose that $\Phi\{f'(x)\}$ exists. Then by definition (1),

$$\Phi\{f'(x)\} = \frac{1}{\sqrt{2\pi}} \int_{-\infty}^{\infty} f'(x) \; e^{ikx} \; dx \; . \qquad (4)$$

Integrating (4) by parts gives

$$\Phi\{f'(x)\} = \frac{1}{\sqrt{2\pi}} \; f(x) \; e^{ikx} \; \Big|_{-\infty}^{\infty} - \frac{ik}{\sqrt{2\pi}} \int_{-\infty}^{\infty} f(x) \; e^{ikx} \; dx \; .$$

$$(5)$$

If the Fourier transform of f(x) exists, this usually
implies that f(x) → 0 as x → ±∞ (this is sometimes not
the case, but then f(x) can be treated as a distribution
so that the derivative formula is then still valid). Thus,
the first term in (5) is zero and

$$\Phi\{f'(x)\} = -ik\ \Phi\{f(x)\}\ \ .\tag{6}$$

If

$$\Phi\{f''(x)\}$$

exists, it is given by

$$\Phi\{f''(x)\} = \frac{1}{\sqrt{2\pi}} \int_{-\infty}^{\infty} f''(x)\ e^{ikx}\ dx\ .$$

Again integrating by parts yields

$$\Phi\{f''(x)\} = \frac{1}{\sqrt{2\pi}}\ f'(x)\ e^{ikx}\ \Big|_{-\infty}^{\infty} - \frac{ik}{\sqrt{2\pi}} \int_{-\infty}^{\infty} f'(x) e^{ikx}\ dx\ .$$

$$\tag{7}$$

For reasons mentioned above, it is expected that

$f'(x) \to 0$ as $x \to \pm\infty$ so that (7) gives

$$\Phi\{f''(x)\} = -ik\ \Phi\{f'(x)\}\ .$$

Using (6) gives

$$\Phi\{f''(x)\} = -k^2\ \Phi\{f(x)\}\ \ .$$

The obvious extension is made by using $f^{(n-1)}(x) = g(x)$ in the formula of (6) to give

$$\Phi\{g'(x)\} = -ik\ \Phi\{g(x)\}$$

or

$$\Phi\{f^{(n)}(x)\} = -ik\ \Phi\{f^{(n-1)}(x)\}\ ,$$

which yields upon iteration

$$\Phi\{f^{(n)}(x)\} = (-ik)^n\ \Phi\{f(x)\}\ .$$

● **PROBLEM 7–33**

a) Prove that if the functions g(x) and F(k) are absolutely integrable on $(-\infty, +\infty)$ and that the Fourier inversion integral for f(x) is valid for all x except possibly at a countably infinite number of points, then

$$\int_{-\infty}^{\infty} F(k)\ G(-k)\,dk = \int_{-\infty}^{\infty} f(x)\ g(x)\,dx\tag{1}$$

where

$$F(k) = \Phi\{f(x)\} \ , \ G(k) = \Phi\{g(x)\} \ .$$

This is known as the second Parseval theorem of Fourier transform theory.

b) From the above equation (1), prove the first Parseval theorem of Fourier transform theory,

$$\int_{-\infty}^{\infty} |F(k)|^2 \ dk = \int_{-\infty}^{\infty} |f(x)|^2 \ dx \ . \tag{2}$$

Solution: a) The Fourier transform of a function f(x) is defined by

$$F(k) = \frac{1}{\sqrt{2\pi}} \int_{-\infty}^{\infty} f(x) \ e^{ikx} \ dx \tag{3}$$

so that by definition

$$G(-k) = \frac{1}{\sqrt{2\pi}} \int_{-\infty}^{\infty} g(x) e^{-ikx} \ dx \ . \tag{4}$$

Therefore,

$$\int_{-\infty}^{\infty} F(k)G(-k)dk = \int_{-\infty}^{\infty} F(k)dk \int_{-\infty}^{\infty} \frac{1}{\sqrt{2\pi}} g(x) e^{-ikx} \ dx \ . \tag{5}$$

Now $F(k)$ and $g(x)$ are absolutely convergent on $(-\infty, +\infty)$, that is, the integrals

$$\int_{-\infty}^{\infty} |F(k)| dk \ , \qquad \int_{-\infty}^{\infty} |g(x)| dx$$

are convergent, so that

$$\int_{-\infty}^{\infty} F(k) e^{-ikx} \ dx \ , \qquad \int_{-\infty}^{\infty} g(x) e^{-ikx} \ dx$$

are absolutely convergent (since

$$|F(k)e^{-ikx}| = |F(k)| \, |e^{-ikx}| = |F(k)|$$

and

$$|g(x)e^{-ikx}| = |g(x)|) \ .$$

Hence, the order of integration in (5) may be interchanged giving

$$\int_{-\infty}^{\infty} F(k)G(-k)dk = \int_{-\infty}^{\infty} g(x)dx \, \frac{1}{\sqrt{2\pi}} \int_{-\infty}^{\infty} F(k)e^{-ikx} \, dk \ . \qquad (6)$$

Since the Fourier inversion integral is valid,

$$\frac{1}{\sqrt{2\pi}} \int_{-\infty}^{\infty} F(k)e^{-ikx} \, dk = f(x) \qquad (7)$$

and using this result in (6) gives the second Parseval theorem:

$$\int_{-\infty}^{\infty} F(k)G(-k)dk = \int_{-\infty}^{\infty} g(x)f(x)dx \ . \qquad (8)$$

The validity of (8) is insured even if the Fourier inversion integral for f(x) has a countably infinite number of discrepancies with f(x) since this will not affect the equality of the integrals

$$\int_{-\infty}^{\infty} g(x)(f(x))dx \quad \text{and} \quad \int_{-\infty}^{\infty} g(x) \, \frac{1}{\sqrt{2\pi}} \int_{-\infty}^{\infty} F(k)e^{-ikx} \, dk \, dx \ .$$

b) The first Parseval theorem is a corollary to the second Parseval theorem stated in equation (8) which follows by letting f(x) = g(x) so that F(k) = G(k) and recalling that (assuming f = g real) G(-k) = G*(k) where G*(k) is the complex conjugate of G(k) where G*(k) is the complex congugate of G(k). Noting that

$$G(k)G* = |G(k)|^2$$

and using these results in (8) gives

$$\int_{-\infty}^{\infty} G(k)G*(k)dk = \int_{-\infty}^{\infty} [g(x)]^2 dx$$

or

$$\int_{-\infty}^{\infty} |G(k)|^2 dk = \int_{-\infty}^{\infty} |g(x)|^2 \, dx \ .$$

Let $F(k) = \Phi\{f(x)\}, G(k) = \Phi\{g(x)\}$ and suppose
$F(k)G(k) = \Phi\{h(x)\}$. Prove the convolution theorem
for Fourier transforms:

If $g(x)$ and $F(k)$ are absolutely integrable on $(-\infty,\infty)$
and if the Fourier inversion integral for $f(x)$ is valid
for all x except possibly a countably infinite number of
points, then

$$h(x) = (f * g),$$

where $(f * g)$ is the convolution of f and g defined by

$$(f * g) = \frac{1}{\sqrt{2\pi}} \int_{-\infty}^{\infty} f(\xi) \, g(x-\xi) d\xi . \tag{1}$$

Solution: Let Φ^{-1} denote the inverse Fourier transform.
Then by definition

$$\Phi^{-1}\{F(k)G(k) = \frac{1}{\sqrt{2\pi}} \int_{-\infty}^{\infty} F(k)G(k) \, e^{-ikx} dk . \tag{2}$$

Using the definition of the Fourier transform of $g(\xi)$,

$$G(k) = \Phi\{g(\xi)\} = \frac{1}{\sqrt{2\pi}} \int_{-\infty}^{\infty} g(\xi) \, e^{ik\xi} d\xi$$

in (2) gives

$$\Phi^{-1}\{F(k)G(k)\} = \frac{1}{2\pi} \int_{-\infty}^{\infty} F(k) e^{-ikx} dk$$

$$\times \int_{-\infty}^{\infty} g(\xi) e^{ik\xi} d\xi . \tag{3}$$

The assumption that $g(x)$ and $F(k)$ are absolutely
integrable on $(-\infty,\infty)$ means that the integrals

$$\int_{-\infty}^{\infty} |F(k)| dk \quad , \quad \int_{-\infty}^{\infty} |g(x)| dx$$

are convergent so that

$$\int_{-\infty}^{\infty} F(k) e^{-ikx} dx \quad , \quad \int_{-\infty}^{\infty} g(x) e^{-ikx} dx$$

are absolutely convergent (since

$$|F(k)e^{-ikx}| = |F(k)|$$

and

$$|g(x)e^{-ikx}| = |g(x)|).$$

Under these conditions the order of integration in (3) may be interchanged giving

$$\Phi^{-1}\{F(k)G(k)\} = \frac{1}{2\pi}\int_{-\infty}^{\infty} g(\xi)d\xi \int_{-\infty}^{\infty} F(k)e^{-ik(x-\xi)}dk . \qquad (4)$$

Since the Fourier inversion integral is valid,

$$\frac{1}{\sqrt{2\pi}}\int_{-\infty}^{\infty} F(k)e^{-ik(x-\xi)}dk = f(x-\xi)$$

and using this result in (4) gives

$$\Phi^{-1}\{F(k)G(k)\} = \frac{1}{\sqrt{2\pi}}\int_{-\infty}^{\infty} g(\xi)f(x-\xi)d\xi . \qquad (5)$$

Noting that

$$\Phi^{-1}\{F(k)G(k)\} = h(x)$$

and using the definition of convolution in (1), (5) becomes

$$h(x) = (f * g)$$

and the theorem is proved. The validity of (5) is assured even if the Fourier inversion integral for f(x) has a countably infinite number of discrepancies with f(x) since this will not affect the equality of the integrals

$$\frac{1}{\sqrt{2\pi}}\int_{-\infty}^{\infty} g(\xi) \ f(x-\xi)d\xi$$

and

$$\frac{1}{\sqrt{2\pi}}\int_{-\infty}^{\infty} g(\xi) \left(\frac{1}{\sqrt{2\pi}}\int_{-\infty}^{\infty} F(k)e^{-ik(x-\xi)}dk\right)d\xi .$$

a)   State conditions under which the Fourier integral formula

$$f(x) = \frac{1}{2\pi} \int_{-\infty}^{\infty} e^{-ikx} \, dk \int_{-\infty}^{\infty} f(\xi) e^{ik\xi} \, d\xi \qquad (1)$$

is valid.   Discuss its validity for the examples in problems 28 and 29 of this chapter.

b)   Recast (1) in real form assuming that f(x) is real.

Solution: The most widely used sufficient conditions for pointwise convergence are as follows:

If f(x) is absolutely integrable and piecewise very smooth on (-∞,∞), then the Fourier integral theorem is valid in the sense that

$$\frac{1}{2\pi} \int_{-\infty}^{\infty} e^{-ikx} \, dk \int_{-\infty}^{\infty} f(\xi) e^{ik\xi} \, d\xi = \frac{1}{2} [f(x+) + f(x-)]$$

where f(x+) and f(x-) are the right and left hand limits of f at x respectively.   Recall that a function f is piecewise very smooth if its second derivative f"(x) is piecewise continuous.

Each of the functions in problems 28 and 29 satisfy these conditions so that the Fourier integral theorem is valid for each.   For example,

$$f(x) = N e^{-\alpha x^2}$$

is absolutely integrable on (-∞,∞) since

$$\int_{-\infty}^{\infty} \left| N e^{-\alpha x^2} \right| \, dx = \int_{-\infty}^{\infty} N e^{-\alpha x^2} \, dx$$

which is a convergent integral.   Also, since f has a continuous second derivative it is piecewise very smooth.

b)   Write (1) as

$$f(x) = \frac{1}{\sqrt{2\pi}} \int_{-\infty}^{0} F(k) e^{-ikx} \, dk + \frac{1}{\sqrt{2\pi}} \int_{0}^{\infty} F(k) e^{-ikx} \, dk \quad .$$

Make the change of variable k' = -k in the first integral to obtain

$$f(x) = \frac{1}{\sqrt{2\pi}} \int_\infty^0 F(-k')e^{ik'x} (-dk') + \frac{1}{\sqrt{2\pi}} \int_0^\infty F(k)e^{-ikx} dk$$

$$= \frac{1}{\sqrt{2\pi}} \int_0^\infty F(-k')e^{ik'x} dk' + \frac{1}{\sqrt{2\pi}} \int_0^\infty F(k)e^{-ikx} dk$$

$$= \frac{1}{\sqrt{2\pi}} \int_0^\infty [F*(k)e^{ikx} + F(k)e^{-ikx}] dk \qquad (2)$$

where the fact that F(-k) = F*(k) has been used in the last step. Now since the Fourier integral formula is assumed valid for f,

$$F(k)e^{-ikx} = \frac{1}{\sqrt{2\pi}} \int_{-\infty}^\infty f(\xi)e^{ik(\xi-x)} d\xi \qquad (3)$$

and taking the complex conjugate of this formula,

$$F*(k)e^{ikx} = \frac{1}{\sqrt{2\pi}} \int_{-\infty}^\infty f(\xi)e^{-ik(\xi-x)} dk \ . \qquad (4)$$

Adding (3) to (4) and recalling that

$$\cos\theta = \frac{e^{i\theta} + e^{-i\theta}}{2}$$

gives

$$F(k)e^{-ikx} + F*(k)e^{ikx} = \frac{1}{\sqrt{2\pi}} \int_{-\infty}^\infty f(\xi)2 \cos k(\xi-x) d\xi$$

which, upon substitution into (2), yields the real form of the Fourier integral formula

$$f(k) = \frac{1}{\pi} \int_0^\infty dk \int_{-\infty}^\infty f(\xi) \cos k (\xi-x) d\xi \qquad .$$

● PROBLEM 7–36

Find the form of the Fourier integral formula if f(x) is an

a) even function, b) odd function;

455

thereby motivating definitions for the Fourier cosine and Fourier sine transforms:

$$\Phi_c\{f(x)\} \qquad \text{and} \qquad \Phi_s\{f(x)\}$$

Solution: The real form of the Fourier integral formula is

$$f(x) = \frac{1}{\pi} \int_0^\infty dk \int_{-\infty}^\infty f(\xi) \cos k(\xi - x) \, d\xi \qquad (1)$$

and using the trigonometric identity

$$\cos(a-b) = \cos a \cos b + \sin a \sin b$$

in (1) yields

$$f(x) = \frac{1}{\pi} \int_0^\infty dk \int_{-\infty}^\infty f(\xi)[\cos kx \cos k\xi + \sin kx \sin k\xi] d\xi$$

$$= \frac{1}{\pi} \int_0^\infty \cos kx \, dk \int_{-\infty}^\infty f(\xi) \cos k\xi \, d\xi + \frac{1}{\pi} \int_0^\infty \sin kx \, dk$$

$$X \int_{-\infty}^\infty f(\xi) \sin k\xi \, d\xi \ . \qquad (2)$$

a) If f is even then $f(\xi) \sin k\xi$ is odd and the second integral is zero. Hence

$$f(x) = \frac{1}{\pi} \int_0^\infty \cos kx \, dk \int_{-\infty}^\infty f(\xi) \cos k\xi \, d\xi , \qquad (3)$$

f even and since $f(\xi) \cos k\xi$ is even, (3) may be written as

$$f(x) = \frac{2}{\pi} \int_0^\infty \cos kx \, dk \int_0^\infty f(\xi) \cos k\xi \, d\xi , \qquad (4)$$

f even.

This formula suggests the definition of a Fourier cosine transform, analogous to that of the regular Fourier transform, by

$$F_c(k) = \Phi_c \{f(x)\} = \sqrt{\frac{2}{\pi}} \int_0^\infty f(x) \cos kx \, dx \qquad (5)$$

with the inverse

$$\Phi_c^{-1} \{F_c(k)\} = \sqrt{\frac{2}{\pi}} \int_0^\infty F_c(k) \cos kx \, dk = \begin{cases} f(x) & x > 0 \\ \\ f(-x) & x < 0 \end{cases} \qquad (6)$$

Note that this definition applies to all functions defined on $(0,\infty)$ even if undefined elsewhere. In any case, the Fourier cosine integral formula (6) will give back $f(x)$ for $x \in (0,\infty)$ and $f(-x)$ for $x < 0$ so

$$\Phi^{-1} \{\Phi_c\{f(x)\}\}$$

represents the even extension of f into the negative real axis.

b) If f is odd then $f(\xi) \cos k\xi$ is odd and the first integral in (2) is zero leaving

$$f(x) = \frac{2}{\pi} \int_0^\infty \sin kx \, dk \int_0^\infty f(\xi) \sin k\xi \, d\xi \ , \ f \text{ odd.} \qquad (7)$$

This formula suggests the definition of a Fourier sine transform by

$$F_s(k) = \Phi_s\{f(x)\} = \sqrt{\frac{2}{\pi}} \int_0^\infty f(x) \sin kx \, dx \qquad (8)$$

with the inverse

$$\Phi_s^{-1} \{F_s(k)\} = \sqrt{\frac{2}{\pi}} \int_0^\infty F_s(k) \sin kx \, dx = \begin{cases} f(x) & x > 0 \\ \\ -f(-x) & x < 0 \end{cases} \qquad (9)$$

As with $\Phi_c\{f(x)\}$ this definition applies to all functions defined on $(0,\infty)$ even if undefined elsewhere. In any case, the Fourier sine integral formula (9) will give back $f(x)$ for $x \in (0,\infty)$ and $-f(-x)$ for $x < 0$. Therefore

$$\Phi^{-1} \{\Phi_s\{f(x)\}\}$$

represents the odd extension of f into the negative real axis.

Define the Dirac delta function, $\delta(x)$, and prove the sifting property of $\delta(x)$ for all functions $f(x)$ which are continuous at $x = 0$,

$$\int_{-\infty}^{\infty} \delta(x) f(x) \, dx = f(0) \; .$$

Solution: The Dirac delta function may be defined in several ways but its most common definition is as the infinitely sharply peaked function given symbolically by

$$\delta(x) = \begin{cases} 0 & x \neq 0 \\[2em] \infty & x = 0 \end{cases} \tag{1}$$

and with the property that

$$\int_{-\infty}^{\infty} \delta(x) \, dx = 1 \; . \tag{2}$$

(i.e., the integral of $\delta(x)$ is normalized to unity).

For this problem start with the integral

$$\int_{-\infty}^{\infty} \delta(x) f(x) \; dx$$

where $f(x)$ is any continuous function. We can evaluate this integral by the following argument:

Since $\delta(x)$ is zero for $x \neq 0$ (by (1)), the limits of integration may be changed to $-\varepsilon$ and $+\varepsilon$, where ε is a small positive number. In addition, since $f(x)$ is continuous at $x = 0$, its values within the interval $(-\varepsilon, +\varepsilon)$ will not differ much from $f(0)$ and we can say, approximately, that

$$\int_{-\infty}^{\infty} \delta(x) f(x) \; dx = \int_{-\varepsilon}^{+\varepsilon} \delta(x) f(x) \; dx \approx f(0) \int_{-\varepsilon}^{+\varepsilon} \delta(x) \, dx$$

where the approximation improves as approaches zero. However, since $\delta(x) = 0$ for $x \neq 0$ and since $\delta(x)$ is normalized we have

458

$$\int_{-\varepsilon}^{+\varepsilon} \delta(x)\,dx = 1$$

for all values of ε .
It appears then that letting $\varepsilon \to 0$, we have exactly

$$\int_{-\varepsilon}^{-\varepsilon} \delta(x)\,f(x)\ dx = f(0). \tag{3}$$

Note that the limits $-\varepsilon$ and $+\varepsilon$ may be replaced by any two numbers a and b provided that a < 0 < b. Now the integral (3) is referred to as the sifting property of the delta function, that is $\delta(x)$ acts as a sieve, selecting from all possible values of $f(x)$ its value at the point x = 0.

APPLICATIONS OF FOURIER TRANSFORMS

• PROBLEM 7-38

Given the current pulse, $i(t) = te^{-bt}$: (a) find the total 1Ω energy associated with this waveform; (b) what fraction of this energy is present in the frequency band from -b to b rad/s?

Solution: Use $j\omega$ as the transfer parameter instead of k. Then the total $1\text{-}\Omega$ energy associated with either a current or voltage waveform can be found by use of Parseval's theorem,

$$W_{1\Omega} = \frac{1}{2\pi} \int_{-\infty}^{\infty} |F(j\omega)|^2 d\omega$$

where $F(j\omega)$ is the Fourier transform of the current or voltage waveform.

The Fourier transform of the current is

$$I(j\omega) = \int_{0}^{\infty} t\,e^{-bt}\,e^{-j\omega t}\ dt$$

$$= \int_0^\infty t\, e^{-t(j\omega+b)}\ dt$$

$$= \left(\frac{-te^{-t(j\omega+b)}}{(j\omega+b)} - \int \frac{e^{-t(j\omega+b)}}{-(j\omega+b)}\ dt \right) \Bigg|_0^\infty$$

$$= \left(-\frac{te^{-t(j\omega+b)}}{j\omega+b} - \frac{e^{-t(j\omega+b)}}{(j\omega+b)^2} \right) \Bigg|_0^\infty$$

$$= \frac{1}{(j\omega+b)^2}$$

$$|I(j\omega)|^2 = \frac{1}{(b^2-\omega^2)^2+4\omega^2 b^2} = \frac{1}{(b^2+\omega^2)^2} \quad .$$

The total energy associated with the current is

$$W = \frac{1}{2\pi} \int_{-\infty}^\infty |I(j\omega)|^2 d\omega \quad .$$

Since

$$W = \frac{1}{\pi} \int_0^\infty |I(j\omega)|^2 d\omega$$

then

$$W = \frac{1}{\pi} \int_0^\infty \frac{1}{(b^2+\omega^2)^2}\ d\omega \quad .$$

If we make the trigonometric substitution,

$$\omega = b\tan\theta,$$

then

$$(b^2+\omega^2)^2 = (b^2\tan^2\theta + b^2)^2 = (b^2\sec^2\theta)^2 \quad .$$

Also,

$$d\omega = b\sec^2\theta d\theta \quad .$$

Hence

$$W = \frac{1}{\pi} \int \frac{b\sec^2\theta}{b^4 \sec^4\theta}\ d\theta$$

$$W = \frac{1}{\pi} \int \frac{1}{b^3 \sec^2 \theta} \, d\theta$$

$$W = \frac{1}{\pi b^3} \int \cos^2 \theta \, d\theta$$

$$W_\tau = \frac{1}{\pi b^3} (\frac{1}{2} \theta + \frac{1}{4} \sin 2\theta) \ .$$

Since

$$\omega = b\tan\theta$$

and

$$\theta = \text{arc tan } \frac{\omega}{b}$$

then

$$W_\tau = \left[\frac{1}{\pi b^3} (\frac{1}{2} \tan^{-1} \frac{\omega}{b} + \frac{1}{4} \sin 2(\tan^{-1} \frac{\omega}{b})) \right] \Bigg|_0^\infty \quad .$$

Since

$$\tan^{-1} \infty = \frac{\pi}{2}$$

then

$$W_\tau = \frac{1}{\pi b^3} (\frac{1}{2} \frac{\pi}{2})$$

$$W_\tau = \frac{1}{4b^3} \quad .$$

b) To find the energy present in the frequency band

$$-b < f < b$$

we use Parseval's theorem and integrate:

$$W_b = \frac{1}{2\pi} \int_{-b}^{b} |I(j\omega)|^2 \, d\omega$$

$$W_b = \frac{1}{\pi} \int_0^b |I(j\omega)|^2 \, d\omega$$

$$W_b = \left[\frac{1}{\pi b^3} (\frac{1}{2} \tan^{-1} \frac{\omega}{b} + \frac{1}{4} \sin 2(\tan^{-1} \frac{\omega}{b})) \right] \Bigg|_0^b$$

461

$$W_b = \left[\frac{1}{\pi b^3} \quad (\frac{1}{2} \tan^{-1} 1 + \frac{1}{4} \sin 2(\tan^{-1} 1))\right.$$
$$\left. - (\frac{1}{2} \tan^{-1} 0 + \frac{1}{4} \sin 2(\tan^{-1} 0))\right]$$

$$W_b = \frac{1}{\pi b^3} \left[\frac{\pi}{8} + \frac{1}{4} - 0 - 0\right]$$

$$W_b = \left(\frac{\pi+2}{8}\right) \frac{1}{\pi b^3} \quad .$$

The fraction of energy present is

$$\frac{W_0}{W_\tau} = \frac{\left[\frac{\pi+2}{8}\right] \frac{1}{\pi b^3}}{\frac{1}{4 b^3}} = \frac{\pi+2}{2\pi} = 0.818 \quad .$$

a)   If a > 0 show that the Fourier transform of the function defined by

$$f(t) = e^{-at} \cos\omega dt \qquad t \geq 0$$

$$= 0 \qquad\qquad t < 0$$

is $(a+j\omega)^2/[(a+j\omega)^2 + \omega_d^2]$.

Then find the total 1Ω energy associated with the function

$$f = e^{-t} \cos t \qquad t \geq 0$$
$$= 0 \qquad\qquad t < 0$$

by using:

b)   time domain integration. That is, find the total energy by integrating

$$W = \int_0^\infty [f(t)]^2 dt \quad .$$

c)   frequency domain integration. That is, find the total energy by integrating

$$W = \frac{1}{2\pi} \int_{-\infty}^{\infty} \left| F(\omega) \right|^2 d\omega$$

where $F(\omega)$ is the Fourier transform of the function $f(t)$.

Solution: Using the relation

$$F(\omega) = \int_{-\infty}^{\infty} f(t) e^{-j\omega t} \, dt$$

we can find the Fourier transform of

$$f(t) = e^{-at} (\cos\omega_d t)$$

Hence,

$$F(\omega) = \int_{0}^{\infty} e^{-at} (\cos\omega_d t) e^{-j\omega t} \, dt$$

$$= \int_{0}^{\infty} e^{-at} \left(\frac{e^{j\omega_d t} + e^{-j\omega_d t}}{2} \right) e^{-j\omega t} \, dt$$

$$= \int_{0}^{\infty} \left[\frac{e^{(j\omega_d - j\omega - a)t}}{2} + \frac{e^{(-j\omega_d - j\omega - a)t}}{2} \right] dt$$

$$= \left[\frac{e^{(j\omega_d - j\omega - a)t}}{2(j\omega_d - j\omega - a)} + \frac{e^{-(j\omega_d + j\omega + a)t}}{2(-j\omega_d - j\omega - a)} \right]_0^{\infty}$$

$$= \left[\frac{-1}{2(j\omega_d - j\omega - a)} - \frac{1}{2(-j\omega_d - j\omega - a)} \right]$$

Hence,

$$F(\omega) = \left[\frac{a + j\omega}{(a + j\omega)^2 + \omega_d^2} \right] \, .$$

463

b) In the time domain, the total energy is found by integrating

$$[f(t)]^2$$

as follows:

$$W = \int_{-\infty}^{\infty} [f(t)]^2 \, dt$$

$$W = \int_{0}^{\infty} \left[e^{-t} (\cos t) \right]^2 \, dt$$

$$= \int_{0}^{\infty} e^{-2t} \left(\frac{1}{2} + \frac{1}{2} \cos 2t \right) dt$$

$$= \int_{0}^{\infty} \left[\frac{e^{-2t}}{4} + \frac{e^{-2t}}{2} \left(\frac{e^{+j2t}}{2} + \frac{e^{-j2t}}{2} \right) \right] dt$$

$$= \left[-\frac{e^{-2t}}{4} + \frac{e^{-2t}e^{j2t}}{4(-2+j2)} - \frac{e^{-2t}e^{-j2t}}{4(2+j2)} \right]\Bigg|_{0}^{\infty}$$

$$W = \frac{8 - (-2-j2) + (2-j2)}{32} = \frac{12}{32} = \frac{3}{8} \quad .$$

c) In the frequency domain the total energy is found by integrating

$$W = \frac{1}{2\pi} \int |F(\omega)|^2 d\omega \quad .$$

We found in (a)

$$F(\omega) = \frac{a + j\omega}{(a+j\omega)^2 + \omega_d^2} \quad .$$

If $a = 1$ and $\omega_d = 1$ then

$$F(\omega) = \frac{1 + j\omega}{(1 + j\omega)^2 + 1}$$

and

$$|F(\omega)|^2 = \frac{1 + \omega^2}{(2-\omega^2)^2 + 4\omega^2} = \frac{1 + \omega^2}{4 + \omega^4} \quad .$$

The energy is

$$W = \frac{1}{2\pi} \int_{-\infty}^{\infty} \frac{1 + \omega^2}{4 + \omega^4} d\omega = \frac{1}{\pi} \int_{0}^{\infty} \frac{1 + \omega^2}{4 + \omega^4} d\omega \quad .$$

Now, by an integral formula,

$$\int_{0}^{\infty} \frac{x^2 + 1}{x^4 + 4} dx = \frac{3\pi}{8} \quad .$$

Hence, we have

$$W = \frac{1}{\pi} \int_{0}^{\infty} \frac{1 + \omega^2}{4 + \omega^2} d\omega = \frac{1}{\pi} \cdot \frac{3\pi}{8} \quad .$$

Thus

$$W = \frac{3}{8}$$

verifying Parseval's identity in this example.

● **PROBLEM 7–40**

Use Fourier transform methods to find the time-domain response of a network having a system function

$$j2\omega/(1 + 2j\omega) \ ,$$

if the input is

$$V(t) = \cos t$$

(For a sinusodial input cos t, the Fourier transform is

$$\pi[\delta(\omega+1) + \delta(\omega-1)]).$$

Solution: The time domain response for a particular input V(t) can be obtained by finding the product of the system function H(jω) and the Fourier transform of the input. The inverse Fourier transform of the resulting function is the time-domain response.

For a sinusoidal input cos t, the Fourier transform pair

$$\cos t \iff \pi[\delta(\omega+1) + \delta(\omega-1)]$$

allows us to find the response

$$f(t) = F^{-1}\left\{\frac{j2\omega}{1+2j}\ \pi(\delta(\omega+1) + \delta(\omega-1))\right\}$$

$$f(t) = F^{-1}\left\{\frac{j2\pi\omega\ \delta(\omega+1)}{1 + 2j\omega} + \frac{j2\pi\omega\ \delta(\omega-1)}{1 + 2j\omega}\right\}$$

$$f(t) = F^{-1}\left\{\frac{j\pi\omega\ \ \delta(\omega+1)}{\frac{1}{2} + j\omega} + \frac{j\pi\omega\ \ \delta(\omega-1)}{\cdot\ \frac{1}{2} + j\omega}\right\}$$

Using the sifting property of the unit impulse, we obtain:

$$f(t) = F^{-1}\left\{-\frac{j\pi\ \delta(\omega+1)}{\frac{1}{2} - j} + \frac{j\pi\ \delta(\omega-1)}{\frac{1}{2} + j}\right\}$$

$$f(t) = F^{-1}\left\{-\frac{j\pi\ \delta(\omega+1)\ (\frac{1}{2}+j)}{\frac{1}{4} + 1} + \frac{j\pi\ \delta(\omega-1)\ (\frac{1}{2}-j)}{\frac{1}{4} + 1}\right\}$$

$$f(t) = F^{-1}\left\{\frac{\pi\ \delta(\omega+1)}{\frac{5}{4}} - \frac{j\frac{1}{2}\pi\ \delta(\omega+1)}{\frac{5}{4}} + \frac{\pi\ \delta(\omega-1)}{\frac{5}{4}}\right.$$

$$\left. + \frac{j\frac{1}{2}\pi\ \delta(\omega-1)}{\frac{5}{4}}\right\}$$

$$f(t) = F^{-1}\left\{\frac{4}{5}\pi(\delta(\omega+1) + \delta(\omega-1)) - \frac{2}{5}\pi(j\ \delta(\omega+1) - j\ \delta(\omega-1))\right\}$$

$$f(t) = \frac{4}{5}\cos t - \frac{2}{5}\sin t .$$

CHAPTER 8

COMPLEX VARIABLES

Basic Attacks and Strategies for Solving Problems in this Chapter. See pages 468 to 529 for step-by-step solutions to problems.

Since there is no real number x which satisfies the polynomial equation $x + 1 = 0$ or similar equations, the set of complex numbers is introduced.

We can consider a complex number as having the form $a + bi$ where a and b are real numbers (called the real and imaginary parts, respectively), and $i = \sqrt{-1}$ is called the imaginary unit. Two complex numbers $a + bi$ and $c + di$ are equal if and only if $a = c$ and $b = d$. We can consider real numbers as the subset of the complex numbers with $b = 0$. The complex number $0 + 0i$ corresponds to the real number 0.

The absolute value or modulus of $a + bi$ is defined as $|a + b_1| = \sqrt{a^2 + b^2}$. The complex conjugate of $a + bi$ is defined as $a - bi$. The complex conjugate of the complex number z is often indicated by z^*.

In performing operations with complex numbers we can operate as in the algebra of real numbers, replacing i^2 by -1 when it occurs. Inequalities for complex numbers are not defined.

From the point of view of an axiomatic foundation of complex numbers, it is desirable to treat a complex number as an ordered pair (a, b) of real numbers, with a and b subject to certain rules which turn out to be equivalent to those above.

DEMOIVRE'S THEOREM

• PROBLEM 8–1

1. Describe the system of polar coordinates.

2. Let z be a complex number $z = x + iy$. Represent z in polar coordinates.

3. Express the following complex numbers in polar form

$$z_1 = 4 + 4\sqrt{3}\ i$$

$$z_2 = -2 + 2i$$

$$z_3 = -5i$$

Solution: 1. We shall introduce in the z plane the polar co-ordinates. The location of each point P of the plane is uniquely determined by two polar coordinates r and θ, as shown in Fig. 1.

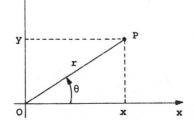

Fig. 1

Here, r is the length of the segment $\overline{OP}$, and θ is the angle be-tween the x-axis and $\overline{OP}$. The relationship between the Carte-sian coordinates (x,y) and the polar coordinates (r,θ) is

$$x = r \cos\theta, \quad y = r \sin\theta \tag{1}$$

2. A complex number $z = x + iy$ can be represented by a point in the z plane, as shown in Fig. 1. We have

$$r = |z| = \sqrt{x^2+y^2}, \quad \tan\theta = \frac{y}{x} \tag{2}$$

Here, θ is called the argument or amplitude of z. We write

$$\theta = \arg z \tag{3}$$

A complex number $z = x + iy$ can be written in the polar form

$$z = r(\cos\theta + i \sin\theta) \tag{4}$$

Note that any integral multiple of 2π can be added to θ, without changing the value of z. The value of arg z which satisfies

$$-\pi < \arg z \leq \pi \tag{5}$$

is called the principal value of arg z, and it is sometimes denoted by Arg z.

3. $z_1 = 4 + 4\sqrt{3}\, i$

$$r = \sqrt{4^2+(4^2 \cdot 3)} = 8 \tag{6}$$

$x = r \cos\theta$ and

$$\cos\theta = \frac{x}{r} = \frac{4}{8} = \frac{1}{2} \tag{7}$$

Therefore $\theta = \dfrac{\pi}{3}$ and

$$4 + 4\sqrt{3}\, i = 8\left[\cos \frac{\pi}{3} + i \sin \frac{\pi}{3}\right] \tag{8}$$

For $z_2 = -2 + 2i$ $\tag{9}$

$$r = \sqrt{4+4} = 2\sqrt{2}$$

$$\cos\theta = \frac{x}{r} = \frac{-2}{2\sqrt{2}} = -\frac{\sqrt{2}}{2}$$

$$\theta = \frac{3\pi}{4}$$

thus $-2 + 2i = 2\sqrt{2}\left[\cos \dfrac{3\pi}{4} + i \sin \dfrac{3\pi}{4}\right]$ $\tag{10}$

For $z_3 = -5i$,

$$r = \sqrt{25} = 5$$

$$\cos\theta = \frac{x}{r} = 0$$

$$\theta = \frac{\pi}{2} \quad \text{or} \quad -\frac{\pi}{2}$$

Since $\sin\theta = \frac{y}{r} = -1$, we choose

$$\theta = -\frac{\pi}{2}$$

Thus

$$-5i = 5\left[\cos\left(-\frac{\pi}{2}\right) + i \sin\left(-\frac{\pi}{2}\right)\right] \tag{11}$$

$$= 5\left(\cos\frac{3\pi}{2} + i \sin\frac{3\pi}{2}\right)$$

● PROBLEM 8–2

Represent graphically the following complex numbers

1. $z_1 = 3(\cos 75^0 + i \sin 75^0)$

2. $z_2 = 5(\cos 210^0 + i \sin 210^0)$

3. $z_3 = 2(\cos\pi + i \sin\pi)$

4. the conjugate of $z_4 = 4(\cos 30^0 + i \sin 30^0)$.

Solution: In general, a complex number can be written in the polar form

$$z = r(\cos\theta + i \sin\theta)$$

the graphical representation of which is shown in Fig. 1.

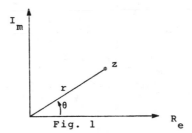

Fig. 1

1. In this case,

$$z_1 = 3(\cos 75^0 + i \sin 75^0)$$

the distance r is 3, and the angle θ is 75^0, thus we can represent this number graphically as shown in Fig. 2.

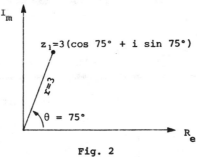

Fig. 2

470

2. In this case, $r = 5$ and $\theta = 210^{\circ}$; the number z_2 is shown in Fig. 3.

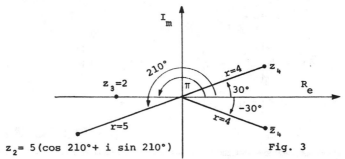

$z_2 = 5(\cos 210^{\circ} + i \sin 210^{\circ})$ Fig. 3

3. Here, the angle is $\theta = \pi$, and $r = 2$. The number z_3 lies on the negative side of the x-axis, as shown in Fig. 3.

4. The conjugate of a number, or point

$$z = a + bi$$

is the reflection of the point about the real axis, and is given by

$$\bar{z} = a - bi$$

We have

$$|\bar{z}| = |z| = r$$

and

$$\arg z = -\arg \bar{z} = -\theta$$

The polar form of the conjugate is therefore

$$\bar{z} = r(\cos(-\theta) + i \sin(-\theta))$$

$$= r(\cos\theta - i \sin\theta)$$

The conjugate of z_4 is thus

$$\bar{z}_4 = 4(\cos 30^{\circ} - i \sin 30^{\circ})$$

$$r = 4 \qquad \theta = -30^{\circ}$$

● **PROBLEM 8–3**

Evaluate the modulus of

$$z = \frac{1 + \cos\theta + i \sin\theta}{1 + \cos\phi + i \sin\phi} \tag{1}$$

Solution: Let us transform the numerator and denominator into polar form. Let

$$z_1 = 1 + \cos\theta + i \sin\theta = r_1(\cos\theta + i \sin\theta)$$

$$z_2 = 1 + \cos\phi + i \sin\phi = r_2(\cos\phi + i \sin\phi)$$

Since r_1 is the modulus of z_1, we obtain

$$r_1 = \sqrt{(1+\cos\theta)^2 + \sin^2\theta} = \sqrt{2(1+\cos\theta)}$$

Likewise

$$r_2 = \sqrt{(1+\cos\phi)^2 + \sin^2\theta} = \sqrt{2(1+\cos\phi)}$$

Now, since

$$z = \frac{z_1}{z_2}$$

then

$$|z| = \left|\frac{z_1}{z_2}\right| = \frac{|z_1|}{|z_2|} = \frac{r_1}{r_2}$$

Thus, the modulus of $z = \frac{z_1}{z_2}$ is

$$|z| = \frac{r_1}{r_2} = \sqrt{\frac{1+\cos\theta}{1+\cos\phi}}$$

● **PROBLEM 8–4**

Compute

$$\frac{\left(\frac{3}{2}\sqrt{3} + \frac{3}{2}i\right)^6}{\left(\sqrt{\frac{5}{2}} + i\sqrt{\frac{5}{2}}\right)^3} \qquad (1)$$

Solution: Let us denote

$$z_1 = \frac{3}{2}\sqrt{3} + \frac{3}{2}i \qquad (2)$$

$$z_2 = \sqrt{\frac{5}{2}} + \sqrt{\frac{5}{2}}i \qquad (3)$$

Then

$$|z_1| = \sqrt{\frac{27}{4} + \frac{9}{4}} = 3 \quad \text{and}$$

$$\cos\theta_1 = \frac{x_1}{|z_1|} = \frac{3\sqrt{3}}{2\cdot3} = \frac{\sqrt{3}}{2} \qquad (4)$$

Thus
$$\theta_1 = \frac{\pi}{6} \tag{5}$$
and
$$z_1 = 3\left(\cos \frac{\pi}{6} + i \sin \frac{\pi}{6}\right) = 3e^{i\frac{\pi}{6}} \tag{6}$$

For z_2, we have
$$|z_2| = \sqrt{\frac{5}{2} + \frac{5}{2}} = \sqrt{5} \quad \text{and}$$
$$\cos\theta_2 = \frac{\sqrt{5}}{\sqrt{2} \cdot \sqrt{5}} = \frac{1}{\sqrt{2}} \tag{7}$$

Hence
$$\theta_2 = \frac{\pi}{4} \quad \text{and}$$
$$z_2 = \sqrt{5}\left(\cos \frac{\pi}{4} + i \sin \frac{\pi}{4}\right) = \sqrt{5}\, e^{\pi i/4} \tag{8}$$

Substituting eqs.(6) and (8) into eq.(1), we find
$$\frac{z_1^{\,6}}{z_2^{\,3}} = \frac{\left(3e^{\pi i/6}\right)^6}{\left(\sqrt{5}\, e^{\pi i/4}\right)^3} = \frac{729}{5\sqrt{5}}\, e^{i\left(\pi - \frac{3\pi}{4}\right)}$$
$$= \frac{729}{5\sqrt{5}}\, e^{\pi i/4} = \frac{729}{5\sqrt{5}}\left(\cos \frac{\pi}{4} + i \sin \frac{\pi}{4}\right) \tag{9}$$
$$= \frac{729}{5\sqrt{5}}\left[\frac{1}{\sqrt{2}} + i\,\frac{1}{\sqrt{2}}\right] = \frac{729}{5\sqrt{5}}(1+i)$$

• PROBLEM 8–5

Let
$$z_1 = r_1 e^{i\theta_1} \quad \text{and} \quad z_2 = r_2 e^{i\theta_2} \tag{1}$$
and
$$z_3 = z_1 + z_2 = r_3 e^{i\theta_3} \tag{2}$$
Evaluate r_3 and θ_3.

Solution: We have
$$z_1 = r_1 e^{i\theta_1} = x_1 + iy_1 = r_1(\cos\theta_1 + i \sin\theta_1) \tag{3}$$
$$z_2 = r_2 e^{i\theta_2} = x_2 + iy_2 = r_2(\cos\theta_2 + i \sin\theta_2) \tag{4}$$

Therefore

$$z_3 = x_3 + iy_3 = r_3 e^{i\theta_3} = (x_1+x_2) + i(y_1+y_2)$$

$$= (r_1\cos\theta_1 + r_2\cos\theta_2) + i(r_1\sin\theta_1 + r_2\sin\theta_2) \tag{5}$$

Thus

$$r_3 = \sqrt{x_3^2 + y_3^2} = \sqrt{(r_1\cos\theta_1 + r_2\cos\theta_2)^2 + (r_1\sin\theta_1 + r_2\sin\theta_2)^2}$$

$$= \left[r_1^2(\cos^2\theta_1 + \sin^2\theta_1) + r_2^2(\cos^2\theta_2 + \sin^2\theta_2) \right.$$

$$\left. + 2r_1 r_2(\cos\theta_1\cos\theta_2 + \sin\theta_1\sin\theta_2) \right]^{\frac{1}{2}} \tag{6}$$

$$= \sqrt{r_1^2 + r_2^2 + 2r_1 r_2 \cos(\theta_1 - \theta_2)}$$

Since

$$z_3 = x_3 + iy_3 = r_3(\cos\theta_3 + i\sin\theta_3) \tag{7}$$

we have

$$\tan\theta_3 = \frac{y_3}{x_3} \tag{8}$$

Substituting x_3 and y_3 into eq.(8) results in

$$\theta_3 = \tan^{-1}\left[\frac{r_1\sin\theta_1 + r_2\sin\theta_2}{r_1\cos\theta_1 + r_2\cos\theta_2} \right] \tag{9}$$

• PROBLEM 8–6

1. Compute the scalar product (also called the dot product) of the two complex vectors

$$z_1 = 3 - 2i$$

$$z_2 = -4 - 6i \tag{1}$$

2. Compute the cross product of z_1 and z_2 as given above.

Solution: 1. Let $z_1 = x_1 + iy_1$ and $z_2 = x_2 + iy_2$ be two complex numbers. The scalar product of z_1 and z_2 is defined by

$$z_1 . z_2 \equiv x_1 x_2 + y_1 y_2 = |z_1||z_2|\cos\theta = Re(\bar{z}_1 z_2) \tag{2}$$

$$= \frac{1}{2}(\bar{z}_1 z_2 + z_1 \bar{z}_2)$$

where θ is the angle between z_1 and z_2, $0 \le \theta < \pi$.

It is easy to show that all parts of eq.(2) are indeed equal. For example

$$Re(\bar{z}_1 z_2) = Re\left[(x_1 - iy_1)(x_2 + iy_2) \right] \tag{3}$$

$$= \text{Re}\left[x_1x_2+y_1y_2+ix_1y_2-ix_2y_1\right] = x_1x_2+y_1y_2$$

For z_1 and z_2 given by eq.(1), we obtain

$$z_1 \cdot z_2 = x_1x_2+y_1y_2 = (3-2i)\cdot(-4-6i)$$

$$= (3)(-4) + (-2)(-6) = 0 \tag{4}$$

2. The cross product of z_1 and z_2 is defined as

$$z_1 \times z_2 \equiv |z_1||z_2|\sin\theta = x_1y_2-y_1x_2$$

$$= \text{Im}(\bar{z}_1z_2) = \frac{1}{2i}(\bar{z}_1z_2-z_1\bar{z}_2) \tag{5}$$

We have

$$z_1 \times z_2 = (3-2i) \times (-4-6i) = x_1y_2-y_1x_2$$

$$= (3)(-6) - (-2)(-4) = -18-8 = -26 \tag{6}$$

• PROBLEM 8–7

Prove that if

$$z_1 = r_1e^{i\theta_1} \quad \text{and} \quad z_2 = r_2e^{i\theta_2} \tag{1}$$

then

1. $z_1 \cdot z_2 = r_1r_2\cos(\theta_2-\theta_1)$ \hfill (2)

$z_1 \times z_2 = r_1r_2\sin(\theta_2-\theta_1)$ \hfill (3)

2. $z_1 \cdot (z_2+z_3) = \left(z_1 \cdot z_2\right)+\left(z_1 \cdot z_3\right)$ \hfill (4)

$z_1 \times (z_2+z_3) = \left(z_1 \times z_2\right)+\left(z_1 \times z_3\right)$ \hfill (5)

Solution: 1. From the definition of the dot and cross product, we obtain

$$z_1 \cdot z_2 = |z_1||z_2|\cos\theta = r_1r_2\cos(\theta_2-\theta_1) \tag{6}$$

$$z_1 \times z_2 = |z_1||z_2|\sin\theta = r_1r_2\sin(\theta_2-\theta_1) \tag{7}$$

2. Let $z_1 = x_1+iy_1$, $z_2 = x_2+iy_2$ and $z_3 = x_3+iy_3$. We have

$$z_1 \cdot (z_2+z_3) = (x_1+iy_1) \cdot \left[(x_2+iy_2) + (x_3+iy_3)\right]$$

$$= (x_1+iy_1) \cdot \left[(x_2+x_3) + i(y_2+y_3)\right] \tag{8}$$

$$= x_1(x_2+x_3) + y_1(y_2+y_3) = x_1x_2 + x_1x_3 + y_1y_2 + y_1y_3$$

$$= (x_1x_2+y_1y_2) + (x_1x_3+y_1y_3)$$

$$= \left(z_1 \cdot z_2\right)+\left(z_1 \cdot z_3\right)$$

$$z_1 \times (z_2+z_3) = (x_1+iy_1) \times \left[(x_2+x_3) + i(y_2+y_3)\right]$$

$$= x_1(y_2+y_3) - y_1(x_2+x_3)$$

$$= (x_1y_2-y_1x_2) + (x_1y_3-y_1x_3) \qquad (9)$$

$$= \left[z_1 \times z_2\right] + \left[z_1 \times z_3\right]$$

• PROBLEM 8–8

Evaluate

$$\frac{(2+i)^{60}}{(3+4i)^{50}} \qquad (1)$$

i.e. , express it in the form x + iy.

Solution: Let $z_1 = 2+i = r_1e^{i\theta_1}$ and $\qquad (2)$

$$z_2 = 3+4i = r_2e^{i\theta_2} \qquad (3)$$

Then

$$r_1 = \sqrt{4+1} = \sqrt{5}$$
$$\qquad (4)$$
$$r_2 = \sqrt{9+16} = 5$$

and

$$\cos\theta_1 = \frac{2}{\sqrt{5}} \qquad \sin\theta_1 = \frac{1}{\sqrt{5}}$$

$$\qquad (5)$$

$$\cos\theta_2 = \frac{3}{5} \qquad \sin\theta_2 = \frac{4}{5}$$

From eq.(5) we evaluate the angles

$$\theta_1 = 27^0$$
$$\qquad (6)$$
$$\theta_2 = 53^0$$

Eq.(1) can be written

$$\frac{(2+i)^{60}}{(3+4i)^{50}} = \frac{[\sqrt{5}\ e^{i\ 27^0}]^{60}}{[\ 5\ e^{i\ 53^0}]^{50}}$$

$$= \frac{(5)^{30}}{(5)^{50}}\ \frac{e^{i\ 1620^0}}{e^{i\ 2650^0}} = \frac{1}{(5)^{20}}\ e^{i(\ 1620^0 - 2650^0)}$$

$$= (5)^{-20}\ e^{-i\ 1030^0} = (5)^{-20}\ e^{i(-310^\circ)}$$

476

$$= 1.0486 \times 10^{-14} \left(\cos 50^\circ + i \, \sin 50^\circ \right)$$

$$= 1.0486 \times 10^{-14} \left(0.6427 + i \, 0.7660 \right)$$

$$= 6.74 \times 10^{-15} + i \, 8.032 \times 10^{-15}$$

● **PROBLEM** 8–9

Find the square roots of

$$z = -4 - 3i \qquad (1)$$

Solution: In the polar form

$$-4 - 3i = 5 \left[\cos(\theta + 2k\pi) + i \, \sin(\theta + 2k\pi) \right] \qquad (2)$$

Thus

$$\cos\theta = -\frac{4}{5} \quad \text{and} \quad \sin\theta = -\frac{3}{5}$$

The square roots of 2 are

$$\sqrt{5} \left[\cos\left(\frac{\theta + 2k\pi}{n} \right) + i \, \sin\left(\frac{\theta + 2k\pi}{n} \right) \right] \qquad (3)$$

Therefore

for k = 0 $\qquad \sqrt{5} \left(\cos\frac{\theta}{2} + i \, \sin\frac{\theta}{2} \right) \qquad (4)$

for k = 1 $\qquad \sqrt{5} \left[\cos\left(\frac{\theta}{2} + \pi \right) + i \, \sin\left(\frac{\theta}{2} + \pi \right) \right] \qquad$

$$= -\sqrt{5} \left(\cos\frac{\theta}{2} + i \, \sin\frac{\theta}{2} \right) \qquad (5)$$

Having $\cos\theta$ and $\sin\theta$, we find

$$\cos\frac{\theta}{2} = \pm\sqrt{\frac{1+\cos\theta}{2}} = \pm\sqrt{\frac{1}{10}} = \pm\frac{1}{\sqrt{10}} \qquad (6)$$

$$\sin\frac{\theta}{2} = \pm\sqrt{\frac{1-\cos\theta}{2}} = \pm\sqrt{\frac{9}{10}} = \pm\frac{3}{\sqrt{10}} \qquad (7)$$

Since both $\cos\theta$ and $\sin\theta$ are negative, θ is an angle in the third quadrant. Therefore, $\frac{\theta}{2}$ is an angle in the second quadrant. In such case, $\cos\frac{\theta}{2} < 0$ and $\sin\frac{\theta}{2} > 0$ and we obtain

$$\cos\frac{\theta}{2} = -\frac{1}{\sqrt{10}}, \quad \sin\frac{\theta}{2} = \frac{3}{\sqrt{10}} \qquad (8)$$

From eqs.(4) and (5), the two roots are found to be

$$\sqrt{5}\left(-\frac{1}{\sqrt{10}} + i\frac{3}{\sqrt{10}}\right) = -\frac{1}{\sqrt{2}} + i\frac{3}{\sqrt{2}} \qquad (9)$$

$$-\sqrt{5}\left(-\frac{1}{\sqrt{10}} + i\frac{3}{\sqrt{10}}\right) = \frac{1}{\sqrt{2}} - i\frac{3}{\sqrt{2}} \qquad (10)$$

There is another method of solving this problem. Let $\alpha + i\beta$ be the solution. Then

$$(\alpha+i\beta)^2 = \alpha^2 - \beta^2 + 2\alpha\beta i = -4 - 3i \qquad (11)$$

or

$$\alpha^2 - \beta^2 = -4$$
$$2\alpha\beta = -3 \qquad (12)$$

This system of equations has two solutions

$$\alpha_1 = -\frac{1}{\sqrt{2}} \qquad \beta_1 = \frac{3}{\sqrt{2}}$$
$$\qquad (13)$$
$$\alpha_2 = \frac{1}{\sqrt{2}} \qquad \beta_2 = -\frac{3}{\sqrt{2}}$$

Thus, the corresponding square roots of 2 are

$$-\frac{1}{\sqrt{2}} + i\frac{3}{\sqrt{2}}$$
$$\qquad (14)$$
$$\frac{1}{\sqrt{2}} - i\frac{3}{\sqrt{2}}$$

● **PROBLEM** 8–10

Apply DeMoivre's formula to prove

1. $\cos2\theta = \cos^2\theta - \sin^2\theta$ $\qquad (1)$

$\sin2\theta = 2\sin\theta\cos\theta$ $\qquad (2)$

2. $\cos3\theta = 4\cos^3\theta - 3\cos\theta$ $\qquad (3)$

$\sin3\theta = 3\sin\theta - 4\sin^3\theta$ $\qquad (4)$

<u>Solution:</u>   1.   From DeMoivre's formula

$$(\cos\theta+i\ \sin\theta)^n = \cos n\theta + i\ \sin n\theta$$

we find

$$(\cos\theta + i\ \sin\theta)^2 = \cos2\theta + i\ \sin2\theta \tag{5}$$

On the other hand

$$(\cos\theta + i\ \sin\theta)^2 = (\cos^2\theta - \sin^2\theta) + 2i\ \sin\theta\cos\theta \tag{6}$$

Eqs.(5) and (6) yield

$$\cos2\theta + i\ \sin2\theta = (\cos^2\theta-\sin^2\theta) + 2i\ \sin\theta\cos\theta \tag{7}$$

Comparing the real and imaginary parts, we find

$$\cos2\theta = \cos^2\theta - \sin^2\theta \tag{8}$$

$$\sin2\theta = 2\sin\theta\cos\theta \tag{9}$$

2. Again, from DeMoivre's formula

$$(\cos\theta+i\ \sin\theta)^3 = \cos3\theta + i\ \sin3\theta \tag{10}$$

But we also have

$$(\cos\theta+i\ \sin\theta)^3 = \cos^3\theta + 3i\ \cos^2\theta\sin\theta - 3\cos\theta\sin^2\theta$$
$$- i\ \sin^3\theta \tag{11}$$
$$= \cos^3\theta-3\cos\theta\sin^2\theta+i(3\cos^2\theta\sin\theta-\sin^3\theta)$$

Hence

$$\cos3\theta = \cos^3\theta-3\cos\theta\sin^2\theta = \cos\theta(\cos^2\theta-3\sin^2\theta)$$
$$= \cos\theta(\cos^2\theta-3+3\cos^2\theta) = 4\cos^3\theta-3\cos\theta \tag{12}$$

and

$$\sin3\theta = 3\cos^2\theta\sin\theta - \sin^3\theta$$
$$= \sin\theta(3\cos^2\theta-\sin^2\theta) = \sin\theta(3-3\sin^2\theta-\sin^2\theta) \tag{13}$$
$$= 3\sin\theta - 4\sin^3\theta$$

The derivation of the above identities gives a clear indication of how useful formulas from the real number system can be obtained from a theorem dealing with complex numbers.

SEQUENCES AND SERIES OF COMPLEX NUMBERS

● **PROBLEM** 8–11

Test the series $\sum\limits_{n=1}^{\infty} \dfrac{i^n}{n}$ and $\sum\limits_{n=1}^{\infty} \dfrac{i^n}{n!}$ for absolute convergence.

Solution: Let us try the ratio test.

$$\left| \frac{z_{n+1}}{z_n} \right| = \left| \frac{i^{n+1} n}{(n+1)i^n} \right| = \frac{n}{n+1} \tag{1}$$

Hence,

$$\lim_{n \to \infty} \left| \frac{z_{n+1}}{z_n} \right| = 1 \tag{2}$$

Thus, the ratio test fails.

We can try to solve this problem by separating z_n into its real and imaginary parts.

Let

$$z_n = x_n + iy_n \tag{3}$$

where

$$z_n = \frac{i^n}{n}$$

Then

$$\sum_{n=1}^{\infty} x_n = \sum_{n=1}^{\infty} \frac{(-1)^n}{2n} \tag{4}$$

and

$$\sum_{n=1}^{\infty} y_n = \sum_{n=1}^{\infty} \frac{(-1)^{n-1}}{2n-1} \tag{5}$$

Each of these series is convergent, hence $\sum\limits_{n=1}^{\infty} z_n$ is convergent. But, neither of the series converges absolutely. Therefore, $\sum\limits_{n=1}^{\infty} z_n$ does not converge absolutely.

$$\left| \frac{z_{n+1}}{z_n} \right| = \left| \frac{i^{n+1} \, n!}{(n+1)! \, i^n} \right| = \frac{1}{n+1} \tag{6}$$

and

480

$$\lim_{n \to \infty} \left| \frac{z_{n+1}}{z_n} \right| = 0 \tag{7}$$

and the series $\sum_{n=1}^{\infty} \frac{i^n}{n}$ converges absolutely.

Determine the convergence of the series

$$1 + \frac{1}{2} + \frac{1}{3} + \frac{1}{4} + \frac{1}{5} + \ldots \tag{1}$$

called harmonic series.

<u>Solution</u>:   Observe that

$$\frac{1}{n+1} + \frac{1}{n+2} + \ldots + \frac{1}{2n} < n\,\frac{1}{2n} = \frac{1}{2} \tag{2}$$

Neglecting the first two terms, we shall assemble the terms of the series in groups consisting of $2, 4, 8, \ldots, 2^{n-1} \ldots$ terms

$$1 + \frac{1}{2} + \underbrace{\frac{1}{3} + \frac{1}{4}}_{2} + \underbrace{\frac{1}{5} + \frac{1}{6} + \frac{1}{7} + \frac{1}{8}}_{2^2} + \underbrace{\frac{1}{9} + \ldots + \frac{1}{16}}_{2^3} \tag{3}$$

$$+ \ldots + \underbrace{\frac{1}{2^{n-1}+1} + \ldots + \frac{1}{2^n}}_{2^{n-1}} + \ldots$$

Each of the partial sums is larger than $\frac{1}{2}$.   Since there are infinitely many groups in (3), the series $\sum_{n=1}^{\infty} \frac{1}{n}$ is divergent.

Determine the convergence of the series

$$\sum_{n=1}^{\infty} \frac{1}{n^s} \tag{1}$$

where s is a real number.

<u>Solution</u>:   Observe that for $s < 1$

$$\frac{1}{n} < \frac{1}{n^s} \tag{2}$$

Thus,

$$\sum_{n=1}^{\infty} \frac{1}{n} \leq \sum_{n=1}^{\infty} \frac{1}{n^s} \qquad (3)$$

Since the harmonic series is divergent (see Problem **8-12**) then, the series on the right hand side in (3) is also divergent.

Thus, for $s \leq 1$ the series $\sum_{n=1}^{\infty} \frac{1}{n^s}$ is divergent.

Consider $s > 1$. Let $s = 1 + \sigma$, $\sigma > 0$. We have

$$\frac{1}{(n+1)^s} + \frac{1}{(n+2)^s} + \ldots + \frac{1}{(2n)^s} < n \cdot \frac{1}{n^s} = \frac{1}{n^\sigma} \qquad (4)$$

Grouping the terms we obtain

$$1 + \frac{1}{2^s} + \underbrace{\frac{1}{3^s} + \frac{1}{4^s}}_{2} + \underbrace{\frac{1}{5^s} + \ldots + \frac{1}{8^s}}_{2^2} + \underbrace{\frac{1}{9^s} + \ldots + \frac{1}{16^s}}_{2^3} + \ldots$$

$$+ \underbrace{\frac{1}{(2^{k-1}+1)^s} + \ldots + \frac{1}{(2^k)^s}}_{2^{k-1}} + \ldots \qquad (5)$$

Each sum is smaller than the respective element of the geometrical sequence

$$\frac{1}{2^\sigma}, \frac{1}{4^\sigma} = \frac{1}{(2^\sigma)^2}, \frac{1}{8^\sigma} = \frac{1}{(2^\sigma)^3}, \ldots, \frac{1}{(2^{k-1})^\sigma} = \frac{1}{(2^\sigma)^{k-1}} \qquad (6)$$

Thus,

$$1 + \frac{1}{2^s} + \frac{1}{3^s} + \frac{1}{4^s} + \ldots < 1 + \frac{1}{2^{1+\sigma}} + \frac{1}{2^\sigma} + \frac{1}{(2^\sigma)^2} + \frac{1}{(2^\sigma)^3} + \ldots$$

$$\qquad (7)$$

$$= 1 + \frac{1}{2^{1+\sigma}} + \frac{1}{1 - \frac{1}{2^\sigma}}$$

Hence, the series (1) is convergent for $s > 1$.

● **PROBLEM** 8–14

Determine the convergence of the series

$$\sum_{n=1}^{\infty} \left(\frac{1}{5n!} + \frac{2i}{5n!} \right) \qquad (1)$$

<u>Solution</u>: Let

$$z_n = \frac{1}{5n!} + \frac{2i}{5n!} \tag{2}$$

Then

$$|z_n| = \sqrt{\frac{1}{25(n!)^2} + \frac{4}{25(n!)^2}} = \frac{1}{\sqrt{5}\, n!} \tag{3}$$

Observe that

$$\frac{1}{\sqrt{5}\, n!} < \frac{1}{n!} \tag{4}$$

Let

$$\omega_n = \frac{1}{n!} \tag{5}$$

The series $\sum\limits_{n=1}^{\infty} \omega_n$ is convergent because

$$e = 1 + \frac{1}{1!} + \frac{1}{2!} + \frac{1}{3!} + \ldots + \frac{1}{n!} + \ldots \tag{6}$$

where e is Euler's constant

$$e = 2.71828182845\ldots \tag{7}$$

Because

$$|z_n| \leq |\omega_n|$$

($\sum\limits_{n=1}^{\infty} |\omega_n|$ is convergent), the series $\sum\limits_{n=1}^{\infty} \left(\frac{1}{5n!} + \frac{2i}{5n!} \right)$ is absolutely convergent

● **PROBLEM 8–15**

Determine the convergence of the series $\sum\limits_{n=1}^{\infty} \frac{1}{n!} \left(\frac{n}{e} \right)^n$.   (1)

<u>Solution</u>: We shall apply Raabe's* test to determine convergence of series (1). If

$$\lim_{n \to \infty} n \left[\frac{|z_n|}{|z_{n+1}|} - 1 \right] = K, \tag{2}$$

then $\sum\limits_{n=1}^{\infty} z_n$ converges absolutely if $K > 1$ and diverges or converges conditionally if $K < 1$.

If $K = 1$, the test fails.

If $\sum\limits_{n=1}^{\infty} z_n$ converges but $\sum\limits_{n=1}^{\infty} |z_n|$ does not converge, then $\sum\limits_{n=1}^{\infty} z_n$ converges conditionally.

483

Let

$$R_n = n \left(\frac{|z_n|}{|z_{n+1}|} - 1 \right) \tag{3}$$

where

$$z_n = \frac{1}{n!} \left(\frac{n}{e} \right)^n \tag{4}$$

then

$$R_n = n \left[\frac{e}{\left(1 + \frac{1}{n} \right)^n} - 1 \right] \tag{5}$$

To evaluate the limit $\lim R_n$ consider a more general expression

$$\frac{1}{x} \left[\frac{e}{(1+x)^{\frac{1}{x}}} - 1 \right] , \quad x \to 0 \tag{6}$$

Applying de l'Hospital's theorem we find

$$- \frac{e}{\left[(1+x)^{\frac{1}{x}} \right]^2} \cdot \left\{ (1+x)^{\frac{1}{x}} + \ln(1+x) \cdot \left(-\frac{1}{x^2} \right) \right.$$

$$\left. + \frac{1}{x}(1+x)^{\frac{1}{x}-1} \right\} = \frac{e}{(1+x)^{\frac{1}{x}}} \cdot \frac{\ln(1+x) - \frac{x}{1+x}}{x^2} \tag{7}$$

Substituting

$$\ln(1+x) = x - \tfrac{1}{2} x^2 + 0(x^2) \tag{8}$$

and

$$\frac{x}{1+x} = x - x^2 + 0(x^2) \tag{9}$$

we find

$$\lim_{n \to \infty} R_n = \frac{1}{2} \tag{10}$$

Hence, the series diverges. It does not converge conditionally because all elements are real and positive.

*J.L. Raabe (1801-1859) of Zürich. His work was almost entirely devoted to the question of convergency of series.

Show that the series

$$\sum_{n=1}^{\infty} \frac{(-1)^n}{n} \qquad (1)$$

converges.

Solution: We shall use the following test called Alternating series test.

If $a_n \geq 0$ and $a_{n+1} \leq a_n$ for $n = 1,2,3\ldots$ and $\lim_{n \to \infty} a_n = 0$, then

$$a_1 - a_2 + a_3 - a_4 \ldots = \Sigma(-1)^{n+1} a_n \qquad (2)$$

converges.

In our case $\qquad\qquad a_n = \frac{1}{n} \qquad (3)$

Hence, $a_n \geq 0$, $a_{n+1} \leq a_n$ and $\lim a_n = \lim \frac{1}{n} = 0$.

We have

$$\sum_{n=1}^{\infty} \frac{(-1)^n}{n} = (-1)\Sigma(-1)^{n+1} a_n \qquad (4)$$

Since $\sum_{n=1}^{\infty} (-1)^{n+1} a_n$ converges also $(-1) \sum_{n=1}^{\infty} (-1)^{n+1} a_n$ converges. Thus, $\sum_{n=1}^{\infty} \frac{(-1)^n}{n}$ is a convergent series.

Find the upper and lower limits of the sequences

1. $n - (-1)^{n+1}n \qquad (1)$

2. $\frac{1}{n} + \sin\left(\frac{n\pi}{2}\right) \qquad (2)$

Solution: 1. Let $x_n = n - (-1)^{n+1}n$

Then

$$x_n = \begin{cases} 0 & \text{for } n = 2k+1 \\ 2n & \text{for } n = 2k \end{cases} \qquad (3)$$

Hence,

$$\overline{\lim_{n \to \infty}} \left[n - (-1)^{n+1} n \right] = +\infty \tag{4}$$

$$\underline{\lim_{n \to \infty}} \left[n - (-1)^{n+1} n \right] = 0 \tag{5}$$

2. Let

$$x_n = \frac{1}{n} + \sin\left(\frac{n\pi}{2}\right) \tag{6}$$

Observe that

$$\sin\left(\frac{n\pi}{2}\right) = \begin{cases} 0 & \text{for} \quad n = 2k \\ 1 & \text{for} \quad n = 4k+1 \\ -1 & \text{for} \quad n = 4k+3 \end{cases} \tag{7}$$

The sequence (6) has three accumulation points $0, 1, -1$.

The sequence $\frac{1}{n}$ converges to zero. Thus, term $\frac{1}{n}$ does not change the situation.

$$\overline{\lim_{n \to \infty}} \left[\frac{1}{n} + \sin\left(\frac{n\pi}{2}\right) \right] = 1$$

$$\underline{\lim_{n \to \infty}} \left[\frac{1}{n} + \sin\left(\frac{n\pi}{2}\right) \right] = -1 \tag{8}$$

• **PROBLEM** 8–18

Show that a sequence of real numbers (x_n) has a limit $\lim_{n \to \infty} x_n = a$, if and only if,

$$\overline{\lim_{n \to \infty}} x_n = \underline{\lim_{n \to \infty}} x_n = a \tag{1}$$

We can denote it as

$$\left(\lim_{n \to \infty} x_n = a \right) \Rightarrow \left(\overline{\lim_{n \to \infty}} x_n = \underline{\lim_{n \to \infty}} x_n = a \right) \tag{2}$$

<u>Solution</u>: $\Rightarrow$ The sequence (x_n) is convergent to a. Thus, it has one accumulation point and

$$\overline{\lim} \, x_n = \underline{\lim} \, x_n \tag{3}$$

Because

$$\bigwedge_{\varepsilon > 0} \bigvee_{N} \bigwedge_{n > N} |x_n - a| < \varepsilon \tag{4}$$

we have

486

$$\overline{\lim} \ x_n = \underline{\lim} \ x_n = \lim x_n = a \qquad (5)$$

$<=$

Let us assume that

$$\overline{\lim_{n \to \infty}} \ x_n = \underline{\lim_{n \to \infty}} \ x_n = a \qquad (6)$$

Since the upper and lower limits of (x_n) are equal the sequence has only one accumulation point and

$$\bigwedge_{\varepsilon} \bigvee_{N \in N} \bigwedge_{n > N} \ |x_n - a| < \varepsilon \qquad (7)$$

Hence,

$$\lim_{n \to \infty} x_n = a \qquad (8)$$

● PROBLEM 8–19

Let (z_n) be a sequence of complex numbers such that

$$\overline{\lim_{n \to \infty}} \ \sqrt[n]{|z_n|} = K \qquad (1)$$

Show that if $K < 1$, $\sum_{n=1}^{\infty} z_n$ is absolutely convergent, if $K > 1$,

$\sum_{n=1}^{\infty} z_n$ is divergent.

<u>Solution</u>: Assume that $K < 1$. A positive number α exists such that

$$K < \alpha < 1 \qquad (2)$$

A number N exists such that

$$\bigwedge_{n > N} \ \sqrt[n]{|z_n|} < \alpha \qquad (3)$$

or

$$|z_n| < \alpha^n \qquad (4)$$

The geometric series $\alpha^N + \alpha^{N+1} + \alpha^{N+2} + \ldots$ converges for $\alpha < 1$. Therefore, the series $|z_N| + |z_{N+1}| + \ldots$ also converges

Consequently $\sum_{n=1}^{\infty} |z_n|$ converges and $\sum_{n=1}^{\infty} z_n$ converges absolutely.

For $K > 1$, there exists an infinite number of z_n such that

$$\sqrt[n]{|z_n|} > 1 \tag{5}$$

or

$$|z_n| > 1 \tag{6}$$

Hence, the series is divergent.

Test for absolute convergence, convergence or divergence.

1. $\displaystyle\sum_{n=1}^{\infty} \frac{n}{2} i^{n+1}$ (1)

2. $\displaystyle\sum_{n=1}^{\infty} \frac{n! i^n}{n^n}$ (2)

3. $\displaystyle\sum_{n=1}^{\infty} \left(\frac{1+i}{2}\right)^n$ (3)

Solution: 1. Let us denote

$$a_n = \frac{n}{2} i^{n+1} \tag{4}$$

Observe that $\lim a_n \neq 0$.

A condition for the series $\displaystyle\sum_{n=1}^{\infty} z_n$ to converge is that $\lim z_n = 0$.

Hence, series (1) does not converge.

2. It appears that d'Alembert's test would be appropriate in this case

$$\left|\frac{z_{n+1}}{z_n}\right| = \left|\frac{(n+1)! i^{n+1}}{(n+1)^{n+1}} \cdot \frac{n^n}{n! i^n}\right|$$

$$= \left|\frac{1 \cdot 2 \cdot \ldots \cdot n(n+1) \cdot n^n i}{1 \cdot 2 \cdot \ldots \cdot n(n+1)^n (n+1)}\right|$$

$$= \frac{n^n}{(n+1)^n} = \frac{1}{\left(\frac{n+1}{n}\right)^n} = \frac{1}{\left(1 + \frac{1}{n}\right)^n} \tag{5}$$

$$\lim_{n \to \infty} \left|\frac{z_{n+1}}{z_n}\right| = \lim_{n \to \infty} \frac{1}{\left(1 + \frac{1}{n}\right)^n} = \frac{1}{e} < 1 \tag{6}$$

Hence, the series is absolutely convergent.

3. Applying the root test we find:

$$z_n := \left(\frac{1+i}{2}\right)^n \tag{7}$$

$$\lim_{n\to\infty} \sqrt[n]{\left|\frac{1+i}{2}\right|^n} = \lim_{n\to\infty} \left|\frac{1+i}{2}\right| = \sqrt{\tfrac{1}{4}+\tfrac{1}{4}} < 1 \tag{8}$$

Hence, the series is absolutely convergent.

MAPPINGS AND CURVES

• PROBLEM 8–21

T maps a set X onto a set Y. Let Y_1 and Y_2 be two disjoint subsets of Y, and let X_1 and X_2 be their respective inverse images under T. Prove X_1 and X_2 are disjoint.

Solution: Definition

Let T be the mapping of X onto Y, T : X → Y. The set

$$T(A) := \{T(x) : x \in A \subset X\} \tag{1}$$

is called the image of a set $A \subset X$.

For any $B \subset Y$, the set

$$T^{-1}(B) := \{x : T(x) \in B\} \tag{2}$$

is called the inverse image of the set B.

To solve the problem, assume that X_1 and X_2 are not disjoint, and that $a \in X_1 \cap X_2$. Then

$$T(a) = b \tag{3}$$

and $b \in Y_1$ and $b \in Y_2$.

This contradicts the hypothesis that Y_1 and Y_2 are disjoint.

> Prove that a composition of continuous mappings is a continuous mapping.

Solution: Let $T_1 : X \to Y$ and $T_2 : Y \to W$, where T_1 and T_2 are continuous. We have

$$T_2 \circ T_1 : X \to W. \qquad (1)$$

However,

$$(T_2 \rho T_1)^{-1}(A) = T_1^{-1}(T_2^{-1}(A)) \qquad (2)$$

where $A \subset W$.

When A is an open set, then $T_2^{-1}(A)$ is an open set in Y. Hence,

$$T_1^{-1}(T_2^{-1}(A)) \text{ is open in X.}$$

> Let $f, g : X \to C$ be continuous functions on X, where C is the set of complex numbers. Show that
>
> 1. $f + g$ (1)
>
> 2. $f - g$ (2)
>
> 3. fg (3)
>
> 4. $\dfrac{f}{g}$ (4)
>
> are continuous functions whenever g does not vanish.

Solution: 1. Let $x_n \to x_0$ and $y_n \to y_0$. Applying the theorem of the sum of the limits, we find

$$\lim(x_n + y_n) = x_0 + y_0. \qquad (5)$$

Since f and g are continuous, we have

$$f(x_n) \to f(x_0)$$

$$g(x_n) \to g(x_0). \qquad (6)$$

Then

$$(f+g)(x_n) = f(x_n) + g(x_n) \to f(x_0) + g(x_0) \qquad (7)$$

$$= (f+g)(x_0).$$

Hence, $f + g$ is continuous.

2. In the same way we can show that f - g is continuous.

3. Using the theorem of the product of the limits of sequences,

if
$$\lim x_n = x_0$$

$$\lim y_n = y_0 \tag{8}$$

then

$$\lim x_n y_n = x_0 y_0 \tag{9}$$

Therefore,

$$\lim (fg)(x_n) = (fg)(x_0). \tag{10}$$

Hence, fg is a continuous function.

4. In the same way, we can show that $\frac{f}{g}$ is continuous whenever $g \neq 0$.

● **PROBLEM 8–24**

Consider the function

$$f : C \to C \tag{1}$$

where
$$f(z) = z^2 + 1. \tag{2}$$

Is this function a homomorphism?

Solution: We shall start with the following definition:

If T is a continuous one-to-one and onto mapping of A to B, such that the inverse T^{-1} mapping of B onto A is also continuous, then T is said to be a homomorphism or a topological mapping of A onto B. The sets A and B are said to be homomorphic.

The function f given by (2) is continuous, but it is not one-to-one. For example,

$$f(1+i) = f(-1-i). \tag{3}$$

Therefore, the inverse function does not exist, and $f(z) = z^2 + 1$ is not a homomorphism.

● **PROBLEM 8–25**

Show that if

$$T_1 : X \to Y \tag{1}$$

$$T_2 : Y \to W$$

where T_1 and T_2 are homomorphisms, then $T_2 \circ T_1 : X \to W$ is also a homomorphism.

Solution: Since both T_1 and T_2 are one-to-one and onto, the product $T_2 \circ T_1$ is also one-to-one and onto. Because both mappings T_1 and T_2 are continuous, $T_2 \circ T_1$ is also continuous (see Problem 6-5). The inverse mapping $(T_2 \circ T_1)^{-1}$ exists, and

$$(T_2 \circ T_1)^{-1} = T_1^{-1} \circ T_2^{-1} \tag{2}$$

Since both T_1^{-1} and T_2^{-1} are continuous, their product $T_1^{-1} \circ T_2^{-1}$ is a continuous mapping.

Hence, both $T_2 \circ T_1$ and $(T_2 \circ T_1)^{-1}$ are continuous.

Thus, $T_2 \circ T_1$ is a homomorphism.

• PROBLEM 8-26

Prove the following theorem:

$$\left. \begin{array}{l} T : X \to Y \\ T \text{ is one-to-one and onto} \\ T \text{ is continuous on } X \\ X \text{ is compact} \end{array} \right\} \Rightarrow \left(\begin{array}{l} T^{-1} : Y \to X \\ T^{-1} \text{ is continu-} \\ \text{ous on } Y \end{array} \right)$$

Solution: Since $T : X \to Y$ is one-to-one and onto, the inverse mapping exists. Let (y_n) be a sequence in Y, such that

$$y_n \to y_0 = T(x_0) \tag{1}$$

Since T is onto and one-to-one,

$$y_n = T(x_n), \quad x_n \in X \tag{2}$$

$$x_n = T^{-1}(T(x_n)). \tag{3}$$

Assume

$$x_n \not\to x_0 \tag{4}$$

Then, either

$$x_n \to x' \neq x_0 \tag{5}$$

or (x_n) is not convergent.

The first possibility is incompatible with the continuity of T because

$$T(x_n) \rightarrow T(x_0) \neq T(x') \qquad (6)$$

If (x_n) is not convergent, we still can select a subsequence convergent to $x \in X$, where $x \neq x_0$, which is incompatible with the continuity of T.

● **PROBLEM** 8–27

Let $I = [t_1, t_2] \subset R'$ be a closed interval. Let C be a continuous mapping of I onto the complex plane. C is said to represent a continuous curve, or path in the plane. We denote

$$z = C(t) \qquad (1)$$

or

$$z = z(t). \qquad (2)$$

Give the "mechanical" interpretation of a curve.

Find the curve for which

$$I = [0,1] \qquad (3)$$

$$z(t) = \cos 2\pi t + i \sin 2\pi t \qquad (4)$$

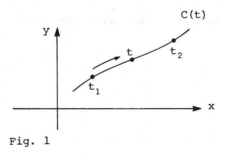

Fig. 1

Solution: Assume that a point z moves continuously in the plane during the time from t_1 to t_2. At each instant $t_1 \leq t \leq t_2$, the point occupies a position in the plane determined by $z = z(t)$. Hence, the mapping $z = z(t)$ gives the motion of the point during the time interval $I = [t_1, t_2]$.

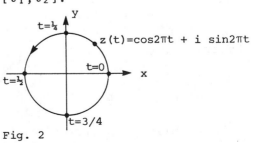

Fig. 2

For t = 0, we have

$$z(0) = \cos 0 + i \sin 0 = 1 \qquad (5)$$

For $t = \frac{1}{4}$,

$$z(\tfrac{1}{4}) = \cos \frac{\pi}{2} + i \sin \frac{\pi}{2} = i$$

$$\qquad (6)$$

$$t = \tfrac{1}{2}, \quad z(\tfrac{1}{2}) = \cos \pi + i \sin \pi = -1$$

$$t = \frac{3}{4}, \quad z = -i$$

$$t = 1, \quad z = 1$$

The point moves once around a unit circle, counterclockwise.

In most cases, the curve $z = z(t)$ has a piecewise smooth parametrization. Italian mathematician Peano found a continuous mapping T

$$T : [0,1] \to [0,1] \times [0,1]$$

i.e., a mapping (continuous) of an interval [0,1] onto a unit square. In Peano's example, the coordinates $x(t)$ and $y(t)$ were continuous but not differentiable functions. $(x(t),y(t))$ is a "space-filling" curve.

• PROBLEM 8–28

Consider an arc of the continuous curve containing the origin

$$y = \begin{cases} x \sin \dfrac{\pi}{x} & \text{for } x \neq 0 \\ 0 & \text{for } x = 0 \end{cases} \qquad (1)$$

Show that this curve is not rectifiable.

Solution: A continuous curve which has no multiple points is called a Jordan* arc.

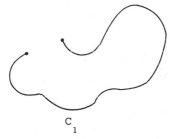

C_1

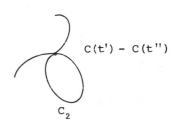

$c(t') - c(t'')$

C_2

Fig. 1

C_1 is a Jordan arc. C_2 is not because two different $t' \neq t''$ are mapped onto the same point of the curve

$$C(t') = C(t'') \qquad (2)$$

This is a multiple point.

If a Jordan arc has a definite length, the arc is said to be rectifiable, and is then called a path segment.

Let $z = z(t)$, $t \in [a,b]$ be a continuous curve. We divide the interval $[a,b]$ into n parts
$$a = t_1 < t_2 < t_3 \ldots < t_n = b \qquad (3)$$

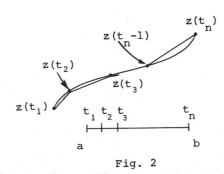

Fig. 2

The points $z(t_1), z(t_2), \ldots, z(t_n)$ are marked on the curve and connected in order by the straight line segments. If the set of the lengths of all such segments is bounded, then the curve is rectifiable. The length of the curve is defined as the least upper bound of this set.

We shall show that a segmental arc of arbitrarily large length can be inscribed in an arc of the curve containing the origin.

Let P_k be the points on the curve, such that

$$y = 0, \quad x = \frac{1}{k} \qquad (4)$$

and Q_k be the points on the curve such that

$$x = \frac{2}{2k-1}, \quad |y| = x \qquad (5)$$

Hence, for Q_k,

$$x = \frac{2}{2k-1}$$

and
$$y = \frac{2}{2k-1} \sin\left[\frac{\pi}{2} \cdot (2k-1)\right] \qquad (6)$$

such that
$$|y| = x.$$

Let the inscribed path be $P_{k-1}Q_kP_k$, for $k = 2,3,4,\ldots$.

The length of $P_{k-1}Q_kP_k$ for any given k exceeds the double ordinate of Q_k.

495

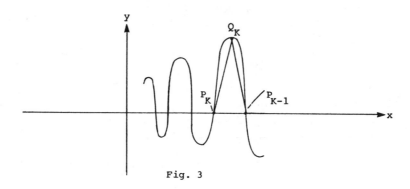

Fig. 3

Hence,

$$\text{Length of } P_{k-1}Q_kP_k > \frac{2}{k} > \frac{1}{k} \tag{7}$$

The series $\sum\limits_{1}^{\infty} \frac{1}{k}$ diverges, thus

$$\underset{M \in R'}{\Lambda} \quad \underset{k' \in N}{V} \quad \underset{k > k'}{\Lambda} \quad \begin{array}{l}\text{length of the}\\ \text{inscribed path}\end{array} > M \tag{8}$$

*M.E.C. Jordan (1838-1922) was a French mathematician. His work was mainly concerned with finite discontinuous substitution groups.

• **PROBLEM** 8–29

{D_ω} is a family of domains, such that

$$\underset{\omega}{\cap} D_\omega = A \tag{1}$$

where A is a non-empty set

Show that

$$\underset{\omega}{\cup} D_\omega \tag{2}$$

is a domain.

Solution: Since D_ω are domains, they are open and connected.

The union of any family of open sets is an open set. Hence,

$$\underset{\omega}{\cup} D_\omega$$

is an open set. We will show that $\underset{\omega}{\cup} D_\omega$ is path-connected. The intersection of {D_ω} is a non-empty set. Assume that

$$\underset{\omega}{\cap} D_\omega = z_0 \tag{3}$$

Taking any two elements of the family $\{D_\omega\}$, for example, D_ω and $D_{\omega'}$, we see that

$$D_\omega \cup D_{\omega'} \tag{4}$$

is path-connected. Therefore, D_ω and $D_{\omega'}$ are path-connected and are not disjoined. Hence, $D_\omega \cup D_{\omega'}$ is path-connected. We conclude that any two points in $\underset{\omega}{\cup} D_\omega$ can be joined by a polygonal line lying in $\underset{\omega}{\cup} D_\omega$. Therefore, $\underset{\omega}{\cup} D_\omega$ is a domain.

• **PROBLEM 8–30**

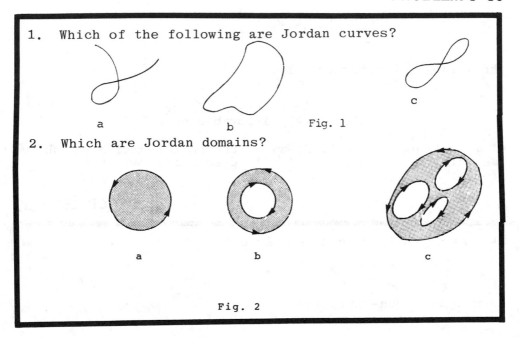

1. Which of the following are Jordan curves?

a   b   Fig. 1   c

2. Which are Jordan domains?

a   b   c

Fig. 2

Solution: 1. A Jordan curve is a simple closed curve (piecewise-smooth). Thus, it is a loop which does not cross itself.

a. is not closed

b. is a Jordan curve

c. crosses itself

The following theorem establishes an important property of Jordan curves:

A Jordan curve decomposes the plane into two separated regions, one lying inside the curve and the other outside the curve.

Intuitively, it is an obvious theorem.

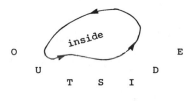

Fig. 3

A rigorous complete proof is difficult (Jordan's proof was not correct), and requires advanced topology.

Note that if the interior lies to the left of the curve, then the curve has positive orientation. Otherwise the orientation is negative.

2. A Jordan domain is a bounded domain D, whose boundary ∂D is the union of a finite number of disjoint Jordan curves, with each curve parametrized so that the boundary curves are positively oriented. a, b and c are Jordan domains.

LIMITS AND CONTINUITY

• **PROBLEM 8–31**

Express the following functions in the form:

$$f(z) = u(x,y) + i\ v(x,y) \qquad (1)$$

1. $f(z) = z + \bar{z}$

2. $f(z) = z + \dfrac{1}{z}$

3. $f(z) = z^{\frac{1}{2}}$

4. $f(z) = \dfrac{1-z}{1+z}$

Solution: 1. $f(z) = x + iy + x - iy = 2x$ $\qquad (2)$

Hence,

$$f(z) = u(x,y) + i\ v(x,y) = 2x$$

$$u(x,y) = 2x$$

$$v(x,y) = 0$$

2. $\qquad f(z) = z + \dfrac{1}{z} = z + \dfrac{\bar{z}}{z\bar{z}}$

$\qquad (3)$

$$= x + iy + \dfrac{x - iy}{x^2 + y^2}$$

498

$$= x + \frac{x}{x^2+y^2} + i\left(y - \frac{y}{x^2+y^2}\right)$$

$$u(x,y) = x + \frac{x}{x^2+y^2}$$

$$v(x,y) = y - \frac{y}{x^2+y^2}$$

3.
$$f(z) = z^{\frac{1}{2}} = (re^{i\theta})^{\frac{1}{2}} = \sqrt{r}\ e^{\frac{i\theta}{2}} \tag{4}$$

$$= r^{\frac{1}{2}}\left(\cos\frac{\theta}{2} + i\sin\frac{\theta}{2}\right)$$

where
$$x = r\cos\theta \tag{5}$$

$$y = r\sin\theta$$

From eq.(5) we can evaluate

$$r = r(x,y) = \sqrt{x^2+y^2} \tag{6}$$

$$\theta = \theta(x,y) = \text{arc cos}\ \frac{x}{\sqrt{x^2+y^2}}$$

Substituting eq.(6) into eq.(4) we find u and v.

4.
$$f(z) = \frac{1-z}{1+z} = \frac{1-x-iy}{1+x+iy}$$

$$= \frac{(1-x-iy)(1+x-iy)}{(1+x+iy)(1+x-iy)} \tag{7}$$

$$= \frac{1+x-iy-x-x^2+ixy-iy-ixy-y^2}{(1+x)^2+y^2}$$

$$= \frac{1-x^2-y^2}{(1+x)^2+y^2} + i\ \frac{-2y}{(1+x)^2+y^2}$$

$$u(x,y) = \frac{1-x^2-y^2}{(1+x)^2+y^2} \tag{8}$$

$$v(x,y) = \frac{-2y}{(1+x)^2+y^2}$$

Definition

Let $f(z) = \omega$ be a single-valued function defined in a domain D. The limit of the function $f(z)$ as z approaches z_0 equals ω_0, we denote

$$\lim_{z \to z_0} f(z) = \omega_0 \qquad (1)$$

if

$$\underset{\varepsilon > 0}{\Lambda} \; \underset{\delta > 0}{V} \; (|z-z_0| < \delta) \Rightarrow (|f(z)-\omega_0| < \varepsilon) \qquad (2)$$

Show that

$$\lim_{z \to 3i} \frac{z^2+6-iz}{z-3i} = 5i \qquad (3)$$

Solution: Note that the function

$$f(z) = \frac{z^2+6-iz}{z-3i} \qquad (4)$$

is defined everywhere except at $z = 3i$. The definition of the limit of $f(z)$ does not require that $f(z)$ be defined at z_0. All that matters is the behavior of $f(z)$ as $z \to z_0$.

$$f(z) = \frac{z^2+6-iz}{z-3i} = \frac{(z-3i)(z+2i)}{z-3i} = z + 2i \qquad (5)$$

for $z \neq 3i$

Thus, for $z \neq 3i$

$$|f(z)-5i| = |z+2i-5i| = |z-3i| \qquad (6)$$

Taking $0 < \varepsilon = \delta$ we have

$$|z-3i| < \delta \Rightarrow \left| \frac{z^2+6-iz}{z-3i} - 5i \right| < \varepsilon \qquad (7)$$

Hence,

$$\lim_{z \to 3i} \frac{z^2+6-iz}{z-3i} = 5i \qquad (8)$$

In general, let $P(z)$ be

$$P(z) = a_n z^n + a_{n-1} z^{n-1} + \ldots + a_1 z + a_0 \tag{9}$$

then

$$\lim_{z \to z_0} P(z) = P(z_0) \tag{10}$$

• PROBLEM 8–33

Show that if a limit of a function $f(z)$ exists, then it is unique.

Solution: Assume the opposite

$$\lim_{z \to z_0} f(z) = \omega_1 \tag{1}$$

and

$$\lim_{z \to z_0} f(z) = \omega_2 \tag{2}$$

Then for any $\varepsilon > 0$ there exist $\delta_1 > 0$ and $\delta_2 > 0$ such that

$$0 < |z - z_0| < \delta_1 \implies |f(z) - \omega_1| < \frac{\varepsilon}{2} \tag{3}$$

and
$$0 < |z - z_0| < \delta_2 \implies |f(z) - \omega_2| < \frac{\varepsilon}{2} \tag{4}$$

Taking
$$\delta = \min(\delta_1, \delta_2) \tag{5}$$

we obtain

$$|\omega_2 - \omega_1| = \left| \left[f(z) - \omega_1 \right] - \left[f(z) - \omega_2 \right] \right| \tag{6}$$

$$\leq |f(z) - \omega_1| + |f(z) - \omega_2| < \frac{\varepsilon}{2} + \frac{\varepsilon}{2} = \varepsilon$$

for $\qquad 0 < |z - z_0| < \delta \tag{7}$

Since ε is an arbitrarily small number, it follows that
$$|\omega_1 - \omega_2| = 0 \tag{8}$$

or
$$\omega_1 = \omega_2$$

That contradicts our assumption.

Evaluate

$$\lim_{z \to 2i} \frac{(3z+1)(z-1)}{z^2+z-1} \qquad (1)$$

using theorems on limits. Show each step.

Solution: We have

$$\lim_{z \to 2i} \frac{(3z+1)(z-1)}{z^2+z-1} = \frac{\lim_{z \to 2i} (3z+1)(z-1)}{\lim_{z \to 2i} (z^2+z-1)}$$

$$= \frac{\lim_{z \to 2i} (3z+1) \lim_{z \to 2i} (z-1)}{\lim_{z \to 2i} z^2 + \lim_{z \to 2i} z - \lim_{z \to 2i} 1}$$

$$= \frac{\left[(3 \lim_{z \to 2i} z) + \lim_{z \to 2i} 1\right]\left[\lim_{z \to 2i} z - \lim_{z \to 2i} 1\right]}{\lim_{z \to 2i} z \cdot \lim_{z \to 2i} z + \lim_{z \to 2i} z - \lim_{z \to 2i} 1}$$

$$(2)$$

$$= \frac{[3 \cdot 2i + 1][2i - 1]}{2i \cdot 2i + 2i - 1}$$

$$= \frac{-12 - 6i + 2i - 1}{-4 - 1 + 2i} = \frac{-13 - 4i}{-5 + 2i}$$

Show that the limit

$$\lim_{z \to 0} \frac{\overline{z}}{z} \qquad (1)$$

does not exist.

Solution: The limit, if it exists, must be independent of the manner in which z approaches the point 0. Let z → 0 along the y-axis. Then,

$$x = 0 \quad \text{and} \quad z = iy, \quad \overline{z} = -iy \qquad (2)$$

so that the limit is

$$\lim_{z \to 0} \frac{\overline{z}}{z} = \lim_{y \to 0} \frac{-iy}{iy} = -1 \qquad (3)$$

Now, let z → 0 along the x-axis. Then,

$$y = 0, \quad z = x, \quad \overline{z} = x \qquad (4)$$

and the limit is

$$\lim_{z \to 0} \frac{\bar{z}}{z} = \lim_{x \to 0} \frac{x}{x} = 1 \qquad (5)$$

Let $f(z)$ be defined on a set A and let z_0 be the accumulation point of A.

Prove that if

$$\lim_{\substack{\alpha \to z_0 \\ \beta \to z_0}} \left[f(\alpha) - f(\beta) \right] = 0 \qquad (1)$$

then, $\lim_{z \to z_0} f(z)$ exists.

Solution: Let $z_n \to z_0$ be a sequence of points in A.

$(f(z_n))$ is a Cauchy sequence, thus it has a limit. We have

$$f(z_n) \to \omega \qquad (2)$$

and

$$\lim_{\substack{\alpha \to z_0 \\ \beta \to z_0}} \left[f(\alpha) - f(\beta) \right] = 0 \qquad (3)$$

Therefore,

$$\bigwedge_{\varepsilon > 0} \quad \bigvee_{N_1 \in \mathbb{N}} \quad \bigvee_{K(z_0, \varepsilon)} \quad \bigwedge_{n > N_1} \text{ we have}$$

$$\begin{aligned}
|f(z) - \omega| &= |f(z) - f(z_n) + f(z_n) - \omega| \\
&\leq |f(z) - f(z_n)| + |f(z_n) - \omega| < \frac{\varepsilon}{2} + \frac{\varepsilon}{2} = \varepsilon
\end{aligned} \qquad (4)$$

Thus,

$$\lim_{z \to z_0} f(z) = \omega \qquad (5)$$

Show that the function

$$\omega = z^n \qquad n = 1, 2, 3, \ldots \qquad (1)$$

is continuous in C.

Solution: Earlier in this book, we discussed in detail continuous mappings on metric spaces. All the definitions and theorems can be applied to the space of complex numbers. Here, we shall only repeat two definitions of continuity. They are equivalent.

Definition I

$f(z)$ is continuous at z_0 when for every $z \to z_0$

$$\lim_{z \to z_0} f(z) = f(z_0) \tag{2}$$

Definition II

$f(z)$ is continuous at z_0 if

$$\underset{\varepsilon > 0}{\wedge} \quad \underset{\delta > 0}{\vee} \quad |z-z_0| < \delta \implies |f(z)-f(z_0)| < \varepsilon \tag{3}$$

Remember that $\delta = \delta(\varepsilon)$.

If $f(z)$ is continuous at every point z_0 of the domain D, we say that $f(z)$ is continuous in D.

Let

$$\omega_0 = z_0^n \quad \text{where } z_0 \in C \tag{4}$$

then

$$\omega - \omega_0 = z^n - z_0^n = (z-z_0)(z^{n-1} + z^{n-2}z_0 + z^{n-3}z_0^2 + \ldots + z_0^{n-1}) \tag{5}$$

Thus

$$|\omega - \omega_0| = |z-z_0| \, |z^{n-1} + z^{n-2}z_0 + \ldots + z_0^{n-1}|$$
$$\leq |z-z_0| (r^{n-1} + r^{n-2}r_0 + \ldots + r_0^{n-1}) \tag{6}$$

where

$$|z| = r, \quad |z_0| = r_0 \tag{7}$$

If $|z-z_0| < \delta$ then,

$$r = |z| = |z_0 + (z-z_0)| \leq |z_0| + |z-z_0| < r_0 + \delta \tag{8}$$

Therefore,

$$|\omega - \omega_0| < |z-z_0| \left[(r_0+\delta)^{n-1} + (r_0+\delta)^{n-2}r_0 + \ldots + r_0^{n-1} \right]$$
$$< n\delta(r_0+\delta)^{n-1} \tag{9}$$

Setting

$$\varepsilon = n\delta(r_0+\delta)^{n-1} \tag{10}$$

we complete the proof that $\omega = z^n$ is continuous in C.

If
$$\lim_{z \to z_0} f(z) = \alpha \neq 0 \qquad (1)$$

show that there exists $\delta > 0$ such that

$$|f(z)| > \tfrac{1}{2}|\alpha| \qquad (2)$$

for

$$0 < |z-z_0| < \delta \qquad (3)$$

Solution: Setting $\varepsilon = \tfrac{1}{2}|\alpha|$ and applying the definition of continuity we find

$$\lim_{z \to z_0} f(z) = \alpha \qquad (4)$$

Hence $\delta > 0$ exists such that

$$|f(z)-\alpha| < \varepsilon = \tfrac{1}{2}|\alpha| \qquad (5)$$

whenever

$$0 < |z-z_0| < \delta$$

Obviously

$$\alpha = \alpha - f(z) + f(z) \qquad (6)$$

then

or

$$|\alpha| \leq |\alpha-f(z)|+|f(z)| < \tfrac{1}{2}|\alpha|+|f(z)| \qquad (7)$$

$$\tfrac{1}{2}|\alpha| < |f(z)|$$

In the same manner, we can show that if a function $f(z)$ is continuous at z_0 in some domain D and $f(z) \neq 0$, then there exists some neighborhood of z_0 throughout which $f(z) \neq 0$.

Prove that
$$f(z) = \frac{1}{1-z} \qquad (1)$$

is not uniformly continuous for $|z| < 1$.

Solution: Suppose $f(z)$ is uniformly continuous. We select $\varepsilon = \frac{1}{10}$. Then $0 < \delta < 1$ can be obtained.

Now, we select z_1 and z_2 such that

$$z_1 = 1 - \delta \tag{2}$$

$$z_2 = 1 - \frac{9}{10} \delta$$

$$|z_1| < 1, \quad |z_2| < 1 \tag{3}$$

We have

$$|z_1 - z_2| = |(1-\delta) - (1 - \frac{9}{10} \delta)| = \frac{\delta}{10} < \delta \tag{4}$$

However,

$$|f(z_1) - f(z_2)| = \left| \frac{1}{1-z_1} - \frac{1}{1-z_2} \right|$$

$$= \left| \frac{1}{1-(1-\delta)} - \frac{1}{1-(1 - \frac{9}{10} \delta)} \right| \tag{5}$$

$$= \left| \frac{1}{\delta} - \frac{1}{\frac{9}{10} \delta} \right| = \frac{1}{9\delta} > \frac{1}{10} = \varepsilon$$

Thus, $f(z) = \frac{1}{1-z}$ is not uniformly continuous in the domain D defined by $|z| < 1$.

• PROBLEM 8–40

Prove that if $f(z)$ is uniformly continuous in a domain D, then $f(z)$ is continuous in D.

Solution: If $f(z)$ is uniformly continuous in D, then

$$\underset{\varepsilon > 0}{\Lambda} \quad \underset{\delta > 0}{V} \quad \underset{z_1, z_2 \in D}{\Lambda} \quad |z_1 - z_2| < \delta \text{ implies that}$$

$$|f(z_1) - f(z_2)| < \varepsilon$$

Hence,

$$\lim_{z \to z_1} f(z) = f(z_1)$$

for any $z_1 \in D$. Function $f(z)$ is continuous.

We see that the condition of uniform continuity is stronger than condition of continuity. There are functions, which are continuous but not uniformly continuous.

• PROBLEM 8–41

Let A be a compact set and $\rho(z)$ be the minimum distance from z to A.

If $z \in A$ then $\rho(z) = 0$.

Show that $\rho(z)$ is a continuous function.

Solution: Let z_1 and z_2 be any two points in the plane and z' and z'' be any two points in A such that

$$\rho(z_1) = |z_1-z'|$$
$$\rho(z_2) = |z_2-z''| \tag{1}$$

as shown in Fig. 1.

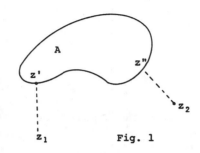

Fig. 1

Then

$$\rho(z_2) \leq |z_2-z'| \tag{2}$$

Hence

$$\rho(z_2)-\rho(z_1) \leq |z_2-z'|-|z_1-z'| \leq |z_2-z_1| \tag{3}$$

In the same way, we show

$$\rho(z_1)-\rho(z_2) \leq |z_1-z_2| \tag{4}$$

Hence

$$|\rho(z_1)-\rho(z_2)| \leq |z_1-z_2| \tag{5}$$

Thus

$$\underset{\varepsilon>0}{\Lambda} \ \underset{\delta>0}{V} \ \ |z_1-z_2| < \delta \Rightarrow |\rho(z_1)-\rho(z_2)| < \varepsilon \tag{6}$$

Setting $\varepsilon = \delta$ we show that $\rho(z)$ is continuous.

DERIVATIVES

Show that

1. $\dfrac{da}{dz} = 0$ (1)

2. $\dfrac{dz}{dz} = 1$ (2)

3. $\dfrac{dz^n}{dz} = nz^{n-1}$ (3)

4. $\dfrac{d\bar{z}}{dz}$ does not exist.

__Solution:__ 1. From the definition of the derivative we have

$$\frac{da}{dz} = \lim_{z \to z_0} \frac{a-a}{z-z_0} = 0 \tag{4}$$

2. Similarly,

$$\frac{dz}{dz} = \lim_{z \to z_0} \frac{z-z_0}{z-z_0} = 1 \tag{5}$$

3. Observe

$$z^n = z \cdot z^{n-1} \tag{6}$$

$$\frac{dz^n}{dz} = \frac{d}{dz}(z \cdot z^{n-1}) = \frac{dz}{dz} z^{n-1} + z \frac{dz^{n-1}}{dz}$$

$$= z^{n-1} + z \frac{dz^{n-1}}{dz} \tag{7}$$

$$= z^{n-1} + z \frac{d}{dz}(z \cdot z^{n-2}) = z^{n-1} + z^{n-1} + z^2 \frac{dz^{n-2}}{dz}$$

$$= 2 z^{n-1} + z^2 \frac{dz^{n-2}}{dz} = \cdots$$

$$= (n-1)z^{n-1} + z^{n-1} \frac{dz}{dz} = nz^{n-1}$$

4. $\dfrac{d\bar{z}}{dz} = \lim\limits_{z \to z_0} \dfrac{\bar{z}-\bar{z}_0}{z-z_0} = \lim\limits_{\Delta z \to 0} \dfrac{\overline{\Delta z}}{\Delta z}$ (8)

This limit does not exist.

If $\Delta z = \Delta x$, then

$$\frac{d\bar{z}}{dz} = 1 \qquad (9)$$

On the other hand, if

$$\Delta z = i\Delta y \qquad (10)$$

then

$$\frac{d\bar{z}}{dz} = -1 \qquad (11)$$

Hence, $\frac{d\bar{z}}{dz}$ does not exist.

● **PROBLEM 8–43**

Compute $\frac{df}{dz}$ for the following functions:

1. $f(z) = z^2 + 4$ $\qquad (1)$

2. $f(z) = 3z^3 + (z+1)^2 + z$ $\qquad (2)$

3. $f(z) = \frac{z-1}{z+1}$ $\qquad (3)$

4. $f(z) = (z + \frac{1}{z})^2$ $\qquad (4)$

<u>Solution</u>: 1.

$$\frac{d}{dz}(z^2+4) = \frac{dz^2}{dz} + \frac{d}{dz}(4)$$

$$= 2z \qquad (5)$$

2.

$$f(z) = 3z^3 + (z+1)^2 + z$$

$$= 3z^3 + z^2 + 3z + 1 \qquad (6)$$

Hence,

$$\frac{df}{dz} = \frac{d}{dz}(3z^3+z^2+3z+1)$$

$$= 9z^2 + 2z + 3 \qquad (7)$$

3. From the definition of the derivative

$$f'(z) = \lim_{\Delta z \to 0} \frac{f(z+\Delta z)-f(z)}{\Delta z}$$

$$= \lim_{\Delta z \to 0} \frac{\frac{z+\Delta z-1}{z+\Delta z+1} - \frac{z-1}{z+1}}{\Delta z} \qquad (8)$$

$$= \lim_{\Delta z \to 0} \frac{(z+\Delta z-1)(z+1)-(z-1)(z+\Delta z+1)}{(z+\Delta z+1)(z+1)\Delta z}$$

$$= \lim_{\Delta z \to 0} \frac{2\Delta z}{\Delta z(z+1)(z+\Delta z+1)} = \frac{2}{(z+1)^2}$$

$$\frac{df}{dz} = \frac{d}{dz}\left(\frac{z-1}{z+1}\right) = \frac{\frac{d}{dz}(z-1)\cdot(z+1)-(z-1)\frac{d}{dz}(z+1)}{(z+1)^2}$$

(9)

$$= \frac{z+1 - z+1}{(z+1)^2} = \frac{2}{(z+1)^2}$$

4. $$f(z) = (z + \frac{1}{z})^2 = z^2 + 2z \cdot \frac{1}{z} + \frac{1}{z^2}$$

(10)

$$= z^2 + \frac{1}{z^2} + 2$$

$$\frac{df}{dz} = \frac{d}{dz}(z^2 + \frac{1}{z^2} + 2) = \frac{d}{dz}(z^2) + \frac{d}{dz}(z^{-2}) + \frac{d}{dz}(2)$$

(11)

$$= 2z - 2z^{-3}$$

We applied the formula

$$\frac{d}{dz}(z^n) = nz^{n-1}$$

(12)

• **PROBLEM 8-44**

Let f(z) be defined by

$$f(z) = \begin{cases} \dfrac{x^3+y^3}{x^2+y^2} + i\,\dfrac{y^3-x^3}{x^2+y^2} & \text{for } x^2+y^2 \neq 0 \\[3mm] 0 & \text{for } x^2+y^2 = 0 \end{cases}$$

(1)

Show that Cauchy-Riemann conditions are satisfied at the origin, but f'(0) does not exist.

<u>Solution</u>:   We have

$$u(x,y) = \frac{x^3+y^3}{x^2+y^2}$$

(2)

$$v(x,y) = \frac{y^3-x^3}{x^2+y^2}$$

(3)

and

$$\left.\frac{\partial u}{\partial x}\right|_{(x,y)=(0,0)} = \lim_{x \to 0} \frac{u(x,0)-u(0,0)}{x} = \lim_{x \to 0} \frac{x^3}{x^3} = 1$$

(4)

$$\left.\frac{\partial u}{\partial y}\right|_{(x,y)=(0,0)} = \lim_{y \to 0} \frac{u(0,y)-u(0,0)}{y} = \lim_{y \to 0} \frac{y^3}{y^3} = 1$$

(5)

$$\frac{\partial v}{\partial x}\bigg|_{(x,y)=(0,0)} = \lim_{x \to 0} \frac{v(x,0)-v(0,0)}{x} = \lim_{x \to 0} \frac{-x^3}{x^3} = -1 \qquad (6)$$

$$\frac{\partial v}{\partial y}\bigg|_{(x,y)=(0,0)} = \lim_{y \to 0} \frac{v(0,y)-v(0,0)}{y} = \lim_{y \to 0} \frac{y^3}{y^3} = 1 \qquad (7)$$

All partial derivatives exist and the Cauchy-Riemann equations are satisfied.

$$\frac{\partial u}{\partial x} = 1 = \frac{\partial v}{\partial y} = 1 \qquad (8)$$

$$\frac{\partial u}{\partial y} = 1 = -\frac{\partial v}{\partial x} = 1 \qquad (9)$$

We can suspect function f(z) of being analytic at z = 0.

Compute f'(0), when z approaches 0 along the x axis, then y = 0 and

$$f'(0) = u_x(0,0) + iv_x(0,0) = 1 - i \qquad (10)$$

When z → 0 along y = x line

$$f(z) = \frac{x^3+x^3}{x^2+x^2} = x \qquad (11)$$

and

$$f'(0) = \lim_{z \to 0} \frac{f(z)-f(0)}{z-0} = \lim_{x \to 0} \frac{x}{x} = 1 \qquad (12)$$

In both cases we obtain different values for f'(0). Hence, f'(0) does not exist and the function f(z) is not analytic at z = 0.

● **PROBLEM 8–45**

Which of the functions are analytic?

1. $f(z) = z^3 + 3i$

2. $f(z) = (\overline{z}+2i)^2 - 1$

3. $f(z) = \dfrac{z+1}{z+4}$

4. $f(z) = \dfrac{z+i}{\overline{z}-i}$

Solution: In the definition of an analytic function, the re-quirement of continuity of the derivative is sometimes included. Actually, that is not necessary.

By theorem of Goursat (sometimes called the Cauchy-Goursat Theorem), if a complex derivative exists, then it is continuous. Therefore, both definitions of analytic functions are equivalent.

1. $f(z) = u+iv = (x+iy)^3 + 3i$
$$= (x^3-3xy^2) + i(3x^2y-y^3+3)$$ (1)

and

$$\frac{\partial u}{\partial x} = 3x^2-3y^2 = \frac{\partial v}{\partial y} = 3x^2-3y^2$$ (2)

$$\frac{\partial u}{\partial y} = -6xy = -\frac{\partial v}{\partial x} = -6xy$$ (3)

$f(z) = z^3+3i$ is analytic everywhere.

2. $f(z) = u+iv = (x-iy+2i)^2 - 1$
$$= (x^2-y^2+4y-5) + i(4x-2xy)$$ (4)

$$\frac{\partial u}{\partial x} = 2x$$

$$\frac{\partial v}{\partial y} = -2x$$ (5)

This function is not analytic, because
$$u_x \neq v_y$$ (6)

3. $f(z) = u+iv = \dfrac{x+iy+1}{x+iy+4}$

$$= \frac{(x+1+iy)(x+4-iy)}{(x+4+iy)(x+4-iy)}$$

$$= \frac{x^2+y^2+5x+4}{x^2+8x+16+y^2} + i\,\frac{3y}{x^2+8x+16+y^2}$$ (7)

$$\frac{\partial u}{\partial x} = \frac{(2x+5)(x^2+8x+16+y^2)-(2x+8)(x^2+y^2+5x+4)}{(x^2+8x+16+y^2)^2}$$ (8)

$$\frac{\partial v}{\partial y} = \frac{3(x^2+8x+16+y^2)-3y\cdot 2y}{(x^2+8x+16+y^2)^2}$$ (9)

and
$$\frac{\partial u}{\partial x} = \frac{\partial v}{\partial y}$$ (10)

$$\frac{\partial u}{\partial y} = \frac{2y(x^2+8x+16+y^2)-2y(x^2+y^2+5x+4)}{(x^2+8x+16+y^2)^2}$$ (11)

$$\frac{\partial v}{\partial x} = \frac{-3y\cdot(2x+8)}{(x^2+8x+16+y^2)^2}$$ (12)

$$\frac{\partial u}{\partial x} = -\frac{\partial v}{\partial x}$$ (13)

$f(z) = \dfrac{z+1}{z+4}$ is analytic everywhere except $z = -4$.

4. $f(z) = u+iv = \dfrac{z+i}{z-i}$

$$= \frac{x+iy+i}{x-iy-i} = \frac{(x+iy+i)(x+i(y+1))}{(x-i(y+1))(x+i(y+1))}$$

$$\qquad (14)$$

$$= \frac{x^2-y^2-2y-1}{x^2+y^2+2y+1} + i \frac{2xy+2x}{x^2+y^2+2y+1}$$

$$\frac{\partial u}{\partial x} = \frac{2x(x^2+y^2+2y+1)-2x(x^2-y^2-2y-1)}{(x^2+y^2+2y+1)^2}$$

$$\qquad (15)$$

$$\frac{\partial v}{\partial y} = \frac{2x(x^2+y^2+2y+1)-(2y+2)(2xy+2x)}{(x^2+y^2+2y+1)^2}$$

Since $\frac{\partial u}{\partial x} \neq \frac{\partial v}{\partial y}$, $f(z) = \frac{z+i}{z-i}$ is not analytic.

E. Goursat, a French mathematician, gave a proof of the integral theorem which requires only that $f'(z)$ exists.

In Paris (1895), E. Goursat, together with P.E. Appell, published a fundamental textbook on the theory of functions entitled, Théorie des Functions Algébriques.

• PROBLEM 8–46

Which of the functions are analytic?

1. $f(z) = \sqrt[n]{z}$ $\qquad (1)$

2. $f(z) = \frac{y-ix}{x^2+y^2}$ $\qquad (2)$

3. $f(z) = \frac{x-iy}{x^2+y^2}$ $\qquad (3)$

Solution: 1. Expressing $f(z) = \sqrt[n]{z}$ in polar coordinates,

$$f(z) = \sqrt[n]{z} = \sqrt[n]{r}\left(\cos \frac{\theta}{n} + i \sin \frac{\theta}{n}\right)$$

$$\qquad (4)$$

$$0 < r, \quad 0 \leq \theta < 2\pi$$

and applying the Cauchy-Riemann conditions,

$$\frac{\partial u}{\partial r} = \frac{1}{r}\frac{\partial v}{\partial \theta}, \quad \frac{\partial v}{\partial r} = -\frac{1}{r}\frac{\partial u}{\partial \theta}$$

$$\qquad (5)$$

we find $\frac{\partial u}{\partial r} = \frac{1}{n} r^{\frac{1}{n}-1} \cdot \cos \frac{\theta}{n}$

$$\qquad (6)$$

$$\frac{\partial v}{\partial r} = \frac{1}{n} r^{\frac{1}{n}-1} \cdot \sin \frac{\theta}{n}$$

$$\qquad (7)$$

$$\frac{\partial u}{\partial \theta} = \frac{-1}{n} \sqrt[n]{r} \sin \frac{\theta}{n}$$

$$\qquad (8)$$

$$\frac{\partial v}{\partial \theta} = \frac{1}{n} \sqrt[n]{r} \, \cos \frac{\theta}{n} \qquad (9)$$

we have

$$\frac{\partial u}{\partial r} = \frac{1}{n} r^{\frac{1}{n}-1} \cos \frac{\theta}{n} = \frac{1}{r} \frac{1}{n} \sqrt[n]{r} \, \cos \frac{\theta}{n} = \frac{1}{r} \frac{\partial v}{\partial \theta} \qquad (10)$$

and

$$\frac{\partial v}{\partial r} = \frac{1}{n} r^{\frac{1}{n}-1} \sin \frac{\theta}{n} = -\frac{1}{r} \cdot \frac{-1}{n} \sqrt[n]{r} \, \sin \frac{\theta}{n} = -\frac{1}{r} \frac{\partial u}{\partial \theta} \qquad (11)$$

Hence, $f(z) = \sqrt[n]{z}$ is analytic.

2. $f(z) = \dfrac{y-ix}{x^2+y^2} = \dfrac{r \sin\theta}{r^2} - i \dfrac{r \cos\theta}{r^2} = \dfrac{\sin\theta}{r} - i \dfrac{\cos\theta}{r}$ $\qquad (12)$

Then, $\dfrac{\partial u}{\partial r} = -\dfrac{\sin\theta}{r^2}, \quad \dfrac{\partial v}{\partial r} = \dfrac{\cos\theta}{r^2}$ $\qquad (13)$

$$\frac{\partial u}{\partial \theta} = \frac{\cos\theta}{r} , \quad \frac{\partial v}{\partial \theta} = \frac{\sin\theta}{r} \qquad (14)$$

$$\frac{\partial u}{\partial r} = -\frac{\sin\theta}{r^2} \neq \frac{1}{r} \cdot \frac{\sin\theta}{r} = \frac{1}{r} \frac{\partial v}{\partial \theta} \qquad (15)$$

Thus, $f(z)$ is not analytic.

3. $f(z) = \dfrac{x-iy}{x^2+y^2} = \dfrac{\cos\theta}{r} - i \dfrac{\sin\theta}{r}, \quad r > 0$ $\qquad (16)$

Here, $\dfrac{\partial u}{\partial r} = \dfrac{1}{r} \dfrac{\partial v}{\partial \theta}$

$$\frac{\partial v}{\partial r} = -\frac{1}{r} \frac{\partial u}{\partial \theta} \qquad (17)$$

and $f(z)$ is an analytic function.

• PROBLEM 8–47

The operators ∇(del) and $\overline{\nabla}$ (del bar) are defined by

$$\nabla := \frac{\partial}{\partial x} + i \frac{\partial}{\partial y} \qquad (1)$$

$$\overline{\nabla} := \frac{\partial}{\partial x} - i \frac{\partial}{\partial y} \qquad (2)$$

1. Prove that

$$\nabla = 2 \frac{\partial}{\partial \overline{z}} \qquad (3)$$

$$\overline{\nabla} = 2 \frac{\partial}{\partial z} \qquad (4)$$

2. The gradient of a real function $f(x,y)$ is defined by

$$\text{grad } f := \nabla f := \frac{\partial f}{\partial x} + i \frac{\partial f}{\partial y} \qquad (5)$$

Express grad f in terms of z and $\overline{z}$.

Solution: 1.

$$\nabla = \frac{\partial}{\partial x} + i \frac{\partial}{\partial y} = \left(\frac{\partial}{\partial z} + \frac{\partial}{\partial \overline{z}} \right) + i^2 \left(\frac{\partial}{\partial z} - \frac{\partial}{\partial \overline{z}} \right)$$

$$= 2 \frac{\partial}{\partial \overline{z}}$$

(6)

Similarly,

$$\overline{\nabla} = \frac{\partial}{\partial x} - i \frac{\partial}{\partial y} = \left(\frac{\partial}{\partial z} + \frac{\partial}{\partial \overline{z}} \right) - i^2 \left(\frac{\partial}{\partial z} - \frac{\partial}{\partial \overline{z}} \right)$$

$$= 2 \frac{\partial}{\partial z}$$

(7)

2. From
$$z = x + iy$$
$$\overline{z} = x - iy$$

(8)

we have

$$x = \frac{z + \overline{z}}{2}$$

$$y = \frac{z - \overline{z}}{2i}$$

(9)

Hence,

$$f(x,y) = f\left(\frac{z + \overline{z}}{2}, \frac{z - \overline{z}}{2i} \right) = g(z, \overline{z})$$

(10)

and

$$\text{grad } f = \frac{\partial f}{\partial x} + i \frac{\partial f}{\partial y} = 2 \frac{\partial g}{\partial \overline{z}}$$

(11)

• PROBLEM 8-48

Let $f(z) = u + iv$ be an analytic function. Show that u and v, when expressed in polar form, satisfy the equation

$$\frac{\partial^2 u}{\partial r^2} + \frac{1}{r} \frac{\partial u}{\partial r} + \frac{1}{r^2} \frac{\partial^2 u}{\partial \theta^2} = 0$$

(1)

Solution: Eq.(1) is Laplace's equation in polar form. Since f(z) is analytic, u and v satisfy the Cauchy-Riemann equations in polar form

$$\frac{\partial u}{\partial r} = \frac{1}{r} \frac{\partial v}{\partial \theta}$$

(2)

$$\frac{\partial v}{\partial r} = - \frac{1}{r} \frac{\partial u}{\partial \theta}$$

(3)

Differentiating eq.(2) with respect to r, and eq.(3) with respect to θ, we find

$$\frac{\partial^2 v}{\partial r \partial \theta} = \frac{\partial}{\partial r}\left(r \frac{\partial u}{\partial r}\right) = r \frac{\partial^2 u}{\partial r^2} + \frac{\partial u}{\partial r} \tag{4}$$

$$\frac{\partial^2 v}{\partial \theta \partial r} = \frac{\partial}{\partial \theta}\left(-\frac{1}{r} \frac{\partial u}{\partial \theta}\right) = -\frac{1}{r} \frac{\partial^2 u}{\partial \theta^2} \tag{5}$$

Since the second partial derivatives are continuous

$$\frac{\partial^2 v}{\partial r \partial \theta} = \frac{\partial^2 v}{\partial \theta \partial r} \tag{6}$$

Hence, from eq.(4) and (5)

$$\frac{\partial^2 u}{\partial r^2} + \frac{1}{r} \frac{\partial u}{\partial r} + \frac{1}{r^2} \frac{\partial^2 u}{\partial \theta^2} = 0 \tag{7}$$

Similarly, we can show that $v(r,\theta)$ satisfies the same equation.

● **PROBLEM 8—49**

Let $u(x,y)$ be a function harmonic in a domain D, such that all partial derivatives exist and are continuous in D.

Show that the functions

$$\frac{\partial u}{\partial x}, \frac{\partial u}{\partial y}, \frac{\partial^2 u}{\partial x^2}, \frac{\partial^2 u}{\partial x \partial y}, \text{ etc.}$$

are harmonic in D.

Solution: $u(x,y)$ is harmonic in D

$$\frac{\partial^2 u}{\partial x^2} + \frac{\partial^2 u}{\partial y^2} = 0 \tag{1}$$

We will show that $\frac{\partial u}{\partial x}$ is also harmonic.

$$\frac{\partial^2}{\partial x^2}\left(\frac{\partial u}{\partial x}\right) + \frac{\partial^2}{\partial y^2}\left(\frac{\partial u}{\partial x}\right) = \frac{\partial}{\partial x}\left(\frac{\partial^2 u}{\partial x^2} + \frac{\partial^2 u}{\partial y^2}\right)$$

$$= \frac{\partial}{\partial x}(0) = 0 \tag{2}$$

Note that since all partial derivatives exist and are continuous, we can always change the order of differentiation.

$$\frac{\partial}{\partial x} \frac{\partial}{\partial y} = \frac{\partial}{\partial y} \frac{\partial}{\partial x} \tag{3}$$

Similarly for $\frac{\partial^2 u}{\partial x^2}$

$$\frac{\partial^2}{\partial x^2}\left(\frac{\partial^2 u}{\partial x^2}\right) + \frac{\partial^2}{\partial y^2}\left(\frac{\partial^2 u}{\partial x^2}\right) = \frac{\partial^2}{\partial x^2}\left(\frac{\partial^2 u}{\partial x^2} + \frac{\partial^2 u}{\partial y^2}\right) = 0 \tag{4}$$

In the same way, we can show that any other derivative of u is a harmonic function in D.

Solve the partial differential equation

$$\frac{\partial^2 f}{\partial x^2} + \frac{\partial^2 f}{\partial y^2} = 2x^2 + y^2 \qquad (1)$$

Solution: We shall replace the independent variables (x,y) by $(z,\bar{z})$ where

$$x = \frac{z+\bar{z}}{2} \qquad (2)$$

$$y = \frac{z-\bar{z}}{2i} \qquad (3)$$

Then,

$$2x^2 + y^2 = 2\left(\frac{z+\bar{z}}{2}\right)^2 + \left(\frac{z-\bar{z}}{2i}\right)^2$$

$$= \frac{z^2}{4} + \frac{3}{2} z\bar{z} + \frac{\bar{z}^2}{4} \qquad (4)$$

$$\frac{\partial^2 f}{\partial x^2} + \frac{\partial^2 f}{\partial x^2} = \nabla^2 f = 4\frac{\partial^2 f}{\partial z \partial \bar{z}} \qquad (5)$$

Hence, eq.(1) becomes

$$\frac{\partial^2 f}{\partial z \partial \bar{z}} = \frac{1}{16} z^2 + \frac{3}{8} z\bar{z} + \frac{1}{16} \bar{z}^2 \qquad (6)$$

Integrating (6) with respect to z, we obtain

$$\frac{\partial f}{\partial \bar{z}} = \frac{1}{48} z^3 + \frac{3}{16} z^2\bar{z} + \frac{1}{16} \bar{z}^2 z + F_1(\bar{z}) \qquad (7)$$

where $F_1(\bar{z})$ is an arbitrary function of $\bar{z}$.

Now, integrating (7) with respect to $\bar{z}$,

$$f(z,\bar{z}) = \frac{1}{48} z^3\bar{z} + \frac{3}{32} z^2\bar{z}^2 + \frac{1}{48} \bar{z}^3 z + F_2(z) + F_3(\bar{z}) \qquad (8)$$

where

$$\int F_1(\bar{z})d\bar{z} = F_3(\bar{z}) \qquad (9)$$

and $F_2(z)$ is an arbitrary function of z.

Returning to (x,y) variables we find

$$f(x,y) = \frac{1}{48}(x+iy)^3(x-iy) + \frac{3}{32}(x+iy)^3(x-iy)^2$$

$$\qquad (10)$$

$$+ \frac{1}{48}(x-iy)^3(x+iy) + F_2(x+iy) + F_3(x-iy)$$

Solve the equation

$$L \frac{d^2q}{dt^2} + R \frac{dq}{dt} + \frac{q}{C} = E_0 \cos \omega t \qquad (1)$$

which appears in the theory of alternating electric currents. L, R, C, E_0 and ω are constants. Replace $E_0 \cos \omega t$ by $E_0 e^{i\omega t}$, and assume that $q(t) = \alpha e^{i\omega t}$ is the solution of the "new" equation.

Solution: Substituting

$$q(t) = \alpha e^{i\omega t} \qquad (2)$$

$$\frac{dq}{dt} = i\omega\alpha e^{i\omega t} \qquad (3)$$

$$\frac{d^2q}{dt^2} = -\omega^2\alpha \, e^{i\omega t} \qquad (4)$$

into

$$L \frac{d^2q}{dt^2} + R \frac{dq}{dt} + \frac{q}{C} = E_0 \, e^{i\omega t} \qquad (5)$$

we find

$$-L\alpha\omega^2 \, e^{i\omega t} + iR\alpha\omega \, e^{i\omega t} + \frac{\alpha}{C} \, e^{i\omega t} \qquad (6)$$

$$= E_0 \, e^{i\omega t}$$

Hence,

$$\alpha = \frac{E_0}{-L\omega^2 + iR\omega + \frac{1}{C}} \qquad (7)$$

and the solution of (5) is

$$q = \frac{E_0}{-L\omega^2 + iR\omega + \frac{1}{C}} \, e^{i\omega t} \qquad (8)$$

Taking the real part of (8),

$$Re(q) = Re \left[\frac{E_0 e^{i\omega t}}{-L\omega^2 + \frac{1}{C} + iR\omega} \right] \qquad (9)$$

we find the solution of (1). Note that

$$e^{i\omega t} = \cos \omega t + i \sin \omega t \qquad (10)$$

INTEGRALS

Compute the value of the integral

$$\int_C (z-z_0)^n dz \tag{1}$$

where n is any integer and C is the circle with center at z_0 and radius r described in the counterclockwise direction.

Solution: The equation of the circle C is

$$z = z(t) = z_0 + re^{it}, \quad 0 \le t \le 2\pi. \tag{2}$$

Note that

$$z' = \frac{dz}{dt} = ir\, e^{it} \tag{3}$$

and

$$\int_C (z-z_0)^n dz = ir^{n+1} \int_0^{2\pi} e^{i(n+1)t}\, dt$$

$$= ir^{n+1} \int_0^{2\pi} \left[\cos(n+1)t + i\sin(n+1)t\right] dt. \tag{4}$$

For n = -1, (4) leads to

$$\int_C (z-z_0)^{-1}\, dz = i\int_0^{2\pi} dt = 2\pi i \tag{5}$$

For $n \ne -1$, we find

$$\int_C (z-z_0)^n dz = \frac{ir^{n+1}}{n+1}\left[\sin(n+1)t - i\cos(n+1)t\right]\Big|_{t=0}^{2\pi} = 0. \tag{6}$$

Hence,

$$\int_C (z-z_0)^n dz = \begin{cases} 2\pi i & \text{for } n = -1 \\ 0 & \text{for } n \ne -1. \end{cases} \tag{7}$$

Let C be the curve

$$y = x^3 + 2x^2 - 2x \qquad (1)$$

joining points (0,0) and (1,1). Evaluate

$$\int_C (4z^2 - 2iz)dz. \qquad (2)$$

Solution: We shall solve this problem using two methods,

I. Since integral (2) is independent of the path joining (0,0) and (1,1), we can choose the path shown in Fig. 1.

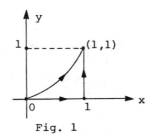

Fig. 1

Thus, $\qquad\qquad\qquad\qquad\qquad\qquad\qquad\qquad\qquad\qquad (3)$

$$\int_C (4z^2-2iz)dz = \int_{(0,0)}^{(1,0)} (4z^2-2iz)dz + \int_{(1,0)}^{(1,1)} (4z^2-2iz)dz$$

Along the path from (0,0) to (1,0) y = 0, dy = 0, dz = dx, z = x.

$$\int_0^1 (4x^2-2ix)dx = \left[\frac{4}{3} x^3 - ix^2 \right]\Big|_0^1$$

$$= \frac{4}{3} - i. \qquad (4)$$

Along the path from (1,0) to (1,1) x = 1, dx = 0, dz = idy, z = 1 + iy.

$$\int_0^1 \left[4(1+iy)^2-2i(1+iy) \right]idy$$

$$= \left[4iy - 4y^2 - \frac{4}{3} y^3 i + 2y + y^2 i \right]\Big|_0^1 \qquad (5)$$

$$= 5i - \frac{4}{3} i - 2.$$

Hence,

$$\int_C (4z^2-2iz)dz = \frac{4}{3} - i + 5i - \frac{4}{3} i - 2$$

$$= -\frac{2}{3} + \frac{8}{3} i. \tag{6}$$

II. The shorter method is shown here.

$$\int_0^{1+i} (4z^2-2iz)dz = \left(\frac{4}{3} z^3 - iz^2\right)\Bigg|_0^{1+i}$$

$$= \frac{4}{3} (1+i)^3 - i(1+i)^2 = -\frac{2}{3} + \frac{8}{3} i \tag{7}$$

● **PROBLEM 8–54**

Evaluate

$$\oint_C \frac{z^2 + \sinh 2z}{z^3-3iz^2+4z-12i} \, dz \tag{1}$$

where C is the circle $|z| = 1$.

Solution: We shall find the singular points of the function

$$f(z) = \frac{z^2 + \sinh 2z}{z^3-3iz^2+4z-12i} \tag{2}$$

The singular points are the solutions of the equation

$$z^3 - 3iz^2 + 4z - 12i = 0. \tag{3}$$

One of the solutions is $z_1 = 3i$. Hence,

$$z^3 - 3iz^2 + 4z - 12i = (z-3i)(z^2+4). \tag{4}$$

Thus, the solutions of eq.(3) are

$$z_1 = 3i, \quad z_2 = 2i, \quad z_3 = -2i \tag{5}$$

These points are not contained within or on C.

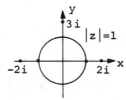

Thus, $f(z)$ is analytic within and on C. Therefore, from the Cauchy theorem we conclude that

521

$$\oint_C \frac{z^2 + \sinh 2z}{z^3 - 3iz^2 + 4z - 12i}\ dz = 0. \tag{6}$$

Evaluate

1. $\displaystyle\int_{z_0}^{z_1} a^z dz$ (1)

2. $\displaystyle\int_{-ki}^{ki} \frac{dz}{z}$,   where k is a positive number   (2)

3. $\displaystyle\int_{z_0}^{z_1} \frac{dz}{z\sqrt{a^2 \pm z^2}}$.   (3)

Solution: 1.   An indefinite integral of the analytic function

$$f(z) = a^z \tag{4}$$

is given by the analytic function

$$F(z) = \frac{a^z}{\ln a}. \tag{5}$$

Thus, for any contour joining z_0 and z_1, we have

$$\int_{z_0}^{z_1} a^z dz = \frac{a^z}{\ln a}\bigg|_{z=z_0}^{z_1} = \frac{a^{z_1} - a^{z_0}}{\ln a}. \tag{6}$$

2.   Integral (2) is evaluated for any curve joining ki and −ki lying in the simply-connected domain D consisting of the complex plane with the nonpositive half of the real axis removed.

An indefinite integral of $f(z) = \frac{1}{z}$, which is analytic in D, is given by the analytic function

$$F(z) = \ln z. \tag{7}$$

Thus,

$$\int_{-ki}^{ki} \frac{dz}{z} = \ln ki - \ln(-ki) = (\ln k + \frac{\pi}{2}i) - (\ln k - \frac{\pi}{2}i)$$

$$= \pi i. \tag{8}$$

3. We have

$$\int \frac{dz}{z\sqrt{a^2 \pm z^2}} = \frac{1}{a} \ln\left(\frac{z}{a+\sqrt{a^2 \pm z^2}}\right) \qquad (9)$$

and

$$\int_{z_0}^{z_1} \frac{dz}{z\sqrt{a^2 \pm z^2}} = \frac{1}{a} \ln \frac{z_1}{a+\sqrt{a^2 \pm z_1^2}} - \frac{1}{a} \ln \frac{z_0}{a+\sqrt{a^2 \pm z_0^2}} \qquad (10)$$

● **PROBLEM 8–56**

Evaluate

$$\oint_C \frac{dz}{1+z^2} \qquad (1)$$

where C is the circle

1. $|z+i| = 1$ (2)

2. $|z-i| = 1$ (3)

3. $|z| = 3.$ (4)

<u>Solution</u>: Function

$$f(z) = \frac{1}{1+z^2} \qquad (5)$$

can be written in the form

$$\frac{1}{z^2+1} = \frac{1}{2i}\left[\frac{1}{z-i} - \frac{1}{z+i}\right]. \qquad (6)$$

1. The circle $|z+i| = 1$ is shown in Fig. 1.

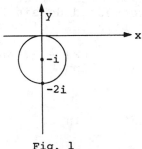

Fig. 1

The singularities of $\frac{1}{z^2+1}$ are located at $z = i$, $z = -i$. Only one singularity $z = -i$ is located inside the circle.

Hence, by Problem 11-20, we obtain

$$\oint_C \frac{1}{z^2+1} \, dz = \frac{1}{2i} \oint_C \frac{1}{z-i} \, dz - \frac{1}{2i} \oint_C \frac{1}{z+i} \, dz$$

$$= -\frac{1}{2i} \oint_C \frac{1}{z+i} \, dz = -\frac{2\pi i}{2i} = -\pi. \tag{7}$$

2. For $|z-i| = 1$, the singularity $z = i$ is inside the circle. Hence,

$$\oint_C \frac{dz}{z^2+1} = \frac{1}{2i} \oint_C \frac{1}{z-i} \, dz = \frac{2\pi i}{2i} = \pi \tag{8}$$

3. The circle $|z| = 3$ contains both singularities

$$\oint_C \frac{dz}{z^2+1} = \frac{1}{2i} \oint_C \frac{1}{z-i} \, dz - \frac{1}{2i} \oint_C \frac{1}{z+i} \, dz \tag{9}$$

$$= \frac{2\pi i}{2i} - \frac{2\pi i}{2i} = 0.$$

● PROBLEM 8–57

Compute the integral

$$\int_C z^2 \, dz \tag{1}$$

where C is the straight-line segment from $z_0 = 0$ to $z_1 = 2 + 2i$.

Solution: We shall use two methods to solve this problem.

I. If C can be parametrized by the variable x, then the points on C have the form

$$z = x + ix = (1+i)x. \tag{2}$$

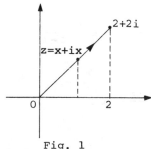

Fig. 1

We find

$$\int_C z^2 dz = \int_{x=0}^{x=2} (1+i)^2 x^2 \, d\left[(1+i)x\right]$$

$$= (1+i)^3 \int_0^2 x^2 dx = (1+i)^3 \cdot \frac{8}{3}. \tag{3}$$

II. In this case, it is easy to find the function $F(z)$ such that

$$F'(z) = z^2 \tag{4}$$

$$F(z) = \frac{z^3}{3}. \tag{5}$$

Then,

$$\int_C z^2 dz = \left(\frac{z^3}{3}\right)\Bigg|_{z_0=0}^{z_1=2+2i} = \frac{(2+2i)^3}{3}$$

$$= \frac{8}{3}(1+i)^3. \tag{6}$$

• **PROBLEM 8–58**

Evaluate

1. $\displaystyle\int_C \frac{\sin\pi(z+1) + \cos\pi z}{(z-1)(z-2)} \, dz$ (1)

   where C is the circle $|z| = 4$.

2. $\displaystyle\int_C \frac{e^{3z}}{(z+2)^{10}} \, dz$ (2)

   where C is the circle $|z| = 3$.

<u>Solution</u>: 1. Note that

$$\frac{1}{(z-1)(z-2)} = \frac{1}{z-2} - \frac{1}{z-1} \tag{3}$$

and we obtain

$$\int_C \frac{\sin\pi(z+1) + \cos\pi z}{(z-1)(z-2)} \, dz$$

$$= \int_C \frac{\sin\pi(z+1) + \cos\pi z}{z-2} \, dz - \int_C \frac{\sin\pi(z+1) + \cos\pi z}{z-1} \, dz. \qquad (4)$$

By Cauchy's integral formula we find

$$\int_C \frac{\sin\pi(z+1) + \cos\pi z}{z-2} \, dz = 2\pi i \left[\sin 3\pi + \cos 2\pi \right]$$

$$= 2\pi i \qquad (5)$$

$$\int_C \frac{\sin\pi(z+1) + \cos\pi z}{z-1} \, dz = 2\pi i (\sin 2\pi + \cos\pi)$$

$$= -2\pi i \qquad (6)$$

Both z=1 and z=2 are inside C and $f(z) = \sin\pi(z+1) + \cos\pi z$ is analytic inside C.

Hence,

$$\int_C \frac{\sin\pi(z+1) + \cos\pi z}{(z-1)(z-2)} \, dz = 2\pi i - (-2\pi i) = 4\pi i \qquad (7)$$

2.   To evaluate $\int_C \frac{e^{3z}}{(z+2)^{10}} \, dz$ where C is the circle $|z| = 3$, we

shall use the expression for the nth derivative of an analytic function

$$f^{(n)}(a) = \frac{n!}{2\pi i} \int_C \frac{f(z)}{(z-a)^{n+1}} \, dz \qquad (8)$$

For $f(z) = e^{3z}$, n = 9 and a = -2 we obtain

$$f^{(9)}(-2) = \frac{9!}{2\pi i} \int_C \frac{e^{3z}}{(z+2)^{10}} \, dz \qquad (9)$$

on the other hand

$$f^{(9)}(z) = 3^9 e^{3z} \qquad (10)$$

Substituting (10) into (9), we obtain

$$3^9 e^{-6} = \frac{9!}{2\pi i} \int_C \frac{e^{3z}}{(z+2)^{10}} \, dz \qquad (11)$$

or

$$\int_C \frac{e^{3z}}{(z+2)^{10}} \, dz = \frac{2\pi i}{9!} \, 3^9 e^{-6} \qquad (12)$$

Prove the formula

$$\int_0^{2\pi} \cos^{2n}\theta\, d\theta = \frac{1\cdot 3\cdot 5\cdot\ldots\cdot(2n-1)}{2\cdot 4\cdot 6\cdot\ldots\cdot(2n)}\, 2\pi \qquad (1)$$

where $n = 1,2,3,4,\ldots$.

Solution: Let $z = e^{i\theta}$, then

$$dz = ie^{i\theta}d\theta = izd\theta \qquad (2)$$

or

$$d\theta = \frac{dz}{iz} \qquad (3)$$

We have

$$\cos\theta = \frac{e^{i\theta} + e^{-i\theta}}{2} = \frac{z + \frac{1}{z}}{2} \qquad (4)$$

Thus, if C is the unit circle $|z| = 1$, Then,

$$\int_0^{2\pi} \cos^{2n}\theta\, d\theta = \int_C \left[\frac{1}{2}\left(z + \frac{1}{z}\right)\right]^{2n} \frac{dz}{iz}$$

$$= \frac{1}{2^{2n}i} \int_C \frac{1}{z}\left[z^{2n} + \binom{2n}{1}z^{2n-1}\frac{1}{z} + \ldots + \binom{2n}{\ell}z^{2n-\ell}\left(\frac{1}{z}\right)^{\ell}\right.$$

$$\left. + \ldots + \left(\frac{1}{z}\right)^{2n}\right]dz \qquad (5)$$

$$= \frac{1}{2^{2n}i} \int_C \left[z^{2n-1} + \binom{2n}{1}z^{2n-3} + \ldots + \binom{2n}{\ell}z^{2n-2\ell-1}\right.$$

$$\left. + \ldots + z^{-2n}\right]dz$$

$$= \frac{1}{2^{2n}i} \cdot 2\pi i \binom{2n}{n} = \frac{1}{2^{2n}} \cdot 2\pi\, \frac{(2n)!}{n!n!}$$

$$= \frac{1\cdot\ldots\cdot n(n-1)\ldots(2n-2)(2n-1)\cdot 2n}{2^{2n}\, n!n!}\, 2\pi$$

$$= \frac{1\cdot 3\cdot 5\cdot\ldots\cdot(2n-1)}{2\cdot 4\cdot 6\cdot\ldots\cdot 2n}\, 2\pi$$

Evaluate

$$\int_C \frac{z^2 - 2z}{(z+1)^2(z^2+4)} \, dz \tag{1}$$

where C is the circle $|z| = 10$.

Solution: The function

$$f(z) = \frac{z^2 - 2z}{(z+1)^2(z^2+4)} \tag{2}$$

has a pole of order two at $z = -1$ and two poles $z = \pm 2i$ of order one.

We shall evaluate the residues. Residue at $z = -1$ is

$$\lim_{z \to -1} \frac{1}{1!} \frac{d}{dz}\left[(z+1)^2 \frac{z^2 - 2z}{(z+1)^2(z^2+4)}\right]$$

$$= \lim_{z \to -1} \frac{(z^2+4)(2z-2)-(z^2-2z)2z}{(z^2+4)^2} = -\frac{14}{25} \tag{3}$$

Residue at $z = 2i$ is

$$\lim_{z \to 2i} \left[(z-2i) \frac{z^2 - 2z}{(z+1)^2(z-2i)(z+2i)}\right]$$

$$= \frac{-4i-4}{4i(2i+1)^2} = \frac{7+i}{25} \tag{4}$$

Remember that $0! = 1$.

Residue at $z = -2i$ is

$$\lim_{z \to -2i} \left[(z+2i) \frac{z^2 - 2z}{(z+1)^2(z-2i)(z+2i)}\right]$$

$$= \frac{4i - 4}{-4i(-2i+1)^2} = \frac{7-i}{25} \tag{5}$$

Since all poles are inside C

$$\int_C \frac{z^2 - 2z}{(z+1)^2(z^2+4)} \, dz = 2\pi i\left(-\frac{14}{25} + \frac{7+i}{25} + \frac{7-i}{25}\right) = 0 \tag{6}$$

Show that all the roots of

$$2z^5 - z^3 + z + 7 = 0 \tag{1}$$

are located between the circles

$$|z| = 1 \quad \text{and} \quad |z| = 2.$$

Solution: First, we shall show that (1) has no solutions inside circle $|z| = 1$. Indeed, let

$$f(z) = 7 \quad \text{and} \quad g(z) = 2z^5 - z^3 + z \tag{2}$$

On $|z| = 1$ we have

$$|g(z)| = |2z^5 - z^3 + z| \leq 2|z^5| + |z^3| + |z|$$
$$= 4 < 7 = |f(z)| \tag{3}$$

Thus, by Rouché's theorem $f(z) + g(z) = 2z^5 - z^3 + z + 7$ has the same number of zeros inside $|z| = 1$ as $f(z) = 7$, that is there are no zeros inside $|z| = 1$.

For the circle $|z| = 2$ let

$$f(z) = 2z^5 \quad \text{and} \quad g(z) = -z^3 + z + 7 \tag{4}$$

Thus,

$$|g(z)| = |-z^3 + z + 7| \leq |z^3| + |z| + 7$$
$$= 8 + 2 + 7 < 2 \cdot 2^5 = |f(z)| \tag{5}$$

Again, by Rouché's theorem $f(z) + g(z) = 2z^5 - z^3 + z + 7$ has the same number of zeros inside $|z| = 2$ as $f(z) = 2z^5$, i.e., five zeros. We conclude that all roots of (1) are located between the circles $|z| = 1$ and $|z| = 2$.

CHAPTER 9

SPECIAL FUNCTIONS

> **Basic Attacks and Strategies for Solving Problems in this Chapter. See pages 531 to 557 for step-by-step solutions to problems.**

The concept of a function of a complex number is a particular case of the general mathematical concept of a function. Many of the applications of the theory of analytic functions are in the areas of physics and mechanics. Examples of such applications are the so-called plane problems of thermal or electrical equilibrium and those of flow past plane contours set up by liquid or gas motions; these lead to the Laplace equation, from different solutions of which all analytic functions can be constructed.

The gamma function Γ has very important applications in various branches in engineering, and is defined as an improper integral. For $x > 0$, it is defined as:

$$\Gamma(x) = \int_{-\infty}^{\infty} t^{x-1} e^{-t} dt.$$

A function of two variables that is ultimately connected to the gamma function is the beta function ß, defined in $\{(x,y):x > 0, y > 0\}$ by

$$B(x,y) = \int_{0}^{1} t^{x-1}(1-t)^{y-1} dt.$$

Note that if $x \geq 1, y \geq 1$, the integral is proper, and if either $x < 1, y < 1$, or both, the integral is a convergent improper integral of the second kind. The beta function is symmetric. That is, for all $x > 0, y > 0$, we have ß (x,y) = ß (y,x).

An important relationship existing between the beta and gamma functions is given by:

$$B(x,y) = \frac{\Gamma(x)\Gamma(y)}{\Gamma(x+y)}$$

GAMMA FUNCTIONS

Define the gamma function $\Gamma(z)$ and show that

$$\Gamma(z + 1) = z\Gamma(z) \tag{1}$$

<u>Solution</u>: The gamma function $\Gamma(z)$ is defined for all z such that for Re z > 0.

$$\Gamma(z) \equiv \int_0^\infty t^{z-1} e^{-t} \, dt \tag{2}$$

Eq.(1) is called the recursion formula.

Integrating by parts we obtain

$$\Gamma(z+1) = \int_0^\infty t^z e^{-t} \, dt = \lim_{p \to \infty} \int_0^P t^z e^{-t} \, dt$$

$$= \lim_{p \to \infty} \left[t^z \cdot (-e^{-t}) \Big|_{t=0}^P - \int_0^P z t^{z-1}(-e^{-t}) dt \right] \tag{3}$$

$$= z \int_0^\infty t^{z-1} e^{-t} \, dt = z\Gamma(z); \quad \text{Re } z > 0$$

531

Note that

$$\lim_{p \to \infty} \frac{p^z}{e^p} = 0 \tag{4}$$

Prove that $\Gamma(z)$ is an analytic function of z in the half-plane Re z > 0.

Solution: We shall show that the limit

$$\lim_{h \to 0} \frac{\Gamma(z+h) - \Gamma(z)}{h} \tag{1}$$

exists. Indeed

$$\frac{\Gamma(z+h) - \Gamma(z)}{h} = \frac{1}{h} \int_0^\infty t^{z+h-1} e^{-t} \, dt - \frac{1}{h} \int_0^\infty t^{z-1} e^{-t} \, dt$$

$$= \int_0^\infty t^{z-1} e^{-t} \left[\frac{t^h - 1}{h} \right] dt \tag{2}$$

The limit is equal to

$$\lim_{h \to 0} \frac{\Gamma(z+h) - \Gamma(z)}{h} = \lim_{h \to 0} \int_0^\infty t^{z-1} e^{-t} \left[\frac{t^h - 1}{h} \right] dt$$

$$= \int_0^\infty t^{z-1} e^{-t} \ln t \, dt \tag{3}$$

To evaluate the limit

$$\lim_{h \to 0} \frac{t^h - 1}{h} = \left[\frac{d}{dh} \left(\frac{t^h - 1}{h} \right) \right] \Bigg|_{h=0} = \ln t \tag{4}$$

we applied l'Hopital's rule. Hence, $\Gamma(z)$ is an analytic function in the half-plane Re z> 0.

Show that $\Gamma(z)$ satisfies the functional equation

$$\Gamma(z) = \frac{\Gamma(z+n)}{z(z+1)(z+2)...(z+n-1)} \qquad (1)$$

where n is a positive integer.

<u>Solution</u>: We have

$$\Gamma(z+1) = z\Gamma(z) \qquad (2)$$

and

$$\Gamma(z+2) = (z+1)\ \Gamma(z+1) \qquad (3)$$

Substituting (2) into (3) we get

$$\Gamma(z+2) = (z+1)z\ \Gamma(z) \qquad (4)$$

Similarly for z + 3

$$\Gamma(z+3) = (z+2)\Gamma(z+2) = (z+2)(z+1)z\Gamma(z) \qquad (5)$$

and, in general

$$\Gamma(z+n) = (z+n-1)(z+n-2)...(z+1)z\Gamma(z) \qquad (6)$$

Show that

$$\Gamma(z) \cdot \Gamma(1-z) = \frac{\pi}{\sin \pi z} \qquad (1)$$

<u>Solution</u>: First, we shall prove (1) for z such that Im z = 0 and 0 < z < 1, and then by analytic continuation we will extend the result to other values of z.

$$\Gamma(m)\Gamma(1-m) = \left[2\int_0^\infty x^{2m-1}e^{-x^2}dx \right]\left[2\int_0^\infty y^{1-2m}e^{-y^2}dy \right]$$

$$= 4\int_0^\infty \int_0^\infty x^{2m-1}y^{1-2m}e^{-(x^2+y^2)}dxdy \qquad (2)$$

In polar coordinates,

$$x = r \cos\theta$$
$$y = r \sin\theta \qquad (3)$$

therefore, (2) becomes

$$4 \int_{\theta=0}^{\frac{\pi}{2}} \int_{r=0}^{\infty} (\tan^{1-2m}\theta)re^{-r^2} \, dr d\theta$$

(4)

$$= 2 \int_{0}^{\frac{\pi}{2}} \tan^{1-2m}\theta \, d\theta = \frac{\pi}{\sin m\pi}$$

Hence,

$$\Gamma(z)\Gamma(1-z) = \frac{\pi}{\sin \pi z}$$

(5)

Here, we used the formula

$$\int_{0}^{\infty} \frac{x^{\alpha-1}}{1+x} \, dx = \frac{\pi}{\sin \alpha\pi}$$

(6)

where $0 < \alpha < 1.$

• **PROBLEM** 9–5

Find the value of $\Gamma(z)$ at the point $z = \frac{1}{2}$.

Solution: We shall apply the following formula

$$\Gamma(p) = 2 \int_{0}^{\infty} x^{2p-1} e^{-x^2} \, dx, \quad p > 0$$

(1)

Setting $p = \frac{1}{2}$ we obtain

$$\Gamma\left(\frac{1}{2}\right) = 2 \int_{0}^{\infty} e^{-x^2} dx = \sqrt{\pi}$$

(2)

because using (5) from Problem 24–18 and setting $z = \frac{1}{2}$

$$\Gamma\left(\frac{1}{2}\right)\Gamma\left(\frac{1}{2}\right) = \Gamma^2\left(\frac{1}{2}\right) = \pi$$

(3)

or

$$\Gamma\left(\frac{1}{2}\right) = \sqrt{\pi}$$

534

Another way of finding the solution is to follow the method shown in Problem 24-18.

$$\left[\Gamma\left(\frac{1}{2}\right)\right]^2 = \left[2 \int_0^\infty e^{-x^2} dx\right]\left[2 \int_0^\infty e^{-y^2} dy\right]$$

$$= 4 \int_0^\infty \int_0^\infty e^{-(x^2+y^2)} dxdy = 4 \int_{\theta=0}^{\frac{\pi}{2}} \int_{r=0}^\infty e^{-r^2} rdrd\theta \qquad (4)$$

$$= \pi$$

Hence,

$$\Gamma\left(\frac{1}{2}\right) = \sqrt{\pi} \qquad (5)$$

● **PROBLEM 9-6**

Evaluate:

1. $\Gamma\left(\frac{5}{2}\right)$

2. $\Gamma\left(-\frac{1}{2}\right)$

Solution: We shall use the formula

$$\Gamma(z + 1) = z\Gamma(z) \qquad (1)$$

From Problem 24-19 we get

$$\Gamma\left(\frac{1}{2}\right) = \sqrt{\pi} \qquad (2)$$

Then

$$\Gamma\left(\frac{5}{2}\right) = \Gamma\left(\frac{3}{2} + 1\right) = \frac{3}{2}\Gamma\left(\frac{3}{2}\right)$$
$$= \frac{3}{2} \Gamma\left(\frac{1}{2} + 1\right) = \frac{3}{2} \cdot \frac{1}{2} \Gamma\left(\frac{1}{2}\right) = \frac{3\sqrt{\pi}}{4} \qquad (3)$$

Similarly,

$$\Gamma\left(\frac{1}{2}\right) = \Gamma\left(-\frac{1}{2} + 1\right) = -\frac{1}{2}\Gamma\left(-\frac{1}{2}\right) \qquad (4)$$

and

$$\Gamma\left(-\frac{1}{2}\right) = -2\sqrt{\pi} \qquad (5)$$

Prove that for k = 1,2,3,...

$$\Gamma\left(\frac{1}{k}\right)\Gamma\left(\frac{2}{k}\right) \cdots \Gamma\left(\frac{k-1}{k}\right) = \frac{(2\pi)^{\frac{k-1}{2}}}{\sqrt{k}} \tag{1}$$

<u>Solution</u>: Let us denote the left-hand side of (1) by A then

$$A = \Gamma\left(\frac{1}{k}\right)\Gamma\left(\frac{2}{k}\right) \cdots \Gamma\left(\frac{k-1}{k}\right) \tag{2}$$

$$= \Gamma\left(1 - \frac{1}{k}\right)\Gamma\left(1 - \frac{2}{k}\right) \cdots \Gamma\left(1 - \frac{k-1}{k}\right)$$

Multiplying term by term and applying

$$\Gamma(k)\,\Gamma(1 - k) = \frac{\pi}{\sin \pi k} \tag{3}$$

we find

$$A^2 = \left[\Gamma\left(\frac{1}{k}\right)\Gamma\left(1 - \frac{1}{k}\right)\right]\left[\Gamma\left(\frac{2}{k}\right)\Gamma\left(1 - \frac{2}{k}\right)\right]\cdots\left[\Gamma\left(1 - \frac{1}{k}\right)\Gamma\left(\frac{1}{k}\right)\right] \tag{4}$$

$$= \frac{\pi}{\sin \frac{\pi}{k}} \cdot \frac{\pi}{\sin \frac{2\pi}{k}} \cdots \cdot \frac{\pi}{\sin \frac{(k-1)\pi}{k}}$$

But, for k = 1,2,3,...

$$\sin \frac{\pi}{k} \cdot \sin \frac{2\pi}{k} \cdot \sin \frac{3\pi}{k} \cdot \cdots \cdot \sin \frac{(k-1)\pi}{k} \tag{5}$$

$$= \frac{k}{2^{k-1}}$$

Substituting (5) into (4) we find

$$\Gamma\left(\frac{1}{k}\right)\,\Gamma\left(\frac{2}{k}\right) \cdots \Gamma\left(\frac{k-1}{k}\right) = \sqrt{\frac{\pi^{k-1}}{k}\,2^{k-1}} \tag{6}$$

$$= \frac{(2\pi)^{\frac{k-1}{2}}}{\sqrt{k}}$$

Evaluate the integrals:

a) $\displaystyle\int_0^\infty y^5\, e^{-2y}\, dy$ \hfill (1)

b) $\displaystyle\int_0^\infty y^{\frac{3}{2}} e^{-3y} dy$ (2)

Solution: a) Substituting $2y = t$ into (1) we find

$$\int_0^\infty y^5 e^{-2y} dy = \int_0^\infty \left(\frac{t}{2}\right)^5 e^{-t} \frac{1}{2} dt$$ (3)

$$= \frac{1}{2^6} \int_0^\infty t^5 e^{-t} dt$$

But,

$$\Gamma(z) = \int_0^\infty t^{z-1} e^{-t} dt$$ (4)

Therefore, (3) is equal to $\frac{1}{2^6} \Gamma(6)$. (5)

Applying $\Gamma(z + 1) = z \Gamma(z)$ we get

$$\int_0^\infty y^5 e^{-2y} dy = \frac{1}{2^6} \cdot 5 \cdot 4 \cdot 3 \cdot 2 = \frac{15}{8}$$ (6)

b) To evaluate integral (2) substitute

$$3y = t$$ (7)

$$\int_0^\infty y^{\frac{3}{2}} e^{-3y} dy = \int_0^\infty \left(\frac{t}{3}\right)^{\frac{3}{2}} e^{-t} \frac{1}{3} dt$$

$$= 3^{-\frac{5}{2}} \int_0^\infty t^{\frac{3}{2}} e^{-t} dt = 3^{-\frac{5}{2}} \Gamma\left(\frac{5}{2}\right)$$ (8)

$$= 3^{-\frac{5}{2}} \cdot \frac{3}{2} \cdot \frac{1}{2} \Gamma\left(\frac{1}{2}\right) = \frac{\sqrt{3\pi}}{36}$$

Prove the functional identity

$$\int_0^\infty x^m e^{-ax^n} dx = \frac{1}{n} a^{-\frac{m+1}{n}} \Gamma\left(\frac{m+1}{n}\right) \tag{1}$$

where m, n, and a are positive constants.

Solution: Substituting

$$ax^n = t \tag{2}$$

we find

$$x = \left(\frac{t}{a}\right)^{\frac{1}{n}}, \quad anx^{n-1} dx = dt \tag{3}$$

and

$$\int_0^\infty x^m e^{-ax^n} dx = \int_0^\infty \left(\frac{t}{a}\right)^{\frac{m}{n}} e^{-t} \frac{1}{an} \cdot \frac{1}{\left(\frac{t}{a}\right)^{\frac{n-1}{n}}} dt$$

$$= \frac{1}{n} \cdot a^{-\frac{m+1}{n}} \int_0^\infty t^{\frac{m-n+1}{n}} e^{-t} dt \tag{4}$$

$$= \frac{1}{n} a^{-\frac{m+1}{n}} \Gamma\left(\frac{m+1}{n}\right)$$

BETA FUNCTION

Define the beta function and show that

$$B(n,m) = B(m,n) \tag{1}$$

Solution: We define the beta function by

$$B(n,m) = \int_0^1 t^{n-1} (1 - t)^{m-1} dt \tag{2}$$

for Re n > 0 and Re m > 0.

Setting

$$t = 1 - u, \tag{3}$$

we obtain

$$B(n,m) = \int_0^1 t^{n-1} (1 - t)^{m-1} dt$$

$$= \int_0^1 (1 - u)^{n-1} u^{m-1} du = B(m,n) \tag{4}$$

● **PROBLEM** 9–11

Show that

$$B(m,n) = 2 \int_0^{\frac{\pi}{2}} \sin^{2m-1} \theta \cos^{2n-1} \theta \, d\theta$$

$$= 2 \int_0^{\frac{\pi}{2}} \cos^{2m-1} \theta \sin^{2n-1} \theta \, d\theta \tag{1}$$

Solution: Let

$$t = \sin^2\theta \tag{2}$$

then, $\qquad dt = 2 \sin\theta \cos\theta \, d\theta \tag{3}$

and

$$B(m,n) = \int_0^1 t^{m-1} (1 - t)^{n-1} dt$$

$$= \int_0^{\frac{\pi}{2}} (\sin^2\theta)^{m-1} (\cos^2\theta)^{n-1} 2\sin\theta\cos\theta d\theta \tag{4}$$

$$= 2 \int_0^{\frac{\pi}{2}} \sin^{2m-1} \theta \, \cos^{2n-1} \theta \, d\theta$$

Since $B(m,n) = B(n,m)$ from (4) we get

$$B(m,n) = 2 \int_0^{\frac{\pi}{2}} \cos^{2m-1} \theta \, \sin^{2n-1} \theta \, d\theta \qquad (5)$$

• PROBLEM 9–12

Show that

$$B(m,n) = \frac{\Gamma(m)\Gamma(n)}{\Gamma(m+n)} \qquad (1)$$

Solution: We shall use the functional identity

$$\Gamma(m) = 2 \int_0^{\infty} x^{2m-1} e^{-x^2} dx, \quad m > 0 \qquad (2)$$

Transforming to polar coordinates we have

$$\Gamma(m)\Gamma(n) = \left[2 \int_0^{\infty} x^{2m-1} e^{-x^2} dx \right] \left[2 \int_0^{\infty} y^{2n-1} e^{-y^2} dy \right]$$

$$= 4 \int_0^{\infty} \int_0^{\infty} x^{2m-1} y^{2n-1} e^{-(x^2+y^2)} dx dy \qquad (3)$$

$$= 4 \int_{\theta=0}^{\frac{\pi}{2}} \int_{r=0}^{\infty} (\cos^{2m-1}\theta \, \sin^{2n-1}\theta)(r^{2m+2n-1} e^{-r^2}) dr d\theta$$

$$= \left[2 \int_0^{\frac{\pi}{2}} \cos^{2m-1}\theta \sin^{2n-1}\theta d\theta \right] \left[2 \int_0^{\infty} r^{2(m+n)-1} e^{-r^2} dr \right]$$

$$= B(m,n) \; \Gamma(m+n)$$

Hence,

$$B(m,n) = \frac{\Gamma(m)\Gamma(n)}{\Gamma(m+n)} \qquad (4)$$

Show that

$$\frac{B(m,n+1)}{B(m+1,n)} = \frac{n}{m} \qquad (1)$$

Solution: We shall utilize the functional identity proved in Problem 24-28.

$$B(m,n) = \frac{\Gamma(m)\Gamma(n)}{\Gamma(m+n)} \qquad (2)$$

Then,

$$\frac{B(m,n+1)}{B(m+1,n)} = \frac{\dfrac{\Gamma(m)\Gamma(n+1)}{\Gamma(m+n+1)}}{\dfrac{\Gamma(m+1)\Gamma(n)}{\Gamma(m+n+1)}}$$

$$= \frac{\Gamma(m)\Gamma(n+1)}{\Gamma(m+1)\Gamma(n)} = \frac{n\Gamma(n)\Gamma(m)}{m\Gamma(m)\Gamma(n)} = \frac{n}{m} \qquad (3)$$

We used the identity

$$\Gamma(z + 1) = z\Gamma(z) \qquad (4)$$

Prove that

$$\frac{B\left(\dfrac{q+1}{2}, \dfrac{1}{2}\right)}{B\left(\dfrac{q+1}{2}, \dfrac{q+1}{2}\right)} = 2^q \qquad (1)$$

Solution: Since,

$$B(m,n) = \frac{\Gamma(m)\Gamma(n)}{\Gamma(m+n)} \qquad (2)$$

the left-hand side of (1) can be written in the form

$$\frac{B\left(\dfrac{q+1}{2}, \dfrac{1}{2}\right)}{B\left(\dfrac{q+1}{2}, \dfrac{q+1}{2}\right)} = \frac{\Gamma\left(\dfrac{q+1}{2}\right)\Gamma\left(\dfrac{1}{2}\right)\Gamma(q+1)}{\Gamma\left(\dfrac{q+2}{2}\right)\Gamma\left(\dfrac{q+1}{2}\right)\Gamma\left(\dfrac{q+1}{2}\right)}$$

$$= \frac{\Gamma\left(\dfrac{1}{2}\right)q\Gamma(q)}{\Gamma\left(\dfrac{q}{2} + 1\right)\Gamma\left(\dfrac{q}{2} + \dfrac{1}{2}\right)} = \frac{2\Gamma\left(\dfrac{1}{2}\right)\Gamma(q)}{\Gamma\left(\dfrac{q}{2}\right)\Gamma\left(\dfrac{q}{2} + \dfrac{1}{2}\right)} \qquad (3)$$

We shall apply the duplication formula

$$\Gamma(z) \ \Gamma\left(z + \frac{1}{2}\right) = \frac{\sqrt{\pi} \ \Gamma(2z)}{2^{2z-1}} \qquad (4)$$

to obtain from (3)

$$\frac{2\Gamma\left(\frac{1}{2}\right)\Gamma(q)}{\Gamma\left(\frac{q}{2}\right)\Gamma\left(\frac{q}{2}+\frac{1}{2}\right)} = \frac{2\sqrt{\pi}\ \Gamma(q)}{\sqrt{\pi}\ \Gamma(q)} \cdot 2^{q-1}$$

$$= 2^q \tag{5}$$

● **PROBLEM 9–15**

Evaluate:

1. $B\left(3, \frac{7}{2}\right)$

2. $B(k,\ell)$, where k, ℓ are positive integers

3. $B\left(\frac{1}{3}, \frac{2}{3}\right)$

<u>Solution</u>: 1.

$$B\left(3, \frac{7}{2}\right) = \frac{\Gamma(3)\Gamma\left(\frac{7}{2}\right)}{\Gamma\left(3 + \frac{7}{2}\right)}$$

$$= \frac{2 \cdot \frac{5}{2} \cdot \frac{3}{2} \cdot \frac{1}{2}\ \Gamma\left(\frac{1}{2}\right)}{\frac{11}{2} \cdot \frac{9}{2} \cdot \frac{7}{2} \cdot \frac{5}{2} \cdot \frac{3}{2} \cdot \frac{1}{2} \cdot \Gamma\left(\frac{1}{2}\right)} \tag{1}$$

$$= \frac{2 \cdot 8}{11 \cdot 9 \cdot 7} = \frac{16}{693}$$

2.

$$B(k,\ell) = \frac{\Gamma(k)\Gamma(\ell)}{\Gamma(k + \ell)} = \frac{(k-1)!\ (\ell-1)!}{(k+\ell-1)!}$$

$$= \frac{(\ell-1)!}{k \cdot (k+1)\ldots(k+\ell-1)} \tag{2}$$

3.

$$B\left(\frac{1}{3}, \frac{2}{3}\right) = \frac{\Gamma\left(\frac{1}{3}\right)\Gamma\left(\frac{2}{3}\right)}{\Gamma\left(\frac{1}{3} + \frac{2}{3}\right)} = \Gamma\left(\frac{1}{3}\right)\Gamma\left(\frac{2}{3}\right) \tag{3}$$

Utilizing the formula

$$\Gamma(z)\Gamma(1-z) = \frac{\pi}{\sin\ \pi z} \tag{4}$$

we obtain

$$B\left(\frac{1}{3}, \frac{2}{3}\right) = \frac{\pi}{\sin\ \frac{\pi}{3}} = \frac{2\pi}{\sqrt{3}} \tag{5}$$

Evaluate the integral

$$\int_0^a \frac{dy}{\sqrt{a^4 - y^4}} \tag{1}$$

Solution: Substituting

$$t = \frac{y^4}{a^4} \tag{2}$$

where

$$dt = \frac{4y^3 dy}{a^4} \tag{3}$$

we obtain

$$\int_0^a \frac{dy}{\sqrt{a^4 - y^4}} = \int_0^1 \frac{1}{a^2 \sqrt{1 - t}} \cdot \frac{a^4}{4a^3 \, t^{\frac{3}{4}}} \, dt$$

$$= \frac{1}{4a} \int_0^1 t^{-\frac{3}{4}} (1 - t)^{-\frac{1}{2}} \, dt$$

$$= \frac{1}{4a} B\left(\frac{1}{4}, \frac{1}{2}\right) \tag{4}$$

We can express the answer in terms of the gamma function

$$\frac{1}{4a} B\left(\frac{1}{4}, \frac{1}{2}\right) = \frac{1}{4a} \frac{\Gamma(\frac{1}{4})\Gamma(\frac{1}{2})}{\Gamma(\frac{1}{4}+\frac{1}{2})} \tag{5}$$

Since,

$$2^{2z-1} \, \Gamma(z)\Gamma(z + \tfrac{1}{2}) = \sqrt{\pi} \, \Gamma(2z) \tag{6}$$

we have

$$\Gamma\left(\frac{1}{4} + \frac{1}{2}\right) = \frac{\pi\sqrt{2}}{\Gamma(\frac{1}{4})} \tag{7}$$

and

$$\int_0^a \frac{dy}{\sqrt{a^4 - y^4}} = \frac{1}{4a} \frac{\Gamma(\frac{1}{4}) \, \sqrt{\pi} \cdot \Gamma(\frac{1}{4})}{\pi\sqrt{2}}$$

$$= \frac{1}{4a} \frac{[\Gamma(\frac{1}{4})]^2}{\sqrt{2\pi}} \tag{8}$$

Evaluate the integrals

1. $\displaystyle\int_0^3 \sqrt{5x(3-x)}\ dx$

2. $\displaystyle\int_0^{\frac{\pi}{2}} \sqrt{\tan\theta}\ d\theta$

3. $\displaystyle\int_0^{\frac{\pi}{2}} \sin^6\theta\ \cos^4\theta\ d\theta$

Solution: 1. Substituting $x = 3t$ we find

$$\int_0^3 \sqrt{5x(3-x)}\ dx = \int_0^1 \sqrt{5\cdot 3t \cdot 3(1-t)}\ 3dt$$

$$= 9\sqrt{5}\int_0^1 t^{\frac{1}{2}}(1-t)^{\frac{1}{2}}\ dt \tag{1}$$

$$= 9\sqrt{5}\ B\left(\frac{3}{2},\frac{3}{2}\right) = 9\sqrt{5}\ \frac{\Gamma\left(\frac{3}{2}\right)\Gamma\left(\frac{3}{2}\right)}{\Gamma(3)}$$

$$= \frac{9\sqrt{5}\cdot\pi}{8}$$

2. $\displaystyle\int_0^{\frac{\pi}{2}}\sqrt{\tan\theta}\ d\theta = \int_0^{\frac{\pi}{2}}\sin^{\frac{1}{2}}\theta\ \cos^{-\frac{1}{2}}\theta\ d\theta \tag{2}$

Using the formula

$$B(m,n) = 2\int_0^{\frac{\pi}{2}}\sin^{2m-1}\theta\ \cos^{2n-1}\theta\ d\theta \tag{3}$$

and because $\Gamma(z)\Gamma(1-z) = \dfrac{n}{\sin nz}$, we find

$$\int_0^{\frac{\pi}{2}} \sqrt{\tan\theta} \; d\theta = \frac{1}{2} B\left(\frac{3}{4}, \frac{1}{4}\right) = \frac{1}{2} \Gamma\left(\frac{3}{4}\right) \Gamma\left(\frac{1}{4}\right)$$

$$\qquad\qquad\qquad = \frac{1}{2} \frac{\pi}{\sin\frac{\pi}{4}} = \frac{\pi\sqrt{2}}{2} \qquad\qquad (4)$$

3. $$\int_0^{\frac{\pi}{2}} \sin^6\theta \; \cos^4\theta \; d\theta = \frac{1}{2} B\left(\frac{7}{2}, \frac{5}{2}\right)$$

$$= \frac{1}{2} \frac{\Gamma\left(\frac{5}{2}\right) \Gamma\left(\frac{7}{2}\right)}{\Gamma(6)} = \frac{\frac{3}{2}\cdot\frac{1}{2}\cdot\frac{5}{2}\cdot\frac{3}{2}\cdot\frac{1}{2}\sqrt{\pi}\cdot\sqrt{\pi}}{2\cdot2\cdot3\cdot4\cdot5} \qquad (5)$$

$$= \frac{3\pi}{512}$$

BESSEL, LEGENDRE, AND OTHER FUNCTIONS

• **PROBLEM** 9–18

Find the general solution of Bessel's differential equation

$$z^2 w'' + zw' + (z^2 - n^2)w = 0 \qquad\qquad (1)$$

where n is not an integer.

Solution: Since z = 0 is a regular singular point the solution
of (1) can be written in the form

$$w = \sum_{k=0}^{\infty} a_k z^{k+\ell} \qquad\qquad (2)$$

Then,

$$w' = \Sigma(k + \ell)a_k z^{k+\ell-1} \qquad\qquad (3)$$

$$w'' = \Sigma(k + \ell)(k + \ell - 1)a_k z^{k+\ell-2} \qquad\qquad (4)$$

Substituting (2), (3), and (4) into (1) we find

$$z^2 w'' + zw' + (z^2 - n^2)w$$
$$= \Sigma\left\{\left[(k+\ell)^2 - n^2\right]a_k + a_{k-2}\right\}z^{k+\ell} = 0 \qquad (5)$$

Hence,

$$\left[(k + \ell)^2 - n^2\right]a_k + a_{k-2} = 0 \tag{6}$$

If $k = 0$ and $a_{-2} = 0$, then

$$(\ell^2 - n^2)a_0 = 0 \tag{7}$$

If $a_0 \neq 0$, then

$$\ell^2 - n^2 = 0 \tag{8}$$

which is the indicial equation with roots $\ell = \pm n$.

For $\ell = n$, (6) yields

$$k(2n + k)a_k + a_{k-2} = 0 \tag{9}$$

Since $a_{-1} = 0$, for $k = 1$ we have $a_1 = 0$ and

$$a_1 = a_3 = a_5 = \ldots = 0 \tag{10}$$

Similarly from (9) if $a_0 \neq 0$

$$a_2 = \frac{-a_0}{2(2n+2)} \tag{11}$$

$$a_4 = \frac{-a_2}{4(2n+4)} = \frac{a_0}{2 \cdot 4(2n+2)(2n+4)} \tag{12}$$

and

$$w_1 = \Sigma a_k z^{k+\ell} = a_0 z^n \left[1 - \frac{z^2}{2(2n+2)} + \frac{z^4}{2 \cdot 4(2n+2)(2n+4)} - \cdots\right]$$

For $\ell = -n$ we get

$$w_2 = a_0 z^{-n} \left[1 - \frac{z^2}{2(2-2n)} + \frac{z^4}{2 \cdot 4(2n+2)(2n+4)} - \cdots\right] \tag{13}$$

The general solution is

$$w = Aw_1 + Bw_2 \tag{14}$$

where A and B are arbitrary constants.

● **PROBLEM** 9–19

Show that

$$z J_{n-1}(z) - 2n J_n(z) + z J_{n+1}(z) = 0 \tag{1}$$

<u>Solution</u>: Differentiating the identity

$$e^{\frac{1}{2}z(u - \frac{1}{u})} = \sum_{n=-\infty}^{\infty} J_n(z)u^n \tag{2}$$

with respect to u we get

$$e^{\frac{1}{2}z(u - \frac{1}{u})} \cdot \frac{1}{2} z \left(1 + \frac{1}{u^2}\right) = \sum_{n=-\infty}^{\infty} \frac{z}{2} \left(1 + \frac{1}{u^2}\right) J_n(z)u^n \tag{3}$$

$$= \sum_{n=-\infty}^{\infty} n \, J_n(z)u^{n-1}$$

Then,

$$\sum_{n=-\infty}^{\infty} z \, J_n(z)u^n + \sum_{n=-\infty}^{\infty} z \, J_n(z)u^{n-2}$$

$$= \sum_{n=-\infty}^{\infty} 2n \, J_n(z)u^{n-1} \tag{4}$$

Equating coefficients of u^n on both sides we have

$$z \, J_n(z) + z \, J_{n+2}(z) = 2(n + 1)J_{n+1}(z) \tag{5}$$

Replacing n by n – 1 we get

$$z \, J_{n-1}(z) + z \, J_{n+1}(z) = 2n \, J_n(z) \tag{6}$$

This result also holds for n which is not an integer.

Eq.(6) is known as the recursion formula for Bessel functions.

● **PROBLEM 9–20**

Let C be a simple closed contour enclosing t = 0. Show that

$$J_n(z) = \frac{1}{2\pi i} \int_C u^{-n-1} \, e^{\frac{1}{2}z(u - \frac{1}{u})} \, du \tag{1}$$

<u>Solution</u>: Multiplying

$$e^{\frac{1}{2}z(u - \frac{1}{u})} = \sum_{m=-\infty}^{\infty} J_m(z)u^m \tag{2}$$

by u^{-n-1} we obtain

$$e^{\frac{1}{2}z(u - \frac{1}{u})} u^{-n-1} = \sum_{m=-\infty}^{\infty} J_m(z)u^{m-n-1} \tag{3}$$

547

Hence,

$$\int_C e^{\frac{1}{2}z(u - \frac{1}{u})} u^{-n-1} \, du = \sum_{m=-\infty}^{\infty} J_m(z) \int_C u^{m-n-1} \, du \qquad (4)$$

But

$$\int_C u^{m-n-1} \, du = \begin{cases} 0 & \text{if } m \neq n \\ 2\pi i & \text{if } m = n \end{cases} \qquad (5)$$

and (4) leads to

$$\int_C e^{\frac{1}{2}z(u - \frac{1}{u})} u^{-n-1} \, du = 2\pi i \, J_n(z) \qquad (6)$$

• **PROBLEM** 9–21

Show that if $\alpha \neq \beta$

$$\int_0^z u J_n(\alpha u) J_n(\beta u) du = z \frac{\alpha J_n(\beta z) J_n'(\alpha z) - \beta J_n(\alpha z) J_n'(\beta z)}{\beta^2 - \alpha^2} \qquad (1)$$

Solution: Both functions $J_n(\alpha u)$ and $J_n(\beta u)$ satisfy the Bessel's differential equation

$$u^2 J_n''(\alpha u) + u J_n'(\alpha u) + (\alpha^2 u^2 - n^2) J_n(\alpha u) = 0 \qquad (2)$$

$$u^2 J_n''(\beta u) + u J_n'(\beta u) + (\beta^2 u^2 - n^2) J_n(\beta u) = 0 \qquad (3)$$

Multiplying (2) by $J_n(\beta u)$ and (3) by $J_n(\alpha u)$ and subtracting we obtain

$$u^2 \left[J_n(\beta u) J_n''(\alpha u) - J_n(\alpha u) J_n''(\beta u) \right]$$
$$+ u \left[J_n'(\alpha u) J_n(\beta u) - J_n'(\beta u) J_n(\alpha u) \right] = (\beta^2 - \alpha^2) u^2 \, J_n(\alpha u) J_n(\beta u) \qquad (4)$$

Eq.(4) can be written in the form

$$\frac{d}{du} \left\{ u \left[J_n(\beta u) J_n'(\alpha u) - J_n(\alpha u) J_n'(\beta u) \right] \right\}$$
$$= (\beta^2 - \alpha^2) u \, J_n(\alpha u) J_n(\beta u) \qquad (5)$$

Integrating (5) with respect to u from 0 to z we find

$$\int_0^z uJ_n(\alpha u)J_n(\beta u)du = \frac{u[J_n(\beta u)J_n'(\alpha u)-J_n(\alpha u)J_n'(\beta u)]}{\beta^2 - \alpha^2}\Bigg|_{u=0}^z$$

$$= \frac{z[\alpha J_n(\beta z)J_n'(\alpha z)-\beta J_n(\alpha z)J_n'(\beta z)]}{\beta^2 - \alpha^2} \tag{6}$$

• PROBLEM 9–22

Show that

$$\int_{-1}^1 P_n(z)P_n(z)dz = \frac{2}{2n+1} \tag{1}$$

Solution:

$$\frac{1}{w^2 - 2zw + 1} = \sum_{m=0}^\infty \sum_{n=0}^\infty P_m(z)P_n(z)w^{m+n} \tag{2}$$

Integrating (2) from –1 to 1 and applying eq.(9) of Problem 24–54 we find

$$\int_{-1}^1 \frac{dz}{w^2-2zw+1} = \sum_{m=0}^\infty \sum_{n=0}^\infty \int_{-1}^1 P_m(z)P_n(z)w^{m+n}\,dz$$

$$= \sum_{n=0}^\infty \left[\int_{-1}^1 P_n(z)P_n(z)dz\right]w^{2n} \tag{3}$$

The left side of (3) is equal to

$$-\frac{1}{2w}\ln(w^2 - 2zw + 1)\Bigg|_{z=-1}^1 = \frac{1}{w}\ln\left(\frac{1+w}{1-w}\right)$$

$$= \sum_{n=0}^\infty \frac{2}{2n+1}w^{2n} \tag{4}$$

Comparing the coefficients of w^{2n} in (3) and (4), we find

$$\int_{-1}^1 P_n(z)P_n(z)dz = \frac{2}{2n+1} \tag{5}$$

Prove the recursion formula for Legendre polynomials

$$z(2n + 1)P_n(z) = nP_{n-1}(z) + (n + 1)P_{n+1}(z) \qquad (1)$$

Solution: Differentiating

$$\frac{1}{\sqrt{w^2 - 2zw + 1}} = \sum_{n=0}^{\infty} P_n(z)w^n \qquad (2)$$

with respect to w, we find

$$\frac{z - w}{(w^2 - 2zw + 1)^{\frac{3}{2}}} = \sum_{n=0}^{\infty} n\, P_n(z)w^{n-1} \qquad (3)$$

Multiplying (3) by $w^2 - 2zw + 1$, we get

$$(z - w)\sum_{n=0}^{\infty} P_n(z)w^n = (w^2 - 2zw + 1)\sum_{n=0}^{\infty} n\, P_n(z)w^{n-1} \qquad (4)$$

or

$$\sum_{n=0}^{\infty} P_n(z)zw^n - \sum_{n=0}^{\infty} P_n(z)w^{n+1}$$

$$= \sum_{n=0}^{\infty} n\, P_n(z)w^{n+1} - \sum_{n=0}^{\infty} 2nz\, P_n(z)w^n + \sum_{n=0}^{\infty} nP_n(z)w^{n-1} \qquad (5)$$

Comparing the coefficients at w^n, we have

$$z\, P_n(z) - P_{n-1}(z) = (n - 1)P_{n-1}(z) - 2nzP_n(z)$$

$$+ (n + 1)P_{n+1}(z) \qquad (6)$$

or

$$z(2n + 1)P_n(z) = n\, P_{n-1}(z) + (n + 1)P_{n+1}(z) \qquad (7)$$

Prove that

$$\ln(1 + z) = z\, F(1,1,2,-z) \qquad (1)$$

Solution: The left side of eq.(1) can be expressed by the power series

$$\ln(1 + z) = z - \frac{z^2}{2} + \frac{z^3}{3} - \frac{z^4}{4} + \ldots + (-1)^{n-1} \frac{z^n}{n} + \ldots \quad (2)$$

Substituting the appropriate numbers for the hypergeometric function and multiplying by z we find

$$z\, F(1,1,2,-z) = z\left[1 + \frac{1 \cdot 1}{1 \cdot 2}(-z) + \frac{1 \cdot 2 \cdot 1 \cdot 2}{1 \cdot 2 \cdot 2 \cdot 3}(-z)^2 \right.$$
$$\left. + \frac{1 \cdot 2 \cdot 3 \cdot 1 \cdot 2 \cdot 3}{1 \cdot 2 \cdot 3 \cdot 2 \cdot 3 \cdot 4}(-z)^3 + \ldots\right] \quad (3)$$

$$= z - \frac{z^2}{2} + \frac{z^3}{3} - \frac{z^4}{4} + \ldots$$

Comparing (2) and (3) we obtain (1).

• PROBLEM 9–25

Show that the zeta function

$$\zeta(z) = \sum_{n=1}^{\infty} \frac{1}{n^z} \quad (1)$$

is analytic in the domain Re z > 1.

Solution: Note that each term $\frac{1}{n^z}$ of (1) is an analytic function. Let ε be any fixed positive number. Then, if

$$x = \text{Re } z \geq 1 + \varepsilon \quad (2)$$

we have

$$\left|\frac{1}{n^z}\right| = \left|\frac{1}{e^{z \ln n}}\right| = \frac{1}{e^{x \ln n}} = \frac{1}{n^x} \leq \frac{1}{n^{1+\varepsilon}} \quad (3)$$

Because $\sum_{n=1}^{\infty} \frac{1}{n^{1+\varepsilon}}$ converges, by the Weierstrass M we conclude that $\sum_{n=1}^{\infty} \frac{1}{n^z}$ converges uniformly for Re z > 1.

Applying the theorem: if $f_n(z)$; $n = 0,1,2,\ldots$ are analytic and $\Sigma f_n(z)$ is uniformly convergent in D; then $\Sigma f_n(z)$ is analytic in D; we conclude that $\zeta(z)$ is analytic for Re z > 1.

Prove that

$$sn(-z) = -sn\ z \qquad (1)$$

$$cn(-z) = cn\ z \qquad (2)$$

$$dn(-z) = dn\ z \qquad (3)$$

Solution: The elliptic integral of the first kind is defined by

$$z = \int_0^w \frac{dt}{\sqrt{(1-t^2)(1-k^2t^2)}}\ , \quad |k| < 1 \qquad (4)$$

Eq.(4) is sometimes written briefly as

$$z = sn^{-1}\ w \qquad (5)$$

That leads to the definition of an elliptic function

$$w = sn\ z \qquad (6)$$

Replacing in (4), t by −t we obtain

$$z = \int_0^{-w} \frac{-dt}{\sqrt{(1-t^2)(1-k^2t^2)}} \qquad (7)$$

That is

$$-z = sn^{-1}(-w) \qquad (8)$$

or

$$sn(-z) = -w = -sn\ z \qquad (9)$$

Relating the elliptic function, w = sn z, with the trigonometric functions, we define

$$cn\ z = \sqrt{1 - sn^2z} \qquad (10)$$

and

$$tn\ z = \frac{sn\ z}{cn\ z} \qquad (11)$$

$$dn\ z = \sqrt{1 - k^2sn^2z} \qquad (12)$$

We have

$$cn(-z) = \sqrt{1 - sin^2(-z)} = \sqrt{1 - sn^2z} = cn\ z \qquad (13)$$

$$dn(-z) = \sqrt{1 - k^2sn^2(-z)} = \sqrt{1 - k^2sn^2z} = dn\ z \qquad (14)$$

Prove that

$$\frac{d}{dz} \, sn \, z = cn \, z \, dn \, z \qquad (1)$$

$$\frac{d}{dz} \, cn \, z = -sn \, z \, dn \, z \qquad (2)$$

<u>Solution</u>: If

$$z = \int_0^w \frac{dt}{\sqrt{(1-t^2)(1-k^2t^2)}} \qquad (3)$$

then

$$w = sn \, z \qquad (4)$$

Thus,

$$\frac{d}{dz} \, sn \, z = \frac{dw}{dz} = \frac{1}{\frac{dz}{dw}} = \sqrt{(1-w^2)(1-k^2w^2)}$$

$$= cn \, z \, dn \, z \qquad (5)$$

$$\frac{d}{dz} \, cn \, z = \frac{d}{dz} \sqrt{1 - sn^2z} = \frac{1}{2\sqrt{1 - sn^2z}} \frac{d}{dz} (-sn^2z)$$

$$= \frac{1}{2\sqrt{1 - sn^2z}} (-2sn \, z)\frac{d}{dz} \, sn \, z \qquad (6)$$

$$= \frac{-cn \, z \, dn \, z \, sn \, z}{cn \, z} = -sn \, z \, dn \, z$$

Prove that

$$sn(z + 2P) = -sn \, z \qquad (1)$$

$$cn(z + 2P) = -cn \, z \qquad (2)$$

<u>Solution</u>: By definition

$$z = \int_0^w \frac{dt}{\sqrt{(1-t^2)(1-k^2t^2)}} \, , \quad |k| < 1 \qquad (3)$$

$$w = sn \, z \qquad (4)$$

Substituting $t = \sin\theta$ and $w = \sin\phi$ we can transform (3) to

$$z = \int_0^\phi \frac{d\theta}{\sqrt{1 - k^2\sin^2\theta}} \tag{5}$$

which is often written in the form

$$\phi = \text{am } z \tag{6}$$

We have

$$w = \sin\phi = \text{sn } z \tag{7}$$

and

$$\cos\phi = \text{cn } z \tag{8}$$

Thus,

$$\int_0^{\phi+\pi} \frac{d\theta}{\sqrt{1-k^2\sin^2\theta}} = \int_0^\pi \frac{d\theta}{\sqrt{1-k^2\sin^2\theta}} + \int_\pi^{\phi+\pi} \frac{d\theta}{\sqrt{1-k^2\sin^2\theta}}$$

$$= 2\int_0^{\frac{\pi}{2}} \frac{d\theta}{\sqrt{1-k^2\sin^2\theta}} + \int_0^\phi \frac{d\lambda}{\sqrt{1-k^2\sin^2\lambda}} \tag{9}$$

where

$$\lambda = \theta - \pi$$

Therefore, denoting

$$P = \int_0^{\frac{\pi}{2}} \frac{d\theta}{\sqrt{1-k^2\sin^2\theta}} \tag{10}$$

we get

$$\phi + \pi = \text{am}(z + 2P) \tag{11}$$

and

$$\text{sn}(z + 2P) = \sin\left[\text{am}(z + 2P)\right] = \sin(\phi + \pi) \tag{12}$$
$$= -\sin\phi = -\text{sn } z$$

$$\text{cn}(z + 2P) = \cos\left[\text{am}(z + 2P)\right] = \cos(\phi + \pi) \tag{13}$$
$$= -\cos\phi = -\text{cn } z$$

Let us denote

$$k' = \sqrt{1 - k^2} \tag{1}$$

$$P' = \int_0^1 \frac{dt}{\sqrt{(1-t^2)(1-k'^2 t^2)}} \tag{2}$$

Prove that

$$sn(P + iP') = \frac{1}{k} \tag{3}$$

$$cn(P + iP') = -\frac{ik'}{k} \tag{4}$$

Solution: We denote

$$s = \frac{1}{\sqrt{1 - k'^2 t^2}} \tag{5}$$

Then, when $t = 0$, $s = 1$, and when $t = 1$, $s = \frac{1}{k}$.

$$\sqrt{1 - t^2} = \frac{-ik's}{\sqrt{1 - k'^2 s^2}} \tag{6}$$

Then,

$$P' = -i \int_1^{\frac{1}{k}} \frac{ds}{\sqrt{(1-s^2)(1-k^2 s^2)}} \tag{7}$$

and we obtain

$$P + iP' = \int_0^1 \frac{ds}{\sqrt{(1-s^2)(1-k^2 s^2)}} + \int_1^{\frac{1}{k}} \frac{ds}{\sqrt{(1-s^2)(1-k^2 s^2)}} \tag{8}$$

$$= \int_0^{\frac{1}{k}} \frac{ds}{\sqrt{(1-s^2)(1-k^2 s^2)}}$$

Therefore,

$$sn(P + iP') = \frac{1}{k} \tag{9}$$

and

$$cn(P + iP') = \sqrt{1 - sn^2(P + iP')}$$

$$=\sqrt{1 - \frac{1}{k^2}} = \frac{-i\sqrt{1-k^2}}{k} = \frac{-ik'}{k} \tag{10}$$

From (3) we also obtain

$$dn(P + iP') = \sqrt{1 - k^2 sn^2(P + iP')} = \sqrt{1 - k^2 \cdot \frac{1}{k^2}} = 0 \tag{11}$$

● **PROBLEM 9–30**

Prove that

$$sn(z_1 + z_2) = \frac{sn z_1 cn z_2 dn z_2 + sn z_2 cn z_1 dn z_1}{1 - k^2 sn^2 z_1 sn^2 z_2} \tag{1}$$

Solution: Let us denote

$$z_1 + z_2 = a \tag{2}$$

where a is an arbitrary constant,

$$A = sn\ z_1$$
$$B = sn\ z_2 \tag{3}$$

Then,

$$\frac{dz_2}{dz_1} = -1 \tag{4}$$

and

$$\frac{dA}{dz_1} = A' = cn\ z_1\ dn\ z_1 \tag{5}$$

$$\frac{dB}{dz_1} = B' = \frac{dB}{dz_2}\frac{dz_2}{dz_1} = -cn\ z_2\ dn\ z_2 \tag{6}$$

Then,

$$A'^2 = (1 - A^2)(1 - k^2 A^2) \tag{7}$$

$$B'^2 = (1 - B^2)(1 - k^2 B^2) \tag{8}$$

Differentiating with respect to z_1 we get

$$A'' = 2k^2 A^3 - (1 + k^2)A \tag{9}$$

$$B'' = 2k^2 B^3 - (1 + k^2)B \tag{10}$$

Multiplying (9) by B and (10) by A and subtracting we find

$$A''B - B''A = 2k^2 AB(A^2 - B^2) \tag{11}$$

and

$$A'B - AB' = \frac{(1 - k^2 A^2 B^2)(B^2 - A^2)}{A'B + AB'} \tag{12}$$

Dividing (11) by (12) we get

$$\frac{A''B - B''A}{A'B - AB'} = \frac{-2k^2AB(A'B + AB')}{1 - k^2A^2B^2} \tag{13}$$

or

$$\frac{(A'B - AB')'}{A'B - AB'} = \frac{(1 - k^2A^2B^2)'}{1 - k^2A^2B^2} \tag{14}$$

Integrating (14) we find

$$\frac{A'B - AB'}{1 - k^2A^2B^2} = \alpha_1 = \text{const} \tag{15}$$

Thus,

$$\frac{snz_1cnz_2dnz_2 + cnz_1snz_2dnz_1}{1 - k^2sn^2z_1sn^2z_2} = \alpha_1 \tag{16}$$

Since $z_1 + z_2 = a$, both constants must be related by

$$\alpha_1 = F(z_1 + z_2) \tag{17}$$

Setting $z_2 = 0$, we have $F(z_1) = sn\ z_1$. Thus

$$F(z_1 + z_2) = sn(z_1 + z_2) \tag{18}$$

From (16), (17), and (18) we obtain (1).

CHAPTER 10

DETERMINANTS AND MATRICES

> **Basic Attacks and Strategies for Solving Problems in this Chapter. See pages 559 to 675 for step-by-step solutions to problems.**

Matrices provide a very powerful tool for dealing with linear models. A matrix is defined to be a rectangular array of numbers arranged into rows and columns. Any matrix which has the same number of rows as columns is called a square matrix. A square matrix with n rows and n columns is often called an n-th order matrix. In general an $m \times n$ matrix has m rows and n columns.

Two matrices are said to be equal, only if their corresponding elements are equal. Two matrices can be added only if they have the same number of rows, and the same number of columns. A matrix can be multiplied only if the number of columns of the first equals the number of rows of the second, the resultant matrix has the number of rows of the first and as many columns as the second. The multiplication of matrices by blocks requires that the partitioning of the columns in the premultiplier be the same as the partitioning of rows in the postmultiplier.

The determinant of an n-th order matrix $A = \|a_{ij}\|$, written $|A|$, is defined to be the number computed from the following sum involving the n^2 elements in A:

$|A|$ = the sum being taken over all permutations of the second subscripts. A term assigned a plus sign if $(i, j, \ldots, \gamma)$ is an even permutation of $(1, 2, \ldots, n)$ and a minus sign if it is an odd permutation.

The definition of a determinant implies that only square matrices have determinants associated with them.

Step-by-Step Solutions to
Problems in this Chapter,
"Determinants and Matrices"

DETERMINANTS

● **PROBLEM** 10–1

Find the product of two permutations σ and φ , where,

(i) $\qquad \sigma = \begin{pmatrix} 1 & 2 & 3 & 4 & 5 \\ 2 & 4 & 1 & 5 & 3 \end{pmatrix} \qquad \varphi = \begin{pmatrix} 1 & 2 & 3 & 4 & 5 \\ 4 & 1 & 2 & 5 & 3 \end{pmatrix}$

(ii) $\qquad \sigma = \begin{pmatrix} 1 & 2 & 3 & 4 & 5 \\ 4 & 1 & 2 & 5 & 3 \end{pmatrix} \qquad \varphi = \begin{pmatrix} 4 & 1 & 2 & 5 & 3 \\ 1 & 2 & 3 & 4 & 5 \end{pmatrix}$

(iii) $\qquad \sigma = \begin{pmatrix} 1 & 2 & 3 & 4 & 5 \\ 5 & 2 & 4 & 3 & 1 \end{pmatrix} \qquad \varphi = \begin{pmatrix} 4 & 1 & 2 & 5 & 3 \\ 1 & 2 & 3 & 4 & 5 \end{pmatrix}$

(iv) $\qquad \sigma = \begin{pmatrix} 1 & 2 & 3 & 4 & 5 \\ 4 & 1 & 2 & 5 & 3 \end{pmatrix} \qquad \varphi = \begin{pmatrix} 1 & 2 & 3 & 4 & 5 \\ 2 & 4 & 1 & 5 & 3 \end{pmatrix}$

Solution: The product of two permutations σ and φ is defined in terms of function composition:

$$(\sigma\varphi)(i) = \sigma[\varphi(i)] , \quad i = 1,\dots,n .$$

The set of all permutations on n objects, together with this operation of multiplication is called the symmetric group of degree n, and is denoted by S_n .

(i) $\qquad (\sigma\varphi)(i) = \sigma[\varphi(i)] , \quad i = 1,\dots,n;$
then

$$(\sigma\varphi)(1) = \sigma[\varphi(1)] = \sigma(4) = 5$$

$$(\sigma\varphi)(2) = \sigma[\varphi(2)] = \sigma(1) = 2$$

$$(\sigma\varphi)(3) = \sigma[\varphi(3)] = \sigma(2) = 4 .$$

Similarly, we find $\sigma\varphi(4) = 3$ and $\sigma\varphi(5) = 1$. Thus,

(i)
$$\sigma\varphi = \begin{pmatrix} 1 & 2 & 3 & 4 & 5 \\ 2 & 4 & 1 & 5 & 3 \end{pmatrix} \begin{pmatrix} 1 & 2 & 3 & 4 & 5 \\ 4 & 1 & 2 & 5 & 3 \end{pmatrix} = \begin{pmatrix} 1 & 2 & 3 & 4 & 5 \\ 5 & 2 & 4 & 3 & 1 \end{pmatrix}$$

(ii)
$$\sigma\varphi = \begin{pmatrix} 1 & 2 & 3 & 4 & 5 \\ 4 & 1 & 2 & 5 & 3 \end{pmatrix} \begin{pmatrix} 4 & 1 & 2 & 5 & 3 \\ 1 & 2 & 3 & 4 & 5 \end{pmatrix} = \begin{pmatrix} 1 & 2 & 3 & 4 & 5 \\ 1 & 2 & 3 & 4 & 5 \end{pmatrix}$$

(iii)
$$\sigma\varphi = \begin{pmatrix} 1 & 2 & 3 & 4 & 5 \\ 5 & 2 & 4 & 3 & 1 \end{pmatrix} \begin{pmatrix} 4 & 1 & 2 & 5 & 3 \\ 1 & 2 & 3 & 4 & 5 \end{pmatrix} = \begin{pmatrix} 1 & 2 & 3 & 4 & 5 \\ 2 & 4 & 1 & 5 & 3 \end{pmatrix}$$

(iv)
$$\sigma\varphi = \begin{pmatrix} 1 & 2 & 3 & 4 & 5 \\ 4 & 1 & 2 & 5 & 3 \end{pmatrix} \begin{pmatrix} 1 & 2 & 3 & 4 & 5 \\ 2 & 4 & 1 & 5 & 3 \end{pmatrix} = \begin{pmatrix} 1 & 2 & 3 & 4 & 5 \\ 1 & 5 & 4 & 3 & 2 \end{pmatrix}$$

Observe that the examples (i) and (iv) show that multiplication of permutations is not always commutative, (i.e., $\sigma\varphi \neq \varphi\sigma$), in general.

● **PROBLEM** 10–2

Define det A and find the determinant of the following matrices:

(a) $[a_{11}]$ (b) $\begin{bmatrix} a_{11} & a_{12} \\ a_{21} & a_{22} \end{bmatrix}$ (c) $\begin{bmatrix} 0 & 0 & 0 \\ 0 & 0 & 0 \\ 0 & 0 & 0 \end{bmatrix}$

(d) $\begin{bmatrix} a_{11} & a_{12} & a_{13} \\ a_{21} & a_{22} & a_{23} \\ a_{31} & a_{32} & a_{33} \end{bmatrix}$

Solution: Determinants are formally defined as:

$$\text{Det}(A) = |A| = \sum_{\sigma} \text{sgn}(\sigma) \prod_{j=1}^{n} a_{\sigma(j)j}$$

$$= \sum_{\sigma} \text{sgn}(\sigma)\, a_{\sigma(1)1}\, a_{\sigma(2)2} \cdots a_{\sigma(n)n}$$

where summation extends over the n!, different permutations, σ, of the n symbols $1,2,\ldots,n$ and $\text{sgn}(\sigma) = +1$, σ even
-1, σ odd

$|A|$ is also known as an $n \times n$ determinant or a determinant of order n.

(a)   det A = a_{11}

(b) $\det A = \begin{vmatrix} a_{11} & a_{12} \\ a_{21} & a_{22} \end{vmatrix} = a_{11}a_{22} - a_{21}a_{12}$

(c) $\det A = |0| = 0$

(d) $\det A = \begin{vmatrix} a_{11} & a_{12} & a_{13} \\ a_{21} & a_{22} & a_{23} \\ a_{31} & a_{32} & a_{33} \end{vmatrix}$

The permutations of S_3 and their signs are:

Permutation	Sign	Permutation	Sign
1 2 3	+	2 1 3	-
1 3 2	-	3 1 2	+
2 3 1	+	3 2 1	-

Then, $\det A = a_{11}a_{22}a_{33} - a_{11}a_{32}a_{23} + a_{21}a_{32}a_{13} - a_{21}a_{12}a_{33} + a_{31}a_{12}a_{23}$
$$- a_{31}a_{22}a_{13} .$$

• PROBLEM 10–3

Find the determinant of the following matrix:

$$A = \begin{bmatrix} 2 & 0 & 3 & 0 \\ 2 & 1 & 1 & 2 \\ 3 & -1 & 1 & -2 \\ 2 & 1 & -2 & 1 \end{bmatrix}$$

Solution: Use the method of expansion by minors.

$$A = \begin{bmatrix} 2 & 0 & 3 & 0 \\ 2 & 1 & 1 & 2 \\ 3 & -1 & 1 & -2 \\ 2 & 1 & -2 & 1 \end{bmatrix}$$

Expanding along the first row:

$$\det A = 2 \begin{vmatrix} 1 & 1 & 2 \\ -1 & 1 & -2 \\ 1 & -2 & 1 \end{vmatrix} + 3 \begin{vmatrix} 2 & 1 & 2 \\ 3 & -1 & -2 \\ 2 & 1 & 1 \end{vmatrix}$$

Note that the minors, whose multiplying factors were zero, have been eliminated. This illustrates the general principle that, when evaluating determinants, expansion along the row (or column) containing the most zeros is the optimal procedure.

Add the second row to the first row for each of the 3 by 3 determinants:

561

$$\det A = 2 \begin{vmatrix} 0 & 2 & 0 \\ -1 & 1 & -2 \\ 1 & -2 & 1 \end{vmatrix} + 3 \begin{vmatrix} 5 & 0 & 0 \\ 3 & -1 & -2 \\ 2 & 1 & 1 \end{vmatrix}$$

Now expand the above determinants by minors using the first row.

$$\det A = 2(-2) \begin{vmatrix} -1 & -2 \\ 1 & 1 \end{vmatrix} + 3(5) \begin{vmatrix} -1 & -2 \\ 1 & 1 \end{vmatrix}$$

$$= (-4)(-1+2) + 15(-1+2)$$

$$= -4+15 = 11 \ .$$

• PROBLEM 10–4

Evaluate the determinant of the matrix A where

$$A = \begin{bmatrix} 4 & 2 & 1 & 3 \\ -1 & 0 & 2 & 8 \\ 5 & -6 & 0 & -1 \\ 0 & 2 & 2 & 3 \end{bmatrix} \ .$$

<u>Solution</u>: The value of the determinant of A is not changed if a multiple of one row (column) is added to a multiple of another. Now, add -4 times the third column to the first column and add -2 times the third column to the second column.

$$\begin{bmatrix} 0 & 0 & 1 & 3 \\ -9 & -4 & 2 & 8 \\ 5 & -6 & 0 & -1 \\ -8 & -2 & 2 & 3 \end{bmatrix} \ .$$

Add -3 times the third column to the fourth column.

$$\begin{bmatrix} 0 & 0 & 1 & 0 \\ -9 & -4 & 2 & 2 \\ 5 & -6 & 0 & -1 \\ -8 & -2 & 2 & -3 \end{bmatrix} \ .$$

Now,

$$\det A = \begin{vmatrix} 0 & 0 & 1 & 0 \\ -9 & -4 & 2 & 2 \\ 5 & -6 & 0 & -1 \\ -8 & -2 & 2 & -3 \end{vmatrix} \ .$$

Expand the above determinant by minors using the first row.

$$\det A = + 0 \begin{vmatrix} -4 & 2 & 2 \\ -6 & 0 & -1 \\ -2 & 2 & -3 \end{vmatrix} - 0 \begin{vmatrix} -9 & 2 & 2 \\ 5 & 0 & -1 \\ 8 & 2 & -3 \end{vmatrix}$$

$$+ 1 \begin{vmatrix} -9 & -4 & 2 \\ 5 & -6 & -1 \\ -8 & -2 & -3 \end{vmatrix} - 0 \begin{vmatrix} -9 & -4 & 2 \\ 5 & -6 & 0 \\ -8 & -2 & 2 \end{vmatrix}$$

Therefore,

$$\det A = 1 \begin{vmatrix} -9 & -4 & 2 \\ 5 & -6 & -1 \\ -8 & -2 & -3 \end{vmatrix} .$$

Add +2 times the second row to the first row and -3 times the second row to the third row.

$$\det A = \begin{vmatrix} 1 & -16 & 0 \\ 5 & -6 & -1 \\ -23 & 16 & 0 \end{vmatrix}$$

Expand the determinant by minors, using column three.

$$\det A = 0 \begin{vmatrix} 5 & -6 \\ -23 & 16 \end{vmatrix} + (-1) \begin{vmatrix} 1 & -16 \\ -23 & 16 \end{vmatrix} + 0 \begin{vmatrix} 1 & -16 \\ 5 & -16 \end{vmatrix}$$

$$= -1 \begin{vmatrix} 1 & -16 \\ -23 & 16 \end{vmatrix}$$

$$= -(16 - 368) = + 352.$$

• PROBLEM 10–5

Find the determinant of the matrix A where:

$$A = \begin{bmatrix} 2 & 7 & -3 & 8 & 3 \\ 0 & -3 & 7 & 5 & 1 \\ 0 & 0 & 6 & 7 & 6 \\ 0 & 0 & 0 & 9 & 8 \\ 0 & 0 & 0 & 0 & 4 \end{bmatrix}$$

Solution: A is an upper triangular matrix. As we know, if A is an nXm triangular matrix (upper or lower), then det A is the product of the entries on the main diagonal.
Hence,

$$\det A = (2) \cdot (-3) \cdot (6) \cdot (9) \cdot (4) = -1296 .$$

Evaluate det A where:

$$A = \begin{bmatrix} 0 & 1 & 5 \\ 3 & -6 & 9 \\ 2 & 6 & 1 \end{bmatrix}$$

<u>Solution:</u>   Interchange the first and second rows of matrix A, obtaining matrix

$$B = \begin{bmatrix} 3 & -6 & 9 \\ 0 & 1 & 5 \\ 2 & 6 & 1 \end{bmatrix} \quad ;$$

and by the properties of the function,

$$\det A = -\det B .$$

$$= -\det \begin{bmatrix} 3 & -6 & 9 \\ 0 & 1 & 5 \\ 2 & 6 & 1 \end{bmatrix}$$

or

$$\det A = -3 \det \begin{bmatrix} 1 & -2 & 3 \\ 0 & 1 & 5 \\ 2 & 6 & 1 \end{bmatrix} .$$

A common factor of 3 from the first row of the matrix B was taken out. Add -2 times the first row to the third row. The value of the determinant of A will remain the same.
Thus,

$$\det A = -3 \det \begin{bmatrix} 1 & -2 & 3 \\ 0 & 1 & 5 \\ 0 & 10 & -5 \end{bmatrix} .$$

Add -10 times the second row to the third row.
Thus,

$$\det A = -3 \det \begin{bmatrix} 1 & -2 & 3 \\ 0 & 1 & 5 \\ 0 & 0 & -55 \end{bmatrix} .$$

As we know, the determinant of a triangular matrix is equal to the product of the diagonal elements.
Thus,

$$\det A = (-3) \cdot (1) \cdot (1) \cdot (-55) = 165 .$$

Compute the determinant of

$$A = \begin{bmatrix} 1 & 0 & 0 & 3 \\ 2 & 7 & 0 & 6 \\ 0 & 6 & 3 & 0 \\ 7 & 3 & 1 & -5 \end{bmatrix}.$$

Solution: First list the basic properties of the determinant. Let A be a square matrix.
(1)   If we interchange two rows (columns) of A, the determinant changes by a factor of (-1).
(2)   By adding to one row (column) of A a multiple of another row (column), the determinant is not changed.
(3)   If A is triangular, i.e., A has zeros above or below the diagonal, the determinant of A is the product of the diagonal elements of A.
(4)   If a row (column) of A is multiplied by a scalar K, but all of the other rows (columns) are left unchanged, then the determinant of the resulting matrix is K times the determinant of A.

Given:

$$A = \begin{bmatrix} 1 & 0 & 0 & 3 \\ 2 & 7 & 0 & 6 \\ 0 & 6 & 3 & 0 \\ 7 & 3 & 1 & -5 \end{bmatrix}.$$

Now add -3 times the first column to the fourth column. But the value of the determinant A will not be changed.
Thus,

$$A = \begin{bmatrix} 1 & 0 & 0 & 0 \\ 2 & 7 & 0 & 0 \\ 0 & 6 & 3 & 0 \\ 7 & 3 & 1 & -26 \end{bmatrix}$$

Now the above matrix A is a lower triangular matrix. Using property (3), yields:

$$\det A = (1) \cdot (3) \cdot (-26) = -546 .$$

Find the cofactors of the matrix A where:

$$A = \begin{bmatrix} 1 & 2 & 3 \\ 3 & 2 & 1 \\ 2 & 0 & 2 \end{bmatrix}.$$

<u>Solution</u>: If A is an n×n matrix, then M_{ij} will denote the (n-1) × (n-1) matrix obtained from A by deleting its ith row and jth column. The determinant $|M_{ij}|$ is called the minor of the element a_{ij} of A, and we define the cofactor of a_{ij}; denoted by A_{ij}, as:

$$A_{ij} = (-1)^{i+j} |M_{ij}| .$$

$$A = \begin{bmatrix} 1 & 2 & 3 \\ 3 & 2 & 1 \\ 2 & 0 & 2 \end{bmatrix} .$$

Then

$$A_{11} = (-1)^2 \begin{vmatrix} 2 & 1 \\ 0 & 2 \end{vmatrix} = 4 - 0 = 4$$

$$A_{12} = (-1)^{1+2} \begin{vmatrix} 3 & 1 \\ 2 & 2 \end{vmatrix} = -(6-2) = -4$$

$$A_{13} = (-1)^{1+3} \begin{vmatrix} 3 & 2 \\ 2 & 0 \end{vmatrix} = +(0-4) = -4$$

$$A_{21} = (-1)^{2+1} \begin{vmatrix} 2 & 3 \\ 0 & 2 \end{vmatrix} = -(4-0) = -4$$

$$A_{22} = (-1)^{2+2} \begin{vmatrix} 1 & 3 \\ 2 & 2 \end{vmatrix} = +(2-6) = -4$$

$$A_{23} = (-1)^{2+3} \begin{vmatrix} 1 & 2 \\ 2 & 0 \end{vmatrix} = -(0-4) = 4$$

$$A_{31} = (-1)^{3+1} \begin{vmatrix} 2 & 3 \\ 2 & 1 \end{vmatrix} = +(2-6) = -4$$

$$A_{32} = (-1)^{3+2} \begin{vmatrix} 1 & 3 \\ 3 & 1 \end{vmatrix} = -1(1-9) = 8$$

$$A_{33} = (-1)^{3+3} \begin{vmatrix} 1 & 2 \\ 3 & 2 \end{vmatrix} = +(2-6) = -4$$

Thus, the cofactors of A are:

$$A_{11} = 4 \qquad A_{12} = -4 \qquad A_{13} = -4$$

$$A_{21} = -4 \qquad A_{22} = -4 \qquad A_{23} = 4$$

$$A_{31} = -4 \qquad A_{32} = 8 \qquad A_{33} = -4 .$$

● PROBLEM 10–9

a) Given:
$$A = \begin{bmatrix} 5 & 2 & -1 \\ 3 & 1 & 2 \\ 2 & 7 & 4 \end{bmatrix} , \text{ find det } A .$$

b) If
$$A = \begin{bmatrix} 2 & 3 & 1 \\ -2 & 4 & 5 \\ 2 & 0 & 7 \end{bmatrix} , \text{ find adj } A .$$

<u>Solution</u>: The definition of the determinant in terms of permutations is a mathematical one. For computational purposes, more efficient evaluative procedures are available.

Define the cofactor a_{ij} of an $n \times n$ determinant $|A|$ as $(-1)^{i+j}|C|$ where $|C|$ is the $(n-1) \times (n-1)$ sub-determinant obtained by deleting the ith row and jth column of $|A|$.

Suppose that $A = (a_{ij})$ is an $n \times n$ matrix, and let A_{ij} denote the cofactor of a_{ij}, $i = 1, 2, \ldots, n$; $j = 1, 2, \ldots, n$. Then,

$$\det A = \sum_{k=1}^{n} a_{kj} A_{kj} \, , \quad j = 1, 2, \ldots, n \, ,$$

and

$$\det A = \sum_{k=1}^{n} a_{ik} A_{ik} \, , \quad i = 1, 2, \ldots, n \, .$$

Note that we sum across k to obtain det A. Since $j = 1, 2, \ldots, n$, there are $2n$ ways of computing det A, one for each row (or column) of elements.

$$A = \begin{bmatrix} 5 & 2 & -1 \\ 3 & 1 & 2 \\ 2 & 7 & 4 \end{bmatrix}$$

The cofactors along the first column are:

$$A_{11} = (-1)^{1+1} \begin{vmatrix} 1 & 2 \\ 7 & 4 \end{vmatrix} = + (4-14) = -10$$

$$A_{21} = (-1)^{2+1} \begin{vmatrix} 2 & -1 \\ 7 & 4 \end{vmatrix} = - (8+7) = -15$$

$$A_{31} = (-1)^{3+1} \begin{vmatrix} 2 & -1 \\ 1 & 2 \end{vmatrix} = + (4+1) = 5 \, .$$

Hence,

$$\det A = \sum_{k=1}^{n} a_{k1} A_{k1} = a_{11}A_{11} + a_{21}A_{21} + a_{31}A_{31}$$

$$= 5(-10) + 3(-15) + 2(5)$$
$$= -50 - 45 + 10$$
$$= -85 \, .$$

Also, expanding along the 2nd row,

$$A_{21} = -15, \ A_{22} = 22, \ A_{23} = -31 \, .$$

Therefore,

$$\det A = a_{21}A_{21} + a_{22}A_{22} + a_{23}A_{23}$$

$$= 3(-15) + 1(22) + 2(-31)$$
$$= -45 + 22 - 62$$
$$= -85 \, .$$

b) Let A be a $n \times n$ matrix. Then the adjoint of A, adj A, is defined as the matrix of transposed cofactors. Let $C_{11}, C_{12}, \ldots, C_{1n}, C_{21}, \ldots, C_{2n}, \ldots, C_{n1}, \ldots, C_{nn}$ denote the cofactors of $a_{11}, a_{12}, \ldots, a_{nn}$, respectively. Then

567

$$\text{adj. } A = \begin{bmatrix} C_{11} & C_{21} & \cdot & \cdot & \cdot & \cdot & \cdot & C_{n1} \\ C_{12} & & & & & & & \\ \cdot & & & & & & & \\ \cdot & & & & & & & \\ \cdot & & & & & & & \\ C_{1n} & \cdot & \cdot & \cdot & \cdot & \cdot & \cdot & \cdot & C_{nn} \end{bmatrix}$$

The adjoint is useful in finding the inverse of a non-singular matrix. The cofactors of A are:

$$A_{11} = + \begin{vmatrix} 4 & 5 \\ 0 & 7 \end{vmatrix} = (28.0) = 28$$

$$A_{12} = - \begin{vmatrix} -2 & 5 \\ 2 & 7 \end{vmatrix} = - (-14-10) = 24$$

$$A_{13} = + \begin{vmatrix} -2 & 4 \\ 2 & 0 \end{vmatrix} = (0-8) = -8$$

$$A_{21} = - \begin{vmatrix} 3 & 1 \\ 0 & 7 \end{vmatrix} = -(21-0) = -21$$

$$A_{22} = + \begin{vmatrix} 2 & 1 \\ 2 & 7 \end{vmatrix} = (14-2) = 12$$

$$A_{23} = - \begin{vmatrix} 2 & 3 \\ 2 & 0 \end{vmatrix} = -(0-6) = 6$$

$$A_{31} = + \begin{vmatrix} 3 & 1 \\ 4 & 5 \end{vmatrix} = (15-4) = 11$$

$$A_{32} = - \begin{vmatrix} 2 & 1 \\ -2 & 5 \end{vmatrix} = - (10+2) = -12$$

$$A_{33} = + \begin{vmatrix} 2 & 3 \\ -2 & 4 \end{vmatrix} = (8+6) = 14 \quad .$$

The matrix of cofactors is:

$$\begin{bmatrix} 28 & 24 & -8 \\ -21 & 12 & 6 \\ 11 & -12 & 14 \end{bmatrix}$$

and the adj of A is:

$$\text{Adj } [A] = \begin{bmatrix} 28 & -21 & 11 \\ 24 & 12 & -12 \\ -8 & 6 & 14 \end{bmatrix} .$$

Given:

$$A = \begin{bmatrix} 3 & 1 & 2 \\ 0 & 1 & 1 \\ -1 & 1 & 0 \end{bmatrix},$$

Show that $(adj\ A) \cdot A = (det\ A) I$ where I is the identity matrix.

Solution: It is known that the classical adjoint, or adj A, is the trans-pose of the matrix of cofactors of the elements a_{ij} of A. The co-factors of the nine elements of the given matrix A are:

$$A_{11} = + \begin{vmatrix} 1 & 1 \\ 1 & 0 \end{vmatrix} = (0-1) = -1$$

$$A_{12} = - \begin{vmatrix} 0 & 1 \\ -1 & 0 \end{vmatrix} = -(0+1) = -1$$

$$A_{13} = + \begin{vmatrix} 0 & 1 \\ -1 & 1 \end{vmatrix} = (0+1) = 1$$

$$A_{21} = - \begin{vmatrix} 1 & 2 \\ 1 & 0 \end{vmatrix} = -(0-2) = +2$$

$$A_{22} = + \begin{vmatrix} 3 & 2 \\ -1 & 0 \end{vmatrix} = (0+2) = 2$$

$$A_{23} = - \begin{vmatrix} 3 & 1 \\ -1 & 1 \end{vmatrix} = -(3+1) = -4$$

$$A_{31} = + \begin{vmatrix} 1 & 2 \\ 1 & 1 \end{vmatrix} = (1-2) = -1$$

$$A_{32} = - \begin{vmatrix} 3 & 2 \\ 0 & 1 \end{vmatrix} = -(3-0) = -3$$

$$A_{33} = + \begin{vmatrix} 3 & 1 \\ 0 & 1 \end{vmatrix} = (3-0) = 3 .$$

Then the matrix of the cofactors is:

$$\begin{bmatrix} -1 & -1 & 1 \\ 2 & 2 & -4 \\ -1 & -3 & 3 \end{bmatrix} .$$

Hence,

$$adj\ A = \begin{bmatrix} -1 & 2 & -1 \\ -1 & 2 & -3 \\ 1 & -4 & 3 \end{bmatrix} .$$

$$(adj\ A) \cdot A = \begin{bmatrix} -1 & 2 & -1 \\ -1 & 2 & -3 \\ 1 & -4 & 3 \end{bmatrix} \begin{bmatrix} 3 & 1 & 2 \\ 0 & 1 & 1 \\ -1 & 1 & 0 \end{bmatrix}$$

$$= \begin{bmatrix} -3+0+1 & -1+2-1 & -2+2+0 \\ -3+0+3 & -1+2-3 & -2+2+0 \\ 3+0-3 & 1-4+3 & 2-4+0 \end{bmatrix}$$

$$= \begin{bmatrix} -2 & 0 & 0 \\ 0 & -2 & 0 \\ 0 & 0 & -2 \end{bmatrix} .$$

$$\text{adj } A \cdot = -2 \begin{bmatrix} 1 & 0 & 0 \\ 0 & 1 & 0 \\ 0 & 0 & 1 \end{bmatrix} = -2I$$

$$\det A = \begin{vmatrix} 3 & 1 & 2 \\ 0 & 1 & 1 \\ -1 & 1 & 0 \end{vmatrix} .$$

$$\det A = 3 \begin{vmatrix} 1 & 1 \\ 1 & 0 \end{vmatrix} - 0 \begin{vmatrix} 1 & 2 \\ 1 & 0 \end{vmatrix} - 1 \begin{vmatrix} 1 & 2 \\ 1 & 1 \end{vmatrix}$$

$$= 3(0-1) -1 (1-2)$$

$$= -3 + 1 = -2 .$$

Hence,

$$\text{adj } A \cdot A = -2I = (\det A)I .$$

• PROBLEM 10–11

Compute the determinants of each of the following matrices and find which of the matrices are invertible.

(a) $\begin{bmatrix} 3 & 1 & 2 \\ 1 & 0 & 6 \\ -1 & 1 & 1 \end{bmatrix}$ (b) $\begin{bmatrix} -1 & 1 & 3 \\ 2 & 1 & 1 \\ 4 & 2 & 2 \end{bmatrix}$ (c) $\begin{bmatrix} 2 & 1 & 1 \\ 0 & 0 & 0 \\ 4 & 3 & 1 \end{bmatrix}$

Solution: We can evaluate determinants by using the basic properties of the determinant function.

Properties of Determinants:
(1)   If each element in a row (or column) is zero, the value of the determinant is zero.
(2)   If two rows (or columns) of a determinant are identical, the value of the determinant is zero.
(3)   The determinant of a matrix A and its transpose A^t are equal: $|A| = |A^t|$.
(4)   The matrix A has an inverse if and only if $\det A \neq 0$.

$$\det A = \begin{vmatrix} 3 & 1 & 2 \\ 1 & 0 & 6 \\ -1 & 1 & 1 \end{vmatrix}$$

$$= 3 \begin{vmatrix} 0 & 6 \\ 1 & 1 \end{vmatrix} - 1 \begin{vmatrix} 1 & 6 \\ -1 & 1 \end{vmatrix} + 2 \begin{vmatrix} 1 & 0 \\ -1 & 1 \end{vmatrix}$$

$$= 3(0-6) - 1(1+6) + 2(1-0)$$

$$= -18-7+2 = -23 .$$

570

Since det A = -23 ≠ 0, this matrix is invertible.

(b)

$$A = \begin{bmatrix} -1 & 1 & 3 \\ 2 & 1 & 1 \\ 4 & 2 & 2 \end{bmatrix}$$

Here det A = 0, since the third row is a multiple of the second row.
Since det A = 0, the matrix is not invertible.

(c)

$$A = \begin{bmatrix} 2 & 1 & 1 \\ 0 & 0 & 0 \\ 4 & 3 & 1 \end{bmatrix}$$

$$\det A = \begin{vmatrix} 2 & 1 & 1 \\ 0 & 0 & 0 \\ 4 & 3 & 1 \end{vmatrix} = 0.$$

Here, each element in the second row is zero, therefore, the value of
the determinant is zero. Since det A = 0, the matrix is not invertible.

• PROBLEM 10–12

Given:
$$A = \begin{bmatrix} 2 & -1 & 1 \\ 4 & 1 & -3 \\ 2 & -1 & 3 \end{bmatrix} ,$$
Evaluate det A and det A^{-1} . What is the relation between det A and
det A^{-1} ?

Solution: First find det A.

$$\det A = \begin{vmatrix} 2 & -1 & 1 \\ 4 & 1 & -3 \\ 2 & -1 & 3 \end{vmatrix}$$

Add the third row to the second row:

$$\det A = \begin{vmatrix} 2 & -1 & 1 \\ 6 & 0 & 0 \\ 2 & -1 & 3 \end{vmatrix}$$

Expand the determinant by minors, using the second row.
$$\det A = -6 \begin{vmatrix} -1 & 1 \\ -1 & 3 \end{vmatrix} + 0 \begin{vmatrix} 2 & 1 \\ 2 & 3 \end{vmatrix} - 0 \begin{vmatrix} 2 & -1 \\ 2 & -1 \end{vmatrix}$$

$$= -6(-3+1) = 12.$$

Thus, det A = 12. Next, find A^{-1} . We know,

$$A^{-1} = \frac{1}{\det A} \cdot adj \ A ,$$

where adj A is the transpose of the matrix of cofactors. Then, the
cofactors are:

$$A_{11} = + \begin{vmatrix} 1 & -3 \\ -1 & 3 \end{vmatrix} = (3-3) = 0$$

$$A_{12} = - \begin{vmatrix} 4 & -3 \\ 2 & 3 \end{vmatrix} = -(12+6) = -18$$

$$A_{13} = + \begin{vmatrix} 4 & 1 \\ 2 & -1 \end{vmatrix} = +(-4-2) = -6$$

$$A_{21} = - \begin{vmatrix} -1 & 1 \\ -1 & 3 \end{vmatrix} = -(-3+1) = 2$$

$$A_{22} = + \begin{vmatrix} 2 & 1 \\ 2 & 3 \end{vmatrix} = (6-2) = 4$$

$$A_{23} = - \begin{vmatrix} 2 & -1 \\ 2 & -1 \end{vmatrix} = -(-2+2) = 0$$

$$A_{31} = + \begin{vmatrix} -1 & 1 \\ 1 & -3 \end{vmatrix} = (3-1) = 2$$

$$A_{32} = - \begin{vmatrix} 2 & 1 \\ 4 & -3 \end{vmatrix} = -(-6-4) = 10$$

$$A_{33} = + \begin{vmatrix} 2 & -1 \\ 4 & 1 \end{vmatrix} = (2+4) = 6 .$$

The matrix of the cofactors is:

$$\begin{bmatrix} 0 & -18 & -6 \\ 2 & 4 & 0 \\ 2 & 10 & 6 \end{bmatrix} .$$

Hence,

$$\text{adj } A = \begin{bmatrix} 0 & 2 & 2 \\ -18 & 4 & 10 \\ -6 & 0 & 6 \end{bmatrix} .$$

Then,

$$A^{-1} = \frac{1}{\det A} (\text{adj } A)$$

$$= \frac{1}{12} \begin{bmatrix} 0 & 2 & 2 \\ -18 & 4 & 10 \\ -6 & 0 & 6 \end{bmatrix}$$

$$= \begin{bmatrix} 0 & 2/12 & 2/12 \\ -18/12 & 4/12 & 10/12 \\ -6/12 & 0 & 6/12 \end{bmatrix} = \begin{bmatrix} 0 & 1/6 & 1/6 \\ -3/2 & 1/3 & 5/6 \\ -1/2 & 0 & 1/2 \end{bmatrix}$$

Now, find $\det A^{-1}$:

$$\det A^{-1} = \begin{vmatrix} 0 & 1/6 & 1/6 \\ -3/2 & 1/3 & 5/6 \\ -1/2 & 0 & 1/2 \end{vmatrix} .$$

Add the third column to the first column.

$$\det A^{-1} = \begin{vmatrix} 1/6 & 1/6 & 1/6 \\ -4/6 & 1/3 & 5/6 \\ 0 & 0 & 1/2 \end{vmatrix} .$$

572

Expand the determinant by minors, using the third row:

$$\det A^{-1} = 0 \begin{vmatrix} 1/6 & 1/6 \\ 1/3 & 5/6 \end{vmatrix} - 0 \begin{vmatrix} 1/6 & 1/6 \\ -4/6 & 5/6 \end{vmatrix} + 1/2 \begin{vmatrix} 1/6 & 1/6 \\ -4/6 & 1/3 \end{vmatrix}$$

$$= 1/2(1/18 + 4/36)$$

$$= 1/2(1/6) = 1/12 .$$

Thus, $\det A = 12$, while $\det A^{-1} = 1/12$. That is,

$$\det A^{-1} = \frac{1}{\det A} .$$

We emphasize that A has an inverse if and only if $\det A \neq 0$.

● **PROBLEM** 10–13

Given:

$$A = \begin{bmatrix} 1 & 1 & 4 \\ -1 & 2 & -3 \\ 2 & -1 & 3 \end{bmatrix} , \quad B = \begin{bmatrix} -1 & 1 & 2 \\ 0 & 2 & 3 \\ 2 & 3 & -1 \end{bmatrix} .$$

Show that the determinant of the product of these two matrices, A and B, is equal to the product of their determinants, i.e., $\det AB = \det A \cdot \det B$.

<u>Solution</u>:

$$\det A = \begin{vmatrix} 1 & 1 & 4 \\ -1 & 2 & -3 \\ 2 & -1 & 3 \end{vmatrix}$$

Add the second row to the first row.

$$\det A = \begin{vmatrix} 0 & 3 & 1 \\ -1 & 2 & -3 \\ 2 & -1 & 3 \end{vmatrix}$$

Add -3 times the third column to the second column.

$$\det A = \begin{vmatrix} 0 & 0 & 1 \\ -1 & 11 & -3 \\ 2 & -10 & 3 \end{vmatrix}$$

Expanding along the first row, two terms drop out, since they have factors of 0 .

$$\text{Det } A = 1 \begin{vmatrix} -1 & 11 \\ 2 & -10 \end{vmatrix} = 10 - 22 = -12$$

$$\det B = \begin{vmatrix} -1 & 1 & 2 \\ 0 & 2 & 3 \\ 2 & 3 & -1 \end{vmatrix}$$

Expand the determinant by minors, using the first column.

$$\det B = -1 \begin{vmatrix} 2 & 3 \\ 3 & -1 \end{vmatrix} - 0 \begin{vmatrix} 1 & 2 \\ 3 & -1 \end{vmatrix} + 2 \begin{vmatrix} 1 & 2 \\ 2 & 3 \end{vmatrix}$$

$$= -1(-2-9) + 2(3-4)$$

$$= + 11 - 2 = 9 .$$

573

$$AB = \begin{bmatrix} 1 & 1 & 4 \\ -1 & 2 & -3 \\ 2 & -1 & 3 \end{bmatrix} \begin{bmatrix} -1 & 1 & 2 \\ 0 & 2 & 3 \\ 2 & 3 & -1 \end{bmatrix}$$

$$= \begin{bmatrix} -1+0+8 & 1+2+12 & 2+3-4 \\ 1+0-6 & -1+4-9 & -2+6+3 \\ -2+0+6 & 2-2+9 & 4-3-3 \end{bmatrix}$$

$$= \begin{bmatrix} 7 & 15 & 1 \\ -5 & -6 & 7 \\ 4 & 9 & -2 \end{bmatrix}$$

Then,

$$\det AB = \begin{vmatrix} 7 & 15 & 1 \\ -5 & -6 & 7 \\ 4 & 9 & -2 \end{vmatrix}$$

$$= 7 \begin{vmatrix} -6 & 7 \\ 9 & -2 \end{vmatrix} - 15 \begin{vmatrix} -5 & 7 \\ 4 & -2 \end{vmatrix} + 1 \begin{vmatrix} -5 & -6 \\ 4 & 9 \end{vmatrix}$$

So,

$\det AB = 7(12-63) -15(10-28) + 1(-45+24) = -357 + 270 - 21 = -108$

and,

$$\det A \cdot \det B = (-12) \times (9)$$
$$= -108 .$$

Hence, $\det AB = \det A \cdot \det B$.

• PROBLEM 10–14

Solve the following linear equations by using Cramer's Rule:

$$-2x_1 + 3x_2 - x_3 = 1$$
$$x_1 + 2x_2 - x_3 = 4$$
$$-2x_1 - x_2 + x_3 = -3 .$$

<u>Solution</u>: Consider a system of n linear equations in n unknowns:

$$a_{11}x_1 + a_{12}x_2 + \dots + a_{1n}x_n = b_1$$

$$a_{21}x_1 + a_{22}x_2 + \dots + a_{2n}x_n = b_2$$

. . . .

$$a_{n1}x_1 + a_{n2}x_2 + \dots + a_{nn}x_n = b_n .$$

Write the above equations in matrix notation.

$$\begin{bmatrix} a_{11} & a_{12} \cdots a_{1n} \\ a_{21} & a_{22} \cdots a_{2n} \\ \vdots & \qquad \vdots \\ a_{n1} & a_{n2} \cdots a_{nn} \end{bmatrix} \begin{bmatrix} x_1 \\ x_2 \\ \vdots \\ x_n \end{bmatrix} = \begin{bmatrix} b_1 \\ b_2 \\ \vdots \\ b_n \end{bmatrix}$$

or, $AX = B$.

Let A be an $n \times n$ matrix over the field F such that $\det A \neq 0$. If $b_1, b_2, \ldots, b_n$ are any scalars in F, the unique solution of the system of equations $AX = B$ is given by:

$$x_i = \frac{\det A_i}{\det A} \qquad i = 1, 2, \ldots, n \ ,$$

where A_i is the $n \times n$ matrix obtained from A by replacing the ith column of A by the column vector

$$\begin{bmatrix} b_1 \\ b_2 \\ \vdots \\ b_n \end{bmatrix}$$

The above theorem is known as "Cramer's Rule" for solving systems of linear equations. Cramer's Rule applies only to systems of n linear equations in n unknowns with non-zero determinants.

Consider the given linear system:

$$-2x_1 + 3x_2 - x_3 = 1$$
$$x_1 + 2x_2 - x_3 = 4$$
$$-2x_1 - x_2 + x_3 = -3$$

or,

$$\begin{bmatrix} -2 & 3 & -1 \\ 1 & 2 & --1 \\ -2 & -1 & 1 \end{bmatrix} \begin{bmatrix} x_1 \\ x_2 \\ x_3 \end{bmatrix} = \begin{bmatrix} 1 \\ 4 \\ 3 \end{bmatrix}$$

$$A \qquad\qquad X \ = \ B$$

$$\text{Det } A = \begin{vmatrix} -2 & 3 & -1 \\ 1 & 2 & -1 \\ -2 & -1 & 1 \end{vmatrix}$$

$$\text{Det } A = -2 \begin{vmatrix} 2 & -1 \\ -1 & 1 \end{vmatrix} - 3 \begin{vmatrix} 1 & -1 \\ -2 & 1 \end{vmatrix} - 1 \begin{vmatrix} 1 & 2 \\ -2 & -1 \end{vmatrix}$$

$$= -2(2-1) - 3(1-2) - (-1+4)$$

$$= -2 + 3 - 3 = -2 \ .$$

Since $\det A \neq 0$, the system has a unique solution. Now,

$$x_1 = \frac{\det A_1}{\det A} \ , \quad x_2 = \frac{\det A_2}{\det A} \ , \quad x_3 = \frac{\det A_3}{\det A} \ .$$

Det A_1 is the determinant of the matrix obtained by replacing the 1st column of A by the column of B. Thus,

$$\det A_1 = \begin{vmatrix} 1 & 3 & -1 \\ 4 & 2 & -1 \\ 3 & -1 & 1 \end{vmatrix}$$

$$= -4 .$$

Then,

$$x_1 = \frac{-4}{-2} = 2 .$$

$$x_2 = \frac{\begin{vmatrix} -2 & 1 & -1 \\ 1 & 4 & -1 \\ -2 & -3 & 1 \end{vmatrix}}{|A|} = \frac{-6}{-2} = 3 .$$

$$x_3 = \frac{\det A_3}{\det A} = \frac{\begin{vmatrix} -2 & 3 & 1 \\ 1 & 2 & 4 \\ -2 & -1 & -3 \end{vmatrix}}{-2} = \frac{-8}{-2} = 4 .$$

Thus,

$$x_1 = 2 , \ x_2 = 3 , \ x_3 = 4 ,$$

is the unique solution to the given system.

• PROBLEM 10–15

Solve the following homogeneous equations:

$$x_1 + 2x_2 + x_3 = 0$$

$$x_2 - 3x_3 = 0$$

$$-x_1 + x_2 - x_3 = 0 .$$

Solution: A homogeneous system of linear equations has either a) the unique trivial solution $x_1 = x_2 = \ldots = x_n = 0$ or b) an infinite number of non-trivial solutions, plus the trivial solution.

First, write the above equations in matrix notation:

$$\begin{bmatrix} 1 & 2 & 1 \\ 0 & 1 & -3 \\ -1 & 1 & -1 \end{bmatrix} \begin{bmatrix} x_1 \\ x_2 \\ x_3 \end{bmatrix} = \begin{bmatrix} 0 \\ 0 \\ 0 \end{bmatrix}$$

or, AX = 0 .

$$\det A = 1 \begin{vmatrix} 1 & -3 \\ 1 & -1 \end{vmatrix} - 2 \begin{vmatrix} 0 & -3 \\ -1 & -1 \end{vmatrix} + 1 \begin{vmatrix} 0 & 1 \\ -1 & 1 \end{vmatrix}$$

$$= \left[(-1+3) - 2(0-3) + (0+1) \right]$$

576

$$= 2 + 6 + 1$$
$$= 9 .$$

Hence, det $A \neq 0$, and according to Cramer's Rule, the above system has a unique solution.

Therefore, $X = 0$, i.e., $x_1 = x_2 = x_3 = 0$ since the homogeneous system $AX = 0$ has a non-zero solution if and only if det $A = 0$.

• PROBLEM 10–16

Solve the system of linear equations:
$$3x + 2y + 4z = 1$$
$$2x - y + z = 0$$
$$x + 2y + 3z = 1 .$$

<u>Solution</u>: Use Cramer's Rule to solve this system. Write the above equations in matrix form:

$$\begin{bmatrix} 3 & 2 & 4 \\ 2 & -1 & 1 \\ 1 & 2 & 3 \end{bmatrix} \begin{bmatrix} x \\ y \\ z \end{bmatrix} = \begin{bmatrix} 1 \\ 0 \\ 1 \end{bmatrix} .$$

Then

$$A = \begin{bmatrix} 3 & 2 & 4 \\ 2 & -1 & 1 \\ 1 & 2 & 3 \end{bmatrix} , \quad B = \begin{bmatrix} 1 \\ 0 \\ 1 \end{bmatrix} .$$

First, check that det $A \neq 0$.

$$\det A = \begin{vmatrix} 3 & 2 & 4 \\ 2 & -1 & 1 \\ 1 & 2 & 3 \end{vmatrix}$$

$$\det A = 3 \begin{vmatrix} -1 & 1 \\ 2 & 3 \end{vmatrix} - 2 \begin{vmatrix} 2 & 1 \\ 1 & 3 \end{vmatrix} + 4 \begin{vmatrix} 2 & -1 \\ 1 & 2 \end{vmatrix}$$

$$= 3(-3-2) - 2(6-1) + 4(4+1)$$
$$= -15x - 10 + 20 = x-5.$$

Since det $A \neq 0$, the system has a unique solution. Then,

$$x = \frac{\det A_1}{\det A}, \quad y = \frac{\det A_2}{\det A}, \quad z = \frac{\det A_3}{\det A} .$$

det A_1 is the determinant of the matrix obtained by replacing the first column of A by the column vector B.

Thus,

$$\det A_1 = \begin{vmatrix} 1 & 2 & 4 \\ 0 & -1 & 1 \\ 1 & 2 & 3 \end{vmatrix} .$$

Expand the determinant by minors, using the first column.

$$\det A_1 = 1 \begin{vmatrix} -1 & 1 \\ 2 & 3 \end{vmatrix} + 1 \begin{vmatrix} 2 & 4 \\ -1 & 1 \end{vmatrix}$$

$$= 1(-3-2) + 1(2+4)$$

$$= -5 + 6 = + 1 .$$

Now, we have $\det A = -5$.

Thus, $x = \dfrac{\det A_1}{\det A} = \dfrac{1}{-5} = -\dfrac{1}{5}$.

$y = \dfrac{\det A_2}{\det A} = \dfrac{\begin{vmatrix} 3 & 1 & 4 \\ 2 & 0 & 1 \\ 1 & 1 & 3 \end{vmatrix}}{-5}$.

Now, expand $\det A_2$ along the second row:

$$\begin{vmatrix} 3 & 1 & 4 \\ 2 & 0 & 1 \\ 1 & 1 & 3 \end{vmatrix} = -2 \begin{vmatrix} 1 & 4 \\ 1 & 3 \end{vmatrix} - 1 \begin{vmatrix} 3 & 1 \\ 1 & 1 \end{vmatrix}$$

$$= -2(3-4) - 1(3-1)$$

$$= 2 - 2 = 0$$

$y = \dfrac{0}{-5} = 0$.

$$z = \dfrac{\det A_3}{\det A} = \dfrac{\begin{vmatrix} 3 & 2 & 1 \\ 2 & -1 & 0 \\ 1 & 2 & 1 \end{vmatrix}}{-5}$$

Expand determinant A_3 by minors, using the third column.

$$\begin{vmatrix} 3 & 2 & 1 \\ 2 & -1 & 0 \\ 1 & 2 & 1 \end{vmatrix} = +1 \begin{vmatrix} 2 & -1 \\ 1 & 2 \end{vmatrix} + 1 \begin{vmatrix} 3 & 2 \\ 2 & -1 \end{vmatrix}$$

$$= (4+1) + (-3-4)$$

$$= 5 - 7 = -2 .$$

Then,

$$z = \dfrac{-2}{-5} = \dfrac{2}{5} .$$

Thus $x = -1/5$, $y = 0$, $z = 2/5$.

MATRIX ARITHMETIC

● **PROBLEM** 10–17

If $A = \begin{bmatrix} 2 & -2 & 4 \\ -1 & 1 & 1 \end{bmatrix}$ and $B = \begin{bmatrix} 0 & 1 & -3 \\ 1 & 3 & 1 \end{bmatrix}$, find $2A + B$.

Solution: For an m×n matrix, $A = (a_{ij})$, we know $cA = (ca_{ij})$. Hence,

$$2A = 2\begin{bmatrix} 2 & -2 & 4 \\ -1 & 1 & 1 \end{bmatrix} = \begin{bmatrix} 2\cdot2 & 2\cdot(-2) & 2\cdot4 \\ 2\cdot(-1) & 2\cdot1 & 2\cdot1 \end{bmatrix} = \begin{bmatrix} 4 & -4 & 8 \\ -2 & 2 & 2 \end{bmatrix}.$$

For two m×n matrices, $A = (\alpha_{ij})$ and $B = (\beta_{ij})$, the ith row of the matrix $A + B$ is given by $e_i \cdot (A+B) = (\alpha_{i1} + \beta_{i1}, \ldots, \alpha_{in} + \beta_{in})$.

Thus,

$$2A + B = \begin{bmatrix} 4 & -4 & 8 \\ -2 & 2 & 2 \end{bmatrix} + \begin{bmatrix} 0 & 1 & -3 \\ 1 & 3 & 1 \end{bmatrix} = \begin{bmatrix} 4+0 & -4+1 & 8-3 \\ -2+1 & 2+3 & 2+1 \end{bmatrix}$$

$$2A + B = \begin{bmatrix} 4 & -3 & 5 \\ -1 & 5 & 3 \end{bmatrix}.$$

● **PROBLEM** 10–18

Show that

a) $A + B = B + A$ where

$$A = \begin{bmatrix} 3 & 1 & 1 \\ 2 & -1 & 1 \end{bmatrix} ; \quad B = \begin{bmatrix} 4 & 2 & -1 \\ 0 & 0 & 2 \end{bmatrix}.$$

b) $(A+B) + C = A + (B+C)$ where

$$A = \begin{bmatrix} -2 & 6 \\ 2 & 1 \end{bmatrix}, \quad B = \begin{bmatrix} 2 & 1 \\ 0 & 3 \end{bmatrix} \quad \text{and} \quad C = \begin{bmatrix} -1 & 0 \\ 7 & 2 \end{bmatrix}.$$

c)   If A and the zero matrix (0_{ij}) have the same size, then $A + 0 = A$ where

$$A = \begin{bmatrix} 2 & 1 \\ 1 & 2 \end{bmatrix}.$$

d) $A + (-A) = 0$ where

$$A = \begin{bmatrix} 2 & 1 \\ 1 & 2 \end{bmatrix}.$$

e) $(ab)A = a(bA)$ where $a = -5$, $b = 3$ and

$$A = \begin{bmatrix} 6 & -1 & 0 \\ 1 & 2 & 1 \end{bmatrix}.$$

f)   Find B if $2A - 3B + C = 0$ where

$$A = \begin{bmatrix} -1 & 3 \\ 0 & 0 \end{bmatrix} \text{ and } C = \begin{bmatrix} -2 & -1 \\ -1 & 1 \end{bmatrix}.$$

<u>Solution</u>: a) By the definition of matrix addition,

$$A + B = \begin{bmatrix} 3 & 1 & 1 \\ 2 & -1 & 1 \end{bmatrix} + \begin{bmatrix} 4 & 2 & -1 \\ 0 & 0 & 2 \end{bmatrix}$$

$$= \begin{bmatrix} 3+4 & 1+2 & 1+(-1) \\ 2+0 & -1+0 & 1+2 \end{bmatrix}$$

$$= \begin{bmatrix} 7 & 3 & 0 \\ 2 & -1 & 3 \end{bmatrix}$$

and

$$B + A = \begin{bmatrix} 4 & 2 & -1 \\ 0 & 0 & 2 \end{bmatrix} + \begin{bmatrix} 3 & 1 & 1 \\ 2 & -1 & 1 \end{bmatrix}$$

$$= \begin{bmatrix} 4+3 & 2+1 & -1+1 \\ 0+2 & 0+(-1) & 2+1 \end{bmatrix} + \begin{bmatrix} 7 & 3 & 0 \\ 2 & -1 & 3 \end{bmatrix}$$

Thus, $A + B = B + A$.

b) $\quad A + B = \begin{bmatrix} -2 & 6 \\ 2 & 1 \end{bmatrix} + \begin{bmatrix} 2 & 1 \\ 0 & 3 \end{bmatrix}$

$$= \begin{bmatrix} -2+2 & 6+1 \\ 2+0 & 1+3 \end{bmatrix} = \begin{bmatrix} 0 & 7 \\ 2 & 4 \end{bmatrix}$$

and

$$(A+B) + C = \begin{bmatrix} 0 & 7 \\ 2 & 4 \end{bmatrix} + \begin{bmatrix} -1 & 0 \\ 7 & 2 \end{bmatrix} = \begin{bmatrix} 0+(-1) & 7+0 \\ 2+7 & 4+2 \end{bmatrix} = \begin{bmatrix} -1 & 7 \\ 9 & 6 \end{bmatrix}.$$

$$B + C = \begin{bmatrix} 2 & 1 \\ 0 & 3 \end{bmatrix} + \begin{bmatrix} -1 & 0 \\ 7 & 2 \end{bmatrix} = \begin{bmatrix} 2+(-1) & 1+0 \\ 0+7 & 3+2 \end{bmatrix} = \begin{bmatrix} 1 & 1 \\ 7 & 5 \end{bmatrix}$$

and

$$A + (B+C) = \begin{bmatrix} -2 & 6 \\ 2 & 1 \end{bmatrix} + \begin{bmatrix} 1 & 1 \\ 7 & 5 \end{bmatrix} = \begin{bmatrix} -2+1 & 6+1 \\ 2+7 & 1+5 \end{bmatrix} = \begin{bmatrix} -1 & 7 \\ 9 & 6 \end{bmatrix}.$$

Thus, $(A+B) + C = A + (B+C)$.

c)   An $m{\times}n$ matrix all of whose elements are zeros is called a zero matrix and is usually denoted by $\underset{m \ n}{0}$.

$$A = \begin{bmatrix} 2 & 1 \\ 1 & 2 \end{bmatrix} \qquad 0 = \begin{bmatrix} 0 & 0 \\ 0 & 0 \end{bmatrix}.$$

Thus,

$$A + 0 = \begin{bmatrix} 2 & 1 \\ 1 & 2 \end{bmatrix} + \begin{bmatrix} 0 & 0 \\ 0 & 0 \end{bmatrix} = \begin{bmatrix} 2+0 & 1+0 \\ 1+0 & 2+0 \end{bmatrix} = \begin{bmatrix} 2 & 1 \\ 1 & 2 \end{bmatrix}.$$

Hence, $A + 0 = A$.

d) $\quad -A = -1 \cdot \begin{bmatrix} 2 & 1 \\ 1 & 2 \end{bmatrix} = \begin{bmatrix} -1 \cdot 2 & -1 \cdot 1 \\ -1 \cdot 1 & -1 \cdot 2 \end{bmatrix} = \begin{bmatrix} -2 & -1 \\ -1 & -2 \end{bmatrix}.$

Thus,

$$A + (-A) = \begin{bmatrix} 2 & 1 \\ 1 & 2 \end{bmatrix} + \begin{bmatrix} -2 & -1 \\ -1 & -2 \end{bmatrix} = \begin{bmatrix} 2+(-2) & 1+(-1) \\ 1+(-1) & 2+(-2) \end{bmatrix} = \begin{bmatrix} 0 & 0 \\ 0 & 0 \end{bmatrix}$$

Therefore, $A + (-A) = 0$.

e)   If $A = \begin{bmatrix} a_1 & b_1 \\ c_1 & d_1 \end{bmatrix}$ and a is any scalar from a field, aA is

defined by
$$aA = \begin{bmatrix} aa_1 & ab_1 \\ ac_1 & ad_1 \end{bmatrix} .$$

So, $bA = 3 \begin{bmatrix} 6 & -1 & 0 \\ 1 & 2 & 1 \end{bmatrix} = \begin{bmatrix} 3 \cdot 6 & 3 \cdot (-1) & 3 \cdot 0 \\ 3 \cdot 1 & 3 \cdot 2 & 3 \cdot 1 \end{bmatrix}$

$$= \begin{bmatrix} 18 & -3 & 0 \\ 3 & 6 & 3 \end{bmatrix}$$

and
$$a(bA) = -5 \begin{bmatrix} 18 & -3 & 0 \\ 3 & 6 & 3 \end{bmatrix} = \begin{bmatrix} -90 & 15 & 0 \\ -15 & -30 & -15 \end{bmatrix}$$

$$(ab)A = ((-5)(3)) \begin{bmatrix} 6 & -1 & 0 \\ 1 & 2 & 1 \end{bmatrix} = -15 \begin{bmatrix} 6 & -1 & 0 \\ 1 & 2 & 1 \end{bmatrix} = \begin{bmatrix} -90 & 15 & 0 \\ -15 & -30 & -15 \end{bmatrix}$$

Thus, $(ab)A = a(bA)$.

f) $2A - 3B + C = 2A + C - 3B = 0$ since matrix addition is commutative.

Now, add 3B to both sides of the equation,
$$2A + C - 3B = 0 ,$$
to obtain $2A + C - 3B + 3B = 0 + 3B$.   (1)

Using the laws we exemplified in parts a) through d), (1) becomes
$2A + C = 3B$. Now,
$$\tfrac{1}{3}(2A + C) = \tfrac{1}{3}(3B)$$
which implies $B = \tfrac{1}{3}(2A + C)$.

$$2A + C = \begin{bmatrix} 2(-1) & 2(3) \\ 2(0) & 2(0) \end{bmatrix} + \begin{bmatrix} -2 & -1 \\ -1 & 1 \end{bmatrix} = \begin{bmatrix} -4 & 5 \\ -1 & 1 \end{bmatrix} .$$
Thus,
$$\tfrac{1}{3}(2A + C) = \tfrac{1}{3} \begin{bmatrix} -4 & 5 \\ -1 & 1 \end{bmatrix} = \begin{bmatrix} -4/3 & 5/3 \\ -1/3 & 1/3 \end{bmatrix} .$$

• PROBLEM 10–19

If $A = \begin{bmatrix} 1 & 2 \\ -1 & 3 \end{bmatrix}$ and $B = \begin{bmatrix} 2 & 1 \\ 0 & 1 \end{bmatrix}$, show $AB \neq BA$.

Solution: A is 2×2 and B is 2×2 ; the product AB is a 2×2
matrix.
$$AB = \begin{bmatrix} 1 & 2 \\ -1 & 3 \end{bmatrix} \begin{bmatrix} 2 & 1 \\ 0 & 1 \end{bmatrix} = \begin{bmatrix} 1 \cdot 2 + 2 \cdot 0 & 1 \cdot 1 + 2 \cdot 1 \\ -1 \cdot 2 + 3 \cdot 0 & -1 \cdot 1 + 3 \cdot 1 \end{bmatrix} = \begin{bmatrix} 2+0 & 1+2 \\ -2+0 & -1+3 \end{bmatrix}$$

$$= \begin{bmatrix} 2 & 3 \\ -2 & 2 \end{bmatrix} .$$

Now,
$$BA = \begin{bmatrix} 2 & 1 \\ 0 & 1 \end{bmatrix} \begin{bmatrix} 1 & 2 \\ -1 & 3 \end{bmatrix} = \begin{bmatrix} 2 \cdot 1 + 1 \cdot (-1) & 2 \cdot 2 + 1 \cdot 3 \\ 0 \cdot 1 + 1 \cdot (-1) & 0 \cdot 2 + 1 \cdot 3 \end{bmatrix}$$

$$= \begin{bmatrix} 2-1 & 4+3 \\ 0-1 & 0+3 \end{bmatrix} = \begin{bmatrix} 1 & 7 \\ -1 & 3 \end{bmatrix}$$

Thus, $AB \neq BA$.

● PROBLEM 10–20

If $A = \begin{bmatrix} 2 & -1 & 0 \\ 1 & 0 & -3 \end{bmatrix}$ and $B = \begin{bmatrix} 1 & -4 & 0 & 1 \\ 2 & -1 & 3 & -1 \\ 4 & 0 & -2 & 0 \end{bmatrix}$

(1) Determine the shape of AB .
(2) If c_{ij} denotes the element in the ith row and jth column of the product matrix AB, find c_{23} c_{14} and c_{21} .

Solution: (1) A matrix with m rows and n columns is called an m by n matrix or an m×n matrix. The pair of numbers (m,n) is called its size or shape. Since A is 2×3 and B is 3×4 , the product AB is a 2×4 matrix. Thus, the shape of AB is (2,4).

(2) Now, c_{23} is the element in the 2nd row and 3rd column of the product matrix AB . To obtain c_{23} , multiply the second row of A by the third column of B . Hence,

$$c_{23} = [1 \cdot 0 \ -3] \begin{bmatrix} 0 \\ 3 \\ -2 \end{bmatrix}$$

$$= 1 \cdot 0 + 0 \cdot 3 + (-3) \cdot (-2) = 0 + 0 + 6 = 6 ,$$

and
$$c_{14} = [2 \ -1 \ 0] \begin{bmatrix} 1 \\ -1 \\ 0 \end{bmatrix}$$

$$= 2 \cdot 1 + (-1) \cdot (-1) + 0 \cdot 0 = 2 + 1 + 0 = 3 .$$

$$c_{21} = [1 \ 0 \ -3] \begin{bmatrix} 1 \\ 2 \\ 4 \end{bmatrix} = 1 \cdot 1 + 0 \cdot 2 + (-3) \cdot 4$$

$$= 1 + 0 - 12 = -11 .$$

● PROBLEM 10–21

Prove   (AB)C = A(BC)   where $A = \begin{bmatrix} 5 & 2 & 3 \\ 2 & -3 & 4 \end{bmatrix}$, $B = \begin{bmatrix} 2 & -1 & 1 & 0 \\ 0 & 2 & 2 & 2 \\ 3 & 0 & -1 & 3 \end{bmatrix}$

and $C = \begin{bmatrix} 1 & 0 & 2 \\ 2 & -3 & 0 \\ 0 & 0 & 3 \\ 2 & 1 & 0 \end{bmatrix}$

<u>Solution</u>: First, find $(AB)C$ and then $A(BC)$.

$$AB = \begin{bmatrix} 5 & 2 & 3 \\ 2 & -3 & 4 \end{bmatrix} \begin{bmatrix} 2 & -1 & 1 & 0 \\ 0 & 2 & 2 & 2 \\ 3 & 0 & -1 & 3 \end{bmatrix}$$

$$= \begin{bmatrix} 10+0+9 & -5+4+0 & 5+4-3 & 0+4+9 \\ 4+0+12 & -2-6+0 & 2-6-4 & 0-6+12 \end{bmatrix}$$

$$= \begin{bmatrix} 19 & -1 & 6 & 13 \\ 16 & -8 & -8 & 6 \end{bmatrix}$$

Now,

$$(AB)C = \begin{bmatrix} 19 & -1 & 6 & 13 \\ 16 & -8 & -8 & 6 \end{bmatrix} \begin{bmatrix} 1 & 0 & 2 \\ 2 & -3 & 0 \\ 0 & 0 & 3 \\ 2 & 1 & 0 \end{bmatrix}$$

$$= \begin{bmatrix} 19-2+0+26 & 0+3+0+13 & 38+0+18+0 \\ 16-16+0+12 & 0+24+0+6 & 32+0-24+0 \end{bmatrix} = \begin{bmatrix} 43 & 16 & 56 \\ 12 & 30 & 8 \end{bmatrix}.$$

Thus,

$$(AB)C = \begin{bmatrix} 43 & 16 & 56 \\ 12 & 30 & 8 \end{bmatrix}.$$

$$BC = \begin{bmatrix} 2 & -1 & 1 & 0 \\ 0 & 2 & 2 & 2 \\ 3 & 0 & -1 & 3 \end{bmatrix} \begin{bmatrix} 1 & 0 & 2 \\ 2 & -3 & 0 \\ 0 & 0 & 3 \\ 2 & 1 & 0 \end{bmatrix}$$

$$= \begin{bmatrix} 2-2+0+0 & 0+3+0+0 & 4+0+3+0 \\ 0+4+0+4 & 0-6+0+2 & 0+0+6+0 \\ 3+0+0+6 & 0+0+0+3 & 6+0-3+0 \end{bmatrix}$$

$$BC = \begin{bmatrix} 0 & 3 & 7 \\ 8 & -4 & 6 \\ 9 & 3 & 3 \end{bmatrix}$$

and

$$A(BC) = \begin{bmatrix} 5 & 2 & 3 \\ 2 & -3 & 4 \end{bmatrix} \begin{bmatrix} 0 & 3 & 7 \\ 8 & -4 & 6 \\ 9 & 3 & 3 \end{bmatrix}$$

$$= \begin{bmatrix} 0+16+27 & 15-8+9 & 35+12+9 \\ 0-24+36 & 6+12+12 & 14-18+12 \end{bmatrix} = \begin{bmatrix} 43 & 16 & 56 \\ 12 & 30 & 8 \end{bmatrix}$$

Thus, $(AB)C = A(BC)$.

Find (i) A^2 (ii) A^3 (iii) A^4 when $A = \begin{bmatrix} 1 & 2 \\ -1 & 1 \end{bmatrix}$.

<u>Solution</u>: $A^2 = AA = \begin{bmatrix} 1 & 2 \\ -1 & 1 \end{bmatrix} \begin{bmatrix} 1 & 2 \\ -1 & 1 \end{bmatrix} = \begin{bmatrix} 1-2 & 2+2 \\ -1-1 & -2+1 \end{bmatrix} = \begin{bmatrix} -1 & 4 \\ -2 & -1 \end{bmatrix}$.

$A^3 = AAA = A^2A = \begin{bmatrix} -1 & 4 \\ -2 & -1 \end{bmatrix} \begin{bmatrix} 1 & 2 \\ -1 & 1 \end{bmatrix} = \begin{bmatrix} -1-4 & -2+4 \\ -2+1 & -4-1 \end{bmatrix} = \begin{bmatrix} -5 & 2 \\ -1 & -5 \end{bmatrix}$.

The usual laws for exponents are $A^m A^n = A^{m+n}$ and $(A^m)^n = A^{mn}$. Thus,
$A^4 = A^3 A$ or, $A^4 = (A^2)^2 = A^2 A^2$.

$\qquad A^4 = A^3 A = \begin{bmatrix} -5 & 2 \\ -1 & -5 \end{bmatrix} \begin{bmatrix} 1 & 2 \\ -1 & 1 \end{bmatrix} = \begin{bmatrix} -5-2 & -10+2 \\ -1+5 & -2-5 \end{bmatrix} = \begin{bmatrix} -7 & -8 \\ 4 & -7 \end{bmatrix}$.

Observe that

$\qquad A^4 = A^2 A^2 = \begin{bmatrix} -1 & 4 \\ -2 & -1 \end{bmatrix} \begin{bmatrix} -1 & 4 \\ -2 & -1 \end{bmatrix} = \begin{bmatrix} 1-8 & -4-4 \\ 2+2 & -8+1 \end{bmatrix}$

$\qquad = \begin{bmatrix} -7 & -8 \\ 4 & -7 \end{bmatrix}$

Compute AB using block multiplication where

$A = \left[\begin{array}{cc:c} 1 & 2 & 1 \\ 3 & 4 & 0 \\ \hdashline 0 & 0 & 2 \end{array} \right]$ and $B = \left[\begin{array}{ccc:c} 1 & 2 & 3 & 1 \\ 4 & 5 & 6 & 1 \\ \hdashline 0 & 0 & 0 & 1 \end{array} \right]$.

<u>Solution</u>: Using a system of horizontal and vertical lines, one can partition a matrix A into smaller "submatrices" of A. The matrix A is then called a block matrix. A given matrix may be divided into blocks in different ways. For example,

$\begin{bmatrix} 1 & -2 & 0 & 1 \\ 2 & 3 & 5 & 7 \\ 3 & 1 & 4 & 5 \end{bmatrix} = \left[\begin{array}{cc:cc} 1 & -2 & 0 & 1 \\ 2 & 3 & 5 & 7 \\ \hdashline 3 & 1 & 4 & 5 \end{array} \right]$

$\qquad\qquad\qquad\quad = \left[\begin{array}{ccc:c} 1 & -2 & 0 & 1 \\ \hdashline 2 & 3 & 5 & 7 \\ \hdashline 3 & 1 & 4 & 5 \end{array} \right]$.

$$A = \begin{bmatrix} 1 & 2 & | & 1 \\ 3 & 4 & | & 0 \\ - & - & - & - & - \\ 0 & 0 & | & 2 \end{bmatrix} .$$

Let $E \begin{bmatrix} 1 & 2 \\ 3 & 4 \end{bmatrix}$, $F = \begin{bmatrix} 1 \\ 0 \end{bmatrix}$ and $G[2]$ then,

$$A = \begin{bmatrix} E & | & F \\ - & -|- & - \\ 0 & | & G \end{bmatrix}$$

$$B = \begin{bmatrix} 1 & 2 & 3 & | & 1 \\ 4 & 5 & 6 & | & 1 \\ - & - & - & - & - & - \\ 0 & 0 & 0 & | & 1 \end{bmatrix} .$$ Let $R = \begin{bmatrix} 1 & 2 & 3 \\ 4 & 5 & 6 \end{bmatrix}$ $S = \begin{bmatrix} 1 \\ 1 \end{bmatrix}$ $T = [1]$.

Then,

$$B = \begin{bmatrix} R & | & S \\ - & -|- & - \\ 0 & | & T \end{bmatrix} .$$

After partitioning the matrices into block matrices, multiplication of the matrices is the usual matrix multiplication with each entire block considered as a unit entry of the matrix. If two matrices can be multiplied, then they can be multiplied as block matrices if they are each partitioned into blocks similarly; that is, into an equal number of blocks so that corresponding blocks have the same size. Suppose

$$A = \begin{bmatrix} A_1 & | & A_2 \\ - & -|- & - \\ A_3 & | & A_4 \end{bmatrix} \quad \text{and} \quad B = \begin{bmatrix} B_1 & | & B_2 \\ - & -|- & - \\ B_3 & | & B_4 \end{bmatrix}$$

where A_1 and B_1, A_2 and B_2, A_3 and B_3, A_4 and B_4 are the same sizes, respectively. Then AB is given by

$$\begin{matrix} A_1 B_1 + A_2 B_3 & A_1 B_2 + A_2 B_4 \\ A_3 B_1 + A_3 B_3 & A_3 B_2 + A_4 B_4 \end{matrix}$$

In the problem,

$$AB = \begin{bmatrix} E & F \\ 0 & G \end{bmatrix} \begin{bmatrix} R & S \\ 0 & T \end{bmatrix} = \begin{bmatrix} ER + F \cdot 0 & ES + FT \\ 0R + G \cdot 0 & 0S + GT \end{bmatrix}$$

$$= \begin{bmatrix} ER & ES + FT \\ 0 & GT \end{bmatrix} =$$

$$= \begin{bmatrix} \begin{bmatrix} 1 & 2 \\ 3 & 4 \end{bmatrix} \begin{bmatrix} 1 & 2 & 3 \\ 4 & 5 & 6 \end{bmatrix} & \begin{bmatrix} 1 & 2 \\ 3 & 4 \end{bmatrix} \begin{bmatrix} 1 \\ 1 \end{bmatrix} + \begin{bmatrix} 1 \\ 0 \end{bmatrix} [1] \\ 0 & [2] \, [1] \end{bmatrix}$$

$$= \begin{bmatrix} \begin{bmatrix} 1+8 & 2+10 & 3+12 \\ 3+16 & 6+20 & 9+24 \end{bmatrix} & \begin{bmatrix} 1+2 \\ 3+4 \end{bmatrix} + \begin{bmatrix} 1 \\ 0 \end{bmatrix} \\ [0 \quad 0 \quad 0] & [2] \end{bmatrix}$$

$$= \begin{bmatrix} 9 & 12 & 15 & 4 \\ 19 & 26 & 33 & 7 \\ 0 & 0 & 0 & 2 \end{bmatrix}$$

If A and B are both diagonal matrices having n rows and n columns, they commute. In other words, show that AB = BA where

$$A = \begin{bmatrix} 2 & 0 & 0 \\ 0 & -1 & 0 \\ 0 & 0 & 3 \end{bmatrix} \quad B = \begin{bmatrix} -2 & 0 & 0 \\ 0 & 4 & 0 \\ 0 & 0 & -6 \end{bmatrix} .$$

Solution: A diagonal matrix is a square matrix whose non-diagonal entries are all zero.

$$AB = \begin{bmatrix} 2 & 0 & 0 \\ 0 & -1 & 0 \\ 0 & 0 & 3 \end{bmatrix} \begin{bmatrix} -2 & 0 & 0 \\ 0 & 4 & 0 \\ 0 & 0 & -6 \end{bmatrix}$$

$$= \begin{bmatrix} 2 \cdot (-2) & 0 & 0 \\ 0 & (-1) \cdot 4 & 0 \\ 0 & 0 & 3 \cdot (-6) \end{bmatrix} = \begin{bmatrix} -4 & 0 & 0 \\ 0 & -4 & 0 \\ 0 & 0 & -18 \end{bmatrix} .$$

$$BA = \begin{bmatrix} -2 & 0 & 0 \\ 0 & 4 & 0 \\ 0 & 0 & -6 \end{bmatrix} \begin{bmatrix} 2 & 0 & 0 \\ 0 & -1 & 0 \\ 0 & 0 & 3 \end{bmatrix}$$

$$= \begin{bmatrix} (-2) \cdot 2 & 0 & 0 \\ 0 & 4 \cdot (-1) & 0 \\ 0 & 0 & (-6) \cdot 3 \end{bmatrix} = \begin{bmatrix} -4 & 0 & 0 \\ 0 & -4 & 0 \\ 0 & 0 & -18 \end{bmatrix}$$

Thus, AB = BA .

Let $A = \begin{bmatrix} 1 & 2 & 0 \\ 3 & -1 & 4 \end{bmatrix}$. Find (i) AA^t , (ii) A^tA .

Solution: The transpose of A, denoted by A^t, is the matrix obtained from A by interchanging the rows and columns of A. For example, if

$$A = \begin{bmatrix} a_{11} & a_{12} \cdots a_{1n} \\ a_{21} & a_{22} \cdots a_{2n} \\ \vdots & \vdots \quad\quad \vdots \end{bmatrix} ,$$

$$\begin{bmatrix} \vdots & \vdots & & \vdots \\ a_{m1} & a_{m2} & & a_{mn} \end{bmatrix}$$

then

$$A^t = \begin{bmatrix} a_{11} & a_{21} \cdots & a_{m1} \\ a_{12} & a_{22} \cdots & a_{m2} \\ \vdots & \vdots & \vdots \\ \vdots & \vdots \cdots \vdots \\ \vdots & \vdots & \vdots \\ a_{1n} & a_{2n} \cdots & a_{mn} \end{bmatrix}$$

Observe that if A is an $m \times n$ matrix, then A^t is an $n \times m$ matrix. Hence, the products AA^t and A^tA are always defined.

$$A = \begin{bmatrix} 1 & 2 & 0 \\ 3 & -1 & 4 \end{bmatrix}$$

then,

$$A^t = \begin{bmatrix} 1 & 3 \\ 2 & -1 \\ 0 & 4 \end{bmatrix} .$$

(i) $AA^t = \begin{bmatrix} 1 & 2 & 0 \\ 3 & -1 & 4 \end{bmatrix} \begin{bmatrix} 1 & 3 \\ 2 & -1 \\ 0 & 4 \end{bmatrix}$

$$= \begin{bmatrix} 1\cdot1+2\cdot2+0\cdot0 & 1\cdot3+2\cdot(-1)+0\cdot4 \\ 3\cdot1+(-1)\cdot2+(4)\cdot0 & 3\cdot3+(-1)\cdot(-1)+4\cdot4 \end{bmatrix}$$

$$= \begin{bmatrix} 1+4+0 & 3-2+0 \\ 3-2+0 & 9+1+16 \end{bmatrix} = \begin{bmatrix} 5 & 1 \\ 1 & 26 \end{bmatrix} .$$

(ii) $A^tA = \begin{bmatrix} 1 & 3 \\ 2 & -1 \\ 0 & 4 \end{bmatrix} \begin{bmatrix} 1 & 2 & 0 \\ 3 & -1 & 4 \end{bmatrix}$

$$= \begin{bmatrix} 1\cdot1+3\cdot3 & 1\cdot2+3\cdot(-1) & 1\cdot0+3\cdot4 \\ 2\cdot1+(-1)\cdot3 & 2\cdot2+(-1)\cdot(-1) & 2\cdot0+(-1)\cdot4 \\ 0\cdot1+4\cdot3 & 0\cdot2+4\cdot(-1) & 0\cdot0+4\cdot4 \end{bmatrix}$$

$$= \begin{bmatrix} 1+9 & 2-3 & 0+12 \\ 2-3 & 4+1 & 0-4 \\ 0+12 & 0-4 & 0+16 \end{bmatrix}$$

$$= \begin{bmatrix} 10 & -1 & 12 \\ -1 & 5 & -4 \\ 12 & -4 & 16 \end{bmatrix}$$

● **PROBLEM** 10–26

Find the inverse of

$$\begin{bmatrix} 1 & 2 \\ 3 & 7 \end{bmatrix}$$

<u>Solution</u>: We call B the inverse of A if $A \cdot B = B \cdot A = I$ where
I is the identity matrix, and $A \cdot B$ represents matrix multiplication. We also say that A is invertible if A has an inverse,
denoted by A^{-1}. Note that

$$A^{-1}A = AA^{-1} = I .$$

We wish to find the inverse of the matrix A.

   If A is invertible, there is a procedure for finding its inverse, which we summarize as follows:

1)   We first form the $n \times 2n$ matrix (A,I); by applying elementary
row operations to (A,I), we

2) reduce the matrix (A,I) to a matrix of the form (I,B) whose
first n columns form the identity matrix, and whose last n columns
give a matrix we call 'B'.

3) We conclude that $A^{-1} = B$.

   We apply the above procedure to the given matrix.

$$A = \begin{bmatrix} 1 & 2 \\ 3 & 7 \end{bmatrix}$$

   We first form the matrix (A,I)

$$A = \left[\begin{array}{cc|cc} 1 & 2 & 1 & 0 \\ 3 & 7 & 0 & 1 \end{array} \right]$$

Reduce it so that the left-hand side, "A" becomes the identity matrix:

$$\left[\begin{array}{cc|cc} 1 & 2 & 1 & 0 \\ 3 & 7 & 0 & 1 \end{array} \right] \rightarrow \left[\begin{array}{cc|cc} 1 & 2 & 1 & 0 \\ 0 & 1 & -3 & 1 \end{array} \right]$$

(We added -3 times the first row to the second row)

$$\left[\begin{array}{cc|cc} 1 & 2 & 1 & 0 \\ 0 & 1 & -3 & 1 \end{array} \right] \rightarrow \left[\begin{array}{cc|cc} 1 & 0 & 7 & -2 \\ 0 & 1 & -3 & 1 \end{array} \right]$$

(We added -2 times the second row to the first row).
   Once the 'left-hand side' is reduced to I, the inverse is given
by the 'right-hand side', which we called "B". Hence the inverse
is

$$\begin{bmatrix} 7 & -2 \\ -3 & 1 \end{bmatrix} .$$

● **PROBLEM** 10–27

Find the inverse of the matrix A where

$$A = \begin{bmatrix} 1 & 1 & 1 & 1 \\ 0 & 1 & 1 & 1 \\ 0 & 0 & 1 & 1 \\ 0 & 0 & 0 & 1 \end{bmatrix}$$

Show that the inverse of a diagonal matrix is obtained by inverting the diagonal entries.

Solution:

$$[A : I] = \begin{bmatrix} 1 & 1 & 1 & 1 & \vdots & 1 & 0 & 0 & 0 \\ 0 & 1 & 1 & 1 & \vdots & 0 & 1 & 0 & 0 \\ 0 & 0 & 1 & 1 & \vdots & 0 & 0 & 1 & 0 \\ 0 & 0 & 0 & 1 & \vdots & 0 & 0 & 0 & 1 \end{bmatrix}$$

Subtract the second row from the first row:

$$\begin{bmatrix} 1 & 0 & 0 & 0 & \vdots & 1 & -1 & 0 & 0 \\ 0 & 1 & 1 & 1 & \vdots & 0 & 1 & 0 & 0 \\ 0 & 0 & 1 & 1 & \vdots & 0 & 0 & 1 & 0 \\ 0 & 0 & 0 & 1 & \vdots & 0 & 0 & 0 & 1 \end{bmatrix}$$

Subtract the third row from the second row, and the fourth row from the third row:

$$\begin{bmatrix} 1 & 0 & 0 & 0 & \vdots & 1 & -1 & 0 & 0 \\ 0 & 1 & 0 & 0 & \vdots & 0 & 1 & -1 & 0 \\ 0 & 0 & 1 & 0 & \vdots & 0 & 0 & 1 & -1 \\ 0 & 0 & 0 & 1 & \vdots & 0 & 0 & 0 & 1 \end{bmatrix}$$

Hence

$$A^{-1} = \begin{bmatrix} 1 & -1 & 0 & 0 \\ 0 & 1 & -1 & 0 \\ 0 & 0 & 1 & -1 \\ 0 & 0 & 0 & 1 \end{bmatrix}$$

A diagonal matrix is a square matrix whose non-diagonal entries are all zero. Let A be a diagonal matrix whose diagonal entries are all non-zero, and let

$$A = \begin{bmatrix} a_{11} & 0 & \cdots & & 0 \\ 0 & a_{22} & \cdots & & 0 \\ \vdots & & & & \vdots \\ & 0 & a_{kk} & & \vdots \\ \vdots & & & & \vdots \\ 0 & \cdots & & & a_{nn} \end{bmatrix}$$

with $a_{ii} \neq 0$, $i = 1,\ldots,n$.

Now apply the procedure for finding the inverse at a matrix.

Then

$$[A : I] = \begin{bmatrix} a_{11} & 0 & \vdots & 1 & 0 & \cdots & 0 \\ 0 & a_{22} & 0 & \vdots & 0 & 1 & \cdots & 0 \\ \cdot & & \vdots & \cdot & & \\ \cdot & & \vdots & \cdot & & \\ \cdot & & \vdots & \cdot & & \\ 0 & \cdots & a_{nn} & \vdots & 0 & \cdots & \cdots & 1 \end{bmatrix}$$

Multiply the first row by $\dfrac{1}{a_{11}}$, the second row by $\dfrac{1}{a_{22}}$ $\cdots$ and the nth row by $\dfrac{1}{a_{nn}}$, to obtain

$$[I : B] = \begin{bmatrix} 1 & 0 & \cdots & 0 & \vdots & 1/a_{11} & 0 & \cdots & 0 \\ 0 & 1 & \cdots & 0 & \vdots & 0 & 1/a_{22} & \cdots & 0 \\ \cdot & & & & \vdots & \cdot & & \\ \cdot & & & & \vdots & \cdot & & \\ \cdot & & & & \vdots & \cdot & & \\ 0 & & \cdots & 1 & \vdots & 0 & \cdots & \cdots & 1/a_{nn} \end{bmatrix}$$

Hence

$$A^{-1} = \begin{bmatrix} 1/a_{11} & 0 & \cdots & 0 \\ 0 & 1/a_{22} & \cdots & 0 \\ \cdot & & & \cdot \\ \cdot & & & \cdot \\ \cdot & & & \cdot \\ 0 & & \cdots & 1/a_{nn} \end{bmatrix}$$

Thus the inverse of a diagonal matrix is obtained by inverting the diagonal entries.

Observe that if one of the diagonal entries is zero, the matrix is not invertible. For example,

$$\begin{bmatrix} 1 & 0 & 0 \\ 0 & 0 & 0 \\ 0 & 0 & 3 \end{bmatrix}$$

is not invertible.

● **PROBLEM** 10–28

Find the inverse of A where

$$A = \begin{bmatrix} 1 & 2 & -1 \\ 2 & 5 & 4 \\ 3 & 7 & 4 \end{bmatrix}$$

<u>Solution</u>: Form the block matrix [A : I] where I is the 3 ✗ 3 identity matrix. Then apply row operations on A to reduce it to

590

the identity matrix. Simultaneously, I will change into the required inverse.
   We have

$$[A,I] \; = \; \begin{bmatrix} 1 & 2 & -1 & 1 & 0 & 0 \\ 2 & 5 & 4 & 0 & 1 & 0 \\ 3 & 7 & 4 & 0 & 0 & 1 \end{bmatrix}$$

Now add −2 times the first row to the second row:

$$\begin{bmatrix} 1 & 2 & -1 & 1 & 0 & 0 \\ 0 & 1 & 6 & -2 & 1 & 0 \\ 3 & 7 & 4 & 0 & 0 & 1 \end{bmatrix}$$

Add −3 times the first row to the third row:

$$\begin{bmatrix} 1 & 2 & -1 & 1 & 0 & 0 \\ 0 & 1 & 6 & -2 & 1 & 0 \\ 0 & 1 & 7 & -3 & 0 & 1 \end{bmatrix}$$

Add −1 times the second row to the third row:

$$\begin{bmatrix} 1 & 2 & -1 & 1 & 0 & 0 \\ 0 & 1 & 6 & -2 & 1 & 0 \\ 0 & 0 & 1 & -1 & -1 & 1 \end{bmatrix}$$

Add the third row to the first row:

$$\begin{bmatrix} 1 & 2 & 0 & 0 & -1 & 1 \\ 0 & 1 & 6 & -2 & 1 & 0 \\ 0 & 0 & 1 & -1 & -1 & 1 \end{bmatrix}$$

Add −6 times the third row to the second row:

$$\begin{bmatrix} 1 & 2 & 0 & 0 & -1 & 1 \\ 0 & 1 & 0 & 4 & 7 & -6 \\ 0 & 0 & 1 & -1 & -1 & 1 \end{bmatrix}$$

Add −2 times the second row to the first row:

$$\begin{bmatrix} 1 & 0 & 0 & -8 & -15 & 13 \\ 0 & 1 & 0 & 4 & 7 & -6 \\ 0 & 0 & 1 & -1 & -1 & 1 \end{bmatrix}$$

Hence

$$A^{-1} \; = \; \begin{bmatrix} -8 & -15 & 13 \\ 4 & 7 & -6 \\ -1 & -1 & 1 \end{bmatrix}$$

Find the inverse of A where

$$A = \begin{bmatrix} 1 & 1 & 1 \\ 0 & 2 & 3 \\ 5 & 5 & 1 \end{bmatrix}$$

Solution: We first form the matrix $[A : I]$

$$\begin{bmatrix} 1 & 1 & 1 & 1 & 0 & 0 \\ 0 & 2 & 3 & 0 & 1 & 0 \\ 5 & 5 & 1 & 0 & 0 & 1 \end{bmatrix}$$

and proceed to reduce it to a matrix of the form $[I : B]$, where A has been reduced to I by elementary row operations. From this we may conclude that $A^{-1} = B$. We arrange our computations as follows:

$$\begin{bmatrix} 1 & 1 & 1 & 1 & 0 & 0 \\ 0 & 2 & 3 & 0 & 1 & 0 \\ 5 & 5 & 1 & 0 & 0 & 1 \end{bmatrix}$$

Subtract 5 times the first row from the third row to obtain:

$$\begin{bmatrix} 1 & 1 & 1 & 1 & 0 & 0 \\ 0 & 2 & 3 & 0 & 1 & 0 \\ 0 & 0 & -4 & -5 & 0 & 1 \end{bmatrix}$$

Divide the second row by 2 to obtain:

$$\begin{bmatrix} 1 & 1 & 1 & 1 & 0 & 0 \\ 0 & 1 & 3/2 & 0 & 1/2 & 0 \\ 0 & 0 & -4 & -5 & 0 & 1 \end{bmatrix}$$

Subtract the second row from the first row to obtain:

$$\begin{bmatrix} 1 & 0 & -1/2 & 1 & -1/2 & 0 \\ 0 & 1 & 3/2 & 0 & 1/2 & 0 \\ 0 & 0 & -4 & -5 & 0 & 1 \end{bmatrix}$$

Divide the third row by −4 to obtain:

$$\begin{bmatrix} 1 & 0 & -1/2 & 1 & -1/2 & 0 \\ 0 & 0 & 3/2 & 0 & 1/2 & 0 \\ 0 & 0 & +1 & +5/4 & 0 & -1/4 \end{bmatrix}$$

Add −3/2 times the third row to the second row to obtain:

$$\begin{bmatrix} 1 & 0 & -1/2 & 1 & -1/2 & 0 \\ 0 & 1 & 0 & -15/8 & 1/2 & 3/8 \\ 0 & 0 & 1 & 5/4 & 0 & -1/4 \end{bmatrix}$$

Add 1/2 times the third row to the first row to obtain:

$$\begin{bmatrix} 1 & 0 & 0 & \vdots & 13/8 & -1/2 & -1/8 \\ 0 & 1 & 0 & \vdots & -15/8 & 1/2 & 3/8 \\ 0 & 0 & 1 & \vdots & 5/4 & 0 & -1/4 \end{bmatrix}.$$

Hence

$$A^{-1} = \begin{bmatrix} 13/8 & -1/2 & -1/8 \\ -15/8 & 1/2 & 3/8 \\ 5/4 & 0 & -1/4 \end{bmatrix}.$$

• PROBLEM 10–30

Find the inverses of the following matrices.

(1)

$$A = \begin{bmatrix} 3 & 1 \\ -1 & 6 \end{bmatrix}$$

(2)

$$A = \begin{bmatrix} 1 & -7 & -14 \\ 2 & 1 & -1 \\ 1 & 3 & 4 \end{bmatrix}$$

(3)

$$A = \begin{bmatrix} 3 & 1 & 0 \\ 1 & -1 & 2 \\ 1 & 1 & 1 \end{bmatrix}.$$

Solution: The method of solution is the same in all three cases, namely, forming the block matrix $[A : I]$ where I is the $n \times n$ identity matrix, and using elementary row operations to reduce it to $[I : A^{-1}]$.

(1) $A = \begin{bmatrix} 3 & 1 \\ -1 & 6 \end{bmatrix}.$

Now $[A : I] = \begin{bmatrix} 3 & 1 & \vdots & 1 & 0 \\ -1 & 6 & \vdots & 0 & 1 \end{bmatrix}.$

Multiply the first row by 6:

$$\begin{bmatrix} 18 & 6 & \vdots & 6 & 0 \\ -1 & 6 & \vdots & 0 & 1 \end{bmatrix}$$

Subtract the second row from the first row:

$$\begin{bmatrix} 19 & 0 & \vdots & 6 & -1 \\ -1 & 6 & \vdots & 0 & 1 \end{bmatrix}$$

593

Multiply the second row by 19:

$$\begin{bmatrix} 19 & 0 & \vdots & 6 & -1 \\ -19 & 114 & \vdots & 0 & 19 \end{bmatrix}$$

Add the first row to the second row:

$$\begin{bmatrix} 19 & 0 & \vdots & 6 & -1 \\ 0 & 114 & \vdots & 6 & 18 \end{bmatrix}$$

Divide the first and second rows by 19:

$$\begin{bmatrix} 1 & 0 & \vdots & 6/19 & -1/19 \\ 0 & 6 & \vdots & 6/19 & 18/19 \end{bmatrix}$$

Divide the second row by 6:

$$\begin{bmatrix} 1 & 0 & \vdots & 6/19 & -1/19 \\ 0 & 1 & \vdots & 1/19 & 3/19 \end{bmatrix}$$

Therefore

$$A^{-1} = \begin{bmatrix} 6/19 & -1/19 \\ 1/19 & 3/19 \end{bmatrix}$$

(2)

$$A = \begin{bmatrix} 1 & -7 & -14 \\ 2 & 1 & -1 \\ 1 & 3 & 4 \end{bmatrix}$$

$$[A : I] = \begin{bmatrix} 1 & -7 & -14 & \vdots & 1 & 0 & 0 \\ 2 & 1 & -1 & \vdots & 0 & 1 & 0 \\ 1 & 3 & 4 & \vdots & 0 & 0 & 1 \end{bmatrix}$$

Subtract the first row from the third row:

$$\begin{bmatrix} 1 & -7 & -14 & \vdots & 1 & 0 & 0 \\ 2 & 1 & -1 & \vdots & 0 & 1 & 0 \\ 0 & 10 & 18 & \vdots & -1 & 0 & 1 \end{bmatrix}$$

Divide the third row by 2:

$$\begin{bmatrix} 1 & -7 & -14 & \vdots & 1 & 0 & 0 \\ 2 & 1 & -1 & \vdots & 0 & 1 & 0 \\ 0 & 5 & 9 & \vdots & -1/2 & 0 & 1/2 \end{bmatrix}$$

Add -2 times the first row to the second row:

$$\begin{bmatrix} 1 & -7 & 9 & \vdots & 1 & 0 & 0 \\ 0 & 15 & 27 & \vdots & -2 & 1 & 0 \\ 0 & 5 & 9 & \vdots & -1/2 & 0 & 1/2 \end{bmatrix}$$

Divide the second row by 3:

$$\begin{bmatrix} 1 & -7 & -14 & \vdots & 1 & 0 & 0 \\ 0 & 5 & 9 & \vdots & -2/3 & 1/3 & 0 \\ 0 & 5 & 9 & \vdots & -1/2 & 0 & 1/2 \end{bmatrix}$$

Subtract the second row from the third row:

$$\begin{bmatrix} 1 & -7 & -14 & \vdots & 1 & 0 & 0 \\ 0 & 5 & 9 & \vdots & -2/3 & 1/3 & 0 \\ 0 & 0 & 0 & \vdots & 1/6 & -1/3 & 1/2 \end{bmatrix}$$

At this point A is row equivalent to

$$F = \begin{bmatrix} 1 & -7 & -14 \\ 0 & 5 & 9 \\ 0 & 0 & 0 \end{bmatrix}$$

The matrix A is singular and therefore A does not have an inverse.

(3)

$$A = \begin{bmatrix} 3 & 1 & 0 \\ 1 & -1 & 2 \\ 1 & 1 & 1 \end{bmatrix}$$

$$[A : I] = \begin{bmatrix} 3 & 1 & 0 & \vdots & 1 & 0 & 0 \\ 1 & -1 & 2 & \vdots & 0 & 1 & 0 \\ 1 & 1 & 1 & \vdots & 0 & 0 & 1 \end{bmatrix}$$

Interchange the first and third rows:

$$\begin{bmatrix} 1 & 1 & 1 & \vdots & 0 & 0 & 1 \\ 1 & -1 & 2 & \vdots & 0 & 1 & 0 \\ 3 & 1 & 0 & \vdots & 1 & 0 & 0 \end{bmatrix}$$

Subtract the first row from the second row and add -3 times the first row to the third row:

$$\begin{bmatrix} 1 & 1 & 1 & \vdots & 0 & 0 & 1 \\ 0 & -2 & 1 & \vdots & 0 & 1 & -1 \\ 0 & -2 & -3 & \vdots & 1 & 0 & -3 \end{bmatrix}$$

Divide the second row by -2:

$$\begin{bmatrix} 1 & 1 & 1 & \vdots & 0 & 0 & 1 \\ 0 & 1 & -1/2 & \vdots & 0 & -1/2 & 1/2 \\ 0 & -2 & -3 & \vdots & 1 & 0 & -3 \end{bmatrix}$$

Subtract the second row from the first row:

$$\begin{bmatrix} 1 & 0 & 3/2 & \vdots & 0 & 1/2 & 1/2 \\ 0 & 1 & -1/2 & \vdots & 0 & -1/2 & 1/2 \\ 0 & -2 & -3 & \vdots & 1 & 0 & -3 \end{bmatrix}$$

Add 2 times the second row to the third row:

$$\begin{bmatrix} 1 & 0 & 3/2 & \vdots & 0 & 1/2 & 1/2 \\ 0 & 1 & -1/2 & \vdots & 0 & -1/2 & 1/2 \\ 0 & 0 & -4 & \vdots & 1 & -1 & -2 \end{bmatrix}$$

Divide the third row by -4:

$$\begin{bmatrix} 1 & 0 & 3/2 & \vdots & 0 & 1/2 & 1/2 \\ 0 & 1 & -1/2 & \vdots & 0 & -1/2 & 1/2 \\ 0 & 0 & 1 & \vdots & -1/4 & 1/4 & +2/4 \end{bmatrix}$$

Add $-3/2$ times the third row to the first row and add $1/2$ times the third row to the second row:

$$\begin{bmatrix} 1 & 0 & 0 & \vdots & 3/8 & +1/8 & -2/8 \\ 0 & 1 & 0 & \vdots & -1/8 & -3/8 & 6/8 \\ 0 & 0 & 1 & \vdots & -1/4 & 1/4 & +2/4 \end{bmatrix}$$

Thus

$$A^{-1} = \begin{bmatrix} 3/8 & 1/8 & -2/8 \\ -1/8 & -3/8 & 6/8 \\ -1/4 & 1/4 & 2/4 \end{bmatrix} .$$

● **PROBLEM** 10–31

Show that A is not invertible where

$$A = \begin{bmatrix} 1 & 6 & 4 \\ 2 & 4 & -1 \\ -1 & 2 & 5 \end{bmatrix}$$

<u>Solution</u>: An $n \times n$ matrix M is said to be invertible if there exists another matrix M^{-1} such that $MM^{-1} = M^{-1}M = I$, the $n \times n$ identity matrix. Matrix inversion corresponds to ordinary division although the rules are quite different.

If a matrix is found to be singular when reduced by elementary row operations, it is not invertible and it cannot be reduced to the identity matrix.

$$[A : I] = \begin{bmatrix} 1 & 6 & 4 & \vdots & 1 & 0 & 0 \\ 2 & 4 & -1 & \vdots & 0 & 1 & 0 \\ -1 & 2 & 5 & \vdots & 0 & 0 & 1 \end{bmatrix}$$

Add -2 times the first row to the second and add the first row to the third:

$$\begin{bmatrix} 1 & 6 & 4 & \vdots & 1 & 0 & 0 \\ 0 & -8 & -9 & \vdots & -2 & 0 & 0 \\ 0 & 8 & 9 & \vdots & 1 & 0 & 1 \end{bmatrix}$$

Add the second row to the third row:

$$\begin{bmatrix} 1 & 6 & 4 & \vdots & 1 & 0 & 0 \\ 0 & -8 & -9 & \vdots & -2 & 0 & 0 \\ 0 & 0 & 0 & \vdots & -1 & 0 & 1 \end{bmatrix}$$

We have obtained a row of zeros on the left side. Therefore, A has been reduced to a singular matrix by elementary row operations, so A is singular and therefore A is not invertible.

• PROBLEM 10–32

Let

$$A = \begin{bmatrix} 1 & 2 \\ 3 & 4 \end{bmatrix}$$

Find the inverse of A directly by solving for the entries of the matrix B which satisfies the equation

$$A \cdot B = I ,$$

where A·B is matrix multiplication.

Solution: The problem asks us to solve for the entries a,b,c, and d of the matrix

$$B = \begin{bmatrix} a & b \\ c & d \end{bmatrix} ,$$

given that A·B = I .

Since A·B = I, we have

$$\begin{bmatrix} 1 & 2 \\ 3 & 4 \end{bmatrix} \cdot \begin{bmatrix} a & b \\ c & d \end{bmatrix} = \begin{bmatrix} 1 & 0 \\ 0 & 1 \end{bmatrix}$$

After performing the multiplication of the matrices, we obtain

$$\begin{bmatrix} a+2c & b+2d \\ 3a+4c & 3b+4d \end{bmatrix} = \begin{bmatrix} 1 & 0 \\ 0 & 1 \end{bmatrix} .$$

Recall that two matrices are equal if, and only, if, their corresponding entries are equal. Thus from the last equation we may conclude that $a+2c = 1$, $b+2d = 0$, $3a+4c = 0$, $3b+4d = 1$.

From these four equations, we can obtain two sets of linear equations from which we can solve for each of a,b,c,d. That is, we have the set:

$$a + 2c = 1$$

$$3a + 4c = 0$$

whose solutions are $a = -2$ and $c = 3/2$, and the set

$$b + 2d = 0$$

$$3b + 4d = 1$$

whose solutions are $b = 1$ and $d = -1/2$. Hence,

$$B = \begin{bmatrix} a & b \\ c & d \end{bmatrix} = \begin{bmatrix} -2 & 1 \\ 3/2 & -1/2 \end{bmatrix}.$$

Since B satisfies $AB = I$, $B = A^{-1}$. Hence,

$$A^{-1} = \begin{bmatrix} -2 & 1 \\ 3/2 & -1/2 \end{bmatrix}.$$

The method used consisted of obtaining sets of linear equations whose unique solutions yielded the required inverse.

• PROBLEM 10–33

Find the inverse of A where
$$A = \begin{bmatrix} 2 & 3 \\ 3 & 5 \end{bmatrix}.$$

<u>Solution</u>: We know that

$$AA^{-1} = I$$

where I is the identity matrix. Let

$$A^{-1} = \begin{bmatrix} a & b \\ c & d \end{bmatrix}.$$

Since $A A^{-1} = I$, we have

$$\begin{bmatrix} 2 & 3 \\ 3 & 5 \end{bmatrix} \begin{bmatrix} a & b \\ c & d \end{bmatrix} = \begin{bmatrix} 1 & 0 \\ 0 & 1 \end{bmatrix}$$

Performing the matrix multiplication, we obtain:

$$\begin{bmatrix} 2a + 3c & 2b + 3d \\ 3a + 5c & 3b + 5d \end{bmatrix} = \begin{bmatrix} 1 & 0 \\ 0 & 1 \end{bmatrix}.$$

Now this matrix equality is equivalent to the following system of equations to be satisfied by a, b, c, d:

$$2a + 3c = 1 \qquad 2b + 3d = 0$$

$$3a + 5c = 0 \qquad 3b + 5d = 1.$$

The pair of equations on the left yields $a = 5$ and $c = -3$, while the pair on the right yields $b = -3$ and $d = 2$. Hence

$$A^{-1} = \begin{bmatrix} 5 & -3 \\ -3 & 2 \end{bmatrix}.$$

The method used in this example reduces a matrix inversion problem to one of solving a system of linear equations, and it may be applied to a square matrix of any order.

● **PROBLEM** 10–34

Find the inverse of the matrix A

$$A = \begin{bmatrix} 2 & 1 & 0 \\ 1 & -1 & 1 \\ 0 & 1 & 3 \end{bmatrix} .$$

Solution: We will apply the method which involves solving directly for the individual entries of the matrix A^{-1} . To do this, we reduce the matrix equation $AXA^{-1} = I$ to systems of linear equations which we can solve for the required entries.

Let

$$A^{-1} = \begin{bmatrix} B_{11} & B_{12} & B_{13} \\ B_{21} & B_{22} & B_{23} \\ B_{31} & B_{32} & B_{33} \end{bmatrix}$$

Since $AA^{-1} = I$, we have

$$\begin{bmatrix} 2 & 1 & 0 \\ 1 & -1 & 1 \\ 0 & 1 & 3 \end{bmatrix} \begin{bmatrix} B_{11} & B_{12} & B_{13} \\ B_{21} & B_{22} & B_{23} \\ B_{31} & B_{32} & B_{33} \end{bmatrix} = \begin{bmatrix} 1 & 0 & 0 \\ 0 & 1 & 0 \\ 0 & 0 & 1 \end{bmatrix}$$

Performing the multiplication of matrices,

$$\begin{bmatrix} 2B_{11}+B_{21} & 2B_{12}+B_{22} & 2B_{13}+B_{23} \\ B_{11}-B_{21}+B_{31} & B_{12}-B_{22}+B_{32} & B_{13}-B_{23}+B_{33} \\ B_{21}+3B_{31} & B_{22}+3B_{32} & B_{23}+3B_{33} \end{bmatrix} = \begin{bmatrix} 1 & 0 & 0 \\ 0 & 1 & 0 \\ 0 & 0 & 1 \end{bmatrix}$$

We now equate corresponding entries in the matrices on either side of the above matrix equation, and obtain:

$$2B_{11}+B_{21} = 1 \qquad 2B_{12}+B_{22} = 0 \qquad 2B_{13}+B_{23} = 0$$

$$B_{11}-B_{21}+B_{31} = 0 \qquad B_{12}-B_{22}+B_{32} = 1 \qquad B_{13}-B_{23}+B_{33} = 0$$

$$B_{21}+3B_{31} = 0 \qquad B_{22}+3B_{32} = 0 \qquad B_{23}+3B_{33} = 1$$

Solving these equations yields:

$$B_{11} = 4/11 \qquad\qquad B_{12} = 3/11 \qquad\qquad B_{13} = -1/11$$

$$B_{21} = 3/11 \qquad\qquad B_{22} = -6/11 \qquad\qquad B_{23} = 2/11$$

$$B_{31} = -1/11 \qquad\qquad B_{32} = 2/11 \qquad\qquad B_{33} = 3/11$$

Thus

$$A^{-1} = \begin{bmatrix} 4/11 & 3/11 & -1/11 \\ 3/11 & -6/11 & 2/11 \\ -1/11 & 2/11 & 3/11 \end{bmatrix} \; .$$

• PROBLEM 10–35

Use the classical adjoint to find A^{-1} where

$$A = \begin{bmatrix} 1 & 0 & -1 \\ 0 & 2 & 2 \\ 1 & 1 & -1 \end{bmatrix}$$

Solution: Recall some definitions: If $A = (a_{ij})$, then a cofactor of an entry a_{ij} is denoted A_{ij} and is given by $(-1)^{i+j}$ times the determinant of the $(n-1) \times (n-1)$ minor matrix obtained from A by deleting its ith row and jth column.

By the matrix of cofactors, we mean the matrix

$$C = \begin{bmatrix} A_{11} & \cdots\cdots & A_{1n} \\ \cdot & \cdots\cdots & \cdot \\ \cdot & & \cdot \\ \cdot & & \cdot \\ A_{n1} & \cdots\cdots & A_{nn} \end{bmatrix} \; .$$

Then the adjoint of A is C^T, i.e.,

$$\text{adj } A = \begin{bmatrix} A_{11} & A_{21} & \cdots & A_{n1} \\ A_{12} & A_{22} & \cdots & \cdot \\ \cdot & \cdot & & \cdot \\ \cdot & \cdot & & \cdot \\ \cdot & \cdot & & \cdot \\ A_{1n} & A_{2n} & \cdots & A_{nn} \end{bmatrix}$$

Recall that A^{-1} exists if and only if $\det A = |A| \neq 0$. The rule for obtaining A^{-1} is then

$$A^{-1} = \frac{1}{|A|} \, [\text{adj } A]$$

600

where $|A| = $ determinant of the $n \times n$ square matrix. Let us first compute the determinant of matrix A

$$A = \begin{bmatrix} 1 & 0 & -1 \\ 0 & 2 & 2 \\ 1 & 1 & -1 \end{bmatrix}$$

$$|A| = \begin{vmatrix} 1 & 0 & -1 \\ 0 & 2 & 2 \\ 1 & 1 & -1 \end{vmatrix}$$

$$= 1 \begin{vmatrix} 2 & 2 \\ 1 & -1 \end{vmatrix} - 0 \begin{vmatrix} 0 & 2 \\ 1 & -1 \end{vmatrix} + (-1) \begin{vmatrix} 0 & 2 \\ 1 & 1 \end{vmatrix}$$

$$= 1(-2-2) - 0(0-2) - 1(0-2)$$

$$= -4-0+2 = -2 \ .$$

We find that $|A| \neq 0$. Therefore A^{-1} exists. The classical adjoint of A is found by replacing each element of A by its cofactor and taking the transpose of the resulting matrix.

Let us now compute the cofactors of the entries of A.

$$A = \begin{bmatrix} 1 & 0 & -1 \\ 0 & 2 & 2 \\ 1 & 1 & -1 \end{bmatrix}$$

To find A_{11}, we delete the first row and first column of A to obtain the matrix

$$\begin{bmatrix} 2 & 2 \\ 1 & -1 \end{bmatrix} \ .$$

The cofactor A_{11} is then $(-1)^{1+1}$ times the determinant of the above matrix, i.e.,

$$A_{11} = (1)^2 \cdot \begin{vmatrix} 2 & 2 \\ 1 & -1 \end{vmatrix} = \begin{vmatrix} 2 & 2 \\ 1 & -1 \end{vmatrix}$$

$$= (-2-2) = -4 \ .$$

We find the cofactors of the remaining elements of A by the same method. The cofactors of the nine elements of A are

$$A_{11} = + \begin{vmatrix} 2 & 2 \\ 1 & -1 \end{vmatrix}, \quad A_{12} = - \begin{vmatrix} 0 & 2 \\ 1 & -1 \end{vmatrix}, \quad A_{13} = + \begin{vmatrix} 0 & 2 \\ 1 & 1 \end{vmatrix}$$

$$\begin{aligned} &= (-2-2) & &= -(0-2) & &= (0-2) \\ &= -4 & &= 2 & &= -2 \end{aligned}$$

$$A_{21} = - \begin{vmatrix} 0 & -1 \\ 1 & -1 \end{vmatrix}, \quad A_{22} = + \begin{vmatrix} 1 & -1 \\ 1 & -1 \end{vmatrix}, \quad A_{23} = - \begin{vmatrix} 1 & 0 \\ 1 & 1 \end{vmatrix}$$

$$\begin{aligned} &= -(0+1) & &= (-1+1) & &= -(1-0) \\ &= -1 & &= 0 & &= -1 \end{aligned}$$

$$A_{31} = + \begin{vmatrix} 0 & -1 \\ 2 & 2 \end{vmatrix}, \quad A_{32} = - \begin{vmatrix} 1 & -1 \\ 0 & 2 \end{vmatrix}, \quad A_{33} = + \begin{vmatrix} 1 & 0 \\ 0 & 2 \end{vmatrix}$$

$$= (0+2) \qquad = -(2-0) \qquad = (2-0)$$
$$= 2 \qquad = -2 \qquad = 2$$

The matrix of cofactors C is given by

$$C = \begin{bmatrix} -4 & 2 & -2 \\ -1 & 0 & -1 \\ 2 & -2 & 2 \end{bmatrix} .$$

We form the transpose of the matrix of cofactors to obtain the classical adjoint of A:

$$\text{Adj } A = \begin{bmatrix} -4 & -1 & 2 \\ 2 & 0 & -2 \\ -2 & -1 & 2 \end{bmatrix}$$

Now,

$$A^{-1} = \frac{1}{|A|} [\text{adj } A]$$

$$= -\frac{1}{2} \begin{bmatrix} -4 & -1 & 2 \\ 2 & 0 & -2 \\ -2 & -1 & 2 \end{bmatrix} ,$$

So

$$A^{-1} = \begin{bmatrix} 2 & 1/2 & -1 \\ -1 & 0 & 1 \\ 1 & 1/2 & -1 \end{bmatrix} .$$

It is easy to check the computation by verifying that

$$AA^{-1} = I$$

$$\begin{bmatrix} 1 & 0 & -1 \\ 0 & 2 & 2 \\ 1 & 1 & -1 \end{bmatrix} \begin{bmatrix} 2 & 1/2 & -1 \\ -1 & 0 & 1 \\ 1 & 1/2 & -1 \end{bmatrix}$$

$$= \begin{bmatrix} 2+0-1 & 1/2+0-1/2 & -1+0+1 \\ 0-2+2 & 0+0+1 & 0+2-2 \\ 2-1-1 & 1/2+0-1/2 & -1+1+1 \end{bmatrix}$$

$$= \begin{bmatrix} 1 & 0 & 0 \\ 0 & 1 & 0 \\ 0 & 0 & 1 \end{bmatrix} = I .$$

Find the rank of the matrix A where $A = \begin{bmatrix} 1 & 3 & 2 \\ 2 & 6 & 1 \end{bmatrix}$.

Solution: If A is a matrix, then the rank of A, written $r(A)$, is the maximum number of linearly independent columns or, equivalently, rows. Since A is 2×3 , the rank must be two or less. First, check the rows for linear independence. Set

$$c_1(1,3,2) + c_3(2,6,1) = 0$$

to obtain the system of equations:

$L_1 : c_1 + 2c_2 = 0$

$L_2 : 3c_1 + 6c_2 = 0$

$L_3 : 2c_1 + c_2 = 0$.

By solving this sytem of equations, we find that it has only a trivial solution:

$-3L_1 : -3c_1 - 6c_2 = 0$

$+L_2 : \underline{3c_1 + 6c_2 = 0}$

$0c_1 + 0c_2 = 0$

$-2L_1 : -2c_1 - 4c_2 = 0$

$+L_3 : \underline{2c_1 + c_2 = 0}$

$0c_1 - 3c_2 = 0$

$c_2 = 0$

From $L_1 : c_1 + 2c_2 = 0$

$c_1 = 0$

Solution: $c_1 = c_2 = 0$.

Thus, the two rows are independent and $r(A) = 2$.

We also could have found $r(A)$ by checking the maximum number of linearly independent columns. The two column vectors

$$\begin{bmatrix} 1 \\ 2 \end{bmatrix} \quad \text{and} \quad \begin{bmatrix} 2 \\ 1 \end{bmatrix}$$

are linearly independent since

$$c_1 \begin{bmatrix} 1 \\ 2 \end{bmatrix} + c_2 \begin{bmatrix} 2 \\ 1 \end{bmatrix} = 0$$

implies $c_1 = c_2 = 0$. Obtaining the system of equations,

$L_1 : c_1 + 2c_2 = 0$

$+L_2 : 2c_1 + c_2 = 0$,

and solving this system of equations, the result is the trivial solution $c_1 = c_2 = 0$.

$$-2L_1 : \quad -2c_1 - 4c_2 = 0$$

$$L_2 : \quad \underline{2c_1 + c_2 = 0}$$

$$-3c_2 = 0$$

$$c_2 = 0$$

From L_1 : $\quad c_1 + 2c_2 = 0$

$$c_1 = 0$$

Solution: $c_1 = c_2 = 0$. Furthermore, since the columns are vectors in R^2 , the maximum number of linearly independent columns can only equal two (dim $R^2 = 2$). Thus, again, $r(A) = 2$.

• PROBLEM 10-37

Find the rank of the matrix A where:

(i)

$$A = \begin{bmatrix} 1 & 3 & 1 & -2 & -3 \\ 1 & 4 & 3 & -1 & -4 \\ 2 & 3 & -4 & -7 & -3 \\ 3 & 8 & 1 & -7 & -8 \end{bmatrix}$$

(ii)

$$A = \begin{bmatrix} 1 & 2 & -3 \\ 2 & 1 & 0 \\ -2 & -1 & 3 \\ -1 & 4 & -2 \end{bmatrix}$$

(iii)

$$A = \begin{bmatrix} 1 & 3 \\ 0 & -2 \\ 5 & -1 \\ -2 & 3 \end{bmatrix}$$

Solution: (i) First, reduce the matrix A to echelon form using the elementary row operations.
(a) Add -1 times the first row to the second row.
(b) Add -2 times the first row to the third row.
(c) Add -3 times the first row to the third row.

$$A = \begin{bmatrix} 1 & 3 & 1 & -2 & -3 \\ 0 & 1 & 2 & 1 & -1 \\ 0 & -3 & -6 & -3 & 3 \\ 0 & -1 & -2 & -1 & 1 \end{bmatrix}$$

Add +3 times the second row to the third row.
Add the second row to the fourth row. Then,

$$A = \begin{bmatrix} 1 & 3 & 1 & -2 & -3 \\ 0 & 1 & 2 & 1 & -1 \\ 0 & 0 & 0 & 0 & 0 \\ 0 & 0 & 0 & 0 & 0 \end{bmatrix}.$$

• PROBLEM 10–38

Is the set $\{1 + t + t^2, 1 - 3t + 2t^2, 3 - t + 4t^2\}$ independent in V?

Solution: The coordinates of $1 + t + t^2$, $1 - 3t + 2t^2$, and $3 - t + 4t^2$, with respect to $(1,t,t^2)$, are $(1,1,1)$, $(1,-3,2)$ and $(3,-1,4)$, respectively.

Form the matrix A whose rows are the above coordinate vectors.

$$A = \begin{bmatrix} 1 & 1 & 1 \\ 1 & -3 & 2 \\ 3 & -1 & 4 \end{bmatrix}$$

Now, reduce the above matrix to echelon form. The non–zero rows (or columns) of an echelon matrix are linearly independent. Since the rank of a matrix is defined as the maximum number of linearly independent rows (or columns) and since equivalent matrices have the same rank, the reduction to echelon form seems appropriate.

Add -1 times the first row to the second row, and add -3 times the first row to the third row.

$$\begin{bmatrix} 1 & 1 & 1 \\ 0 & -4 & 1 \\ 0 & -4 & 1 \end{bmatrix}$$

Add -1 times the second row to the third row.

$$\begin{bmatrix} 1 & 1 & 1 \\ 0 & -4 & 1 \\ 0 & 0 & 0 \end{bmatrix}$$

Since the echelon matrix A has two non–zero row vectors, its rank is 2.

The coordinate vectors generate a space of dimension two. We conclude that $\{(1,1,1), (1,-3,2), (3,-1,4)\}$ is independent, and, hence, that $\{1 + t + t^2, 1 - 3t + 2t^2, 3 - t + 4t^2\}$ is dependent.

Consider the system of equations

$$\begin{bmatrix} 2 & 1 & 3 \\ 1 & -2 & 2 \\ 0 & 1 & 3 \end{bmatrix} \begin{bmatrix} x_1 \\ x_2 \\ x_3 \end{bmatrix} = \begin{bmatrix} 1 \\ 2 \\ 3 \end{bmatrix} .$$

Show that the system has a solution without actually computing a solution.

Solution: Let $AX = B$ be a system of m linear equations in n unknowns, where $A = [a_{ij}]$ is an $m \times n$ matrix, X is an n-dimensional column vector and B is an m-dimensional column vector. Thus, the system has the form

$$a_{11}x_1 + a_{12}x_2 + \cdots + a_{1n}x_n = b_1$$
$$a_{21}x_1 + a_{22}x_2 + \cdots + a_{2n}x_n = b_2$$
$$\vdots \qquad \vdots \qquad \qquad \vdots \qquad \qquad (1)$$
$$a_{m1}x_1 + a_{n2}x_2 + \cdots + a_{mn}x_n = b_m \quad ,$$

The system (1) may also be written as the vector equation

$$x_1 \begin{bmatrix} a_{11} \\ a_{21} \\ \cdot \\ \cdot \\ \cdot \\ a_{m1} \end{bmatrix} + x_2 \begin{bmatrix} a_{12} \\ a_{22} \\ \cdot \\ \cdot \\ \cdot \\ a_{m2} \end{bmatrix} + \cdots + x_n \begin{bmatrix} a_{1n} \\ a_{2n} \\ \cdot \\ \cdot \\ \cdot \\ a_{mn} \end{bmatrix} = \begin{bmatrix} b_1 \\ b_2 \\ \cdot \\ \cdot \\ \cdot \\ b_m \end{bmatrix}$$

Thus, $AX = B$ has a solution when B is a linear combination of the columns of A; that is, if B belongs to the column space of A. But the dimension of the column space of A is the rank of A. Since B is in this space, the rank of A is also the rank of the augmented matrix $[A \,|\, B]$. Thus, a necessary and sufficient condition for $AX = B$ to have a solution is that rank $A = \text{rank}[A \,|\, B]$.

Applying elementary rwo operations to the given matrix, it can be seen that

$$\begin{bmatrix} 2 & 1 & 3 \\ 1 & -2 & 2 \\ 0 & 1 & 3 \end{bmatrix}$$

is equivalent to

$$\begin{bmatrix} 1 & 0 & 8 \\ 0 & 1 & 3 \\ 0 & 0 & -16 \end{bmatrix} \qquad (2)$$

Since the columns of the matrix (2) form a basis for R^3, rank A = 3. Next, form the augmented matrix

$$\begin{bmatrix} 2 & 1 & 3 & | & 1 \\ 1 & -2 & 2 & | & 2 \\ 0 & 1 & 3 & | & 3 \end{bmatrix}$$

and reduce to echelon form:

$$\begin{bmatrix} 1 & 0 & 0 & | & -1 \\ 0 & 1 & 0 & | & -3/8 \\ 0 & 0 & 1 & | & 9/8 \end{bmatrix}$$

Thus, rank $[A \vdots B] = 3$ and the given system of equations has a solution.

• PROBLEM 10-40

Let A be the matrix

$$\begin{bmatrix} 0 & 1 & 3 & -2 & -1 & 2 \\ 0 & 2 & 6 & -4 & -2 & 4 \\ 0 & 1 & 3 & -2 & 1 & 4 \\ 0 & 2 & 6 & 1 & -1 & 0 \end{bmatrix}$$

Find the determinant rank of A.

Solution: If A is an m×n matrix, the determinant rank of A is defined as follows: The order of the largest non-zero determinant which is obtainable by the possible deletion of rows and columns from the matrix.

The standard method of computing the determinant rank is the one shown below. First, use elementary row operations to reduce the matrix to echelon form (the leading coefficient of each equation equals one). Then, from the echelon matrix, select the largest upper triangular matrix which has one's along the main diagonal. The determinant of this matrix is the product of the diagonal elements, and, hence, the determinant rank is the order of this determinant. Applying the three elementary row operations on A, we obtain the equivalent matrix:

$$\begin{bmatrix} 0 & 1 & 3 & -2 & -1 & 2 \\ 0 & 0 & 0 & 1 & 1/5 & -4/5 \\ 0 & 0 & 0 & 0 & 1 & 1 \\ 0 & 0 & 0 & 0 & 0 & 0 \end{bmatrix} \qquad (1)$$

Examining (1), it can be seen that the second, fourth and fifth columns form the largest possible upper triangular matrix. Thus,

$$\begin{bmatrix} 1 & -2 & -1 \\ 0 & 1 & 1/5 \\ 0 & 0 & 1 \end{bmatrix}$$

has determinant equal to one, and the determinant rank of A is three. Since (1) contains three non-zero rows, the row-rank of A is also three, i.e., determinant rank = row rank. The last statement is always true.

Show that the matrix

$$A = \begin{bmatrix} 0 & 1 & 2 \\ 2 & 3 & 4 \\ 4 & 7 & 10 \end{bmatrix}$$

is equivalent to D_2^{33} where $D_r^{m,n}$ denotes the canonical form under equivalence of A. $D_r^{m,n}$ is the echelon form that has one's along the diagonal and zeros elsewhere, and where all the zero rows are consigned to the depths of the matrix.

Solution: An $m \times n$ matrix A is equivalent to $D_r^{m,n}$ if and only if it has rank r. The matrices $D_r^{m,n}$ are called the canonical forms under equivalence. Each $m \times n$ matrix is equivalent to exactly one canonical form under equivalence.

The given matrix A has rank 2, and may be converted to $D_2^{3,3}$ by the following sequence of matrices, each of which is obtained from the preceding matrix by a row or column operation.

$$A = \begin{bmatrix} 0 & 1 & 2 \\ 2 & 3 & 4 \\ 4 & 7 & 10 \end{bmatrix}$$

Interchange the first and the second rows.

$$A_1 = \begin{bmatrix} 2 & 3 & 4 \\ 0 & 1 & 2 \\ 4 & 7 & 10 \end{bmatrix}$$

Add −2 times the first row to the third row.

$$A_2 = \begin{bmatrix} 2 & 3 & 4 \\ 0 & 1 & 2 \\ 0 & 1 & 2 \end{bmatrix}$$

Add −3 times the second row to the first row, and add −1 times the second row to the third row.

$$A_3 = \begin{bmatrix} 2 & 0 & -2 \\ 0 & 1 & 2 \\ 0 & 0 & 0 \end{bmatrix}$$

Add the first column to the third column.

$$A_4 = \begin{bmatrix} 2 & 0 & 0 \\ 0 & 1 & 2 \\ 0 & 0 & 0 \end{bmatrix}$$

Add −2 times the second column to the third column.

$$A_5 = \begin{bmatrix} 2 & 0 & 0 \\ 0 & 1 & 0 \\ 0 & 0 & 0 \end{bmatrix}$$

Divide the first row by 1/2 . Then,

$$D_2^{3,3} = \begin{bmatrix} 1 & 0 & 0 \\ 0 & 1 & 0 \\ 0 & 0 & 0 \end{bmatrix}$$

SYSTEMS OF LINEAR EQUATIONS

• **PROBLEM** 10–42

Reduce the following matrices to echelon form and then to row reduced echelon form.

(a)
$$A = \begin{bmatrix} 0 & 1 & 3 & -2 \\ 2 & 1 & -4 & 3 \\ 2 & 3 & 2 & -1 \end{bmatrix}$$

(b)
$$A = \begin{bmatrix} 6 & 3 & -4 \\ -4 & 1 & -6 \\ 1 & 2 & -5 \end{bmatrix}$$

<u>Solution:</u> In echelon form, the first non-zero entry of any row is a 1, and any row of zeros lies below the rows with non-zero entries. Furthermore, the first non-zero entry of any row is in a column to the left of the first non-zero entry in the next row. In addition, in reduced echelon form, the entire column containing the first non-zero entry of any row is all zeros except for that entry.

(a) Perform the following row operations: Interchange the first and the second rows.

$$\begin{bmatrix} 2 & 1 & -4 & 3 \\ 0 & 1 & 3 & -2 \\ 2 & 3 & 2 & -1 \end{bmatrix}$$

Add −1 times the first row to the third row.

$$\begin{bmatrix} 2 & 1 & -4 & 3 \\ 0 & 1 & 3 & -2 \\ 0 & 2 & 6 & -4 \end{bmatrix}$$

Add −2 times the second row to the third row.

$$\begin{bmatrix} 2 & 1 & -4 & 3 \\ 0 & 1 & 3 & -2 \\ 0 & 0 & 0 & 0 \end{bmatrix}$$

Finally, to obtain the echelon form, multiply the first column by $\frac{1}{2}$. Hence,

$$\begin{bmatrix} 1 & \frac{1}{2} & 2 & 3/2 \\ 0 & 1 & 3 & -2 \\ 0 & 0 & 0 & 0 \end{bmatrix}.$$

Now add $-\frac{1}{2}$ times the second row to the first row, to obtain the row reduced echelon form

$$\begin{bmatrix} 1 & 0 & -7/2 & 5/2 \\ 0 & 1 & 3 & -2 \\ 0 & 0 & 0 & 0 \end{bmatrix}.$$

(b) First interchange the first and third rows.

$$\begin{bmatrix} 1 & 2 & -5 \\ -4 & 1 & -6 \\ 6 & 3 & -4 \end{bmatrix}.$$

Add 4 times the first row to the second row and -6 times the first row to the third row.

$$\begin{bmatrix} 1 & 2 & -5 \\ 0 & 9 & -26 \\ 0 & -9 & 26 \end{bmatrix}$$

Now add the second row to the third row.

$$\begin{bmatrix} 1 & 2 & -5 \\ 0 & 9 & -26 \\ 0 & 0 & 0 \end{bmatrix}$$

Divide the second row by 9 to obtain the echelon form.

$$\begin{bmatrix} 1 & 2 & -5 \\ 0 & 1 & -26/9 \\ 0 & 0 & 0 \end{bmatrix}$$

Add -2 times the second row to the first row to obtain the row-reduced echelon form.

$$\begin{bmatrix} 1 & 0 & 7/9 \\ 0 & 1 & -26/9 \\ 0 & 0 & 0 \end{bmatrix}.$$

● **PROBLEM** 10–43

Given

$$A = \begin{bmatrix} 1 & -2 & 3 & -1 \\ 2 & -1 & 2 & 2 \\ 3 & 1 & 2 & 3 \end{bmatrix},$$

(i) Reduce A to echelon form.
(ii) Reduce A to row reduced echelon form.

<u>Solution:</u> To obtain the echelon form, the first non-zero entry of any row must be contained in a column to the left of the first non-zero entry in the next row. Also, the first non-zero entry must be a 1, and any row of all zeros lies below all the rows that have non-zero entries. In reduced echelon form, the column containing the first non-zero entry of the ith row is $\vec{e}_i$.

(i)   To reduce A to echelon form, apply the following row operations:

add -2 times the first row to the second row and -3 times the first row to the third row.

$$\begin{bmatrix} 1 & -2 & 3 & -1 \\ 0 & 3 & -4 & 4 \\ 0 & 7 & -7 & 6 \end{bmatrix} .$$

Multiply the second row by 7 and the third row by 3.

$$\begin{bmatrix} 1 & -2 & 3 & -1 \\ 0 & 21 & -28 & 28 \\ 0 & 21 & -21 & 18 \end{bmatrix} .$$

Then, add -1 times the second row to the third row, to obtain

$$\begin{bmatrix} 1 & -2 & 3 & -1 \\ 0 & 3 & -4 & 4 \\ 0 & 0 & 7 & -10 \end{bmatrix} .$$

Divide the second row by 3 and the third row by 7 to obtain the echelon form.

$$\begin{bmatrix} 1 & -2 & 3 & -1 \\ 0 & 1 & -4/3 & 4/3 \\ 0 & 0 & 1 & -10/7 \end{bmatrix} .$$

(ii)   To obtain the reduced echelon form, add 2 times the second row to the first row.

$$\begin{bmatrix} 1 & 0 & 1/3 & 5/3 \\ 0 & 1 & -4/3 & 4/3 \\ 0 & 0 & 1 & -10/7 \end{bmatrix} .$$

Add -1/3 times the third row to the first row and 4/3 times the third row to the second row, resulting in the row reduced echelon form.

$$\begin{bmatrix} 1 & 0 & 0 & 15/7 \\ 0 & 1 & 0 & -4/7 \\ 0 & 0 & 1 & -10/7 \end{bmatrix} .$$

Let

$$A = \begin{bmatrix} 0 & 0 & 0 & 0 & 1 & 1 & 1 \\ 0 & 2 & 6 & 2 & 0 & 0 & 4 \\ 0 & 1 & 3 & 1 & 1 & 0 & 1 \\ 0 & 1 & 3 & 1 & 2 & 1 & 2 \end{bmatrix}.$$

Reduce A to the Hermite normal form.

Solution: The row-reduced echelon matrix is also known as the Hermite normal form. Perform the following elementary row operations to obtain this form. Interchange rows one and three:

$$\begin{bmatrix} 0 & 1 & 3 & 1 & 1 & 0 & 1 \\ 0 & 2 & 6 & 2 & 0 & 0 & 4 \\ 0 & 0 & 0 & 0 & 1 & 1 & 1 \\ 0 & 1 & 3 & 1 & 2 & 1 & 2 \end{bmatrix}.$$

Add -1 times the first row to the fourth row and -2 times the first row to the second row.

$$\begin{bmatrix} 0 & 1 & 3 & 1 & 1 & 0 & 1 \\ 0 & 0 & 0 & 0 & -2 & 0 & 2 \\ 0 & 0 & 0 & 0 & 1 & 1 & 1 \\ 0 & 0 & 0 & 0 & 1 & 1 & 1 \end{bmatrix}.$$

Divide the second row by -2. Add -1 times the third row to the fourth row.

$$\begin{bmatrix} 0 & 1 & 3 & 1 & 1 & 0 & 1 \\ 0 & 0 & 0 & 0 & 1 & 0 & -1 \\ 0 & 0 & 0 & 0 & 1 & 1 & 1 \\ 0 & 0 & 0 & 0 & 0 & 0 & 0 \end{bmatrix}.$$

Add -1 times the second row to the first row and the third row.

$$\begin{bmatrix} 0 & 1 & 3 & 1 & 0 & 0 & 2 \\ 0 & 0 & 0 & 0 & 1 & 0 & -1 \\ 0 & 0 & 0 & 0 & 0 & 1 & 2 \\ 0 & 0 & 0 & 0 & 0 & 0 & 0 \end{bmatrix}.$$

The above matrix is the Hermite normal form of A.

Given
$$A = \begin{bmatrix} 0 & 0 & 0 & 0 \\ 0 & 0 & 0 & 0 \\ 5 & 7 & 8 & 0 \\ 10 & 9 & 6 & 0 \\ 0 & 10 & 5 & 0 \end{bmatrix} \quad ,$$

reduce   A   to column-reduced echelon form.

Solution: To reduce a matrix to a form that specifies the arrangement of its columns, it is necessary to perform column operations. These are merely the column analogs of the basic row operations; i.e.:
1.   interchanging the ith and jth column.
2.   multiplying the ith column by a scalar k.
3.   adding the ith column to the jth column.

To reduce A to column-reduced echelon form, apply the following column operations. Divide the first column by 5.

$$\begin{bmatrix} 0 & 0 & 0 & 0 \\ 0 & 0 & 0 & 0 \\ 1 & 7 & 8 & 0 \\ 2 & 9 & 6 & 0 \\ 0 & 10 & 5 & 0 \end{bmatrix} \quad .$$

Add –7 times the first column to the second column and –8 times the first column to the third column.

$$\begin{bmatrix} 0 & 0 & 0 & 0 \\ 0 & 0 & 0 & 0 \\ 1 & 0 & 0 & 0 \\ 2 & -5 & -10 & 0 \\ 0 & 10 & 5 & 0 \end{bmatrix} \quad .$$

Divide the second column by −5.

$$\begin{bmatrix} 0 & 0 & 0 & 0 \\ 0 & 0 & 0 & 0 \\ 1 & 0 & 0 & 0 \\ 2 & 1 & -10 & 0 \\ 0 & -2 & 5 & 0 \end{bmatrix}.$$

Add −2 times the second column to the first column and 10 times the second column to the third column.

$$\begin{bmatrix} 0 & 0 & 0 & 0 \\ 0 & 0 & 0 & 0 \\ 1 & 0 & 0 & 0 \\ 0 & 1 & 0 & 0 \\ 4 & -2 & -15 & 0 \end{bmatrix}.$$

Divide the third column by −15.

$$\begin{bmatrix} 0 & 0 & 0 & 0 \\ 0 & 0 & 0 & 0 \\ 1 & 0 & 0 & 0 \\ 0 & 1 & 0 & 0 \\ 4 & -2 & 1 & 0 \end{bmatrix}.$$

Add −4 times the third column to the first column and 2 times the third column to the second column to obtain the column-reduced echelon form.

$$\begin{bmatrix} 0 & 0 & 0 & 0 \\ 0 & 0 & 0 & 0 \\ 1 & 0 & 0 & 0 \\ 0 & 1 & 0 & 0 \\ 0 & 0 & 1 & 0 \end{bmatrix}.$$

Find the dimension of the vector space spanned by:

(i) $(1,-2,3,-1)$ and $(1,1,-2,3)$

(ii) $(3,-6,3,-9)$ and $(-2,4,-2,6)$

(iii) $t^3 + 2t^2 + 3t + 1$ and $2t^3 + 4t^2 + 6t + 2$

(iv) $t^3 - 2t^2 + 5$ and $t^2 + 3t - 4$

(v) $\begin{bmatrix} 1 & 2 \\ 1 & 2 \end{bmatrix}$ and $\begin{bmatrix} 1 & 1 \\ 2 & 2 \end{bmatrix}$

(vi) $\begin{bmatrix} 1 & 1 \\ -1 & -1 \end{bmatrix}$ $\begin{bmatrix} -3 & -3 \\ 3 & 3 \end{bmatrix}$

(vii) 3 and -3.

Solution: Two non-zero vectors span a space W of dimension 2 if they are independent, and of dimension 1 if they are dependent. Two vectors are dependent if and only if one is a scalar multiple of the other. Now, using the above facts, the dimension of the given vector space can be found. Hence, the dimensions of the subspaces spanned by the given sets of vectors are, respectively:

(1) 2, (ii) 1, (iii) 1, (iv) 2, (v) 2, (vi) 1, and (vii) 1.

Note that in (i) and (ii) the subspace spanned is a subspace of the vector space R^4 ; in (iii) and (iv) it is a subspace of the real vector space of polynomials in T over the field R; in (v) and (vi) it is a subspace of the real vector space of 2×2 matrices with entries in R; and in (vii) it is just R considered as a real vector space.

Let V be the real vector space of 2 by 2 symmetric matrices with entries in R. Show that dim V = 3.

Solution: Recall that $A = (a_{ij})$ is symmetric if $A = A^t$ or, equivalently, $a_{ij} = a_{ji}$. Let $A = \begin{bmatrix} a & b \\ b & c \end{bmatrix}$ be an arbitrary 2 by 2 symmetric matrix where $a,b,c \in R$. Set (i) a = 1, b = 0, c = 0; (ii) a = 0, b = 1, c = 0; (iii) a = 0, b = 0, c = 1. Then, the respective matrices are:

$$E_1 = \begin{bmatrix} 1 & 0 \\ 0 & 0 \end{bmatrix}, \quad E_2 = \begin{bmatrix} 0 & 1 \\ 1 & 0 \end{bmatrix}, \quad E_3 = \begin{bmatrix} 0 & 0 \\ 0 & 1 \end{bmatrix}.$$

We shall show that $\{E_1, E_2, E_3\}$ is a basis of V, i.e., it (1) spans V and (2) is a linearly independent set.

For the above arbitrary matrix A in V, we have,

$$A = \begin{bmatrix} a & b \\ b & c \end{bmatrix} = aE_1 + bE_2 + cE_3 .$$

Thus, $\{E_1, E_2, E_3\}$ generates V, i.e., spans V. (2) By definition, the vectors $v_1, v_2, \ldots, v_m$ are linearly independent if there exist scalars

$a_1, a_2, \ldots, a_m \in R$, all of them zero, such that $a_1 v_1 + a_2 v_2 + \ldots + a_m v_m = 0$. If v_1, v_2 and v_3 are linearly independent, the set $\{v_1, v_2, v_3\}$ is called linearly independent. Let $a_1 E_1 + a_2 E_2 + a_3 E_3 = 0$ where a_1, a_2, a_3 are unknown scalars. Note that 0 is the 0 element of V, i.e., the 0 matrix. Then,

$$a_1 \begin{bmatrix} 1 & 0 \\ 0 & 0 \end{bmatrix} + a_2 \begin{bmatrix} 0 & 1 \\ 1 & 0 \end{bmatrix} + a_3 \begin{bmatrix} 0 & 0 \\ 0 & 1 \end{bmatrix} = \begin{bmatrix} 0 & 0 \\ 0 & 0 \end{bmatrix}$$

or,

$$\begin{bmatrix} a_1 & a_2 \\ a_2 & a_3 \end{bmatrix} = \begin{bmatrix} 0 & 0 \\ 0 & 0 \end{bmatrix} .$$

Thus, the result is

$$a_1 = 0, \ a_2 = 0, \ a_3 = 0 .$$

Therefore, $\{E_1, E_2, E_3\}$ is independent. Thus, $\{E_1, E_2, E_3\}$ is a basis of V and so the dimension of V is 3.

● **PROBLEM** 10–48

For the system

$$2x - y + z = 0$$
$$-7x + 7/2 \ y - 7/2 \ z = 0$$
$$4x + \quad y - 2z = 0$$

form the matrix of coefficients. Find the form of the solutions to the system by reducing the coefficient matrix to a reduced matrix.

Solution: The coefficient matrix is

$$\begin{bmatrix} 2 & -1 & 1 \\ -7 & 7/2 & -7/2 \\ 4 & 1 & -2 \end{bmatrix} .$$

Add $7/2$ times the first row to the second row and add -2 times the first row to the third row.

$$\begin{bmatrix} 2 & -1 & 1 \\ 0 & 0 & 0 \\ 0 & 3 & -4 \end{bmatrix} \qquad (2)$$

Add $1/3$ times the third row to the first row.

$$\begin{bmatrix} 2 & 0 & -1/3 \\ 0 & 0 & 0 \\ 0 & 3 & -4 \end{bmatrix}$$

Note the row of zeros which indicates that the second equation has been entirely eliminated, being superfluous. Divide the first row by 2 and the third row by 3. Then interchange the second row and the third row.

616

$$\begin{bmatrix} 1 & 0 & -1/6 \\ 0 & 1 & -4/3 \\ 0 & 0 & -0 \end{bmatrix}$$

This matrix is reduced. It is the coefficient matrix (after dropping the third row) for the system

$$x \quad - \: 1/6 \: z = 0$$
$$y \quad - \: 4/3 \: z = 0 \: .$$

(1)

The dependence of the solutions on z can be expressed as follows:

$$x = 1/6 \: z$$
$$y = 4/3 \: z$$
$$z = z \quad .$$

To obtain another form for the solution, reduce the matrix (2) in a different manner. Add 1/4 times the third row of matrix (2) to the first row, obtaining

$$\begin{bmatrix} 2 & -1/4 & 0 \\ 0 & 0 & 0 \\ 0 & 3 & -4 \end{bmatrix} \quad .$$

Divide the first row by 2 and the third row by 3, and interchange the second and third rows

$$\begin{bmatrix} 1 & -1/8 & 0 \\ 0 & 1 & -4/3 \\ 0 & 0 & 0 \end{bmatrix} \quad .$$

This gives

$$x - 1/8 \: y = 0$$
$$y - 4/3 \: z = 0$$

(3)

The dependence of the solutions on y can be expressed by rewriting system (3) as

$$x = 1/8 \: y$$
$$y = y$$
$$z = 3/4 \: y \quad .$$

● **PROBLEM** 10–49

Solve the following system of equations:

$$x + 3y = 0$$
$$2x + 6y + 4z = 0$$

(1)

Solution: To solve the given system of equations, first form the matrix of the coefficients. Then reduce this matrix to echelon form. The matrix of the coefficients is

$$\begin{bmatrix} 1 & 3 & 0 \\ 2 & 6 & 4 \end{bmatrix} \quad .$$

Now add −2 times the first row to the second row

$$\begin{bmatrix} 1 & 3 & 0 \\ 0 & 0 & 4 \end{bmatrix} .$$

Divide the second row by 4

$$\begin{bmatrix} 1 & 3 & 0 \\ 0 & 0 & 1 \end{bmatrix} .$$

The above is the matrix of coefficients for

$$\begin{aligned} x + 3y &= 0 \\ z &= 0 . \end{aligned} \qquad (2)$$

This system is easy to solve. We have $z = 0$ and can assign y any value. Then compute x from (2). This gives a solution to (1).

● **PROBLEM** 10–50

Solve the following homogeneous system of linear equations.

$$\begin{aligned} 2x_1 + 2x_2 - x_3 \quad + x_5 &= 0 \\ -x_1 - x_2 + 2x_3 - 3x_4 + x_5 &= 0 \\ x_1 + x_3 - 2x_3 - x_5 &= 0 \\ x_3 + x_4 + x_5 &= 0 \end{aligned} \qquad (1)$$

Solution: The system (1) has five unknowns but only four equations. We know that a homogeneous system of linear equations with more unknowns than equations has a non-zero (non-trivial) solution. Now, to solve the system (1), form the matrix of the coefficients. Then reduce this matrix to reduced row-echelon form. The coefficient matrix is

$$A = \begin{bmatrix} 2 & 2 & -1 & 0 & 1 \\ -1 & -1 & 2 & -3 & 1 \\ 1 & 1 & -2 & 0 & -1 \\ 0 & 0 & 1 & 1 & 1 \end{bmatrix} . \qquad (1)$$

Add the fourth row to the first row and the third row to the second row.

$$\begin{bmatrix} 2 & 2 & 0 & 1 & 2 \\ 0 & 0 & 0 & -3 & 0 \\ 1 & 1 & -2 & 0 & -1 \\ 0 & 0 & 1 & 1 & 1 \end{bmatrix} . \qquad (2)$$

Divide the second row by −3. Then add −1 times the second row to the first row and to the fourth row.

$$\begin{bmatrix} 2 & 2 & 0 & 0 & 2 \\ 0 & 0 & 0 & 1 & 0 \\ 1 & 1 & -2 & 0 & -1 \\ 0 & 0 & 1 & 0 & 1 \end{bmatrix} \qquad (3)$$

Divide the first row by 2. Then add −1 times the first row to the third row.

$$\begin{bmatrix} 1 & 1 & 0 & 0 & 1 \\ 0 & 0 & 0 & 1 & 0 \\ 0 & 0 & -2 & 0 & -2 \\ 0 & 0 & 1 & 0 & 1 \end{bmatrix} . \qquad (4)$$

Add 2 times the fourth row to the third row

$$\begin{bmatrix} 1 & 1 & 0 & 0 & 1 \\ 0 & 0 & 0 & 1 & 0 \\ 0 & 0 & 0 & 0 & 0 \\ 0 & 0 & 1 & 0 & 1 \end{bmatrix} \qquad (5)$$

Interchange the second and fourth rows. Next, interchange the third and fourth rows.

$$\begin{bmatrix} 1 & 1 & 0 & 0 & 1 \\ 0 & 0 & 1 & 0 & 1 \\ 0 & 0 & 0 & 1 & 0 \\ 0 & 0 & 0 & 0 & 0 \end{bmatrix} \qquad (6)$$

This matrix is in row reduced echelon form. The corresponding system of equations is

$$x_1 + x_2 \qquad + x_5 = 0$$
$$x_3 \qquad + x_5 = 0$$
$$x_4 \qquad = 0 \quad .$$

Solving for the leading variables yields

$$x_1 = -x_2 - x_5$$
$$x_3 = -x_5$$
$$x_4 = 0 \ .$$

The solution set is, therefore, given by

$$x_1 = -s - t, \ x_2 = s, \ x_3 = -t, \ x_4 = 0, \ x_5 = t.$$

That is, we have chosen x_2 and x_5 to be free variables and, hence, set them equal to the parameters s and t, respectively. The dependent variables are then x_1 and x_3, and their dependence on s and t are given by the reduced form of the system. Thus, any solution vector is of the form $(-s-t, s, -t, 0, t)$. Recall that a basis for the subspace of vectors of this form (i.e., the solution space) would be the vectors

(1) $(-0-1,0,-1,0,1) = (-1,0,-1,0,1)$

(2) $(-1-0,1,-0,0,0) = (-1,1,0,0,0,)$

• **PROBLEM** 10–51

By forming the augmented matrix and row reducing, determine the solutions of the following system

$$\begin{aligned} 2x - y + 3z &= 4 \\ 3x + 2z &= 5 \\ -2x + y + 4z &= 6 \ . \end{aligned} \qquad (1)$$

Solution: The system of equations

$$a_{11}x_1 + a_{12}x_2 + \cdots + a_{1n}x_n = c_1$$
$$a_{21}x_1 + a_{22}x_2 + \cdots + a_{2n}x_n = c_2$$
$$\cdot$$
$$\cdot$$
$$\cdot$$
$$a_{m1}x_1 + a_{m2}x_2 + \cdots + a_{mn}x_n = c_m$$

(2)

is called a non-homogeneous linear system if the constants $c_1, c_2, \ldots, c_m$ are not all zero.

We form the $m \times (n+1)$ matrix A' defined by

$$A' = \begin{bmatrix} a_{11} & a_{12} & \cdots & a_{1n} & c_1 \\ a_{21} & a_{22} & \cdots & a_{2n} & c_2 \\ \cdot & & & & \\ \cdot & & & & \\ \cdot & & & & \\ a_{m1} & a_{m2} & \cdots & a_{mn} & c_m \end{bmatrix} .$$

This matrix is called the augmented matrix of the system (2). The first n columns of A' consist of the coefficient matrix of (2), and the last column of A' consists of the corresponding constants.

To solve the non-homogeneous linear system, form the augmented matrix A'. Apply row operations to A' to reduce it to echelon form. Now, the augmented matrix of the system (1) is

$$\begin{bmatrix} 2 & -1 & 3 & 4 \\ 3 & 0 & 2 & 5 \\ -2 & 1 & 4 & 6 \end{bmatrix} .$$

Add the first row to the third row

$$\begin{bmatrix} 2 & -1 & 3 & 4 \\ 3 & 0 & 2 & 5 \\ 0 & 0 & 7 & 10 \end{bmatrix} .$$

This is the augmented matrix of

$$2x - y + 3z = 4$$
$$3x + 2z = 5$$
$$7z = 10 .$$

The system has been sufficiently simplified now so that the solution can be found.

From the last equation we have $z = 10/7$. Substituting this value into the second equation and solving for x gives $x = 5/7$. Substituting $x = 5/7$ and $z = 10/7$ into the first equation and solving for y yields $y = 12/7$. The solution to system (1) is, therefore,
$$x = 5/7 , y = 12/7 , z = 10/7 .$$

Note: We could have further reduced the matrix to row-reduced echelon form and solved the system directly from the reduced matrix. That is, by adding $-2/3$ times the second row to the first row, we have

$$\begin{bmatrix} 0 & -1 & 5/3 & 2/3 \\ 3 & 0 & 2 & 5 \\ 0 & 0 & 7 & 10 \end{bmatrix} .$$

Multiplying the second row by 1/3 and the first row by −1 and inter-changing the two, then multiplying the third row by 1/7 results in

$$\begin{bmatrix} 1 & 0 & 2/3 & 5/3 \\ 0 & 1 & -5/3 & -2/3 \\ 0 & 0 & 1 & 10/7 \end{bmatrix} .$$

Then, adding 5/3 times the third row to the second and −2/3 times the third row to the first gives

$$\begin{bmatrix} 1 & 0 & 0 & 5/7 \\ 0 & 1 & 0 & 12/7 \\ 0 & 0 & 1 & 10/7 \end{bmatrix} .$$

The solution to a non-homogeneous system found in the above man-ner is called the particular solution. The non-homogeneous system will be satisfied by any sum of the particular solution and a solution to the corresponding homogeneous system. In this case, the only solution to the homogeneous system is the trivial solution. Therefore, the only solution to the non-homogeneous problem is the particular solution.

• PROBLEM 10–52

Solve the following linear system of equations:

$$\begin{aligned} 2x + 3y - 4z &= 5 \\ -2x + z &= 7 \\ 3x + 2y + 2z &= 3 . \end{aligned} \qquad (1)$$

Solution: The matrix of coefficients is

$$\begin{bmatrix} 2 & 3 & -4 \\ -2 & 0 & 1 \\ 3 & 2 & 2 \end{bmatrix} .$$

Let

$$X = \begin{bmatrix} x \\ y \\ z \end{bmatrix} \quad \text{and} \quad B = \begin{bmatrix} 5 \\ 7 \\ 3 \end{bmatrix} .$$

The given linear system can be written in matrix form as

$$AX = B .$$

Since the system (1) is non-homogeneous, the solutions to the system are completely determined by its augmented matrix.

The augmented matrix for the system (1) is

$$\left[\begin{array}{ccc|c} 2 & 3 & -4 & 5 \\ -2 & 0 & 1 & 7 \\ 3 & 2 & 2 & 3 \end{array} \right]$$

which can be reduced by using the following sequence of row operations: Add the first row to the second row

$$\left[\begin{array}{ccc|c} 2 & 3 & -4 & 5 \\ 0 & 3 & -3 & 12 \\ 3 & 2 & 2 & 3 \end{array} \right] .$$

Divide the first row by 2 and the second row by 3

$$\begin{bmatrix} 1 & 3/2 & -2 & | & 5/2 \\ 0 & 1 & -1 & | & 4 \\ 3 & 2 & 2 & | & 3 \end{bmatrix} .$$

Add -3 times the first row to the third row

$$\begin{bmatrix} 1 & 3/2 & -2 & | & 5/2 \\ 0 & 1 & -1 & | & 4 \\ 0 & -5/2 & 8 & | & -9/2 \end{bmatrix} .$$

Add 5/2 times the second row to the third row

$$\begin{bmatrix} 1 & 3/2 & -2 & | & 5/2 \\ 0 & 1 & -1 & | & 4 \\ 0 & 0 & 11/2 & | & 11/2 \end{bmatrix} .$$

This is the augmented matrix for the system

$$\begin{array}{rcl} x + 3/2\,y - 2z &=& 5/2 \\ y - z &=& 4 \\ 11/2\,z &=& 11/2 . \end{array}$$

Now the solution to this system can be easily found. From the last equation we have $z = 1$. Substituting $z = 1$ in the second equation gives $y = 5$. Next, substitute $y = 5$ and $z = 1$ into the first equation. This gives $x = -3$. Therefore, the solution to system (1) is $x = -3$, $y = 5$, $z = 1$.

This is the particular solution. A solution to the corresponding homogeneous problem plus the particular solution also solves the inhomogeneous system and is known as a general solution.

● **PROBLEM** 10–53

Solve the following system

$$\begin{array}{rcl} x + y + 2z &=& 9 \\ 2x + 4y - 3z &=& 1 \\ 3x + 6y - 5z &=& 0 \end{array} \qquad (1)$$

Solution: The augmented matrix for the system (1) is

$$\begin{bmatrix} 1 & 1 & 2 & | & 9 \\ 2 & 4 & -3 & | & 1 \\ 3 & 6 & -5 & | & 0 \end{bmatrix} .$$

It can be reduced to echelon form by elementary row operations. Add -2 times the first row to the second row and -3 times the first row to the third row

$$\begin{bmatrix} 1 & 1 & 2 & | & 9 \\ 0 & 2 & -7 & | & -17 \\ 0 & 3 & -11 & | & -27 \end{bmatrix} .$$

Multiply the second row by 1/2

$$\begin{bmatrix} 1 & 1 & 2 & \vdots & 9 \\ 0 & 1 & -7/2 & \vdots & -17/2 \\ 0 & 3 & -11 & \vdots & -27 \end{bmatrix} .$$

Add -3 times the second row to the third row

$$\begin{bmatrix} 1 & 1 & 2 & \vdots & 9 \\ 0 & 1 & -7/2 & \vdots & -17/2 \\ 0 & 0 & -1/2 & \vdots & -3/2 \end{bmatrix} .$$

Multiply the third row by -2 to obtain

$$\begin{bmatrix} 1 & 1 & 2 & \vdots & 9 \\ 0 & 1 & -7/2 & \vdots & -17/2 \\ 0 & 0 & 1 & \vdots & 3 \end{bmatrix} .$$

This is the augmented matrix for the system

$$\begin{aligned} x + y + 2z &= 9 \\ y - 7/2\, z &= -17/2 \\ z &= 3 \end{aligned} .$$

Solving this system gives $x = 1$, $y = 2$, $z = 3$.

● **PROBLEM** 10–54

Solve the following system
$$\begin{aligned} 2x + y - 2z &= 10 \\ 3x + 2y + 2z &= 1 \\ 5x + 4y + 3z &= 4 \end{aligned} \qquad (1)$$

Solution: The augmented matrix is

$$\begin{bmatrix} 2 & 1 & -2 & \vdots & 10 \\ 3 & 2 & 2 & \vdots & 1 \\ 5 & 4 & 3 & \vdots & 4 \end{bmatrix} .$$

Reduce this matrix to echelon form. Add -1 times the second row to the first row

$$\begin{bmatrix} 1 & -1 & -4 & \vdots & 9 \\ 3 & 2 & 2 & \vdots & 1 \\ 5 & 4 & 3 & \vdots & 4 \end{bmatrix} .$$

Add 3 times the first row to the second row and 5 times the first row to the third row.

$$\begin{bmatrix} -1 & -1 & -4 & \vdots & 9 \\ 0 & -1 & -10 & \vdots & 28 \\ 0 & -1 & -17 & \vdots & 49 \end{bmatrix} .$$

Add -1 times the second row to the third row.

$$\begin{bmatrix} -1 & -1 & -4 & \vdots & 9 \\ 0 & -1 & -10 & \vdots & 28 \\ 0 & 0 & -7 & \vdots & 21 \end{bmatrix} \, .$$

This augmented matrix can be represented in equation form as

$$\begin{aligned} -x - y - 4z &= 9 \\ - y - 10z &= 28 \\ - 7z &= 21 \, . \end{aligned}$$

Solving this system yields $z = -3$, $y = 2$, $x = 1$. Thus, the solution of the system (1) is $x = 1$, $y = 2$ and $z = 3$.

• PROBLEM 10–55

Solve the following system by Gauss–Jordan elimination

$$\begin{aligned} x_1 + 3x_2 - 2x_3 + 2x_5 &= 0 \\ 2x_1 + 6x_2 - 5x_3 - 2x_4 + 4x_5 - 3x_6 &= -1 \\ 5x_3 + 10x_4 + 15x_6 &= 5 \\ 2x_1 + 6x_2 + 8x_4 + 4x_5 + 18x_6 &= 6 \, . \end{aligned}$$

Solution: The augmented matrix for the system is

$$\begin{bmatrix} 1 & 3 & -2 & 0 & 2 & 0 & \vdots & 0 \\ 2 & 6 & -5 & -2 & 4 & -3 & \vdots & -1 \\ 0 & 0 & 5 & 10 & 0 & 15 & \vdots & 5 \\ 2 & 6 & 0 & 8 & 4 & 18 & \vdots & 6 \end{bmatrix} \, .$$

Reduce this matrix to row-reduced echelon form. Add -2 times the first row to the second and fourth rows

$$\begin{bmatrix} 1 & 3 & -2 & 0 & 2 & 0 & \vdots & 0 \\ 0 & 0 & -1 & -2 & 0 & -3 & \vdots & -1 \\ 0 & 0 & 5 & 10 & 0 & 15 & \vdots & 5 \\ 0 & 0 & 4 & 8 & 0 & 18 & \vdots & 6 \end{bmatrix} \, .$$

Add 5 times the second row to the third row and 4 times the second row to the fourth row.

$$\begin{bmatrix} 1 & 3 & -2 & 0 & 2 & 0 & \vdots & 0 \\ 0 & 0 & -1 & -2 & 0 & -3 & \vdots & -1 \\ 0 & 0 & 0 & 0 & 0 & 0 & \vdots & 0 \\ 0 & 0 & 0 & 0 & 0 & 6 & \vdots & 2 \end{bmatrix} \, .$$

Multiply the second row by -1 and the fourth row by 1/6. Then, interchange the third and fourth rows.

$$\begin{bmatrix} 1 & 3 & -2 & 0 & 2 & 0 & \vdots & 0 \\ 0 & 0 & 1 & 2 & 0 & 3 & \vdots & 1 \\ 0 & 0 & 0 & 0 & 0 & 1 & \vdots & 1/3 \\ 0 & 0 & 0 & 0 & 0 & 0 & \vdots & 0 \end{bmatrix} \, .$$

Add -3 times the third row to the second row. Then add 2 times the second row to the first row.

$$\begin{bmatrix} 1 & 3 & 0 & 4 & 2 & 0 & | & 0 \\ 0 & 0 & 1 & 2 & 0 & 0 & | & 0 \\ 0 & 0 & 0 & 0 & 0 & 1 & | & 1/3 \\ 0 & 0 & 0 & 0 & 0 & 0 & | & 0 \end{bmatrix} .$$

Now the corresponding system of equations is

$$x_1 + 3x_2 + 4x_4 + 2x_5 = 0$$
$$x_3 + 2x_4 = 0$$
$$x_6 = 1/3 .$$

Then, solving for the leading variables results in

$$x_1 = -3x_2 - 4x_4 - 2x_5$$
$$x_3 = -2x_4$$
$$x_6 = 1/3 .$$

If we assign x_2, x_4 and x_5 the arbitrary values r,s, and t, respectively, the solution set is given by the formulas,

$$x_1 = -3r - 4s - 2t, \quad x_2 = r, \quad x_3 = -2s, \quad x_4 = s$$
$$x_5 = t, \quad x_6 = 1/3 .$$

• PROBLEM 10–56

Find the necessary and sufficient conditions for the existence of a solution to the following system.

$$x + y + 2z = a_1$$
$$-2x \quad\quad - z = a_2 \quad\quad\quad (1)$$
$$x + 3y + 5z = a_3 .$$

Solution: The procedure for solving nonhomogeneous systems is to form the augmented matrix of the system and reduce this matrix to a row-echelon matrix. Suppose, during this reduction, we obtain a matrix in which there is a row where the first non-zero entry appears in the last column. This matrix is the augmented matrix of a system with no solution.

If we obtain a reduced matrix with the property that no row has its first non-zero entry in the last column then the system has a solution. Now, the augmented matrix of the system (1) is

$$\begin{bmatrix} 1 & 1 & 2 & | & a_1 \\ -2 & 0 & -1 & | & a_2 \\ 1 & 3 & 5 & | & a_3 \end{bmatrix} .$$

Add 2 times the first row to the second row and -1 times the first row to the third row

625

$$\begin{bmatrix} 1 & 1 & 2 & | & a_1 \\ 0 & 2 & 3 & | & a_2+2a_1 \\ 0 & 2 & 3 & | & a_3-a_1 \end{bmatrix} .$$

Add -1 times the second row to the third row

$$\begin{bmatrix} 1 & 1 & 2 & | & a_1 \\ 0 & 2 & 3 & | & a_2+2a_1 \\ 0 & 0 & 0 & | & a_3-a_2-3a_1 \end{bmatrix} .$$

Suppose $a_3 - a_2 - 3a_1 \neq 0$. Then the reduced matrix has a row in which the first non-zero entry appears in the last column. Thus, the system has no solution. Therefore, a necessary condition for a solution to exist is that $a_3 - a_2 - 3a_1$ must be equal to zero.

Let $a_3 - a_2 - 3a_1 = 0$. Then, we have

$$\begin{bmatrix} 1 & 1 & 2 & | & a_1 \\ 0 & 2 & 3 & | & a_2+2a_1 \\ 0 & 0 & 0 & | & 0 \end{bmatrix}$$

which is the augmented matrix for the system

$$x + y + 2z = a_1$$
$$2y + 3z = a_2 + 2a_1 .$$

The above system can be written as

$$x + \tfrac{1}{2}z = -a_2/2$$
$$y + 3/2z = a_1 + a_2/2 .$$

Assign to z any arbitrary value, then compute x and y from the above equations to obtain a solution to system (1). Thus, we can conclude that $a_3 - a_2 - 3a_1 = 0$ is a necessary and sufficient condition for the existence of a solution to a system (1).

● **PROBLEM** 10–57

Determine the values of a so that the following system of equations has: (a) no solution, (b) more than one solution, (c) a unique solution.

$$x + y - z = 1$$
$$2x + 3y + az = 3$$
$$x + ay + 3z = 2$$

Solution: First form the augmented matrix for the system and then reduce it to echelon form.

The augmented matrix is

$$\begin{bmatrix} 1 & 1 & -1 & | & 1 \\ 2 & 3 & a & | & 3 \\ 1 & a & 3 & | & 2 \end{bmatrix} .$$

Add -2 times the first row to the second row and -1 times the first row to the third row.

$$\begin{bmatrix} 1 & 1 & -1 & \vdots & 1 \\ 0 & 1 & a+2 & \vdots & 1 \\ 0 & a-1 & 4 & \vdots & 1 \end{bmatrix} .$$

Add $-(a-1)$ times the second row to the third row

$$\begin{bmatrix} 1 & 1 & -1 & \vdots & 1 \\ 0 & 1 & a+2 & \vdots & 1 \\ 0 & 0 & 4-(a+2)(a-1) & \vdots & 1-(a-1) \end{bmatrix} .$$

The above matrix can be written as

$$\begin{bmatrix} 1 & 1 & -1 & \vdots & 1 \\ 0 & 1 & a+2 & \vdots & 1 \\ 0 & 0 & (3+a)(2-a) & \vdots & (2-a) \end{bmatrix} \qquad (1)$$

since $4-(a+2)(a-1) = 4-(a^2+a-2) = 6 - a - a^2 = (3+a)(2-a)$. Suppose $(3+a) = 0$. Then $(2-a) \neq 0$, and we have a reduced matrix with the property that a row has its first non-zero entry in the last column. In this case, the system has no solution. Thus, the system has a solution only if $(3+a) \neq 0$, that is, if $a \neq -3$. If $a= 3$, then the last row of the matrix (1) becomes $[0\ 0\ 0\ 5]$, and the system has no solution. Suppose $a = -2$. Then the last row of the matrix (1) is $[0\ 0\ 0\ 0]$.

In this case, the system has an infinite number of solutions. We can summarize our results as follows: If
(a) a = -3, then the system has no solution.
(b) if a = 2, the system has more than one solution.
(c) if $a \neq 2$ and $a \neq -3$, the system has a unique solution.

• PROBLEM 10–58

Solve the following system
$$x_1 - 2x_2 - 3x_3 = 3$$
$$2x_1 - x_2 - 4x_3 = 7 \qquad (1)$$
$$3x_1 - 3x_2 - 5x_3 = 8 .$$

<u>Solution</u>: The matrix of coefficients for the system (1) is

$$A = \begin{bmatrix} 1 & -2 & -3 \\ 2 & -1 & -4 \\ 3 & -3 & -5 \end{bmatrix} .$$

The system (1) may be written in matrix form as

$$\begin{bmatrix} 1 & -2 & -3 & x_1 \\ 2 & -1 & -4 & x_2 \\ 3 & -3 & -5 & x_3 \end{bmatrix} = \begin{bmatrix} 3 \\ 7 \\ 8 \end{bmatrix} . \qquad (2)$$

Let

$$X = \begin{bmatrix} x_1 \\ x_2 \\ x_3 \end{bmatrix}, \qquad b = \begin{bmatrix} 3 \\ 7 \\ 8 \end{bmatrix} .$$

Then equation (2) is written

$$\vec{AX} = \vec{b} . \qquad (3)$$

A solution vector $\vec{x}$ can be found by multiplying both sides of equation (3) by A^{-1}. Then we have

$$A^{-1}AX = A^{-1}b ,$$

but $A^{-1}A = I$. Hence,

$$IX = A^{-1}b$$

or

$$X = A^{-1}b . \qquad (4)$$

Thus, the solutions of a system of linear equations can be obtained by finding the inverse matrix of the coefficient matrix of the system and then solving equation (4). To find A^{-1}, first form the matrix $[A : I]$, and reduce this matrix, by applying row operations, to the form $[I : B]$. Then, $B = A^{-1}$. Now,

$$[A : I] = \begin{bmatrix} 1 & -2 & -3 & | & 1 & 0 & 0 \\ 2 & -1 & -4 & | & 0 & 1 & 0 \\ 3 & -3 & -5 & | & 0 & 0 & 1 \end{bmatrix} .$$

Add -2 times the first row to the second row and -3 times the first row to the third row

$$\begin{bmatrix} 1 & -2 & -3 & | & 1 & 0 & 0 \\ 0 & 3 & 2 & | & -2 & 1 & 0 \\ 0 & 3 & 4 & | & -3 & 0 & 1 \end{bmatrix} .$$

Now add -1 times the second row to the third row

$$\begin{bmatrix} 1 & -2 & -3 & | & 1 & 0 & 0 \\ 0 & 3 & 2 & | & -2 & 1 & 0 \\ 0 & 0 & 2 & | & -1 & -1 & 0 \end{bmatrix} .$$

Divide the third row by 2

$$\begin{bmatrix} 1 & -2 & -3 & | & 1 & 0 & 0 \\ 0 & 3 & 2 & | & -2 & 1 & 0 \\ 0 & 0 & 1 & | & -\frac{1}{2} & -\frac{1}{2} & \frac{1}{2} \end{bmatrix} .$$

Add -2 times the third row to the second row and 3 times the third row to the first row

$$\begin{bmatrix} 1 & -2 & 0 & | & -\frac{1}{2} & -3/2 & 3/2 \\ 0 & 3 & 0 & | & -1 & 2 & -1 \\ 0 & 0 & 1 & | & -\frac{1}{2} & -\frac{1}{2} & \frac{1}{2} \end{bmatrix} .$$

Divide the second row by 3; then add 2 times the resulting second row to the first row.

$$\begin{bmatrix} 1 & 0 & 0 & | & -7/6 & -1/6 & 5/6 \\ 0 & 1 & 0 & | & -1/3 & 2/3 & -1/3 \\ 0 & 0 & 1 & | & -\frac{1}{2} & -\frac{1}{2} & \frac{1}{2} \end{bmatrix} .$$

628

Thus,

$$A^{-1} = \begin{bmatrix} -7/6 & -1/6 & 5/6 \\ -1/3 & 2/3 & -1/3 \\ -\frac{1}{2} & -\frac{1}{2} & \frac{1}{2} \end{bmatrix} .$$

Then equation (4) becomes

$$\begin{bmatrix} x_1 \\ x_2 \\ x_3 \end{bmatrix} = \begin{bmatrix} -7/6 & -1/6 & 5/6 \\ -1/3 & 2/3 & -1/3 \\ -\frac{1}{2} & -\frac{1}{2} & \frac{1}{2} \end{bmatrix} \begin{bmatrix} 3 \\ 7 \\ 8 \end{bmatrix} .$$

Multiplying, we have

$$\begin{bmatrix} x_1 \\ x_2 \\ x_3 \end{bmatrix} = \begin{bmatrix} -21/6 - 7/6 + 40/6 \\ -1 + 14/3 - 8/3 \\ -3/2 - 7/2 + 4 \end{bmatrix} = \begin{bmatrix} 2 \\ 1 \\ -1 \end{bmatrix} .$$

Thus,

$$x_1 = 2, \ x_2 = 1, \ x_3 = -1 .$$

It is interesting to note that the calculation of the inverse matrix is closely related to the solution of simultaneous equations. Indeed, the two processes are essentially the same. To show this, we can use the method of successive elimination. Eliminating x_1 from the second and third equations,

$$x_1 - 2x_2 - 3x_3 = 3$$
$$3x_2 + 2x_3 = 1$$
$$3x_2 + 4x_3 = -1 .$$

Then, eliminating x_2 from the third equation,

$$x_1 - 2x_2 - 3x_3 = 3$$
$$3x_2 + 2x_3 = 1$$
$$2x_3 = -2 .$$

We obtain

$$x_1 - 2x_2 - 3x_3 = 3$$
$$3x_2 + 2x_3 = 1$$
$$x_3 = -1 ,$$

$$x_1 - 2x_2 = 0$$
$$3x_2 = 3$$
$$x_3 = -1$$

and, finally,

$$x_1 = 2$$
$$x_2 = 1$$
$$x_3 = -1 .$$

It can be seen that the solution to the system $A\vec{x} = \vec{b}$, calculated directly, is the same solution we obtained by calculating A^{-1} and finding $A^{-1}\vec{b}$. Observe that the inverse matrix can also be found by using the formula

$$A^{-1} = \frac{1}{\det A}\, \text{adj } A \, .$$

• **PROBLEM** 10–59

Find the solution set of the following system of equations:

$$\begin{aligned}
2x_1 + x_2 - 4x_3 &= 8 \\
3x_1 - x_2 + 2x_3 &= -1 \, .
\end{aligned} \qquad (1)$$

Solution: If $v = <c_1, c_2, \ldots, c_n>$ is a particular solution of a given system of linear equations and S is the solution set of its associated homogeneous system, then the solution set of the given system is $P = V + S$. The system (1) has more than one particular solution since it has more unknowns than equations. The system (1) has a particular solution, $x = 1$, $x_2 = 2$, $x_3 = -1$, as may be easily verified.

Now, the associated homogeneous system is

$$\begin{aligned}
2x_1 + x_2 - 4x_3 &= 0 \\
3x_1 - x_2 + 2x_3 &= 0 \, ,
\end{aligned}$$

and its solution set consists of all vectors orthogonal to both $<2,1,-4>$ and $<3,-1,2>$. This is the set of scalar multiples of

$$<2,1,-4> \ \times\ <3,-1,2> = <-2,-16,-5> \ .$$

Hence, the solution set of the given system is $<1,2,1> + \text{Sp}\{<-2,-16,-5>\}$.

POLYNOMIAL ALGEBRA

• **PROBLEM** 10–60

(B) Let

$$A = \begin{bmatrix} 1 & 2 & -1 \\ 1 & 0 & 1 \\ 4 & -4 & 5 \end{bmatrix} \, .$$

Find the characteristic polynomial of A.

(B) From A, as given, compute the matrix

$$\lambda I - A = \begin{bmatrix} \lambda-1 & -2 & 1 \\ -1 & \lambda & -1 \\ -4 & 4 & \lambda-5 \end{bmatrix} \, .$$

Then, the characteristic polynomial of A is:

$$f(\lambda) = \det[\lambda I - A] = \begin{vmatrix} \lambda-1 & -2 & 1 \\ -1 & \lambda & -1 \\ -4 & 4 & \lambda-5 \end{vmatrix}$$

$$= (\lambda-1) \begin{vmatrix} \lambda & -1 \\ 4 & \lambda-5 \end{vmatrix} -(-2) \begin{vmatrix} -1 & -1 \\ -4 & \lambda-5 \end{vmatrix} +1 \begin{vmatrix} -1 & \lambda \\ -4 & 4 \end{vmatrix}$$

$$= (\lambda-1) \left[\lambda(\lambda-5) + 2[-1(\lambda-5)-4] + 1[-4+4\lambda] \right.$$

$$f(\lambda) = (\lambda-1)[\lambda^2 - 5\lambda + 4] - 2(\lambda-1) + 4(\lambda-1) = (\lambda-1)[\lambda^2 - 5\lambda + 4 - 2 + 4]$$

$$= (\lambda-1)(\lambda-3)(\lambda-2)$$

or

$$f(\lambda) = \lambda^3 - 6\lambda^2 + 11\lambda - 6 .$$

It is worthwhile to note that $f(\lambda) = 0$ has roots 1, 3 and 2 which are called the characteristic roots or eigenvalues of A.

• **PROBLEM** 10–61

Find the characteristic polynomials and the eigenvalues of the matrices.

(i) $A = \begin{bmatrix} 2 & 3 \\ 1 & 4 \end{bmatrix}$; (ii) $B = \begin{bmatrix} \cos \alpha & \sin \alpha \\ -\sin \alpha & \cos \alpha \end{bmatrix}$;

(iii) $C = \begin{bmatrix} 1 & 2 & 3 \\ 2 & 1 & 3 \\ 3 & 3 & 6 \end{bmatrix}$

Solution: The characteristic polynomial of A is

$$f(\lambda) = \det(\lambda I - A) = \det \begin{bmatrix} \lambda-2 & -3 \\ -1 & \lambda-4 \end{bmatrix}$$

Therefore, $f(\lambda) = (\lambda-2)(\lambda-4) - 3 = \lambda^2 - 6\lambda + 5 = (\lambda-1)(\lambda-5)$.

The characteristic equation is $f(\lambda) = (\lambda-1)(\lambda-5) = 0$. Then the characteristic values are $\lambda = 1$ and $\lambda = 5$.

The zeros of the characteristic polynomial of a matrix are also called characteristic numbers or, proper values.

(ii) The characteristic polynomial of B is:

$$f(\lambda) = \det(\lambda I - B) = \det \begin{bmatrix} \lambda - \cos \alpha & -\sin \alpha \\ \sin \alpha & \lambda-\cos \alpha \end{bmatrix}$$

$$f(\lambda) = (\lambda - \cos \alpha)(\lambda - \cos \alpha) + \sin^2\alpha$$

$$= \lambda^2 - 2\lambda \cos \alpha + \cos^2\alpha + \sin^2\alpha .$$

But, $\sin^2\alpha + \cos^2\alpha = 1$. Therefore, $f(\lambda) = \lambda^2 - 2\lambda\cos\alpha + 1$. The characteristic equation is

$$f(\lambda) = \lambda^2 - 2\cos\alpha\,\lambda + 1 = 0.$$

We know the root of an equation, $ax^2 + bx + c = 0$ is

$$x = \frac{-b \pm \sqrt{b^2 - 4ac}}{2a}.$$

Thus,

$$\lambda = \frac{2\cos\alpha \pm \sqrt{4\cos^2\alpha - 4}}{2}$$

or,

$$\lambda = \frac{2\cos\alpha \pm 2\sqrt{\cos^2\alpha - 1}}{2}$$

$$= \cos\alpha \pm \sqrt{\cos^2\alpha - 1}$$

But, $\cos^2\alpha - 1 = -\sin^2\alpha$; therefore, $\lambda = \cos\alpha \pm \sqrt{-\sin^2\alpha}$ or,

$$\lambda = \cos\alpha \pm i\sin\alpha.$$

(iii) The characteristic polynomial of C is:

$$f(\lambda) = \det(\lambda I - C) = \det \begin{bmatrix} \lambda-1 & -2 & -3 \\ -2 & \lambda-1 & -3 \\ -3 & -3 & \lambda-6 \end{bmatrix}$$

$$= (\lambda-1)\begin{vmatrix} \lambda-1 & -3 \\ -3 & \lambda-6 \end{vmatrix} -(-2)\begin{vmatrix} -2 & -3 \\ -3 & \lambda-6 \end{vmatrix} + (-3)\begin{vmatrix} -2 & \lambda-1 \\ -3 & -3 \end{vmatrix}$$

$f(\lambda) = (\lambda-1)[(\lambda-1)(\lambda-6) -9] + 2[-2(\lambda-6)-9] - 3[6 + 3(\lambda-1)]$

$\quad = (\lambda-1)[\lambda^2 - 7\lambda - 3] + 2[-2\lambda + 3] - 3[3 + 3\lambda]$

$\quad = \lambda^3 - 8\lambda^2 + 4\lambda + 3 - 4\lambda + 6 - 9 - 9\lambda$

$\quad = \lambda^3 - 8\lambda^2 - 9\lambda = \lambda(\lambda-9)(\lambda+1)$.

The characteristic equation is $f(\lambda) = \lambda(\lambda-9)(\lambda+1) = 0$. Then, the characteristic values are $\lambda = 0$, $\lambda = -1$ and $\lambda = 9$.

• **PROBLEM** 10-62

Find the minimal polynomials of the following matrices:

(i) $\begin{bmatrix} 3 & 1 \\ 0 & 3 \end{bmatrix}$; (ii) $\begin{bmatrix} 3 & 1 & 0 \\ 0 & 3 & 0 \\ 0 & 0 & 3 \end{bmatrix}$; (iii) $\begin{bmatrix} 2 & 0 & 0 \\ 0 & 3 & 1 \\ 0 & 0 & 3 \end{bmatrix}$;

(iv) $\begin{bmatrix} 2 & 0 & 0 & 0 \\ 0 & 2 & 0 & 0 \\ 0 & 0 & 3 & 0 \\ 0 & 0 & 0 & 3 \end{bmatrix}$

Solution: (i) The characteristic polynomial of the matrix A is

$$f(\lambda) = \det(\lambda I - A) = \det \begin{bmatrix} \lambda-3 & -1 \\ 0 & \lambda-3 \end{bmatrix}$$

$$= (\lambda-3)(\lambda-3) \ .$$

The minimum polynomial $m(\lambda)$ must divide $f(\lambda)$. Thus, $m(\lambda)$ is exactly one of the following:

$$m_1(\lambda) = (\lambda-3) \qquad m_2(\lambda) = (\lambda-3)(\lambda-3) \ .$$

Now, $\qquad m_1(A) = (A-3I) \neq 0 \ .$

$\qquad m_2(A) = f(A) = 0$ by the Cayley-Hamilton Theorem.

Therefore, the characteristic polynomial, $(\lambda-3)(\lambda-3)$, is the minimal polynomial of A.

(ii)
$$A = \begin{bmatrix} 3 & 1 & 0 \\ 0 & 3 & 0 \\ 0 & 0 & 3 \end{bmatrix}$$

The characteristic polynomial of A is

$$f(\lambda) = \det \begin{bmatrix} \lambda-3 & -1 & 0 \\ 0 & \lambda-3 & 0 \\ 0 & 0 & \lambda-3 \end{bmatrix}$$

$$= (\lambda-3)(\lambda-3)(\lambda-3),$$

since the determinant of an upper or lower triangular matrix is the product of the main diagonal elements. Now, $(\lambda-3)$ must be a factor of $m(\lambda)$, the minimal polynomial. Also, $m(\lambda)$ must divide $f(\lambda)$; hence, $m(\lambda)$ must be one of the following three polynomials:

$$m_1(\lambda) = (\lambda-3), \ m_2(\lambda) = (\lambda-3)(\lambda-3) \quad \text{or,}$$
$$m_3(\lambda) = (\lambda-3)(\lambda-3)(\lambda-3).$$

We see that $m_1(A) = (A-3I) \neq 0$, since

$$A-3I = \begin{bmatrix} 0 & 1 & 0 \\ 0 & 0 & 0 \\ 0 & 0 & 0 \end{bmatrix}$$

Therefore, $m_1(\lambda)$ is not the minimal polynomial. But,

$$m_2(A) = (A-3I)^2 = 0 \ .$$

Thus, $(\lambda-3)(\lambda-3)$ is a minimal polynomial of A.

(iii)
$$A = \begin{bmatrix} 2 & 0 & 0 \\ 0 & 3 & 1 \\ 0 & 0 & 3 \end{bmatrix}$$

The characteristic polynomial of A is

$$f(\lambda) = \det(\lambda I - A) = \det \begin{bmatrix} \lambda-2 & 0 & 0 \\ 0 & \lambda-3 & -1 \\ 0 & 0 & \lambda-3 \end{bmatrix}$$

$$= (\lambda-2)(\lambda-3)(\lambda-3) \ .$$

As we know, the minimum polynomial $m(\lambda)$ must be one of the following two polynomials:

$$m_1(\lambda) = (\lambda-2)(\lambda-3), \quad m_2(\lambda) = (\lambda-2)(\lambda-3)^2$$

$$m_1(A) = (A-2I)(A-3I)$$

$$= \begin{bmatrix} 0 & 0 & 0 \\ 0 & 1 & 1 \\ 0 & 0 & 1 \end{bmatrix} \begin{bmatrix} -1 & 0 & 0 \\ 0 & 0 & 1 \\ 0 & 0 & 0 \end{bmatrix} \neq 0 \; .$$

But, $m_2(A) = f(A) = 0$ (by the Cayley-Hamilton Theorem). Thus, $m_2(\lambda) = (\lambda-2)(\lambda-3)^2$ is the minimum polymial of A since $m_2(\lambda)$ is of least degree among the $m_i(\lambda)$'s and since $m_2(A) = 0$.

(iv)

$$A = \begin{bmatrix} 2 & 0 & 0 & 0 \\ 0 & 2 & 0 & 0 \\ 0 & 0 & 3 & 0 \\ 0 & 0 & 0 & 3 \end{bmatrix}$$

The determinant of an upper triangular matrix is the product of the diagonal entries. Therefore,

$$f(\lambda) = \det(\lambda I-A) = (\lambda-2)(\lambda-2)(\lambda-3)(\lambda-3).$$

Let $m_1(\lambda) = (\lambda-2)(\lambda-3)$. Then,

$$m_1(A) = (A-2I)(A-3I) = 0 \; .$$

Therefore, the minimum polynomial of A is $(\lambda-2)(\lambda-3)$.

• PROBLEM 10–63

Find the characteristic and minimum polynomials of each of the following matrices

(a) $\begin{bmatrix} 3 & -1 \\ -1 & 3 \end{bmatrix}$ (b) $\begin{bmatrix} 1 & 1 \\ 0 & 2 \end{bmatrix}$ (c) $\begin{bmatrix} 1 & -2 \\ 1 & -1 \end{bmatrix}$ (d) $\begin{bmatrix} 1 & 1 \\ 0 & 1 \end{bmatrix}$

(e) $\begin{bmatrix} 0 & 1 & 0 & 0 \\ 0 & 0 & 0 & 0 \\ 0 & 0 & 1 & -2 \\ 0 & 0 & 1 & -1 \end{bmatrix}$ (f) $\begin{bmatrix} 3 & 1 & 0 & 0 \\ 0 & 3 & 0 & 0 \\ 0 & 0 & 2 & 1 \\ 0 & 0 & 1 & 2 \end{bmatrix}$

Solution: (a) The characteristic polynomial is

$$f(\lambda) = \det(\lambda I - A)$$

$$f(\lambda) = \det \begin{bmatrix} \lambda-3 & +1 \\ 1 & \lambda-3 \end{bmatrix}$$

$$= (\lambda-3)(\lambda-3) - 1$$

$$= \lambda^2 - 6\lambda + 8 = (\lambda-4)(\lambda-2) \; . \tag{1}$$

Then the characteristic equation is

$$f(\lambda) = \det(\lambda I - A) = 0 \; .$$

Thus,

$$(\lambda-4)(\lambda-2) = 0$$

and the characteristic values are $\lambda = 4$, and $\lambda = 2$.

Let $m(\lambda)$ be the minimum polynomial of A. Then, λ is a character-

istic value of A if and only if $m(\lambda) = 0$. Now, for the characteristic values, $\lambda = 4$ and $\lambda = 2$, we have

$$(\lambda-4)(\lambda-2) = 0 .$$

Therefore, the minimum polynomial $m(\lambda) = (\lambda-4)(\lambda-2)$.

(b) The characteristic polynomial is

$$f(\lambda) = \det \begin{bmatrix} \lambda-1 & -1 \\ 0 & \lambda-2 \end{bmatrix} = (\lambda-1)(\lambda-2)$$

and the characteristic values are $\lambda = 1$ and $\lambda = 2$. For these values,
$(\lambda-4)(\lambda-2)$ is equal to zero. Therefore, the minimum polynomial is
$(\lambda-4)(\lambda-2)$.

(c)
$$f(\lambda) = \det \begin{bmatrix} \lambda-1 & 2 \\ -1 & \lambda+1 \end{bmatrix} = (\lambda-1)(\lambda+1) + 2 = \lambda^2 + 1 .$$

$f(\lambda) = \lambda^2 + 1$ has no real roots (roots of $\lambda^2 + 1$ are $\pm\, i$). Therefore, it is irreducible over R (that is, it cannot be factored into
any lesser degree polynomials in $R[X]$, the set of polynomials with coefficients in the real numbers). The matrix A, however is a root of
$\lambda^2 + 1$ by the Cayley-Hamilton Theorem. Therefore, $f(\lambda) = \lambda^2 + 1$ is
the polynomial of least degree over the reals having A as a root. So,
$m(\lambda) = \lambda^2 + 1$.

(d)
$$A = \begin{bmatrix} 1 & 1 \\ 0 & 1 \end{bmatrix}$$

Then,

$$f(\lambda) = \det \begin{bmatrix} \lambda-1 & -1 \\ 0 & \lambda-1 \end{bmatrix} = (\lambda-1)^2 .$$

The minimum polynomial must divide $f(\lambda)$ and each irreducible factor
of $f(\lambda)$ must be a factor of $m(\lambda)$. Therefore, the minimal polynomial
is either $m_1(\lambda) = (\lambda-1)$ or $m_2(\lambda) = (\lambda-1)^2$. But, $m_1(A) = (A - 1I) \neq 0$.
$m_2(A) = (A - 1I)^2 = 0$ (by the Cayley-Hamilton Theorem). Therefore,
$m(\lambda) = (\lambda-1)^2$ is the minimum polynomial.

(e)
$$f(\lambda) = \det \begin{bmatrix} \lambda & -1 & 0 & 0 \\ 0 & \lambda & 0 & 0 \\ 0 & 0 & \lambda-1 & 2 \\ 0 & 0 & -1 & \lambda+1 \end{bmatrix} = \lambda \begin{vmatrix} \lambda & 0 & 0 \\ 0 & \lambda-1 & 2 \\ 0 & -1 & \lambda+1 \end{vmatrix}$$

$$= \lambda[\, \lambda\{(\lambda-1)(\lambda+1) + 2\}]$$
$$= \lambda^2(\lambda^2 + 1) .$$

Then, $m(\lambda)$ is exactly one of the following:

$$m_1(\lambda) = \lambda(\lambda^2 + 1) , \quad m_2(\lambda) = \lambda^3(\lambda^3 + 1).$$

We have:

$$m_1(A) = A(A^2 + 1I)$$

$$= \begin{bmatrix} 0 & 1 & 0 & 0 \\ 0 & 0 & 0 & 0 \\ 0 & 0 & 1 & -2 \\ 0 & 0 & 1 & -1 \end{bmatrix} \begin{bmatrix} 1 & 0 & 0 & 0 \\ 0 & 1 & 0 & 0 \\ 0 & 0 & 0 & 0 \\ 0 & 0 & 0 & 0 \end{bmatrix}$$

$$= \begin{bmatrix} 0 & 1 & 0 & 0 \\ 0 & 0 & 0 & 0 \\ 0 & 0 & 0 & 0 \\ 0 & 0 & 0 & 0 \end{bmatrix} \neq 0$$

By the Cayley-Hamilton Theorem, $m_2(A) = f(A) = 0$. Therefore, $\lambda^2(\lambda^2 + 1)$ is a minimal polynomial of A.

(f)

$$f(\lambda) = \det \begin{bmatrix} \lambda-3 & -1 & 0 & 0 \\ 0 & \lambda-3 & 0 & 0 \\ 0 & 0 & \lambda-2 & -1 \\ 0 & 0 & -1 & \lambda-2 \end{bmatrix}$$

$$= \lambda-3 \begin{vmatrix} \lambda-3 & 0 & 0 \\ 0 & \lambda-2 & -1 \\ 0 & -1 & \lambda-2 \end{vmatrix}$$

$$= (\lambda-3)[(\lambda-3)\{(\lambda-2)^2 - 1\}]$$

$$= (\lambda-3)[(\lambda-3)(\lambda^2-4\lambda+3)]$$

$$= (\lambda-3)^3(\lambda-1) .$$

Therefore, $m(\lambda)$ must be one of the following:

$$m_1(\lambda) = (\lambda-3)(\lambda-1) , \quad m_2(\lambda) = (\lambda-3)^2(\lambda-1)$$

and

$$m_3(\lambda) = (\lambda-3)^3(\lambda-1) .$$

$m_1(A) = (A - 3I)(A - I)$

$$= \begin{bmatrix} 0 & 1 & 0 & 0 \\ 0 & 0 & 0 & 0 \\ 0 & 0 & -1 & 1 \\ 0 & 0 & 1 & -1 \end{bmatrix} \begin{bmatrix} 2 & 1 & 0 & 0 \\ 0 & 2 & 0 & 0 \\ 0 & 0 & 1 & 1 \\ 0 & 0 & 1 & 1 \end{bmatrix} \neq 0 .$$

$m_2(A) = (A - 3I)^2(A - I)$

$$= \begin{bmatrix} 0 & 1 & 0 & 0 \\ 0 & 0 & 0 & 0 \\ 0 & 0 & -1 & 1 \\ 0 & 0 & 1 & -1 \end{bmatrix}^2 \begin{bmatrix} 2 & 1 & 0 & 0 \\ 0 & 2 & 0 & 0 \\ 0 & 0 & 1 & 1 \\ 0 & 0 & 1 & 1 \end{bmatrix}$$

$$= \begin{bmatrix} 0 & 0 & 0 & 0 \\ 0 & 0 & 0 & 0 \\ 0 & 0 & 2 & -2 \\ 0 & 0 & -2 & 2 \end{bmatrix} \begin{bmatrix} 2 & 1 & 0 & 0 \\ 0 & 2 & 0 & 0 \\ 0 & 0 & 1 & 1 \\ 0 & 0 & 1 & 1 \end{bmatrix} = 0$$

We know that $m_3(A) = f(A) = 0$ by the Cayley-Hamilton Theorem. But, the degree of $m_2(\lambda)$ is less than the degree of $m_3(\lambda)$. Therefore, $m_2(\lambda) = (\lambda-3)^2(\lambda-1)$ is the minimum polynomial of A.

636

If $F(X) = \begin{bmatrix} 1 & 0 & 1 \\ 2 & 1 & 1 \\ 1 & 1 & 1 \end{bmatrix} - \begin{bmatrix} 2 & 1 & 0 \\ -1 & 1 & 1 \\ 0 & 1 & 0 \end{bmatrix} X + \begin{bmatrix} 1 & 1 & 1 \\ 1 & 0 & 1 \\ 0 & 1 & 0 \end{bmatrix} X^2 ,$

and $B = \begin{bmatrix} 1 & 1 & 1 \\ 1 & 0 & 1 \\ 0 & 1 & 1 \end{bmatrix}$, find $F_L(B)$ and $F_R(B)$.

Solution: A matrix polynomial of order n is defined as:

$$F(X) = A_0 + A_1 X + A_2 X^2 + \ldots + A_m X^m \tag{1}$$

where $A_0, A_1, \ldots, A_m$ are $n \times n$ matrices. The degree of the matrix polynomial (1) is the largest integer K for which $A_K \neq 0$. The coefficient A_K of a polynomial $F(X)$ of degree K is called the leading coefficient of $F(X)$. Two matrix polynomials of order n,

$$F(X) = A_0 + A_1 X + A_2 X^2 + \ldots + A_m X^m ,$$

$$G(X) = B_0 + B_1 X + B_2 X^2 + \ldots + B_m X^m ,$$

are said to be equal if $A_i = B_i$; $i = 0,1,2,\ldots,m$.

If $F(X)$ is the matrix polynomial of order n given by (1), and B is an $n \times n$ matrix, then $F_L(B)$ and $F_R(B)$ are defined as follows:

$$F_L(B) = A_0 + B A_1 + B^2 A_2 + \ldots + B^m A_m$$

$$F_R(B) = A_0 + A_1 B + A_2 B^2 + \ldots + A_m B^m .$$

Now, from the given example, we have

$$A_0 = \begin{bmatrix} 1 & 0 & 1 \\ 2 & 1 & 1 \\ 1 & 1 & 1 \end{bmatrix}, \quad A_1 = -\begin{bmatrix} 2 & 1 & 0 \\ -1 & 1 & 1 \\ 0 & 1 & 0 \end{bmatrix}$$

and $A_2 = \begin{bmatrix} 1 & 1 & 1 \\ 1 & 0 & 1 \\ 0 & 1 & 0 \end{bmatrix}$.

Then,

$$F_L(B) = \begin{bmatrix} 1 & 0 & 1 \\ 2 & 1 & 1 \\ 1 & 1 & 1 \end{bmatrix} - \begin{bmatrix} 1 & 1 & 1 \\ 1 & 0 & 1 \\ 0 & 1 & 1 \end{bmatrix}\begin{bmatrix} 2 & 1 & 0 \\ -1 & 1 & 1 \\ 0 & 1 & 0 \end{bmatrix}$$

$$+ \begin{bmatrix} 1 & 1 & 1 \\ 1 & 0 & 1 \\ 0 & 1 & 1 \end{bmatrix}^2 \begin{bmatrix} 1 & 1 & 1 \\ 1 & 0 & 1 \\ 0 & 1 & 0 \end{bmatrix} .$$

Performing the matrix multiplication yields

$$F_L(B) = \begin{bmatrix} 1 & 0 & 1 \\ 2 & 1 & 1 \\ 1 & 1 & 1 \end{bmatrix} - \begin{bmatrix} 1 & 3 & 1 \\ 2 & 2 & 0 \\ -1 & 2 & 1 \end{bmatrix} + \begin{bmatrix} 4 & 5 & 4 \\ 3 & 3 & 3 \\ 2 & 3 & 2 \end{bmatrix}$$

$$= \begin{bmatrix} 4 & 2 & 4 \\ 3 & 2 & 4 \\ 4 & 2 & 2 \end{bmatrix} .$$

Then,

$$F_R(B) = \begin{bmatrix} 1 & 0 & 1 \\ 2 & 1 & 1 \\ 1 & 1 & 1 \end{bmatrix} - \begin{bmatrix} 2 & 1 & 0 \\ -1 & 1 & 1 \\ 0 & 1 & 0 \end{bmatrix} \begin{bmatrix} 1 & 1 & 1 \\ 1 & 0 & 1 \\ 0 & 1 & 1 \end{bmatrix}$$

$$+ \begin{bmatrix} 1 & 1 & 1 \\ 1 & 0 & 1 \\ 0 & 1 & 0 \end{bmatrix} \begin{bmatrix} 1 & 1 & 1 \\ 1 & 0 & 1 \\ 0 & 1 & 1 \end{bmatrix}^2$$

$$= \begin{bmatrix} 1 & 0 & 1 \\ 2 & 1 & 1 \\ 1 & 1 & 1 \end{bmatrix} - \begin{bmatrix} 3 & 2 & 3 \\ 0 & 0 & 1 \\ 1 & 0 & 1 \end{bmatrix} + \begin{bmatrix} 4 & 5 & 7 \\ 3 & 3 & 5 \\ 1 & 2 & 2 \end{bmatrix}$$

$$= \begin{bmatrix} 2 & 3 & 5 \\ 5 & 4 & 5 \\ 1 & 3 & 2 \end{bmatrix} .$$

● **PROBLEM** 10–65

Let
$$\varphi(\lambda) = -2 - 5\lambda + 3\lambda^2 , \quad A = \begin{bmatrix} 1 & 2 \\ 3 & 1 \end{bmatrix} . \quad \text{Show that}$$

$$\varphi(A) = \begin{bmatrix} 14 & 2 \\ 3 & 14 \end{bmatrix} .$$

<u>Solution</u>: Consider a polynomial in λ,
$$\Phi(\lambda) = a_0 + a_1\lambda + \ldots + a_n\lambda^n ,$$
where $a_0, a_1, \ldots, a_n$ are elements from a field R. The expression
$$a_0 I + a_1 A + \ldots + a_n A^n ,$$
where A is an arbitrary square matrix with elements from R and I is an identity matrix, is called a polynomial in A and is denoted by $\varphi(A)$. The expression $\varphi(A)$ is also called the value of $\varphi(\lambda)$ for $\lambda = A$. For example, consider the equality
$$\lambda^2 - 1 = (\lambda-1)(\lambda+1) .$$
Evaluate the right and left sides for $\lambda = A$. Then,

$$A^2 - I = (A - I)(A + I) \ .$$

Similarly, for $\lambda^3 + 1 = (\lambda+1)(\lambda^2-\lambda+1)$ we have

$$A^3 + I = (A+I)(A^2-A+I) \ .$$

In general, from such relations between polynomials in λ, one obtains a matrix identity. Given

$$\varphi(\lambda) = -2-5\lambda + 3\lambda^2 \ ,$$

then

$$\varphi(\lambda) = 3\lambda^2 - 5\lambda - 2 = (\lambda-2)(3\lambda+1) \ .$$

Therefore,

$$\varphi(A) = (A - 2I)(3A + I) \tag{1}$$

$$A = \begin{bmatrix} 1 & 2 \\ 3 & 1 \end{bmatrix}$$

$$A - 2I = \begin{bmatrix} 1 & 2 \\ 3 & 1 \end{bmatrix} - 2 \begin{bmatrix} 1 & 0 \\ 0 & 1 \end{bmatrix} = \begin{bmatrix} -1 & 2 \\ 3 & -1 \end{bmatrix}$$

and

$$3A + I = 3 \begin{bmatrix} 1 & 2 \\ 3 & 1 \end{bmatrix} + \begin{bmatrix} 1 & 0 \\ 0 & 1 \end{bmatrix} = \begin{bmatrix} 4 & 6 \\ 9 & 4 \end{bmatrix}$$

Substitute this into (1). Then,

$$\varphi(A) = \begin{bmatrix} -1 & 2 \\ 3 & -1 \end{bmatrix} \begin{bmatrix} 4 & 6 \\ 9 & 4 \end{bmatrix} = \begin{bmatrix} -4+18 & -6+8 \\ 12-9 & 18-4 \end{bmatrix}$$

$$= \begin{bmatrix} 14 & 2 \\ 3 & 14 \end{bmatrix} \ .$$

• PROBLEM 10–66

Find the minimal polynomial of
$$A = \begin{bmatrix} 9 & -2 & 2 \\ -8 & 3 & -2 \\ -48 & 12 & -11 \end{bmatrix}$$

Solution: The method that is normally used for finding the minimal polynomial of a given matrix A is to find the characteristic polynomial $f(\lambda)$, to factor $f(\lambda)$ into linear factors, and then to find $m(\lambda)$, the monic divisor of $f(\lambda)$ of least degree for which $m(\lambda) = m(A) = 0$. This method does not always work, however, because sometimes it is difficult to carry out the needed factorization of $f(\lambda)$. Now, the following theorem describes another method for finding $m(\lambda)$.

Theorem: For each $n \times n$ matrix A, $m(\lambda) = f(\lambda)/d(\lambda)$, where $d(\lambda)$ is the monic greatest common divisor of the polynomial entries of $adj(\lambda I - A)$. Given

$$A = \begin{bmatrix} 9 & -2 & 2 \\ -8 & 3 & -2 \\ -48 & 12 & -11 \end{bmatrix}$$

The characteristic polynomial $f(\lambda)$ is

$$f(\lambda) = \det(\lambda I - A) = \det \begin{bmatrix} \lambda-9 & 2 & -2 \\ 8 & \lambda-3 & 2 \\ 48 & -12 & \lambda+11 \end{bmatrix}$$

$f(\lambda) = (\lambda-9)[(\lambda-3)(\lambda+11) + 24] - 2[8(\lambda+11) - 96] - 2[-96 - 48(\lambda-3)]$

$\qquad = (\lambda-9)[(\lambda+9)(\lambda-1)] - 16[(\lambda-1)] + 96[(\lambda-1)]$

$\qquad = (\lambda-1)[(\lambda-9)(\lambda+9) - 16 + 96]$

$\qquad = (\lambda-1)[\lambda^2 - 81 + 80] = (\lambda-1)(\lambda^2-1)$.

Now, the transpose of the matrix of cofactors of A is called the adjoint of A.

$$\lambda I - A = \begin{bmatrix} \lambda-9 & 2 & -2 \\ 8 & \lambda-3 & 2 \\ 48 & -12 & \lambda+11 \end{bmatrix} .$$

Then,

$$\text{adj}(\lambda I - A) = \begin{bmatrix} \lambda^2+8\lambda-9 & -2\lambda+2 & 2\lambda-2 \\ -8\lambda+8 & \lambda^2+2\lambda-3 & 2\lambda-2 \\ -48\lambda+48 & 12\lambda-12 & \lambda^2-12\lambda+11 \end{bmatrix}$$

or,

$$\text{adj}(\lambda I - A) = \begin{bmatrix} (\lambda+9)(\lambda-1) & -2(\lambda-1) & 2(\lambda-1) \\ -8(\lambda-1) & (\lambda+3)(\lambda-1) & 2(\lambda-1) \\ -48(\lambda-1) & 12(\lambda-1) & (\lambda-11)(\lambda-1) \end{bmatrix}$$

The greatest common divisor of the entries of $\text{adj}(\lambda I - A)$ is $d(\lambda) = \lambda-1$. Therefore,

$$m(\lambda) = f(\lambda)/d(\lambda) = (\lambda-1)(\lambda^2-1)/(\lambda-1)$$

$$= (\lambda^2-1) .$$

The above method for finding minimal polynomial $m(\lambda)$ always works, but it requires long computations. Thus, it would be used only when needed, as a general method, or when the characteristic polynomial $f(\lambda)$ is unfactorable.

• PROBLEM 10–67

Find the companion matrix of the following polynomials.

(i) $(\lambda-1)(\lambda-2)$, (ii) $(\lambda-1)^2$, (iii) $(\lambda-1)(\lambda-2)(\lambda-3)$.

<u>Solution</u>: Let

$$f(\lambda) = a_0 + a_1\lambda + \ldots + a_{n-1}\lambda^{n-1} + \lambda^n$$

be an arbitrary nonconstant polynomial with leading coefficient 1. Let the coefficients of this polynomial lie in a field R. The matrix

$$C = \begin{bmatrix} 0 & 0 & 0 & \ldots & 0 & -a_0 \\ 1 & 0 & 0 & \ldots & 0 & -a_1 \\ 0 & 1 & 0 & \ldots & 0 & -a_2 \\ . & . & . & . & . & . \\ 0 & 0 & 0 & \ldots & 0 & -a_{n-2} \\ 0 & 0 & 0 & \ldots & 1 & -a_{n-1} \end{bmatrix}$$

is called the companion matrix of the polynomial $f(\lambda)$.

(i) $f(\lambda) = (\lambda-1)(\lambda-2) = \lambda^2 - 3\lambda + 2$. Here, $a_0 = 2$, $a_1 = -3$ and $n = 2$. Thus, we have the 2×2 matrix,

$$C = \begin{bmatrix} 0 & -2 \\ 1 & 3 \end{bmatrix}$$

This is the companion matrix.

(ii) $f(\lambda) = (\lambda-1)^2 = \lambda^2 - 2\lambda + 1$. Here, $a_0 = 1$, $a_1 = -2$. Then, the companion matrix is

$$C = \begin{bmatrix} 0 & -1 \\ 1 & 2 \end{bmatrix} \quad .$$

(iii) $f(\lambda) = (\lambda-1)(\lambda-2)(\lambda-3) = \lambda^3 - 6\lambda^2 + 11\lambda - 6$. Here, $a_0 = -6$, $a_1 = 11$, $a_2 = -6$. The companion matrix is

$$C = \begin{bmatrix} 0 & 0 & 6 \\ 1 & 0 & -11 \\ 0 & 1 & 6 \end{bmatrix}$$

• **PROBLEM** 10–68

Let $A = \begin{bmatrix} 9 & 5 & -4 \\ -8 & -4 & 4 \\ 2 & 2 & 0 \end{bmatrix}$. Find the square root of A.

Solution: Let A be an $n\times n$ matrix which is similar to a diagonal matrix. Suppose that the characteristic values of A are all positive real numbers. Then, there is one and only one $n\times n$ matrix B such that $B^2 = A$. Also, there is a polynomial $f(x)$ with real coefficients such that $B = f(A)$. In other words, one can find the square root B of any matrix A which satisfies this hypotheses.

Now, the characteristic polynomial of A is

$$f(\lambda) = \det(\lambda I - A) = \det \begin{bmatrix} \lambda-9 & -5 & 4 \\ 8 & \lambda+4 & -4 \\ -2 & -2 & \lambda \end{bmatrix}$$

$$= (\lambda-9) \begin{vmatrix} \lambda+4 & -4 \\ -2 & \lambda \end{vmatrix} - (-5) \begin{vmatrix} 8 & -4 \\ -2 & \lambda \end{vmatrix} + 4 \begin{vmatrix} 8 & \lambda+4 \\ -2 & -2 \end{vmatrix}$$

$$= (\lambda-9)[\lambda^2 + 4\lambda - 8] + 5[8\lambda - 8] + 4[2\lambda - 8]$$

$$= \lambda^3 - 5\lambda^2 + 4\lambda = \lambda(\lambda-1)(\lambda-4) \quad .$$

Then, the characteristic equation of A is $\lambda(\lambda-1)(\lambda-4) = 0$. Hence, the characteristic values are $0, 1$, and 4. Since A has three distinct characteristic values, A is similar to a diagonal matrix. Then, from the given hypotheses, $B^2 = A$ and $B = f(A)$.

Let $f(x)$ be a scalar polynomial for which $f(\lambda_i) = \sqrt{\lambda_i}$, $i = 1,2,\ldots,n$ Thus, $f(0) = 0$, $f(1) = 1$, and $f(4) = 2$. If $f(x) = ax^2 + bx + c$, then

641

$$f(0) = c = 0$$
$$f(1) = a + b + c = 1$$

and

$$f(4) = 16a + 4b + c = 2 .$$

Solving the above equations yields

$$a = -1/6, \quad b = 7/6 , \quad \text{and} \quad c = 0 .$$

Thus,
$$f(x) = -1/6x^2 + 7/16x .$$

Then,

$$B = f(A) = -1/6\,A^2 + 7/6\,A$$

$$= -1/6 \begin{bmatrix} 9 & 5 & -4 \\ -8 & -4 & 4 \\ 2 & 2 & 0 \end{bmatrix}^2 + 7/6 \begin{bmatrix} 9 & 5 & -4 \\ -8 & -4 & 4 \\ 2 & 2 & 0 \end{bmatrix}$$

$$= -1/6 \begin{bmatrix} 33 & 17 & -16 \\ -32 & -16 & 16 \\ 2 & 2 & 0 \end{bmatrix} + 7/6 \begin{bmatrix} 9 & 5 & -4 \\ -8 & -4 & 4 \\ 2 & 2 & 0 \end{bmatrix}$$

$$= +1/6 \begin{bmatrix} -33 & -17 & 16 \\ 32 & 16 & -16 \\ -2 & -2 & 0 \end{bmatrix} + 1/6 \begin{bmatrix} 63 & 35 & -28 \\ -56 & -28 & 28 \\ 14 & 14 & 0 \end{bmatrix}$$

$$B = f(A) = 1/6 \begin{bmatrix} 30 & 18 & -12 \\ -24 & -12 & 12 \\ 12 & 12 & 0 \end{bmatrix} = \begin{bmatrix} 5 & 3 & -2 \\ -4 & -2 & 2 \\ 2 & 2 & 0 \end{bmatrix}$$

Hence,

$$B = f(A) = \begin{bmatrix} 5 & 3 & -2 \\ -4 & -2 & 2 \\ 2 & 2 & 0 \end{bmatrix} \quad \text{is a square root of } A.$$

• PROBLEM 10–69

Let V be the vector space R^2 and let T be the operator defined by
$$T(xy) = (2x-y, \ x+y).$$
Let $f(x) = 2 + 3x$ and $g(x) = x + x^2$. Find $f(T)$ and $g(T)$.

<u>Solution</u>: Consider a polynomial $f(x)$ over a field R:
$$f(x) = a_0 + a_1 x + a_2 x^2 + \ldots + a_n x^n .$$
Now, suppose $T: V \to V$ is a linear operator on a vector space V over R . Then, define
$$f(T) = a_0 I + a_1 T + a_2 T^2 + \ldots + a_n T^n$$
where I is the identity mapping. Now, $f(x) = 2 + 3x$. Then, $f(T) = 2I + 3T$ or,
$$f(T)(x,y) = 2(x,y) + 3T(x,y).$$
We know that $\alpha T(x,y) = T(\alpha x, \alpha y)$ where α is any real scalar, and
$$(S+T)(x,y) = S(x,y) + T(x,y)$$

since T is a linear operator. Therefore,

$$f(T)(x,y) = 2(x,y) + (6x - 3y, 3x + 3y)$$
$$= (8x - 3y, 3x + 5y) .$$

Now,

$$g(x) = x + x^2 , \quad g(T) = T + T^2 .$$
$$T^2(x,y) = T(T(x,y)) = T(2x - y, x + y)$$
$$= (2(2x - y) - x - y, (2x - y) + (x + y)),$$

so

$$T^2(x,y) = (3x - 3y, 3x) .$$

Then,

$$g(T)(x,y) = T(x,y) + T^2(x,y)$$
$$= (2x - y, x + y) + (3x - 3y, 3x)$$
$$= (5x - 4y, 4x + y).$$

EIGENVALUES

• **PROBLEM** 10-70

Find the real eigenvalues of the matrix,
$$A = \begin{bmatrix} -2 & -1 \\ 5 & 2 \end{bmatrix}$$

<u>Solution:</u> Let T be a linear transformation and A its matrix with respect to a given basis. Then λ is an eigenvalue if

$$AX = \lambda X ,\tag{1}$$

where X is a non-zero vector. Choosing R^n as the underlying vector space, $T: R^n \to R^n$, (1) becomes

$$\begin{bmatrix} a_{11} & a_{12} & \cdots & a_{1n} \\ a_{21} & & & \cdot \\ \cdot & & & \cdot \\ \cdot & & & \cdot \\ \cdot & & & \cdot \\ a_{n1} & a_{n2} & \cdots & a_{nn} \end{bmatrix} \begin{bmatrix} x_1 \\ x_2 \\ \cdot \\ \cdot \\ \cdot \\ x_n \end{bmatrix} = \lambda \begin{bmatrix} x_1 \\ x_2 \\ \cdot \\ \cdot \\ \cdot \\ x_n \end{bmatrix} .\tag{2}$$

Expanding (2),

$$a_{11}x_1 + a_{12}x_2 + \cdots + a_{1n}x_n = \lambda x_1 \tag{3}$$
$$a_{21}x_1 + a_{22}x_2 + \cdots + a_{2n}x_n = \lambda x_2$$
$$\vdots \qquad \vdots \qquad \qquad \vdots \qquad \vdots$$
$$a_{n1}x_1 + a_{n2}x_2 + \cdots + a_{nn}x_n = \lambda x_n$$

Rewriting (3),

$$(\lambda - a_{11})x_1 - a_{12}x_2 - \cdots - a_{1n}x_n = 0 \qquad (4)$$

$$-a_{21}x_1 + (\lambda - a_{22})x_2 - \cdots - a_{2n}x_n = 0$$

$$\vdots$$

$$-a_{n1}x_1 - a_{n2}x_2 - \cdots + (\lambda - a_{nn})x_n = 0 \, .$$

The set of linear homogeneous equations (4) can be expressed in matrix form as:

$$\begin{bmatrix} \lambda & 0 & \cdots & 0 \\ 0 & \lambda & \cdots & 0 \\ \vdots & & & \\ \vdots & & & \\ 0 & 0 & \cdots & \lambda \end{bmatrix} \begin{bmatrix} x_1 \\ x_2 \\ \vdots \\ x_n \end{bmatrix} - \begin{bmatrix} a_{11} & a_{12} & \cdots & a_{1n} \\ a_{21} & & & \\ \vdots & & & \\ a_{n1} & \cdots & & a_{nn} \end{bmatrix} \begin{bmatrix} x_1 \\ x_2 \\ \vdots \\ x_n \end{bmatrix} = \begin{bmatrix} 0 \\ 0 \\ \vdots \\ 0 \end{bmatrix}$$

or,

$$[\lambda I - A][X] = [0] \, . \qquad (5)$$

Recall now that a set of n linear homogeneous equations in n unknowns can have a non-trivial solution only if $\det[\lambda I - A] = 0$. The equation $\det[\lambda I - A] = 0$ is an nth degree polynomial in λ and its roots provide the eigenvalues of A and, thus, of T. By the Fundamental Theorem of Algebra there are n such roots in the complex field.

Form the matrix

$$\lambda I - A = \lambda \begin{bmatrix} 1 & 0 \\ 0 & 1 \end{bmatrix} - \begin{bmatrix} -2 & -1 \\ 5 & 2 \end{bmatrix}$$

$$= \begin{bmatrix} \lambda + 2 & 1 \\ -5 & \lambda - 2 \end{bmatrix}$$

Take its determinant to obtain the characteristic polynomial of A:

$$f(\lambda) = \det \ (\lambda I - A)$$

$$= \det \begin{vmatrix} \lambda + 2 & 1 \\ -5 & \lambda - 2 \end{vmatrix}$$

$$= (\lambda + 2)(\lambda - 2) + 5$$

$$= \lambda^2 + 1$$

The eigenvalues of A must, therefore, satisfy the quadratic equation $\lambda^2 + 1 = 0$. Since the only solutions to this equation are the imaginary numbers $\lambda = i$ and $\lambda = -i$.

A has no real eigenvalues.

Find the real eigenvalues of A and their associated eigenvectors when

$$A = \begin{bmatrix} 1 & 1 \\ -2 & 4 \end{bmatrix} .$$

Solution: We wish to find all real numbers λ and all non-zero vectors $X = \begin{bmatrix} x_1 \\ x_2 \end{bmatrix}$ such that $AX = \lambda X$:

$$\begin{bmatrix} 1 & 1 \\ -2 & 4 \end{bmatrix} \begin{bmatrix} x_1 \\ x_2 \end{bmatrix} = \lambda \begin{bmatrix} x_1 \\ x_2 \end{bmatrix}$$

The above matrix equation is equivalent to the homogeneous system,

$$x_1 + x_2 = \lambda x_1$$

$$-2x_1 + 4x_2 = \lambda x_2$$

or,

$$(\lambda - 1)x_1 - x_2 = 0 \qquad (1)$$
$$+2x_1 + (\lambda - 4)x_2 = 0 \quad .$$

Recall that a homogeneous system has a non-zero solution if and only if the determinant of the matrix of coefficients is zero. Thus,

$$\begin{vmatrix} \lambda - 1 & -1 \\ 2 & \lambda - 4 \end{vmatrix} = 0$$

or,

$$(\lambda - 1)(\lambda - 4) + 2 = 0 .$$

Therefore,

$$\lambda^2 - 5\lambda + 6 = 0$$

or,

$$(\lambda - 3)(\lambda - 2) = 0 .$$

Hence, $\lambda_1 = 2$ and $\lambda_2 = 3$ are the eigenvalues of A. To find an eigenvector of A associated with $\lambda_1 = 2$, form the linear system:

$$AX = 2X$$

or,

$$\begin{bmatrix} 1 & 1 \\ -2 & 4 \end{bmatrix} \begin{bmatrix} x_1 \\ x_2 \end{bmatrix} = 2 \begin{bmatrix} x_1 \\ x_2 \end{bmatrix}$$

This gives

$$x_1 + x_2 = 2x_1 \qquad \qquad (2-1)x_1 - x_2 = 0$$
$$\qquad \qquad \text{or}$$
$$-2x_1 + 4x_2 = 2x_2 \qquad \qquad 2x_1 + (2-4)x_2 = 0$$

or,

$$x_1 - x_2 = 0$$
$$2x_1 - 2x_2 = 0 \qquad \text{or, simply, } x_1 - x_2 = 0 .$$

Observe that we could have obtained this last linear system by substituting $\lambda = 2$ in (1). It can be seen that any vector in R^2 of the form $x = k \begin{bmatrix} 1 \\ 1 \end{bmatrix}$, k a scalar, is an eigenvector of A associated with $\lambda_1 = 2$. Thus, $x_1 = \begin{bmatrix} 1 \\ 1 \end{bmatrix}$ is an eigenvector of A associated with

$\lambda_1 = 2$. Similarly, for $\lambda_2 = 3$, we obtain from (1):

$$(3-1)x_1 - x_2 = 0 \qquad\qquad 2x_1 = x_2 = 0$$
$$\text{or,}$$
$$2x_1 + (3-4)x_2 = 0 \qquad\qquad 2x_1 - x_2 = 0 .$$

Thus, $x_2 = \begin{bmatrix} 1 \\ 2 \end{bmatrix}$ is an eigenvector of A associated with $\lambda_2 = 3$.

• PROBLEM 10-72

Find the eigenvalues and the corresponding eigenvectors of A where

$$A = \begin{bmatrix} 0 & \frac{1}{2} \\ \frac{1}{2} & 0 \end{bmatrix}$$

Solution: An eigenvalue of A is a scalar λ such that $AX = \lambda X$ for some non-zero vectors X. This may be converted to $(\lambda I - A)X = 0$ which implies $\det(\lambda I - A) = 0$, the characteristic equation. The roots of this equation yield the required eigenvalues.

$$\lambda I - A = \begin{bmatrix} \lambda & 0 \\ 0 & \lambda \end{bmatrix} - \begin{bmatrix} 0 & \frac{1}{2} \\ \frac{1}{2} & 0 \end{bmatrix} = \begin{bmatrix} \lambda & -\frac{1}{2} \\ -\frac{1}{2} & \lambda \end{bmatrix}$$

$\det(\lambda I - A) = \lambda^2 - \frac{1}{4}$.

Then, the characteristic equation is $\lambda^2 - \frac{1}{4} = 0$ and the eigenvalues are $\lambda_1 = \frac{1}{2}$ and $\lambda_2 = -\frac{1}{2}$. Substitute $\lambda = \frac{1}{2}$ in the equation $(\lambda I - A)x = 0$ to obtain the corresponding eigenvectors. $(\frac{1}{2}I - A)x = 0$

$$\begin{bmatrix} \frac{1}{2} & -\frac{1}{2} \\ -\frac{1}{2} & \frac{1}{2} \end{bmatrix} \begin{bmatrix} x_1 \\ x_2 \end{bmatrix} = \begin{bmatrix} 0 \\ 0 \end{bmatrix}$$

or,

$$\frac{1}{2}x_1 - \frac{1}{2}x_2 = 0 \qquad \text{or, } x_1 - x_2 = 0 .$$
$$-\frac{1}{2}x_1 + \frac{1}{2}x_2 = 0$$

Thus,

$$x_1 = \begin{bmatrix} 1 \\ 1 \end{bmatrix}$$ is an eigenvector of A associated with the eigen-

value $\lambda_1 = \frac{1}{2}$. Now, let $\lambda = -\frac{1}{2}$. Then,

$$\begin{bmatrix} -\frac{1}{2} & -\frac{1}{2} \\ -\frac{1}{2} & -\frac{1}{2} \end{bmatrix} \begin{bmatrix} x_1 \\ x_2 \end{bmatrix} = \begin{bmatrix} 0 \\ 0 \end{bmatrix}$$

therefore,

$$-\frac{1}{2}x_1 - \frac{1}{2}x_2 = 0 ;$$
$$\text{or, } x_1 + x_2 = 0$$
$$-\frac{1}{2}x_1 - \frac{1}{2}x_2 = 0$$

Then,

$$x_2 = \begin{bmatrix} 1 \\ -1 \end{bmatrix}$$ is an eigenvector of A associated with the eigen-

value $\lambda_2 = -\frac{1}{2}$. If we let $L: R^2 \rightarrow R^2$ be defined by

$$L(X) = AX = \begin{bmatrix} 0 & \frac{1}{2} \\ \frac{1}{2} & 0 \end{bmatrix} \begin{bmatrix} x_1 \\ x_2 \end{bmatrix}$$

then Figure 1 shows that X_1 and $L(X_1)$ are parallel and that X_2 and $L(X_2)$ are parallel also. This illustrates the fact that if X is an eigenvector of A, then X and AX are parallel.

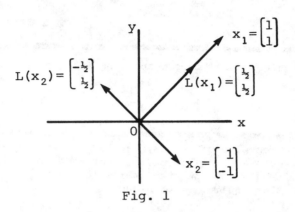

Fig. 1

● **PROBLEM** 10–73

Find the eigenvalues and the eigenvectors of the matrix A, where

$$A = \begin{bmatrix} 1 & 2 & -1 \\ 1 & 0 & 1 \\ 4 & -4 & 5 \end{bmatrix} .$$

Solution: First form the matrix $[\lambda I - A]$.

$$[\lambda I - A] = \begin{bmatrix} \lambda & 0 & 0 \\ 0 & \lambda & 0 \\ 0 & 0 & \lambda \end{bmatrix} - \begin{bmatrix} 1 & 2 & -1 \\ 1 & 0 & 1 \\ 4 & -4 & 5 \end{bmatrix}$$

$$= \begin{bmatrix} \lambda-1 & -2 & 1 \\ -1 & \lambda & -1 \\ -4 & 4 & \lambda-5 \end{bmatrix} .$$

Then,

$$\det(\lambda I - A) = \lambda-1 \begin{vmatrix} \lambda & -1 \\ 4 & \lambda-5 \end{vmatrix} -(-2) \begin{vmatrix} -1 & -1 \\ -4 & \lambda-5 \end{vmatrix} +1 \begin{vmatrix} -1 & \lambda \\ -4 & 4 \end{vmatrix}$$

$$\det(\lambda I - A) = (\lambda-1)[\lambda(\lambda-5) + 4] + 2[-\lambda+5 - 4] + 1[-4 + 4\lambda]$$

$$= (\lambda-1)[\lambda^2-5\lambda + 4] - 2\lambda + 2 - 4 + 4\lambda$$

$$= (\lambda-1)[\lambda^2-5\lambda + 4] + 2\lambda - 2$$

$$= (\lambda-1)[\lambda^2-5\lambda + 4 + 2]$$

$$= (\lambda-1)(\lambda-3)(\lambda-2) .$$

The characteristic equation of A is $(\lambda-1)(\lambda-3)(\lambda-2) = 0$. The eigenvalues of A are $\lambda = 1$, $\lambda = 2$, $\lambda = 3$. To find an eigenvector x_1 associated with $\lambda_1 = 1$, form the system:

$$(\lambda_1 I - A)X = 0, \text{ with } \lambda = 1:$$

$$\begin{bmatrix} 1-1 & -2 & 1 \\ -1 & 1 & -1 \\ -4 & 4 & 1-5 \end{bmatrix} \begin{bmatrix} x_1 \\ x_2 \\ x_3 \end{bmatrix} = \begin{bmatrix} 0 \\ 0 \\ 0 \end{bmatrix}$$

or,

$$\begin{bmatrix} 0 & -2 & 1 \\ -1 & 1 & -1 \\ -4 & 4 & -4 \end{bmatrix} \begin{bmatrix} x_1 \\ x_2 \\ x_3 \end{bmatrix} = \begin{bmatrix} 0 \\ 0 \\ 0 \end{bmatrix}$$

or,

$$-2x_2 + x_3 = 0 \qquad\qquad 2x_2 - x_3 = 0$$
$$-x_1 + x_2 - x_3 = 0 \qquad \text{or,} \qquad x_1 - x_2 + x_3 = 0 .$$
$$-4x_1 + 4x_2 - 4x_3 = 0$$

One solution to this system is
$$x_1 = -1 , \; x_2 = 1 , \; x_3 = 2 . \quad \text{Thus,} \quad X = \begin{bmatrix} -1 \\ 1 \\ 2 \end{bmatrix} \text{ is an eigen-}$$

vector of A associated with $\lambda = 1$.

To find an eigenvector X_2 associated with $\lambda = 2$, form the system:

$$(\lambda I - A)X = 0 \text{ , with } \lambda = 2.$$

$$\begin{bmatrix} 2-1 & -2 & 1 \\ -1 & 2 & -1 \\ -4 & 4 & 2-5 \end{bmatrix} \begin{bmatrix} x_1 \\ x_2 \\ x_3 \end{bmatrix} = \begin{bmatrix} 0 \\ 0 \\ 0 \end{bmatrix}$$

or,

$$x_1 - 2x_2 + x_3 = 0 \qquad\qquad x_1 - 2x_2 + x_3 = 0$$
$$-x_1 + 2x_2 - x_3 = 0 \qquad \text{or,} \qquad 4x_1 - 4x_2 + 3x_3 = 0 .$$
$$-4x_1 + 4x_2 - 3x_3 = 0$$

Solving this system gives $x_1 = -2$, $x_2 = 1$, and $x_3 = 4$ as one solution.
Thus,

$$x_2 = \begin{bmatrix} -2 \\ 1 \\ 4 \end{bmatrix} \text{ is an eigenvector of } A \text{ associated}$$

with $\lambda = 2$.

To find an eigenvector, X for $\lambda = 3$, solve the system $(\lambda I - A)X = 0$ for X with $\lambda = 3$.

$\lambda = 3$
$$[3I - A]X = 0$$

or,

$$\begin{bmatrix} 3-1 & -2 & 1 \\ -1 & 3 & -1 \\ -4 & 4 & 3-5 \end{bmatrix} \begin{bmatrix} x_1 \\ x_2 \\ x_3 \end{bmatrix} = \begin{bmatrix} 0 \\ 0 \\ 0 \end{bmatrix} .$$

This is equivalent to

$$2x_1 - 2x_2 + x_3 = 0 \qquad\qquad 2x_1 - 2x_2 + x_3 = 0$$

$$-x_1 + 3x_2 - x_3 = 0 \qquad \text{or,} \qquad x_1 - 3x_2 + x_3 = 0 \quad.$$

$$-4x_1 + 4x_2 - 2x_3 = 0$$

Solving this system yields $x_1 = -1$, $x_2 = 1$, $x_3 = 4$. Thus, $X_3 = \begin{bmatrix} -1 \\ 1 \\ 4 \end{bmatrix}$

is an eigenvector of A associated with $\lambda = 3$.

● **PROBLEM** 10–74

The characteristic values of the matrix

$$A = \begin{bmatrix} 8 & 2 & -2 \\ 3 & 3 & -1 \\ 24 & 8 & -6 \end{bmatrix}$$

are 2 and 1. Find the characteristic vectors.

Solution: Let W_1 denote the set of characteristic vectors that belong to 2. A vector $V = (x_1, x_2, x_3)$ is in W_1 if and only if $VA = 2V$; that is, $V[A - 2I] = 0$ or,

$$[x_1, x_2, x_3] \begin{bmatrix} 6 & 2 & -2 \\ 3 & 1 & -1 \\ 24 & 8 & -8 \end{bmatrix} = [0 \ 0 \ 0]$$

or,

$$6x_1 + 3x_2 + 24x_3 = 0$$

$$2x_1 + x_2 + 8x_3 = 0 \qquad\qquad (1)$$

$$-2x_1 - x_2 - 8x_3 = 0 \quad.$$

Thus, W_1 is the solution space of (1). The system (1) is clearly equivalent to $2x_1 + x_2 + 8x_3 = 0$. Since $x_2 + 8x_3 = -2x_1$ and $2x_1 + x_2 = -8x_3$, it follows that $v_1 = (1, -2, 0)$ and $v_2 = (0, -8, 1)$; $[v_1 v_2]$ is a basis of W_1 . The vectors v_1 and v_2 are characteristic vectors belonging to 2, and every characteristic vector belonging to 2 is a linear combination of v_1 and v_2 .

Let W_2 denote the set of characteristic vectors that belong to 1. A vector $V = (x_1, x_2, x_3)$ belongs to W_2 if and only if $VA = V$; that is,

$$v[A - 1I] = 0 \ ,$$

or,

$$[x_1 x_2 x_3] \begin{bmatrix} 7 & 2 & -2 \\ 3 & 2 & -1 \\ 24 & 8 & -7 \end{bmatrix} = [0 \ 0 \ 0]$$

or,

$$7x_1 + 3x_2 + 24x_3 = 0$$
$$2x_1 + 2x_2 + 8x_3 = 0$$
$$-2x_1 - x_2 - 7x_3 = 0$$

Solving this system gives $x_1 = 3$, $x_2 = 1$, $x_3 = -1$ as one uncomplicated solution. Thus, $v_3 = (3,1,-1)$ is a basis of w_2. The vector v_3 is a characteristic vector belonging to 1, and every characteristic vector belonging to 1 is a scalar multiple of v_3.

• PROBLEM 10-75

Find the eigenvalues of m'm:

(1) $\quad m = \begin{bmatrix} 1 & 0 \\ 0 & 0 \end{bmatrix}$

(2) $\quad m = \begin{bmatrix} 0 & 1 \\ -1 & 0 \end{bmatrix}$

(3) $\quad m = \begin{bmatrix} 1 & 1 \\ 0 & 1 \end{bmatrix}.$

Solution: The map $X \rightarrow mX = Y$ maps the unit sphere in x-space onto an ellipsoid in y space, and the squares of the semi-axes of the image are equal to the eigenvalues of m'm. If we have the map $X \rightarrow mX$ and put $mX = Y$, then the x's which are mapped into $Y'Y = 1$ satisfy $X'm'mX = 1$. They lie on a quadric surface in the original space.

When the eigenvalues are all positive, the locus is an ellipsoid.

(1) $\quad m = \begin{bmatrix} 1 & 0 \\ 0 & 0 \end{bmatrix}$

The locus is $X'm'mX = 1$ (in this case, the pair of lines $x_1 = \pm 1$). These map onto the two points $(\pm 1, 0)$. m'm has eigenvalues 1 and 0.

(2) $\quad m = \begin{bmatrix} 0 & 1 \\ -1 & 0 \end{bmatrix}$

This is a rotation. The unit circle is rotated through a right angle. m'm = I. The unit circle is invariant.

(3) $m = \begin{bmatrix} 1 & 1 \\ 0 & 1 \end{bmatrix}$

This is a shear. The unit circle becomes the ellipse

$$x^2 - 2x_1x_2 + 2x_2^2 = 1.$$

The major axis is in the direction $(2,-1 + \sqrt{5})$ and is of length $1 + \sqrt{5}$. The minor axis is in the direction $(2,-1 - \sqrt{5})$ and is of length $-1 + \sqrt{5}$.

$mm' = \begin{bmatrix} 1 & 1 \\ 1 & 2 \end{bmatrix}$. The eigenvalues are $\frac{1}{2}(3 \pm \sqrt{5})$. The ellipse

$x_1^2 + 2x_1x_2 + 2x_2^2 = 1$ is mapped onto the unit circle.

● **PROBLEM** 10–76

Find the eigenvalues of A and a basis for each eigenspace, where

$$A = \begin{bmatrix} 2 & 2 \\ -1 & 5 \end{bmatrix} .$$

Solution: Each eigenvalue of A has associated with it a set of eigen-vectors, i.e., vectors X such that
$$AX = \lambda X \tag{1}$$
where λ is the eigenvalue. The set of eigenvectors forms a subspace of R^n called the eigenspace. From (1) we obtain the characteristic equation as follows:

$$(\lambda I - A) = \lambda \begin{bmatrix} 1 & 0 \\ 0 & 1 \end{bmatrix} - \begin{bmatrix} 2 & 2 \\ -1 & 5 \end{bmatrix} = \begin{bmatrix} \lambda-2 & -2 \\ 1 & \lambda-5 \end{bmatrix} .$$

$$\det(\lambda I - A) = \det \begin{vmatrix} \lambda-2 & -2 \\ 1 & \lambda-5 \end{vmatrix} = [(\lambda-2)(\lambda-5) + 2]$$
$$= \lambda^2 - 7\lambda + 12$$
$$= (\lambda-4)(\lambda-3) .$$

The characteristic equation of A is: $(\lambda-4)(\lambda-3) = 0$. Then, eigen-values of A are $\lambda = 4$ and $\lambda = 3$. By definition, the eigenspace of A corresponding to λ is $(\lambda I - A)X = 0$. If $\lambda = 4$ then

$$(4I - A)X = \begin{bmatrix} 4-2 & -2 \\ 1 & 4-5 \end{bmatrix} \begin{bmatrix} x_1 \\ x_2 \end{bmatrix} = \begin{bmatrix} 0 \\ 0 \end{bmatrix}$$

or,

$$\begin{bmatrix} 2 & -2 \\ 1 & -1 \end{bmatrix} \begin{bmatrix} x_1 \\ x_2 \end{bmatrix} = \begin{bmatrix} 0 \\ 0 \end{bmatrix}$$

or, $2x_1 - 2x_2 = 0$

$x_1 - x_2 = 0$

or, $x_1 - x_2 = 0$.

The system has only one independent solution, i.e., $x_1 = 1$, $x_2 = 1$.

Thus, the eigenvectors corresponding to $\lambda = 4$ are the non-zero vectors of the form

$$X = \alpha \begin{bmatrix} 1 \\ 1 \end{bmatrix} ,$$

where α is a scalar, so that $\begin{bmatrix} 1 \\ 1 \end{bmatrix}$ is a basis for the eigenspace corresponding to $\lambda = 4$. If $\lambda = 3$,

$$\begin{bmatrix} 3-2 & -2 \\ 1 & 3-5 \end{bmatrix} \begin{bmatrix} x_1 \\ x_2 \end{bmatrix} = \begin{bmatrix} 0 \\ 0 \end{bmatrix}$$

or,

$$\begin{bmatrix} 1 & -2 \\ 1 & -2 \end{bmatrix} \begin{bmatrix} x_1 \\ x_2 \end{bmatrix} = \begin{bmatrix} 0 \\ 0 \end{bmatrix} .$$

This gives

$x_1 - 2x_2 = 0$

$x_1 - 2x_2 = 0$

or, $x_1 - 2x_2 = 0$.

Thus, $X = \begin{bmatrix} 2 \\ 1 \end{bmatrix}$ is an eigenvector which generates and forms a basis of the eigenspace of 3.

● PROBLEM 10–77

Find a basis for the eigenspace of

$$A = \begin{bmatrix} 3 & -2 & 0 \\ -2 & 3 & 0 \\ 0 & 0 & 5 \end{bmatrix} .$$

Solution: If λ is an eigenvalue of A, then the solution space for the system of equations $(\lambda I - A)X = 0$ is called the eigenspace of A corresponding to λ, and the non-zero vectors in the eigenspace are called the eigenvectors of A corresponding to λ .

Form the matrix

$$\lambda I - A = \lambda \begin{bmatrix} 1 & 0 & 0 \\ 0 & 1 & 0 \\ 0 & 0 & 1 \end{bmatrix} - \begin{bmatrix} 3 & -2 & 0 \\ -2 & 3 & 0 \\ 0 & 0 & 5 \end{bmatrix}$$

$$= \begin{bmatrix} \lambda-3 & 2 & 0 \\ 2 & \lambda-3 & 0 \\ 0 & 0 & \lambda-5 \end{bmatrix} .$$

$$\det(\lambda I - A) = \det \begin{vmatrix} \lambda-3 & 2 & 0 \\ 2 & \lambda-3 & 0 \\ 0 & 0 & \lambda-5 \end{vmatrix}$$

$$= \lambda{-}5 \begin{vmatrix} \lambda{-}3 & 2 \\ 2 & \lambda{-}3 \end{vmatrix}$$

$$= \lambda{-}5[\,(\lambda{-}3)^2 - 4\,]$$

$$= \lambda{-}5[\,\lambda^2 - 6\lambda + 9 - 4\,]$$

$$= (\lambda{-}5)[\,\lambda^2 - 6\lambda + 5\,]$$

$$= (\lambda{-}5)(\lambda{-}5)(\lambda{-}1)$$

$$= (\lambda{-}5)^2(\lambda{-}1)\ .$$

The characteristic equation of A is $(\lambda{-}5)^2(\lambda{-}1) = 0$, so that the eigenvalues of A are $\lambda = 1$ and $\lambda = 5$.

By definition,

$$X = \begin{bmatrix} x_1 \\ x_2 \\ x_3 \end{bmatrix}$$

is an eigenvector of A corresponding to λ if and only if x is a non-trivial solution of $(\lambda I - A)X = 0$. Thus,

$$\begin{bmatrix} \lambda{-}3 & 2 & 0 \\ 2 & \lambda{-}3 & 0 \\ 0 & 0 & \lambda{-}5 \end{bmatrix} \begin{bmatrix} x_1 \\ x_2 \\ x_3 \end{bmatrix} = \begin{bmatrix} 0 \\ 0 \\ 0 \end{bmatrix}. \tag{1}$$

If $\lambda = 5$, then equation (1) becomes

$$\begin{bmatrix} 2 & 2 & 0 \\ 2 & 2 & 0 \\ 0 & 0 & 0 \end{bmatrix} \begin{bmatrix} x_1 \\ x_2 \\ x_3 \end{bmatrix} = \begin{bmatrix} 0 \\ 0 \\ 0 \end{bmatrix}.$$

Solving this system yields $x_1 = -s$, $x_2 = s$, $x_3 = t$, where s and t are any scalars. Thus, the eigenvectors of A corresponding to $\lambda = 5$ are the non-zero vectors of the form

$$X = \begin{bmatrix} -s \\ s \\ t \end{bmatrix} = \begin{bmatrix} -s \\ s \\ 0 \end{bmatrix} + \begin{bmatrix} 0 \\ 0 \\ t \end{bmatrix} = s \begin{bmatrix} -1 \\ 1 \\ 0 \end{bmatrix} + t \begin{bmatrix} 0 \\ 0 \\ 1 \end{bmatrix}$$

Since $\begin{bmatrix} -1 \\ 1 \\ 0 \end{bmatrix}$ and $\begin{bmatrix} 0 \\ 0 \\ 1 \end{bmatrix}$ are linearly independent, they form a basis for the eigenspace corresponding to $\lambda = 5$. If $\lambda = 1$, then equation (1) becomes

$$\begin{bmatrix} -2 & 2 & 0 \\ 2 & 2 & 0 \\ 0 & 0 & -4 \end{bmatrix} \begin{bmatrix} x_1 \\ x_2 \\ x_3 \end{bmatrix} = \begin{bmatrix} 0 \\ 0 \\ 0 \end{bmatrix}.$$

Solving this system yields $x_1 = t$, $x_2 = t$, $x_3 = 0$; t is any scalar. Thus, the eigenvectors corresponding to $\lambda = 1$ are non-zero vectors of the form:

$$X = \begin{bmatrix} t \\ t \\ 0 \end{bmatrix} = t \begin{bmatrix} 1 \\ 1 \\ 0 \end{bmatrix} \text{ so that } \begin{bmatrix} 1 \\ 1 \\ 0 \end{bmatrix}$$

is a basis for the eigenspace corresponding to $\lambda = 1$.

Find the eigenvalues and an orthonormal basis for the eigenspace of A where,

$$A = \begin{bmatrix} 1 & 2 & 0 \\ 2 & 1 & 0 \\ 0 & 0 & 3 \end{bmatrix} .$$

Solution: A is a symmetric matrix. An important result concerning symmetric matrices is that all eigenvalues of a symmetric matrix are real. Form the matrix

$$(\lambda I - A) = \begin{bmatrix} \lambda & 0 & 0 \\ 0 & \lambda & 0 \\ 0 & 0 & \lambda \end{bmatrix} - \begin{bmatrix} 1 & 2 & 0 \\ 2 & 1 & 0 \\ 0 & 0 & 3 \end{bmatrix}$$

$$= \begin{bmatrix} \lambda-1 & -2 & 0 \\ -2 & \lambda-1 & 0 \\ 0 & 0 & \lambda-3 \end{bmatrix} .$$

Now, expanding along the third column yields

$$\det(\lambda I - A) = (\lambda-3) \begin{vmatrix} \lambda-1 & -2 \\ -2 & \lambda-1 \end{vmatrix}$$

$$= (\lambda-3)(\lambda^2 - 2\lambda + 1 - 4)$$

$$= (\lambda-3)(\lambda-3)(\lambda+1) .$$

The characteristic equation of A is $(\lambda-3)^2(\lambda+1) = 0$ and, therefore, the eigenvalues of A are $\lambda = 3$ and $\lambda = -1$.

Now find eigenvectors corresponding to $\lambda_1 = 3$. To do this, solve $(\lambda I - A)X = 0$ for X with $\lambda = 3$.

$$(3I - A)X = 0$$

or,

$$\begin{bmatrix} 2 & -2 & 0 \\ -2 & 2 & 0 \\ 0 & 0 & 0 \end{bmatrix} \begin{bmatrix} x_1 \\ x_2 \\ x_3 \end{bmatrix} = \begin{bmatrix} 0 \\ 0 \\ 0 \end{bmatrix} .$$

This is equivalent to

$$\begin{matrix} 2x_1 - 2x_2 = 0 \\ -2x_1 + 2x_2 = 0 \end{matrix} \quad \text{or,} \quad x_1 - x_2 = 0 .$$

Solving this system gives $x_1 = s$, $x_2 = s$, $x_3 = t$. Therefore,

$$X = \begin{bmatrix} s \\ s \\ t \end{bmatrix} = \begin{bmatrix} s \\ s \\ 0 \end{bmatrix} + \begin{bmatrix} 0 \\ 0 \\ t \end{bmatrix} = s\begin{bmatrix} 1 \\ 1 \\ 0 \end{bmatrix} + \begin{bmatrix} 0 \\ 0 \\ 1 \end{bmatrix}$$

or,

$$x_1 = \begin{bmatrix} 1 \\ 1 \\ 0 \end{bmatrix} \quad x_2 = \begin{bmatrix} 0 \\ 0 \\ 1 \end{bmatrix} .$$

Note that x_1 and x_2 are orthogonal to each other since $x_1 \cdot x_2 = 0$. Next, normalize x_1 and x_2 to obtain the unit orthogonal solutions by replacing x_i with

$$\frac{x_i}{|x_i|} \quad .$$

Since $|x_1| = \sqrt{2}$ and $|x_2| = 1$,

$$u_1 = \begin{bmatrix} 1/\sqrt{2} \\ 1/\sqrt{2} \\ 0 \end{bmatrix}; \quad u_2 = \begin{bmatrix} 0 \\ 0 \\ 1 \end{bmatrix}$$

and they form a basis for the eigenspace corresponding to $\lambda = 3$. To find the eigenvectors corresponding to $\lambda = -1$, solve $(\lambda I - A)X = 0$ for X with $\lambda = -1$.

$(-1I - A)X = 0$ or,

$$\begin{bmatrix} -2 & -2 & 0 \\ -2 & -2 & 0 \\ 0 & 0 & -4 \end{bmatrix} \begin{bmatrix} x_1 \\ x_2 \\ x_3 \end{bmatrix} = 0 \quad .$$

Carrying out the indicated matrix multiplication,

$$\begin{array}{ll} -2x_1 - 2x_2 = 0 & \quad x_1 + x_2 = 0 \\ -2x_1 - 2x_2 = 0 \quad \text{or,} & \quad x_3 = 0 \quad . \\ \quad\quad\;\; - 4x_3 = 0 & \end{array}$$

Solving this system gives

$$x_3 = \begin{bmatrix} 1 \\ -1 \\ 0 \end{bmatrix} .$$

Now, normalize x_3 to obtain the unit orthogonal solution. Thus,

$$u_3 = \begin{bmatrix} 1/\sqrt{2} \\ -1\sqrt{2} \\ 0 \end{bmatrix}$$

forms a basis for the eigenspace corresponding to $\lambda = -1$. Since $u_1 \cdot u_3 = 0$ and $u_2 \cdot u_3 = 0$, $\{u_1, u_2, u_3\}$ is an orthonormal basis of R^3. In general, if A is a symmetric $n \times n$ matrix, then the eigenvectors of A contain an orthonormal basis of R^n.

● **PROBLEM** 10–79

Find an orthogonal matrix P such that $P^{-1}AP$ is a diagonal matrix B where,

$$A = \begin{bmatrix} 3 & 1 \\ 1 & 3 \end{bmatrix} .$$

Solution: Recall that the transpose A^t of A is the matrix obtained from A by interchanging the rows and columns of A. We say that A is symmetric if $A^t = A$.

For symmetric matrices there is the following theorem:

If A is symmetric, there is an invertible matrix P such that $B = P^{-1}AP$ is a diagonal matrix.

This theorem is usually called the Spectral Theorem; it tells us, in particular, that if a matrix is symmetric, its characteristic polynomial must have only real roots. If A is symmetric, one may actually find an orthogonal matrix P such that $P^{-1}AP$ is diagonal. Recall that an orthogonal matrix is a matrix whose columns are orthonormal.

Now, consider the given matrix

$$A = \begin{bmatrix} 3 & 1 \\ 1 & 3 \end{bmatrix} \, .$$

A is symmetric. Form the matrix

$$(\lambda I - A) = \begin{bmatrix} \lambda & 0 \\ 0 & \lambda \end{bmatrix} - \begin{bmatrix} 3 & 1 \\ 1 & 3 \end{bmatrix}$$

$$= \begin{bmatrix} \lambda-3 & -1 \\ -1 & \lambda-3 \end{bmatrix} \, .$$

Then,

$$\det(\lambda I - A) = \det \begin{vmatrix} \lambda-3 & -1 \\ -1 & \lambda-3 \end{vmatrix}$$

$$= (\lambda-3)^2 - 1$$

$$= \lambda^2 - 6\lambda + 9 - 1$$

$$= \lambda^2 - 6\lambda + 8$$

$$= (\lambda-4)(\lambda-2) \, .$$

The characteristic equation of A is $(\lambda-4)(\lambda-2) = 0$ so that the characteristic values are $\lambda = 4$ and $\lambda = 2$. If $\lambda = 4$, then

$$4I - A = \begin{bmatrix} 4 & 0 \\ 0 & 4 \end{bmatrix} - \begin{bmatrix} 3 & 1 \\ 1 & 3 \end{bmatrix} = \begin{bmatrix} 1 & -1 \\ -1 & 1 \end{bmatrix} \, .$$

Now find the characteristic vectors (or eigenvectors).

$$(4I - A)X = 0$$

$$\begin{bmatrix} 1 & -1 \\ -1 & 1 \end{bmatrix} \begin{bmatrix} x_1 \\ x_2 \end{bmatrix} = 0$$

or, $x_1 - x_2 = 0$

$-x_1 + x_2 = 0$ or, $x_1 - x_2 = 0$. Thus, $x_1 = \begin{bmatrix} t \\ t \end{bmatrix}$ where $t \in R$ are

the eigenvectors. Clearly, $\begin{bmatrix} 1 \\ 1 \end{bmatrix}$ is a basis for the space. Since the norm of $[1,1]$ is $([1,1]\cdot[1,1])^{\frac{1}{2}} = \sqrt{2}$, normalize $[1,1]$ to obtain

$$x_1 = \begin{bmatrix} 1/\sqrt{2} \\ 1/\sqrt{2} \end{bmatrix} \, .$$

This is an orthonormal basis for the null space of $4I - A$. Now let $\lambda = 2$. Then,

$$2I - A = \begin{bmatrix} 2 & 0 \\ 0 & 2 \end{bmatrix} - \begin{bmatrix} 3 & 1 \\ 1 & 3 \end{bmatrix} = \begin{bmatrix} -1 & -1 \\ -1 & -1 \end{bmatrix} \, .$$

Thus, $(4I - A)X = 0$,

$$\begin{bmatrix} -1 & -1 \\ -1 & -1 \end{bmatrix} \begin{bmatrix} x_1 \\ x_2 \end{bmatrix} = \begin{bmatrix} 0 \\ 0 \end{bmatrix}$$

or,

$$\begin{array}{c} -x_1 - x_2 = 0 \\ -x_1 - x_2 = 0 \end{array} \quad \text{or, } x_1 + x_2 = 0 .$$

Clearly, $\begin{bmatrix} -1 \\ 1 \end{bmatrix}$ is a basis for the space. If we normalize, $u_2 = \begin{bmatrix} -1/\sqrt{2} \\ 1/\sqrt{2} \end{bmatrix}$

is an orthonormal basis for the null space of $2I - A$.

Observe that $u_1 \cdot u_2 = 0$ so that u_1 and u_2 are an orthonormal basis for R^2. Thus, in general, if A is symmetric and λ_1 and λ_2 are distinct characteristic values of A, the corresponding characteristic vectors x_1, x_2 must be orthogonal. Now construct an orthogonal matrix P whose columns are orthonormal. Thus,

$$P = \begin{bmatrix} 1/\sqrt{2} & -1/\sqrt{2} \\ 1/\sqrt{2} & 1/\sqrt{2} \end{bmatrix} .$$

Since P is an orthogonal matrix, $P^{-1} = P^t$. Therefore,

$$P^{-1} = \begin{bmatrix} 1/\sqrt{2} & 1/\sqrt{2} \\ -1/\sqrt{2} & 1/\sqrt{2} \end{bmatrix} .$$

Then, $B = P^{-1}AP$ is a diagonal matrix.

$$B = \begin{bmatrix} 1/\sqrt{2} & 1/\sqrt{2} \\ -1/\sqrt{2} & 1/\sqrt{2} \end{bmatrix} \begin{bmatrix} 3 & 1 \\ 1 & 3 \end{bmatrix} \begin{bmatrix} 1/\sqrt{2} & -1/\sqrt{2} \\ 1/\sqrt{2} & 1/\sqrt{2} \end{bmatrix}$$

$$= \begin{bmatrix} 1/\sqrt{2} & 1/\sqrt{2} \\ -1/\sqrt{2} & 1/\sqrt{2} \end{bmatrix} \begin{bmatrix} 4/\sqrt{2} & -2/\sqrt{2} \\ 4/\sqrt{2} & 2/\sqrt{2} \end{bmatrix}$$

$$= \begin{bmatrix} 2+2 & -1+1 \\ -2+2 & 1+1 \end{bmatrix}$$

$$= \begin{bmatrix} 4 & 0 \\ 0 & 2 \end{bmatrix} .$$

• PROBLEM 10–80

Find an orthogonal matrix P such that $P^{-1}AP$ is a diagonal matrix B where,

$$A = \begin{bmatrix} 1 & 1 & 0 \\ 1 & 1 & 0 \\ 0 & 0 & 2 \end{bmatrix} .$$

Solution: A is a symmetric matrix.

$$\lambda I - A = \begin{bmatrix} \lambda & 0 & 0 \\ 0 & \lambda & 0 \\ 0 & 0 & \lambda \end{bmatrix} - \begin{bmatrix} 1 & 1 & 0 \\ 1 & 1 & 0 \\ 0 & 0 & 2 \end{bmatrix}$$

$$= \begin{bmatrix} \lambda-1 & -1 & 0 \\ -1 & \lambda-1 & 0 \\ 0 & 0 & \lambda-2 \end{bmatrix} .$$

To find $\det(\lambda I - A)$, expand along the third column.

$$\det(\lambda I - A) = (\lambda-2) \begin{vmatrix} \lambda-1 & -1 \\ -1 & \lambda-1 \end{vmatrix}$$

$$= (\lambda-2)[(\lambda-1)^2 - 1]$$

$$= (\lambda-2)[\lambda^2 - 2\lambda + 1 - 1]$$

$$= (\lambda-2)(\lambda^2 - 2\lambda)$$

$$= \lambda(\lambda-2)^2 .$$

Thus, the characteristic equation of A is $\lambda(\lambda-2)^2 = 0$ so that characteristic values are $\lambda = 0$ and $\lambda = 2$. If $\lambda = 0$, then

$$0I - A = \begin{bmatrix} -1 & -1 & 0 \\ -1 & -1 & 0 \\ 0 & 0 & -2 \end{bmatrix} ;$$

$(0I - A)X = 0$

$$\begin{bmatrix} -1 & -1 & 0 \\ -1 & -1 & 0 \\ 0 & 0 & -2 \end{bmatrix} \begin{bmatrix} x_1 \\ x_2 \\ x_3 \end{bmatrix} = \begin{bmatrix} 0 \\ 0 \\ 0 \end{bmatrix}$$

or,

$$\begin{array}{ll} -x_1 - x_2 = 0 & x_1 + x_2 = 0 \\ -x_1 - x_2 = 0 \quad \text{or,} & x_3 = 0 . \\ -2x_3 = 0 & \end{array}$$

Therefore, the null space consists of vectors of the form

$$\alpha \begin{bmatrix} -1 \\ 1 \\ 0 \end{bmatrix} , \ \alpha \text{ a scalar, so } x_1 = \begin{bmatrix} 1/\sqrt{2} \\ -1/\sqrt{2} \\ 0 \end{bmatrix}$$

is an orthonormal basis for the null space of $0I - A$. Now let $\lambda = 2$. Then,

$$2I - A = \begin{bmatrix} 2 & 0 & 0 \\ 0 & 2 & 0 \\ 0 & 0 & 2 \end{bmatrix} - \begin{bmatrix} 1 & 1 & 0 \\ 1 & 1 & 0 \\ 0 & 0 & 2 \end{bmatrix}$$

$$= \begin{bmatrix} 1 & -1 & 0 \\ -1 & 1 & 0 \\ 0 & 0 & 0 \end{bmatrix} .$$

Then, $(2I - A)X = 0$;

$$\begin{bmatrix} 1 & -1 & 0 \\ -1 & 1 & 0 \\ 0 & 0 & 0 \end{bmatrix} \begin{bmatrix} x_1 \\ x_2 \\ x_3 \end{bmatrix} = \begin{bmatrix} 0 \\ 0 \\ 0 \end{bmatrix}$$

Solving this system yields $x_1 = s$, $x_2 = s$, $x_3 = t$. Thus, the null space of $2I - A$ consists of vectors of the form:

$$\begin{bmatrix} s \\ s \\ t \end{bmatrix} = s\begin{bmatrix} s \\ s \\ 0 \end{bmatrix} + \begin{bmatrix} 0 \\ 0 \\ t \end{bmatrix} = s\begin{bmatrix} 1 \\ 1 \\ 0 \end{bmatrix} + t\begin{bmatrix} 0 \\ 0 \\ 1 \end{bmatrix} .$$

Then,

$$\begin{bmatrix} 1 \\ 1 \\ 0 \end{bmatrix} \text{ and } \begin{bmatrix} 0 \\ 0 \\ 1 \end{bmatrix}$$

are an orthogonal basis for this null space. Normalizing shows that

$$x_2 = \begin{bmatrix} 1/\sqrt{2} \\ 1/\sqrt{2} \\ 0 \end{bmatrix} \quad \text{and} \quad x_3 = \begin{bmatrix} 0 \\ 0 \\ 1 \end{bmatrix}$$

are an orthonormal basis for the null space of $2I - A$. Using the inner product function, $x_1 \cdot x_2 = 0$, $x_1 \cdot x_3 = 0$. Therefore, $\{x_1, x_2, x_3\}$ is a orthonormal basis consisting of characteristic vectors for A. Now, we construct an orthogonal matrix P whose columns are orthonormal.

$$P = \begin{bmatrix} 1/\sqrt{2} & 1/\sqrt{2} & 0 \\ -1/\sqrt{2} & 1/\sqrt{2} & 0 \\ 0 & 0 & 1 \end{bmatrix}$$

Then, since this is an orthogonal matrix,

$$P^{-1} = P^t = \begin{bmatrix} 1/\sqrt{2} & -1/\sqrt{2} & 0 \\ 1/\sqrt{2} & 1/\sqrt{2} & 0 \\ 0 & 0 & 1 \end{bmatrix} .$$

Recall that $B = P^{-1}AP$ is diagonal, and its diagonal entries are the characteristic values of A corresponding to the columns of P. Thus,

$$B = \begin{bmatrix} 0 & 0 & 0 \\ 0 & 2 & 0 \\ 0 & 0 & 2 \end{bmatrix}$$

which can be checked by calculating $P^{-1}AP$ directly.

The matrix of the transformation, $(x,y)T = (y,x)$, is

$$A = \begin{bmatrix} 0 & 1 \\ 1 & 0 \end{bmatrix}.$$

Find the eigenvalues, the eigenvectors, and also the diagonal matrix of T.

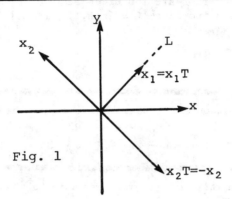

Fig. 1

Solution: To find eigenvalues, set $AX = \lambda X$. Then,

$$[A - \lambda I] = \begin{bmatrix} 0 & 1 \\ 1 & 0 \end{bmatrix} - \begin{bmatrix} \lambda & 0 \\ 0 & \lambda \end{bmatrix}$$

$$= \begin{bmatrix} -\lambda & 1 \\ 1 & -\lambda \end{bmatrix} .$$

Therefore, the characteristic equation $\det(A - \lambda I) = 0$ is $\lambda^2 - 1 = 0$, and the eigenvalues are $\lambda = 1$ and $\lambda = -1$. To find an eigenvector corresponding to $\lambda = 1$, solve $X(A - 1I) = 0$ for X. Thus,

$$\begin{bmatrix} x_1 \\ x_2 \end{bmatrix} - \begin{bmatrix} -1 & 1 \\ 1 & -1 \end{bmatrix} = \begin{bmatrix} 0 \\ 0 \end{bmatrix}$$

or,

$$-x_1 + x_2 = 0 \quad \text{or,} \quad x_1 - x_2 = 0 .$$
$$x_1 - x_2 = 0$$

Thus,

$$X_1 = \begin{bmatrix} 1 \\ 1 \end{bmatrix}$$ is an eigenvector corresponding to $\lambda = 1$.

To find an eigenvector corresponding to $\lambda = -1$, solve $X[A - (-1)I] = 0$ for X. Hence,

$$\begin{bmatrix} x_1 \\ x_2 \end{bmatrix} \begin{bmatrix} 1 & 1 \\ 1 & 1 \end{bmatrix} = \begin{bmatrix} 0 \\ 0 \end{bmatrix}$$

or,

$$x_1 + x_2 = 0 \quad \text{or,} \quad x_1 + x_2 = 0 .$$
$$x_1 + x_2 = 0$$

So, in general, $x_1 = -x_2$. One specific solution is: $x_1 = -1$, $x_2 = +1$.

Therefore,

$$X_2 = \begin{bmatrix} -1 \\ 1 \end{bmatrix}$$ is an eigenvector corresponding to $\lambda = -1$.

Since we have two distinct eigenvalues and a two-dimensional vector space, the eigenvectors form a basis for R^2. Relative to this basis, T has the diagonal matrix,

$$B = \begin{bmatrix} 1 & 0 \\ 0 & -1 \end{bmatrix},$$

whose diagonal entries are the eigenvalues. T is now seen to be the reflection across the line $L = \{a(1,1): a \in R\}$ (Figure 1).

• PROBLEM 10–82

Find the eigenvalues and a basis for each of the eigenspaces of the linear operator $T: P_2 \rightarrow P_2$ defined by:

$T(a + bx + cx^2) = (3a - 2b) + (-2a + 3b)x + (5c)x^2$.

Solution: The matrix of T with respect to the standard basis $\overline{B = \{1,x,x^2\}}$ is

$$A = \begin{bmatrix} 3 & -2 & 0 \\ -2 & 3 & 0 \\ 0 & 0 & 5 \end{bmatrix}.$$

The eigenvalues of T are the eigenvalues of A. We form the matrix

$$\lambda I - A = \begin{bmatrix} \lambda & 0 & 0 \\ 0 & \lambda & 0 \\ 0 & 0 & \lambda \end{bmatrix} - \begin{bmatrix} 3 & -2 & 0 \\ -2 & 3 & 0 \\ 0 & 0 & 5 \end{bmatrix}$$

$$= \begin{bmatrix} \lambda-3 & 2 & 0 \\ 2 & \lambda-3 & 0 \\ 0 & 0 & \lambda-5 \end{bmatrix}.$$

Expanding along the third column yields

$$\det(\lambda I - A) = (\lambda-5) \begin{vmatrix} \lambda-3 & 2 \\ 2 & \lambda-3 \end{vmatrix}$$

$$= (\lambda-5) [(\lambda-3)^2 - 4]$$

$$= (\lambda-5)[\lambda^2 - 6\lambda + 5]$$

$$= (\lambda-5)(\lambda-5)(\lambda-1)$$

$$= (\lambda-5)^2(\lambda-1) .$$

The characteristic equation of A is $(\lambda-5)^2(\lambda-1) = 0$ so that the eigenvalues of A are $\lambda = 1$ and $\lambda = 5,5$. If $\lambda = 1$, then

$(1I - A)u = 0$

or,

$$\begin{bmatrix} -2 & 2 & 0 \\ 2 & 2 & 0 \\ 0 & 0 & -4 \end{bmatrix} \begin{bmatrix} u_1 \\ u_2 \\ u_3 \end{bmatrix} = \begin{bmatrix} 0 \\ 0 \\ 0 \end{bmatrix}.$$

Solving this system yields $u_1 = t$, $u_2 = t$, $u_3 = 0$. Thus, the eigen-vectors corresponding to $\lambda = 1$ are the non-zero vectors of the form

$$u = \begin{bmatrix} t \\ t \\ 0 \end{bmatrix} = t \begin{bmatrix} 1 \\ 1 \\ 0 \end{bmatrix}, \text{ so that } \begin{bmatrix} 1 \\ 1 \\ 0 \end{bmatrix} \text{ is a basis for the}$$

eigenspace corresponding to $\lambda = 1$. If $\lambda = 5$, then

$$(\lambda I - A)u = 0$$

$$\text{or,} \quad \begin{bmatrix} 2 & 2 & 0 \\ 2 & 2 & 0 \\ 0 & 0 & 0 \end{bmatrix} \begin{bmatrix} u_1 \\ u_2 \\ u_3 \end{bmatrix} = \begin{bmatrix} 0 \\ 0 \\ 0 \end{bmatrix}.$$

Solving this system gives $u_1 = -s$, $u_2 = s$, $u_3 = t$. Thus, the eigen-vectors of A corresponding to $\lambda = 5$ are the non-zero vectors of the form:

$$u = \begin{bmatrix} -s \\ s \\ t \end{bmatrix} = \begin{bmatrix} -s \\ s \\ 0 \end{bmatrix} + \begin{bmatrix} 0 \\ 0 \\ t \end{bmatrix} = s \begin{bmatrix} -1 \\ 1 \\ 0 \end{bmatrix} + t \begin{bmatrix} 0 \\ 0 \\ 1 \end{bmatrix}$$

Hence, the eigenspace of A corresponding to $\lambda = 5$ has the basis $\{u_1, u_2\}$ and that corresponding to $\lambda = 1$ has the basis $\{u_3\}$ where

$$u_1 = \begin{bmatrix} -1 \\ 1 \\ 0 \end{bmatrix} \quad u_2 = \begin{bmatrix} 0 \\ 0 \\ 1 \end{bmatrix} \quad u_3 = \begin{bmatrix} 1 \\ 1 \\ 0 \end{bmatrix}.$$

These matrices are the coordinate matrices with respect to B of

$$P_1 = -1 + x, \ P_2 = x^2, \ P_3 = 1 + x.$$

Thus, $\{-1 + x, \ x^2\}$ is a basis for the eigenspace of T corresponding to $\lambda = 5$, and $\{1 + x\}$ is a basis for the eigenspace corresponding to $\lambda = 1$.

● **PROBLEM** 10–83

T is given by the rule $(x \ y)T = (-4x + y, \ -5x + 2y)$. Find the eigen-values and eigenvectors for T.

Solution: The usual matrix A for T is

$$A = \begin{bmatrix} -4 & -5 \\ 1 & 2 \end{bmatrix}.$$

Then,

$$\lambda I - A = \begin{bmatrix} \lambda & 0 \\ 0 & \lambda \end{bmatrix} - \begin{bmatrix} -4 & -5 \\ 1 & 2 \end{bmatrix} = \begin{bmatrix} \lambda+4 & 5 \\ -1 & \lambda-2 \end{bmatrix}$$

$$\det(\lambda I - A) = (\lambda+4)(\lambda-2) + 5$$

$$= \lambda^2 + 2\lambda - 8 + 5 = \lambda^2 + 2\lambda - 3$$

$$= (\lambda+3)(\lambda-1).$$

Therefore, the characteristic equation of A is $(\lambda+3)(\lambda-1) = 0$. Also, the eigenvalues are $\lambda = -3$ and $\lambda = 1$.

Now, find an eigenvector corresponding to $\lambda_1 = -3$. To do this, solve $X(A - \lambda I) = 0$ for X with $\lambda = -3$.

$$A - \lambda I = \begin{bmatrix} -4-\lambda & -5 \\ 1 & 2-\lambda \end{bmatrix}$$

or,

$$X \begin{bmatrix} -4-\lambda & -5 \\ 1 & 2-\lambda \end{bmatrix} = 0 \; ;$$

that is,

$$X \begin{bmatrix} -1 & -5 \\ 1 & 5 \end{bmatrix} = 0$$

or,

$$(x_1 \; x_2) \begin{bmatrix} -1 & -5 \\ 1 & 5 \end{bmatrix} = (0 \; 0) \; .$$

This is the system of equations,

$$-1x_1 + 1x_2 = 0$$
$$-5x_1 + 5x_2 = 0 \; ,$$

whose matrix is

$$\begin{bmatrix} -1 & 1 \\ -5 & 5 \end{bmatrix}$$ which gives $X_1 = X_2$.

Therefore, $(1,1)$ is an eigenvector corresponding to $\lambda_1 = -3$. Similarly, to find an eigenvector corresponding to $\lambda_2 = 1$, solve $X(A - 1I) = 0$ for X. Thus,

$$X \begin{bmatrix} -5 & -5 \\ 1 & 1 \end{bmatrix} = 0 \; .$$

This gives a system of equations with matrix $\begin{bmatrix} -5 & 1 \\ -5 & 1 \end{bmatrix}$. That is,

$-5x_1 + x_2 = 0 \; ; \; -5x_1 + x_2 = 0$ which shows that every eigenvector of $\lambda_2 = 1$ is of the form $(k,5k) = k(1,5)$ with k a scalar. Thus, $(1,5)$ is an eigenvector corresponding to $\lambda_2 = 1$. Therefore, $x_1 = (1,1)$, $x_2 = (1,5)$. Since we have two distinct eigenvalues and a two-dimensional vector space, the eigenvectors form a basis for R^2; relative to this basis, T has the diagonal matrix

$$B = \begin{bmatrix} -3 & 0 \\ 0 & 1 \end{bmatrix}$$

whose diagonal entries are the eigenvalues. In general, a linear operator $T : V \to V$ can be represented by a diagonal matrix B if and only if V has a basis consisting of eigenvectors of T. In this case, the diagonal elements of B are the corresponding eigenvalues.

Reduce the matrix A of the linear transformation to a diagonal form
where

$$A = \begin{bmatrix} 1 & 3 & 1 & 2 \\ 0 & -1 & 1 & 3 \\ 0 & 0 & 2 & 5 \\ 0 & 0 & 0 & -2 \end{bmatrix}$$

Solution: If the characteristic polynomial of a linear transformation
of an n–dimensional linear space has n distinct real roots, then the
matrix of the transformation reduces to diagonal form in an appropriate
coordinate system. The characteristic values of A are the diagonal entries since A is an upper triangular matrix. Thus,

$$\lambda_1 = 1 \ , \ \lambda_2 = -1 \ , \ \lambda_3 = 2 \ , \ \lambda_4 = -2 \ .$$

To find an eigenvector corresponding to $\lambda = 1$, solve the system
$X(A - 1I) = 0$ for X . Thus,

$$[x_1 \ x_2 \ x_3 \ x_4] \begin{bmatrix} 0 & 3 & 1 & 2 \\ 0 & -2 & 1 & 3 \\ 0 & 0 & 1 & 5 \\ 0 & 0 & 0 & -3 \end{bmatrix} = 0$$

or,

$$\begin{aligned} 3x_1 - 2x_2 \qquad\qquad &= 0 \\ x_1 + x_2 + x_3 \qquad &= 0 \\ 2x_1 + 3x_2 + 5x_3 - 3x_4 &= 0 \ . \end{aligned}$$

Solving these equations yields: $x_1 = 2$, $x_2 = 3$, $x_3 = -5$, $x_4 = -4$.

Thus,

$$X_1 = (2,3,-5,-4) \quad \text{is an eigenvector associated with} \quad \lambda = 1.$$

Similarly,

$$X_2 = (0,-3,1,-4)$$
$$X_3 = (0,0,4,5)$$
$$X_4 = (0,0,0,1)$$

are eigenvectors associated with $\lambda_2 = -1$; $\lambda_3 = 2$; $\lambda_4 = -2$. Since we have
four linearly independent eigenvalues and a four–dimensional vector space,
the eigenvectors form a basis for R^4 . The matrix of the transformation
A reduces to diagonal form,

$$B = \begin{bmatrix} 1 & 0 & 0 & 0 \\ 0 & -1 & 0 & 0 \\ 0 & 0 & 2 & 0 \\ 0 & 0 & 0 & -2 \end{bmatrix}$$

whose diagonal entries are characteristic values of A.

MATRIX DIAGONALIZATION

Let (1) $E = [e_1, e_2]$, the usual basis for R^2. (2) $T \overset{E}{\Leftrightarrow} A = \begin{bmatrix} 2 & 5 \\ 3 & 7 \end{bmatrix}$

is a linear transformation. (3) $F = [f_1, f_2]$ is another basis for R^2, where $f_1 = (1,1) = e_1 + e_2$; $f_2 = (-1,1) = -e_1 + e_2$.
Find the matrix of T relative to F.

Solution: Suppose T is a linear transformation relative to the usual basis. A is then the matrix of T with respect to a given basis. If B is the matrix of T with respect to another basis then $B = P^{-1}AP$. In other words, B is similar to A. The matrix P is the conversion matrix from the old to the new basis, or simply the conversion matrix. Thus, the matrix of T relative to the new basis = (conversion matrix)$^{-1}$ · (matrix of T relative to old basis) · (conversion matrix).

$F = PE$ and $T \overset{E}{\Leftrightarrow} A$ implies that the matrix of T under the basis f is given by $P^{-1}AP$. We must find P and P^{-1}.

$$F = [f_1, f_2] \quad f_1 = 1e_1 + 1e_2$$
$$f_2 = -1e_1 + 1e_2 .$$

Then,

$$P = \begin{bmatrix} 1 & -1 \\ 1 & 1 \end{bmatrix}$$

The adjoint method can be used to find P^{-1}. Thus,

$$P^{-1} = \frac{1}{\det P} \text{ adj } P = \begin{bmatrix} \frac{1}{2} & \frac{1}{2} \\ -\frac{1}{2} & \frac{1}{2} \end{bmatrix} = \frac{1}{2} \begin{bmatrix} 1 & 1 \\ -1 & 1 \end{bmatrix} .$$

Therefore,

$$T \overset{F}{\Leftrightarrow} P^{-1}AP = \frac{1}{2} \begin{bmatrix} 1 & 1 \\ -1 & 1 \end{bmatrix} \begin{bmatrix} 2 & 5 \\ 3 & 7 \end{bmatrix} \begin{bmatrix} 1 & -1 \\ 1 & 1 \end{bmatrix}$$

$$= \frac{1}{2} \begin{bmatrix} 1 & 1 \\ -1 & 1 \end{bmatrix} \begin{bmatrix} 7 & 3 \\ 10 & 4 \end{bmatrix}$$

$$= \frac{1}{2} \begin{bmatrix} 17 & 7 \\ 3 & 1 \end{bmatrix} = \begin{bmatrix} 17/2 & 7/2 \\ 3/2 & 1/2 \end{bmatrix} .$$

This is the matrix of T relative to the basis $\{f_1, f_2\}$.

Show that the matrices M' and M are similar, where $M = \begin{bmatrix} 1 & 1 \\ 0 & 1 \end{bmatrix}$ and

$M' = \begin{bmatrix} 1 & 0 \\ 1 & 1 \end{bmatrix}$.

Solution: Two nxn matrices M and M' are similar if $A^{-1}(M'A) = M$, where A is an invertible nxn matrix.

To see that M and M' are similar, let

$$A = \begin{bmatrix} 0 & 1 \\ 1 & 0 \end{bmatrix}.$$

Then

$$A^{-1} = \begin{bmatrix} 0 & 1 \\ 1 & 0 \end{bmatrix} \quad \text{since } A^{-1} = \frac{1}{\det A} \text{ Adj A , and}$$

$$A^{-1}(m'A) = \begin{bmatrix} 0 & 1 \\ 1 & 0 \end{bmatrix} \begin{bmatrix} 1 & 0 \\ 1 & 1 \end{bmatrix} \begin{bmatrix} 0 & 1 \\ 1 & 0 \end{bmatrix}$$

$$= \begin{bmatrix} 0 & 1 \\ 1 & 0 \end{bmatrix} \begin{bmatrix} 0 & 1 \\ 1 & 1 \end{bmatrix}$$

$$= \begin{bmatrix} 1 & 1 \\ 0 & 1 \end{bmatrix} = M .$$

Given that
$$A = \begin{bmatrix} 1 & 1 \\ -2 & 4 \end{bmatrix} ,$$
find an invertible matrix P such that $P^{-1}AP$ is a diagonal matrix D.

Solution: The characteristic equation of the matrix A is $\det(\lambda I - A) = 0$. Thus,

$$\det \begin{vmatrix} \lambda-1 & -1 \\ 2 & \lambda-4 \end{vmatrix} = 0 ,$$

or

$$(\lambda-1)(\lambda-4) + 2 = 0 ,$$

$$\lambda^2 - 5\lambda + 6 = 0$$

or

$$(\lambda-3)(\lambda-2) = 0 .$$

The eigenvalues are therefore, $\lambda_1 = 2$ and $\lambda_2 = 3$. To obtain the eigenvector corresponding to the eigenvalue $\lambda_1 = 2$, solve the equation $(2I - A)X = 0$ for x:

$$\begin{bmatrix} 1 & -1 \\ 2 & -2 \end{bmatrix} \begin{bmatrix} x_1 \\ x_2 \end{bmatrix} = \begin{bmatrix} 0 \\ 0 \end{bmatrix},$$

or

$$x_1 - x_2 = 0$$

which gives $x_1 - x_2 = 0$ or

$$2x_1 - 2x_2 = 0$$

$$x_1 = x_2 .$$

Thus, $x_1 = \begin{bmatrix} 1 \\ 1 \end{bmatrix}$ is an eigenvector of 2. Similarly, an eigenvector x_2 corresponding to the eigenvalue $\lambda = 3$ is $\begin{bmatrix} 1 \\ 2 \end{bmatrix}$. Since the eigenvectors x_1 and x_2 are linearly independent A is diagonalizable. It is possible to see that x_1 and x_2 are independent because $a_1 x_1 + a_2 x_2 = 0$ has only the trivial solution, $a_1 = a_2 = 0$. Let P be the matrix whose columns are x_1 and x_2:

$$P = \begin{bmatrix} 1 & 1 \\ 1 & 2 \end{bmatrix}$$

and

$$P^{-1} = \begin{bmatrix} 2 & -1 \\ -1 & 1 \end{bmatrix}$$

(using the adjoint method). Thus,

$$P^{-1}AP = \begin{bmatrix} 2 & -1 \\ -1 & 1 \end{bmatrix} \begin{bmatrix} 1 & 1 \\ -2 & 4 \end{bmatrix} \begin{bmatrix} 1 & 1 \\ 1 & 2 \end{bmatrix} = \begin{bmatrix} 2 & 0 \\ 0 & 3 \end{bmatrix}$$

On the other hand, if we let $\lambda_1 = 3$ and $\lambda_2 = 2$, then $x_1 = \begin{bmatrix} 1 \\ 2 \end{bmatrix}$ and $x_2 = \begin{bmatrix} 1 \\ 1 \end{bmatrix}$. In that case,

$$P = \begin{bmatrix} 1 & 1 \\ 2 & 1 \end{bmatrix} \quad \text{and} \quad P^{-1} = \begin{bmatrix} -1 & 1 \\ 2 & -1 \end{bmatrix},$$

and

$$P^{-1}AP = \begin{bmatrix} -1 & 1 \\ 2 & -1 \end{bmatrix} \begin{bmatrix} 1 & 1 \\ -2 & 4 \end{bmatrix} \begin{bmatrix} 1 & 1 \\ 2 & 1 \end{bmatrix} = \begin{bmatrix} 3 & 0 \\ 0 & 2 \end{bmatrix} .$$

• PROBLEM 10–88

Diagonalize the matrix $A = \begin{bmatrix} 1 & 2 \\ -1 & 4 \end{bmatrix}$ and find a diagonalizer for it.

Solution: If A is diagonalizable, there exists a matrix P such that PAP^{-1} is a diagonal matrix. P is called the diagonalizer of A. The diagonal elements of PAP^{-1} are the eigenvalues of A. The characteristic equation of A is $\det(\lambda I - A) = 0$. Then

$$\det \begin{vmatrix} \lambda - 1 & -2 \\ 1 & \lambda - 4 \end{vmatrix} = 0 ,$$

667

or
$$(\lambda-1)(\lambda-4) + 2 = 0 \text{ ,}$$
or
$$\lambda^2 - 5\lambda + 6 = (\lambda-3)(\lambda-2) = 0 \text{ .}$$

It follows that the eigenvalues of A are $\lambda_1 = 2$ and $\lambda_2 = 3$.

To find eigenvectors for A, form the vector $\begin{bmatrix} x_1 \\ x_2 \end{bmatrix}$ and set

$$(\lambda I - A)\begin{bmatrix} x_1 \\ x_2 \end{bmatrix} = 0 \text{ for } \lambda = 2,3 \text{ . For } \lambda = 2,$$

$$\begin{bmatrix} 1 & -2 \\ 1 & -2 \end{bmatrix}\begin{bmatrix} x_1 \\ x_2 \end{bmatrix} = \begin{bmatrix} 0 \\ 0 \end{bmatrix} \text{ ,}$$

or $x_1 - 2x_2 = 0$. Thus, $x_1 = \begin{bmatrix} 2 \\ 1 \end{bmatrix}$ is an eigenvector corresponding to $\lambda = 2$. Similarly, for $\lambda = 3$, we find $\begin{bmatrix} 1 \\ 1 \end{bmatrix}$ is an eigenvector.

Therefore,

$$P = \begin{bmatrix} 2 & 1 \\ 1 & 1 \end{bmatrix} \text{ diagonalizes } A = \begin{bmatrix} 1 & 2 \\ -1 & 4 \end{bmatrix} \text{ ,}$$

yielding the diagonal form

$$D = \begin{bmatrix} 2 & 0 \\ 0 & 3 \end{bmatrix} \text{ .}$$

• PROBLEM 10–89

Find an invertible matrix P such that $P^{-1}AP$ is a diagonal matrix B where

$$A = \begin{bmatrix} 1 & 0 & -2 \\ 0 & 0 & 0 \\ -2 & 0 & 4 \end{bmatrix} \text{ .}$$

Solution: The matrix P has as its columns linearly independent eigenvectors each one of which belongs to an eigenvalue of A. Thus, we form the matrix $[\lambda I - A]$:

$$[\lambda I - A] = \begin{bmatrix} \lambda & 0 & 0 \\ 0 & \lambda & 0 \\ 0 & 0 & \lambda \end{bmatrix} - \begin{bmatrix} 1 & 0 & -2 \\ 0 & 0 & 0 \\ -2 & 0 & 4 \end{bmatrix} = \begin{bmatrix} \lambda-1 & 0 & 2 \\ 0 & \lambda & 0 \\ 2 & 0 & \lambda-4 \end{bmatrix}$$

Since the characteristic equation of A is $\det(\lambda I - A) = 0$, the result is:

$$\det \begin{vmatrix} \lambda-1 & 0 & 2 \\ 0 & \lambda & 0 \\ 2 & 0 & \lambda-4 \end{vmatrix} = (\lambda-1)[\lambda(\lambda-4)] + 2[0 - 2\lambda] = 0 \text{ ,}$$

or
$$\lambda[\lambda^2 - 5\lambda + 4 - 4] = 0 \text{ ,}$$
$$\lambda^2 (\lambda-5) = 0 \text{ ,}$$

and the characteristic values of A are $\lambda = 0, \lambda = 0$ and $\lambda = 5$.

The characteristic vector associated with $\lambda = 0$ is obtained by solving the equation $(0I - A)x = 0$. Thus,

$$\begin{bmatrix} -1 & 0 & 2 \\ 0 & 0 & 0 \\ 2 & 0 & -4 \end{bmatrix} \begin{bmatrix} x_1 \\ x_2 \\ x_3 \end{bmatrix} = \begin{bmatrix} 0 \\ 0 \\ 0 \end{bmatrix} ,$$

or
$$-x_1 + 2x_3 = 0$$
$$2x_1 - 4x_3 = 0 ,$$

or
$$X = \begin{bmatrix} 2a \\ b \\ a \end{bmatrix}$$ with a and b arbitrary scalars(not both zero).

Two such vectors which are also linearly independent are $\begin{bmatrix} 2 \\ 0 \\ 1 \end{bmatrix}$ and $\begin{bmatrix} 0 \\ 1 \\ 0 \end{bmatrix}$.

To find the eigenvector corresponding to $\lambda = 5$, solve the equation $(5I - A)x = 0$:

$$\begin{bmatrix} 4 & 0 & 2 \\ 0 & 5 & 0 \\ 2 & 0 & 1 \end{bmatrix} \begin{bmatrix} x_1 \\ x_2 \\ x_3 \end{bmatrix} = \begin{bmatrix} 0 \\ 0 \\ 0 \end{bmatrix} ,$$

and, therefore,

(1) $4x_1 + 2x_3 = 0$
(2) $5x_2 = 0$
(3) $2x_1 + x_3 = 0$.

Equation (2) yields $x_2 = 0$. Equations (1) and (3) are dependent, so, taking equation (3) yields $x_3 = -2x_1$. One vector which satisfies these conditions is $x = \begin{bmatrix} 1 \\ 0 \\ -2 \end{bmatrix}$. We have obtained

$$\begin{bmatrix} 2 \\ 0 \\ 1 \end{bmatrix} , \begin{bmatrix} 0 \\ 1 \\ 0 \end{bmatrix} , \text{ and } \begin{bmatrix} 1 \\ 0 \\ -2 \end{bmatrix}$$

as a basis for R^3 that consists of characteristic vectors of A. If we let P be the matrix with these basis vectors as columns, P is invertible and $P^{-1}AP$ is a diagonal matrix whose diagonal entries are the characteristic values of A. Thus,

$$P = \begin{bmatrix} 2 & 0 & 1 \\ 0 & 1 & 0 \\ 1 & 0 & -2 \end{bmatrix} , \quad P^{-1} = \frac{1}{\det P}(\text{adj } P) =$$

$$-\frac{1}{5} \cdot \begin{bmatrix} -2 & 3 & -1 \\ 0 & -5 & 0 \\ -1 & 0 & 2 \end{bmatrix} \text{ and}$$

$$P^{-1}AP = \begin{bmatrix} 0 & 0 & 0 \\ 0 & 0 & 0 \\ 0 & 0 & 5 \end{bmatrix} .$$

We know this to be true, but in order to check do the matrix arithmatic:

$$P^{-1}AP = -\frac{1}{5} \begin{bmatrix} -2 & 3 & -1 \\ 0 & -5 & 0 \\ -1 & 0 & 2 \end{bmatrix} \begin{bmatrix} 1 & 0 & -2 \\ 0 & 0 & 0 \\ -2 & 0 & 4 \end{bmatrix} \begin{bmatrix} 2 & 0 & 1 \\ 0 & 1 & 0 \\ 1 & 0 & -2 \end{bmatrix}$$

$$= -\frac{1}{5} \begin{bmatrix} 0 & 0 & 0 \\ 0 & 0 & 0 \\ -5 & 0 & 10 \end{bmatrix} \begin{bmatrix} 2 & 0 & 1 \\ 0 & 1 & 0 \\ 1 & 0 & -2 \end{bmatrix}$$

$$= -\frac{1}{5} \begin{bmatrix} 0 & 0 & 0 \\ 0 & 0 & 0 \\ 0 & 0 & -25 \end{bmatrix} = \begin{bmatrix} 0 & 0 & 0 \\ 0 & 0 & 0 \\ 0 & 0 & 5 \end{bmatrix} \quad .$$

• PROBLEM 10–90

Show that the matrix A is not diagonalizable where
$$A = \begin{bmatrix} -3 & 2 \\ -2 & 1 \end{bmatrix} \quad .$$

Solution: The characteristic equation of A is $\det(\lambda I - A) = 0$; therefore,

$$\begin{vmatrix} \lambda+3 & -2 \\ +2 & \lambda-1 \end{vmatrix} = 0 ,$$

or

$$(\lambda+3)(\lambda-1) + 4 = 0$$
$$(\lambda+1)^2 = 0 .$$

Thus, $\lambda = -1$ is the only eigenvalue of A; the eigenvectors corresponding to $\lambda = -1$ are the solutions of $(-I-A)x = 0$. Thus,

$$\begin{bmatrix} 2 & -2 \\ 2 & -2 \end{bmatrix} \begin{bmatrix} x_1 \\ x_2 \end{bmatrix} = \begin{bmatrix} 0 \\ 0 \end{bmatrix} ,$$

or

$$2x_1 - 2x_2 = 0$$
$$2x_1 - 2x_2 = 0 .$$

The solutions of this system are $x_1 = t$, $x_2 = t$. Hence, the eigenspace consists of all vectors of the form
$$\begin{bmatrix} t \\ t \end{bmatrix} = t \begin{bmatrix} 1 \\ 1 \end{bmatrix} \quad .$$

Since this space is 1-dimensional, A does not have two linearly independent eigenvectors and, therefore is not diagonalizable.

Given:
$$A = \begin{bmatrix} 1 & 1 \\ 0 & 1 \end{bmatrix} \quad \text{and} \quad P = \begin{bmatrix} 1 & 1 \\ 1 & -1 \end{bmatrix} .$$

(a) Find P^{-1} .

(b) Find $P^{-1}AP$.

(c) Verify that, if B is similar to A, then A is similar to B.

(d) Show that $B^k = P^{-1}A^kP$ if $B = P^{-1}AP$ where k is any positive integer.

<u>Solution:</u> (a) It is known that $P^{-1} = \dfrac{1}{\det P} \text{adj } P.$

$$\det P = \begin{vmatrix} 1 & 1 \\ 1 & -1 \end{vmatrix} = -1-1 = -2 .$$

$$\text{adj } P = \begin{bmatrix} -1 & -1 \\ -1 & 1 \end{bmatrix} .$$

Therefore,

$$P^{-1} = \begin{bmatrix} \frac{1}{2} & \frac{1}{2} \\ \frac{1}{2} & -\frac{1}{2} \end{bmatrix} .$$

(b)

$$P^{-1}AP = \begin{bmatrix} \frac{1}{2} & \frac{1}{2} \\ \frac{1}{2} & -\frac{1}{2} \end{bmatrix} \begin{bmatrix} 1 & 1 \\ 0 & 1 \end{bmatrix} \begin{bmatrix} 1 & 1 \\ 1 & -1 \end{bmatrix}$$

$$= \begin{bmatrix} \frac{1}{2} & \frac{1}{2} \\ \frac{1}{2} & -\frac{1}{2} \end{bmatrix} \begin{bmatrix} 2 & 0 \\ 1 & -1 \end{bmatrix}$$

$$= \begin{bmatrix} 3/2 & -\frac{1}{2} \\ \frac{1}{2} & \frac{1}{2} \end{bmatrix} .$$

We say that the matrix B is similar to the matrix A if there is an invertible matrix P such that

$$B = P^{-1}AP .$$

Therefore, let

$$B = \begin{bmatrix} 3/2 & -\frac{1}{2} \\ \frac{1}{2} & \frac{1}{2} \end{bmatrix}$$

and then B is similar to A.

(c) If P is invertible and $B = P^{-1}AP$, then

$$PBP^{-1} = P(P^{-1}AP)P^{-1}$$

$$= (PP^{-1})A(PP^{-1})$$
$$= A \quad \text{since} \quad PP^{-1} = I \ .$$

Let $Q = P^{-1}$ so that $Q^{-1} = P$. Then $Q^{-1}BQ = A$.

$$Q = P^{-1} = \begin{bmatrix} \frac{1}{2} & \frac{1}{2} \\ \frac{1}{2} & -\frac{1}{2} \end{bmatrix} \ .$$

Thus,

$$Q^{-1} = \frac{1}{\det Q} \ \text{adj} \ Q$$

$$Q^{-1} = \frac{1}{-\frac{1}{2}} \begin{bmatrix} -\frac{1}{2} & -\frac{1}{2} \\ -\frac{1}{2} & \frac{1}{2} \end{bmatrix} = \begin{bmatrix} 1 & 1 \\ 1 & -1 \end{bmatrix}$$

$$Q^{-1}BQ = \begin{bmatrix} 1 & 1 \\ 1 & -1 \end{bmatrix} \begin{bmatrix} 3/2 & -\frac{1}{2} \\ \frac{1}{2} & \frac{1}{2} \end{bmatrix} \begin{bmatrix} \frac{1}{2} & \frac{1}{2} \\ \frac{1}{2} & -\frac{1}{2} \end{bmatrix}$$

$$= \begin{bmatrix} 1 & 1 \\ 1 & -1 \end{bmatrix} \begin{bmatrix} \frac{1}{2} & 1 \\ \frac{1}{2} & 0 \end{bmatrix}$$

$$= \begin{bmatrix} 1 & 1 \\ 0 & 1 \end{bmatrix}$$

$$= A \ .$$

Thus, if B is similar to A, then A is similar to B.

(d) Let $K = 2$; check that $P^{-1}A^2 P = B^2$. Then

$$B^2 = \begin{bmatrix} 3/2 & -\frac{1}{2} \\ \frac{1}{2} & \frac{1}{2} \end{bmatrix} \begin{bmatrix} 3/2 & -\frac{1}{2} \\ \frac{1}{2} & \frac{1}{2} \end{bmatrix}$$

$$= \begin{bmatrix} 2 & -1 \\ 1 & 0 \end{bmatrix} \ .$$

$$A^2 = \begin{bmatrix} 1 & 1 \\ 0 & 1 \end{bmatrix} \begin{bmatrix} 1 & 1 \\ 0 & 1 \end{bmatrix} = \begin{bmatrix} 1 & 2 \\ 0 & 1 \end{bmatrix} \ .$$

Then,

$$P^{-1}A^2 P = \begin{bmatrix} \frac{1}{2} & \frac{1}{2} \\ \frac{1}{2} & -\frac{1}{2} \end{bmatrix} \begin{bmatrix} 1 & 2 \\ 0 & 1 \end{bmatrix} \begin{bmatrix} 1 & 1 \\ 1 & -1 \end{bmatrix}$$

$$= \begin{bmatrix} \frac{1}{2} & \frac{1}{2} \\ \frac{1}{2} & -\frac{1}{2} \end{bmatrix} \begin{bmatrix} 3 & -1 \\ 1 & -1 \end{bmatrix}$$

$$= \begin{bmatrix} 2 & -1 \\ 1 & 0 \end{bmatrix}$$

$$= B^2 \ .$$

Suppose $K = 3$:

$$A^3 = A^2 A = \begin{bmatrix} 1 & 2 \\ 0 & 1 \end{bmatrix} \begin{bmatrix} 1 & 1 \\ 0 & 1 \end{bmatrix} = \begin{bmatrix} 1 & 3 \\ 0 & 1 \end{bmatrix} ,$$

$$B^3 = B^2 B = \begin{bmatrix} 2 & -1 \\ 1 & 0 \end{bmatrix} \begin{bmatrix} 3/2 & -\frac{1}{2} \\ \frac{1}{2} & \frac{1}{2} \end{bmatrix} = \begin{bmatrix} 5/2 & -3/2 \\ 3/2 & -\frac{1}{2} \end{bmatrix} .$$

Then,

$$P^{-1} A^3 P = \begin{bmatrix} \frac{1}{2} & \frac{1}{2} \\ \frac{1}{2} & -\frac{1}{2} \end{bmatrix} \begin{bmatrix} 1 & 3 \\ 0 & 1 \end{bmatrix} \begin{bmatrix} 1 & 1 \\ 1 & -1 \end{bmatrix}$$

$$= \begin{bmatrix} \frac{1}{2} & \frac{1}{2} \\ \frac{1}{2} & -\frac{1}{2} \end{bmatrix} \begin{bmatrix} 4 & -2 \\ 1 & -1 \end{bmatrix}$$

$$= \begin{bmatrix} 5/2 & -3/2 \\ 3/2 & -\frac{1}{2} \end{bmatrix}$$

$$= B^3 .$$

In general for any positive integer k, $B^k = P^{-1} A^k P$ if $B = P^{-1} A P$. To prove this rigorously for any matrices A and B and an invertible matrix P, use an inductive argument. Given $P^{-1} A P = B$, show $P^{-1} A^k P = B^k$. Take $n = 1$; $P^{-1} A P = B$. When $n = 2$; $P^{-1} A P = B$ gives $B^2 = P^{-1} A P P^{-1} A P$ so $B^2 = P^{-1} A^2 P$ since $P P^{-1} = I$.

Assume $P^{-1} A^k P = B^k$ is true for $k = n$; show that it is true for $k = n+1$. $P^{-1} A^n P = B^n$ so, since $B^{n+1} = B^n \cdot B = P^{-1} A^n P B = P^{-1} A^n P P^{-1} A P = P^{-1} A^{n+1} P$, $B^{n+1} = P^{-1} A^{n+1} P$.

From this it follows that if B is similar to A, then B^k is similar to A^k. Observe that the powers of A are easy to find. Direct calculation gives:

$$A^3 = \begin{bmatrix} 1 & 3 \\ 0 & 1 \end{bmatrix} ; \quad A^4 = \begin{bmatrix} 1 & 4 \\ 0 & 1 \end{bmatrix} .$$

In general, we obtain the formula

$$A^k = \begin{bmatrix} 1 & k \\ 0 & 1 \end{bmatrix} .$$

Again, to be rigorous, one would need to use an inductive argument. To find B^k, use the formula $B = P^{-1} A^k B$. Thus,

$$B^k = P^{-1} \begin{bmatrix} 1 & k \\ 0 & 1 \end{bmatrix} \begin{bmatrix} 1 & 1 \\ 1 & -1 \end{bmatrix}$$

$$= P^{-1} \begin{bmatrix} 1+k & 1-k \\ 1 & -1 \end{bmatrix}$$

$$= \begin{bmatrix} \frac{1}{2} & \frac{1}{2} \\ \frac{1}{2} & -\frac{1}{2} \end{bmatrix} \begin{bmatrix} 1+k & 1-k \\ 1 & -1 \end{bmatrix}$$

$$= \begin{bmatrix} 1+k/2 & -k/2 \\ k/2 & 1-k/2 \end{bmatrix} .$$

Let $A = \begin{bmatrix} 2 & 1 & 1 \\ 1 & 2 & 1 \\ 1 & 1 & 2 \end{bmatrix}$. Find a (real) orthogonal matrix P such that $P^t AP$ is diagonal.

<u>Solution</u>: A is a symmetric matrix. Hence, it has real eigenvalues. First, find the characteristic polynomial of A:

$$f(\lambda) = \det(\lambda I - A) = \begin{vmatrix} \lambda-2 & -1 & -1 \\ -1 & \lambda-2 & -1 \\ -1 & -1 & \lambda-2 \end{vmatrix}$$

$$= (\lambda-2) \begin{vmatrix} \lambda-2 & -1 \\ -1 & \lambda-2 \end{vmatrix} - (-1) \begin{vmatrix} -1 & -1 \\ -1 & \lambda-2 \end{vmatrix} + (-1) \begin{vmatrix} -1 & \lambda-2 \\ -1 & -1 \end{vmatrix}$$

$$= (\lambda-2)[(\lambda-2)(\lambda-2) - 1] + [-(\lambda-2) - 1] - [1 + (\lambda-2)]$$

$$= (\lambda-2)(\lambda-3)(\lambda-1) - (\lambda-1) - (\lambda-1)$$

$$= (\lambda-1)[\lambda^2 - 5\lambda + 6 - 1 - 1]$$

$$= (\lambda-1)(\lambda-1)(\lambda-4).$$

Therefore, the characteristic equation of A is $(\lambda-1)^2(\lambda-2) = 0$, and the eigenvalues are $\lambda = 1$ and $\lambda = 4$. To obtain eigenvectors corresponding to $\lambda = 1$, solve the equation $(1I - A)x = 0$:

$$\begin{bmatrix} -1 & -1 & -1 \\ -1 & -1 & -1 \\ -1 & -1 & -1 \end{bmatrix} \begin{bmatrix} x_1 \\ x_2 \\ x_3 \end{bmatrix} = \begin{bmatrix} 0 \\ 0 \\ 0 \end{bmatrix},$$

or $x_1 + x_2 + x_3 = 0$. Thus,

$$X_1 = \begin{bmatrix} 1 \\ -1 \\ 0 \end{bmatrix} \text{ and } X_2 = \begin{bmatrix} 1 \\ 1 \\ -2 \end{bmatrix}$$

are the eigenvectors corresponding to the eigenvalue $\lambda = 1$. For $\lambda = 4$, $(4I - A)x = 0$, or

$$\begin{bmatrix} +2 & -1 & -1 \\ -1 & 2 & -1 \\ -1 & -1 & 2 \end{bmatrix} \begin{bmatrix} x_1 \\ x_2 \\ x_3 \end{bmatrix} = \begin{bmatrix} 0 \\ 0 \\ 0 \end{bmatrix},$$

or

$$2x_1 - x_2 - x_3 = 0$$
$$-x_1 + 2x_2 - x_3 = 0$$
$$-x_1 - x_2 + 2x_3 = 0 \; .$$

Thus,

$$X_3 = \begin{bmatrix} 1 \\ 1 \\ 1 \end{bmatrix} \text{ is an eigenvector associated with } \lambda = 4.$$

Next, normalize X_1, X_2, X_3 to obtain the unit orthogonal solutions,

$$u_1 = \frac{X_1}{|X_1|} \; , \; u_2 = \frac{X_2}{|X_2|} \; , \; u_3 = \frac{X_3}{|X_3|} \; .$$

Thus,

$$u_1 = \begin{bmatrix} 1/\sqrt{2} \\ -1/\sqrt{2} \\ 0 \end{bmatrix} \qquad u_2 = \begin{bmatrix} 1/\sqrt{6} \\ 1/\sqrt{6} \\ -2/\sqrt{6} \end{bmatrix} \qquad u_3 = \begin{bmatrix} 1/\sqrt{3} \\ 1/\sqrt{3} \\ 1/\sqrt{3} \end{bmatrix} \; .$$

If P is the matrix whose columns are the u_i respectively,

$$P = \begin{bmatrix} 1/\sqrt{2} & 1/\sqrt{6} & 1/\sqrt{3} \\ -1/\sqrt{2} & 1/\sqrt{6} & 1/\sqrt{3} \\ 0 & -2/\sqrt{6} & 1/\sqrt{3} \end{bmatrix}$$

and

$$P^t A P = \begin{bmatrix} 1 & 0 & 0 \\ 0 & 1 & 0 \\ 0 & 0 & 4 \end{bmatrix} \; .$$

CHAPTER 11

PROBABILITY

> **Basic Attacks and Strategies for Solving Problems in this Chapter. See pages 677 to 744 for step-by-step solutions to problems.**

Probability models have assumed important roles in the study and practical utilization of natural phenomena. The success of probability models in so many diverse contexts clearly demonstrates a need for a unified approach to a study of chance phenomena. Several such approaches were proposed. Among these, the axiomatic approach, based on "the stability properties of relative frequencies," proved to be the most popular.

The applicability of probability theory is a consequence of the observed fact that chance phenomena exhibit statistical irregularities. A basic concept for all discussions of chance phenomena is the notion of a random experiment. A random experiment is a procedure that involves certain actions under specified conditions and that has as its outcome one (and only one) of a collection of possible simple results. The particular simple result that occurs cannot be predicted with certainty before the experiment is performed.

Corresponding to simple results of a random experiment, we conceive of basic mathematical entities in our model as outcomes. An event is defined as a collection of outcomes. Corresponding to every composite result E of a random experiment, there is an event defined in the model for that experiment—namely, the event consisting of the outcomes identified with the simple results of E. For every event E, there exists a number $P(E)$, called the probability of the event E, that is the idealization of the long-run relative frequency of the composite result corresponding to E.

When every outcome is equally probable, then the probabilities of events can be determined by counting outcomes. Any finite model for which the simple events S_i have equal probabilities is called a uniform probability model.

There are many situations in which we obtain partial knowledge concerning the outcome of a random experiment before the complete result becomes known. The availability of this partial knowledge suggests that we might improve our predictions concerning the outcome of the experiment if we could re-evaluate the probabilities of events of interest taking into account such partial knowledge.

Step-by-Step Solutions to
Problems in this Chapter,
"Probability"

ELEMENTARY PROBABILITY

• PROBLEM 11–1

How many different numbers of 3 digits can be formed from
the numbers 1, 2, 3, 4, 5 (a) If repetitions are allowed?
(b) If repetitions are not allowed? How many of these
numbers are even in either case?

Solution: (a) If repetitions are allowed, there are
5 choices for the first digit, 5 choices for the second
and 5 choices for the third. By the Fundamental Principle
of counting, there are 5 × 5 × 5 = 125 possible three digit
numbers. If the number is even, the final digit must be
either 2 or 4, 2 choices. So there will be 5 choices for
the first digit, 5 choices for the second and 2 for the
third, 5 × 5 × 2 = 50 such numbers.

     (b) If repetitions are not allowed, there are 5
choices for the first digit. After this has been picked,
there will be 4 choices for the second digit, and 3
choices for the third. Hence, 5 × 4 × 3 = 60 such numbers
can be selected.

     If the number must be even, then there are 2 choices
for the final digits, 2 or 4. This leaves 4 choices for
the next digit and 3 choices for the first digit. Hence
there are 4 × 3 × 2 = 24 possible even numbers that can be
selected in this way.

Calculate the number of permutations of the letters
a,b,c,d taken four at a time.

Solution: The number of permutations of the four letters
taken four at a time equals the number of ways the four
letters can be arranged or ordered. Consider four places
to be filled by the four letters. The first place can be
filled in four ways choosing from the four letters. The
second place may be filled in three ways selecting one of
the three remaining letters. The third place may be filled
in two ways with one of the two still remaining. The fourth
place is filled one way with the last letter. By the
fundamental principle, the total number of ways of ordering
the letters equals the product of the number of ways of
filling each ordered place, or $4 \cdot 3 \cdot 2 \cdot 1 = 24 = P(4,4) = 4!$
(read 'four factorial').

In general, for n objects taken r at a time,

$$P(n,r) = n(n-1)(n-2)\ldots(n-r+1) = \frac{n!}{(n-r)!} \quad (r < n).$$

For the special case where $r = n$,

$$P(n,n) = n(n-1)(n-2)\ldots(3)(2)(1) = n!,$$

since $(n-r)! = 0!$ which $= 1$ by definition.

Determine the number of permutations of the letters in
the word BANANA.

Solution: In solving this problem we use the fact that
the number of permutations P of n things taken all at a
time $[P(n,n)]$, of which n_1 are alike, n_2 others are alike,
n_3 others are alike, etc. is

$$P = \frac{n!}{n_1!n_2!n_3!\ldots} \quad , \text{ with } n_1 + n_2 + n_3 + \ldots = n.$$

In the given problem there are six letters ($n = 6$),
of which two are alike, (there are two N's so that
$n_1 = 2$), three others are alike (there are three A's,
so that $n_2 = 3$), and one is left (there is one B, so
$n_3 = 1$). Notice that $n_1 + n_2 + n_3 = 2 + 3 + 1 = 6 = n$;
thus,

$$P = \frac{6!}{2!3!1!} = \frac{6 \cdot 5 \cdot \overset{2}{\cancel{4}} \cdot 3!}{\cancel{2} \cdot 1 \cdot \cancel{3}! \cdot 1} = 60.$$

Thus, there are 60 permutations of the letters
in the word BANANA.

How many different sums of money can be obtained by choosing two coins from a box containing a penny, a nickel, a dime, a quarter, and a half dollar?

Solution: The order makes no difference here, since a selection of a penny and a dime is the same as a selection of a dime and a penny, insofar as a sum of money is concerned. This is a case of combinations, then, rather than permutations. Then the number of combinations of n different objects taken r at a time is equal to:

$$\frac{n(n-1)\ldots(n-r+1)}{1\cdot 2\cdots r} \; .$$

In this example, n = 5, r = 2, therefore

$$C(5,2) = \frac{5\cdot 4}{1\cdot 2} = 10.$$

As in the problem of selecting four committee members from a group of seven people, a distinct two coins can be selected from five coins in

$$\frac{5\cdot 4}{1\cdot 2} = 10 \text{ ways (applying the fundamental principle).}$$

How many baseball teams of nine members can be chosen from among twelve boys, without regard to the position played by each member?

Solution: Since there is no regard to position, this is a combinations problem (if order or arrangement had been important it would have been a permutations problem). The general formula for the number of combinations of n things taken r at a time is

$$C(n,r) = \frac{n!}{r!(n-r)!} \; .$$

We have to find the number of combinations of 12 things taken 9 at a time. Hence we have

$$C(12,9) = \frac{12!}{9!(12-9)!} = \frac{12!}{9!3!} = \frac{12\cdot 11\cdot 10\cdot \cancel{9!}}{3\cdot 2\cdot 1\cdot \cancel{9!}} = 220$$

Therefore, there are 220 possible teams.

What is the probability of throwing a "six" with a single die?

Solution: The die may land in any of 6 ways:

1, 2, 3, 4, 5, 6

The probability of throwing a six,

$$P(6) = \frac{\text{number of ways to get a six}}{\text{number of ways the die may land}}$$

Thus $P(6) = \frac{1}{6}$.

● **PROBLEM** 11–7

A box contains 7 red, 5 white, and 4 black balls. What is the probability of your drawing at random one red ball? One black ball?

Solution: There are 7 + 5 + 4 = 16 balls in the box. The probability of drawing one red ball,

$$P(R) = \frac{\text{number of possible ways of drawing a red ball}}{\text{number of ways of drawing any ball}}$$

$$P(R) = \frac{7}{16}.$$

Similarly, the probability of drawing one black ball

$$P(B) = \frac{\text{number of possible ways of drawing a black ball}}{\text{number of ways of drawing any ball}}$$

Thus,
$$P(B) = \frac{4}{16} = \frac{1}{4}$$

● **PROBLEM** 11–8

If a card is drawn from a deck of playing cards, what is the probability that it will be a jack or a ten?

Solution: The probability that an event A or B occurs, but not both at the same time, is $P(A \cup B) = P(A) + P(B)$. Here the symbol "$\cup$" stands for "or."
    In this particular example, we only select one card at a time. Thus, we either choose a jack "or" a ten. $P(\text{a jack or a ten}) = P(\text{a jack}) + P(\text{a ten})$.

$$P(\text{a jack}) = \frac{\text{number of ways to select a jack}}{\text{number of ways to choose a card}} = \frac{4}{52} = \frac{1}{13}.$$

$$P(\text{a ten}) = \frac{\text{number of ways to choose a ten}}{\text{number of ways to choose a card}} = \frac{4}{52} = \frac{1}{13}.$$

$$P(\text{a jack or a ten}) = P(\text{a jack}) + P(\text{a ten}) = \frac{1}{13} + \frac{1}{13} = \frac{2}{13}.$$

A bag contains 4 white balls, 6 black balls, 3 red balls, and 8 green balls. If one ball is drawn from the bag, find the probability that it will be either white or green.

Solution: The probability that it will be either white or green is:
P(a white ball or a green ball) = P(a white ball) + P(a green ball).
This is true because if we are given two mutually exclusive events A or B, then P(A or B) = P(A) + P(B). Note that two events, A and B, are mutually exclusive events if their intersection is the null or empty set. In this case the intersection of choosing a white ball and of choosing a green ball is the empty set. There are no elements in common.

P (a white ball) = $\dfrac{\text{number of ways to choose a white ball}}{\text{number of ways to select a ball}}$

$= \dfrac{4}{21}$

P(a green ball) = $\dfrac{\text{number of ways to choose a green ball}}{\text{number of ways to select a ball}}$

$= \dfrac{8}{21}$

Thus,
P(a white ball or a green ball) = $\dfrac{4}{21} + \dfrac{8}{21} = \dfrac{12}{21} = \dfrac{4}{7}$.

A coin is tossed nine times. What is the total number of possible outcomes of the nine-toss experiment? How many elements are in the subset "6 heads and 3 tails"? What is the probability of getting exactly 6 heads and 3 tails in nine tosses of thus unbiased coin?

Solution: There are 2 possible outcomes for each toss, that is, $\underbrace{2 \cdot 2 \cdot \ldots \cdot 2}_{\text{nine terms}} = 2^9$ possible outcomes in 9 tosses, or 512 outcomes.

To count the number of elements in the subset "6 heads and 3 tails" is equivalent to counting the number of ways 6 objects can be selected from 9. These objects will then be labeled "heads" and the remaining 3 objects will be labeled tails. There are

$$\binom{9}{6} = \frac{9!}{6! \ 3!} = 84 \qquad \text{ways to do this and hence the}$$

probability of observing this configuration is

$$\frac{\text{the number of ways 6 heads and 3 tails can occur}}{\text{the total possible outcomes}}$$

$$= \frac{\binom{9}{6}}{2^9} = \frac{84}{512} = .164 .$$

A card is drawn at random from a deck of cards. Find the probability that at least one of the following three events will occur:

Event A : a heart is drawn.
Event B: a card which is not a face card is drawn.
Event C: the number of spots (if any) on the drawn card is divisible by 3.

Solution: Let $A \cup B \cup C$ = the event that at least one of the three events above will occur. We wish to find $P(A \cup B \cup C)$, the probability of the event $A \cup B \cup C$. Let us count the number of ways that at least A, B or C will occur. There are 13 hearts, 40 non-face cards, and 12 cards such that the number of spots is divisible by 3. (Cards numbered 3, 6, or 9 are all divisible by 3 and there are 4 suits each with 3 such cards, $3 \times 4 = 12$). If we add $40 + 13 + 12$ we will have counted too many times. There are 10 cards which are hearts and non-face cards. 3 cards divisible by 3 and hearts, 12 cards which are non-face cards and divisible by 3. We must subtract each of these from our total of $40 + 13 + 12$ giving $40 + 13 + 12 - 10 - 3 - 12$. But we have subtracted too much; we have subtracted the 3 cards which are hearts and non-face cards and divisible by 3. We must add these cards to our total making

$$P(A \cup B \cup C) = \frac{40 + 13 + 12 - 10 - 3 - 12 + 3}{52} = \frac{43}{52} \quad .$$

Our counting technique used was called the principle of inclusion/exclusion and is useful for problems of this sort. Also look again at our answer,

$$P(A \cup B \cup C) = \frac{13 + 40 + 12 - 10 - 3 - 12 + 3}{52}$$

$$= \frac{13}{52} + \frac{40}{52} + \frac{12}{52} - \frac{10}{52} - \frac{3}{52} - \frac{12}{52} + \frac{3}{52} \quad .$$

Note that $P(A) = \dfrac{\text{number of hearts}}{\text{number of cards}} = \dfrac{13}{52}$

$P(B) = \dfrac{\text{number of non-face cards}}{\text{number of cards}} = \dfrac{40}{52}$

$P(C) = \dfrac{\text{number of cards divisible by 3}}{\text{number of cards}} = \dfrac{12}{52}$

$P(AB) = \dfrac{\text{number of hearts and non-face cards}}{\text{number of cards}} = \dfrac{10}{52}$

$$P(AC) = \frac{\text{number of hearts and cards divisible by 3}}{\text{number of cards}}$$

$$= \frac{3}{52} \; .$$

• PROBLEM 11–12

A penny is to be tossed 3 times. What is the probability there will be 2 heads and 1 tail?

Solution: We start this problem by constructing a set of all possible outcomes:

We can have heads on all 3 tosses: (HHH)
head on first 2 tosses, tail on the third: (HHT)   (1)
head on first toss, tail on next two: (HTT)
•   (HTH)   (2)
•   (THH)   (3)
•   (THT)
                                             (TTH)
                                             (TTT)

Hence there are eight possible outcomes (2 possibilities on first toss x 2 on second x 2 on third = 2 x 2 x 2 = 8).

We assume that these outcomes are all equally likely and assign the probability 1/8 to each. Now we look for the set of outcomes that produce 2 heads and 1 tail. We see there are 3 such outcomes out of the 8 possibilities (numbered (1), (2), (3) in our listing). Hence the probability of 2 heads and 1 tail is 3/8.

• PROBLEM 11–13

A survey was made of 100 customers in a department store. Sixty of the 100 indicated they visited the store because of a newspaper advertisement. The remainder had not seen the ad. A total of 40 customers made purchases; of these customers, 30 had seen the ad. What is the probability that a person who did not see the ad made a purchase? What is the probability that a person who saw the ad made a purchase?

Solution: In these two questions we have to deal with conditional probability, the probability that an event occurred given that another event occurred. In symbols, $P(A|B)$ means "the probability of A given B". This is defined as the probability of A and B, divided by the probability of B. Symbolically,

$$P(A|B) = \frac{P(A \cap B)}{P(B)} \; .$$

683

In the problem, we are told that only 40 customers made purchases. Of these 40, only 30 had seen the ad. Thus, 10 of 100 customers made purchases without seeing the ad. The probability of selecting such a customer at random is

$$\frac{10}{100} = \frac{1}{10} .$$

Let A represent the event of "a purchase", B the event of "having seen the ad", and $\overline{B}$ the event of "not having seen the ad."

Symbolically, $P(A \cap \overline{B}) = \frac{1}{10}$. We are told that 40 of the customers did not see the ad. Thus $P(\overline{B}) = \frac{40}{100} = \frac{4}{10}$.

Dividing, we obtain $\frac{1/10}{4/10} = \frac{1}{4}$, and, by definition of conditional probability, $P(A|\overline{B}) = \frac{1}{4}$. Thus the probability that a customer purchased given they did not see the ad is $\frac{1}{4}$.

To solve the second problem, note that 30 purchasers saw the ad. The probability that a randomly selected customer saw the ad and made a purchase is $\frac{30}{100} = \frac{3}{10}$. Since 60 of the 100 customers saw the ad, the probability that a randomly-picked customer saw the ad is $\frac{60}{100} = \frac{6}{10}$. Dividing we obtain

$$P(A|B) = \frac{P(A \cap B)}{P(B)} = \frac{\frac{3}{10}}{\frac{6}{10}} = \frac{3}{6} = \frac{1}{2} .$$

• PROBLEM 11-14

A coin is tossed 3 times. Find the probability that all 3 are heads,
    (a) if it is known that the first is heads,
    (b) if it is known that the first 2 are heads,
    (c) if it is known that 2 of them are heads.

Solution: This problem is one of conditional probability. If we have two events, A and B, the probability of event A given that event B has occurred is

684

$$P(A/B) = \frac{P(AB)}{P(B)}.$$

(a) We are asked to find the probability that all three tosses are heads given that the first toss is heads. The first event is A and the second is B.

P(AB) = probability that all three tosses are heads given that the first toss is heads

$$= \frac{\text{the number of ways that all three tosses are heads given that the first toss is a head}}{\text{the number of possibilities resulting from 3 tosses}}$$

$$= \frac{\{H, \ HH\}}{\{\{H,H,H\}, \ \{H,H,T\}, \ \{H,T,H\}, \ \{H,T,T\}, \ \{T,T,T\}, \{T,T,H\}, \ \{T,H,T\}, \ \{T,H,H\}}$$

$$= \frac{1}{8}.$$

P(B) = P(first toss is a head)

$$= \frac{\text{the number of ways to obtain a head on the first toss}}{\text{the number of ways to obtain a head or a tail on the first of 3 tosses}}$$

$$= \frac{\{H,H,H\}, \ \{H,H,T\}, \ \{H,T,H\}, \ \{H,T,T\}}{8}$$

$$= \frac{4}{8} \quad = \frac{1}{2}.$$

$$P(A/B) = \frac{P(AB)}{P(B)} = \frac{\frac{1}{8}}{\frac{1}{2}} = \frac{1}{8} \quad \frac{2}{1} = \frac{1}{4}.$$

To see what happens, in detail, we note that if the first toss is heads, the logical possibilities are HHH, HHT, HTH, HTT. There is only one of these for which the second and third are heads. Hence,

$$P(A/B) = \frac{1}{4}.$$

(b) The problem here is to find the probability that all 3 tosses are heads given that the first two tosses are heads.

P(AB) = the probability that all three tosees are heads given that the first two are heads

$$= \frac{\text{the number of ways to obtain 3 heads given that the first two tosses are heads}}{\text{the number of possibilities resulting from 3 tosses}}$$

$$= \frac{1}{8}.$$

685

P(B) = the probability that the first two are heads

$$= \frac{\text{number of ways to obtain heads on the first two tosses}}{\text{number of possibilities resulting from three tosses}}$$

$$= \frac{\{H,H,H\}, \ \{H,H,T\}}{8} = \frac{2}{8} = \frac{1}{4}.$$

$$P(A/B) = \frac{P(AB)}{P(B)} = \frac{\frac{1}{8}}{\frac{1}{4}} = \frac{4}{8} = \frac{1}{2}.$$

(c) In this last part, we are asked to find the probability that all 3 are heads on the condition that any 2 of them are heads.
Define:

A = the event that all three are heads

B = the event that two of them are heads

P(AB) = the probability that all three tosses are heads knowing that two of them are heads

$$= \frac{1}{8}.$$

P(B) = the probability that two tosses are heads

$$= \frac{\text{number of ways to obtain at least two heads out of three tosses}}{\text{number of possibilities resulting from 3 tosses}}$$

$$= \frac{\{H,H,T\}, \ \{H,H,H\}, \ \{H,T,H\}, \ \{T,H,H\}}{8}$$

$$= \frac{4}{8} = \frac{1}{2}.$$

$$P(A/B) = \frac{P(AB)}{P(B)} = \frac{\frac{1}{8}}{\frac{1}{2}} = \frac{2}{8} = \frac{1}{4}.$$

• PROBLEM 11–15

A hand of five cards is to be dealt at random and without replacement from an ordinary deck of 52 playing cards. Find the conditional probability of an all spade hand given that there will be at least 4 spades in the hand.

Solution: Let C_1 be the event that there are at least 4 spades in the hand and C_2 that there are five. We want $P(C_2|C_1)$.

$C_1 \cap C_2$ is the intersection of the events that there are at least 4 and there are five spades. Since C_2 is contained in C_1, $C_1 \cap C_2 = C_2$. Therefore

$$P(C_2|C_1) = \frac{P(C_1 \cap C_2)}{P(C_1)} = \frac{P(C_2)}{P(C_1)} \; ;$$

$$P(C_2) = P \text{ (5 spades)} = \frac{\text{number of possible 5 spade hands}}{\text{number of total hands}} \; .$$

The denominator is $\binom{52}{5}$ since we can choose any 5 out of 52 cards. For the numerator we can have only spades, of which there are 13. We must choose 5, hence we have $\binom{13}{5}$ and $P(C_2) = \binom{13}{5} / \binom{52}{5}$.

$$P(C_1) = P(4 \text{ or } 5 \text{ spades}) = \frac{\text{\# of possible 4 or 5 spades}}{\text{\# of total hands}}$$

The denominator is still $\binom{52}{5}$. The numerator is $\binom{13}{5}$ + (number of 4 spade hands). To obtain a hand with 4 spades we can choose any 4 of the 13, $\binom{13}{4}$. We must also choose one of the 39 other cards, $\binom{39}{1}$. By the Fundamental Principle of Counting, the number of four spade hands is $\binom{13}{4}\binom{39}{1}$. Hence the numerator is $\binom{13}{5} + \binom{13}{4}\binom{39}{1}$ and

$$P(C_1) = \frac{\binom{13}{5} + \binom{13}{4}\binom{39}{1}}{\binom{52}{5}} \; . \text{ Thus}$$

$$P(C_2|C_1) = \frac{P(C_2)}{P(C_1)} = \frac{\dfrac{\binom{13}{5}}{\binom{52}{5}}}{\dfrac{\binom{13}{5} + \binom{13}{4}\binom{39}{1}}{\binom{52}{5}}}$$

$$= \frac{\binom{13}{5}}{\binom{13}{5} + \binom{13}{4}\binom{39}{1}} = .044 \; .$$

From an ordinary deck of playing cards, cards are drawn successively at random and without replacement. Compute the probability that the third spade appears on the sixth draw.

Solution: Recall the following form of the multiplication rule: $P(C_1 \cap C_2) = P(C_1) \, P(C_2|C_1)$.

Let C_1 be the event of 2 spades in the first five draws and let C_2 be the event of a spade on the sixth draw. Thus the probability that we wish to compute is $P(C_1 \cap C_2)$.

After 5 cards have been picked there are $52 - 5 = 47$ cards left. We also have $13 - 2 = 11$ spades left after 2 spades have been picked in the first 5 cards. Thus, by the classical model of probability,

$$P(C_2|C_1) = \frac{\text{favorable outcomes}}{\text{total possibilities}} = \frac{11}{47} \, .$$

To compute $P(C_1)$, use the classical model of probability. $P(C_1) = \dfrac{\text{ways of drawing 2 spades in 5}}{\text{All ways of drawing 5}}$.

The number of ways to choose 5 cards from 52 is $\binom{52}{5}$. Now count how many ways one can select two spades in five draws. We can take any 2 of 13 spades, $\binom{13}{2}$. The other 3 cards can be chosen from any of the 39 non-spades, there are $\binom{39}{3}$ ways to choose 3 from 39.

To determine the total number of ways of drawing 2 spades and 3 non-spades we invoke the basic principle of counting and obtain $\binom{13}{2}\binom{39}{3}$. Hence

$$P(C_1) = \frac{\binom{13}{2}\binom{39}{3}}{\binom{52}{5}} \, .$$

$$P(C_1 \cap C_2) = P(C_1)P(C_2|C_1) = \frac{\binom{13}{2}\binom{39}{3}}{\binom{52}{5}} \cdot \frac{11}{47} = 0.274$$

More generally, suppose X is the number of draws required to produce the 3rd spade. Let C_1 be the event of 2 spades in the first X - 1 draws and let C_2 be the event that a spade is drawn on the Xth draw. Again we want to compute the probability $P(C_1 \cap C_2)$. To find $P(C_2|C_1)$ note that after X-1 cards have been picked, 2 of which were spades, 11 of the remaining 52-(X-1) cards are spades. The classical model of probability gives

$$P(C_2|C_1) = \frac{11}{52 - (x - 1)} .$$

Again by the classical model,

$$P(C_1) = \frac{\text{ways of 2 spaced in X-1}}{\text{All ways of X-1 cards}}$$. The denominator is the number of ways to choose X-1 from 52 or $\binom{52}{X-1}$. Now determine the number of ways of choosing 2 spades in X-1 cards.

There are still only 13 spades in the deck, 2 of which we must choose. Hence we still have a $\binom{13}{2}$ term. The other (X-1) - 2 = X - 3 cards must be non-spades. Thus we must choose X - 3 out of 39 possibilities. This is $\binom{39}{X-3}$. The basic principle of counting says that to get the number of ways of choosing 2 spades and X - 3 non-spades we must multiply the two terms, $\binom{13}{2} \times \binom{39}{X-3}$.

Therefore, $P(C_1) = \dfrac{\binom{13}{2}\binom{39}{X-3}}{\binom{52}{X-1}}$ and the probability of

drawing the third spade on the Xth card is

$$P(C_1 \cap C_2) = P(C_1) \times P(C_2|C_1) = \frac{\binom{13}{2}\binom{39}{X-3}}{\binom{52}{X-1}} \times \frac{11}{52-(X-1)}$$

If 4 cards are drawn at random and without replacement from a deck of 52 playing cards, what is the chance of drawing the 4 aces as the first 4 cards?

Solution: We will do this problem in two ways. First we will use the classical model of probability which tells us

$$\text{Probability} = \frac{\text{Number of favorable outcomes}}{\text{All possible outcomes}} \text{ , assuming all}$$

outcomes are equally likely.

There are four aces we can draw first. Once that is gone any one of 3 can be taken second. We have 2 choices for third and only one for fourth. Using the Fundamental Principle of Counting we see that there are $4 \times 3 \times 2 \times 1$ possible favorable outcomes. Also we can choose any one of 52 cards first. There are 51 possibilities for second, etc. The Fundamental Principle of Counting tells us that there are $52 \times 51 \times 50 \times 49$ possible outcomes in the drawing of four cards. Thus,

$$\text{Probability} = \frac{4 \times 3 \times 2 \times 1}{52 \times 51 \times 50 \times 49} = \frac{1}{270,725} = .0000037.$$

Our second method of solution involves the multi-plication rule and shows some insights into its origin and its relation to conditional probability.

The formula for conditional probability $P(A|B) = \frac{P(A \cap B)}{P(B)}$ can be extended as follows:

$$P(A|B \cap C \cap D) = \frac{P(A \cap B \cap C \cap D)}{P(B \cap C \cap D)} \text{ ;} \qquad \text{thus}$$

$$P(A \cap B \cap C \cap D) = P(A|B \cap C \cap D) \ P(B \cap C \cap D) \text{ but}$$

$$P(B \cap C \cap D) = P(B|C \cap D) \ P(C \cap D) \qquad \text{therefore}$$

$$P(A \cap B \cap C \cap D) = P(A|B \cap C \cap D) \ P(B|C \cap D) \ P(C \cap D) \qquad \text{but}$$

$$P(C \cap D) = P(C|D) \ P(D) \qquad \text{hence}$$

$$P(A \cap B \cap C \cap D) = P(A|B \cap C \cap D) \ P(B|C \cap D) \ P(C|D) \ P(D) \ .$$

Let event D = drawing an ace on the first card

C = drawing an ace on second card

B = ace on third draw

A = ace on fourth card .

Our conditional probability extension becomes

$$P \text{ (4 aces)} = P \text{ (on 4th}|\text{first 3)} \times P\text{(3rd}|\text{first 2)} \times$$

$$P\text{(2nd}|\text{on first)} \times P\text{(on first)}.$$

Assuming all outcomes are equally likely;

$P \text{ (on 1st draw)} = \frac{4}{52}$. There are 4 ways of success in 52 possibilities. Once we pick an ace there are 51 remaining cards, 3 of which are aces. This leaves a probability of $\frac{3}{51}$ for picking a second ace once we have chosen the first. Once we have 2 aces there are 50 remaining cards, 2 of which are aces, thus $P\text{(on 3rd}|\text{first 2)} = \frac{2}{50}$. Similarly $P\text{(4th ace}|\text{first 3)} = \frac{1}{49}$. According to our formula above

$$P \text{ (4 aces)} = \frac{1}{49} \times \frac{2}{50} \times \frac{3}{51} \times \frac{4}{52} = .000037.$$

• **PROBLEM** 11–18

Twenty percent of the employees of a company are college graduates. Of these, 75% are in supervisory position. Of those who did not attend college, 20% are in supervisory positions. What is the probability that a randomly selected supervisor is a college graduate?

Solution: Let the events be as followed:

E : The person selected is a supervisor
E_1 : The person is a college graduate.

E_2 : The person is not a college graduate.

We are searching for $P(E_1|E)$.

By the definition of conditional probability

$$P(E_1|E) = \frac{P(E_1 \cap E)}{P(E)} \quad .$$

But also by conditional probability $P(E_1 \cap E) = P(E|E_1) \cdot P(E_1)$. Since, E is composed of mutually exclusive events, E_1 and E_2, $P(E) = P(E_1 \cap E) + P(E_2 \cap E)$. Furthermore, $P(E_2 \cap E) = P(E|E_2) P(E_2)$, by conditional probability. Inserting these expressions into $\dfrac{P(E_1 \cap E)}{P(E)}$, we obtain

$$P(E_1|E) = \frac{P(E_1)P(E|E_1)}{P(E_1)P(E|E_1) + P(E_2)P(E|E_2)} .$$

This formula is a special case of the well-known Bayes' Theorem. The general formula is

$$P(E_1|E) = \frac{P(E_1)\ P(E|E_1)}{\sum\limits_{1}^{n} P(E_n)P(E|E_n)} .$$

In our problem,

$P(E_1) = P(\text{College graduate}) = 20\% = .20$

$P(E_2) = P(\text{Not graduate}) = 1 - P(\text{Graduate}) = 1 - .2 = .80$

$P(E|E_1) = P(\text{Supervisor}|\text{Graduate}) = 75\% = .75.$

$P(E|E_2) = P(\text{Supervisor}|\text{Not a graduate}) = 20\% = .20.$

Substituting,

$$P(E_1|E) = \frac{(.20)(.75)}{(.20)(.75) + (.80)(.20)} = \frac{.15}{.15 + .16}$$

$$= \frac{15}{31} .$$

PROBABILITY DISTRIBUTIONS

If $f(x) = 1/4$, $x = 0,1,2,3$ is a probability mass function, find $F(t)$, the cumulative distribution function and sketch its graph.

<u>Solution</u>: $F(t) = \sum\limits_{x=0}^{t} f(x) = Pr(X \leq t)$. $F(t)$ changes for integer values of t. We have:

$$F(t) = 0 \qquad\qquad t < 0$$
$$F(t) = f(0) = 1/4, \quad 0 \leq t < 1$$
$$F(t) = f(0) + f(1) = 1/4 + 1/4 = 1/2 ,$$
$$1 \leq t < 2$$

$$F(t) = f(0) + f(1) + f(2) \qquad 2 \leq t < 3$$
$$= \tfrac{1}{4} + \tfrac{1}{4} + \tfrac{1}{4} = 3/4 .$$

$$F(t) = \sum\limits_{x=0}^{t} f(x) = 1 \qquad\qquad\qquad 3 \leq t .$$

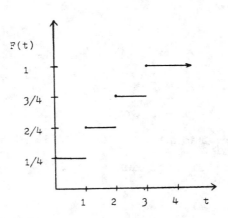

Given that the discrete random variable X has mass function,

$$Pr(X = x) = \begin{cases} x/6 & \text{where } x = 1,2,3 \\ 0 & \text{elsewhere,} \end{cases}$$

describe and graph its cumulative distribution function, $F(x)$.

<u>Solution</u>: The cumulative distribution function $F(x)$ is defined to be $Pr(X \leq x)$. $Pr(X = x) = 0$ except when $x = 1,2$ or 3, hence $P(X < 1) = 0$. Since the events "X = 1" and "1 < x < 2" are exclusive, we can add their probabilities: $P(X = 1) + P(1 < x < 2) = P(1 \leq x < 2)$.

P(X = 1) = 1/6 and P(1 < x < 2) = 0 because F(x) = 0 for
1 < x < 2. Thus P(1 ≤ x < 2) = 1/6 + 0 = 1/6 .

By the same reasoning, P(2 ≤ x < 3) = P(x = 2) + P(2 < x < 3) =
2/6 + 0 = 2/6. For 2 ≤ x < 3, F(x) = P(x < 1) + P(1 ≤ x < 2) +
P(2 ≤ x < 3) = 0 + 1/6 + 2/6 = 3/6 = 1/2.

Finally, Pr(x = 3) = 3/6 = 1/2. For 3 ≤ x, F(x) = 1/2 + 1/2 = 1.

Now we summarize and graph

$$F(x) = \begin{cases} 0 & x < 1 \\ 1/6 & 1 \le x < 2 \\ 3/6 & 2 \le x < 3 \\ 1 & 3 \le x \end{cases}$$

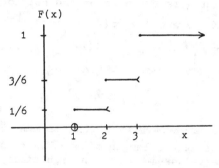

The graph shows that F(x) is a step function with steps of
heights 1/6, 2/6 and 3/6. It is constant and continuous in every
interval not containing 1,2 or 3.

● PROBLEM 11-21

X is a discrete random variable with probability distribution
$$f(x) = (\tfrac{1}{2})^x \quad , \ x = 1,2,\ldots$$
$$= 0 \qquad \text{otherwise.}$$
Find the Pr(X is even).

Solution: X has an infinite but countable number of values with
positive probability.

$$\text{Pr}(X \text{ is even}) = \text{Pr}(X = 2) + \text{Pr}(X = 4) + \ldots + \text{Pr}(X = 2n) + \ldots$$

$$= \sum_{n=1}^{\infty} \text{Pr}(X = 2n)$$

$$= \sum_{n=1}^{\infty} (\tfrac{1}{2})^{2n} = \sum_{n=1}^{\infty} (\tfrac{1}{4})^n$$

$$= \tfrac{1}{4} \sum_{n=1}^{\infty} (\tfrac{1}{4})^{n-1}$$

$$= (\tfrac{1}{2}) \sum_{n-1=0}^{\infty} (\tfrac{1}{4})^{n-1} .$$

Let $n-1 = m$ then

$$\Pr(X \text{ is even}) = (\tfrac{1}{2}) \sum_{m=0}^{\infty} (\tfrac{1}{4})^m$$

but this is now $\tfrac{1}{4}$ times the familiar geometric series. In general

$$\sum_{m=0}^{\infty} r^m = \frac{1}{1-r} \text{ if } |r| < 1 .$$

In our case $r = \tfrac{1}{4} < 1$, thus

$$\Pr(X \text{ is even}) = (\tfrac{1}{2}) \sum_{m=0}^{\infty} (\tfrac{1}{4})^m = \tfrac{1}{2}\left(\frac{1}{1-\tfrac{1}{4}}\right)$$

$$= \tfrac{1}{2}\left(\frac{1}{3/4}\right) = 1/3 .$$

Since all positive integers are even or odd but not both

$$\Pr(X \text{ is even}) + \Pr(X \text{ is odd}) = 1$$

and

$$\Pr(X \text{ is odd}) = 1 - \Pr(X \text{ is even})$$

$$= 1 - 1/3 = 2/3 .$$

● PROBLEM 11–22

Defects occur along the length of a cable at an average of 6 defects per 4000 feet. Assume that the probability of k defects in t feet of cable is given by the probability mass function:

$$\Pr(k \text{ defects}) = \frac{e^{-\frac{6t}{4000}} \left(\frac{6t}{4000}\right)^k}{k!}$$

for $k = 0,1,2,\ldots$. Find the probability that a 3000-foot cable will have at most two defects.

Solution: The probability of exactly k defects in 3000 feet is determined by the given discrete probability distribution as

Pr(k defects in 3000 ft.)

$$= \frac{e^{-\frac{6(3000)}{4000}} \left(\frac{6(3000)}{4000}\right)^k}{k!}$$

$$= \frac{e^{-4.5}(4.5)^k}{k!} , \quad k = 0,1,2,\ldots .$$

We use the probability distribution to find the probability of at most two defects.

Pr(at most two defects) = Pr(0,1 or 2 defects).

The events "0 defects", "1 defect" and "2 defects" are all mutually exclusive, thus,

Pr(at most two defects) = Pr(0 defects) + Pr(1 defect) + Pr(2 defects)

$$= \frac{e^{-4.5}(4.5)^0}{0!} + \frac{e^{-4.5}(4.5)^1}{1!} + \frac{e^{-4.5}(4.5)^2}{2!}$$

$$= e^{-4.5}(1 + 4.5 + \frac{(4.5)^2}{2!})$$

$$= .1736 .$$

● **PROBLEM** 11–23

Given that the random variable X has density function

$$f(x) = \begin{cases} 2x & 0 < x < 1 \\ 0 & \text{otherwise} \end{cases}$$

Find $\Pr(\frac{1}{2} < x < 3/4)$ and $\Pr(-\frac{1}{2} < x < \frac{1}{2})$.

Solution: Since f(x) = 2x is the density function of a continuous random variable, $\Pr(\frac{1}{2} < x < 3/4)$ = area under f(x) from $\frac{1}{2}$ to 3/4.

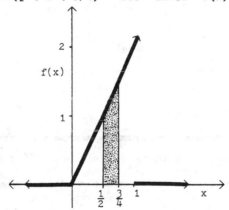

f(x) is indicated by the heavy line.

The area under f(x) is the area of the triangle with vertices at (0,0), (1,0) and (1,2).

The area of this triangle is A = $\frac{1}{2}$ bh where b = base of the triangle and h is the altitude.

Thus, A = $\frac{1}{2}$(1) x 2 = $\frac{2}{2}$ = 1

proving that f(x) is a proper probability density function.

To find the probability that $\frac{1}{2} < x < 3/4$ we find the area of the shaded region in the diagram. This shaded region is the difference in areas of the   right triangle with vertices (0,0), ($\frac{1}{2}$,0) and ($\frac{1}{2}$,f($\frac{1}{2}$)) and the area of the triangle with vertices (0,0), (3/4,0) and (3/4, f(3/4)).

This difference is $\Pr(\frac{1}{2} < x < 3/4) = \frac{1}{2}(3/4)f(3/4) - \frac{1}{2}(\frac{1}{2})f(\frac{1}{2})$

$$= \frac{1}{2}[\frac{3}{4} \cdot \frac{6}{4} - \frac{1}{2} \cdot 1]$$

$$= \frac{1}{2}[\frac{9}{8} - \frac{1}{2}] = \frac{1}{2} \cdot \frac{5}{8} = \frac{5}{16} \, .$$

The probability that $-\frac{1}{2} < x < \frac{1}{2}$ is

$\Pr(-\frac{1}{2} < x < \frac{1}{2}) =$ Area under $f(x)$ from $-\frac{1}{2}$ to $\frac{1}{2}$.

Because $f(x) = 0$ from $-\frac{1}{2}$ to 0, the area under $f(x)$ from $-\frac{1}{2}$ to 0 is 0. Thus

$\Pr(-\frac{1}{2} < x < \frac{1}{2}) = \Pr(0 < x < \frac{1}{2}) =$ area under $f(x)$ from 0 to $\frac{1}{2}$

$$= \frac{1}{2}(\frac{1}{2}) \, f(\frac{1}{2})$$

$$= \frac{1}{2}(\frac{1}{2}) \cdot 1 = \frac{1}{4} \, .$$

• PROBLEM 11-24

Let X be a continuous random variable where $f(x) = cx^2$; $0 < x < 1$. Find c.

Solution: We want the total probability to sum or integrate to one.

$$1 = \int_{-\infty}^{\infty} f(x) \, dx = \int_{0}^{1} cx^2 \, dx = \frac{cx^3}{3} \Big|_{x=0}^{1}$$

$$= c(1/3 - 0) = c/3 \, .$$

Hence $c = 3$.

697

Let the probability density function for X be given as $f(x) = 1/k$ for $0 < x < k$ and 0 otherwise. Find $\Pr(X \leq t) = F(t)$ and sketch $F(t)$.

Solution: $F(t) = \Pr(X \leq t)$ is defined as the cumulative distribution function, $F(t) = \int_{-\infty}^{t} f(x) \, dx$.

In this case $F(t) = \int_{-\infty}^{t} 1/k \, dx$ and

$$F(t) = \begin{cases} 0 & t \leq 0 \\ \int_{0}^{t} 1/k \, dx, & 0 < t < k \\ 1 & k \leq t \end{cases}$$

$$F(t) = \begin{cases} 0 & t \leq 0 \\ t/k & 0 < t < k \\ 1 & k \leq t \, . \end{cases}$$

This function is graphed below:

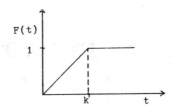

Let x be the random variable representing the length of a telephone conversation. Let $f(x) = \lambda e^{-\lambda x}$, $0 \leq x < \infty$. Find the c.d.f., $F(x)$, and find $\Pr(5 < x \leq 10)$.

Solution: The c.d.f. is the probability that an observation of a random variable will be less than or equal to x. The probability is equal to the following shaded area.

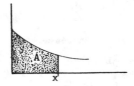

Area can be found by integration:

$$A = \int_0^x \lambda e^{-\lambda t} \, dt = - \left. e^{-\lambda t} \right|_{t=0}^x = 1 - e^{-\lambda x} = F(x).$$

$\Pr(5 < x \leq 10)$ is represented by the following shaded area.

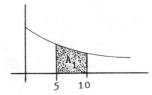

We find A_1 by integration,

$$A_1 = \int_{x=5}^{10} \lambda e^{-\lambda t} \, dt = -e^{-\lambda t} \Big|_{x=5}^{x=10} = e^{-5\lambda} - e^{10\lambda}.$$

THE BINOMIAL AND JOINT DISTRIBUTIONS

• **PROBLEM** 11-27

John Reeves completed 60% of his passes one season. Assuming he is as good a quarterback the next fall, what is the probability that he will complete 80 of his first 100 passes?

Solution: Since we are working with sequences of independent successes and failures, the binomial distribution is the correct one to use.

The probability of 80% success is given by substitution in the formula $\binom{n}{x} p^x (1 - p)^{n - x}$ where n is the number of trials. In this case $n = 100$, x is the number of desired successes ($X=80$ here) and p the probability of success in any trial ($p = \frac{6}{10}$ here).

$\binom{n}{x}$ is the number of combinations of n trials taken x at a time. This is defined to be $\frac{n!}{x!(n-x)!}$.

The distribution $p(x) = \binom{n}{x} p^x (1-p)^{n-x}$ is called binomial since for each integer value of x, the probability of x corresponds to a term in the binomial expansion

$$(q + p)^n = q^n + \binom{n}{1} q^{n-1} p + \binom{n}{2} q^{n-2} p + \ldots + p^n.$$

We interpret p to be the probability of success in any trial and $q = 1 - p$ the probability of failure because $p + q = p + 1 - p = (p - p) + 1 = 1$.

We have, therefore,

$$p(x) = \binom{n}{x} p^x (1-p)^{n-x} = \binom{100}{80} \left(\frac{60}{100}\right)^{80} \left(\frac{40}{100}\right)^{20}$$

$$= \frac{100!}{80!20!} \left(\frac{6}{10}\right)^{80} \left(\frac{4}{10}\right)^{20} = 1.053(10^{-5}).$$

Thus the probability that Reeves completes Exactly 80 of his first 100 passes in the upcoming football season is about 10^{-5}.

The binomial theorem provides a convenient way to compute binomial probabilities. The following three examples provide the statement of this theorem and examples of its use.

• PROBLEM 11-28

Expand $(x + 2y)^5$.

Solution: Apply the binomial theorem. If n is a positive integer, then

$$(a + b)^n = \binom{n}{0} a^n b^0 + \binom{n}{1} a^{n-1} b + \binom{n}{2} a^{n-2} b^2 + \ldots + \binom{n}{r} a^{n-r} b^r$$

$$+ \ldots + \binom{n}{n} b^n.$$

Note that $\binom{n}{r} = \frac{n!}{r!(n-r)!}$ and that $0! = 1$. Then, we obtain:

$$(x + 2y)^5 = \binom{5}{0} x^5 (2y)^0 + \binom{5}{1} x^4 (2y)^1 + \binom{5}{2} x^3 (2y)^2$$

$$+ \binom{5}{3} x^2 (2y)^3 + \binom{5}{4} x^1 (2y)^4 + \binom{5}{5} x^0 (2y)^5$$

$$= \frac{5!}{0!5!} x^5 + \frac{5!}{1!4!} x^4 2y + \frac{5!}{2!3!} x^3 \left(4y^2\right)$$

$$+ \frac{5!}{3!2!} x^2 \left(8y^3\right) + \frac{5!}{4!1!} x \left(16y^4\right) + \frac{5!}{5!0!} 1 \left(32y^5\right)$$

$$= x^5 + \frac{5 \cdot 4!}{4!} x^4 2y + \frac{5 \cdot 4 \cdot 3!}{2 \cdot 1 \cdot 3!} x^3 \left(4y^2\right)$$

$$+ \frac{5 \cdot 4 \cdot 3!}{3! \cdot 2 \cdot 1} x^2 \left(8y^3\right) + \frac{5 \cdot 4!}{4!1!} x \left(16y^4\right) + \frac{5!}{5!0!} \left(32y^5\right)$$

$$= x^5 + 10x^4 y + 40x^3 y^2 + 80x^2 y^3 + 80xy^4 + 32y^5 .$$

• PROBLEM 11–29

Find the expansion of $(x + y)^6$.

Solution: Use the Binomial Theorem which states that

$$(a+b)^n = \frac{1}{0!} a^n + \frac{n}{1!} a^{n-1} b + \frac{n(n-1)}{2!} a^{n-2} b^2 + \dots + nab^{n-1} + b^n .$$

Replacing a by x and b by y:

$$(x+y)^6 = \frac{1}{0!} x^6 + \frac{6}{1!} x^5 y + \frac{6 \cdot 5}{2!} x^4 y^2 + \frac{6 \cdot 5 \cdot 4}{3!} x^3 y^3 + \frac{6 \cdot 5 \cdot 4 \cdot 3}{4!} x^2 y^4$$

$$+ \frac{6 \cdot 5 \cdot 4 \cdot 3 \cdot 2}{5!} x^1 y^5 + \frac{6 \cdot 5 \cdot 4 \cdot 3 \cdot 2 \cdot 1}{6!} x^0 y^6$$

$$= \frac{1}{1} x^6 + \frac{6}{1} x^5 y + \frac{6 \cdot 5}{2 \cdot 1} x^4 y^2 + \frac{6 \cdot 5 \cdot 4}{3 \cdot 2 \cdot 1} x^3 y^3 + \frac{6 \cdot 5 \cdot 4 \cdot 3}{4 \cdot 3 \cdot 2 \cdot 1} x^2 y^4$$

$$+ \frac{6 \cdot 5 \cdot 4 \cdot 3 \cdot 2}{5 \cdot 4 \cdot 3 \cdot 2 \cdot 1} xy^5 + \frac{6 \cdot 5 \cdot 4 \cdot 3 \cdot 2 \cdot 1}{6 \cdot 5 \cdot 4 \cdot 3 \cdot 2 \cdot 1} y^6$$

$$(x+y)^6 = x^6 + 6x^5 y + 15x^4 y^2 + 20x^3 y^3 + 15x^2 y^4 + 6xy^5 + y^6 .$$

• PROBLEM 11–30

What is the probability of getting exactly 3 heads in 5 flips of a balanced coin?

Solution: We have here the situation often referred to
as a Bernoulli trial. There are two possible outcomes,
head or tail, each with a finite probability. Each flip
is independent. This is the type of situation to which the
binomial distribution,

$$P(X = k) = \binom{n}{k} p^k (1 - p)^{n-k},$$

applies. The a priori probability of tossing a head is
$p = \frac{1}{2}$. the probability of a tail is $q = 1 - p =$
$1 - \frac{1}{2} = \frac{1}{2}$. Also n = 5 and k = 3 (number of heads required).
we have

$$P(X = 3) = \binom{5}{3} \left(\frac{1}{2}\right)^3 \left(1 - \frac{1}{2}\right)^2 = \frac{5!}{3!2!} \left(\frac{1}{2}\right)^3 \left(\frac{1}{2}\right)^2$$

$$= \frac{5 \cdot 4 \cdot 3 \cdot 2 \cdot 1}{3 \cdot 2 \cdot 1 \cdot 2 \cdot 1} \left(\frac{1}{2}\right)^5 = \frac{10}{2^5} = \frac{10}{32} = \frac{5}{16} .$$

● **PROBLEM** 11–31

Find the probability that in three rolls of a pair of dice,
exactly one total of 7 is rolled.

Solution: Consider each of the three rolls of the 3
pairs as a trial and the probability of rolling a total of s
seven as a "success." Assume that each roll is independent o
of the others. If X = the number of successes, then we want
to find

$$Pr(X = 1) = \binom{3}{1} p^1 (1 - p)^2$$

where p = probability of rolling a total of 7 on a single
roll. We have earlier found in the Probability Chapter this
probability to be $\frac{6}{36} = \frac{1}{6}$ and hence the probability of
rolling a total of 7 exactly once in three rolls is

$$Pr(X = 1) = \binom{3}{1} \left(\frac{1}{6}\right)^1 \left(\frac{5}{6}\right)^2 = 3 \cdot \frac{1}{6} \cdot \frac{25}{36} = \frac{25}{72} .$$

● **PROBLEM** 11–32

A deck of cards can be dichotomized into black cards and
red cards. If p is the probability of a black card on a
single draw and q the probability of a red card, p = 1/2
and q = 1/2. Six cards are sampled with replacement. What
is the probability on six draws of getting 4 black and 2
red cards? Of getting all black cards?

Solution: Let X = the number of black cards observed in six draws from this deck.

Pr(X = 4) = the probability that 4 black cards and 2 red cards are in this sample of 6. The probability of drawing a black card on a single draw is p = 1/2 and since each draw is independent, X is distributed binomially with parameters 6 and 1/2.

Thus $\Pr(X = 4) = \binom{6}{4}\left(\frac{1}{2}\right)^4\left(\frac{1}{2}\right)^2 = 15\left(\frac{1}{2}\right)^6 = \frac{15}{64}$

and $\Pr(X = 6) = \binom{6}{6}\left(\frac{1}{2}\right)^6\left(\frac{1}{2}\right)^0 = \left(\frac{1}{2}\right)^6 = \frac{1}{64}$.

● **PROBLEM** 11–33

Suppose that the probability of parents to have a child with blond hair is $\frac{1}{4}$. If there are four children in the family, what is the probability that exactly half of them have blond hair?

Solution: We assume that the probability of parents having a blond child is $\frac{1}{4}$. In order to compute the probability that 2 of 4 children have blond hair we must make another assumption. We must assume that the event consisting of a child being blond when it is born is independent of whether any of the other children are blond. The genetic determination of each child's hair color can be considered one of four independent trials with the probability of success, observing a blond child, equal to $\frac{1}{4}$.

If X = the number of children in the family with blond hair, we are interested in finding Pr(X = 2). By our assumptions X is binomially distributed with n = 4 and p = $\frac{1}{4}$.

Thus Pr(X = 2) = Pr(exactly half the children are blond)

$$= \binom{4}{2}\left(\frac{1}{4}\right)^2\left(\frac{3}{4}\right)^2 = \frac{4!}{2!2!} \cdot \left(\frac{1}{4}\right)^2\left(\frac{3}{4}\right)^2$$

$$= \frac{4 \cdot 3}{2 \cdot 1} \cdot \left(\frac{1}{16}\right)\left(\frac{9}{16}\right) = \frac{27}{128} = .21.$$

If a fair coin is tossed four times, what is the probability of at least two heads?

Solution: Let X = the number of heads observed in 4 tosses of a fair coin. X is binomially distributed if we assume that each toss is independent. If the coin is fair, p = Pr (a head is observed on a single toss) = $\frac{1}{2}$.

Thus Pr (at least two heads in 4 tosses)

$$= Pr(X \overset{\geq}{=} 2) = \sum_{x=2}^{4} \binom{4}{x} \left(\frac{1}{2}\right)^x \left(\frac{1}{2}\right)^{4-x}$$

$$= \binom{4}{2} \left(\frac{1}{2}\right)^2 \left(\frac{1}{2}\right)^2 + \binom{4}{3} \left(\frac{1}{2}\right)^3 \left(\frac{1}{2}\right) + \binom{4}{4} \left(\frac{1}{2}\right)^4 \left(\frac{1}{2}\right)^0$$

$$= \frac{6}{16} + \frac{4}{16} + \frac{1}{16} = \frac{11}{16} .$$

Given the following cumulative binomial distribution, find (a) P(X = 1); (b) P(X = 4); (c) P(X = 5).

$$(n = 5, p = 0.31)$$

x	P(X≥x)
0	1.0000
1	0.8436
2	0.4923
3	0.1766
4	0.0347
5	0.0029

Solution: (a) The probabilities on the table, P(X≥x), give the values of the probabilities that a specific occurrence of a random variable will be at least as great as the given x. For example, 1 is the only value at least as great as 1, but not at least as great as 2. Therefore the probability that X = 1 is the probability that X is at least 1 but NOT at least 2.

The addition rule for mutually exclusive probabilities therefore says, P(X≥1) = P(X=1) + P(X≥2).

Equivalently, P(X=1) = P(X≥1) − P(X≥2)

$$= .8436 - .4923 = .3513.$$

(b) $P(X=4) = P(X\geq4) - P(X\geq5) = .0347 - .0029 = .0318.$

(c) 5 is the only possible value at least as large as 5, hence $P(X=5) = P(X\geq5) = .0029.$

• **PROBLEM** 11–36

Given the following binomial distribution, find (a) $P(X=4)$; (b) $P(X=1)$; (c) $P(X=0)$.

$$(n = 5, \ p = 0.69)$$

x	$P(X\leq x)$
0	0.0029
1	0.0347
2	0.1766
3	0.4923
4	0.8436
5	1.0000

Solution: (a) The events $P(X\leq3)$ and $P(X=4)$ are disjoint therefore $P(X=4 \cup X\leq3) = P(X\leq3) + P(X=4)$ by the addition rule or

$$P(X\leq4) = P(X\leq3) + P(X=4)$$

equivalently, $P(X=4) = P(X\leq4) - P(X\leq3)$

$$= .8437 - .4923 = .3513.$$

(b) Similarly, $P(X=1) = P(X\leq1) - P(X\leq0)$

$$= .0347 - .0029 = .0318.$$

(c) $P(X=0) = P(X\leq0) = .0029.$

You might notice that the previous problem had the same solution. The similarity can be explained if the binomial formula,

$$P(X=k) = \binom{n}{k} p^k (1 - p)^{n-k}$$

is used to solve both examples. In the first one where $n = 5$ and $p = .31$, $1 - p = .69$ and

$$P(X=1) = \binom{5}{1} (.31)^1 (.69)^4 = \frac{5!}{1!4!} (.31)^1 (.69)^4 = .3513.$$

In this example where n = 5 and p = 0.69, 1 - p = .31 and

$$P(X=4) = \binom{5}{4} (.69)^4 (.31)^1 = .3513.$$

Similar operations can be performed on (b) and (c).

● **PROBLEM** 11–37

The probability of hitting a target on a shot is $\frac{2}{3}$. If a person fires 8 shots at a target, Let X denote the number of times he hits the target, and find:

(a) P(X = 3)   (b) P(1 < X ≤ 6)   (c) P(X > 3) .

Solution: If we assume that each shot is independent of any other shot then X is a binomially distributed random variable with parameters n = 8 and $\pi = \frac{2}{3}$. (π equals the probability of hitting the target on any particular shot and n = the number of shots.)

Thus, $Pr(X=3) = \binom{8}{3} \left(\frac{2}{3}\right)^3 \left(\frac{1}{8}\right)^{8-3} = \frac{8!}{3!5!} \left(\frac{2}{3}\right)^3 \left(\frac{1}{3}\right)^5$

$$= \frac{8 \cdot 7 \cdot 6}{3 \cdot 2 \cdot 1} \left(\frac{8}{27}\right) \left(\frac{1}{243}\right) = \frac{448}{6561} = .06828$$

$Pr(1 < X \leq 6) = Pr(X = 2, 3, 4, 5 \text{ or } 6)$.

Each of these events is mutually exclusive and thus,

$Pr(X = 2, 3, 4, 5, \text{ or } 6) = Pr(X=2) + Pr(X=3) + Pr(X=4)$

$+ Pr(X=5) + Pr(X=6)$

$$= \sum_{n=2}^{6} Pr(X=n) .$$

$$Pr(1 < X \leq 6) = \sum_{n=2}^{6} \binom{8}{n} \left(\frac{2}{3}\right)^n \left(\frac{1}{3}\right)^{8-n} .$$

Using tables of cumulative probabilities and the fact that $Pr(1 < X < 6) = Pr(X \leq 6) - Pr(X \leq 1)$, or calculating single probabilities and adding, we see that $Pr(1 < X \leq 6) = .8023.$

$Pr(X > 3) = Pr(X = 4, 5, 6, 7 \text{ or } 8)$

$= Pr(X=4) + Pr(X=5) + Pr(X=6) + Pr(X=7)$

$$+ \operatorname{Pr}(X=8)$$

$$= \sum_{n=4}^{8} \operatorname{Pr}(X = n)$$

$$= \sum_{n=4}^{8} \binom{8}{n} \left(\frac{2}{3}\right)^{n} \left(\frac{1}{3}\right)^{8-n} .$$

Again, using a table of cumulative probabilities or calculating each single probability we see that

$$\operatorname{Pr}(X > 3) = .912.$$

• **PROBLEM** 11-38

If a bag contains three white two black, and four red balls and four balls are drawn at random with replacement, calculate the probabilities that

(a) The sample contains just one white ball.
(b) The sample contains just one white ball given that it contains just one red ball.

Solution: Since there are nine balls and we are sampling with replacement and choosing the balls at random, on each draw

$$\operatorname{Pr}(\text{white ball}) = \frac{3}{9} = \frac{1}{3} .$$

$$\operatorname{Pr}(\text{black ball}) = \frac{2}{9} .$$

$$\operatorname{Pr}(\text{red ball}) = \frac{4}{9} .$$

(a) On each draw, $\operatorname{Pr}(\text{white}) + \operatorname{Pr}(\text{black or red}) = 1$. Let X = number of white balls. Then X is distributed binomially with $n = 4$ trials and $\operatorname{Pr}(\text{white ball}) = \frac{1}{3}$. Thus

$$\operatorname{Pr}(\text{just one white}) = \operatorname{Pr}(X=1) = \binom{4}{1} \left(\frac{1}{3}\right)^{1} \left(1 - \frac{1}{3}\right)^{4-1}$$

$$= 4 \left(\frac{1}{3}\right) \left(\frac{2}{3}\right)^{3} = \frac{32}{81} .$$

(b) $\operatorname{Pr}(\text{just 1 white} | \text{just 1 red})$

$$= \frac{\operatorname{Pr}(\text{just 1 white and just 1 red})}{\operatorname{Pr}(\text{just 1 red})} .$$

If Y = number of red balls then Y is distributed binomially with parameters $n = 4$ and $p = \frac{4}{9}$.

Thus $\Pr(\text{just 1 red}) = \Pr(Y = 1) = \binom{4}{1}\left(\frac{4}{9}\right)^1 \left(1 - \frac{4}{9}\right)^{4-1}$

$$= 4 \binom{4}{9}\left(\frac{5}{9}\right)^3 .$$

$\Pr(\text{just 1 white and just 1 red})$

$= \Pr(\text{1 white, 1 red and 2 blacks}).$

Any particular sequence of outcomes in which 1 white ball is chosen, 1 red ball is chosen and 2 black balls are chosen has probability $\left(\frac{3}{9}\right)^1 \left(\frac{2}{9}\right)^2 \left(\frac{4}{9}\right)^1$. We now must find the number of such distinguishable arrangements. There are $\binom{4}{1}$ ways to select the position of the white ball. There are now three positions available to select the position of the red ball and $\binom{3}{1}$ ways to do this. The position of the black balls are now fixed. There are thus

$$\binom{4}{1}\binom{3}{1} = \frac{4!}{1!3!}\,\frac{3!}{1!2!} = \frac{4!}{1!2!1!} \quad \text{distinguishable arrange-}$$

ments.

Thus the $\Pr(\text{1 red ball, 1 white ball and 2 black balls})$

$$= \frac{4!}{1!2!1!} \cdot \left(\frac{3}{9}\right)\left(\frac{2}{9}\right)^2\left(\frac{4}{9}\right)^1 = \frac{4 \cdot 3 \cdot 3 \cdot 4 \cdot 4}{9^4} ,$$

$$\Pr(\text{just 1 white}\,|\,\text{just 1 red}) = \frac{\dfrac{4 \cdot 3 \cdot 3 \cdot 4 \cdot 4}{9^4}}{4 \binom{4}{9}\left(\frac{5}{9}\right)^3}$$

$$= \frac{4 \cdot 3 \cdot 3 \cdot 4 \cdot 4}{4 \cdot 4 \cdot 5 \cdot 5 \cdot 5} = \frac{36}{125} .$$

• PROBLEM 11–39

Consider the joint distribution of X and Y given in the form of a table below. The cell (i,j) corresponds to the joint probability that $X = i$, $Y = j$, for $i = 1,2,3$, $j = 1,2,3$.

Y \ X	1	2	3
1	0	1/6	1/6
2	1/6	0	1/6
3	1/6	1/6	0

Check that this is a proper probability distribution. What is the marginal distribution of X? What is the marginal distribution of Y?

Solution: A joint probability mass function gives the probabilities of events. These events are composed of the results of two (or more) experiments. An example might be the toss of two dice. In this case, each event or outcome has two numbers associated with it. The numbers are the outcomes from the toss of each die. The probability distribution of the pair (X,Y) is

$$Pr(X = i,\ X = j) = \frac{1}{36} \qquad\qquad i = 1,2,3,4,5,6$$
$$j = 1,2,3,4,5,6 \ .$$

Another example is the toss of two dice where X = number observed on first die ; Y = the larger of the two numbers.

In order for $f(x,y) = Pr(X = x,\ Y = y)$ to be a proper joint probability, the sum of $Pr(X = x,\ Y = y)$ over all (x,y), over all points in the sample space must equal 1.

In the case of the pair of tossed dice,

$$\sum_x \sum_y Pr(X = x,\ Y = y) = \sum_{i=1}^{6} \sum_{j=1}^{6} \frac{1}{36} = \sum_{i=1}^{6} \frac{6}{36} = \frac{6 \cdot 6}{36} = 1 \ .$$

Thus, this is a proper probability distribution.

In our original example,

$$\sum_{i=1}^{3} \sum_{j=1}^{3} Pr(X = i,\ Y = j) = \sum_{i=1}^{3} [Pr(X=i,Y=1) + Pr(X=i,Y=2) + Pr(X=i,Y=3)]$$

$$= \sum_{i=1}^{3} Pr(X=i,Y=1) + \sum_{i=1}^{3} Pr(X=i,Y=2) + \sum_{i=1}^{3} Pr(X=i,Y=3)$$

$$= (0 + 1/6 + 1/6) + (1/6 + 0 + 1/6) + (1/6 + 1/6 + 0)$$

$$= 1/3 + 1/3 + 1/3 = 1 \ .$$

Thus, the probability distribution specified in the table is a proper distribution.

We can compute the individual probability distributions of X and Y. These are called the marginal distributions of X and Y and are calculated in the following way.

We wish to find the probability that $X = 1,2,3$,

$$Pr(X = 1) = Pr(X = 1,\ Y = 1,2,\ or\ 3) \ .$$

Because the events $"X = 1,\ Y = 1"$, $"X = 1,\ Y = 2"$, $"X = 1,\ Y = 3"$ are mutually exclusive,

$$Pr(X=1) = Pr(X=1,Y=1) + Pr(X=1,Y=2) + Pr(X=1,Y=3)$$

$$= \sum_{i=1}^{3} Pr(X=1,Y=i) \ .$$

Thus, $\qquad Pr(X=1) = 0 + 1/6 + 1/6 = 1/3 \ .$

Similarly, $\quad Pr(X=2) = \sum_{i=1}^{3} Pr(X=2,Y=i) = 1/6 + 0 + 1/6 = 1/3$

and $\qquad\qquad Pr(X=3) = \sum_{i=1}^{3} Pr(X=3,Y=i) = 1/6 + 1/6 + 0 = 1/3 \ .$

We compute the marginal probabilities of Y in a similar way.

$$Pr(Y=1) = Pr(X=1,Y=1) + Pr(X=2,Y=1) + Pr(X=3,Y=1)$$

$$= 0 + 1/6 + 1/6 = 2/6 = 1/3$$

$$Pr(Y=2) = \sum_{j=1}^{3} Pr(X=j,Y=2) = 1/6 + 0 + 1/6 = 1/3 .$$

$$Pr(Y=3) = \sum_{j=1}^{3} Pr(X=j,Y=3) = 1/6 + 1/6 + 0 .$$

To see why these are called marginal probabilities we examine the way they were computed.

The marginal probabilities of X were found by summing along the rows of the table of the joint distribution. The marginal probabilities of Y were found by summing along the columns of the table of the joint distribution.

The probabilities resulting from these summations are often placed in the margins, as in the table below, hence the name marginal probabilities.

● **PROBLEM** 11–40

Consider the table representing the joint distribution between X' and Y'.

X'＼Y'	1	2	3
1	1/9	1/9	1/9
2	1/9	1/9	1/9
3	1/9	1/9	1/9

Find the marginal distributions of X' and Y'. Are X' and Y' independent? In the previous problem, were X and Y independent?

Solution: The marginal distributions of X' and Y' are found by summing across the rows and columns of the table above.

$$Pr(X'=1) = \sum_{i=1}^{3} Pr(X'=1,Y'=i) = 1/9 + 1/9 + 1/9 = 1/3 .$$

$$Pr(X'=2) = \sum_{i=1}^{3} Pr(X'=2,Y'=i) = 1/9 + 1/9 + 1/9 = 1/3 .$$

$$Pr(X'=3) = \sum_{i=1}^{3} Pr(X'=3,Y'=i) = 1/9 + 1/9 + 1/9 = 1/3 .$$

Similarly, $$Pr(Y'=1) = \sum_{j=1}^{3} Pr(X'=j,Y'=i) = 1/9 + 1/9 + 1/9 = 1/3 .$$

$$Pr(Y'=2) = \sum_{j=1}^{3} Pr(X'=j,Y'=2) = 1/9 + 1/9 + 1/9 = 1/3.$$

$$Pr(Y'=3) = \sum_{j=1}^{3} Pr(X'=j, Y=3) = 1/3.$$

The marginal distributions are hence

$$Pr(X'=x) = \begin{cases} 1/3 & x = 1,2,3 \\ 0 & \text{otherwise} \end{cases}$$

and

$$Pr(Y'=y) = \begin{cases} 1/3 & y = 1,2,3 \\ 0 & \text{otherwise} . \end{cases}$$

Two random variables, X and Y, will be independent if and only if $Pr(X=x, Y=y) = Pr(X=x) Pr(Y=y)$, for all x and y.

Checking X' and Y' we see that for all x and y,

$$Pr(X'=x, Y'=y) = 1/9 = Pr(X'=x)Pr(Y'=y) = 1/3 \cdot 1/3 = 1/9 .$$

In the previous problem, we see that X and Y are not independent but dependent. To see this, consider $Pr(X=i, Y=i)$ for i = 1,2,3.

$$Pr(X=i)Pr(Y=i) = 1/3 \cdot 1/3 = 1/9$$

for i = 1,2,3 but the joint probability function specifies that $Pr(X=i, Y=i) = 0$ for i = 1,2, and 3 . Thus

$$Pr(X=i)Pr(Y=i) \neq Pr(X=i, Y=i)$$

and X and Y are not independent.

• PROBLEM 11–41

Consider a bag containing two white and 4 black balls. If two balls are drawn at random without replacement from the bag, let X and Y be random variables representing the results of these two drawings. Let 0 correspond to drawing a black ball and 1 correspond to drawing a white ball. Find the joint, marginal, and conditional distributions of X and Y.

Solution: $Pr(X=0)$ = Pr(a black ball is drawn first)

$$= \frac{\text{number of black balls}}{\text{total number of balls}} = \frac{4}{6} = \frac{2}{3} .$$

$Pr(X=1)$ = Pr(a white ball is drawn first)

$$= \frac{\text{number of white balls}}{\text{total number of balls}} = \frac{2}{6} = \frac{1}{3} .$$

We may use the notion of conditional probability to find the probability of a particular event on the second draw given a particular event on the first. We may talk about conditional distribution for the variable Y given X. A conditional distribution is defined in terms of conditional probability.

$Pr(Y=y|X=x) = f(y|x) =$ conditional distribution of Y given X, $f(x,y) = Pr(X=x, Y=y) =$ joint distribution of X and Y and $f(x) = Pr(X=x)$ is the marginal distribution of X.

The conditional probabilities may be calculated directly from the problem. We then use the conditional probabilities and marginal distribution of X to find the joint distribution of X and Y.

$Pr(Y=0|X=0) = Pr($black ball is second$|$black ball first$)$

$$= \frac{\text{number of black balls } - 1}{\text{total number of balls } - 1} = \frac{4-1}{6-1} = \frac{3}{5}$$

$Pr(Y=1|X=0) = Pr($white second$|$black first$)$

$$= \frac{\text{number of white balls}}{\text{total number of balls } - 1} = \frac{2}{6-1} = \frac{2}{5}$$

$Pr(Y=0|X=1) = Pr($black ball second$|$white ball first$)$

$$= \frac{\text{number of black balls}}{\text{total number of balls } - 1} = \frac{4}{6-1} = \frac{4}{5}$$

$Pr(Y=1|X=1) = Pr($white ball second$|$white first$)$

$$= \frac{\text{number of white balls } -1}{\text{number of balls } - 1} = \frac{2-1}{6-1} = \frac{1}{5} .$$

We now calculate the joint probabilities of X and Y.

$$\frac{Pr(X=0, Y=0)}{Pr(X=0)} = Pr(Y=0|X=0) , \text{ or,}$$

$Pr(X=0,Y=0) = Pr(X=0)Pr(Y=0|X=0) = \frac{2}{3} \cdot \frac{3}{5} = \frac{6}{15}$. Similarly,

$Pr(X=1,Y=0) = Pr(Y=0|X=1)Pr(X=1) = \frac{4}{5} \cdot \frac{1}{3} = \frac{4}{15}$

$Pr(X=0,Y=1) = Pr(Y=1|X=0)Pr(X=0) = \frac{2}{5} \cdot \frac{2}{3} = \frac{4}{15}$

$Pr(X=1,Y=1) = Pr(Y=1|X=1)Pr(X=1) = \frac{1}{5} \cdot \frac{1}{3} = \frac{1}{15}$.

Summarizing these results in the following table, we see:

X Y	0	1	Pr(Y = y)
0	6/15	4/15	10/15
1	4/15	1/15	5/15
Pr(X = x)	10/15	5/15	

This is the joint distribution of X and Y, with the marginal distribution indicated.

The conditional distribution of Y given X is

$$\Pr(Y=y \mid X=x) = \frac{\Pr(Y=y, X=x)}{\Pr(X=x)} = \begin{cases} \dfrac{6/15}{10/15} = 3/5 & x = y = 0 \\[2mm] \dfrac{4/15}{5/15} = 4/5 & x = 1; \ y = 0 \\[2mm] \dfrac{4/15}{10/15} = 2/5 & x = 0; \ y = 1 \\[2mm] \dfrac{1/15}{5/15} = 1/5 & x = y = 1 . \end{cases}$$

For any fixed value of X, we see that $f(y \mid x) = \Pr(Y=y \mid X=x)$ is a proper probability distribution for Y.

$$\sum_{y=0}^{1} \Pr(Y=y \mid X=0) = \Pr(Y=0 \mid X=0) + \Pr(Y=1 \mid X=0) = 3/5 + 2/5 = 1$$

and

$$\sum_{y=0}^{1} \Pr(Y=y \mid X=1) = \Pr(Y=0 \mid X=1) + \Pr(Y=1 \mid X=1) = 1/5 + 4/5 = 1 .$$

● **PROBLEM** 11–42

Use

$$f(x,y) = \begin{cases} e^{-x} e^{-y} & \begin{array}{l} x > 0 \\ y > 0 \end{array} \\ 0 & \text{otherwise} \end{cases}$$

to find the probability that $\{ 1 < X < 2 \quad \text{and} \quad 0 < Y < 2 \}$.

Solution: $\Pr(1 < X < 2 \quad \text{and} \quad 0 < Y < 2)$ is the volume over the shaded rectangle:

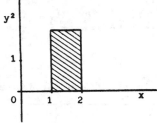

This volume over the rectangle and under $f(x,y)$ is pictured below:

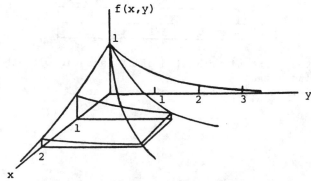

713

To find this volume we integrate X from 1 to 2 and Y from 0 to 2. Thus,

$$\Pr(1 < X < 2 \text{ and } 0 < Y < 2) = \int_0^2 \int_1^2 f(x,y)\ dx\ dy$$

$$= \int_0^2 e^{-y}\left(\int_1^2 e^{-x}\ dx\right) dy$$

$$= \int_0^2 e^{-y} dy\left[-e^{-x}\Big|_1^2\right] = \int_0^2 e^{-y} dy\,(e^{-1} - e^{-2})$$

$$= -e^{-y}\Big]_0^2\ (e^{-1} - e^{-2})$$

$$= (e^0 - e^{-2})(e^{-1} - e^{-2}) = (1 - e^{-2})(e^{-1} - e^{-2})$$

$$= (.865)(.233) = .20 \ .$$

• **PROBLEM** 11–43

Given that the joint density function of X and Y is

$$f(x,y) = 6x^2 y \quad \text{where } \begin{cases} 0 < x < 1 \text{ and} \\ 0 < y < 1, \end{cases}$$

$$= 0 \quad \text{elsewhere,}$$

find $\Pr(0 < x < 3/4,\ 1/3 < y < 2)$.

Solution: The joint density function of two random variables represents a surface over the region on which it is defined. The volume under the surface is always 1. We need to find the volume over the region bounded by the given limits. We construct a double integral:

$$\int_{y=1/3}^{y=2} \int_{x=0}^{x=3/4} f(x,y)dx\ dy = \int_{y=1/3}^{y=1} \int_{x=0}^{x=3/4} 6x^2 y\ dx\ dy$$

$$+ \int_{y=1}^{2} \int_{x=0}^{x=3/4} 0\ dx\ dy\ ;$$

because $f(x,y) = 0$ where $y \geq 1$), this becomes

$$\int_{y=1/3}^{y=2} \int_{x=0}^{x=3/4} f(x,y)dxdy = 6\int_{y=1/3}^{y=1} \frac{x^3}{3} y\Big]_0^{3/4} dy = 6\int_{y=1/3}^{y=1} \frac{9}{64} y\ dy$$

$$= \frac{54}{64} \frac{y^2}{2}\Big]_{1/3}^{1} = \frac{27}{32}\left(\frac{1}{2} - \frac{1}{18}\right)$$

$$= \frac{27}{32}\left(\frac{8}{18}\right) = \frac{27}{32}\left(\frac{4}{9}\right)$$

$$= \frac{3}{8} = \Pr(0 < x < 3/4,\ 1/3 < y < 2).$$

Given that the continuous random variables X and Y have joint probability density function $f(x,y) = x + y$ when $0 < x < 1$ and $0 < y < 1$, and $f(x,y) = 0$, otherwise, use the marginal probability functions to decide whether or not X and Y are stochastically independent.

Solution: The marginal probability density function of X is found by integrating the joint density function, over the entire domain of Y, with respect to y. This produces a function of x:

$$f_x(x) = \int_{-\infty}^{\infty} f(x,y)dy = \int_0^1 (x+y)dy = \int_0^1 xdy + \int_0^1 ydy$$

$$= (x)y \Big]_0^1 + \frac{y^2}{2}\Big]_0^1 = x(1-0) + (\tfrac{1}{2}-0) = \begin{cases} x + \tfrac{1}{2}, & 0 < x < 1 \\ 0 & \text{elsewhere} \end{cases}$$

Similarly,
$$f_y(y) = \int_{-\infty}^{\infty} f(x,y)dx = \int_0^1 (x+y)dx = \int_0^1 xdx + \int_0^1 ydx$$

$$= \frac{x^2}{2}\Big]_0^1 + (y)x\Big]_0^1 = (\tfrac{1}{2}-0) + y(1-0) = \tfrac{1}{2} + y, \quad 0 < y < 1$$
$$= 0 \qquad \text{elsewhere.}$$

By definition, X and Y are stochastically independent if and only if $f(x,y) = f(x) \cdot f(y)$. If $f(x,y) \neq f_x(x) \cdot f_y(y)$, X and Y are said to be stochastically dependent.

$$f_x(x) \cdot f_y(y) = (x+\tfrac{1}{2})(y+\tfrac{1}{2}) = xy + \tfrac{1}{2}y + \tfrac{1}{2}x + \tfrac{1}{4} \neq x+y = f(x,y).$$

Therefore, X and Y are stochastically dependent.

Given that the joint density function of the random variables X, Y and Z is $f(x,y,z) = e^{-(x+y+z)}$ when $0 < x < \infty$, $0 < y < \infty$, $0 < z < \infty$, find the cumulative distribution function $F(x,y,z)$ of X, Y, and Z.

Solution: $F(x,y,z)$ is defined to be $\Pr(X \leq x, Y \leq y, Z \leq z)$. Here we have extended the concept of cumulative probability from one random variable to three. The cumulative distribution function of one variable gives the area under the density curve to the left of the particular value of x. The distribution function of two variables gives the volume under the surface represented by the bivariate density function over the region bounded by the specified values of x and y.

This is obtained by integration with respect to each variable. The number given by a distribution function of three variables can be interpreted as a 4-dimensional volume. It is obtained by constructing an iterated triple integral and integrating with respect to each variable.

$$F(x,y,z) = \int_0^z \int_0^y \int_0^x e^{-u-v-w} \, du\,dv\,dw$$

$$= \int_0^z \int_0^y \int_0^x e^{-u}(e^{-v-w}) \, du \, dv \, dw$$

$$= \int_0^z \int_0^y (e^{-v-w})(1-e^{-x}) \, dv \, dw = (1-e^{-x}) \int_0^z e^{-w} \int_0^y e^{-v} \, dv \, dw$$

$$= (1-e^{-x}) \int_0^z e^{-w}(1-e^{-y}) \, dw$$

$$= (1-e^{-x})(1-e^{-y}) \int_0^z e^{-w} \, dw = (1-e^{-x})(1-e^{-y})(1-e^{-z})$$

for $x, y, z > 0$.

● **PROBLEM** 11–46

Let (X, Y) have the distribution defined by the joint density function,

$$f(x,y) = \begin{cases} e^{-x-y} & \begin{array}{l} x > 0 \\ y > 0 \end{array} \\ 0 & \text{otherwise} . \end{cases}$$

Find the marginal and conditional densities of Y and X. Are X and Y independent?

Solution: The marginal distributions of X and Y are:

$$g(x) = \int_0^\infty f(x,y) \, dy = \int_0^\infty e^{-x-y} \, dy$$

and

$$h(y) = \int_0^\infty f(x,y) \, dx = \int_0^\infty e^{-x-y} \, dx .$$

Thus

$$g(x) = e^{-x}(-e^{-y}) \Big|_0^\infty = e^{-x} \qquad x > 0$$

and

$$h(y) = e^{-y}(-e^{-x}) \Big|_0^\infty = e^{-y} \qquad y > 0 .$$

We see that $f(x,y) = g(x)h(y)$ because $e^{-x-y} = e^{-x}e^{-y}$. Thus X and Y are independent.

The conditional densities of X and Y are defined by analogy to conditional probability.

$f(x|y)$ = conditional density of x given y is by definition

$$= \frac{f(x,y)}{h(y)}$$

for fixed y and

$f(y|x)$ = conditional density of y given x is

$$= \frac{f(x,y)}{g(x)}$$

for fixed y.

The conditional densities in this example are

$$f(x|y) = \frac{e^{-x-y}}{e^{-y}} \qquad x > 0$$

and

$$f(y|x) = \frac{e^{-x-y}}{e^{-x}} \qquad y > 0 .$$

Hence

$$f(x|y) = g(x) = e^{-x} \quad \text{and} \quad f(y|x) = h(y) = e^{-y} .$$

Another alternate condition for independence is that the conditional density function is equal to the marginal density function.

● **PROBLEM** 11–47

A random vector (X,Y) has a density,

$$f(x,y) = \begin{cases} 2 & x + y < 1, \ x \geq 0 \\ & \qquad\qquad y > 0 \\ 0 & \text{otherwise.} \end{cases}$$

Find the marginal density of X and the conditional density of Y given X.

Solution: The marginal density of X is the area under the cross-section at $X = x$ as a function of x:

$$f(x) = \int_{-\infty}^{\infty} f(x,y) dy = \int_0^{1-x} 2dy = 2(1-x)$$

$$f(y|x) = \frac{f(x,y)}{f(x)} = \frac{2}{2(1-x)} \quad \text{for} \quad \begin{matrix} x + y < 1 \\ 1 > x \geq 0 \\ y > 0 \end{matrix}$$

or for fixed x, $\qquad\qquad 0 < x < 1$,

$$f(y|x) = \begin{cases} \dfrac{1}{1-x} & 0 < y < 1-x \\ 0 & \text{otherwise .} \end{cases}$$

Note that for some fixed x such that $0 < x < 1$, $f(y|x)$ is a constant density for $0 < y < 1-x$.

● **PROBLEM** 11–48

Let

$$f(x,y) = \begin{cases} 2 - x - y & \begin{matrix} 0 < x < 1 \\ 0 < y < 1 \end{matrix} \\ 0 & \text{otherwise .} \end{cases}$$

Find the conditional distribution of Y given X.

Solution: The conditional distribution of Y given $X = x$ is defined to be $\qquad f(y|X = x) = \dfrac{f(x,y)}{g(x)}$, where g(x) is the marginal distribution of X and f(x,y) is the joint distribution of X and Y.

The marginal distribution of X is

$$g(x) = \int_{-\infty}^{\infty} f(x,y)dy = \int_{0}^{1} (2-x-y)dy$$

$$= (2-x)y - \frac{y^2}{2}\Bigg]_{0}^{1} = 2 - x - \tfrac{1}{2} = \frac{3}{2} - x \quad \text{for} \quad 0 < x < 1 .$$

Thus the conditional density is

$$f(y|x) = \frac{2-y-x}{3/2 - x} , \quad 0 < y < 1 \quad \text{and} \quad x \text{ fixed.}$$

EXPECTED VALUE

● PROBLEM 11–49

Let X be the random variable defined as the number of dots observed on the upturned face of a fair die after a single toss. Find the expected value of X.

Solution: X can take on the values 1, 2, 3, 4, 5, or 6. Since the die is fair we assume that each value is observed with equal probability. Thus,

$$\Pr(X = 1) = \Pr(X = 2) = \ldots = \Pr(X = 6)$$

$$= \frac{1}{6} .$$

The expected value of X is

$$E(X) = \Sigma x \Pr(X = x) . \qquad \text{Hence}$$

$$E(X) = 1 \cdot \frac{1}{6} + 2 \cdot \frac{1}{6} + 3 \cdot \frac{1}{6} + 4 \cdot \frac{1}{6} + 5 \cdot \frac{1}{6} + 6 \cdot \frac{1}{6}$$

$$= \frac{1}{6} (1 + 2 + 3 + 4 + 5 + 6)$$

$$= \frac{21}{6} = 3 \frac{1}{2} .$$

● PROBLEM 11–50

Suppose the earnings of a laborer, denoted by X, are given by the following probability function.

X	0	8	12	16
Pr(X = x)	0.3	0.2	0.3	0.2

Find the laborer's expected earnings.

Solution: The laborer's expected earnings are denoted by E(X), the expected value of the random variable X.
The expected value of X is defined to be,

E(X) = (0) Pr(X = 0) + (8) Pr(X = 8)

        + (12) Pr(X = 12) + (16) Pr(X = 16)

    = (0)(.3) + (8)(.2) + (12)(.3) + (16)(.2)

    = 0 + 1.6 + 3.6 + 3.2

    = 8.4 .

Thus the expected earnings are 8.4.

● **PROBLEM** 11–51

Find the expected number of boys on a committee of 3 selected at random from 4 boys and 3 girls.

Solution: Let X represent the number of boys on the committee. X can be equal to 0, 1, 2 or 3. The sampling procedure is without replacement.

To find the probability distribution we calculate the probabilities that X = 0, 1, 2 or 3. The probability that there are zero boys on the committee is

Pr(X = 0)

$= \dfrac{\text{number of ways 0 boys can be picked}}{\text{number of ways a committee of 3 can be chosen}}$.

The number of ways a committee of 3 can be chosen from the 4 boys and 3 girls is the number of ways 3 can be selected from 7 or $\binom{7}{3}$.

The number of ways a committee of 3 can be chosen that contains 0 boys is $\binom{4}{0} \cdot \binom{3}{3}$, the number of ways 0 boys are chosen from 4 multiplied by the number of ways 3 girls are chosen from the 3 available.

Thus, $\Pr(X = 0) = \dfrac{\binom{4}{0}\binom{3}{3}}{\binom{7}{3}}$.

Similarly,

$$\Pr(X = 1) = \frac{\binom{4}{1}\binom{3}{2}}{\binom{7}{3}} ,$$

$$\Pr(X = 2) = \frac{\binom{4}{2}\binom{3}{1}}{\binom{7}{3}}$$

and

$$\Pr(X = 3) = \frac{\binom{4}{3}\binom{3}{0}}{\binom{7}{3}} .$$

By definition, the expected number of boys is

$$E(X) = (0)\,\Pr(X = 0) + (1)\,\Pr(X = 1) + (2)\,\Pr(X = 2)$$
$$+ (3)\,\Pr(X = 3)$$

$$= (0)\,\frac{\binom{4}{0}\binom{3}{3}}{\binom{7}{3}} + (1)\,\frac{\binom{4}{1}\binom{3}{2}}{\binom{7}{3}} + (2)\,\frac{\binom{4}{2}\binom{3}{1}}{\binom{7}{3}} + (3)\,\frac{\binom{4}{3}\binom{3}{0}}{\binom{7}{3}}$$

$$= (1)\,\frac{\frac{4!}{3!1!} \cdot \frac{3!}{2!1!}}{\binom{7}{3}} + (2)\,\frac{\frac{4!}{2!2!} \cdot \frac{3!}{2!1!}}{\binom{7}{3}}$$

$$+ (3)\,\frac{\frac{4!}{3!1!} \cdot \frac{3!}{0!3!}}{\binom{7}{3}}$$

$$= \frac{1}{\binom{7}{3}}\ [(4)(3) + (2)(6)(3) + (3)(4)(1)]$$

$$= \frac{12 + 36 + 4}{\binom{7}{3}} = \frac{52}{\frac{7!}{3!4!}}$$

$$= \frac{52}{\frac{7 \cdot 6 \cdot 5}{3 \cdot 2 \cdot 1}} = \frac{52}{35} = 1.5 .$$

Thus, if a comittee of 3 is selected at random over and over it would contain on the average 1.5 boys.

Let X be a random variable denoting the hours of life in an electric light bulb. Suppose X is distributed with density function

$$f(x) = \frac{1}{1,000} e^{-x/1000} \qquad \text{for } x > 0$$

Find the expected lifetime of such a bulb.

Solution: The expected value of a continuous random variable is the "sum" of all the values of the random variable multiplied by their probabilities. In the continuous case, this "summing" necessitates integration. Thus

$$E(X) = \int_{\text{all } x} x f(x) dx .$$

In our problem, x can take all positive values; thus

$$E(X) = \int_{0}^{\infty} x f(x) dx$$

$$= \int_{0}^{\infty} \frac{x}{1000} e^{-\frac{x}{1000}} dx .$$

Integrating by parts, let u = x and

$$dv = \frac{1}{1000} e^{-\frac{x}{1000}} dx; \text{ then } du = dx \text{ and } v = - e^{-\frac{x}{1000}} .$$

Thus

$$E(X) = uv \Big|_{0}^{\infty} - \int_{0}^{\infty} v du$$

$$E(X) = - xe^{-\frac{x}{1000}} \Big|_{0}^{\infty} - \int_{0}^{\infty} - e^{-\frac{x}{1000}} dx$$

$$= 0 - 1000 \int_0^\infty -\frac{1}{1000} e^{-\frac{x}{1000}} \, dx$$

$$= -1000 \cdot e^{-\frac{x}{1000}} \Bigg]_0^\infty$$

$$= -1000 \, [0 - 1] = 1000 \, .$$

Thus, the expected lifetime of the bulb is 1000 hours.

• PROBLEM 11–53

Find E(X) for the continuous random variables with probability density functions;

a) $f(x) = 2x$, $0 < x < 1$.

b) $f(x) = \frac{1}{(2\sqrt{x})}$, $0 < x < 1$.

c) $f(x) = 6x(1 - x)$, $0 < x < 1$.

d) $f(x) = \frac{1}{2}x^2 e^{-x}$, $0 < x < \infty$.

e) $f(x) = \frac{1}{x^2}$, $1 \leq x < \infty$.

f) $f(x) = 1 - |1 - x|$, $0 \leq x \leq 2$.

<u>Solution:</u>   For a continuous random variable, X,

$$E(X) = \int_{-\infty}^{\infty} x \, f(x) \, dx.$$

It is possible that $f(x) = 0$ for large portions of the real line reducing E(X) to a proper integral.

(a) $E(X) = \int_0^1 x \cdot 2x \, dx = \int_0^1 2x^2 \, dx = \frac{2}{3} x^3 \Bigg|_0^1$

$$= \frac{2}{3} \, [1 - 0] = \frac{2}{3} \, .$$

(b) $E(X) = \int_0^1 x \, \frac{1}{2\sqrt{x}} \, dx = \frac{1}{2} \int_0^1 \sqrt{x} \, dx$

722

$$= \frac{1}{2} \left. \frac{x^{\frac{1}{2} + 1}}{1 + \frac{1}{2}} \right]_0^1 \quad \frac{1}{2} \cdot \frac{2}{3} \left. x^{\frac{3}{2}} \right]_0^1 = \frac{1}{3} \cdot$$

(c) $\quad E(X) = \displaystyle\int_0^1 x(6x(1-x))\, dx = 6 \int_0^1 (x^2 - x^3)\, dx$

$$= 6 \left[\frac{x^3}{3} - \frac{x^4}{4} \right]_0^1 = 6 \left(\frac{1}{3} - \frac{1}{4} \right) = \frac{6}{12} = \frac{1}{2} \cdot$$

(d) $\quad E(X) = \displaystyle\int_0^\infty x \cdot f(x)\, dx = \int_0^\infty \frac{1}{2} x^3 e^{-x}\, dx \, .$

Using integration by parts,

let $\quad\quad u = x^3 \quad\quad\quad\quad$ thus $\quad\quad du = 3x^2\, dx$

$\quad\quad\quad\quad dv = e^{-x}\, dx \quad\quad\quad\quad\quad\quad v = -e^{-x}$

and we see that

$$E(X) = \frac{1}{2} \left[\left. -x^3 e^{-x} \right|_0^\infty - \int_0^\infty -e^{-x}\, 3x^2\, dx \right]$$

$$= \frac{1}{2} \left[0 + 0 + 3 \int_0^\infty x^2 e^{-x}\, dx \right]$$

$$= \frac{3}{2} \int_0^\infty x^2 e^{-x}\, dx \quad\quad = 3 \int_0^\infty \frac{1}{2} x^2 e^{-x}\, dx \, ,$$

but the integrand is $f(x) = \frac{1}{2} x^2 e^{-x}$ our original density function and by definition a density function is a positive-valued function $f(x)$ such that

$$\int_0^\infty f(x)\, dx = 1; \quad\quad\quad\quad\quad\quad \text{thus}$$

$$E(X) = 3 \int_0^\infty f(x)\, dx = 3.$$

(e) $\quad E(X) = \displaystyle\int_1^\infty x \cdot \frac{1}{x^2}\, dx = \int_1^\infty \frac{1}{x}\, dx = \lim_{b \to \infty} \int_1^b \frac{dx}{x}$

$$= \lim_{b\to\infty} \; [\log b - \log 1]$$

$$= \lim_{b\to\infty} \log b = \infty;$$

thus the expected value of x does not exist.

(f) $\quad E(X) = \displaystyle\int_0^2 x\,f(x)\,dx$

$$= \int_0^2 x(1 - |1 - x|)\,dx$$

$$= \int_0^2 [x - x\,|1 - x|]\,dx$$

$$= \frac{1}{2}\,x^2\,\Big|_0^2 - \int_0^2 x\,|1 - x|\,dx$$

$$= 2 - \int_0^2 x\,|1 - x|\,dx;$$

but $\quad x\,|1 - x| = \begin{cases} x(1 - x) & \text{for } 0 \le x \le 1 \\ x(x - 1) & \text{for } 1 \le x \le 2 \end{cases}.$

Thus

$$E(X) = 2 - \left[\int_0^1 x(1 - x)\,dx + \int_1^2 x\,(x - 1)\,dx \right]$$

$$= 2 - \int_0^1 x\,dx + \int_0^1 x^2\,dx - \int_1^2 x^2 + \int_1^2 x\,dx$$

$$= 2 - \frac{1}{2}\,x^2\,\Big|_0^1 + \frac{1}{3}\,x^3\,\Big|_0^1 - \frac{1}{3}\,x^3\,\Big|_1^2 + \frac{1}{2}\,x^2\,\Big|_1^2$$

$$= 2 - \frac{1}{2} + \frac{1}{3} - \frac{8}{3} + \frac{1}{3} + \frac{4}{2} - \frac{1}{2}$$

$$= 1.$$

Find the expected value of the random variable X if X is distributed with probability density function

$$f(x) = \lambda e^{-\lambda x} \qquad \text{for} \quad 0 < X < \infty .$$

Solution: To find this expected value we will use another method.

This new method computes the expected value from $F(x) = \Pr(X \leq x)$. For our random variable,

$$\Pr(X \leq x) = \int_0^x f(t)\, dt = \int_0^x \lambda e^{-\lambda t}\, dt$$

$$= -\int_0^x (-\lambda) e^{-\lambda t}\, dt$$

$$= -e^{-\lambda t} \Big|_0^x = -e^{-\lambda x} - \left(-e^{-\lambda \cdot 0}\right)$$

$$= 1 - e^{-\lambda x}$$

We have defined $E(X) = \int_0^\infty x\, f(x)\, dx = \int_0^\infty x\, \lambda e^{-\lambda x}\, dx;$

but $x = \int_0^x dt.$ Thus substituting,

$$E(X) = \int_0^\infty f(x) \left[\int_0^x dt \right] dx .$$

This is an iterated integration over the shaded region,

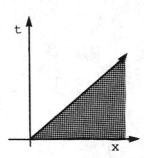

Reversing the order of integration, we integrate with respect to x first. The variable x is integrated from t to ∞ and then t is integrated from 0 to ∞. Thus,

$$\int_0^\infty f(x) \left[\int_0^x dt \right] dx = \int_0^\infty \left[\int_t^\infty f(x)\, dx \right] dt .$$

But $\displaystyle\int_t^\infty f(x)\ dx = Pr(X \geq t) = 1 - Pr(X < t) = 1 - F(t)$

or $\quad E(X) = \displaystyle\int_0^\infty [1 - F(t)]\ dt$.

Thus, $\quad E(X) = \displaystyle\int_0^\infty \left[1 - \left(1 - e^{-\lambda t}\right)\right]\ dt$

$$= \int_0^\infty e^{-\lambda t} dt = -\frac{1}{\lambda} e^{-\lambda t}\ \Big|_0^\infty = -\frac{1}{\lambda}[0 - 1] = \frac{1}{\lambda}\ .$$

● **PROBLEM** 11–55

Find the expected value of the random variable
Y = f(X), when X is a discrete random variable with
probability mass function g(x). Let $f(X) = X^2 + X + 1$

and $Pr(X = x) = g(x) = \begin{cases} \dfrac{1}{2} & x = 1 \\ \dfrac{1}{3} & x = 2 \\ \dfrac{1}{6} & x = 3\ . \end{cases}$

Solution: To find the expected value of a function of
a random variable, we define

$$E(Y) = E(f(X)) = \sum_x f(X)\ g(x) = \sum_x f(X)\ Pr(X = x).$$

As an example, we consider the above problem.

$$E(Y) = f(1)\ Pr(X = 1) + f(2)\ Pr(X = 2) + f(3)\ Pr(X = 3).$$

But $\quad f(1) = 1^2 + 1 + 1 = 3$

$f(2) = 2^2 + 2 + 1 = 7$

$f(3) = 3^2 + 3 + 1 = 13.$

Substituting we see that,

$$E(Y) = 3\ Pr(X = 1) + 7\ Pr(X = 2) + 13 Pr(X = 3)$$

$$= 3 \cdot \frac{1}{2} + 7 \cdot \frac{1}{3} + 13 \cdot \frac{1}{6}$$

$$= \frac{3}{2} + \frac{7}{3} + \frac{13}{6} = \frac{9}{6} + \frac{14}{6} + \frac{13}{6}$$

$$= \frac{36}{6} = 6.$$

Suppose the random vector (X, Y) is distributed with probability density,

$$f(x, y) = \begin{cases} x + y & \begin{array}{l} 0 < x < 1 \\ 0 < y < 1 \end{array} \\ 0 & \text{otherwise.} \end{cases}$$

Find $E[XY]$, $E[X + Y]$ and $E(X)$.

Solution: By definition,

$$E(g(x, y)) = \int \int_{(x, y)} g(x, y) f(x, y) \, dx \, dy$$

Thus, if $g(x,y) = xy$, we have: $E(xy) = \int_0^1 \int_0^1 xy \, (x + y) \, dx \, dy$

$$= \int_0^1 \left[\int_0^1 (x^2 y + xy^2) \, dx \right] dy = \int_0^1 \left[\frac{x^3 y}{3} + \frac{x^2 y^2}{2} \right]_0^1 dy$$

$$= \int_0^1 \left(\frac{y}{3} + \frac{y^2}{2} \right) dy = \left[\frac{y^2}{6} + \frac{y^3}{6} \right]_0^1 = \frac{2}{6} = \frac{1}{3} .$$

$$E(X + Y) = \int_0^1 \int_0^1 (x + y)(x + y) \, dx \, dy$$

$$= \int_0^1 \int_0^1 (x^2 + 2xy + y^2) \, dx \, dy$$

$$= \int_0^1 \left[\frac{x^3}{3} + \frac{2 \, x^2 y}{2} + y^2 x \right]_0^1 dy$$

$$= \int_0^1 \left[\frac{1}{3} + y + y^2 \right] dy$$

$$= \left[\frac{y}{3} + \frac{y^2}{2} + \frac{y^3}{3} \right]_0^1 = \frac{1}{3} + \frac{1}{2} + \frac{1}{3} = \frac{7}{6} .$$

$$E(X) = \int_0^1 \int_0^1 x(x + y) \, dx \, dy = \int_0^1 \left[\frac{x^3}{3} + \frac{x^2 y}{2} \right]_0^1 dy$$

$$= \int_0^1 \left[\frac{1}{3} + \frac{y}{2} \right] dy = \left[\frac{y}{3} + \frac{y^2}{4} \right]_0^1$$

$$= \frac{1}{3} + \frac{1}{4} = \frac{7}{12} \; .$$

• **PROBLEM** 11-57

A population consists of the measurements 2, 3, 3, 4, 4, 4, 5, 5, 5, 6, 6, 7. Compute: (a) μ, (b) σ^2.

Solution: Because the entire population is known, we may calculate μ and σ^2 directly. This is only possible when the entire population is known. If we have a sample from the entire population, we can only calculate estimates of μ and σ^2.

To find μ, we multiply each value in the population by its frequency of occurrence. Thus,

$$E(X) = \mu = 2\left(\frac{1}{12}\right) + 3\left(\frac{2}{12}\right) + 4\left(\frac{3}{12}\right) + 5\left(\frac{3}{12}\right) + 6\left(\frac{2}{12}\right) + 7\left(\frac{1}{12}\right)$$

$$\mu = \frac{2 + 6 + 12 + 15 + 12 + 7}{12} = \frac{54}{12} = 4.5 \; .$$

We also could have found μ by adding the population values and dividing by the number of values in the population.

By definition $E(X - \mu)^2 = \sum_i (X_i - \mu)^2 \Pr(X_i - \mu)$

$$= \sum_i (X_i - \mu)^2 \Pr(X_i) \text{ since } \mu \text{ is invariant}$$

$$= \sum_i (X_i - \mu)^2 \frac{1}{n} \text{ since all } X_i \text{ are equally}$$
likely to be chosen

$$= \frac{1}{n} \sum_i (X_i - \mu)^2, \text{ the average squared}$$
deviation from the mean μ.

To find σ^2, we calculate the average squared deviation from the mean μ. Thus,

$$\sigma^2 = \frac{\sum_{i=1}^{n} (X_i - \mu)^2}{n}$$

$$= \frac{(2 - 4.5)^2 + 2(3 - 4.5)^2 + 3(4 - 4.5)^2}{12}$$

$$+ \frac{3(5 - 4.5)^2 + 2(6 - 4.5)^2 + (7 - 4.5)^2}{12}$$

$$= \frac{(-2.5)^2 + 2(-1.5)^2 + 3(-.5)^2 + 3(.5)^2}{12}$$

728

$$+ \frac{2(1.5)^2 + (2.5)^2}{12}$$

$$= \frac{2(2.5)^2 + 4(1.5)^2 + 6(.5)^2}{12}$$

$$= \frac{12.5 + 4(2.25) + 1.5}{12} = \frac{12.5 + 10.5}{12} = \frac{23}{12} = 1.9 \ .$$

● **PROBLEM** 11–58

Given that the random variable X has density function

$$f(x) = \frac{1}{2} (x + 1) \text{ when } - 1 < x < 1$$

and $f(x) = 0$ \qquad elsewhere ;

calculate the mean value or expected value of X and the variance of X.

Solution: Recall that when X is a discrete random variable, its mean value

$$\mu = E(X) = \Sigma x \ f(x), \text{ where } f(x) = Pr(X = x).$$

This sum of products is a weighted average of the values of X.

In this problem X is a continuous random variable. Therefore we must integrate xf(x) from -1 to $+1$, since $f(x) = 0$ when $x \leq -1$ or $x \geq 1$.

$$\mu = \int_{-\infty}^{\infty} xf(x) \ dx = \int_{-1}^{1} x \ \frac{x + 1}{2} \ dx$$

(we have substituted $\frac{x + 1}{2}$ for $f(x)$)

$$= \int_{-1}^{1} \left(\frac{x^2}{2} + \frac{x}{2} \right) dx = \frac{1}{2} \int_{-1}^{1} x^2 \ dx + \frac{1}{2} \int_{-1}^{1} x dx$$

$$= \frac{1}{2} \left. \frac{x^3}{3} \right]_{-1}^{1} + \frac{1}{2} \left. \frac{x^2}{2} \right]_{-1}^{1}$$

$$= \frac{1}{2} \left(\frac{1}{3} + \frac{1}{3} \right) + \frac{1}{2} \left(\frac{1}{2} - \frac{1}{2} \right) = \frac{1}{2} \left(\frac{2}{3} \right) = \frac{1}{3} \ .$$

The variance of X is defined to be

$$\sigma^2 = E[(X - \mu)^2]. \qquad \text{Since}$$

$$(X - \mu)^2 = (X - \mu)(X - \mu) = X^2 - 2\mu X + \mu^2,$$

we can write,

$$\sigma^2 = E[(X - \mu)^2] = E(X^2 - 2\mu X + \mu^2)$$

$$= \int_{-1}^{1} (X^2 - 2\mu X + \mu^2) \, f(x) \, dx$$

$$= \int_{-1}^{1} X^2 \, f(x) \, dx - 2\mu \int_{-1}^{1} x f(x) \, dx + \mu^2 \int_{-1}^{1} 1 \, f(x) \, dx$$

$$= E(X^2) - 2\mu E(X) + \mu^2 E(1)$$

$$= E(X^2) - 2\mu^2 + \mu^2 \quad [E(1) = 1 \text{ because } f(x) \text{ is a density}$$

function and must satisfy the condition that $\int_{-\infty}^{\infty} f(x) \, dx = 1$].

$$E(X^2) - 2\mu^2 + \mu^2 = E(X^2) - \mu^2$$

$$= \int_{-\infty}^{\infty} X^2 \, f(x) \, dx - \left(\frac{1}{3}\right)^2$$

$$= \int_{-1}^{1} X^2 \left(\frac{x + 1}{2}\right) \, dx - \frac{1}{9}$$

$$= \int_{-1}^{1} \frac{x^3}{2} \, dx + \int_{-1}^{1} \frac{x^2}{2} \, dx - \frac{1}{9}$$

$$= \frac{x^4}{8} \Big|_{-1}^{1} + \frac{x^3}{6} \Big|_{-1}^{1} - \frac{1}{9}$$

$$= \left(\frac{1}{8} - \frac{1}{8}\right) + \frac{1}{6} - \left(\frac{-1}{6}\right) - \frac{1}{9} = \frac{1}{3} - \frac{1}{9} = \frac{2}{9} \; .$$

THE MOMENT GENERATING FUNCTION AND CHEBYSHEV'S INEQUALITY

• **PROBLEM** 11–59

Given that the probability density function of a discrete random variable X is

$$f(x) = \frac{6}{\pi^2 x^2} \quad , \quad x = 1, 2, \ldots$$

find its moment generating function $M(t)$.

Solution: The mathematical expectation of a function $u(X)$ of a discrete random variable X, with density function $f(x)$, is defined to be

$$E[u(X)] = \sum_{X}^{\infty} u(x) f(x).$$

(If X is continuous, $E(u(X)] = \int_{-\infty}^{\infty} u(x) f(x) dx$.) The moment generating function $M(t)$ is defined by $M(t) = E(e^{tX})$. This is obtained by setting $u(X) = e^{tX}$. $E[u(X)] = E(e^{tX}) = \sum_{X} e^{tX} f(x)$. Given that

$$f(x) = \frac{6}{\pi^2 x^2}, \quad \sum_{x=1}^{\infty} e^{tX} f(x) = \sum_{x=1}^{\infty} \frac{6e^{tx}}{\pi^2 x^2} = M(t) .$$

If this series does not converge, then $M(t)$ does not exist. By the ratio test,

$$\frac{\frac{6e^{t(X+1)}}{\pi^2 (X+1)^2}}{\frac{6e^{tX}}{\pi^2 x^2}} = \frac{6e^{t(X+1)}}{6e^{tX}} \frac{\pi^2 x^2}{\pi^2 (X+1)^2} = \frac{e^{t(x+1)}}{e^{tX}} \cdot \frac{x^2}{(X+1)^2} .$$

Observe that: $\quad \dfrac{e^{t(X+1)}}{e^{tX}} = e^{t(X+1)-tX} = e^{tX+t-tX} = e^{t} .$

Substitution yields: $\quad \dfrac{e^{t(X+1)}}{e^{tX}} \cdot \dfrac{x^2}{(X+1)^2} = \dfrac{e^{t} x^2}{(X+1)^2} .$

For any positive integer value of X, $X + 1 > X$. Thus $(X+1)^2 > X^2$ and $\dfrac{x^2}{(X+1)^2} < 1$. As X increases without bound, the ratio $\dfrac{X}{X+1}$ approaches 1 and $\dfrac{x^2}{(X+1)^2} = \left(\dfrac{X}{X+1}\right)^2$ also approaches 1. For $t = 0$, $e^{t} = e^{0} = 1$. For $t > 0$, $e^{t} > 1$. It follows that for large values of x, the ratio of the $(X+1)$st to the Xth term is greater than 1. Thus the series $\displaystyle\sum_{X=1}^{\infty} \frac{6e^{tX}}{\pi^2 x^2}$ diverges.

Therefore, $\quad f(x) = \dfrac{6}{\pi^2 x^2}$

does not have a moment generating function.

● **PROBLEM** 11–60

Given that the random variable X has moment generating function

$$M(t) = e^{t^2/2} ,$$

find $E(X^{2k})$ and $E(X^{2k-1})$.

Solution: We first represent M(t) as a MacLaurin's series. Recall that for all x,

$$e^x = \sum_{n=0}^{\infty} \frac{x^n}{n!} .$$

Thus,

$$e^{t^2/2} = \sum_{n=0}^{\infty} \frac{(t^2/2)^n}{n!} = \sum_{n=0}^{\infty} \frac{t^{2n}}{2^n n!}$$

$$= 1 + \frac{1}{2!} t^2 + \frac{t^4}{4 \cdot 2!} + \ldots + \frac{t^{2k}}{2^k \cdot k!} + \ldots$$

but

$$\frac{1}{4 \cdot 2!} = \frac{3 \cdot 1}{4 \cdot 3 \cdot 2 \cdot 1} = \frac{3 \cdot 1}{4!}$$

$$\frac{1}{8 \cdot 3!} = \frac{5 \cdot 3}{(3 \cdot 2) \cdot 5 \cdot 4 \cdot 3!} = \frac{5 \cdot 3}{6!}$$

$$\frac{1}{16 \cdot 4!} = \frac{7 \cdot 5 \cdot 3 \cdot 1}{8 \cdot 2 \cdot 7 \cdot 5 \cdot 3 \cdot 4!} = \frac{7 \cdot 5 \cdot 3 \cdot 1}{8 \cdot 7 \cdot 6 \cdot 5 \cdot 4!} = \frac{7 \cdot 5 \cdot 3 \cdot 1}{8!} .$$

There is a pattern here and it can be shown by induction that

$$\frac{1}{2^k \cdot k!} = \frac{(2k-1)(2k-3)\ldots 3 \cdot 1}{(2k)!} .$$

Thus,

$$M(t) = 1 + 0 \cdot t + \frac{3 \cdot 1}{4!} t^2 + 0 \cdot t^3 + \frac{5 \cdot 3 \cdot 1}{6!} t^4 + \ldots + 0 \cdot t^{(2k-1)}$$

$$+ \frac{(2k-1)(2k-3)\ldots 3 \cdot 1}{(2k)!} t^{2k} + \ldots$$

Thus,

$$E(X^{2k}) = \frac{d^{(2k)} M(0)}{dt^{(2k)}} = 0 + 0 + 0 + \ldots + \frac{[(2k-1)(2k-3)\ldots (3)(1)]}{(2k)!} (2k)!$$

because

$$\frac{d^{(2k)}[t^{2k}]}{dt^{(2k)}} = 2k!$$

Thus,

$$E(X^{2k}) = (2k-1)(2k-3)(2k-5)\ldots(3) \cdot (1)!$$

Also since every odd-powered term is 0, $E(X^{2k-1}) = 0$.

● **PROBLEM** 11-61

Let X be a continuous random variable with probability density function,

$$f(x) = \begin{cases} \lambda e^{-\lambda x} & \infty > x > 0 \\ 0 & \text{otherwise} . \end{cases}$$

Find the moment generating function of X.

Solution: $\qquad M_X(t) = E(e^{Xt}) = \int_0^{\infty} e^{xt} f(x) dx$

$$= \int_0^\infty \lambda e^{xt-\lambda x} \, dx$$

$$= \lambda \int_0^\infty e^{-x(\lambda-t)} \, dx$$

$$= \frac{-\lambda}{\lambda-t} \left[\lim_{x \to \infty} e^{-x(\lambda-t)} - 1 \right];$$

if $\lambda > t$ then $-x(\lambda-t) < 0$ and $\lim_{x \to \infty} e^{-x(\lambda-t)} = 0$. Thus,

$$M_X(t) = \frac{-\lambda}{\lambda-t} [0 - 1] = \frac{\lambda}{\lambda-t} \qquad \lambda > t \,.$$

• PROBLEM 11–62

Use Chebyshev's inequality to find a lower bound on $\Pr(-4 < X < 20)$ where the random variable X has a mean $\mu = 8$ and variance $\sigma^2 = 9$.

Solution: Chebyshev's inequality gives

$\Pr(\mu - k\sigma < X < \mu + k\sigma) \geq 1 - \frac{1}{k^2}$. We wish to find k.

Let $\mu - k\sigma = -4$ and $\mu + k\sigma = 20$

$\mu = 8$ and $\sigma = \sqrt{\sigma^2} = \sqrt{9} = 3$. Thus,   k satisfies either

$8 - 3k = -4$ or $8 + 3k = 20$ Hence $k = 4$.

Then $\Pr(\mu - k\sigma < X < \mu + k\sigma) \geq 1 - \frac{1}{k^2}$

$$\Rightarrow \Pr(-4 < X < 20) \geq 1 - \frac{1}{(4)^2} \quad = 1 - \frac{1}{(4)^2} = 1 - \frac{1}{16} = \frac{15}{16} \,.$$

Thus a lower bound on $\Pr(-4 < X < 20)$ is $\frac{15}{16}$.

• PROBLEM 11–63

Given that the discrete random variable X has density function $f(x)$ given by $f(-1) = \frac{1}{8}$, $f(0) = \frac{6}{8}$, $f(1) = \frac{1}{8}$, use Chebyshev's inequality,

$$\Pr(|X - \mu| \geq k\sigma) \leq \frac{1}{k^2} \,,$$

to find the upper bound when k = 2. What does this tell us about the possibility of improving the inequality to make the upper bound closer to the exact probability?

Solution: In order to use the inequality, we need to know the mean and variance of X.

$$\mu = E(x) = \sum_x x\, f(x) = (-1)\left(\frac{1}{8}\right) + \left[0 \cdot \frac{6}{8}\right] + \left(1 \cdot \frac{1}{8}\right)$$

$$= \frac{1}{8} - \frac{1}{8} = 0 \,. \quad \sigma^2 = E[(X - \mu)^2] = E[(X - 0)^2] = E(X^2)$$

$$= \sum_x x^2 f(x) = (-1)^2\left(\frac{1}{8}\right) + 0 + (1^2)\left(\frac{1}{8}\right) = \frac{1}{8} + \frac{1}{8} = \frac{1}{4} \,.$$

When $k = 2$, $\Pr(|X - \mu| \geq k\sigma) =$

$$\Pr\left(|X| \geq 2\sqrt{\frac{1}{4}}\right) = \Pr(|X| \geq 1) \leq \frac{1}{2^2} = \frac{1}{4} \,.$$

The exact $\Pr(|X| \geq 1) = \Pr(X \leq -1 \text{ or } X \geq 1)$.

$\Pr(X < 1) = 0 = \Pr(X > 1)$, because the sum of the probabilities for $x = -1$, $x = 0$, and $x = 1$ is

$\frac{1}{8} + \frac{6}{8} + \frac{1}{8} = \frac{8}{8} = 1$. Therefore we need to consider only

$$\Pr(X = -1) = \frac{1}{8} \quad \text{and} \quad \Pr(X = 1) = \frac{1}{8} \,.$$

Since $x = -1$ and $x = 1$ are mutually exclusive events, we can add their probabilities:

$$\frac{1}{8} + \frac{1}{8} = \frac{2}{8} = \frac{1}{4} \,. \quad \text{Therefore, the exact}$$

$\Pr(|X| \geq 1) = \frac{1}{4}$ equals the upper bound given by Chebyshev's inequality, so that we cannot improve the inequality for a random variable having finite variance σ^2.

● **PROBLEM** 11–64

Suppose that X assumes the values 1 and -1, each with probability .5. Find and compare the lower bound on $\Pr(-1 < X < 1)$ given by Chebyshev's inequality and the actual probability that $-1 < X < 1$.

Solution: Chebyshev's inequality gives

$\Pr(\mu - k\sigma < X < \mu + k\sigma) \geq 1 - \frac{1}{k^2}$. For this random variable,

$\mu = E(X) = \Pr(X = 1) \cdot (1) + \Pr(X = -1) \cdot (-1)$
$= .5 - .5 = 0 \quad$ and $\quad \text{Var } X = E(X^2) - [E(X)]^2 = E(X^2)$

$E(X^2) = \Pr(X = 1)(1)^2 + \Pr(X = -1)(-1)^2 = .5 + .5 = 1$.

Thus, $\text{Var } X = \sigma^2 = 1$ and $\sigma = 1$.

Now $\Pr(0 - k < X < 0 + k) \geq 1 - \frac{1}{k^2}$.

If $k = 1$, then $\Pr(-1 < X < 1) \geq 1 - \frac{1}{1} = 0$.

Thus $\Pr(-1 < X < 1) > 0$.

The actual probability that X is between 1 and -1 is found by

$$\Pr(-1 \leq X \leq 1) = \Pr(X = 1) + \Pr(-1 < X < 1) + \Pr(X$$
$$+\Pr(X = 1) .$$

But $\Pr(-1 \leq X \leq 1) = 1$ because -1 and 1 are the values that X assumes with positive probability. That is,

$$1 = .5 + \Pr(-1 < X < 1) + .5 \quad \text{or} \quad \Pr(-1 < X < 1) = 0 .$$

Thus, in this case the lower bound on this probability equals the true probability when the distribution of X is known.

● **PROBLEM 11–65**

Find a lower bound on $\Pr(-3 < X < 3)$ where
$\mu = E(X) = 0$ and $\text{Var } X = \sigma^2 = 1$.

Solution: From Chebyshev's inequality,

$$\Pr(\mu - k\sigma < X < \mu + k\sigma) \geq 1 - \frac{1}{k^2} .$$

We know that $\mu = 0$ and $\sigma = 1$. Thus $\mu - k\sigma = -3$ implies

$0 - k(1) = -3$ or $-k = -3$ $\qquad k = 3$.

Thus $\Pr(0 - 3 < X < 0 + 3) \geq 1 - \frac{1}{3^2}$

$\Pr(-3 < X < 3) > 1 - \frac{1}{9} = \frac{8}{9}$. The lower bound is thus $\frac{8}{9}$.

SPECIAL DISCRETE AND OTHER DISTRIBUTIONS

● **PROBLEM 11–66**

If X follows a discrete uniform distribution, i.e.
$F(x) = \frac{1}{N}$ for $x = 1, 2, \ldots, N$, find $E(x)$ and $\text{Var}(x)$.

Solution: By definition $E(x) = \sum\limits_{x} x\, F(x)$.

$$E(x) = \sum\limits_{x=1}^{N} x\, \frac{1}{N} = \frac{1}{N} \sum\limits_{x=1}^{N} x, \text{ but}$$

$\sum\limits_{x=1}^{N} x = \frac{N\,(N+1)}{2}$ algebraically. Therefore

$$E(x) = \frac{1}{N}\, \frac{N\,(N+1)}{2} = \frac{N+1}{2}.$$

For the variance, first find

$$E(x^2) = \sum\limits_{x} x^2\, F(x) = \sum\limits_{x=1}^{N} \frac{1}{N}\, x^2 = \frac{1}{N} \sum\limits_{x=1}^{N} x^2$$

But by a known algebraic formula,

$\sum\limits_{x=1}^{N} x^2 = \frac{N\,(2N+1)\,(N+1)}{6}$. Hence

$$E(x^2) = \frac{N\,(N+1)\,(2N+1)}{6N} = \frac{(N+1)\,(2N+1)}{6}. \text{ But}$$

$$Var(x) = E(x^2) - (E(x))^2$$

$$= \frac{(N+1)\,(2N+1)}{6} - \left(\frac{N+1}{2}\right)^2$$

$$= \frac{2N^2 + 3N + 1}{6} - \frac{N^2 + 2N + 1}{4}$$

$$= \frac{4N^2 + 6N + 2}{12} - \frac{3N^2 + 6N + 3}{12}$$

$$= \frac{N^2 - 1}{12}.$$

• **PROBLEM** 11–67

In order to attract customers, a grocery store has started a SAVE game. Any person who collects all four letters of the word SAVE gets a prize. A diligent Mrs. Y who has three letters S, A, and E keeps going to the store until she gets the fourth letter V. The probability that she gets the letter V on any visit is 0.002 and remains the same from visit to visit. Let X denote the number of times she visits the store until she gets the letter V for the first time. Find:

(a) the probability function of X

(b) the probability that she gets the letter V for the first time on the twentieth visit

(c) the probability that she will not have to visit more than three times.

Solution: The process consists of a number of failures before a success, the obtaining of a V. The distribution is therefore geometric and

$$F(x) = (1 - p)^{x - 1} p; \text{ for } x = 1, 2, \ldots.$$

In this case $p = .002$ and $F(x) = (1 - .002)^{x - 1} (.002)$

$$= (.002)(.998)^{x-1}.$$

(a) $F(x) = (.002)(.998)^{x-1}.$

(b) We want $F(20) = (.002)(.998)^{19} = .002 \times .963 = .0019.$

(c) $Pr(x \leq 3) = Pr(x = 1) + Pr(x = 2) + Pr(x = 3)$

$$= F(1) + F(2) + F(3)$$

$$= (.002)(.998)^{1-1} + (.002)(.998)^{2-1}$$

$$+ (.002)(.998)^{3-1}$$

$$= .002 + (.002)(.998) + (.002)(.998)^2$$

$$= .002 + .001996 + .001992$$

$$= .005988.$$

● **PROBLEM** 11-68

Given that the random variable X has a Poisson distribution with mean $\mu = 2$, find the variance σ^2 and compute $Pr(1 \leq x)$.

Solution: The density function for a random variable X with Poisson distribution is

$$F(x) = \frac{\lambda^x e^{-\lambda}}{x!} \text{ when } x = 0, 1, 2, \ldots,$$

$F(x) = 0$ when $x \neq 0, 1, 2, \ldots$ λ is a constant that is specified for the particular circumstances. We are given that the mean $\mu = 2$, but we are not given σ^2. But recall that a Poisson random variable has the unique property that the expectation equals the variance. Hence $\sigma^2 = 2$ also. Note that λ now must be 2. Also

737

$$Pr(1 \leq x) = 1 - Pr(x = 0)$$

$$= 1 - \frac{e^{-\lambda} \lambda^0}{0!} = 1 - \frac{e^{-2} \cdot 1}{1} = 1 - \frac{1}{e^2}$$

$$= 1 - .135 = .865.$$

● **PROBLEM** 11–69

Suppose X has a Poisson distribution with parameter λ.

(a) Show that $p(k + 1) = \frac{\lambda}{k + 1} p(k)$, where $p(k) = P(X = k)$

(b) If $\lambda = 2$, compute $p(0)$, and then use the recursive relation in (a) to compute $p(1)$, $p(2)$, $p(3)$, and $p(4)$.

Solution: (a) Note that $P(X = k) = p(k) = \dfrac{e^{-\lambda} \lambda^k}{k!}$

But $p(k + 1) = \dfrac{e^{-\lambda} \lambda^{k + 1}}{(k + 1)!}$. Factor out $\dfrac{\lambda}{k + 1}$;

$$p(k + 1) = \frac{e^{-\lambda} \lambda^k}{k!} \cdot \frac{\lambda}{k + 1} = \frac{\lambda}{k + 1}\ p(k)$$

(b) Since $\lambda = 2$, $p(k) = \dfrac{e^{-2} 2^k}{k!}$. In particular

$$p(0) = \frac{e^{-2} 2^0}{0!} = e^{-2} .$$

$$p(1) = \frac{\lambda}{k + 1}\ p(0) = \frac{2}{0 + 1}\ p(0) = 2e^{-2},$$

$$p(2) = \frac{\lambda}{1 + 1}\ p(1) = \frac{2}{2}\ p(1) = p(1) = 2\ e^{-2},$$

$$p(3) = \frac{\lambda}{2 + 1}\ p(2) = \frac{2}{3}\ p(2) = \frac{4}{3}\ e^{-2},\quad \text{and}$$

$$p(4) = \frac{\lambda}{3 + 1}\ p(3) = \frac{2}{4}\ p(3) = \frac{2}{3}\ e^{-2} .$$

● **PROBLEM** 11–70

Find $Pr(- .47 < Z < .94)$.

Solution:

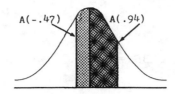

A(-.47)   A(.94)

        Pr(-.47 < Z < .94) is equal to the shaded
area above. To find the value of the shaded area we add
the areas labeled A(- .47) and A(.94).

    Pr(-.47 ≤ Z ≤ .94) = A(- .47) + A(.94).

    By the symmetry of the normal distribution,
A(- .47) = A(.47) = .18082 from the table.

Also   A(.94) = .32639 so

    Pr(-.47 < Z < .94) = .18082 + .32639 = .50721 .

• PROBLEM 11-71

Find Pr(-.47 < Z < .94) using Φ(- .47) and Φ(.94).

Solution: Φ(- .47) = .5000 - A(- .47)

                    = .5000 - A(.47)

and   Φ(.94) = .5000 + A(.94). Hence

    Pr(- .47 < Z < .94) = Φ(.94) - Φ(- .47)

      = [.5000 + A(.94)] - [.5000 - A(.47)]

      = .82639 - .31918 = .50721.

• PROBLEM 11-72

Let X be a normally distributed random variable representing
the hourly wage in a certain craft. The mean of the hourly
wage is $4.25 and the standard deviation is $.75.

    (a) What percentage of workers receive hourly wages
between $3.50 and $4.90?

    (b) What hourly wage represents the 95th percentile?

Solution: (a) We seek,

    Pr(3.50 ≤ X ≤ 4.90).

Converting to Z-scores we see that

$$\Pr(3.50 \leq X \leq 4.90) = \Pr\left(\frac{3.50 - \mu}{\sigma} \leq Z \leq \frac{4.90 - \mu}{\sigma}\right)$$

$$= \Pr\left(\frac{3.50 - 4.25}{.75} \leq Z \leq \frac{4.90 - 4.25}{.75}\right)$$

$$= \Pr\left(\frac{-.75}{.75} \leq Z \leq \frac{.65}{.75}\right) = \Pr(-1 \leq Z \leq .87)$$

$$= \Pr(Z \leq .87) - \Pr(Z \leq -1) = \Pr(Z \leq .87) - (1 - \Pr(Z \geq -1))$$

$$= \Pr(Z \leq .87) - [1 - \Pr(Z \leq 1)] = .809 - [1 - .841]$$
$$= .650 .$$

Thus 65% of the hourly wages are between \$3.50 and \$4.90.

(b) The 95th percentile is that number Z_α such that

$\Pr(X \leq K) = .95.$

To find Z_α, we first convert to Z-scores. Thus

$$\Pr(X \leq Z_\alpha) = \Pr\left(\frac{X - 4.25}{.75} \leq \frac{K - 4.25}{.75}\right) = .95$$

$$= \Pr\left(Z \leq \frac{K - 4.25}{.75}\right) = .95 .$$

But $\Pr(Z \leq 1.645) = .95$ thus $\frac{K - 4.25}{.75} = 1.645,$

$K = 4.25 + (.75)(1.645) = 5.48.$

Thus 95% of the craftsmen have hourly wages less than \$5.48.

• PROBLEM 11-73

The simplest continuous random variable is the one whose distribution is constant over some interval (a, b) and zero elsewhere. This is the uniform distribution.

$$f(x) = \begin{cases} \dfrac{1}{b - a}, & a \leq X \leq b \\ 0, & \text{elsewhere} \end{cases}$$

Find the mean and variance of this distribution.

Solution: By definition,

$$E(x) = \int_{-\infty}^{\infty} x\, f(x)\, dx = \int_{a}^{b} \frac{1}{b-a}\, x\, dx$$

$$= \frac{1}{b-a} \int_{a}^{b} x\, dx = \frac{1}{b-a}\, \frac{x^2}{2}\, \Big|_{a}^{b}$$

$$= \frac{b^2 - a^2}{2}\left(\frac{1}{b-a}\right) = \frac{a+b}{2} \ .$$

For the variance we must first find $E(X^2)$. By definition

$$E(X^2) = \int_{-\infty}^{\infty} x^2\, f(x)\, dx$$

$$= \int_{a}^{b} x^2\, \frac{1}{b-a}\, dx$$

$$= \frac{1}{b-a} \int_{a}^{b} x^2\, dx$$

$$= \frac{1}{b-a}\, \frac{x^3}{3}\, \Big|_{a}^{b}$$

$$= \frac{b^3 - a^3}{3(b-a)} \ .$$

But $\text{Var}(X) = E(X^2) - (E(x))^2$

$$= \frac{b^3 - a^3}{3(b-a)} - \left(\frac{b+a}{2}\right)^2$$

$$= \frac{b^3 - a^3}{3(b-a)} - \frac{(a^2 + 2ab + b^2)}{4}$$

$$= \frac{(b^2 + ab + a^2)(b-a)}{3(b-a)} - \frac{(a^2 + 2ab + b^2)}{4}$$

$$= \frac{(b^2 + ab + a^2)}{3} - \frac{(a^2 + 2ab + b^2)}{4}$$

$$= \frac{(4b^2 + 4ab + 4a^2)}{12} - \frac{(3a^2 + 6ab + 3b^2)}{12}$$

$$= \frac{b^2 - 2ab + a^2}{12} = \frac{(b-a)^2}{12} \ .$$

Consider the exponential distribution $f(x) = \lambda e^{-\lambda x}$ for x > 0. Find the moment generating function and from it, the mean and variance of the exponential distribution.

<u>Solution:</u>   By definition $M_x(t) = E(e^{tx})$

$$= \int_{-\infty}^{\infty} e^{tx} f(x) \, dx$$

$$= \int_{x=0}^{\infty} e^{tx} \lambda e^{-\lambda x} \, dx$$

$$= \int_{0}^{\infty} \lambda e^{(t - \lambda)x} \, dx = \lambda \int_{0}^{\infty} e^{(t - \lambda)x} \, dx$$

$$= \lambda \left[\frac{-1}{t - \lambda} e^{(t - \lambda)x} \right]_{0}^{\infty}$$

$$= \frac{\lambda}{\lambda - t} \left[e^{(t - \lambda)x} \right]_{0}^{\infty}.$$

Consider $t < \lambda$. Then $\lambda - t > 0$ and $t - \lambda < 0$. Hence $e^{(t - \lambda)x} = e^{-kx}$ and $M_x(t) = \frac{\lambda}{\lambda - t} (0 - (-1)) = \frac{\lambda}{\lambda - t}$

for $t < \lambda$.

The mean is

$$E(x) = M_x'(t) \Big|_{t=0}$$

$$M_x'(t) = \frac{d}{dt} \left[\frac{\lambda}{\lambda - t} \right] = \lambda \left[\frac{d}{dt} \frac{1}{\lambda - t} \right]$$

$$= \lambda \left[\frac{-1}{(\lambda - t)^2} \right] \frac{d}{dt} (\lambda - t) = \frac{\lambda}{(\lambda - t)^2}$$

$$M_x'(0) = E(x) = \frac{\lambda}{(\lambda - 0)^2} = \frac{\lambda}{\lambda^2} = \frac{1}{\lambda} .$$

Also by the moment generating function's properties

$$E(x^2) = M_x''(t) \Big|_{t=0}$$

$$M_x''(t) = \frac{d}{dt} \frac{\lambda}{(\lambda - t)^2} = \lambda \frac{d}{dt} \frac{1}{(\lambda - t)^2} .$$

$$= \lambda \; \frac{-2}{(\lambda - t)^3} \; \frac{d}{dt} \; (\lambda - t) = \frac{2\lambda}{(\lambda - t)^3}$$

$$M_x''(0) = E(x^2) = \frac{2\lambda}{\lambda^3} = \frac{2}{\lambda^2} \; ,$$

Now $\quad Var(X) = E(x^2) - (E(x))^2$

$$= \frac{2}{\lambda^2} - \left(\frac{1}{\lambda}\right)^2$$

$$= \frac{1}{\lambda^2} \; .$$

• PROBLEM 11–75

Engineers determine that the lifespans of electric light bulbs manufactured by their company have the exponential distribution $f(x) = \frac{1}{1000} e^{-x/1000}$ when $x \geq 0$ and $f(x) = 0$, $x < 0$.

Compute the probability that a randomly selected light bulb has a lifespan of less than 1,000 hours. Graph the density function.

<u>Solution:</u>   The cumulative distribution function $F(a) = P(X < a)$ is used to compute $P(X < 1000)$. We find the area under the graph of $f(x)$ from 0 to a (the area from $-\infty$ to 0 is 0 because $f(x) = 0$ when $x < 0$).

$$P(X \leq a) = F(a) = \int_0^a f(x) \; dx = \int_0^a \frac{1}{1000} e^{-x/1000} \; dx$$

$$= \int_0^a \frac{1}{1000} e^{-x/1000} dx = \frac{1}{1000} \left[-1000 \; e^{-x/1000} \right]_0^a$$

$$= -e^{-x/1000} \Big]_0^a = - \left[e^{-a/1000} - e^0 \right]$$

$$= 1 - e^{-a/1000} = F(a)$$

$$P(X \leq 1000) = F(1000) = 1 - e^{-1000/1000} = 1 - e^{-1}$$

$$= 1 - \frac{1}{e} \; .$$

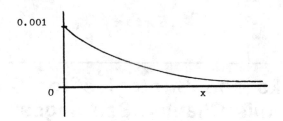

Graph of $f(x) = 0.001\ e^{-0.001\ x}$

When $x = 0$, $f(x) = 0.001$. As $x \to \infty$, $f(x) \to 0$.

CHAPTER 12

STATISTICS

> **Basic Attacks and Strategies for Solving Problems in this Chapter. See pages 746 to 808 for step-by-step solutions to problems.**

Probability theory directly leads to the discussion of statistical inference. By using the concept of a sampling distribution of a statistic, a simple way can be found to calculate the probability of selecting a sample from a population.

A point estimate of a parameter is not very meaningful without some measure of the possible error in the estimate confidence intervals and regions provide good illustrations of uncertain inferences.

In many problems, the population parameters are unknown and the main concern is to estimate them. A statistician may select a sample, find the sample average, and use that as an estimator of the average income for the whole community.

The second major area of statistical inference, besides the estimation of parameters, is the testing of hypothesis. General methods are developed for testing hypotheses and to apply those methods to some common problems.

One of the most frequently used techniques in economics and business research, to find a relation between two or more variables that are related causally, is regression analysis. This analysis is applied in several linear and non-linear models to understand the behavior of firms and industry as a whole.

Step-by-Step Solutions to
Problems in this Chapter,
"Statistics"

SAMPLING THEORY

A population consists of the number of defective transistors in shipments received by an assembly plant. The number of defectives is 2 in the first, 4 in the second, 6 in the third, and 8 in the fourth.

(a) Find the mean $\bar{x}$ and the standard deviation s'_x of the given population.

(b) List all random samples, with replacement, of size 2 that can be formed from the population and find the distribution of the sample mean.

(c) Find the mean and the standard deviation of the sample mean.

Solution: The population is 2,4,6,8.

a) $\quad \bar{x} = \dfrac{\sum\limits_{i=1}^{n} x_i}{n} = \dfrac{\sum\limits_{i=1}^{4} x_i}{4} = \dfrac{2 + 4 + 6 + 8}{4} = \dfrac{20}{4} = 5$.

We will compute

$$s'_x = \sqrt{\dfrac{\sum\limits_{i=1}^{n} (x - \bar{x})^2}{n}} = \sqrt{\dfrac{(2-5)^2 + (4-5)^2 + (6-5)^2 + (8-5)^2}{4}}$$

$$= \sqrt{\dfrac{9 + 1 + 1 + 9}{4}} = \sqrt{\dfrac{20}{4}} = \sqrt{5}$$.

b) The following table should prove useful.

Sample		Sample Mean
1.	2,2	2
2.	2,4	3
3.	2,6	4

4.	2,8	5
5.	4,2	3
6.	4,4	4
7.	4,6	5
8.	4,8	6
9.	6,2	4
10.	6,4	5
11.	6,6	6
12.	6,8	7
13.	8,2	5
14.	8,4	6
15.	8,6	7
16.	8,8	8

Collating the data we have

x = Sample Mean	N(x) = Number of times x occurs	F(x) = N(x)/n = N(x)/16
2	1	1/16
3	2	1/8
4	3	3/16
5	4	1/4
6	3	3/16
7	2	1/8
8	1	1/16

c) $\text{Sample Mean} = \dfrac{\sum\limits_{i=1}^{n} i^{th}\ \text{Sample Mean}}{n}$

$= \dfrac{2+3+4+5+3+4+5+6+4+5+6+7+5+6+7+8}{16}$

$= \dfrac{80}{16} = 5.$

$s'_{\bar{x}} = \sqrt{\dfrac{\sum(sm - \overline{sm})^2}{n}}$

$= \sqrt{\dfrac{(2-5)^2+2(3-5)^2+3(4-5)^2+4(5-5)^2+3(6-5)^2+2(7-5)^2+(8-5)^2}{16}}$

$= \sqrt{\dfrac{9+8+3+0+3+8+9}{16}} = \sqrt{\dfrac{40}{16}} = \sqrt{\dfrac{5}{2}}.$

● **PROBLEM** 12–2

A population of Australian Koala bears has a mean height of 20 inches and a standard deviation of 4 inches. You plan to choose a a sample of 64 bears at random. What is the probability of a sample mean between 20 and 21?

<u>Solution</u>: Our method of attack will be to transform 20 and 21 into standard normal statistics. The sample is large enough, 64, so that the Central Limit Theorem will apply and $\sqrt{n}(\bar{X} - \mu)/\sigma$ will approximate a standard normal statistic.

We want to know

$$Pr(20 < \bar{X} < 21).$$

Equivalently we want to know

$$Pr\left(\frac{(20 - \mu)\sqrt{n}}{\sigma} < \frac{\sqrt{n}(\bar{X} - \mu)}{\sigma} < \sqrt{n}(21 - \mu)\right)$$

or

$$Pr\left(\frac{(\sqrt{n}(20 - \mu)}{\sigma} < Z < \frac{\sqrt{n}(21 - \mu)}{\sigma}\right).$$

Substituting the values $n = 64$, $\mu = 20$, and $\sigma = 4$, we obtain

$$Pr\left(\frac{\sqrt{64}(20 - 20)}{4} < Z < \frac{\sqrt{64}(21 - 20)}{4}\right)$$

$$= Pr(0 < Z < 2).$$

From the standard normal table, this is .4772.

• PROBLEM 12-3

Random samples of size 100 are drawn, with replacement, from two populations, P_1 and P_2, and their means, $\bar{X}_1$ and $\bar{X}_2$, computed. If

$\mu_1 = 10$, $\sigma_1 = 2$, $\mu_2 = 8$, and $\sigma_2 = 1$, find

(a) $E(\bar{X}_1 - \bar{X}_2)$;

(b) $\sigma_{(\bar{X}_1 - \bar{X}_2)}$;

(c) the probability that the difference between a given pair of sample means is less than 1.5;

(d) the probability that the difference between a given pair of sample means is greater than 1.75 but less than 2.5.

<u>Solution</u>: a) By the linearity properties of the expectation operator

$$E(\bar{X}_1 - \bar{X}_2) = E(\bar{X}_1) - E(\bar{X}_2) = \mu_1 - \mu_2 = 10 - 8 = 2.$$

b) $\bar{X}_1$ and $\bar{X}_2$ are independent. In light of this we can say

$$Var(a\bar{X}_1 + b\bar{X}_2) = a^2 Var(\bar{X}_1) + b^2 Var(\bar{X}_2).$$

In our case $a = 1$ and $b = -1$. $Var(\bar{X}_1 - \bar{X}_2) = 1^2 Var(\bar{X}_1) + (-1)^2 Var(\bar{X}_2)$

$$= Var(\bar{X}_1) + Var(\bar{X}_2) = \sigma_1^2/n_1 + \sigma_2^2/n_2. \quad \text{Finally}$$

$$\sigma_{(\bar{X}_1 - \bar{X}_2)} = \sqrt{Var(\bar{X}_1 - \bar{X}_2)} = \sqrt{\sigma_1^2/n_1 + \sigma_2^2/n_2} =$$

$$= \sqrt{\frac{2^2}{100} + \frac{1^2}{100}} = \sqrt{\frac{5}{100}} = \frac{\sqrt{5}}{10}.$$

c) We want $\Pr(|\bar{X}_1 - \bar{X}_2| < 1.5)$ or equivalently

$$\Pr(-1.5 < \bar{X}_1 - \bar{X}_2 < 1.5) \ .$$

Subtract $E(\bar{X}_1 - \bar{X}_2) = 2$ and divide by $\sigma_{(\bar{X}_1-\bar{X}_2)} = \frac{\sqrt{5}}{10}$ to obtain

$$\Pr\left(\frac{-1.5-2}{\sqrt{5}/10} < \frac{(\bar{X}_1 - \bar{X}_2)-E(\bar{X}_1 - \bar{X}_2)}{\sigma_{\bar{X}_1-\bar{X}_2}} < \frac{1.5-2}{\sqrt{5}/10}\right)$$

$$= \Pr\left(-15.652 < \frac{(\bar{X}_1 - \bar{X}_2) -E(\bar{X}_1 - \bar{X}_2)}{\sigma_{\bar{X}_1-\bar{X}_2}} < -2.236\right) \ .$$

The Central Limit Theorem tells us that since we have large samples

$$\frac{(\bar{X}_1 - \bar{X}_2) - E(\bar{X}_1 - \bar{X}_2)}{\sigma_{\bar{X}_1-\bar{X}_2}}$$ is approximately standard normal. We then have

$\Pr(-15.652 < \text{Standard Normal} < -2.236) \cong \Pr(\text{Standard Normal} < -2.236)$, since the area under the standard normal curve to the left of -15.652 is negligible.

From the standard normal tables
$$\Pr(Z \le -2.236) \cong .0127.$$

d) To solve this, we will follow exactly the method of part c). We want $\Pr(-1.75 < \bar{X} - \bar{X}_2 < 2.5)$. As in c), this is equivalent to

$$\Pr\left(\frac{1.75-2}{5/10} < \frac{(\bar{X}_1 - \bar{X}_2)- E(\bar{X}_1 - \bar{X}_2)}{\sigma_{\bar{X}_1 - \bar{X}_2}} < \frac{2.5-2}{5/10}\right)$$

$$= \Pr(-1.118 < \text{Standard Normal} < 2.236)$$

$$= \Pr(-1.12 < z < 0) + \Pr(0 < z < 2.24)$$

(by the symmetry of the standard normal curve)

$$= .3686 + .4875 = .8561.$$

• PROBLEM 12-4

The chi-square density function is the special case of a gamma density with parameters $\alpha = K/2$ and $\lambda = \frac{1}{2}$. Find the mean, variance and moment-generating function of a chi-square random variable.

Solution: Recall the gamma distribution

$$f(x) = \frac{\lambda}{\Gamma(\alpha)} (\lambda x)^{\alpha-1} e^{-\lambda x} \ , \quad x \ge 0 \ .$$

With $\alpha = K/2$ and $\lambda = \frac{1}{2}$,

$$f(x) = \frac{1}{2\Gamma(K/2)} \left(\frac{x}{2}\right)^{K/2-1} e^{-X/2} = \frac{1}{\Gamma(K/2)}\left(\frac{1}{2}\right)^{K/2} X^{K-2/2} e^{-X/2},$$

for $X \geq 0$. Earlier for the gamma distribution we found

$$M_x(t) = \left(\frac{\lambda}{\lambda-t}\right)^{\alpha} \text{ for } t < \lambda, \ E(x) = \frac{\alpha}{\lambda}, \text{ and } Var(X) = \frac{\alpha}{\lambda^2}.$$

Making the substitution $\alpha = K/2$ and $\lambda = \frac{1}{2}$ into the formulae for the gamma distribution, we see that the chi-square distribution has

$$M_x(t) = \left[\frac{\frac{1}{2}}{\frac{1}{2}-t}\right]^{K/2} = \left[\frac{1}{1-2t}\right]^{K/2}, \ t < \frac{1}{2},$$

and

$$E(x) = \frac{K/2}{\frac{1}{2}} = \frac{K}{2} \cdot \frac{2}{1} = K,$$

$$Var(X) = \frac{K/2}{(\frac{1}{2})^2} = \frac{K}{2} \frac{4}{1} = 2K.$$

• PROBLEM 12-5

A manufacturer of kitchen clocks claims that a certain model will last 5 years on the average with a standard deviation of 1.2 years. A random sample of six of the clocks lasted 6, 5.5, 4, 5.2, 5, and 4.3 years. Compute

$$\frac{(n-1)S^2}{\sigma^2}$$

and use the chi-square tables to find the probability of a X^2 value this high.

Solution: First we must calculate the sample mean.

$$\bar{X} = \frac{\sum\limits_{i=1}^{6} X_i}{n} = \frac{6+5.5+4+5.2+5+4.3}{6} = \frac{30}{6} = 5.$$

$$S^2 = \frac{1}{n-1} \sum\limits_{i=1}^{6} (X_i - \bar{X})^2 = \frac{1}{6-1}((6-5)^2 + (5.5-5)^2 + (4-5)^2 + (5.2-5)^2$$

$$+ (4.3-5)^2)$$

$$= \frac{1}{5}(1^2 + (.5)^2 + (-1)^2 + (.2)^2 + 0^2 + (-.7)^2)$$

$$= \frac{1}{5}(1 + .25 + 1 + .04 + .49) = \frac{1}{5} \cdot 2.78$$

$$= .556$$

The standard deviation, σ, is 1.2. Hence $\sigma^2 = (1.2)^2 = 1.44$.

Finally

$$U = \frac{(n-1)S^2}{\sigma^2} = \frac{(6-1)(.556)}{1.44} = 1.931.$$

In this case U has a chi-square distribution with $n-1 = 5$ degrees of freedom. On a chi-square table, look down the left side for the row with 5 degrees of freedom. The probabilities on the top row are $Pr(X^2_{(n-1)} \geq U)$. Looking across the rwo for 5 d.o.f., we see that

the value for Pr = .80 is 2.343 and that for Pr = .90 is 1.610.
To find $Pr(\chi^2(5) \geq 1.931)$, we use linear interpolation,

$$\frac{1.931 - 1.610}{2.343 - 1.610} = \frac{X - .90}{.80 - .90}$$

Thus, $X = .90 + (.80 - .90)\left(\frac{1.931 - 1.610}{2.343 - 1.610}\right)$

$$= .90 - .1\left(\frac{.321}{.743}\right) = .90 - .1(.432)$$

$$= .857.$$

Hence $Pr(\chi^2_{(5)} \geq 1.931) = .857.$

• PROBLEM 12-6

Find the probability that a t-distribution has a t score

(a) greater than 1.740 when d.o.f. = 17

(b) less than - 1.323 when d.o.f. = 21.

Solution: This problem involves reading the t-table.

a) This part involves a t-distribution with 17 degrees of freedom. We want to know $Pr(t(17) > 1.74)$. We look at the row on the t-table which has 17 degrees of freedom. The number 1.74 appears under the column headed by .95. These column headings are the probabilities that $t(n) \leq$ number in the column below. Hence $Pr(t(17) < 1.74) = 0.95$. Finally $Pr(t(17) > 1.74) = 1 - Pr(t(17) < 1.74) = 1 - 0.95 = .05.$

b) We have 21 degrees of freedom. In the row marked 21, 1.323 is under the column headed .90. Hence $Pr(t(21) < 1.323) = .90$. By the symmetry of the t-distribution,

$$Pr(t(21) > -1.323) = Pr(t(21) < 1.323) = .90 .$$

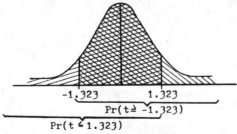

Also note that $Pr(t(21) < -1.323) = 1 - Pr(t(21) > 1.323) = 1 - .90 = .10.$

• PROBLEM 12-7

Find $t_{.975}$ (5).

Solution:

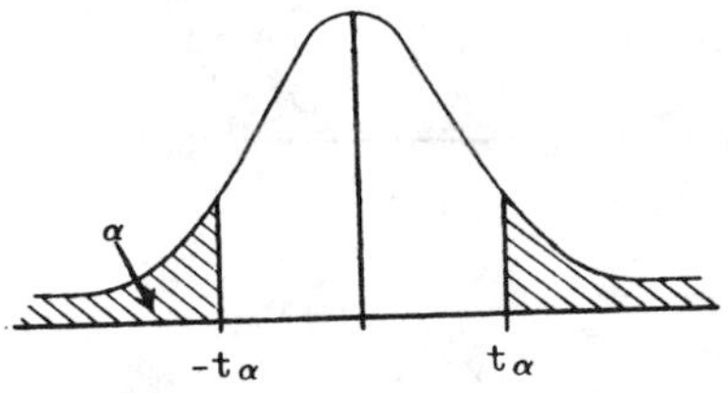

t_α is the number such that $\Pr(t(n) > t_\alpha) = \alpha$. Let $\alpha = .975$. By the symmetry of the t-distribution $t_\alpha = -t_{1-\alpha}$ and $t_{.975} = -t_{.025}$. Using the t-table we see that $t_{.025}(5) = 2.57$ and $t_{.975} = -2.57$.

• PROBLEM 12-8

Find the probability that X is greater than 3.28 if X has an F distribution with 12 and 8 degrees of freedom.

Solution: We look on the following cumulative F-table. Look across the top for 12, the degrees of freedom of the numerator. Now we look along the side for 8, the degrees of freedom of the denomonator. Find the block that is in row 8 and column 12. In it the value 3.28 corresponds to the G-value .95 in the extreme left hand column. This signifies $G(3.28) = \Pr(X < 3.28) = .95$. Hence

$$\Pr(X > 3.28) = 1 - \Pr(X < 3.28) = 1 - .95 = .05.$$

• PROBLEM 12-9

Find the tenth quantile point of an F distribution with 15 and seven degrees of freedom.

Solution: We know that if the pth quantile ξ_p is given for an F-distribution with m and n degrees of freedom, then the quantile ξ'_{1-p} for an F-distribution with n and m degrees of freedom is given by $1/\xi_p$.

Following this, we see

$$\xi_{.10}(15,7) = \frac{1}{\xi_{.90}(7,15)} = \frac{1}{2.16} = .463 .$$

• PROBLEM 12-10

Let X_1, X_2 be a random independent sample from N(0,1).

a) What is the distribution of

$$\frac{X_2 - X_1}{\sqrt{2}} ?$$

b)   What is the distribution of

$$\frac{(X_1 + X_2)^2}{(X_2 - X_1)^2} \quad ?$$

c)   What is the distribution of

$$\frac{(X_2 + X_1)}{\sqrt{(X_2 - X_1)^2}} \quad ?$$

d)   What is the distribution of $1/Z$ if $Z = X_1^2/X_2^2$?

Solution: First we will establish the distributions of $X_1 + X_2$ and $X_2 - X_1$. $X_1 + X_2$ is the sum of two normal random variables. Hence it is normally distributed with $E(X_1+X_2) = E(X_1) + E(X_2) = 0 + 0 = 0$, and $Var(X_1+X_2) = Var(X_1) + Var(X_2) = 1 + 1 = 2$, since X_1, X_2, are independent.

Similarly, $X_2 - X_1$ will also be normally distributed with

$$E(X_2 - X_1) = E(X_2) - E(X_1) = 0 - 0 = 0$$

and

$$Var(X_2 - X_1) = Var(X_2) + Var(X_1) = 1 + 1 = 2.$$

a)   We have now $1/\sqrt{2} \; N(0,2)$. A constant times a normal distribution will still be normally distributed. $E(ax) = a \, E(x)$. Therefore the new mean will be $1/\sqrt{2} \cdot 0 = 0$. $Var(ax) = a^2 Var(X)$; hence the new variance will be

$$(1/\sqrt{2})^2(2) = 1.$$

The new distribution is $N(0,1)$.

b)   We start with

$$\frac{(X_1 + X_2)^2}{(X_2 - X_1)^2} \; .$$

Multiply numerator and denominator by $\frac{1}{2}$ to obtain

$$\frac{\frac{1}{2}(X_1 + X_2)^2}{\frac{1}{2}(X_2 - X_1)^2} \; = \; \frac{\left(\frac{X_1 + X_2}{\sqrt{2}}\right)^2}{\left(\frac{X_2 - X_1}{\sqrt{2}}\right)^2}$$

From part a) we know the denomonator is $[N(0,1)]^2$. For similar reasons the numerator is also $[N(0,1)]^2$. But we know that a $[N(0,1)]^2$ is $\chi^2(1)$. We now have

$$\frac{\chi^2(1)}{\chi^2(1)} = \frac{\chi^2(1)/1}{\chi^2(1)/1} \; .$$

By definition this is an $F(1,1)$ random variable.

c)   We start with

$$\frac{(X_1 + X_2)}{\sqrt{(X_2 - X_1)^2}} \; .$$

Multiply numerator and denomonator by $1/\sqrt{2}$ to obtain

753

$$\frac{\frac{1}{\sqrt{2}}(X_1 + X_2)}{\frac{1}{\sqrt{2}}\sqrt{(X_2 - X_1)^2}} = \frac{\left(\dfrac{X_1 + X_2}{\sqrt{2}}\right)}{\sqrt{\left(\dfrac{X_2 - X_1}{\sqrt{2}}\right)^2}} .$$

From previous experience $X_1 + X_2/\sqrt{2}$ and $X_2 - X_1/\sqrt{2}$ are both $N(0,1)$. We now have

$$\frac{N(0,1)}{\sqrt{[N(0,1)]^2}} = \frac{N(0,1)}{\sqrt{\chi^2(1)}} , \text{ since } [N(0,1)]^2 \text{ is } \chi^2(1) ,$$

$$= \frac{N(0,1)}{\sqrt{\dfrac{\chi^2(1)}{1}}} = t(1) \text{ by definition.}$$

d) If $Z = X_1^2/X_2^2$, then $1/Z = X_2^2/X_1^2$. X_1^2 and X_2^2 are both $[N(0,1)]^2$ and thereby $\chi^2(1)$, $1/Z$ is therefore

$$\frac{\chi^2(1)}{\chi^2(1)} = \frac{\chi^2(1)/1}{\chi^2(1)/1}$$

which is $F(1,1)$ by definition.

• PROBLEM 12–11

> For the density $f(x) = 2x; \; 0 \le x \le 2$, find the cumulative distribution for the twelfth order statistic in a sample of 13.

Solution: We know that

$$F_{y_\alpha}(y) = \sum_{j=\alpha}^{n} \binom{n}{j} [F(y)]^j [1 - F(y)]^{n-j} .$$

In our problem $\alpha = 12$, $n = 13$, and

$$F(y) = \int_{-\infty}^{y} f(x)dx = \int_{0}^{y} 2x \, dx; \; 0 \le y \le 2$$

$$= y^2 ; \; 0 \le y \le 2 .$$

$$F_{y_{12}}(y) = \sum_{j=12}^{13} \binom{13}{j}\left(y^2\right)^j (1 - y^2)^{n-j}$$

$$= \binom{13}{12}\left(y^2\right)^{12} (1 - y^2)^{13-12} + \binom{13}{13}\left(y^2\right)^{13}(1-y)^{13-13}$$

$$= 13 \, y^{24}(1 - y^2)^1 + y^{26} = 13 \, y^{24} - 13y^{26} + y^{26}$$

$$= 13 \, y^{24} - 12 \, y^{26} .$$

CONFIDENCE INTERVALS

● **PROBLEM** 12–12

Find a 95 per cent confidence interval for μ, the true mean of a normal population which has variance $\sigma^2 = 100$. Consider a sample of size 25 with a mean of 67.53.

Solution: We have a sample mean $\overline{X} = 67.53$. We want to transform that into a standard normal quantity, for we know from the standard normal tables that

Pr (– 1.96 < Standard Normal Quantity < 1.96) = .95 .

$$\frac{\overline{X} - E(\overline{X})}{\sqrt{Var\ (\overline{X})}}$$ is a standard normal quantity.

Recall now that the expectation of a sample mean is μ, the true mean of a population. Also recall that the variance of a sample mean is $\frac{\sigma^2}{n}$ where σ^2 is the true variance of the population and n is the size of our sample. Applying this to our case, $E(\overline{X}) = \mu$ and $\sqrt{Var(\overline{X})} = \sqrt{\frac{\sigma^2}{n}} = \sqrt{\frac{100}{25}} = 2$.

For our sample: $Pr\left(- 1.96 < \frac{\overline{X} - \mu}{2} < 1.96\right) = .95$.

Multiplying by 2: $Pr(- 3.92 < \overline{X} - \mu < 3.92) = .95$.

Transposing: $Pr(\overline{X} - 3.92 < \mu < \overline{X} + 3.92) = .95$.

$\overline{X} - 3.92 < \mu < \overline{X} + 3.92$ is our required confidence interval. If we insert our given sample mean, we come up with $67.53 - 3.92 < \mu < 67.53 + 3.92$

or $63.61 < \mu < 71.45$.

● **PROBLEM** 12–13

A group of experts feel that polishing would have a positive effect on the average endurance limit of steel. Eight specimens of polished steel were measured and their average, $\overline{X}$, was computed to be 86,375 psi. Similarly, 8 specimens of unpolished steel were measured and their average, $\overline{Y}$, was computed to be 79,838 psi. Assume all measurements to be normally distributed with standard deviation 4,000 psi (both samples). Find a 90% lower one-sided confidence limit for $\mu_X - \mu_Y$, the difference in means.

<u>Solution</u>:   Our plan of attack will be a common one. We
will try to construct a standard normal pivotal quantity
involving $\mu_X - \mu_Y$ and use the fact, obtainable from the
standard normal table, that

Pr (Standard Normal Quantity < 1.28) = .90.

The sample mean, $\overline{X}$, must be normally distributed, since
each measurement is. Also $\overline{X}$ is distributed with $E(\overline{X}) = \mu_X$

and $\sqrt{\text{Var } \overline{X}} = \dfrac{\sigma_x}{\sqrt{n_x}}$. (μ_x, σ_x, and n_x are the mean of popula-

tion X, its standard deviation, and the size of the sample
we draw from X.) A similar discussion applies to population

Y. $E(\overline{Y}) = \mu_y$; $\sqrt{\text{Var } \overline{Y}} = \dfrac{\sigma_y}{\sqrt{n_y}}$.

Consider the random variable $\overline{X} - \overline{Y}$.

$E(\overline{X} - \overline{Y}) = E(\overline{X} + (-\overline{Y}) = E(\overline{X}) + E(-\overline{Y}) = E(\overline{X}) - E(\overline{Y})$.

This conclusion follows from the linearity properties
of expectation (i.e. $E(X + Y) = E(X) + E(Y)$ and $E(aX) =$
$aE(x)$ when a is a constant). Also note that $E(\overline{X} - \overline{y}) =$
$E(\overline{X}) - E(\overline{Y}) = \mu_x - \mu_y$. Furthermore,

$\sigma_{(\overline{X}-\overline{Y})} = \sigma \sqrt{\dfrac{1}{n_x} + \dfrac{1}{n_y}}$. This results from the rule

$\text{Var}(ax + by) = a^2 \text{Var}(x) + b^2 \text{Var}(y)$. In our case a = 1 and b =
- 1 therefore

$\sigma_{(\overline{X}-\overline{Y})} = \sqrt{\text{Var}(\overline{X} - \overline{Y})} = \sqrt{\text{Var}(x) + \text{Var}(\overline{y})}$

$= \sqrt{\dfrac{\sigma^2}{n_x} + \dfrac{\sigma^2}{n_y}} = \sigma \sqrt{\dfrac{1}{n_x} + \dfrac{1}{n_y}}$.

$\dfrac{(\overline{X} - \overline{Y}) - E(\overline{X} - \overline{Y})}{\sigma_{(\overline{X}-\overline{Y})}}$ must be standard normal since

$\overline{X} - \overline{Y}$, the difference of 2 normal distributions, is normal.

$\dfrac{(\overline{X} - \overline{Y}) - E(\overline{X} - \overline{Y})}{\sigma_{(\overline{X}-\overline{Y})}} = \dfrac{(\overline{X} - \overline{Y}) - (\mu_x - \mu_y)}{\sigma \left(\sqrt{\dfrac{1}{n_x} + \dfrac{1}{n_y}} \right)}$

by inserting our derived results.

We obtain:
$$\Pr\left[\frac{(\overline{X} - \overline{Y}) - (\mu_x - \mu_y)}{\sigma \sqrt{\dfrac{1}{n_x} + \dfrac{1}{n_y}}} < 1.28\right] = .90.$$

We are given:

$\overline{X} = 86,375;\ \overline{Y} = 79,838;\ n_x = n_y = 8;\ \sigma = 4000.$

Substituting, the inequality becomes

$$\frac{(86,375 - 79,838) - (\mu_x - \mu_y)}{4000 \sqrt{\dfrac{1}{8} + \dfrac{1}{8}}} < 1.28.$$

Combining:
$$\frac{6537 - (\mu_x - \mu_y)}{2000} < 1.28.$$

Multiplying through by 2000: $6537 - (\mu_x - \mu_y) < 2560.$

Subtracting 6537: $- (\mu_x - \mu_y) < - 3977.$

Multiplying by - 1: $\mu_x - \mu_y > 3\ 977$

Our lower limit is 3 977.

• **PROBLEM** 12–14

A survey was conducted in 1970 to determine the average hourly earnings of a female sales clerk employed by a department store in metropolitan Los Angeles. A simple random sample of 225 female clerks was selected and the following information obtained:

X = hourly wage rate earned by female sales clerk,

$\Sigma x = \$450.00,\quad \Sigma(x - \overline{x})^2 = \$2016.00$.

What is the .99 confidence interval estimate of the average hourly wage rate?

Solution: We have a large sample problem here. Two major theorems facilitate such problems. The Law of Large Numbers will allow us to approximate σ, the true standard deviation by s the sample deviation. In addition the Central Limit Theorem tells us that $\dfrac{(\overline{X} - \mu)\ \sqrt{n}}{\sigma}$ is approxi-

mately standard normal. Hence we will assume $\dfrac{(\overline{X} - \mu)\ \sqrt{n}}{s}$

757

is standard normal. One look at the tables will tell us that,

$$\Pr\left(-2.58 < \frac{(\overline{X} - \mu) \sqrt{n}}{s} < 2.58\right) = .99 \ .$$

We have to compute $\overline{X}$ and s .

$$\overline{X} = \frac{\Sigma X}{n} = \frac{\$450.00}{225} = \$2.00 \ \text{(the sample mean)}.$$

$$s = \sqrt{\frac{\Sigma(X_i - \overline{X})^2}{n - 1}} = \sqrt{\frac{2016.00}{224}} = \sqrt{9.00} = 3.00$$

(the standard deviation of the sample).

Now we can insert these values into the inequality

$$-2.58 < \frac{(\overline{X} - \mu) \sqrt{n}}{s} < 2.58 \ \text{to obtain}$$

$$-2.58 < \frac{(2.00 - \mu) \sqrt{225}}{3.00} < 2.58 \ .$$

Multiply by $\dfrac{3.00}{\sqrt{225}}$: $-0.516 < 2.00 - \mu < 0.516$.

Subtracting 2.00: $-2.516 < -\mu < -1.484$.

Multiply by -1: $1.484 < \mu < 2.516$.

Thus the average hourly earnings of a female sales clerk is 99% interval estimated to be between $ 1.48 and $2.52.

• PROBLEM 12-15

Mr. Greenberg owns a gas station in Philadelphia. His busiest hour is from 11 to noon. On May 18th, Mr. Greenberg's daughter Beth surveyed 36 customers. She found that the 36 people bought an average of 12 gallons of gasoline with a standard deviation of 4 gallons.

(a) Find a point estimate for μ, the mean number of gallons of gas people buy.
(b) Establish a .95 confidence interval for μ.
(c) Establish a .99 confidence interval for μ.

Solution: (a) 36 people constitutes a large sample. The law of large numbers applies. We have taken a large sample and therefore can estimate μ by $\overline{X} = 12$ gallons.

(b) Note that we have a sample standard deviation and not the true one. In a large sample, however, we can use the sample deviation as an estimate of σ, the real one and use the Central Limit Theorem which tells us $\dfrac{\overline{X} - \mu}{S/\sqrt{n}}$ will approach standard normal. We can justify this fairly easily by one look at the t table. To form an inequality using the t-statistic we would consider

$$Pr(- 2.03 < t_{(35)} < 2.03) = .95$$

while using the standard normal we would use

$$Pr(- 1.96 < \text{Standard Normal} < 1.96) = .95$$

The accuracy of 2.03 to 1.96 is certainly no less than the accuracy of most of our measurements anyway. To continue we will use the inequality $- 1.96 < \dfrac{\overline{X} - \mu}{\sigma/\sqrt{n}} < 1.96$. Substituting the known quantities $\overline{X} = 12$, $\sigma = 4$, $\sqrt{n} = \sqrt{36} = 6$, we obtain $- 1.96 < \dfrac{12 - \mu}{4/6} < 1.96$.

Multiplying by $\frac{4}{6}$: $- 1.31 < 12 - \mu < 1.31$.

Subtracting 12: $- 13.31 < - \mu < - 10.69$.

Multiplying through by $- 1$: $10.69 < \mu < 13.31$.

(c) To get a 99 percent confidence interval note

$$Pr\left[- 2.58 < \frac{\overline{X} - \mu}{\sigma/\sqrt{n}} < 2.58\right] = .99 \quad \text{for the same}$$

reasons as in (b). Substituting into the central inequality we obtain; $- 2.58 < \dfrac{12 - \mu}{4/6} < 2.58$.

Multiplying by $\frac{4}{6}$: $- 1.72 < 12 - \mu < 1.72$.

Subtracting 12: $- 13.72 < - \mu < - 10.28$.

Multiplying by $- 1$: $10.28 < \mu < 13.72$.

Our answer is $10.28 < \mu < 13.72$.

The interval (10.28, 13.72) is a 99% confidence interval for the true mean number of gallons of gasoline purchased.

The nicotine contents of five cigarettes of a certain brand, measured in milligrams, are 21, 19, 23, 19, 23. Establish a .99 confidence interval estimate of the average nicotine content of this brand of cigarette.

Solution: In this problem we are given specific measurements and must determine the sample mean and standard deviation for ourselves.

For the mean: $\bar{X} = \dfrac{\Sigma X}{n}$

$$= \frac{21 + 19 + 23 + 19 + 23}{5} = \frac{105}{5}$$

$$= 21 \text{ milligrams.}$$

For the sample standard deviation:

$$S = \sqrt{\frac{\Sigma (X - \bar{X})^2}{n - 1}}$$

X	$(X - \bar{X})$	$(X - \bar{X})^2$
21	0	0
19	− 2	4
23	+ 2	4
19	− 2	4
23	+ 2	4

$$\Sigma (X - \bar{X})^2 = 16$$

$$S = \sqrt{\frac{\Sigma (X - \bar{X})^2}{n - 1}} = \sqrt{\frac{16}{5 - 1}} = 2 \text{ milligrams.}$$

We have a small sample with an unknown variance.

We search for the quotient of a standard normal random variable and the square root of a chi-square random variable divided by its degrees of freedom.

Since $\dfrac{\bar{X} - \mu}{\sigma/\sqrt{n}}$ is standard normal and $\dfrac{\Sigma (X_i - \bar{X})^2}{\sigma^2}$ is

χ^2_{n-1} $\dfrac{(\bar{X} - \mu)/(\sigma/\sqrt{n})}{\sqrt{\Sigma (X_i - \bar{X})^2/(n-1)\sigma^2}} = \dfrac{\bar{X} - \mu}{s/\sqrt{n}}$ is such a random

variable and as we have seen in previous problems is t distributed with n − 1 d.o.f. In our case n − 1 = 5 − 1 so $\dfrac{\bar{X} - \mu}{s/\sqrt{n}}$ is $t_{(4)}$. Looking at the t table we see

$$Pr(- 4.604 < t_{(4)} < 4.604) = .99 .$$

Substituting our t random variable,

$$Pr \left[- 4.604 < \frac{\overline{X} - \mu}{s / \sqrt{n}} < 4.604 \right] = 0.99.$$

Since we know $\overline{X} = 21$, $s = 2$, and $n = 5$, we only need to use the inequality

$$- 4.604 < \frac{21 - \mu}{2/\sqrt{5}} < 4.604 .$$

Multiplying through by $\frac{2}{\sqrt{5}}$: $- 4.118 < 21 - \mu < 4.118 .$

Subtracting 21 yields: $- 25.118 < - \mu < - 16.882.$

Multiplying by $- 1$ gives the 99% confidence interval for μ:

$$16.882 < \mu < 25.118 .$$

• PROBLEM 12–17

In the previous problem, find a 95% interval using the data n = 9, $S^2 = 7.63$.

Solution: Since n = 9, we are dealing with a chi-square statistic with n - 1 = 8 degrees of freedom.

We want a 95% (= .95) confidence interval, therefore, .95 = 1 - α or α = .05 and $\alpha/2$ = .025. Hence

$$P(\chi^2_{(8)} < 2.18) = .025 \text{ and } P(\chi^2_{(8)} > 17.5) = .025$$

which implies a = 2.18 and b = 17.5.

Making the required substutitions into

$$\left(\frac{(n - 1)S^2}{b} , \frac{(n - 1)S^2}{a} \right) \text{ yields } \left(\frac{8(7.63)}{17.5} , \frac{8(7.63)}{2.18} \right)$$

or   (3.49, 28).

• PROBLEM 12–18

The lengths of a random sample of 10 staples have a sample variance of .32 centimeters squared. Find a .95 confidence interval estimate for the variance of all staple lengths.

Solution: We have a small sample with an unknown
variance. We are looking for a confidence interval for the
variance. The statistic of concern here is

$\frac{(n - 1)S^2}{\sigma^2}$ which is chi-square distributed with n - 1

degrees of freedom. Here n - 1 = 10 - 1 = 9.   Thus

$\frac{(n - 1)S^2}{\sigma^2}$ is $\chi^2_{(9)}$.

We want values a, b such that $Pr(a < \chi^2_{(9)} < b) = .95$.

We choose a such that $P(\chi^2_{(9)} < a) = .025$, and b such

that $P(\chi^2_{(9)} > b) = .025$. From the chi-square tables

a = 2.7 and b = 19. Since $\frac{(n - 1)S^2}{\sigma^2}$ is $\chi^2_{(9)}$,

$$Pr\left[2.7 < \frac{(n - 1)S^2}{\sigma^2} < 19\right] = .95.$$

Substitute n = 10 and s^2 = .32 in the central inequality. Thus

$$2.7 < \frac{(10 - 1)(.32)}{\sigma^2} < 19 = 2.7 < \frac{2.88}{\sigma^2} < 19.$$

Divide by 2.88 and obtain $\frac{2.7}{2.88} < \frac{1}{\sigma} < \frac{19}{2.88}$

Taking the reciprocal produces $\frac{2.88}{19} < \sigma^2 < \frac{2.88}{2.7}$

or $.15 < \sigma^2 < 1.06.$

● **PROBLEM** 12–19

Eight scholars are working on a book. They are scheduled for
an eight hour day, but no one works exactly eight hours.
Yesterday the totals were 7.9, 7.8, 8.0, 8.1, 8.2, 7.9, 7.7,
and 8.3 hours. Find a .95 confidence interval estimate for
the variance of all 8 hour days these scholars will put in
before their work is published.

Solution: We have a small sample with an unknown mean.
We want a confidence interval for σ^2. The statistic we
resort to is $\frac{(n - 1)S^2}{\sigma^2}$ which is $\chi^2_{(n - 1)}$ or $\chi^2_{(7)}$ here.

We must find $(n - 1)S^2$. Since $S^2 = \frac{\Sigma(X - \bar{X})^2}{n - 1}$,

$$(n - 1)S^2 = \Sigma(X - \overline{X})^2.$$

We must find $\overline{X}$.

$$\overline{X} = \frac{\Sigma X}{n} = \frac{7.9+7.8+8+8.1-8.2+7.9+7.7+8.3}{8} = \frac{63.9}{8} = 7.99.$$

The following table will help us find $\Sigma(X - \overline{X})^2$.

X	$X - \overline{X}$	$(X - \overline{X})^2$
7.9	− .09	.0081
7.8	− .19	.0361
8.0	.01	.0001
8.1	.11	.0121
8.2	.21	.0441
7.9	− .09	.0081
7.7	− .29	.0841
8.3	.31	.0961

$$\Sigma(X - \overline{X})^2 = 0.2888 = (n - 1)S^2$$

We know that there exists a and b so that

$$Pr\left[a < \frac{(n - 1)S^2}{\sigma^2} < b\right].$$ Since $\frac{(n - 1)S^2}{\sigma^2}$ is $\chi^2_{(7)}$, we choose a and b such that

$$P(\chi^2_{(7)} < a) = .025 \quad \text{and} \quad P(\chi^2_{(7)} > b) = .025.$$

The chi-square tables tell us a = 1.69 and b = 16.0.

We will construct our confidence interval from the

inequality $\quad a < \frac{(n - 1)S^2}{\sigma^2} < b \quad$ which equals

$$1.69 < \frac{.2888}{\sigma^2} < 16.$$

Dividing by .2888 yields: $\quad \frac{1.69}{.2888} < \frac{1}{\sigma^2} < \frac{16}{.2888}.$

Taking the reciprocal produces the result:

$$\frac{.2888}{16} < \sigma^2 < \frac{.2888}{1.69} \quad \text{or} \quad .018 < \sigma^2 < .171.$$

One night New York City was victimized by a power
failure. Most homes were without power 12 hours or more.
Twenty-five homes were polled as to how long they were
without power. It was observed that the variance of the
sample was $S^2 = 4$. Find a .98 confidence interval estimate for the variance for all homes in New York.

Solution: As in the previous problems concerning
confidence intervals of variances when the mean is unknown
we use the statistic $\dfrac{(n-1)S^2}{\sigma^2}$ which has a chi-square distribution with n - 1 degrees of freedom. We have n - 1 =
25 - 1 = 24. We want a and b so that

$$\Pr(a < \chi^2_{(24)} < b) = .98.$$

To obtain this a and b, we set .98 = 1 - α so that
α = .02 and $\frac{\alpha}{2}$ = .01. a is the value such that

$$P\left(\chi^2_{(24)} < a = .01 \right. \text{and b is the value such that}$$

$$P\chi^2_{(24)} > b = .01. \text{ From chi-square tables } a = 10.9$$
and b = 43.0. Since $\dfrac{(n-1)S^2}{\sigma^2}$ is $\chi^2_{(24)}$,

$$P\left(10.9 < \dfrac{(n-1)S^2}{\sigma^2} < 43\right) = .98.$$

Substituting n = 25, $S^2 = 4$ yields the inequality

$$10.9 < \dfrac{(25-1)\ 4}{\sigma^2} < 43$$

Simplification produces: $10.9 < \dfrac{96}{\sigma^2} < 43$.

Divide by 96: $\dfrac{10.9}{96} < \dfrac{1}{\sigma^2} < \dfrac{43}{96}$.

Take the reciprocal: $\dfrac{96}{43} < \sigma^2 < \dfrac{96}{10.9}$

or $(2.233 < \sigma^2 < 8.807)$

Thus (8.807, 8.772) is our 98% confidence interval for σ^2.

A random sample of 10 salt-waterfish had variance, S_1^2,
in girth of 7.2 inches^2, while a random sample of 8 fresh-
water fish had a variance S_2^2 in girth of 3.6 in^2. Find a
.90 confidence interval for the ratio between the two

variances $\frac{\sigma_2^2}{\sigma_1^2}$. Assume normal populations.

Solution: From the last problem. we know the interval is of the form: $a \frac{S_2^2}{S_1^2} < \frac{\sigma_2^2}{\sigma_1^2} < b \frac{S_2^2}{S_1^2}$.

Since $1 - \alpha = .90$, $\alpha = .10$ and $\alpha/2 = .05$. In this case $m = 8$ and $n = 10$. Thus $n - 1 = 9$ and $m - 1 = 7$. a is the value such that $Pr(F_{(9,7)} < a) = .05$. b is the value such that $Pr(F_{(9,7)} > b) = .050$.

From the F tables $a = .304$ and $b = 3.68$.

Since $S_1^2 = 7.2$ and $S_2^2 = 3.6$, the interval becomes

$.304 \left(\frac{3.6}{7.2} \right) < \frac{\sigma_2^2}{\sigma_1^2} < 3.68 \left(\frac{3.6}{7.2} \right)$ or $.152 < \frac{\sigma_2^2}{\sigma_1^2} < 1.84$.

• PROBLEM 12–22

A random sample of 225 students at a college showed that 135 had used and benefited from   problem solvers.
(a) Make a point estimate for the proportion of college students who were helped by   problem solvers. (b) Make a .95 confidence interval estimate of the exact proportion.

Solution: (a) We will use as our point estimate, for p, $\hat{p} = \frac{x}{n}$. One reason for this is that the Weak Law of Large Numbers tells us that $\lim_{n \to \infty} \left| \frac{x}{n} - p \right| \to 0$.

Therefore $\hat{p} = \frac{x}{n} = \frac{135}{225} = .6$, $\hat{q} = 1 - \hat{p} = 1 - .6 = .4$.

(b) We will approximate a binomial probability with the normal distribution. We know $\dfrac{\frac{x}{n} - p}{\sqrt{\frac{pq}{n}}}$ can be considered a standard normal distribution for a large sample. Our sample is large, $n = 225$. The size of the sample will afford us another luxury. We do not know p or q, but our sample size allows us to use point estimates for them in computing $\sqrt{\frac{pq}{n}}$.

The rationale is that as the sample size increases, $\sqrt{\frac{\hat{p}\hat{q}}{n}}$ will be very close to $\sqrt{\frac{pq}{n}}$.

Consider $\dfrac{\dfrac{x}{n} - p}{\sqrt{\dfrac{\hat{p}\hat{q}}{n}}}$ to be a standard normal random

variable. This implies $\Pr\left(- 1.96 < \dfrac{\dfrac{x}{n} - p}{\sqrt{\dfrac{\hat{p}\hat{q}}{n}}} < 1.96\right) = .95.$

Let us clear the parentheses and multiply by $\sqrt{\dfrac{\hat{p}\hat{q}}{n}}$,

$$- 1.96 \sqrt{\dfrac{\hat{p}\hat{q}}{n}} < \dfrac{x}{n} - p < 1.96 \sqrt{\dfrac{\hat{p}\hat{q}}{n}}.$$

Subtract $\dfrac{x}{n}$: $\quad - \dfrac{x}{n} - 1.96 \sqrt{\dfrac{\hat{p}\hat{q}}{n}} < - p < - \dfrac{x}{n} + 1.96 \sqrt{\dfrac{\hat{p}\hat{q}}{n}}.$

Multiply by $- 1$: $\quad \dfrac{x}{n} - 1.96 \sqrt{\dfrac{\hat{p}\hat{q}}{n}} < p < \dfrac{x}{n} + 1.96 \sqrt{\dfrac{\hat{p}\hat{q}}{n}}.$

Substitute our given values:

$$.6 - 1.96 \sqrt{\dfrac{(.6)(.4)}{225}} < p < .6 + 1.96 \sqrt{\dfrac{(.6)(.4)}{225}}$$

or $\quad .6 - .064 < p < .6 + .064,$

which reduces to $\quad .536 < p < .664.$

POINT ESTIMATION

• PROBLEM 12–23

A psychologist wishes to determine the variation in I.Q.s of the population in his city. He takes many random samples of size 64. The standard error of the mean is found to be equal to 2. What is the population standard deviation?

Solution: The standard error of the mean is defined to be

$$\sigma_{\bar{x}} = \dfrac{\sigma}{\sqrt{n}} \tag{1}$$

where σ is the positive square root of the population variance and n is the size of the sample. Formula (1) is valid when sampling occurs with replacement or when the population is infinite.

We are given $n = 64$ and $\sigma_{\bar{x}} = 2$.

Substituting into (1),

$$2 = \frac{\sigma}{\sqrt{64}} \quad \text{or,} \quad \sigma = 16. \tag{2}$$

Thus, the standard deviation of the distribution of I.Q.s in the city is 16.

If we assume that I.Q.s are normally distributed with mean 100, then a standard deviation of 16 tells us that approximately 68% of the population have I.Q.s between 84 and 116.

• PROBLEM 12–24

An investigator collected 50 different samples; each sample contained 17 scores. He studied the 50 means and estimated $\sigma_{\bar{x}}^2$ to be 2.9. Estimate σ^2 of the original population.

Solution: Consider a population containing a finite number of elements N. Suppose we wish to take a sample of size n from this population. How many different samples can we take provided sampling is done with re-placement? The answer is N^n possible samples. For there are N ways of choosing the first element of the sample, N ways of choosing the second element, and pro-ceeding, N ways of choosing the last element.

For each sample of size n, there is a sample mean, $\bar{x}$. We can imagine the means of the samples forming a distri-bution with a mean value and a variance. The variance of this distribution is given by the formula

$$\sigma_{\bar{x}}^2 = \frac{\sigma^2}{n} , \tag{1}$$

where the subscript $\bar{x}$ indicates that we are dealing with the distribution of sample means and n is the size of the sample.

Substituting the given data into (1),

$$2.9 = \frac{\sigma^2}{17} \quad \text{or,} \quad \sigma^2 = 17(2.9) = 49.3 .$$

Note that we did not use 50, the number of samples taken, for the problem at hand. This was irrelevant data.

An urn contains a number of black and a number of white balls, the ratio of the numbers being 3 : 1. It is not known, however, which color ball is more numerous. From a random sample of three elements drawn with replacement from the urn, estimate the probability of drawing a black ball.

Solution: Let p be the probability of drawing a black ball, and n the number of balls drawn. Then p is either $\frac{1}{4}$ or $\frac{3}{4}$. Since a drawn ball is either black or white, the number of black balls is given by the binomial distribution

$$f(x;\ p) = \binom{n}{x}\ p^x\ (1-p)^{n-x}\ ;\quad x = 0, 1, \ldots, n. \quad (1)$$

Letting $p = \frac{1}{4}$ and then $p = \frac{3}{4}$ we obtain the following table from (1).

Outcome: x	0	1	2	3
f(x; 1/4)	27/64	27/64	9/64	1/64
f(x; 3/4)	1/64	9/64	27/64	27/64

Now assume that we draw a sample and find x = 2. Then, it is more likely that black balls are more numerous in the urn. If, on the other hand, no black balls were drawn, i.e., x = 0, then it is more likely that the white balls are three times more numerous than the black balls.

In general,

$$\hat{p} = \hat{p}(x) = \begin{array}{ll} .25 & \text{for}\ \ x = 0, 1 \\ .75 & \text{for}\ \ x = 2, 3. \end{array}$$

is the estimator of the parameter p. For given sample outcomes it yields the most likely values of the parameter.

Eight trials are conducted of a given system with the following results: S, F, S, F, S, S, S, S (S = success, F = failure). What is the maximum likelihood estimate of p, the probability of successful operation?

Solution: If we assume that the trials are independent of each other, the sequence above has the probability

$$L(p) = p^6 (1 - p)^2.\tag{1}$$

We wish to find that value of p which maximizes (1), the likelihood of this particular sequence. Differentiating (1) with respect to p and setting the derivative equal to zero:

$$L'(p) = 8p^7 - 14p^6 + 6p^5 = 0,$$

or, $2p^5 (4p^2 - 7p + 3) = 0.$ \hfill (2)

From (2) either $\hat{p} = 0$ or

$$\hat{p} = \frac{7 \pm \sqrt{49 - 4(4)(3)}}{8}$$

$$= \frac{7 \pm 1}{8} = 1 \text{ or } \frac{3}{4}.$$

The values $\hat{p} = 0$ and $\hat{p} = 1$ when substituted into (1) yield $L(p) = 0$. Hence the likelihood is maximized when $\hat{p} = \frac{3}{4}$. This is the maximum likelihood estimate of p.

• PROBLEM 12–27

For a certain sample the sample variance

$$s^2 = \sum_{i=1}^{n} \frac{(x_i - \bar{x})^2}{n}$$

is calculated to be 16. The sample contains 9 elements. Using an unbiased estimator, estimate σ^2.

Solution: We must first examine whether s^2 is an unbiased estimator of σ^2. An estimator $\hat{\theta}$ of a population parameter θ is said to be unbiased if

$$E(\hat{\theta}) = \theta.$$

Now, $E(s^2) = E\left\{ \frac{1}{n} \sum_{i=1}^{n} (x_i - \bar{x})^2 \right\}.$ \hfill (1)

But we may express $\sum_{i=1}^{n} (x_i - \bar{x})^2$ as

$$\left[\sum_{i=1}^{n} (x_i - \mu)^2 \right] - n(\bar{x} - \mu)^2.\tag{2}$$

Substituting (2) into (1),

$$E(s^2) = E \left\{ \frac{1}{n} \left[\sum_{i=1}^{n} (x_i - \mu)^2 - n(\overline{x} - \mu)^2 \right] \right\}. \quad (3)$$

But $\quad E \left[\frac{1}{n} \sum_{i=1}^{n} (x_i - \mu)^2 \right] = \sigma^2$ (the population variance)

and $\quad E \left[\frac{1}{n} (n(\overline{x} - \mu)^2) \right] = \sigma_{\overline{x}}^2 \quad$ (the variance of the sampling mean distribution). Hence,

$$E(s^2) = \sigma^2 - \sigma_{\overline{v}}^2$$

$$= \sigma^2 - \frac{\sigma^2}{n} \qquad (4)$$

$$= \sigma^2 \left(1 - \frac{1}{n} \right) = \sigma^2 \left(\frac{n-1}{n} \right). \qquad (5)$$

Hence s^2 is a biased estimator of σ^2, the amount of bias being $- \frac{\sigma^2}{n}$. We use (5) to find an unbiased estimator of σ^2. Define the statistic

$$\overline{s}^2 = \frac{1}{n-1} \sum_{i=1}^{n} (x_i - \overline{x})^2. \qquad (6)$$

Note that

$$E(\overline{s}^2) = E \left[\sum_{i=1}^{n} \frac{(x_i - \overline{x})^2}{n-1} \right] = \sigma^2. \qquad (7)$$

Hence $\overline{s}^2$ is an unbiased estimator of σ^2.

We now are ready to estimate σ in the given problem. Since $\overline{s}^2 = \frac{n}{n-1} s^2$,

$$\overline{s}^2 = \frac{9}{8} (16) = 18$$

Therefore, an unbiased estimate of σ^2 for this set of data is $\overline{s}^2 = 18$

• **PROBLEM** 12–28

An investigator is interested in the distribution of weights at a convention of runners. He draws three samples, each sample containing 16 runners. The mean weights were $\overline{x}_1 = 152$, $\overline{x}_2 = 156$ and $\overline{x}_3 = 157$.

(a) Compute the variance using the three sample means and then find the best estimate of $\sigma_{\bar{x}}$. (b) Use this estimate of $\sigma_{\bar{x}}$ to estimate the population variance σ^2.

Solution: The variance of a finite set of numbers can be defined as

$$\sigma^2 = \sum_{i=1}^{n} \frac{(x_i - \mu)^2}{n} \,. \tag{1}$$

For a sample of numbers, from this set, the corresponding sample variance is

$$s^2 = \sum_{i=1}^{n} \frac{(x_i - \bar{x})^2}{n} \,.$$

In the given problem, the x_i are the different sample means and the $\bar{x}$ represents the mean of the sample means. For the above set of data,

$$\bar{x} = \frac{\bar{x}_1 + \bar{x}_2 + \bar{x}_3}{3} = \frac{152 + 156 + 157}{3}$$

$$= 155.$$

Hence, $\quad s_{\bar{x}}^2 = \dfrac{(x_1 - \bar{x}) + (x_2 - \bar{x}) + (x_3 - \bar{x})}{n}$

$$= \frac{[(152 - 155)^2 + (156 - 155)^2 + (157 - 155)^2]}{3}$$

$$= \frac{14}{3} \,.$$

We know from theoretical statistics that in general, s^2 is a biased estimator of σ^2, the amount of bias being $-\dfrac{\sigma^2}{n}$. In fact,

$$E(s_{\bar{x}}^2) = \frac{n-1}{n} \, \sigma_{\bar{x}}^2 \,. \tag{2}$$

Using (2), we can construct an unbiased estimator of $\sigma_{\bar{x}}^2$,

$$\hat{s}_{\bar{x}}^2 = \frac{1}{n-1} \sum_{i=1}^{n} (x_i - \bar{x})^2. \tag{3}$$

(3) represents the best estimator of $\sigma_{\bar{x}}^2$ for the following reasons (1) it is unbiased, (2) it is consistent and (3) it has the minimum variance amongst the class of estimators of σ^2 that are unbiased.

Studying (3), we see that it is equal to

$$\frac{n}{n-1} \; s_{\overline{x}}^2 \; . \qquad \text{Hence,}$$

$$\text{est } \sigma_{\overline{x}}^2 = \frac{n}{n-1} \; \sum_{i=1}^{n} \frac{(x_i - \overline{x})^2}{n}$$

$$= \frac{3}{2} \left(\frac{14}{3}\right) = 7.0 \; .$$

Hence, the best estimate of $\sigma_{\overline{x}}$ is

$$\sqrt{7.0} = 2.6. \qquad\qquad\qquad (4)$$

(b)   We now wish to use (4) to estimate the population variance σ^2. First we must derive the relationship between $\sigma_{\overline{x}}^2$, (the variance of the sampling mean distribution) and σ^2, the population variance.

Now,

$$\text{Var } (\overline{x}) = \text{Var } \left(\sum_{i=1}^{n} \frac{x_i}{n}\right)$$

$$= \frac{1}{n^2} \text{Var } \left(\sum_{i=1}^{n} x_i\right) \; .$$

But since the x_i are independent of each other,

$$\text{Var } \left(\sum_{i=1}^{n} x_i\right) = \sum_{i=1}^{n} \text{Var}(x_i).$$

By definition $\text{Var } (x_i) = \sigma^2$.

Hence, since the x_i are identically distributed,

$$\sum_{i=1}^{n} \text{Var } (x_i) = n\sigma^2.$$

Thus, $\text{Var } (\overline{x}) = \dfrac{1}{n^2} n\sigma^2 = \dfrac{\sigma^2}{n} \; .$

This implies that

$$\sigma^2 = n \text{ Var } (\overline{x}) = n\sigma_{\overline{x}}^2 \; .$$

Since $n = 16$ (the size of the samples), and $\sigma_{\overline{x}}^2$ has been estimated to be 7, the estimate of the population variance is

$\sigma^2 = 16(7) = 112.$

From this, the estimated standard deviation is

$\sqrt{112} = 10.6.$

● **PROBLEM** 12–29

Let x_1, x_2, x_3 be a sample of size 3 from the Bernoulli distribution. Show that the statistic

$S_1 = u(x_1, x_2, x_3) = x_1 + x_2 + x_3$

is a sufficient statistic but that the statistic

$S_2 = v(x_1, x_2, x_3) = x_1 x_2 + x_3$

is not a sufficient statistic.

<u>Solution:</u>   A random variable x has the Bernoulli distribution if it has the frequency function

$$f(x) = p^x (1 - p)^{1-x}; \qquad x = 0, 1,$$

where p is the probability of success. The sample space for the above problem consists of

$$\{\{0, 0, 0\} \ \{0, 0, 1\}\{0, 1, 0\}\{0, 1, 1\}\{1, 0, 0\}$$

$$\{1, 0, 1\}\{1, 1, 0\}\{1, 1, 1\}\} \quad .$$

For each sample point we can find the values of the statistics S_1 and S_2. Thus we may construct the following table:

(1)	(2) Values of S_1	(3) Values of S_2	(4) $f_{x_1,x_2,x_3/S_1}$	(5) $f_{x_1,x_2,x_3/S_2}$
(0, 0, 0)	0	0	1	1 - p/1 + p
(0, 0, 1)	1	1	1/3	1 - p/1 + 2p
(0, 1, 0)	1	0	1/3	p/1 + p
(1, 0, 0)	1	0	1/3	p/1 + p
(0, 1, 1)	2	1	1/3	p/1 + 2p
(1, 0, 1)	2	1	1/3	p/1 + 2p
(1, 1, 0)	2	1	1/3	p/1 + 2p
(1, 1, 1)	3	2	1	1

Since $S_1 = x_1 + x_2 + x_3$, column (2) is found by adding each coordinate of the triple (x_1, x_2, x_3). For example, $(0, 1, 1) = 0 + 1 + 1 = 2$. Similarly, $S_2 = x_1 x_2 + x_3$ and an example is $(1, 0, 0) = (1)(0) + (0) = 0$.

The results of column (4) are calculated as follows. Consider the point $(0, 0, 0)$. The conditional density of $x_1, x_2, x_3 = (0, 0, 0)$ given $S_1 = 0$ is

$$f_{x_1, x_2, x_3/S_1 = 0} (0, 0, 0/0)$$

$$= \frac{P[x_1 = 0; x_2 = 0; x_3 = 0; S_1 = 0]}{P[S_1 = 0]}$$

$$= \frac{(1 - p)^3}{\binom{3}{0}(1 - p)^3} = 1.$$

Similarly,

$$f_{x_1, x_2, x_3/S_1 = 1} (0, 0, 1/S_1 = 1)$$

$$= \frac{P[x_1 = 0; x_2 = 0; x_3 = 1; S_1 = 1]}{P[S_1 = 1]}$$

$$= \frac{(1 - p)^2 p}{\binom{3}{1}(1 - p)^2 p} = \frac{1}{3}.$$

Continuing in this way we obtain column (4). Note that the conditional distribution of the sample given the values of S_1 is independent of p. This is in accordance with the definition of a sufficient statistic.

Now consider the construction of column (5). We find that,

$$f_{x_1, x_2, x_3 | S_2 = 0} (0, 0, 0/0) = \frac{P[x_1 = 0; x_2 = 0; x_3 = 0; S_2 = 0]}{P[S_2 = 0]}$$

$$= \frac{(1 - p)^3}{(1 - p)^3 + 2(1 - p)^2 p}$$

$$= \frac{(1 - p)^3}{(1 - p)^2 [1 - p + 2p]} = \frac{1 - p}{1 + p}.$$

The rest of column (5) is constructed in a similar manner. Note that the conditional distribution of the sample given S_2 is not independent of the parameter p. Hence S_2 is not a sufficient statistic.

Consider a probability distribution having mean μ and variance σ^2. Show that

$$\bar{x}_n = \frac{1}{n} \sum_{i=1}^{n} x_i \quad \text{and} \quad s_n^2 = \frac{1}{n-1} \sum_{i=1}^{n} (x_i - \bar{x}_n)^2$$

are consistent sequences of estimators of μ and σ^2.

Solution: The sample mean and the sample estimate of the distribution variance are unbiased estimates of the population mean μ and population variance σ^2. Recall that an estimate $\hat{\theta}$ of a parameter θ is consistent if

$$\lim_{n\to\infty} E(\hat{\theta}_n) = \theta \quad \text{and} \quad \lim_{n\to\infty} \text{Var}(\hat{\theta}_n) = 0$$

where $\hat{\theta}_n$ represents a sequence of estimators that depend on the sample size n.

Hence, we must show that $\text{Var}[\bar{x}_n]$ and $\text{Var}[s_n^2]$ both tend to zero as $n \to \infty$.

The variance of a random variable is defined as

$$\text{Var}(\theta) = E[(\theta - E(\theta))^2] \tag{1}$$

From this definition it follows that, for independent random variables x_1, x_2 and constants a, b,

$$\text{Var}(aX_1 + bX_2) = a^2 \text{Var}(x_1) + b^2 \text{Var}(x_2).$$

In general,

$$\text{Var}\left[\sum_{i=1}^{n} a_i x_i\right] = \sum_{i=1}^{n} a_i^2 \text{Var}(x_i)$$

Now, $$\text{Var}(\bar{x}) = \text{Var}\left[\frac{1}{n} \sum_{i=1}^{n} x_i\right] = \frac{1}{n^2} \text{Var}\left[\sum_{i=1}^{n} x_i\right]. \tag{2}$$

But since the sample is randomly chosen, the x_i are independent. That is,

$$\text{Var}[x_1 + x_2 + \dots + x_n] = \text{Var}\left[\sum_{i=1}^{n} x_i\right]$$

$$= \sum_{i=1}^{n} \text{Var}(x_i).$$

Hence, (2) becomes

$$\frac{1}{n^2} \sum_{i=1}^{n} \text{Var} (x_i) = \frac{1}{n^2} n\sigma^2 = \frac{\sigma^2}{n} .$$

Hence, $\lim_{n\to\infty} \text{Var} (\bar{x}) = \lim_{n\to\infty} \frac{\sigma^2}{n} = 0$

and $\bar{x}$ is a consistent estimator of μ.

We now consider $\text{Var} (s_n^2)$. Using (1)

$$\text{Var} (s_n^2) = E [(s_n^2 - \sigma^2)^2]$$

$$= \frac{1}{n} \left(\mu_4 - \frac{n-3}{n-1} \sigma^4 \right) \qquad (3)$$

where μ_4 denotes the fourth moment about the origin. Letting $n \to \infty$ in (3) we find

$$\lim_{n\to\infty} \text{Var} (s_n^2) = \lim_{n\to\infty} \frac{1}{n} \left(\mu_4 - \frac{n-3}{n-1} \sigma^4 \right) = 0.$$

Hence s_n^2 is a consistent estimator of σ^2.

• PROBLEM 12–31

Let $x_1, \ldots, x_n$ be a random sample from the uniform density

$$f(x; a, b) = \frac{1}{b-a} . \qquad (1)$$

Show that the order statistics

$$y_1 = \min [x_1, \ldots, x_n] \quad \text{and} \quad y_n = \max [x_1, \ldots, x_n]$$

are jointly complete.

Solution: We must first define what jointly complete means. The two statistics y_1 and y_n are jointly complete when $E [g(y_1, y_n)] \equiv 0$ for all values of a and b implies $P[g(y_1, y_n) = 0] \equiv 1$, where $g(y_1, y_n)$ is a statistic.

Assume $g(y_1, y_n)$ is an unbiased estimator of 0, i.e.,

$$F [g(y_1, y_n)] \equiv 0. \qquad (1)$$

By the definition of expectation,

$$E[g(y_1, y_n)] \equiv \iint\limits_{R} g(y_1, y_n)\, f(y_1, y_n)\, dy_1\, dy_n$$

$$= \int_a^b \left[\int_0^{y_n} g(y_1, y_n)\, n(n-1) \left(\frac{y_n - a}{b - a} - \frac{y_1 - a}{b - a} \right)^{n-2} \right.$$

$$\left. \left(\frac{1}{b - a} \cdot \frac{1}{b - a} \right) dy_1 \right] dy_n.$$

Using the uniform density function for $f(y_1, y_n)$ equation (2) may be rewritten

$$E[g(y_1, y_n)]$$

$$\equiv \frac{n(n-1)}{(b-a)^n} \int_a^b \int_0^{y_n} g(y_1, y_n)(y_n - y_1)^{n-2}\, dy_1\, dy_n; \quad (3)$$

(3) is equal to zero only if

$$\int_a^b \int_0^{y_n} g(y_1, y_n)(y_n - y_1)^{n-2}\, dy_1\, dy_n \equiv 0. \quad (4)$$

We differentiate (4) with respect to b using Leibnitz's rule. Thus, we obtain

$$\int_a^b g(y_1, y_n)(b - y_1)^{n-2}\, dy_1 \equiv 0. \quad (5)$$

Differentiating (5) with respect to the parameter a, (again using Leibnitz's rule for the derivative of an integral)

$$- g(y_1, y_n)(b - a)^{n-2} \equiv 0 \quad \text{for all } a < b.$$

But $a < b$ implies $(b - a)^{n-2} \not\equiv 0$. Hence, we must have $g(y_1, y_n) \equiv 0$, i.e.

$$P[g(y_1, y_n) = 0] \equiv 1.$$

Therefore y_1 and y_n are jointly complete.

● **PROBLEM 12–32**

Consider the exponential density $f(x; \theta) = \theta e^{-\theta x}$. Let $x_1, \ldots, x_n$ denote a random sample from this density.

Show that the sample mean, $\bar{x}$, is a minimum variance estimator of the mean of the distribution $\frac{1}{\theta}$.

Solution: We use the Cramér-Rao inequality to solve the problem. Let T be an unbiased estimator of the parameter $\frac{1}{\theta}$. Then, by the Cramér-Rao inequality,

$$\text{Var } [T] \geq \frac{\left[\left(\frac{1}{\theta}\right)' \right]^2}{nE\left[\left[\frac{\partial}{\partial\theta} \ln f(x;\ \theta)\right]^2 \right]} \tag{1}$$

where $\left(\frac{1}{\theta}\right)' = \frac{d}{d\theta} \left(\frac{1}{\theta}\right)$ and $\ln z \equiv \log_e z$.

Verbally, (1) states that the variance of any unbiased estimator of a parameter of a distribution is always greater than the expression on the right side of (1). Equality prevails only when there exists a function $K(\theta, n)$ such that

$$\sum_{i=1}^{n} \frac{\partial}{\partial\theta} \ln f(x_i;\ \theta)$$

$$= K(\theta,\ n) \left[u_1(x_1,\ \ldots,\ x_n) - \left(\frac{1}{\theta}\right) \right] \tag{2}$$

When (2) is possible, T is called a minimum variance unbiased estimator.

We first find $\left[\frac{\partial}{\partial\theta} \ln f(x;\ \theta) \right]^2$. Thus,

$$f(x;\ \theta) = \theta e^{-\theta x}$$

$$\ln f(x;\ \theta) = \ln \theta - \theta x$$

$$\frac{\partial}{\partial\theta} \ln f(x;\ \theta) = \frac{1}{\theta} - x. \quad \text{Hence,}$$

$$E\left[\left[\frac{\partial}{\partial\theta} \ln f(x;\ \theta)\right]^2 \right] = E\left[\left(\frac{1}{\theta} - x\right)^2 \right]$$

$$= E\left[\left(x - \frac{1}{\theta}\right)^2 \right] = \text{Var } [x] = \frac{1}{\theta^2}, \tag{3}$$

since the variance of a random variable which has the negative exponential density is $\frac{1}{\theta^2}$.

Next, we find $\left[\left(\frac{1}{\theta}\right)' \right]^2$. $\frac{d}{d\theta}\left(\frac{1}{\theta}\right) = -\frac{1}{\theta^2}$.

Hence, $\left[\left(\frac{1}{\theta}\right)' \right]^2 = \frac{1}{\theta^4}$. \tag{4}

Substituting (3) and (4) into (1)

$$\text{Var } [T] \geq \frac{\frac{1}{\theta^4}}{n \frac{1}{\theta^2}} = \frac{1}{n \theta^2} \; . \tag{5}$$

Changing the inequality in (5) to equality,

$$\text{Var } [T] = \frac{1}{n \theta^2} \; .$$

Let us try to put the exponential density into the form (2). Thus

$$\sum_{i=1}^{n} \frac{\partial}{\partial \theta} \ln f(x; \theta) = \sum_{i=1}^{n} \frac{\partial}{\partial \theta} \ln \theta - \theta x_i$$

$$= \sum_{i=1}^{n} \frac{1}{\theta} - x_i = - n \left(\bar{x} - \frac{1}{\theta} \right).$$

Letting $K(\theta, n) = - n$ and $u_1(x_1, \ldots, x_n) = \bar{x}$,
we see that (2) is indeed possible for the negative exponential density.

This is a sufficient condition to show that $\bar{x}$ is an unbiased estimator of $\left(\frac{1}{\theta} \right)$ with minimum variance (equal to the lower bound).

• PROBLEM 12–33

A machine manufactures metal screws in a given lot size. Among the lots there is variation in the percentage of defective metal screws, with the distribution below:

Fraction Defective p	Prior Probability Function $P(p)$
.01	.25
.02	.35
.03	.20
.06	.15
.10	.04
.20	.01
	1.00

A random sample of 10 screws is chosen from the lots and after testing, it is found that one of the screws is defective. Use the prior distribution and the sample evidence to find the posterior probability function of fraction defective.

Solution: We are given the probability distribution of the random variable p. Let us use the binomial probability density function to find the various likelihoods. That is, we ask "what is the likelihood of obtaining one defective in a sample of 10 with the different probabilities in the table?". A random variable x is said to be binomially distributed if it has the frequency function

$$f(x) = \binom{n}{x} p^x q^{n-x}. \tag{1}$$

Letting n = 10, x = 1 and p the different probabilities we obtain the likelihood values.

$$\binom{10}{1} \quad (.01)^1 \ (.99)^9 \quad = .0091$$

$$\binom{10}{1} \quad (.02)^1 \ (.98)^9 \quad = .167$$

$$\binom{10}{1} \quad (.03)^1 \ (.97)^9 \quad = .228 \tag{2}$$

$$\binom{10}{1} \quad (.06)^1 \ (.94)^9 \quad = .344$$

$$\binom{10}{1} \quad (.10)^1 \ (.90)^9 \quad = .387$$

$$\binom{10}{1} \quad (.20)^1 \ (.80)^9 \quad = .268$$

Examining the table (2) we see that the likelihood function achieves its maximum when p = .10.

To find the posterior probabilities we use Bayes' formula:

$$P(A_i|A) = \frac{P(A_i) \ P(A|A_i)}{\sum\limits_{i=1}^{n} P(A_i) \ P(A|A_i)}. \tag{3}$$

In the present problem, the A_i are the various values that p can take and hence $P(A_i)$ is given by the prior probability function. Consider the term $P(A|A_i)$. The likelihood function defined by (2) gives the probability that $A = \frac{1}{10}$ given the various prior probabilities. Thus, it may be substituted for $P(A|A_i)$. We construct the following table:

P(p) = P(A_i)	Likelihood (= P(A|A_i)	Joint Probability P(A_i)P(A|A_i)	Posterior Probability $\dfrac{P(A_i)P(A\|A_i)}{\Sigma\ P(A_i)P(A\|A_i)}$
1. .25	.0091	.023	.117
2. .35	.167	.058	.294
3. .20	.229	.046	.234
4. .15	.344	.052	.264
5. .04	.387	.016	.076
6. .01	.268	.003	.015

The entries in the last column of the above table, when substituted into (3) yield $P(A_i|A)$, the posterior probability function.

HYPOTHESIS TESTING

• **PROBLEM** 12–34

A sample of size 49 yielded the values $\bar{x} = 87.3$ and $s^2 = 162$. Test the hypothesis that $\mu = 95$ versus the alternative that it is less. Let $\alpha = .01$.

<u>Solution</u>: The null and alternative hypotheses are given respectively by

$$H_0 : \mu = 95; \quad H_1 : \mu < 95 .$$

$\alpha = .01$ is the given level of significance.

Because the sample size is quite large (≥ 30), we can assume that the distribution of $\bar{X}$ is approximately normal. We are using the sample variance s^2 as an estimate of the true but unknown population variance and if the sample were not as large we would use a t-test.

The critical region consists of all z-scores that are less than $z_{.01} = -2.33$. The observed z-score is

$$z = \frac{\bar{X} - \mu}{\sqrt{s^2/n}} = \frac{87.3 - 95}{\sqrt{162/49}} = \frac{(-7.7)(7)}{\sqrt{162}} = -4.23 .$$

This observed score is in the critial region; thus we reject the null hypothesis and accept the alternative that $\mu < 95$.

For the following given information, find the critical region, compute t, and decide whether the results are significant or not significant.

Sample mean $\bar{X} = 26$

Sample standard deviation $s = 6$

Sample size $n = 25$

Null hypothesis $\mu = 30$

Alternate hypothesis $\mu < 30$

Significance level $\alpha = .01$.

Solution: The critical region consists of those values of t for which we will reject the null hypothesis, H_0 . For $\alpha = .01$,

df = n-1 = 24, and a one-tailed test, we will reject H_0 if

$t < -2.492$. This is because the test statistic, $t = (\bar{X} - \mu)/s_{\bar{X}}$ has

a t-distribution with a mean of 0 and a standard deviation of 1, and for 24 degrees of freedom, 1% of scores will have a t-value less than -2.492. We will accept H_0 if $t_c \geq -2.492$. We must calculate

$$t = \frac{\bar{X} - \mu}{s_{\bar{X}}} \quad \text{where} \quad s_{\bar{X}} = \frac{s}{\sqrt{n}} .$$

For the data given in this problem,

$$s_{\bar{X}} = \frac{6}{\sqrt{25}} = \frac{6}{5} = 1.2$$

and

$$t_c = \frac{26 - 30}{1.2} = \frac{-4}{1.2} = -3.33.$$

Since $-3.33 < -2.492$ we reject H_0 and conclude the results are significant in this problem.

A certain printing press is known to turn out an average of 45 copies a minute. In an attempt to increase its output, an alteration is made to the machine, and then in 3 short test runs it turns out 46, 47, and 48 copies a minute. Is this increase statistically significant, or is it likely to be simply the result of chance variation? Use a significance level of .05.

Solution: Since the sample size is small (n = 3), we can use a t-test for this problem to determine whether the sample mean number of copies is far enough from 45 for us to conclude that the population mean is greater than 45. We therefore set up our null and alternate hypotheses as follows:

$$H_0 : \mu = 45 \; ; \quad H_1 : \mu > 45 .$$

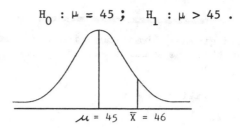

$\mu = 45 \quad \bar{X} = 46$

Our next step in problems of this type is to calculate

$$t = \frac{\bar{X} - \mu}{S_{\bar{X}}} \quad \text{where} \quad S_{\bar{X}} = \frac{S}{\sqrt{n}} .$$

We use the t-distribution because when the sample size is small (n < 30) the statistic $(\bar{X} - \mu)/S_{\bar{X}}$ has a t-statistic with a mean of 0 and a standard deviation of 1. But since $\bar{X}$ and S are not given we must calculate them.

$$\bar{X} = \frac{46+47+48}{3} = 47 \; ; \quad S = \sqrt{\frac{\Sigma(X - \bar{X})^2}{n-1}}$$

X	X - $\bar{X}$	$(X - \bar{X})^2$
46	-1	1
47	0	0
48	1	1

$2 = \Sigma(X - \bar{X})^2$

$$S = \sqrt{2/2} = \sqrt{1} = 1.$$

Now we may calculate $S_{\bar{X}} = \frac{1}{\sqrt{3}} = \frac{1}{1.732} = .577$

and

$$t = \frac{47 - 45}{.577} = \frac{2}{.577} = 3.47 .$$

For this problem where we have $\alpha = .05$ and a one-tailed test, our critical value of t is 2.92 since for the t-distribution with mean of 0 and standard deviation of 1, 5% of scores will have a t-value above 2.92. Therefore our decision rule is: reject H_0 if t > 2.92 and accept H_0 if t ≤ 2.92. Since our calculated t of 3.47 > 2.92 we would reject H_0 and conclude that the increase in copies in our sample is statistically significant. If however, we were to choose a 1% level of significance we would obtain for our decision rule: Reject H_0 if t > 6.95 and accept H_0 if t ≤ 6.965 because 1% of scores will have a t-value above 6.965. Since 3.47 ≤ 6.965, we would accept H_0 and conclude that the increase observed was simply the result of chance variation.

● **PROBLEM** 12–37

Suppose that you want to decide which of two equally-priced brands of light bulbs lasts longer. You choose a random sample of 100 bulbs of each brand and find that brand A has sample mean of 1180 hours and sample standard deviation of 120 hours, and that brand B has sample mean of 1160 hours

and sample standard deviation of 40 hours. What decision
should you make at the 5% significance level?

Solution: Arrange the data into a table:

	n	$\overline{X}$	s
Brand A	100	1180	120
Brand B	100	1160	40

Establish two hypotheses: H_0 asserts that A and B last
the same, on the average, and H_1 asserts that A and B have
different average lifespans. Thus:

$$H_0 : \mu_A = \mu_B , \text{ or, } \quad H_1 : \mu_A \neq \mu_B ;$$

equivalently, $H_0 : \mu_A - \mu_B = 0; \qquad H_1 : \mu_A - \mu_B \neq 0.$

Now define the acceptance region and rejection region
for this test. We can use the standard normal curve to
determine these regions because of the theorem that if two
populations from which two independent random samples are
taken are normally distributed or if $n_1 + n_2 \geq 30$, then
the sampling distribution of the difference between the
sample means is normal or approximately normal, and its
standard error is

$$\sqrt{\frac{\sigma_1^2}{n_1} + \frac{\sigma_2^2}{n_2}}$$

when σ_1^2 and σ_2^2 are the variances of populations 1 and 2
respectively.   In this problem, $n_1 + n_2 = 200 > 30$, so
that the sampling distribution of $\overline{X}_1 - \overline{X}_2$ is approximately
normal.   The acceptance region is the interval which lies
under 95% of the area under the standard normal curve,
because your decision will be made at the 5% level of
significance.   The acceptance region is therefore
$|Z| \leq 1.96$ and the rejection region is $|Z| > 1.96$.

$$\text{Now } \quad Z = \frac{(\overline{X}_A - \overline{X}_B) - (\mu_A - \mu_B)}{\sqrt{\frac{\sigma_A^2}{n_A} + \frac{\sigma_B^2}{n_B}}} = \frac{(\overline{X}_A - \overline{X}_B)}{\sqrt{\frac{\sigma_A^2}{n_A} + \frac{\sigma_B^2}{n_B}}} \text{ by } H_0.$$

$(S_A)^2 = (120)^2$ and $(S_B)^2 = (40)^2$ can be used as
estimates for σ_A^2 and σ_B^2. Substituting known values in
the formula for Z, we have

$$Z = \frac{1180 - 1160}{\sqrt{\frac{14400}{100} + \frac{1600}{100}}} = \frac{20}{\sqrt{144 + 16}} = \frac{20}{\sqrt{160}} = \frac{20}{\sqrt{12.65}} = 1.58 .$$

784

Since $-1.96 < (Z = 1.58) < 1.96$, Z is in the acceptance region. Therefore, we accept the hypothesis that there is no difference between the average lifespans of the two brands, at the 5% significance level.

• PROBLEM 12–38

A reading test is given to an elementary school class that consists of 12 Anglo-American children and 10 Mexican-American children. The results of the test are - Anglo-American children: $\overline{X}_1 = 74$, $S_1 = 8$; Mexican-American children: $\overline{X}_2 = 70$, $S_2 = 10$. Is the difference between the means of the two groups significant at the .05 level of significance?

Solution: Assuming the test scores are normally distributed, we may use the t-test with $n_1 + n_2 - 2$ degrees of freedom to test the significance of the difference between the means because the statistic

$$\frac{(\overline{X}_2 - \overline{X}_1) - (\mu_2 - \mu_1)}{S_{\overline{X}_2 - \overline{X}_1}}$$

has a t-distribution when $n_1 + n_2 \leq 30$. The figure below depicts this problem.

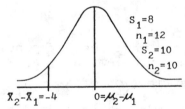

We have for our null and alternate hypotheses

$H_0: \mu_2 - \mu_1 = 0$ $\qquad$ $H_1: \mu_2 - \mu_1 \neq 0$.

We must calculate $\qquad t = \dfrac{(\overline{X}_2 - \overline{X}_1) - 0}{S_{\overline{X}_2 - \overline{X}_1}}$ $\qquad$ where

$$S_{\overline{X}_2 - \overline{X}_1} = \sqrt{\frac{(n_1 - 1)S_1^2 + (n_2 - 1)S_2^2}{n_1 + n_2 - 2}} \sqrt{\frac{1}{n_1} + \frac{1}{n_2}}.$$

The critical t for $n_1 + n_2 - 2 = 20$ df's and a two-tailed test at $\alpha = .05$ is 2.09 because for the t-distribution with mean of 0 and standard deviation of 1, 2.5% scores will have a t-value greater than 2.09 and 2.5% of scores will have a t-value less than -2.09. Therefore, our decision rule is: reject H_0 if $t > 2.09$ or $t < -2.09$; accept H_0 if $-2.09 \leq t \leq 2.09$.

For the data of this problem,

$$S_{\overline{X}_2 - \overline{X}_1} = \sqrt{\frac{11(8)^2 + 9(10)^2}{12 + 10 - 2}} \sqrt{\frac{1}{12} + \frac{1}{10}}$$

$$= \sqrt{\frac{704 + 900}{20}} \sqrt{.0833 + .1}$$

$$= \sqrt{80.2} \sqrt{.1833} = 3.834 \qquad \text{and}$$

$$t = \frac{(70 - 74) - 0}{3.834} = \frac{-4}{3.834} = -1.04.$$

Since $-2.09 < -1.04 < 2.09$, we accept H_0 and conclude that the difference between the means of the two groups is not significant.

• PROBLEM 12–39

For the following samples of data, compute t and determine whether μ_1 is significantly less than μ_2. For your test use a level of significance of .10. Sample 1: $n = 10$, $\overline{X} = 10.0$, $S = 5.2$; sample 2: $n = 10$, $\overline{X} = 13.3$, $S = 5.7$.

<u>Solution:</u>   For this problem, we have for our hypotheses

$H_0: \mu_1 - \mu_2 \geq 0$ $\qquad\qquad$ $H_1: \mu_1 - \mu_2 < 0$.

This problem can be depicted by the following diagram:

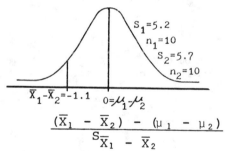

The statistic $\dfrac{(\overline{X}_1 - \overline{X}_2) - (\mu_1 - \mu_2)}{S_{\overline{X}_1 - \overline{X}_2}}$

has a t-distribution because $n_1 + n_2 \leq 30$.

We have $n_1 + n_2 - 2 = 18$ df's and our decision rule for $\alpha = .10$ is: reject H_0 if $t < -1.330$; accept H_0 if $t \geq -1.330$.

We must calculate

$$t = \frac{(\overline{X}_1 - \overline{X}_2) - (\mu_1 - \mu_2)}{S_{\overline{X}_1 - \overline{X}_2}} \qquad \text{where}$$

$$S_{\overline{X}_1 - \overline{X}_2} = \sqrt{\frac{(n_1 - 1)S_1^2 + (n_2 - 1)S_2^2}{n_1 + n_2 - 2}} \cdot \sqrt{\frac{1}{n_1} + \frac{1}{n_2}}.$$

For the data of this problem we have

$$S_{\overline{X}_1 - \overline{X}_2} = \sqrt{\frac{9(5.2)^2 + 9(5.7)^2}{18}} \cdot \sqrt{\frac{1}{10} + \frac{1}{10}}$$

$$= \sqrt{\frac{9(27.04) + 9(32.49)}{18}} \cdot \sqrt{\frac{1}{5}}$$

$$= \sqrt{\frac{243.36 + 292.41}{18}} \cdot \sqrt{\frac{1}{5}}$$

$$= \sqrt{29.765} \cdot \sqrt{\frac{1}{5}} = 2.44 \qquad \text{and}$$

$$t = \frac{(10.0 - 13.3) - 0}{2.44} = \frac{-3.3}{2.44} = -1.35 .$$

Since $-1.35 < -1.330$, we reject H_0 and conclude that μ_2 is significantly less than μ_1 at a 10% level of significance.

• PROBLEM 12-40

A sports magazine reports that the people who watch Monday night football games on television are evenly divided between men and women. Out of a random sample of 400 people who regularly watch the Monday night game, 220 are men. Using a .10 level of significance, can be conclude that the report is false?

Solution: We have for this problem as our hypotheses:

H_0: $p = .50$, where p is the true population proportion.

H_1: $p \neq .50$.

The following diagram depicts the data of this problem, where $\overline{p}$ is the sample proportion of men who watch Monday night football.

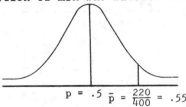

$$p = .5 \quad \overline{p} = \frac{220}{400} = .55$$

The statistic $(\overline{p} - p)/\sigma_p$ is approximately normally distributed with a mean of 0 and a standard deviation of 1. We calculate this value, which is called Z,

$$Z = \frac{\overline{p} - p}{\sigma_{\overline{p}}} \quad \text{where}$$

$$\sigma_{\overline{p}} = \sqrt{\frac{pq}{n}}$$

and compare the value of Z to a critical value. If Z lies beyond this critical value, we will reject H_0. For this problem, where we

787

have $\alpha = 10\%$ and a two-tailed test, our critical value is 1.645, since for the normal distribution with mean of 0 and standard deviation of 1, 5% of scores will have a Z-value above 1.645 and 5% of scores will have a Z-value below -1.645. Therefore our decision rule is: reject H_0 if $|Z| > 1.645$, accept H_0 if $|Z| \leq 1.645$.

For the data of this problem,

$$\sigma_p = \sqrt{\frac{(.50)(.50)}{400}} \approx .025 \quad \text{and}$$

$$Z = \frac{.55 - .50}{.025} = \frac{.05}{.025} = 2.0 .$$

Since $2.0 > 1.645$, we reject H_0 and conclude that the report of the sports magazine is incorrect at a 10% level of significance.

• PROBLEM 12–41

The makers of a certain brand of car mufflers claim that the life of the mufflers has a variance of .8 year. A random sample of 16 of these mufflers showed a variance of 1 year. Using a 5% level of significance, test whether the variance of all the mufflers of this manufacturer exceeds .8 year.

<u>Solution:</u>   Our hypotheses for this problem are:

$H_0: \sigma^2 = .8$ $H_1: \sigma^2 > .8$.

The statistic $\dfrac{(n - 1)s^2}{\sigma^2}$ has a χ^2 distribution with $n - 1$ degrees of freedom.

For $\alpha = .05$, a one-tailed test, and $n - 1 = 15$ degrees of freedom, we will have for our decision rule: reject H_0 if $\chi^2 > 24.996$; accept H_0 if $\chi^2 \leq 24.996$.

We calculate χ^2.

$$\chi^2 = \frac{(n - 1)s^2}{\sigma^2} = \frac{15(1)^2}{(.8)} = 18.75 .$$

Since $18.75 < 24.996$, we accept H_0 and conclude that the variance of all the mufflers of this manufacturer does not exceed .8 year.

• PROBLEM 12–42

A plant manager claims that on the average no more than 5 service calls per hour are made by the plant's workers. Suppose in one particular hour, 9 service calls were required. At a 5% level of significance, could we now reject the plant manager's claim?

<u>Solution:</u>   The number of accidents, claims, errors, or other such occurrences in a fixed time interval has a Poisson distribution.

We may use the Poisson distribution for this problem, where the variable in question is the number of service calls.

Our hypotheses in this problem are

$H_0: \lambda = 5$ $\qquad\qquad$ $H_1: \lambda > 5$

where λ is the average number of service calls in the fixed time interval of one hour.

For the hypothesized value of $\lambda = 5$, the probability of obtaining 9 or more service calls in one hour is, using a table of Poisson probabilities, given by

.0363 + .0181 + .0082 + .0034 + .0013 + .0005 + .0002 = .068

which equals Pr(9 service calls) + Pr(10 service calls) + Pr(11 service calls) + Pr(12 service calls) + Pr(13 service calls) + Pr(14 service calls) + Pr(15 service calls). (Note: the probability of higher numbers of service calls is virtually 0).

Since this value of .068 is greater than .05, the probability of an hour occurring where 9 or more service calls are required is greater than 5%. Hence at a 5% level of significance, we would not reject H_0.

● PROBLEM 12–43

A study in which infants were fed baked beans showed that such infants tended to gain more weight than a control group. The Z-value calculated was 1.96, which is equivalent to significance at p = .05. In a second independent study, the Z-value calculated was 1.64, significant only at p = .10. What conclusion can be reached if we combine the results of these 2 studies?

<u>Solution:</u>   The Z-values cannot be added together like χ^2 values but we may convert each p-value associated with a Z-value to a χ^2 value by using the following formula

$$\chi^2 \text{ (for 2 d.f.)} = -2 \log_e \left(\frac{P\%}{100}\right)$$

$$-2 \log_e 10 \left[\log_{10} \frac{P\%}{100}\right]$$

$$= -2(2.3)(\log_{10} P\% - \log_{10} 100)$$

789

$$= -4.6 \ (\log_{10} P \% - 2)$$

$$= 4.6 \ (2 - \log_{10} P \%).$$

For p = 5

$$\chi^2 = 4.6(2 - .6990) = 4.6(1.3010) = 5.98.$$

For p = 10, we obtain

$$\chi^2 = 4.6 \ (2 - 1) = 4.60.$$

In each case df = 2, so our combined

$$\chi^2 = 5.98 + 4.60 = 10.58$$

with df = 4, which is significant for p < .05. Therefore, combining the results of the 2 studies, we <u>can</u> conclude that a diet of baked beans for a group of infants does cause a greater weight gain than occurs for a control group.

● PROBLEM 12–44

Suppose we have a binomial distribution for which H_0 is $p = \frac{1}{2}$ where p is the probability of success on a single trial. Suppose the type I error, $\alpha = .05$ and n = 100. Calculate the power of this test for each of the following alternate hypotheses, H_1: p = .55, p = .60, p = .65, p = .70, and p = .75. Do the same when $\alpha = .01$.

<u>Solution</u>: Since N(p) and N(q) are both greater than 5, we may use the normal approximation to the binomial distribution for this problem. For $p = \frac{1}{2}$, the mean, μ, and standard deviation, σ, for this data are $p = \frac{1}{2}$, $\mu = np = 100(\frac{1}{2}) = 50$ and $\sigma = \sqrt{npq} = \sqrt{100(\frac{1}{2})(\frac{1}{2})} = 5$.

Since $\alpha = .05$, we will reject H_0 when Z > 1.65. Our formula for Z is

$$Z = \frac{X - \mu}{\sigma}.$$

Substituting the given values for Z, μ, and σ, we obtain

$$\frac{X - 50}{5} > 1.65.$$

Multiplying both sides of this equation by 5 yields

$$X - 50 > 8.25.$$

Adding 50 to both sides gives

$$X > 58.25.$$

So we will reject H_0 when X > 58.

The power of a test is given by the probability of accepting H_1 when H_1 is true. We must therefore calculate the probability of $X > 58$ for each of the specified H_1's.

We use the formula
$$Z = \frac{X - \mu}{\sigma}$$

where $X = 57.5$ ($58.5 - .5$ because the binomial distribution is discrete and the normal distribution is continuous). Also, for each case $\mu = np$ and $\sigma = \sqrt{npq}$.

The table below gives the values of μ, σ, Z, and the power of the test for each specified H_1.

H_1	$\mu = np$	$\sigma = \sqrt{npq}$	$\dfrac{X-\mu}{\sigma} = Z$	Power
p = .55	55	4.97	.5	.308
p = .60	60	4.90	-.51	.695
p = .65	65	4.77	-1.57	.942
p = .70	70	4.58	-2.73	.997
p = .75	75	4.33	-4.04	1.000

The power is obtained by using a table for the normal distribution and finding the probability that Z is greater than the value obtained in the prior column.

For $\alpha = .01$, we will reject H_0 when $Z > 2.33$. To find the value of X corresponding to $Z = 2.33$, we use

$$Z = \frac{X - \mu}{\sigma}$$

where $Z = 2.33$, $\mu = 50$, and $\sigma = 5$. Substituting $\dfrac{X - 50}{5} > 2.33$.

Multiplying both sides by 5 yields $X - 50 > 11.65$.

Adding 50 to each side yields $X > 61.65$.

Since 61.65 is in the interval 61.5 to 62.5, we will reject H_0 when X is greater than 61.5 for an $\alpha = .01$. Since the binomial distribution is discrete we use the value of X as 61.5, and we may now construct a table to give the power of the test for each specified H_1 when $\alpha = .01$.

H_1	$\mu = np$	$\sigma = \sqrt{npq}$	$\dfrac{X - \mu}{\sigma} = Z$	Power
p = .55	55	4.97	1.31	.095
p = .60	60	4.90	.31	.378
p = .65	65	4.77	-.73	.767
p = .70	70	4.58	-1.86	.967
p = .75	75	4.33	-3.12	.999

Note that a decrease in the size of the critical region from .05 to .01 uniformly reduced the power of the test for all H_1.

REGRESSION AND CORRELATION ANALYSIS

● **PROBLEM** 12-45

Given the following pairs of measurements for the two
variables:

X	5	8	3	9	10	12
Y	9	12	5	15	18	20

(a) Construct a scattergram and draw a calculated
     regression line.

(b) Using the regression line in part (a) estimate
     the values of Y when:

  (1) X = 4, (2) X = 1, and (3) X = 15.

Solution:

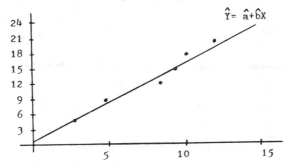

$$\hat{Y} = \hat{a} + \hat{b}X$$

From the data we compute the regression coefficients $\hat{a}$ and $\hat{b}$.

X	Y	X^2	XY
5	9	25	45
8	12	64	96
3	5	9	15
9	15	81	135
10	18	100	180
12	20	144	240
$\Sigma X=47$	$\Sigma Y=79$	$\Sigma X^2=423$	$\Sigma XY=711$

$$\hat{b} = \frac{\Sigma XY - \frac{\Sigma X \ \Sigma Y}{n}}{\Sigma X^2 - n\bar{X}^2}$$

$$= \frac{711 - \frac{(47)(79)}{6}}{423 - 6\left(\frac{47}{6}\right)^2}$$

$$= \frac{711 - 618.833}{423 - 368.16}$$

$$= \frac{92.167}{54.84}$$

$$= 1.68 .$$

The first normal equation is

$$\Sigma Y = n\hat{a} + \hat{b}\Sigma X$$

Divide by n.

$$\frac{\Sigma Y}{n} = \hat{a} + \hat{b}\,\frac{\Sigma X}{n} \quad \text{or equivalently}$$

$$\overline{Y} = \hat{a} + \hat{b}\,\overline{X}.$$

Thus $\quad \hat{a} = \overline{Y} - \hat{b}\overline{X} = \dfrac{79}{6} - 1.68\left(\dfrac{47}{6}\right)$

$$= 13.166 - 13.16 = .006 .$$

(b) $\quad \hat{Y} = \hat{a} + \hat{b}X = .006 + 1.68X .$

When $\quad X = 4, \quad \hat{Y} = .006 + 1.68(4) = 6.73$

$\qquad X = 1, \quad \hat{Y} = .006 + 1.68(1) = 1.69$

$\qquad X = 15, \quad \hat{Y} = .006 + 1.68(16) = 25.21.$

● PROBLEM 12–46

Given 4 pairs of observations

X	6	7	4	3
Y	8	10	4	2

compute and graph the least-squares regression line.

Solution: We use the table to compute the summary statistics needed.

X	Y	X^2	XY
6	8	36	48
7	10	49	70
4	4	16	16
3	2	9	6
$\Sigma X=20$	$\Sigma Y=24$	$\Sigma X^2=110$	$\Sigma XY=140$

$n = 4$

$$\hat{b} = \frac{\Sigma XY - n\overline{X}\,\overline{Y}}{\Sigma X^2 - n\overline{X}^2} = \frac{140 - 4\left(\dfrac{20}{4}\right)\left(\dfrac{24}{4}\right)}{110 - 4\left(\dfrac{20}{4}\right)^2}$$

$$= \frac{140 - 120}{110 - 100} = \frac{20}{10} = 2.0 ,$$

$$\hat{a} = \overline{Y} - \hat{b}\overline{X} = \frac{24}{4} - (2.0)\left(\frac{20}{4}\right) = 6 - 10$$

$$= -4,$$

$$\hat{Y} = \hat{a} + \hat{b}\,X$$

$$= -4 + 2X.$$

Y = a + bX = -4 + 2X

• **PROBLEM** 12–47

Three arctic zoologists spent 5 winters in the Yukon Territory. Their hypothesis was that the number of days on which the temperature dropped below - 50° Fahrenheit affected the length of moose horns. Find the correlation coefficient based on their data.

X, number of days - 50°F or less	Y, average length of moose Horns (in meters)
30	.9
20	.8
10	.5
30	1.0
10	.8

Solution: We compute the correlation coefficient with the aid of the table below.

X	Y	X^2	Y^2	XY
30	.9	900	.81	27
20	.8	400	.64	16
10	.5	100	.25	5
30	1.0	900	1.0	30
10	.8	100	.64	8

$$\Sigma X = 100 \qquad \Sigma Y = 4.0$$

$$\Sigma Y^2 = 3.34 \qquad \Sigma X^2 = 2400$$

$$\Sigma XY = 86$$

$$r = \frac{n\Sigma XY - (\Sigma X)(\Sigma Y)}{[\sqrt{n\Sigma X^2 - (\Sigma X)^2}][\sqrt{n\Sigma Y^2 - (\Sigma Y)^2}]}$$

$$= \frac{5(86) - (100)(4.0)}{[\sqrt{5(2400) - (100)^2}][\sqrt{5(3.34) - 4^2}]}$$

$$r = \frac{430 - 400}{[\sqrt{2000}][\sqrt{16.7 - 16}]} = \frac{30}{(44.721)(.8366)} = .80 .$$

It seems there is a high positive correlation between number of days when the temperature drops to - 50° F or less and the average length of moose horns.

• **PROBLEM** 12–48

A researcher suspects there is a correlation between the number of promises a political candidate makes and the number of promises that are fulfilled once the candidate is elected. He keeps track of several prominent politicians and records the following data:

Promises made, X	20	30	30	40	50	50	60
Promises kept, Y	7	6	5	4	3	2	1

What is the correlation between promises made and promises kept?

Solution: The coefficient of correlation measures the strength of a linear relation between the two variables X and Y.

The sign on the correlation coefficient indicates the direction of the relationship.

X	X^2	Y	Y^2	XY
20	400	7	49	140
30	900	6	36	180
30	900	5	25	150
40	1600	4	16	160
50	2500	3	9	150
50	2500	2	4	100
60	3600	1	1	60

n = 7

$$\Sigma X \;=\; 280 \qquad\qquad \Sigma Y \;=\; 28 \qquad\qquad \Sigma XY \;=\; 940 \;.$$

$$\Sigma X^2 \;=\; 12400 \qquad\qquad \Sigma Y^2 \;=\; 140$$

The coefficient of correlation is

$$r \;=\; \frac{n\,\Sigma XY \;-\; (\Sigma X)(\Sigma Y)}{[\sqrt{n\Sigma X^2 \;-\; (\Sigma X)^2}\,][\sqrt{n\Sigma Y^2 \;-\; (\Sigma Y)^2}\,]}$$

$$=\; \frac{7(940) \;-\; (280)(28)}{[\sqrt{7(12400) \;-\; (280)^2}\,][\sqrt{7(140) \;-\; (28)^2}\,]}$$

$$=\; \frac{6580 \;-\; 7840}{\sqrt{8400}\ \ \sqrt{196}} \;=\; \frac{-\,1260}{(91.65)(14)} \;=\; -\,.98 \;\;.$$

The coefficient of correlation between promises made and promises kept is $-\,.98$, very close to $-\,1$, indicating a very strong negative correlation. On the basis of this data it seems that the politicians who promise the most deliver the least. However it is important to remember that correlation does not imply causality. It seems unlikely that the number of promises made induces or causes a certain number of promises to be kept. Promises made and promises kept are probably both due to the politician's character and hence highly correlated.

● PROBLEM 12–49

Plot the following points.

X	- 1	- 2	0	1	1.5
Y	1	3	1	2	4

Use linear regression to estimate the relationship $Y = \alpha + \beta X$. Compute r^2. How well does the least squares equation "fit" the data? Now transform each X by squaring. Use linear regression to estimate the relationship between Y and X^2. What is r^2 for this new regression.

Solution:

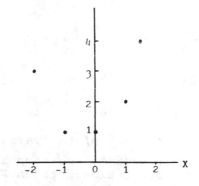

The following table is useful in computing the regression coefficients and r .

X	Y	X^2	Y^2	XY
- 1	1	1	1	- 1
- 2	3	4	9	- 6
0	1	0	1	0
1	2	1	4	2
1.5	4	2.25	16	6

$$\Sigma X = - .5 \quad \Sigma Y = 11 \quad \Sigma X^2 = 8.25$$

$$\Sigma Y^2 = 31 \quad \Sigma XY = 1 , \quad n = 5$$

$$\hat{\beta} = \frac{\Sigma XY - n \bar{X} \bar{Y}}{\Sigma X^2 - n\bar{X}^2} = \frac{1 - 5\left(\frac{-.5}{5}\right)\left(\frac{11}{5}\right)}{8.25 - 5\left(\frac{-.5}{5}\right)^2}$$

$$= \frac{2.1}{8.2} = .256 .$$

$$\hat{\alpha} = \bar{Y} - \hat{\beta}\bar{X} = \frac{11}{5} - (.256)\left(\frac{-.5}{5}\right) = 2.226 .$$

Thus, the best fitting linear equation is

$$\hat{Y} = \hat{\alpha} + \hat{\beta}X = 2.226 + .256X.$$

$$r = \frac{\Sigma (X - \bar{X})(Y - \bar{Y})}{\sqrt{\Sigma X^2 - n\bar{X}^2} \sqrt{\Sigma Y^2 - n\bar{Y}^2}}$$

$$= \frac{\Sigma XY - n\bar{X} \bar{Y}}{\sqrt{\Sigma X^2 - n\bar{X}^2} \sqrt{\Sigma Y^2 - n\bar{Y}^2}}$$

$$= \frac{2.1}{\sqrt{8.2} \sqrt{31 - 5\left(\frac{11}{5}\right)^2}}$$

$$= \frac{2.1}{\sqrt{8.2} \sqrt{6.8}}$$

$$= \frac{2.1}{7.47} = .28$$

$$r^2 = (r)^2 = (.28)^2 = .079 .$$

Approximately 8 percent of the variation in Y is explained by the least squares equation $\hat{Y}$ = 2.226 + .256X. This is quite a low correlation between X and Y.

$$= \frac{5.85}{\sqrt{6.8} \quad \sqrt{9.45}}$$

$$= \frac{5.85}{(2.61)(3.07)}$$

$$= .730$$

and $r^2 = (r)^2 = .53$.

   Thus 53 percent of the variation in Y is explained by the regression line, $\hat{Y} = 3.22 + .619X$.

   The functional form of $Y = a + bX^2$ is more appropriate in describing this data than the form $Y = a + bX$. It is necessary that the researcher specify an appropriate functional form before linear regression is used.

● **PROBLEM** 12–50

   The table below lists the ranks assigned by two securities analysts to 12 investment opportunities in terms of the degree of investor risk involved.

Investment	Rank by analyst 1	Rank by analyst 2
A	7	6
B	8	4
C	2	1
D	1	3
E	9	11
F	3	2
G	12	12
H	11	10
I	4	5
J	10	9
K	6	7
L	5	8

   Find the correlation between the two rankings. What is the relationship between the two rankings?

Solution: We compute r_s, the coefficient of rank correlation between the rankings given by Analyst 1 and Analyst 2.

Investment	Analyst 1	Analyst 2	$X_i - Y_i$	$(X_i - Y_i)^2$
A	7	6	1	1
B	8	4	4	16
C	2	1	1	1
D	1	3	- 2	4
E	9	11	- 2	4
F	3	2	1	1
G	12	12	0	0
H	11	10	1	1
I	4	5	- 1	1
J	10	9	1	1
K	6	7	- 1	1
L	5	8	- 3	9

$$r_s = 1 - \frac{6 \sum\limits_{i=1}^{12} (X_i - Y_i)^2}{n(n^2 - 1)} \qquad \sum\limits_{i=1}^{12} (X_i - Y_i)^2 = 40$$

$$n = 12 .$$

$$r_s = 1 - \frac{6(40)}{12(144 - 1)} = 1 - \frac{240}{1716}$$

$$= 1 - .14$$

$$r_s = .86 .$$

The correlation between the rankings given by Analyst 1 and Analyst 2 is .86. This is a very strong, positive correlation between the two rankings. It seems to imply that these two analysts have very similar ideas about the degree of investor risk involved with these twelve securities.

● **PROBLEM** 12–51

Find the z_f that corresponds to r = - .26.

Solution: z_f, the Fischer-z transform, is defined to be $z_f = (1.1513) \log \frac{1 + r}{1 - r}$, where r = the sample

correlation coefficient. The sample correlation coefficient can vary from + 1 to - 1. We have seen how to
solve for z_f given r > 0 using the table of Fis her-z
values. To find z_f when r < 0, find the z_f value for - r;
the z_f value of r < 0 is then - (the z_f value for - r).

Thus to find the z_f value of r = - .26, we find

- (the z_f value for - (- .26))

= - (the z_f for value for .26)

= - (.26611) = - .26611.

● **PROBLEM** 12-52

Twenty samples of sediment from the ocean floor were
analyzed for the presence of the metal uranium and the
mineral feldspar. The twenty pairs of observations (given
in micrograms) are below.

X, feldspar	Y, Uranium
10	12
19	15
17	12
8	8
5	10
4	7
8	5
16	20
21	19
26	20
5	15
9	5
12	25
8	20
9	12
26	15
19	19
20	24
2	8
3	15

You are a statistician brought in to analyze this
data. What is the equation of the least square regression?
Is there a significant relation between feldspar and
uranium?

Solution: The following table is useful for computing
$\hat{\alpha}$ and $\hat{\beta}$.

X	Y	X^2	Y^2	XY
10	12	100	144	120
19	15	361	225	285
17	12	289	224	204
8	8	64	64	64
5	10	25	100	50
4	7	16	49	28
8	5	64	25	40
16	20	256	400	320
21	19	441	361	399
26	20	676	400	520
5	15	25	225	75
9	5	81	25	45
12	25	144	625	300
8	20	64	400	160
9	12	81	144	108
26	15	676	225	390
19	19	361	361	361
20	24	400	576	480
2	8	4	64	16
3	15	9	225	45

$\Sigma X = 247 \quad \Sigma Y = 286 \quad \Sigma X^2 = 4137$

$\Sigma Y^2 = 4862 \quad\quad \Sigma XY = 4010 \quad\quad n = 20.$

$$\hat{\beta} = \frac{\Sigma\, XY - n\overline{X}\,\overline{Y}}{\Sigma X^2 - n\overline{X}^2}$$

$$= \frac{4010 - 20\,(12.35)(14.3)}{4137 - 20\,(12.35)^2}$$

$$= \frac{4010 - 3532.1}{4137 - 3050.45} = \frac{477.9}{1086.55}$$

$$= .4398 .$$

$$\hat{\alpha} = \overline{Y} - \hat{\beta}\overline{X} = 14.3 - (.4398)(12.35)$$

$$= 8.868 .$$

Thus the regression equation is

$$\hat{Y} = 8.868 + (.4398)X .$$

To test for the significance of this relationship, we test the hypothesis that

$H_0 : \beta = 0 \quad\quad$ versus

$H_A : \beta \neq 0 .$

If $\beta = 0$ is accepted at a .05 level of significance there is no significant relationship between feldspar and uranium.

We calculate the previously derived t-statistic,

$$T = \frac{\hat{\beta} - \beta}{\hat{\sigma}} \sqrt{\Sigma (X_i - \overline{X})^2}$$

and compare it with the values $\pm t_{.025}$ $(20 - 2)$.

To calculate T, first calculate $\hat{\sigma}^2$.

$$\hat{\sigma}^2 = \frac{1}{n - 2} \left[\Sigma (Y_i - \overline{Y})^2 - [\Sigma (X_i - \overline{X})(Y_i - \overline{Y})] \hat{\beta} \right]$$

$$= \frac{1}{18} [4862 - 20(14.3)^2 - (477.9)(.4398)]$$

$$= \frac{1}{18} [562] = 31.22 .$$

Thus,

$$\hat{\sigma} = \sqrt{\hat{\sigma}^2} = \sqrt{31.22} = 5.58 .$$

$$\sqrt{\Sigma (X_i - \overline{X})^2} = \sqrt{\Sigma X_i^2 - n\overline{X}^2}$$

$$= \sqrt{4137 - 20 (12.35)^2}$$

$$= \sqrt{4137 - 3050.45}$$

$$= \sqrt{1086.55} = 32.96 .$$

Under the null hypothesis, $\beta = 0$. Thus the T statistic is

$$T = \frac{\hat{\beta}}{\hat{\sigma}} \sqrt{\Sigma (X_i - \overline{X})^2} = \frac{(.4398)(32.96)}{5.58} = 2.597 .$$

The critical values from the table of the t-distribution are $\pm t_{.025}$ $(18) = \pm 2.101$. The calculated t-statistic is greater than $+ 2.101$, thus we reject the null hypothesis, that $\beta = 0$ in favor of $\beta \neq 0$. There is a significant relation between feldspar and uranium.

A data set relates proportional limit and tensile strength in certain alloys of gold collected for presentation at a Dentistry Convention. (Proportional limit is the load in psi at which the elongation of a sample no longer obeys Hooke's Law.)

Let (X_i, Y_i) be an observed ordered pair consisting of X_i, an observed tensile strength, and Y_i, an observed proportional limit, each measured in pounds per square inch (psi). After 25 observations of this sort the following summary statistics are:

$$\Sigma X_i = 2{,}991{,}300 , \qquad \overline{X} = 119{,}652$$

$$\Sigma X_i^2 = 372{,}419{,}750{,}000.$$

$$\Sigma Y_i = 2{,}131{,}200 , \qquad \overline{Y} = 85{,}248$$

$$\Sigma Y_i^2 = 196{,}195{,}960{,}000$$

$$\Sigma X_i Y_i = 269{,}069{,}420{,}000$$

Compute the regression coefficients relating proportional limit and tensile strength.

Solution: If Y is the proportional limit and X the tensile strength we hypothesize a relationship of the form $Y = \alpha + \beta X$.

The regression coefficients are estimated in the usual way

$$\hat{\alpha} = \overline{Y} - \hat{\beta}\overline{X} \qquad \text{and}$$

$$\hat{\beta} = \frac{\Sigma XY - n\overline{X}\,\overline{Y}}{\Sigma X^2 - n\overline{X}^2} ,$$

Thus,

$$\hat{\beta} = \frac{269{,}069{,}420{,}000 - 25(119{,}652)(85{,}248)}{372{,}419{,}750{,}000 - 25(119{,}652)^2}$$

$$= \frac{1406708 \times 10^4}{14504.73 \times 10^6}$$

$$= 96.9827 \times 10^{-2} = .9698$$

and

$$\hat{\alpha} = \overline{Y} - \hat{\beta}\overline{X} = 85{,}248 - (.9698)(119652) = -30{,}790 .$$

Thus our regression equation is

$$\hat{Y} = -30{,}790 + .9698X .$$

Test the hypothesis that $\underline{\beta}$, the true but unknown regression coefficient in the equation relating tensile strength to proportional limit, is equal to 1 versus $\beta \neq 1$.

Solution: Our hypotheses are

$$H_0: \beta = 1 \qquad \text{vs.}$$

$$H_A: \beta \neq 1 .$$

This is a two-tailed rejection region. We set the level of significance to be $\alpha = .05$. The test statistic will be a t-statistic with $n - 2 = 23$ degrees of freedom.

The critical values are those values $-t_{\alpha/2}(23)$, $t_{\alpha/2}(23)$ such that

$$\Pr(-t_{\alpha/2}(23) \leq T \leq t_{\alpha/2}(23)) = 1 - \alpha$$
$$= .95 ,$$

From the tables we see that

$$-t_{.025}(23) = -2.069 \quad \text{and}$$

$$t_{.025}(23) = +2.069 .$$

If the computed t-statistic, T, is greater than 2.069 or less than -2.069 we will reject H_0 in favor of H_A. It has been shown that

$$T = \frac{Z}{\sqrt{\dfrac{U}{n-2}}} \qquad \text{where Z is a standard normal}$$

random variable and U is a chi-square random variable with $n - 2$ degrees of freedom. We have seen that

$$U = \frac{n-2}{\sigma^2}\hat{\sigma}^2 . \qquad \text{What is Z in this case?}$$

Under the null hypothesis, $\beta = 1$. The expected value of $\hat{\beta}$ is $E(\hat{\beta}) = \beta$ and $\text{Var }\hat{\beta} = \dfrac{\sigma^2}{\Sigma(X_i - \overline{X})^2}$. Thus

$$Z = \frac{\hat{\beta} - E(\hat{\beta})}{\sqrt{\text{Var }\hat{\beta}}} = \frac{\hat{\beta} - \beta}{\dfrac{\sigma}{\sqrt{\Sigma(X_i - \overline{X})^2}}} = \frac{\hat{\beta} - 1}{\dfrac{\sigma}{\sqrt{\Sigma(X_i - \overline{X})^2}}}$$

Also

$$T = \frac{Z}{\sqrt{\frac{U}{n-2}}} = \frac{\frac{(\hat{\beta}-1)}{\sigma}\left(\sqrt{\Sigma(X_i-\bar{X})^2}\right)}{\frac{\hat{\sigma}}{\sigma}}$$

or

$$T = \frac{(\hat{\beta}-1)\sqrt{\Sigma(X_i-\bar{X})^2}}{\hat{\sigma}} \quad .$$

We compute T by substitution of the numerical values into this formula

$$T = \frac{(.9698-1)\sqrt{(14,504,722,400)}}{6160} = \frac{-3637}{6160} = -.59 \quad .$$

$-.59$ is greater than -2.069 and less than 2.069 thus the null hypothesis is accepted. That is, $\beta = 1$.

● **PROBLEM** 12–55

The following is a two variable, one common factor model:

$$X_1 = .8F + .6U_1 \qquad (1)$$

$$X_2 = .6F + .8U_2 \qquad (2)$$

a) Draw a path model for the system.
b) Find the covariance and correlation between X_1, X_2 and F, U_1, U_2 .

Solution: This is an example of a basic theoretical structure in Factor Analysis. The model (1)-(2) asserts that the observed variables X_1, X_2 are linear combinations of an unobserved common factor F and two unobserved unique factors, U_1 (uniquely affecting X_1) and U_2 (uniquely affecting X_2). The general form of the model is:

$$X_1 = b_1F + d_1U_1 \qquad (3)$$

$$X_2 = b_2F + d_2U_2 \qquad (4)$$

a) A path model for the system (3)-(4) consists of the observed and unobserved variables connected by lines showing the causal structure.

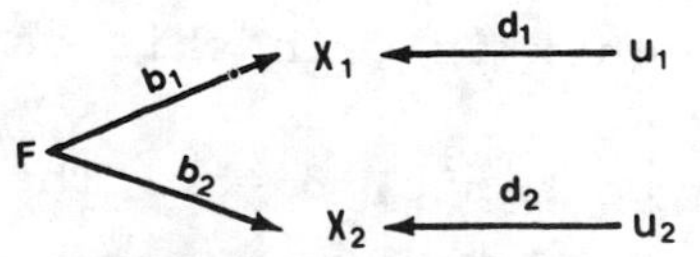

Fig. 1

Applying Fig. 1 to the given model (1)-(2) yields

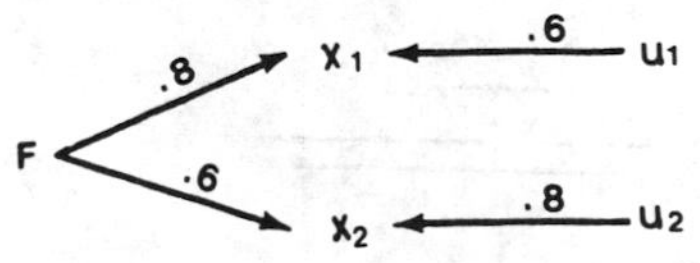

b) The covariance and correlation between the given variables is determined by the assumptions made involving their statistical nature. A convenient set of assumptions is the following:

i) $\text{Cov}(F, U_1) = \text{Cov}(F, U_2) = \text{Cov}(U_1, U_2) = 0$ where

$\text{Cov}(Y, Z) = E[(Y - \bar{Y})(Z - \bar{Z})]$.

ii) All the variables are normally distributed with mean zero and variance one. In fact, they are standard normal variables.
Note that i) is implicit in the path diagram (Fig. 1) since there are no causal connections between F and U_1 or U_2 and between U_1 and U_2. Using the general model (3)-(4):

$$\text{Var}(X_1) = E(X_1 - \bar{X})^2 = E(X_1)^2$$
$$= E[b_1 F + d_1 U_1]^2 = E[b_1^2 F^2 + d_1^2 U_1^2 + 2b_1 d_1 F U_1]$$
$$= b_1^2 E[F^2] + d_1^2 E[U_1^2] + 2b_1 d_1 E[F U_1]$$
$$= b_1^2 \text{Var}(F) + d_1^2 \text{Var}(U_1) + 2b_1 d_1 \text{Cov}(F, U_1)$$

$$\left(\text{since} \quad \text{Var}(F) = E(F - \bar{F})^2 = E(F - 0)^2 = E(F^2) \quad \text{and} \quad \text{Var}(U_1) = E(U_1 - \bar{U}_1)^2 \right.$$
$$\left. = E(U_1^2)\right)$$

$$= b_1^2 \text{Var}(F) + d_1^2 \text{Var}(U_1) .$$

(Since $\text{Cov}(FU_1) = 0$ by assumption). (The property of the expectation operator $E[aX + bY] = aE(X) + bE(Y)$ was used in the above derivation).
But $\text{Var}(F) = \text{Var}(U_1) = 1$ (by assumption) and thus $\text{Var}(X_1) = 1 = b_1^2 + d_1^2$. Similarly $\text{Var}(X_2) = 1 = b_2^2 + d_2^2$.
Applying these results to the given model (1)-(2),

$$\text{Var}(X_1) = (.8)^2 + (.6)^2 = 1$$
$$\text{Var}(X_2) = (.6)^2 + (.8)^2 = 1 .$$

Thus the proportion of variance in X_1 determined by F is .64 and the proportion determined by U_1 is .36 . Next, the covariance between a factor and an observed variable is given by

$$\text{Cov}(F, X_1) = E[(F - \bar{F})(X_1 - \bar{X})] = E[(F - 0)(X_1 - 0)] = E(F X_1)$$
$$= E[F(b_1 F + d_1 U_1)] = b_1 E(F^2) + d_1 E(F U_1)$$

$$= b_1 \text{Var}(F) + d_1 \text{Cov}(F U_1) = b_1 \text{Var}(F) .$$

But $\text{Var}(F) = 1$ by assumption and thus $\text{Cov}(F, X_1) = b_1$.
Similarly $\text{Cov}(F, X_2) = b_2$. Now r_{xy} (the correlation coefficient)

$$= \frac{E(X - \bar{X})(Y - \bar{Y})}{\sigma_x \ \sigma_y} = \frac{\text{Cov}(X, Y)}{\sigma_x \ \sigma_y} .$$

But here, $\sigma_x, \sigma_y = 1$ and thus $\text{Cov}(F, X_1) = b_1 = r_{F, X_1}$.
The covariance equals the correlation coefficient. For the given model, $\text{Cov}(F, X_1) = .8$ and $\text{Cov}(F, X_2) = .6$.

Also, since $Cov(X_1, U_1) = r_{X_1 U_1} = d_1$ and $Cov(X_2, U_2) = r_{X_2 U_2} = d_2$,

$Cov(X_1, U_1) = .6$ and $Cov(X_2, U_2) = .8$.

Finally, the covariance between X_1 and X_2 is:

$$
\begin{aligned}
Cov(X_1, X_2) &= E[(X_1 - \bar{X}_1)(X_2 - \bar{X}_2)] \\
&= E[(b_1 F + d_1 U_1)(b_2 F + d_2 U_2)] \\
&= E[b_1 b_2 F^2 + b_1 d_2 FU_2 + b_2 d_1 FU_1 + d_1 d_2 U_1 U_2] \\
&= b_1 b_2 Var(F) + b_1 d_2 Cov(FU_2) + b_2 d_1 Cov(F, U_1) \\
&\qquad + d_1 d_2 Cov(U_1, U_2) \\
&= b_1 b_2 Var(F) = b_1 b_2 .
\end{aligned}
$$

Thus the covariance between two observed variables sharing one common factor is equivalent to the variance of the factor times the two respective linear factors involved.

$$
\begin{aligned}
\Big(\text{since} \quad Var(F) &= E(F - \bar{F})^2 = E(F - 0)^2 = E(F^2) \quad \text{and} \quad Var(U_1) = E(U_1 - \bar{U}_1)^2 \\
&= E(U_1^2)\Big) \\
&= b_1^2 \, Var(F) + d_1^2 \, Var(U_1) .
\end{aligned}
$$

(Since $Cov(FU_1) = 0$ by assumption). (The property of the expectation operator $E[aX + bY] = aE(X) + bE(Y)$ was used in the above derivation).
But $Var(F) = Var(U_1) = 1$ (by assumption) and thus $Var(X_1) = 1 = b_1^2 + d_1^2$. Similarly $Var(X_2) = 1 = b_2^2 + d_2^2$.
Applying these results to the given model (1)–(2),

$$
\begin{aligned}
Var(X_1) &= (.8)^2 + (.6)^2 = 1 \\
Var(X_2) &= (.6)^2 + (.8)^2 = 1 .
\end{aligned}
$$

Thus the proportion of variance in X_1 determined by F is .64 and the proportion determined by U_1 is .36. Next, the covariance between a factor and an observed variable is given by

$$
\begin{aligned}
Cov(F, X_1) &= E[(F - \bar{F})(X_1 - \bar{X})] = E[(F - 0)(X_1 - 0)] = E(FX_1) \\
&= E[F(b_1 F + d_1 U_1)] = b_1 E(F^2) + d_1 E(FU_1) \\
&= b_1 \, Var(F) + d_1 \, Cov(FU_1) = b_1 \, Var(F) .
\end{aligned}
$$

But $Var(F) = 1$ by assumption and thus $Cov(F, X_1) = b_1$.
Similarly $Cov(F, X_2) = b_2$. Now r_{xy} (the correlation coefficient)

$$
= \frac{E(X - \bar{X})(Y - \bar{Y})}{\sigma_x \sigma_y} = \frac{Cov(X, Y)}{\sigma_x \sigma_y} .
$$

But here, $\sigma_x, \sigma_y = 1$ and thus $Cov(F, X_1) = b_1 = r_{F, X_1}$.

The covariance equals the correlation coefficient. For the given model, $Cov(F,X_1) = .8$ and $Cov(F,X_2) = .6$.

Also, since $Cov(X_1,U_1) = r_{X_1U_1} = d_1$ and $Cov(X_2,U_2) = r_{X_2U_2} = d_2$,

$Cov(X_1,U_1) = .6$ and $Cov(X_2,U_2) = .8$.

Finally, the covariance between X_1 and X_2 is:

$$
\begin{aligned}
Cov(X_1,X_2) &= E[(X_1-\bar{X}_1)(X_2-\bar{X}_2)] \\
&= E[(b_1F + d_1U_1)(b_2F + d_2U_2)] \\
&= E[b_1b_2F^2 + b_1d_2FU_2 + b_2d_1FU_1 + d_1d_2U_1U_2] \\
&= b_1b_2Var(F) + b_1d_2Cov(FU_2) + b_2d_1Cov(F,U_1) \\
&\quad\quad + d_1d_2Cov(U_1,U_2) \\
&= b_1b_2Var(F) = b_1b_2 \ .
\end{aligned}
$$

Thus the covariance between two observed variables sharing one common factor is equivalent to the variance of the factor times the two respective linear factors involved.

In the given model, $Cov(X_1,X_2) = (.8)(.6) = .48$. This shows that the covariation between the observed variables is completely determined by the common factor; if F were removed, there would be no correlation between X_1 and X_2 .

INDEX

Numbers on this page refer to <u>PROBLEM NUMBERS</u>, not page numbers

REA's **Problem Solvers**

The "PROBLEM SOLVERS" are comprehensive supplemental text-books designed to save time in finding solutions to problems. Each "PROBLEM SOLVER" is the first of its kind ever produced in its field. It is the product of a massive effort to illustrate almost any imaginable problem in exceptional depth, detail, and clarity. Each problem is worked out in detail with a step-by-step solution, and the problems are arranged in order of complexity from elementary to advanced. Each book is fully indexed for locating problems rapidly.

ADVANCED CALCULUS
ALGEBRA & TRIGONOMETRY
AUTOMATIC CONTROL
   SYSTEMS/ROBOTICS
BIOLOGY
BUSINESS, ACCOUNTING, & FINANCE
CALCULUS
CHEMISTRY
COMPLEX VARIABLES
COMPUTER SCIENCE
DIFFERENTIAL EQUATIONS
ECONOMICS
ELECTRICAL MACHINES
ELECTRIC CIRCUITS
ELECTROMAGNETICS
ELECTRONIC COMMUNICATIONS
ELECTRONICS
FINITE & DISCRETE MATH
FLUID MECHANICS/DYNAMICS
GENETICS
GEOMETRY

HEAT TRANSFER
LINEAR ALGEBRA
MACHINE DESIGN
MATHEMATICS for ENGINEERS
MECHANICS
NUMERICAL ANALYSIS
OPERATIONS RESEARCH
OPTICS
ORGANIC CHEMISTRY
PHYSICAL CHEMISTRY
PHYSICS
PRE-CALCULUS
PSYCHOLOGY
STATISTICS
STRENGTH OF MATERIALS &
   MECHANICS OF SOLIDS
TECHNICAL DESIGN GRAPHICS
THERMODYNAMICS
TOPOLOGY
TRANSPORT PHENOMENA
VECTOR ANALYSIS

If you would like more information about any of these books,
complete the coupon below and return it to us or visit your local bookstore.

RESEARCH & EDUCATION ASSOCIATION
61 Ethel Road W. • Piscataway, New Jersey 08854
Phone: (908) 819-8880

Please send me more information about your Problem Solver Books

Name _____

Address _____

City _____ State _____ Zip _____

REA's Test Preps
The Best in Test Preparations

The REA "Test Preps" are far more comprehensive than any other test series. They contain more tests with much more extensive explanations than others on the market. Each book provides several complete practice exams, based on the most recent tests given in the particular field. Every type of question likely to be given on the exams is included. Each individual test is followed by a complete answer key. **The answers are accompanied by full and detailed explanations.** By studying each test and the pertinent explanations, students will become well-prepared for the actual exam.

REA has published 40 Test Preparation volumes in several series. They include:

Advanced Placement Exams (APs)
Biology
Calculus AB & Calculus BC
Chemistry
Computer Science
English Literature & Composition
European History
Government & Politics
Physics
Psychology
United States History

College Board Achievement Tests (CBATs)
American History
Biology
Chemistry
English Composition

French
German
Literature
Mathematics Level I, II & IIC
Physics
Spanish

Graduate Record Exams (GREs)
Biology
Chemistry
Computer Science
Economics
Engineering
General
History
Literature in English
Mathematics
Physics
Political Science
Psychology

CBEST - California Basic Educational Skills Test
CDL - Commercial Drivers License Exam
ExCET - Exam for Certification of Educators in Texas
FE (EIT) - Fundamentals of Engineering Exam
GED - High School Equivalency Diploma Exam
GMAT - Graduate Management Admission Test
LSAT - Law School Admission Test
MCAT - Medical College Admission Test
NTE - National Teachers Exam
SAT - Scholastic Aptitude Test
TOEFL - Test of English as a Foreign Language

RESEARCH & EDUCATION ASSOCIATION
61 Ethel Road W. • Piscataway, New Jersey 08854
Phone: (908) 819-8880

Please send me more information about your Test Prep Books

Name _____

Address _____

City _____ State _____ Zip _____

"The ESSENTIALS"
of Math & Science

Each book in the ESSENTIALS series offers all essential information of the field it covers. It summarizes what every textbook in the particular field must include, and is designed to help students in preparing for exams and doing homework. The ESSENTIALS are excellent supplements to any class text.

The ESSENTIALS are complete, concise, with quick access to needed information, and provide a handy reference source at all times. The ESSENTIALS are prepared with REA's customary concern for high professional quality and student needs.

Available in the following titles:

Advanced Calculus I & II
Algebra & Trigonometry I & II
Anthropology
Automatic Control Systems /
   Robotics I & II
Biology I & II
Boolean Algebra
Calculus I, II & III
Chemistry
Complex Variables I & II
Differential Equations I & II
Electric Circuits I & II
Electromagnetics I & II

Electronic Communications I & II
Electronics I & II
Finite & Discrete Math
Fluid Mechanics /
   Dynamics I & II
Fourier Analysis
Geometry I & II
Group Theory I & II
Heat Transfer I & II
LaPlace Transforms
Linear Algebra
Math for Engineers I & II
Mechanics I, II & III

Modern Algebra
Numerical Analysis I & II
Organic Chemistry I & II
Physical Chemistry I & II
Physics I & II
Set Theory
Statistics I & II
Strength of Materials &
   Mechanics of Solids I & II
Thermodynamics I & II
Topology
Transport Phenomena I & II
Vector Analysis

If you would like more information about any of these books,
complete the coupon below and return it to us or go to your local bookstore.

RESEARCH & EDUCATION ASSOCIATION
61 Ethel Road W. • Piscataway, New Jersey 08854
Phone: (908) 819-8880

Please send me more information about your Essentials Books

Name _____

Address _____

City _____ State _____ Zip _____

HANDBOOK of MATHEMATICAL, SCIENTIFIC, & ENGINEERING FORMULAS, FUNCTIONS, GRAPHS, TABLES, & TRANSFORMS

REA RESEARCH & EDUCATION ASSOCIATION

A particularly useful reference for those in math, science, engineering and other technical fields. Includes the most-often used formulas, tables, transforms, functions, and graphs which are needed as tools in solving problems. The entire field of special functions is also covered. A large amount of scientific data which is often of interest to scientists and engineers has been included.

Available at your local bookstore or order directly from us by sending in coupon below.